Pearson New International Edition

Biostatistical Analysis

Jerrold H. Zar
Fifth Edition

PEARSON

Pearson Education Limited
Edinburgh Gate
Harlow
Essex CM20 2JE
England and Associated Companies throughout the world

Visit us on the World Wide Web at: www.pearsoned.co.uk

© Pearson Education Limited 2014

ISBN 10: 1-292-02404-6
ISBN 13: 978-1-292-02404-2

British Library Cataloguing-in-Publication Data
A catalogue record for this book is available from the British Library

Printed in the United States of America

Table of Contents

Data: Types and Presentation

Scientific study involves the systematic collection, organization, analysis, and presentation of knowledge. Many investigations in the biological sciences are quantitative, where knowledge is in the form of numerical observations called *data*. (One numerical observation is a *datum*.*) In order for the presentation and analysis of data to be valid and useful, we must use methods appropriate to the type of data obtained, to the design of the data collection, and to the questions asked of the data; and the limitations of the data, of the data collection, and of the data analysis should be appreciated when formulating conclusions.

The word *statistics* is derived from the Latin for "state," indicating the historical importance of governmental data gathering, which related principally to demographic information (including census data and "vital statistics") and often to their use in military recruitment and tax collecting.[†]

The term *statistics* is often encountered as a synonym for *data*: One hears of college enrollment statistics (such as the numbers of newly admitted students, numbers of senior students, numbers of students from various geographic locations), statistics of a basketball game (such as how many points were scored by each player, how many fouls were committed), labor statistics (such as numbers of workers unemployed, numbers employed in various occupations), and so on. Hereafter, this use of the word *statistics* will not appear in this text. Instead, it will be used in its other common manner: to refer to the *orderly collection, analysis, and interpretation of data with a view to objective evaluation of conclusions based on the data*.

Statistics applied to biological problems is simply called *biostatistics* or, sometimes, *biometry*[‡] (the latter term literally meaning "biological measurement"). Although

---

*The term *data* is sometimes seen as a singular noun meaning "numerical information." This book refrains from that use.

[†]Peters (1987: 79) and Walker (1929: 32) attribute the first use of the term *statistics* to a German professor, Gottfried Achenwall (1719–1772), who used the German word *Statistik* in 1749, and the first published use of the English word to John Sinclair (1754–1835) in 1791.

[‡]The word *biometry*, which literally means "biological measurement," had, since the nineteenth century, been found in several contexts (such as demographics and, later, quantitative genetics; Armitage, 1985; Stigler, 2000), but using it to mean the application of statistical methods to biological information apparently was conceived between 1892 and 1901 by Karl Pearson, along with the name *Biometrika* for the still-important English journal he helped found; and it was first published in the inaugural issue of this journal in 1901 (Snedecor, 1954). The Biometrics Section of the American

the field of statistics has roots extending back hundreds of years, its development began in earnest in the late nineteenth century, and a major impetus from early in this development has been the need to examine biological data.

Statistical considerations can aid in the design of experiments intended to collect data and in the setting up of hypotheses to be tested. Many biologists attempt the analysis of their research data only to find that too few data were collected to enable reliable conclusions to be drawn, or that much extra effort was expended in collecting data that cannot be of ready use in the analysis of the experiment. Thus, a knowledge of basic statistical principles and procedures is important as research questions are formulated *before* an experiment and data collection are begun.

Once data have been obtained, we may organize and summarize them in such a way as to arrive at their orderly and informative presentation. Such procedures are often termed *descriptive statistics*. For example, measurements might be made of the heights of all 13-year-old children in a school district, perhaps determining an average height for each sex. However, perhaps it is desired to make some generalizations from these data. We might, for example, wish to make a reasonable estimate of the heights of all 13-year-olds in the state. Or we might wish to conclude whether the 13-year-old boys in the state are on the average taller than the girls of that age. The ability to make such generalized conclusions, inferring characteristics of the whole from characteristics of its parts, lies within the realm of *inferential statistics*.

1 TYPES OF BIOLOGICAL DATA

A characteristic (for example, size, color, number, chemical composition) that may differ from one biological entity to another is termed a *variable* (or, sometimes, a *variate**), and several different kinds of variables may be encountered by biologists. Because the appropriateness of descriptive or inferential statistical procedures depends upon the properties of the data obtained, it is desirable to distinguish among the principal kinds of data. The classification used here is that which is commonly employed (Senders, 1958; Siegel, 1956; Stevens, 1946, 1968). However, not all data fit neatly into these categories and some data may be treated differently depending upon the questions asked of them.

(a) Data on a Ratio Scale. Imagine that we are studying a group of plants, that the heights of the plants constitute a variable of interest, and that the number of leaves per plant is another variable under study. It is possible to assign a numerical value to the height of each plant, and counting the leaves allows a numerical value to be recorded for the number of leaves on each plant. Regardless of whether the height measurements are recorded in centimeters, inches, or other units, and regardless of whether the leaves are counted in a number system using base 10 or any other base, there are two fundamentally important characteristics of these data.

First, there is a constant size interval between adjacent units on the measurement scale. That is, the difference in height between a 36-cm and a 37-cm plant is the same

---

Statistical Association was established in 1938, successor to the Committee on Biometrics of that organization, and began publishing the *Biometrics Bulletin* in 1945, which transformed in 1947 into the journal *Biometrics*, a journal retaining major importance today. More recently, the term *biometrics* has become widely used to refer to the study of human physical characteristics (including facial and hand characteristics, fingerprints, DNA profiles, and retinal patterns) for identification purposes.

*"Variate" was first used by R. A. Fisher (1925: 5; David, 1995).

as the difference between a 39-cm and a 40-cm plant, and the difference between eight and ten leaves is equal to the difference between nine and eleven leaves.

Second, it is important that there exists a zero point on the measurement scale and that there is a physical significance to this zero. This enables us to say something meaningful about the ratio of measurements. We can say that a 30-cm (11.8-in.) tall plant is half as tall as a 60-cm (23.6-in.) plant, and that a plant with forty-five leaves has three times as many leaves as a plant with fifteen.

Measurement scales having a constant interval size and a true zero point are said to be *ratio scales* of measurement. Besides lengths and numbers of items, ratio scales include weights (mg, lb, etc.), volumes (cc, cu ft, etc.), capacities (ml, qt, etc.), rates (cm/sec, mph, mg/min, etc.), and lengths of time (hr, yr, etc.).

(b) Data on an Interval Scale. Some measurement scales possess a constant interval size but not a true zero; they are called *interval scales*. A common example is that of the two common temperature scales: Celsius (C) and Fahrenheit (F). We can see that the same difference exists between 20°C (68°F) and 25°C (77°F) as between 5°C (41°F) and 10°C (50°F); that is, the measurement scale is composed of equal-sized intervals. But it cannot be said that a temperature of 40°C (104°F) is twice as hot as a temperature of 20°C (68°F); that is, the zero point is arbitrary.* (Temperature measurements on the absolute, or Kelvin [K], scale can be referred to a physically meaningful zero and thus constitute a ratio scale.)

Some interval scales encountered in biological data collection are *circular scales*. Time of day and time of the year are examples of such scales. The interval between 2:00 P.M. (i.e., 1400 hr) and 3:30 P.M. (1530 hr) is the same as the interval between 8:00 A.M. (0800 hr) and 9:30 A.M. (0930 hr). But one cannot speak of ratios of times of day because the zero point (midnight) on the scale is arbitrary, in that one could just as well set up a scale for time of day which would have noon, or 3:00 P.M., or any other time as the zero point. Circular biological data are occasionally compass points, as if one records the compass direction in which an animal or plant is oriented. As the designation of north as 0° is arbitrary, this circular scale is a form of interval scale of measurement.

(c) Data on an Ordinal Scale. The preceding paragraphs on ratio and interval scales of measurement discussed data between which we know numerical differences. For example, if man *A* weighs 90 kg and man *B* weighs 80 kg, then man *A* is known to weigh 10 kg more than *B*. But our data may, instead, be a record only of the fact that man *A* weighs more than man *B* (with no indication of how much more). Thus, we may be dealing with relative differences rather than quantitative differences. Such data consist of an ordering or ranking of measurements and are said to be on an *ordinal* scale of measurement (*ordinal* being from the Latin word for "order"). We may speak of one biological entity being shorter, darker, faster, or more active than another; the sizes of five cell types might be labeled 1, 2, 3, 4, and 5, to denote

---

*The German-Dutch physicist Gabriel Daniel Fahrenheit (1686–1736) invented the thermometer in 1714 and in 1724 employed a scale on which salt water froze at zero degrees, pure water froze at 32 degrees, and pure water boiled at 212 degrees. In 1742 the Swedish astronomer Anders Celsius (1701–1744) devised a temperature scale with 100 degrees between the freezing and boiling points of water (the so-called "centigrade" scale), first by referring to zero degrees as boiling and 100 degrees as freezing, and later (perhaps at the suggestion of Swedish botanist and taxonomist Carolus Linnaeus [1707–1778]) reversing these two reference points (Asimov, 1982: 177).

their magnitudes relative to each other; or success in learning to run a maze may be recorded as *A*, *B*, or *C*.

It is often true that biological data expressed on the ordinal scale could have been expressed on the interval or ratio scale had exact measurements been obtained (or obtainable). Sometimes data that were originally on interval or ratio scales will be changed to ranks; for example, examination grades of 99, 85, 73, and 66% (ratio scale) might be recorded as A, B, C, and D (ordinal scale), respectively.

Ordinal-scale data contain and convey less information than ratio or interval data, for only relative magnitudes are known. Consequently, quantitative comparisons are impossible (e.g., we cannot speak of a grade of C being half as good as a grade of A, or of the difference between cell sizes 1 and 2 being the same as the difference between sizes 3 and 4). However, we will see that many useful statistical procedures are, in fact, applicable to ordinal data.

(d) Data in Nominal Categories.   Sometimes the variable being studied is classified by some qualitative measure it possesses rather than by a numerical measurement. In such cases the variable may be called an *attribute*, and we are said to be dealing with *nominal*, or *categorical*, data. Genetic phenotypes are commonly encountered biological attributes: The possible manifestations of an animal's eye color might be brown or blue; and if human hair color were the attribute of interest, we might record black, brown, blond, or red. As other examples of nominal data (*nominal* is from the Latin word for "name"), people might be classified as male or female, or right-handed or left-handed. Or, plants might be classified as dead or alive, or as with or without fertilizer application. Taxonomic categories also form a nominal classi-fication scheme (for example, plants in a study might be classified as pine, spruce, or fir).

Sometimes, data that might have been expressed on an ordinal, interval, or ratio scale of measurement may be recorded in nominal categories. For example, heights might be recorded as tall or short, or performance on an examination as pass or fail, where there is an arbitrary cut-off point on the measurement scale to separate tall from short and pass from fail.

As will be seen, statistical methods useful with ratio, interval, or ordinal data gen-erally are not applicable to nominal data, and we must, therefore, be able to identify such situations when they occur.

(e) Continuous and Discrete Data.   When we spoke previously of plant heights, we were dealing with a variable that could be any conceivable value within any observed range; this is referred to as a *continuous variable*. That is, if we measure a height of 35 cm and a height of 36 cm, an infinite number of heights is possible in the range from 35 to 36 cm: a plant might be 35.07 cm tall or 35.988 cm tall, or 35.3263 cm tall, and so on, although, of course, we do not have devices sensitive enough to detect this infinity of heights. A continuous variable is one for which there is a possible value between any other two values.

However, when speaking of the number of leaves on a plant, we are dealing with a variable that can take on only certain values. It might be possible to observe 27 leaves, or 28 leaves, but 27.43 leaves and 27.9 leaves are values of the variable that are impossible to obtain. Such a variable is termed a *discrete* or *discontinuous variable* (also known as a *meristic variable*). The number of white blood cells in 1 mm³ of blood, the number of giraffes visiting a water hole, and the number of eggs laid by a grasshopper are all discrete variables. The possible values of a discrete variable generally are consecutive integers, but this is not necessarily so. If the leaves on our plants are always formed in pairs, then only even integers are possible values of the variable. And the ratio of number of wings to number of legs of insects is a discrete variable that may only have the value of 0, 0.3333..., or 0.6666... (i.e., $\frac{0}{6}$, $\frac{2}{6}$, or $\frac{4}{6}$, respectively).*

Ratio-, interval-, and ordinal-scale data may be either continuous or discrete. Nominal-scale data by their nature are discrete.

2 ACCURACY AND SIGNIFICANT FIGURES

Accuracy is the nearness of a measurement to the true value of the variable being measured. *Precision* is not a synonymous term but refers to the closeness to each other of repeated measurements of the same quantity. Figure 1 illustrates the difference between accuracy and precision of measurements.

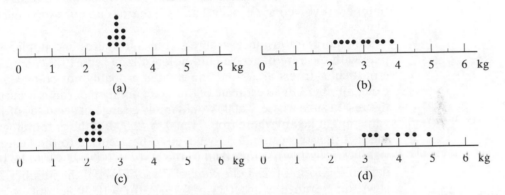

FIGURE 1: Accuracy and precision of measurements. A 3-kilogram animal is weighed 10 times. The 10 measurements shown in sample (a) are relatively accurate and precise; those in sample (b) are relatively accurate but not precise; those of sample (c) are relatively precise but not accurate; and those of sample (d) are relatively inaccurate and imprecise.

Human error may exist in the recording of data. For example, a person may miscount the number of birds in a tract of land or misread the numbers on a heart-rate monitor. Or, a person might obtain correct data but record them in such a way (perhaps with poor handwriting) that a subsequent data analyst makes an error in reading them. We shall assume that such errors have not occurred, but there are other aspects of accuracy that should be considered.

Accuracy of measurement can be expressed in numerical reporting. If we report that the hind leg of a frog is 8 cm long, we are stating the number 8 (a value of a continuous variable) as an estimate of the frog's true leg length. This estimate was made using some sort of a measuring device. Had the device been capable of more accuracy, we might have declared that the leg was 8.3 cm long, or perhaps 8.32 cm long. When recording values of continuous variables, it is important to designate the accuracy with which the measurements have been made. By convention, the value 8 denotes a measurement in the range of 7.50000... to 8.49999..., the value 8.3 designates a range of 8.25000... to 8.34999..., and the value 8.32 implies that the true value lies within the range of 8.31500... to 8.32499.... That is, the reported value is the midpoint of the implied range, and the size of this range is designated by the last decimal place in the measurement. The value of 8 cm implies an ability to

---

*The ellipsis marks (...) may be read as "and so on." Here, they indicate that $\frac{2}{6}$ and $\frac{4}{6}$ are repeating decimal fractions, which could just as well have been written as 0.3333333333333... and 0.6666666666666..., respectively.

determine length within a range of 1 cm, 8.3 cm implies a range of 0.1 cm, and 8.32 cm implies a range of 0.01 cm. Thus, to record a value of 8.0 implies greater accuracy of measurement than does the recording of a value of 8, for in the first instance the true value is said to lie between 7.95000 ... and 8.049999... (i.e., within a range of 0.1 cm), whereas 8 implies a value between 7.50000 ... and 8.49999 ... (i.e., within a range of 1 cm). To state 8.00 cm implies a measurement that ascertains the frog's limb length to be between 7.99500 ... and 8.00499 ... cm (i.e., within a range of 0.01 cm). Those digits in a number that denote the accuracy of the measurement are referred to as *significant figures*. Thus, 8 has one significant figure, 8.0 and 8.3 each have two significant figures, and 8.00 and 8.32 each have three.

In working with exact values of discrete variables, the preceding considerations do not apply. That is, it is sufficient to state that our frog has four limbs or that its left lung contains thirteen flukes. The use of 4.0 or 13.00 would be inappropriate, for as the numbers involved are exactly 4 and 13, there is no question of accuracy or significant figures.

But there are instances where significant figures and implied accuracy come into play with discrete data. An entomologist may report that there are 72,000 moths in a particular forest area. In doing so, it is probably not being claimed that this is the exact number but an estimate of the exact number, perhaps accurate to two significant figures. In such a case, 72,000 would imply a range of accuracy of 1000, so that the true value might lie anywhere from 71,500 to 72,500. If the entomologist wished to convey the fact that this estimate is believed to be accurate to the nearest 100 (i.e., to three significant figures), rather than to the nearest 1000, it would be better to present the data in the form of *scientific notation*,* as follows: If the number $7.2 \times 10^4 (= 72,000)$ is written, a range of accuracy of $0.1 \times 10^4 (= 1000)$ is implied, and the true value is assumed to lie between 71,500 and 72,500. But if 7.20×10^4 were written, a range of accuracy of $0.01 \times 10^4 (= 100)$ would be implied, and the true value would be assumed to be in the range of 71,950 to 72,050. Thus, the accuracy of large values (and this applies to continuous as well as discrete variables) can be expressed succinctly using scientific notation.

Calculators and computers typically yield results with more significant figures than are justified by the data. However, it is good practice—to avoid rounding error—to retain many significant figures until the last step in a sequence of calculations, and on attaining the result of the final step to round off to the appropriate number of figures.

3 FREQUENCY DISTRIBUTIONS

When collecting and summarizing large amounts of data, it is often helpful to record the data in the form of a *frequency table*. Such a table simply involves a listing of all the observed values of the variable being studied and how many times each value is observed. Consider the tabulation of the frequency of occurrence of sparrow nests in each of several different locations. This is illustrated in Example 1, where the observed kinds of nest sites are listed, and for each kind the number of nests observed is recorded. The distribution of the total number of observations among the various categories is termed a *frequency distribution*. Example 1 is a frequency table for nominal data, and these data may also be presented graphically by means of a *bar graph* (Figure 2), where the height of each bar is proportional to the frequency in the class represented. The widths of all bars in a bar graph should be equal so

---

*The use of scientific notation—by physicists—can be traced back to at least the 1860s (Miller, 2004b).

EXAMPLE 1   The Location of Sparrow Nests: A Frequency Table of Nominal Data

The variable is nest site, and there are four recorded categories of this variable. The numbers recorded in these categories constitute the frequency distribution.

Nest Site	Number of Nests Observed
A. Vines	56
B. Building eaves	60
C. Low tree branches	46
D. Tree and building cavities	49

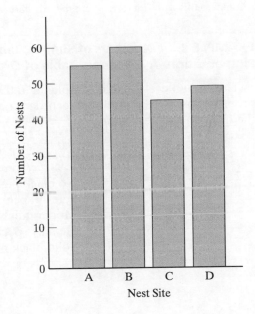

FIGURE 2: A bar graph of the sparrow nest data of Example 1. An example of a bar graph for nominal data.

that the eye of the reader is not distracted from the differences in bar heights; this also makes the area of each bar proportional to the frequency it represents. Also, the frequency scale on the vertical axis should begin at zero to avoid the apparent differences among bars. If, for example, a bar graph of the data of Example 1 were constructed with the vertical axis representing frequencies of 45 to 60 rather than 0 to 60, the results would appear as in Figure 3. Huff (1954) illustrates other techniques that can mislead the readers of graphs. It is good practice to leave space between the bars of a bar graph of nominal data, to emphasize the distinctness among the categories represented.

A frequency tabulation of ordinal data might appear as in Example 2, which presents the observed numbers of sunfish collected in each of five categories, each category being a degree of skin pigmentation. A bar graph (Figure 4) can be prepared for this frequency distribution just as for nominal data.

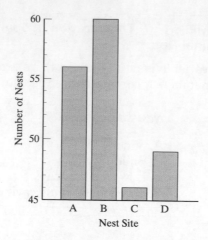

FIGURE 3: A bar graph of the sparrow nest data of Example 1, drawn with the vertical axis starting at 45. Compare this with Figure 1, where the axis starts at 0.

EXAMPLE 2   Numbers of Sunfish, Tabulated According to Amount of Black Pigmentation: A Frequency Table of Ordinal Data

The variable is amount of pigmentation, which is expressed by numerically ordered classes. The numbers recorded for the five pigmentation classes compose the frequency distribution.

Pigmentation Class	Amount of Pigmentation	Number of Fish
0	No black pigmentation	13
1	Faintly speckled	68
2	Moderately speckled	44
3	Heavily speckled	21
4	Solid black pigmentation	8

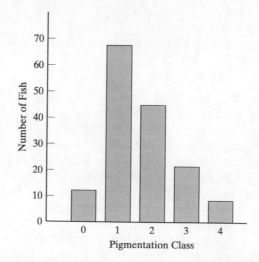

FIGURE 4: A bar graph of the sunfish pigmentation data of Example 2. An example of a bar graph for ordinal data.

In preparing frequency tables of interval- and ratio-scale data, we can make a procedural distinction between discrete and continuous data. Example 3 shows discrete data that are frequencies of litter sizes in foxes, and Figure 5 presents this frequency distribution graphically.

EXAMPLE 3   Frequency of Occurrence of Various Litter Sizes in Foxes: A Frequency Table of Discrete, Ratio-Scale Data

The variable is litter size, and the numbers recorded for the five litter sizes make up frequency distribution.

Litter Size	Frequency
3	10
4	27
5	22
6	4
7	1

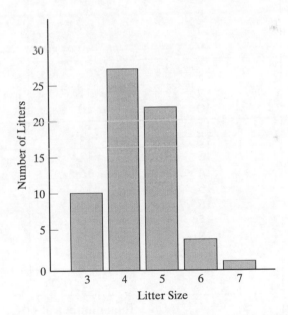

FIGURE 5: A bar graph of the fox litter data of Example 3. An example of a bar graph for discrete, ratio-scale data.

Example 4a shows discrete data that are the numbers of aphids found per clover plant. These data create quite a lengthy frequency table, and it is not difficult to imagine sets of data whose tabulation would result in an even longer list of frequencies. Thus, for purposes of preparing bar graphs, we often cast data into a frequency table by grouping them.

Example 4b is a table of the data from Example 4a arranged by grouping the data into size classes. The bar graph for this distribution appears as Figure 6. Such grouping results in the loss of some information and is generally utilized only to make frequency tables and bar graphs easier to read, and not for calculations performed on the data. There have been several "rules of thumb" proposed to aid in deciding into how many classes data might reasonably be grouped, for the use of too few groups will obscure the general shape of the distribution. But such "rules" or recommendations are only rough guides, and the choice is generally left to good judgment, bearing in mind that from 10 to 20 groups are useful for most biological work. (See also Doane, 1976.) In general, groups should be established that are equal in the size interval of the variable being measured. (For example, the group size interval in Example 4b is four aphids per plant.)

EXAMPLE 4a   Number of Aphids Observed per Clover Plant: A Frequency Table of Discrete, Ratio-Scale Data

Number of Aphids on a Plant	Number of Plants Observed	Number of Aphids on a Plant	Number of Plants Observed
0	3	20	17
1	1	21	18
2	1	22	23
3	1	23	17
4	2	24	19
5	3	25	18
6	5	26	19
7	7	27	21
8	8	28	18
9	11	29	13
10	10	30	10
11	11	31	14
12	13	32	9
13	12	33	10
14	16	34	8
15	13	35	5
16	14	36	4
17	16	37	1
18	15	38	2
19	14	39	1
		40	0
		41	1

Total number of observations = 424

Because continuous data, contrary to discrete data, can take on an infinity of values, one is essentially always dealing with a frequency distribution tabulated by groups. If the variable of interest were a weight, measured to the nearest 0.1 mg, a frequency table entry of the number of weights measured to be 48.6 mg would be interpreted to mean the number of weights grouped between 48.5500 . . . and 48.6499 . . . mg (although in a frequency table this class interval is usually written as 48.55–48.65). Example 5 presents a tabulation of 130 determinations of the amount of phosphorus, in milligrams per gram, in dried leaves. (Ignore the last two columns of this table until Section 4.)

EXAMPLE 4b **Number of Aphids Observed per Clover Plant: A Frequency Table Grouping the Discrete, Ratio-Scale Data of Example 4a**

Number of Aphids on a Plant	Number of Plants Observed
0–3	6
4–7	17
8–11	40
12–15	54
16–19	59
20–23	75
24–27	77
28–31	55
32–35	32
36–39	8
40–43	1

Total number of observations = 424

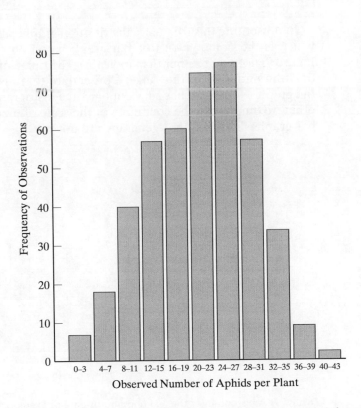

FIGURE 6: A bar graph of the aphid data of Example 4b. An example of a bar graph for grouped discrete, ratio-scale data.

EXAMPLE 5   Determinations of the Amount of Phosphorus in Leaves: A Frequency Table of Continuous Data

Phosphorus (mg/g of leaf)	Frequency (i.e., number of determinations)	Cumulative frequency	
		Starting with Low Values	Starting with High Values
8.15–8.25	2	2	130
8.25–8.35	6	8	128
8.35–8.45	8	16	122
8.45–8.55	11	27	114
8.55–8.65	17	44	103
8.65–8.75	17	61	86
8.75–8.85	24	85	69
8.85–8.95	18	103	45
8.95–9.05	13	116	27
9.05–9.15	10	126	14
9.15–9.25	4	130	4

Total frequency = 130 = n

In presenting this frequency distribution graphically, one can prepare a *histogram*,* which is the name given to a bar graph based on continuous data. This is done in Figure 7; note that rather than indicating the range on the horizontal axis, we indicate only the midpoint of the range, a procedure that results in less crowded printing on the graph. Note also that adjacent bars in a histogram are often drawn touching each other, to emphasize the continuity of the scale of measurement, whereas in the other bar graphs discussed they generally are not.

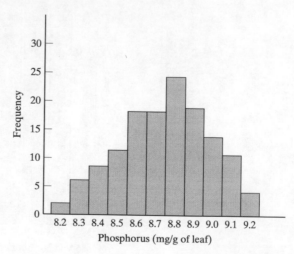

FIGURE 7: A histogram of the leaf phosphorus data of Example 5. An example of a histogram for continuous data.

*The term *histogram* is from Greek roots (referring to a pole-shaped drawing) and was first published by Karl Pearson in 1895 (David 1995).

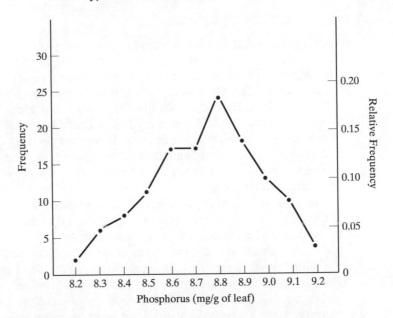

FIGURE 8: A frequency polygon for the leaf phosphorus data of Example 5.

Often a *frequency polygon* is drawn instead of a histogram. This is done by plotting the frequency of each class as a dot (or other symbol) at the class midpoint and then connecting each adjacent pair of dots by a straight line (Figure 8). It is, of course, the same as if the midpoints of the tops of the histogram bars were connected by straight lines. Instead of plotting frequencies on the vertical axis, one can plot *relative frequencies*, or proportions of the total frequency. This enables different distributions to be readily compared and even plotted on the same axes. Sometimes, as in Figure 8, frequency is indicated on one vertical axis and the corresponding relative frequency on the other. (Using the data of Example 5, the relative frequency for 8.2 mg/g is $2/130 = 0.015$, that for 8.3 mg/g is $6/130 = 0.046$, that for 9.2 mg/g is $4/130 = 0.030$, and so on. The total of all the frequencies is n, and the total of all the relative frequencies is 1.)

Frequency polygons are also commonly used for discrete distributions, but one can argue against their use when dealing with ordinal data, as the polygon implies to the reader a constant size interval horizontally between points on the polygon. Frequency polygons should not be employed for nominal-scale data.

If we have a frequency distribution of values of a continuous variable that falls into a large number of class intervals, the data may be grouped as was demonstrated with discrete variables. This results in fewer intervals, but each interval is, of course, larger. The midpoints of these intervals may then be used in the preparation of a histogram or frequency polygon. The user of frequency polygons is cautioned that such a graph is simply an aid to the eye in following trends in frequency distributions, and one should not attempt to read frequencies between points on the polygon. Also note that the method presented for the construction of histograms and frequency polygons requires that the class intervals be equal. Lastly, the vertical axis (e.g., the frequency scale) on frequency polygons and bar graphs generally should begin with zero, especially if graphs are to be compared with one another. If this is not done, the eye may be misled by the appearance of the graph (as shown for nominal-scale data in Figures 2 and 3).

4 CUMULATIVE FREQUENCY DISTRIBUTIONS

A frequency distribution informs us how many observations occurred for each value (or group of values) of a variable. That is, examination of the frequency table of Example 3 (or its corresponding bar graph or frequency polygon) would yield information such as, "How many fox litters of four were observed?", the answer being 27. But if it is desired to ask questions such as, "How many litters of four or more were observed?", or "How many fox litters of five or fewer were observed?", we are speaking of *cumulative frequencies*. To answer the first question, we sum all frequencies for litter sizes four and up, and for the second question, we sum all frequencies from the smallest litter size up through a size of five. We arrive at answers of 54 and 59, respectively.

In Example 5, the phosphorus concentration data are cast into two cumulative frequency distributions, one with cumulation commencing at the low end of the measurement scale and one with cumulation being performed from the high values toward the low values. The choice of the direction of cumulation is immaterial, as can be demonstrated. If one desired to calculate the number of phosphorus determinations less than 8.55 mg/g, namely 27, a cumulation starting at the low end might be used, whereas the knowledge of the frequency of determinations greater than 8.55 mg/g, namely 103, can be readily obtained from the cumulation commencing from the high end of the scale. But one can easily calculate any frequency from a low-to-high cumulation (e.g., 27) from its complementary frequency from a high-to-low cumulation (e.g., 103), simply by knowing that the sum of these two frequencies is the total frequency (i.e., $n = 130$); therefore, in practice it is not necessary to calculate both sets of cumulations.

Cumulative frequency distributions are useful in determining medians, percentiles, and other quantiles. They are not often presented in bar graphs, but *cumulative frequency polygons* (sometimes called *ogives*) are not uncommon. (See Figures 9 and 10.)

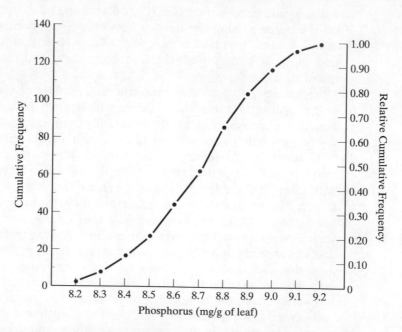

FIGURE 9: Cumulative frequency polygon of the leaf phosphorus data of Example 5, with cumulation commencing from the lowest to the highest values of the variable.

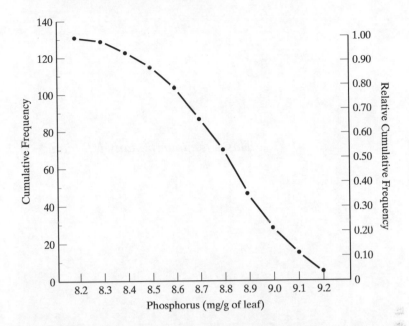

FIGURE 10: Cumulative frequency polygon of the leaf phosphorus data of Example 5, with cumulation commencing from the highest to the lowest values of the variable.

Relative frequencies (proportions of the total frequency) can be plotted instead of (or, as in Figures 9 and 10, in addition to) frequencies on the vertical axis of a cumulative frequency polygon. This enables different distributions to be readily compared and even plotted on the same axes. (Using the data of Example 5 for Figure 9, the relative cumulative frequency for 8.2 mg/g is 2/130 = 0.015, that for 8.3 mg/g is 8/130 = 0.062, and so on. For Figure 10, the relative cumulative frequency for 8.2 mg/g is 130/130 = 1.000, that for 8.3 mg/g is 128/130 = 0.985, and so on.)

This page intentionally left blank

Populations and Samples

From Chapter 2 of *Biostatistical Analysis*, Fifth Edition, Jerrold H. Zar. Copyright © 2010 by
Pearson Education, Inc. Publishing as Pearson Prentice Hall. All rights reserved.

Populations and Samples

The primary objective of a statistical analysis is to infer characteristics of a group of data by analyzing the characteristics of a small sampling of the group. This generalization from the part to the whole requires the consideration of such important concepts as population, sample, parameter, statistic, and random sampling. These topics are discussed in this chapter.

1 POPULATIONS

Basic to statistical analysis is the desire to draw conclusions about a group of measurements of a variable being studied. Biologists often speak of a "population" as a defined group of humans or of another species of organisms. Statisticians speak of a *population* (also called a *universe*) as a group of measurements (not organisms) about which one wishes to draw conclusions. It is the latter definition, the statistical definition of *population*, that will be used throughout this text. For example, an investigator may desire to draw conclusions about the tail lengths of bobcats in Montana. All Montana bobcat tail lengths are, therefore, the population under consideration. If a study is concerned with the blood-glucose concentration in three-year-old children, then the blood-glucose levels in all children of that age are the population of interest.

Populations are often very large, such as the body weights of all grasshoppers in Kansas or the eye colors of all female New Zealanders, but occasionally populations of interest may be relatively small, such as the ages of men who have traveled to the moon or the heights of women who have swum the English Channel.

2 SAMPLES FROM POPULATIONS

If the population under study is very small, it might be practical to obtain all the measurements in the population. If one wishes to draw conclusions about the ages of all men who have traveled to the moon, it would not be unreasonable to attempt to collect all the ages of the small number of individuals under consideration. Generally, however, populations of interest are so large that obtaining all the measurements is unfeasible. For example, we could not reasonably expect to determine the body weight of every grasshopper in Kansas. What can be done in such cases is to obtain a subset of all the measurements in the population. This subset of measurements constitutes a *sample*, and from the characteristics of samples we can draw conclusions about the characteristics of the populations from which the samples came.*

Biologists may sample a population that does not physically exist. Suppose an experiment is performed in which a food supplement is administered to 40 guinea pigs, and the sample data consist of the growth rates of these 40 animals. Then the population about which conclusions might be drawn is the growth rates of all the guinea pigs that conceivably might have been administered the same food supplement under identical conditions. Such a population is said to be "imaginary" and is also referred to as "hypothetical" or "potential."

3 RANDOM SAMPLING

Samples from populations can be obtained in a number of ways; however, for a sample to be representative of the population from which it came, and to reach valid conclusions about populations by induction from samples, statistical procedures typically assume that the samples are obtained in a *random* fashion. To sample a population randomly requires that each member of the population has an equal and independent chance of being selected. That is, not only must each measurement in the population have an equal chance of being chosen as a member of the sample, but the selection of any member of the population must in no way influence the selection of any other member. Throughout this text, "sample" will always imply "random sample."[†]

It is sometimes possible to assign each member of a population a unique number and to draw a sample by choosing a set of such numbers at random. This is equivalent to having all members of a population in a hat and drawing a sample from them while blindfolded. Table 41 from *Appendix: Statistical Tables and Graphs* provides 10,000 random digits for this purpose. In this table, each digit from 0 to 9 has an equal and independent chance of appearing anywhere in the table. Similarly, each combination of two digits, from 00 to 99, is found at random in the table, as is each three-digit combination, from 000 to 999, and so on.

Assume that a random sample of 200 names is desired from a telephone directory having 274 pages, three columns of names per page, and 98 names per column. Entering Table 41 from *Appendix: Statistical Tables and Graphs* at random (i.e., do not always enter the table at the same place), one might decide first to arrive at a random combination of three digits. If this three-digit number is 001 to 274, it can be taken as a randomly chosen page number (if it is 000 or larger than 274, simply skip it and choose another three-digit number, e.g., the next one on the table). Then one might examine the next digit in the table; if it is a 1, 2, or 3, let it denote a page column (if a digit other than 1, 2, or 3 is encountered, it is ignored, passing to the next digit that is 1, 2, or 3). Then one could look at the next two-digit number in the table; if it is from 01 to 98, let it represent a randomly selected name within that column. This three-step procedure would be performed a total of 200 times to obtain the desired random sample. One can proceed in any direction in the random number table: left to right, right to left, upward, downward, or diagonally; but the direction should be decided on before looking at the table. Computers are capable of quickly generating random numbers (sometimes called "pseudorandom" numbers because the number generation is not perfectly random), and this is how Table 41 from *Appendix: Statistical Tables and Graphs* was derived.

---

*This use of the terms *population* and *sample* was established by Karl Pearson (1903).

[†]This concept of random sampling was established by Karl Pearson between 1897 and 1903 (Miller, 2004a).

Very often it is not possible to assign a number to each member of a population, and random sampling then involves biological, rather than simply mathematical, considerations. That is, the techniques for sampling Montana bobcats or Kansas grasshoppers require knowledge about the particular organism to ensure that the sampling is random. Researchers consult relevant books, periodical articles, or reports that address the specific kind of biological measurement to be obtained.

4 PARAMETERS AND STATISTICS

Several measures help to describe or characterize a population. For example, generally a preponderance of measurements occurs somewhere around the middle of the range of a population of measurements. Thus, some indication of a population "average" would express a useful bit of descriptive information. Such information is called a *measure of central tendency* (also called a *measure of location*).

It is also important to describe how dispersed the measurements are around the "average." That is, we can ask whether there is a wide spread of values in the population or whether the values are rather concentrated around the middle. Such a descriptive property is called a *measure of variability* (or a *measure of dispersion*).

A quantity such as a measure of central tendency or a measure of dispersion is called a *parameter* when it describes or characterizes a population, and we shall be very interested in discussing parameters and drawing conclusions about them. Section 2 pointed out, however, that one seldom has data for entire populations, but nearly always has to rely on samples to arrive at conclusions about populations. Thus, one rarely is able to calculate parameters. However, by random sampling of populations, parameters can be estimated well. An estimate of a population parameter is called a *statistic*.* It is statistical convention to represent population parameters by Greek letters and sample statistics by Latin letters; will demonstrate this custom for specific examples.

The statistics one calculates will vary from sample to sample for samples taken from the same population. Because one uses sample statistics as estimates of population parameters, it behooves the researcher to arrive at the "best" estimates possible. As for what properties to desire in a "good" estimate, consider the following.

First, it is desirable that if we take an indefinitely large number of samples from a population, the long-run average of the statistics obtained will equal the parameter being estimated. That is, for some samples a statistic may underestimate the parameter of interest, and for others it may overestimate that parameter; but in the long run the estimates that are too low and those that are too high will "average out." If such a property is exhibited by a statistic, we say that we have an *unbiased* statistic or an unbiased estimator.

Second, it is desirable that a statistic obtained from any single sample from a population be very close to the value of the parameter being estimated. This property of a statistic is referred to as *precision*,[†] *efficiency*, or *reliability*. As we commonly secure only one sample from a population, it is important to arrive at a close estimate of a parameter from a single sample.

---

*This use of the terms *parameter* and *statistic* was defined by R. A. Fisher as early as 1922 (Miller, 2004a; Savage, 1976).

[†]The precision of a sample statistic, as defined here, should not be confused with the precision of a measurement.

Third, consider that one can take larger and larger samples from a population (the largest sample being the entire population). As the sample size increases, a *consistent* statistic will become a better estimate of the parameter it is estimating. Indeed, if the sample were the size of the population, then the best estimate would be obtained: the parameter itself.

5 OUTLIERS

Occasionally, a set of data will have one or more observations that are so different, relative to the other data in the sample, that we doubt they should be part of the sample. For example, suppose a researcher collected a sample consisting of the body weights of nineteen 20-week-old mallard ducks raised in individual laboratory cages, for which the following 19 data were recorded:

1.87, 3.75, 3.79, 3.82, 3.85, 3.87, 3.90, 3.94, 3.96, 3.99,

3.99, 4.00, 4.03, 4.04, 4.05, 4.06, 4.09, 8.97, and 39.8 kilograms.

Visual inspection of these 19 recorded data casts doubt upon the smallest datum (1.87 kg) and the two largest data (8.97 kg and 39.8 kg) because they differ so greatly from the rest of the weights in the sample. Data in striking disagreement with nearly all the other data in a sample are often called *outliers* or *discordant data*, and the occurrence of such observations generally calls for closer examination.

Sometimes it is clear that an outlier is the result of incorrect recording of data. In the preceding example, a mallard duck weight of 39.8 kg is highly unlikely (to say the least!), for that is about the weight of a 12-year-old boy or girl (and such a duck would probably not fit in one of the laboratory cages). In this case, inspection of the data records might lead us to conclude that this body weight was recorded with a careless placement of the decimal point and should have been 3.98 kg instead of 39.8 kg. And, upon interrogation, the research assistant may admit to weighing the eighteenth duck with the scale set to pounds instead of kilograms, so the metric weight of that animal should have been recorded as 4.07 (not 8.97) kg.

Also, upon further examination of the data-collection process, we may find that the 1.87-kg duck was taken from a wrong cage and was, in fact, only 4 weeks old, not 20 weeks old, and therefore did not belong in this sample. Or, perhaps we find that it was not a mallard duck, but some other bird species (and, therefore, did not belong in this sample). Statisticians say a sample is *contaminated* if it contains a datum that does not conform to the characteristics of the population being sampled. So the weight of a 4-week-old duck, or of a bird of a different species, would be a statistical contaminant and should be deleted from this sample.

There are also instances where it is known that a measurement was faulty—for example, when a laboratory technician spills coffee onto an electronic measuring device or into a blood sample to be analyzed. In such a case, the measurements known to be erroneous should be eliminated from the sample.

However, outlying data can also be correct observations taken from an intended population, collected purely by chance. As we shall see, when drawing a random sample from a population, it is relatively likely that a datum in the sample will be around the average of the population and very unlikely that a sample datum will be dramatically far from the average. But sample data very far from the average still may be possible.

It should also be noted that in some situations the examination of an outlier may reveal the effect of a previously unsuspected factor. For example, the 1.87-kg duck might, indeed, have been a 20-week-old mallard but suffering from a genetic mutation or a growth-impeding disease deserving of further consideration in additional research.

In summary, it is not appropriate to discard data simply because they appear (to someone) to be unreasonably extreme. However, if there is a very obvious reason for correcting or eliminating a datum, such as the situations described previously, the incorrect data should be corrected or eliminated. In some other cases questionable data can be *accommodated* in statistical analysis, perhaps by employing statistical procedures that give them less weight or analytical techniques that are *robust* in that they are resistant to effects of discrepant data. And in situations when this cannot be done, dubious data will have to remain in the sample (perhaps encouraging the researcher to repeat the experiment with a new set of data).

The idea of rejecting erroneous data dates back over 200 years; and recommendations for formal, objective methods for such rejection began to appear about 150 years ago. Major discussions of outliers, their origin, and treatment (rejection or accommodation) are those of Barnett and Lewis (1994), Beckman and Cook (1983), and Thode (2002: 123–142).

Measures of Central Tendency

1   THE ARITHMETIC MEAN
2   THE MEDIAN
3   THE MODE
4   OTHER MEASURES OF CENTRAL TENDENCY
5   CODING DATA

In samples, as well as in populations, one generally finds a preponderance of values somewhere around the middle of the range of observed values. The description of this concentration near the middle is an *average*, or a *measure of central tendency* to the statistician. It is also termed a *measure of location*, for it indicates where, along the measurement scale, the sample or population is located. Various measures of central tendency are useful population parameters, in that they describe an important property of populations. This chapter discusses the characteristics of these parameters and the sample statistics that are good estimates of them.

1   THE ARITHMETIC MEAN

The most widely used measure of central tendency is the *arithmetic mean*,[*] usually referred to simply as the *mean*,[†] which is the measure most commonly called an "average."

Each measurement in a population may be referred to as an X_i (read "X sub i") value. Thus, one measurement might be denoted as X_1, another as X_2, another as X_3, and so on. The subscript i might be any integer value up through N, the total number of X values in the population.[‡] The mean of the population is denoted by the Greek letter μ (lowercase mu) and is calculated as the sum of all the X_i values divided by the size of the population.

The calculation of the population mean can be abbreviated concisely by the formula

$$\mu = \frac{\sum\limits_{i=1}^{N} X_i}{N}.$$ (1)

---

[*]As an adjective, *arithmetic* is pronounced with the accent on the third syllable. In early literature on the subject, the adjective *arithmetical* was employed.

[†]The term *mean* (as applied to the arithmetic mean, as well as to the geometric and harmonic means of Section 4) dates from ancient Greece (Walker, 1929: 183), with its current statistical meaning in use by 1755 (Miller, 2004a; Walker, 1929: 176); *central tendency* appeared by the late 1920s (Miller, 2004a).

[‡]Charles Babbage (1791–1871) (O'Connor and Robertson, 1998) was an English mathematician and inventor who conceived principles used by modern computers—well before the advent of electronics—and who, in 1832, proposed the modern convention of italicizing Latin (also called Roman) letters to denote quantities; nonitalicized letters had already been employed for this purpose for more than six centuries (Miller, 2001).

The Greek letter Σ (capital sigma) means "summation"* and $\sum_{i=}^{N} X$ means "summation of all X_i values from X_1 through X_N." Thus, for example, $\sum_{i=1}^{4} X_i = X_1 + X_2 + X_3 + X_4$ and $\sum_{i=3}^{5} X_i = X_3 + X_4 + X_5$. Since, in statistical computations, summations are nearly always performed over the entire set of X_i values, this text will assume $\sum X_i$ to mean "sum X_i's over all values of i," simply as a matter of printing convenience, and $\mu = \sum X_i/N$ would therefore designate the same calculation as would $\mu = \sum_{i=1}^{N} X_i/N$.

The most efficient, unbiased, and consistent estimate of the population mean, μ, is the sample mean, denoted as $\overline{X}$ (read as "X bar"). Whereas the size of the population (which we generally do not know) is denoted as N, the size of a sample is indicated by n, and $\overline{X}$ is calculated as

$$\overline{X} = \frac{\sum_{i=1}^{n} X_i}{n} \quad \text{or} \quad \overline{X} = \frac{\sum X_i}{n}, \tag{2}$$

which is read "the sample mean equals the sum of all measurements in the sample divided by the number of measurements in the sample."[†] Example 1 demonstrates the calculation of the sample mean. Note that the mean has the same units of measurement as do the individual observations.

EXAMPLE 1   A Sample of 24 from a Population of Butterfly Wing Lengths

X_i (in centimeters): 3.3, 3.5, 3.6, 3.6, 3.7, 3.8, 3.8, 3.8, 3.9, 3.9, 3.9, 4.0, 4.0, 4.0, 4.0, 4.1, 4.1, 4.1, 4.2, 4.2, 4.3, 4.3, 4.4, 4.5.

$$\sum X_i = 95.0 \text{ cm}$$
$$n = 24$$
$$\overline{X} = \frac{\sum X_i}{n} = \frac{95.0 \text{ cm}}{24} = 3.96 \text{ cm}$$

---

*Mathematician Leonhard Euler (1707–1783; born in Switzerland, worked mostly in Russia), in 1755, was the first to use Σ to denote summation (Cajori, 1928/9, Vol. II: 61).

[†]The modern symbols for plus and minus ("+" and "−") appear to have first appeared in a 1456 unpublished manuscript by German mathematician and astronomer Regiomontanus (Johannes Müller, 1436–1476), with Bohemia-born Johann (Johannes) Widman (1562–1498) the first, in 1489, to use them in print (Cajori, 1928/9, Vol. I: 128, 231–232). The modern equal sign ("=") was invented by Welsh physician and mathematician Robert Recorde (1510–1558), who published it in 1557 (though its use then disappeared in print until 1618), and it was well recognized starting in 1631 (Cajori, ibid.: 298; Gullberg, 1997: 107). Recorde also was the first to use the plus and minus symbols in an English work (Miller, 2004b). Using a horizontal line to express division derives from its use, in denoting fractions, by Arabic author Al-Ḥaṣṣâr in the twelfth century, though it was not consistently employed for several more centuries (Cajori, ibid. I: 269, 310). The slash mark ("/"; also known as a solidus, virgule, or diagonal) was recommended to denote division by the English logician and mathematician Augustus De Morgan (1806–1871) in 1845 (ibid. I: 312–313), and the India-born Swiss author Johann Heinhirch Rahn (1622–1676) proposed, in 1659, denoting division by the symbol "÷", which previously was often used by authors as a minus sign (ibid.: 211, 270; Gullberg, 1997: 105). Many other symbols were used for mathematical operations, before and after these introductions (e.g., Cajori, ibid.: 229–245).

If, as in Example 1, a sample contains multiple identical data for several values of the variable, then it may be convenient to record the data in the form of a frequency table, as in Example 2. Then X_i can be said to denote each of k different measurements and f_i can denote the frequency with which that X_i occurs in the sample. The sample mean may then be calculated, using the sums of the products of f_i and X_i, as*

$$\overline{X} = \frac{\sum\limits_{i=1}^{k} f_i X_i}{n}. \tag{3}$$

Example 2 demonstrates this calculation for the same data as in Example 1.

EXAMPLE 2	**The Data from Example 1 Recorded as a Frequency Table**		

X_i (cm)	f_i	$f_i X_i$ (cm)	
3.3	1	3.3	$k = 13$
3.4	0	0	
3.5	1	3.5	$\sum\limits_{i=1}^{k} f_i = n = 24$
3.6	2	7.2	
3.7	1	3.7	
3.8	3	11.4	
3.9	3	11.7	$\overline{X} = \dfrac{\sum\limits_{i=1}^{k} f_i X_i}{n} = \dfrac{95.0 \text{ cm}}{24} = 3.96 \text{ cm}$
4.0	4	16.0	
4.1	3	12.3	$\text{median} = 3.95 \text{ cm} + \left(\dfrac{1}{4}\right)(0.1 \text{ cm})$
4.2	2	8.4	
4.3	2	8.6	$= 3.95 \text{ cm} + 0.025 \text{ cm}$
4.4	1	4.4	$= 3.975 \text{ cm}$
4.5	1	4.5	
	$\sum f_i = 24$	$\sum f_i X_i = 95.0 \text{ cm}$	

A similar procedure is computing what is called a *weighted mean*, an expression of the average of several means. For example, we may wish to combine the mean of 3.96 cm from the sample of 24 measurements in Example 1 with a mean of 3.78 cm from a sample of 30 measurements and a mean of 4.02 cm from a sample of 15. These three means would be from a total of $24 + 30 + 15 = 69$ data; and if we had all 69 of the data we could sum them and divide the sum by 69 to obtain the overall mean length. However, that overall mean can be obtained without knowing the 69

---

*Denoting the multiplication of two quantities (e.g., a and b) by their adjacent placement (i.e., ab) derives from practices in Hindu manuscripts of the seventh century (Cajori, 1928/9, Vol. I: 77, 250). Modern multiplication symbols include a raised dot (as in $a \cdot b$), which was suggested in a 1631 posthumous publication of Thomas Harriot (1560?–1621) and prominently adopted in 1698 by the outstanding mathematician Gottfried Wilhelm Leibniz (1646–1716, in what is now Germany); the St. Andrew's cross (as in $a \times b$), which was used in 1631 by English mathematician William Oughtred (1574–1660) though it was not in general use until more than 200 years later; and the letter X, which was used, perhaps by Oughtred, as early as 1618 (Cajori, ibid.: 251; Gullberg, 1997: 104; Miller 2004b). Johann Rahn's 1659 use of an asterisk-like symbol (as in $a * b$) (Cajori, ibid: 212–213) did not persist but resurfaced in electronic computer languages of the latter half of the twentieth century.

individual measurements, by employing Equation 3 with $f_1 = 24$, $X_1 = 3.96$ cm, $f_2 = 30$, $X_2 = 3.78$ cm, $f_3 = 15$, $X_3 = 4.02$ cm, and $n = 69$. This would yield a weighted mean of $\overline{X} = [(24)(3.96 \text{ cm}) + (30)(3.78 \text{ cm}) + (15)(4.02 \text{ cm})]/69 = (268.74 \text{ cm})/69 = 3.89$ cm.

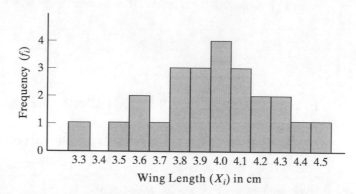

FIGURE 1: A histogram of the data in Example 2. The mean (3.96 cm) is the center of gravity of the histogram, and the median (3.975 cm) divides the histogram into two equal areas.

If data are plotted as a histogram (Figure 1), the mean is the *center of gravity* of the histogram.* That is, if the histogram were made of a solid material, it would balance horizontally with the fulcrum at $\overline{X}$. The mean is applicable to both ratio- and interval-scale data; it should not be used for ordinal data and cannot be used for nominal data.

2 THE MEDIAN

The median is typically defined as the middle measurement in an ordered set of data.[†] That is, there are just as many observations larger than the median as there are smaller. The sample median is the best estimate of the population median. In a symmetrical distribution (such as Figures 2a and 2b) the sample median is also an unbiased and consistent estimate of μ, but it is not as efficient a statistic as $\overline{X}$ and should not be used as a substitute for $\overline{X}$. If the frequency distribution is asymmetrical, the median is a poor estimate of the mean.

The median of a sample of data may be found by first arranging the measurements in order of magnitude. The order may be either ascending or descending, but ascending order is most commonly used as is done with the samples in Examples 1, 2, and 3. Then, we define the sample median as

$$\text{sample median} = X_{(n+1)/2}. \tag{4}$$

---

*The concept of the mean as the center of gravity was used by L. A. J. Quetelet in 1846 (Walker, 1929: 73).

[†]The concept of the median was conceived as early as 1816, by K. F. Gauss; enunciated and reinforced by others, including F. Galton in 1869 and 1874; and independently discovered and promoted by G. T. Fechner beginning in 1874 (Walker, 1929: 83–88, 184). It received its name, in English, from F. Galton in 1882 (David, 1995) and, in French, from A. A. Cournot in 1843 (David, 1998a).

Measures of Central Tendency

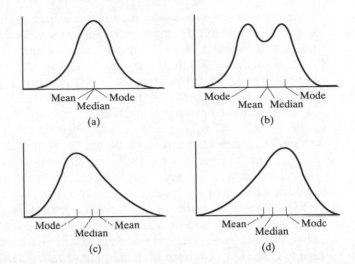

FIGURE 2: Frequency distributions showing measures of central tendency. Values of the variable are along the abscissa (horizontal axis), and the frequencies are along the ordinate (vertical axis). Distributions (a) and (b) are symmetrical, (c) is asymmetrical and said to be positively skewed, and (d) is asymmetrical and said to be negatively skewed. Distributions (a), (c), and (d) are unimodal, and distribution b is bimodal. In a unimodal asymmetric distribution, the median lies about one-third the distance between the mean and the mode.

EXAMPLE 3   Life Span for Two Species of Birds in Captivity

The data for each species are arranged in order of magnitude

X_i (mo)	X_i (mo)
16	34
32	36
37	38
39	45
40	50
41	54
42	56
50	59
82	69
	91

$n = 9$	$n = 10$
median $= X_{(n+1)/2} = X_{(9+1)/2}$	median $= X_{(n+1)/2} = X_{(10+1)/2}$
$= X_5 = 40$ mo	$= X_{5.5} = 52$ mo
$\overline{X} = 42.11$ mo	$\overline{X} = 53.20$ mo

If the sample size (n) is odd, then the subscript in Equation 4 will be an integer and will indicate which datum is the middle measurement in the ordered sample. For the data of species A in Example 3, $n = 9$ and the sample median is $X_{(n+1)/2} = X_{(9+1)/2} = X_5 = 40$ mo. If n is even, then the subscript in Equation 4 will be a number midway between two integers. This indicates that there is not a middle value in the ordered list of data; instead, there are two middle values, and the median is defined as the midpoint between them. For the species B data in Example 3, $n = 10$ and $X(n + 1)/2 = X_{(10+1)/2} = X_{5.5}$, which signifies that the median is midway between X_5 and X_6, namely a median of $(50 \text{ mo} + 54 \text{ mo})/2 = 52$ mo.

Note that the median has the same units as each individual measurement. If data are plotted as a frequency histogram (e.g., Figure 1), the median is the value of X that divides the area of the histogram into two equal parts. In general, the sample median is a more efficient estimate of the population median when the sample size is large.

If we find the middle value(s) in an ordered set of data to be among identical observations (referred to as *tied* values), as in Example 1 or 2, a difficulty arises. If we apply Equation 4 to these 24 data, then we conclude the median to be $X_{12.5} = 4.0$ cm. But four data are tied at 4.0 cm, and eleven measurements are less than 4.0 cm and nine are greater. Thus, 4.0 cm does not fit the definition above of the median as that value for which there is the same number of data larger and smaller. Therefore, a better definition of the median of a set of data is that value for which no more than half the data are smaller and no more than half are larger.

When the sample median falls among tied observations, we may interpolate to better estimate the population median. Using the data of Example 2, we desire to estimate a value below which 50% of the observations in the population lie. Fifty percent of the observations in the sample would be 12 observations. As the first 7 classes in the frequency table include 11 observations and 4 observations are in class 4.0 cm, we know that the desired sample median lies within the range of 3.95 to 4.05 cm. Assuming that the four observations in class 4.0 cm are distributed evenly within the 0.1-cm range of 3.95 to 4.05 cm, then the median will be $\left(\frac{1}{4}\right)$ (0.1 cm) = 0.025 cm into this class. Thus, the median = 3.95 cm + 0.025 cm = 3.975 cm. In general, for the sample median within a class interval containing tied observations,

$$\text{median} = \binom{\text{lower limit}}{\text{of interval}} + \left(\frac{0.5n - \text{cum. freq.}}{\text{no. of observations in interval}}\right)\binom{\text{interval}}{\text{size}}, \quad (5)$$

where "cum. freq." refers to the cumulative frequency of the previous classes.* By using this procedure, the calculated median will be the value of X that divides the area of the histogram of the sample into two equal parts.

The median expresses less information than does the mean, for it does not take into account the actual value of each measurement, but only considers the rank of each measurement. Still, it offers advantages in some situations. For example, extremely high or extremely low measurements ("outliers") do not affect the median as much as they affect the mean (causing the sample median to be called a "resistant" statistic). Distributions that are not symmetrical around the mean (such as in Figures 2c and 2d) are said to be *skewed*.† When we deal with skewed

---

*This procedure was enunciated in 1878 by the German psychologist Gustav Theodor Fechner (1801–1887) (Walker, 1929: 86).

†This term, applied to a distribution and to a curve, was used as early as 1895 by Karl Pearson (Miller, 2004a).

populations and do not want the strong influence of outliers, we may prefer the median to the mean to express central tendency.

Note that in Example 3 the researcher would have to wait 82 months to compute a mean life expectancy for species *A* and 91 months for species *B*, whereas the median for species *A* could be determined in only 40 months and in only 52 months for species *B*. Also, to calculate a median one does not need to have accurate data for all members of the sample. If, for example, we did not have the first three data for species *A* accurately recorded, but could state them as "less than 39 months," then the median could have been determined just as readily as if we had all 9 data fully recorded, while calculation of the mean would not have been possible.

The expression "LD fifty" (LD_{50}), used in some areas of biological research, is simply the median lethal dose (and is so named because the median is the 50th percentile).

The median can be determined not only for interval-scale and ratio-scale data, but also for data on an ordinal scale, data for which the use of the mean usually would not be considered appropriate. But neither the median nor the mean is applicable to nominal data.

3 THE MODE

The *mode* is commonly defined as the most frequently occurring measurement in a set of data.* In Example 2, the mode is 4.0 cm. But it is perhaps better to define a mode as a measurement of relatively great concentration, for some frequency distributions may have more than one such point of concentration, even though these concentrations might not contain precisely the same frequencies. Thus, a sample consisting of the data 6, 7, 7, 8, 8, 8, 8, 8, 8, 9, 9, 10, 11, 12, 12, 12, 12, 12, 13, 13, and 14 mm would be said to have two modes: at 8 mm and 12 mm. (Some authors would refer to 8 mm as the "major mode" and call 12 mm the "minor mode.") A distribution in which each different measurement occurs with equal frequency is said to have no mode. If two consecutive values of *X* have frequencies great enough to declare the *X* values modes, the mode of the distribution may be said to be the midpoint of these two *X*'s; for example, the mode of 3, 5, 7, 7, 7, 8, 8, 8, and 10 liters is 7.5 liters. A distribution with two modes is said to be *bimodal* (e.g., Figure 2b) and may indicate a combination of two distributions with different modes (e.g., heights of men and women). Modes are often discerned from histograms or frequency polygons; but we should be aware that the shape of such graphs and therefore the appearance of modes, may be influenced by the measurement intervals on the horizontal axis.

The sample mode is the best estimate of the population mode. When we sample a symmetrical unimodal population, the mode is an unbiased and consistent estimate of the mean and median (Figure 2a), but it is relatively inefficient and should not be so used. As a measure of central tendency, the mode is affected by skewness less than is the mean or the median, but it is more affected by sampling and grouping than these other two measures. The mode, but neither the median nor the mean, may be used for data on the nominal, as well as the ordinal, interval, and ratio scales of measurement. In a unimodal asymmetric distribution (Figures 2c and 2d), the median lies about one-third the distance between the mean and the mode.

The mode is not often used in biological research, although it is often interesting to report the number of modes detected in a population, if there are more than one.

---

*The term *mode* was introduced by Karl Pearson in 1895 (David, 1995).

4 OTHER MEASURES OF CENTRAL TENDENCY

(a) The Geometric Mean. The *geometric mean* is the nth root of the product of the n data:

$$\overline{X}_G = \sqrt[n]{X_1 X_2 \ _3 \dots X_n} = \sqrt[n]{\prod_{i=1}^{n} X_i} .$$ (6)

Capital Greek pi, Π, means "take the product"[*] in an analogous fashion as Σ indicates "take the sum." The geometric mean may also be calculated as the antilogarithm of the arithmetic mean of the logarithms of the data (where the logarithms may be in any base); this is often more feasible computationally:

$$\overline{X}_G = \text{antilog}\left(\frac{\log X_1 + \log X_2 + \cdots + \log X_n}{n}\right) = \text{antilog}\frac{\sum_{i=1}^{n} \log X_i}{n}.$$ (7)

The geometric mean is appropriate to use only for ratio-scale data and only when all of the data are positive (that is, greater than zero). If the data are all equal, then the geometric mean, $\overline{X}_G$, is equal to the arithmetic mean, $\overline{X}$ (and also equal to the harmonic mean described below); if the data are not all equal, then[†] $\overline{X}_G < \overline{X}$.

$\overline{X}_G$ is sometimes used as a measure of location when the data are highly skewed to the right (i.e., when there are many more data larger than the arithmetic mean than there are data smaller than the arithmetic mean).

$\overline{X}_G$ is also useful when dealing with data that represent ratios of change. As an illustration of this, Example 4 considers changes in the size of a population of organisms over four decades. Each of the original data (population size at the end of a decade) is expressed as a ratio, X_i, of the population size to the population size of the previous decade. The geometric mean of those ratios is computed and may be thought of as representing the average rate of growth per decade (which is the same as a constant rate of compound interest). This example demonstrates that the arithmetic mean of those ratios is $\overline{X} = 1.1650$ (i.e., 16.50% growth) per decade. But over the four decades of population change, this mean would have us calculate a final population size of $(10,000)(1.1650)(1.1650)(1.1650)(1.1650) = 18,421$, which is *not* the population size recorded at the end of the fourth decade. However, using the geometric mean, $\overline{X}_G$, to indicate the average rate of growth, the final population size would be computed to be $(10,000)(1.608)(1.608)(1.608)(1.608) = 18,156$, which is the fourth-decade population size that was observed.

---

[*]Use of this symbol to indicate taking the product was introduced by René Descartes (Gullberg, 1997: 105).

[†]The symbols "$<$" and "$>$" (meaning "less than" and "greater than") were inserted by someone else into a 1631 posthumous publication by the English mathematician and astronomer Thomas Harriot (1560?–1621), (Cajori, 1928/9, Vol. I: 199; Gullberg, 1997: 109; Miller, 2004b). The symbols for "less than or equal to" ($\leqq$) and "greater than or equal to" ($\geqq$) were written as $\leqq$ and $\geqq$ when introduced by the French scientist Pierre Bouguere (1698–1758) in 1734. (Gullberg, 1997: 109).

Measures of Central Tendency

EXAMPLE 4   The Geometric Mean of Ratios of Change

Decade	Population Size	Ratio of Change X_i
0	10,000	
1	10,500	$\dfrac{10,500}{10,000} = 1.05$
2	11,550	$\dfrac{11,550}{10,500} = 1.10$
3	13,860	$\dfrac{13,860}{11,550} = 1.20$
4	18,156	$\dfrac{18,156}{13,860} = 1.31$

$$\overline{X} = \frac{1.05 + 1.10 + 1.20 + 1.31}{4} = \frac{4.66}{4} = 1.1650$$

and $(10,000)(0.1650)(1.650)(1.650)(1.650) = 18,421$

But,

$$\overline{X}_G = \sqrt[4]{(1.05)(1.10)(1.20)(1.31)} = \sqrt[4]{1.8157} = 1.1608$$

or

$$\overline{X}_G = \text{antilog}\left[\frac{\log(1.05) + \log(1.10) + \log(1.20) + \log(1.31)}{4}\right]$$

$$= \frac{\text{antilog}(0.0212 + 0.0414 + 0.0792 + 0.1173)}{4} = \frac{\text{antilog}(0.2591)}{4}$$

$$= \text{antilog } 0.0648 = 1.1608$$

and $(10,000)(1.1608)(1.1608)(1.1608)(1.1608) = 18,156$

(b) The Harmonic Mean. The *harmonic mean* is the reciprocal of the arithmetic mean of the reciprocals of the data:

$$\overline{X}_H = \frac{1}{\dfrac{1}{n}\sum\dfrac{1}{X_i}} = \frac{n}{\sum\dfrac{1}{X_i}}. \tag{8}$$

It may be used for ratio-scale data when no datum is zero. If all of the data are identical, then the harmonic mean, $\overline{X}_H$, is equal to the arithmetic mean, $\overline{X}$ (and equal to the geometric mean, $\overline{X}_G$). If the data are all positive and not identical, then $\overline{X}_H < \overline{X}_G < \overline{X}$.

$\overline{X}_H$ finds use when desiring an average of rates, as described by Croxton, Cowden, and Klein (1967: 182–188). For example, consider that a flock of birds flies from a roosting area to a feeding area 20 km away, flying at a speed of 40 km/hr (which

31

takes 0.5 hr). The flock returns to the roosting area along the same route (20 km), flying at 20 km/hr (requiring 1 hr of flying time). To ask what the average flying speed was, we might employ Equation 2 and calculate the arithmetic mean as $\overline{X} = $ (40 km/hr + 20 km/hr)/2 = 30 km/hr. However, this answer may not be satisfying, because a total of 40 km was traveled in 1.5 hr, indicating a speed of (40 km)/(1.5 hr) = 26. $\overline{X}_G$) is 26.7 km/hr.

EXAMPLE 5   The Harmonic Mean of Rates

$$X_1 = 40 \text{ km/hr}, \ X_2 = 20 \text{ km/hr}$$

$$\overline{X} = \frac{40 \text{ km/hr} + 20 \text{ km/hr}}{2} = \frac{60 \text{ km/hr}}{2} = 30 \text{ km/hr}$$

But

$$\overline{X}_H = \frac{2}{\dfrac{1}{40 \text{ km/hr}} + \dfrac{1}{20 \text{ km/hr}}} = \frac{2}{0.0250 \text{ hr/km} + 0.0500 \text{ hr/km}}$$

$$= \frac{2}{0.075 \text{ hr/km}} = 26.67 \text{ km/hr}$$

(c) The Range Midpoint. The *range midpoint*, or *midrange*, is a measure of location defined as the point halfway between the minimum and the maximum values in the set of data. It may be used with data measured on the ratio, interval, or ordinal scale; but it is not generally a good estimate of location, for it utilizes relatively little information from the data. (However, the so-called mean daily temperature is often reported as the mean of the minimum and maximum and is, therefore, a range midpoint.)

The midpoint of any two symmetrically located percentiles, such as the point midway between the first and third quartiles (i.e., the 25th and 75th percentiles), may be used as a location measure in the same fashion as the range midpoint is used (see Dixon and Massey, 1969: 133–134). Such measures are not as adversely affected by aberrantly extreme values as is the range midpoint, and they may be applied to ratio or interval data. If used with ordinal data, they (and the range midpoint) would be the same as the median.

5 CODING DATA

Often in the manipulation of data, considerable time and effort can be saved if *coding* is employed. Coding is the conversion of the original measurements into easier-to-work-with values by simple arithmetic operations. Generally coding employs a *linear transformation* of the data, such as multiplying (or dividing) or adding (or subtracting) a constant. The addition or subtraction of a constant is sometimes termed a translation of the data (i.e., changing the origin), whereas the multiplication or division by a constant causes an expansion or contraction of the scale of measurement.

EXAMPLE 6   **Coding Data to Facilitate Calculations**

Sample 1 (Coding by Subtraction: $A = -840$ g)		Sample 2 (Coding by Division: $M = 0\ 001$ liters/ml)	
X_i (g)	coded $X_i = X_i - 840$ g	X_i (ml)	coded $X_i = (X_i)(0.001$ liters/ml$)$ $= X_i$ liters
842	2	8,000	8.000
844	4	9,000	9.000
846	6	9,500	9.500
846	6	11,000	11.000
847	7	12,500	12.500
848	8	13,000	13.000
849	9		

$$\sum X_i = 5922\text{ g} \quad \text{coded} \sum X_i = 42\text{ g} \qquad \sum X_i = 63,000\text{ ml} \quad \text{coded} \sum X_i$$

$$\overline{X} = \frac{5922\text{ g}}{7} \qquad \text{coded } \overline{X} = \frac{42\text{ g}}{7} \qquad\qquad = 63.000 \text{ liters}$$

$$= 846\text{ g} \qquad\qquad = 6\text{ g} \qquad \overline{X} = 10,500\text{ ml} \quad \text{coded } \overline{X}$$

$$= 10.500 \text{ liters}$$

$$\overline{X} = \text{coded } \overline{X} - A \qquad\qquad \overline{X} = \text{coded } \frac{\overline{X}}{M}$$

$$= 6\text{ g} - (-840\text{ g}) \qquad\qquad = \frac{10.500 \text{ liters}}{0.001 \text{ liters/ml}}$$

$$= 846\text{ g} \qquad\qquad\qquad = 10,500 \text{ ml}$$

The first set of data in Example 6 are coded by subtracting a constant value of 840 g. Not only is each coded value equal to $X_i - 840$ g, but the mean of the coded values is equal to $\overline{X} - 840$ g. Thus, the easier-to-work-with coded values may be used to calculate a mean that then is readily converted to the mean of the original data, simply by adding back the coding constant.

In Sample 2 of Example 6, the observed data are coded by dividing each observation by 1000 (i.e., by multiplying by 0.001).* The resultant mean only needs to be multiplied by the coding factor of 1000 (i.e., divided by 0.001) to arrive at the mean of the original data. As the other measures of central tendency have the same units as the mean, they are affected by coding in exactly the same fashion.

Coding affects the median and mode in the same way as the mean is affected. The widespread use of computers has greatly diminished the need for researchers to utilize coding (although computer software may use it).

---

*In 1593, mathematician Christopher Clavius (1538–1612, born in what is now Germany but spent most of his life in what is now Italy; also credited with proposing the currently used Gregorian calendar rules regarding leap years; O'Connor and Robertson, 1996) became the first to use a decimal point to separate units from tenths; in 1617, the Scottish mathematician John Napier (1550–1617) used both points and commas for this purpose (Cajori, 1928/9. Vol. I: 322–323), and the comma is still so used in some parts of the world. In some countries a raised dot has been used—a symbol Americans sometimes employ to denote multiplication.

EXERCISES

1. If $X_1 = 3.1$ kg, $X_2 = 3.4$ kg, $X_3 = 3.6$ kg, $X_4 = 7$ kg, and $X_5 = 4.0$ kg, calculate the value of

(a) $\displaystyle\sum_{i=1}^{4} X_i,$

(b) $\displaystyle\sum_{i=2}^{4} X_i,$

(c) $\displaystyle\sum_{i=1}^{5} X_i,$

(d) $\displaystyle\sum X_i.$

2. (a) Calculate the mean of the five weights in Exercise 1.

(b) Calculate the median of those weights.

3. The ages, in years, of the faculty members of a university biology department are 32.2, 37.5, 41.7, 53.8, 50.2, 48.2, 46.3, 65.0, and 44.8.

(a) Calculate the mean age of these nine faculty members.

(b) Calculate the median of the ages.

(c) If the person 65.0 years of age retires and is replaced on the faculty with a person 46.5 years old, what is the new mean age?

(d) What is the new median age?

4. Consider the following frequency tabulation of leaf weights (in grams):

X_i	f_i
1.85–1.95	2
1.95–2.05	1
2.05–2.15	2
2.15–2.25	3
2.25–2.35	5
2.35–2.45	6
2.45–2.55	4
2.55–2.65	3
2.65–2.75	1

Using the midpoints of the indicated ranges of X_i,
(a) Calculate the mean leaf weight using Equation 2, and
(b) Calculate the mean leaf weight using Equation 3.
(c) Calculate the median leaf weight using Equation 4, and
(d) Calculate the median using Equation 5.
(e) Determine the mode of the frequency distribution.

5. A fruit was collected from each of eight lemon trees, with the intent of measuring the calcium concentration in the rind (grams of calcium per 100 grams of dry rind). The analytical method used could only detect a concentration of at least 0.80 g/100 g of dry weight. Six of the eight concentrations were measured to be 1.02, 0.98, 0.91, 0.84, 0.87, 1.04 g/100 g of dry weight, and two of the concentrations were known to be less than 0.80 g/100 g of dry weight. What is the median of this sample of eight data?

ANSWERS TO EXERCISES

1. (a) 13.8 kg; **(b)** 10.7 kg; **(c)** 17.8 kg; **(d)** 17.8 kg.

2. (a) 3.56 kg; **(b)** 3.6 kg.

3. (a) 46.63 yr; **(b)** 46.3 yr; **(c)** 44.58 yr; **(d)** 46.3 yr.

4. (a) 2.33 g; **(b)** 2.33 g; **(c)** 2.4 g; **(d)** 2.358 g; **(e)** 2.4 g.

5. 0.89 g/100 g.

Measures of Variability and Dispersion

In addition to a description of the central tendency of a set of data, it is generally desirable to have a description of the *variability*, or of the *dispersion*,* of the data. A measure of variability (or measure of dispersion, as it is often called) is an indication of the spread of measurements around the center of the distribution. Measurements that are concentrated around the center of a distribution of data have low variability (low dispersion), whereas data that are very spread out along the measurement scale have high variability (high dispersion). Measures of variability of a population are population parameters, and sample measures of variability are statistics that estimate those parameters.

1 THE RANGE

The difference between the highest and lowest measurements in a group of data is termed the *range*.[†] If sample measurements are arranged in increasing order of magnitude, as if the median were about to be determined, then

$$\text{sample range} = X_n - X_1,\qquad(1)$$

which is

$$\text{sample range} = \text{largest } X - \text{smallest } X.$$

Sample 1 in Example 1 is a hypothetical set of ordered data in which $X_1 = 1.2$ g and $X_n = 2.4$ g. Thus, the range may be expressed as 1.2 to 2.4 g, or as 2.4 g $-$ 1.2 g $=$ 1.2 g. Note that the range has the same units as the individual measurements. Sample 2 in Example 1 has the same range as Sample 1.

---

*The statistical use of this term first appeared in an 1876 publication by Francis Galton (David, 1998a).

   [†]This statistical term dates from an 1848 paper by H. Lloyd (David, 1995). It was already used by the Greek astronomer Hipparchus as a measure of dispersion in the second century B.C.E. (David, 1998b).

EXAMPLE 1   Calculation of Measures of Dispersion for Two Hypothetical Samples of 7 Insect Body Weights

Sample 1

X_i (g)	$X_i - \overline{X}$ (g)	$\lvert X_i - \overline{X}\rvert$ (g)	$(X_i - \overline{X})^2$ (g²)
1.2	−0.6	0.6	0.36
1.4	−0.4	0.4	0.16
1.6	−0.2	0.2	0.04
1.8	0.0	0.0	0.00
2.0	0.2	0.2	0.04
2.2	0.4	0.4	0.16
2.4	0.6	0.6	0.36

$$\sum X_i = 12.6\text{ g} \qquad \sum(X_i - \overline{X}) = 0.0\text{ g} \qquad \sum \lvert X_i - \overline{X}\rvert = 2.4\text{ g} \qquad \sum(X_i - \overline{X})^2 = 1.12\text{ g}^2$$

= sum of squared deviations
from the mean
= "sum of squares"

$$n = 7; \quad \overline{X} = \frac{\sum X_i}{n} = \frac{12.6\text{ g}}{7} = 1.8\text{ g}$$

$$\text{range} = X_7 - X_1 = 2.4\text{ g} - 1.2\text{ g} = 1.2\text{ g}$$

$$\text{interquartile range} = Q_3 - Q_1 = 2.2\text{ g} - 1.4\text{ g} = 0.8\text{ g}$$

$$\text{mean deviation} = \frac{\sum \lvert X_i - \overline{X}\rvert}{n} = \frac{2.4\text{ g}}{7} = 0.34\text{ g}$$

$$\text{variance} = s^2 = \frac{\sum(X_i - \overline{X})^2}{n - 1} = \frac{1.12\text{ g}^2}{6} = 0.1867\text{ g}^2$$

$$\text{standard deviation} = s = \sqrt{0.1867\text{ g}^2} = 0.43\text{ g}$$

Sample 2

X_i (g)	$X_i - \overline{X}$ (g)	$\lvert X_i - \overline{X}\rvert$ (g)	$(X_i - \overline{X})^2$ (g²)
1.2	−0.6	0.6	0.36
1.6	−0.2	0.2	0.04
1.7	−0.1	0.1	0.01
1.8	0.0	0.0	0.00
1.9	0.1	0.1	0.01
2.0	0.2	0.2	0.04
2.4	0.6	0.6	0.36

$$\sum X_i = 12.6\text{ g} \qquad \sum(X_i - \overline{X}) = 0.0\text{ g} \qquad \sum \lvert X_i - \overline{X}\rvert = 1.8\text{ g} \qquad \sum(X_i - \overline{X})^2 = 0.82\text{ g}^2$$

= sum of squared deviations
from the mean
= "sum of squares"

$$n = 7; \quad \overline{X} = \frac{\sum X_i}{n} = \frac{12.6\text{ g}}{7} = 1.8\text{ g}$$

$$\text{range} = X_7 - X_1 = 2.4\text{ g} - 1.2\text{ g} = 1.2\text{ g}$$

$$\text{interquartile range } = Q_3 - Q_1 = 2.0\,\text{g} - 1.6\,\text{g} = 0.4\,\text{g}$$

$$\text{mean deviation } = \frac{\sum |X_i - \overline{X}|}{n} = \frac{1.8\,\text{g}}{7} = 0.26\,\text{g}$$

$$\text{variance } = s^2 = \frac{\sum (X_i - \overline{X})^2}{n - 1} = \frac{0.82\,\text{g}^2}{6} = 0.1367\,\text{g}^2$$

$$\text{standard deviation } = s = \sqrt{0.1367\,\text{g}^2} = 0.37\,\text{g}$$

The range is a relatively crude measure of dispersion, inasmuch as it does not take into account any measurements except the highest and the lowest. Furthermore, it is unlikely that a sample will contain both the highest and lowest values in the population, so the sample range usually underestimates the population range; therefore, it is a biased and inefficient estimator. Nonetheless, it is considered useful by some to present the sample range as an estimate (although a poor one) of the population range. For example, taxonomists are often concerned with having an estimate of what the highest and lowest values in a population are expected to be. Whenever the range is specified in reporting data, however, it is usually a good practice to report another measure of dispersion as well. The range is applicable to ordinal-, interval-, and ratio-scale data.

2 DISPERSION MEASURED WITH QUANTILES

Because the sample range is a biased and inefficient estimate of the population range, being sensitive to extremely large and small measurements, alternative measures of dispersion may be desired. Just as the median is the value above and below which lies half the set of data, one can define measures, called *quantiles*, above or below which lie other fractional portions of the data.

For example, if the data are divided into four equal parts, we speak of *quartiles*. One-fourth of all the ranked observations are smaller than the first quartile, one-fourth lie between the first and second quartiles, one-fourth lie between the second and third quartiles, and one-fourth are larger than the third quartile. The second quartile is identical to the median. As with the median, the first and third quartiles might be one of the data or the midpoint between two of the data. The first quartile, Q_1, is

$$Q_1 = X_{(n+1)/4}; \tag{2}$$

if the subscript, $(n + 1)/4$, is not an integer or half-integer, then it is rounded up to the nearest integer or half-integer. The second quartile is the median, and the subscript on X for the third quartile, Q_3, is

$$n + 1 - (\text{subscript on } X \text{ for } Q_1, \text{ after any rounding}). \tag{3}$$

For species A, $n = 9$, $(n + 1)/4 = 2.5$, and $Q_1 = X_{2.5} = 34.5$ mo; and $Q_3 = X_{10-2.5} = X_{7.5} = 46$ mo. For species B, $n = 10$, $(n + 1)/4 = 2.75$ (which we round up to 3), and $Q_1 = X_3 = 38$ mo, and $Q_3 = X_{11-3} = X_8 = 59$ mo.

The distance between Q_1 and Q_3, the first and third quartiles (i.e., the 25th and 75th percentiles), is known as the *interquartile range* (or *semiquartile range*):

$$\text{interquartile range } = Q_3 - Q_1. \tag{4}$$

One may also encounter the *semi-interquartile range*:

$$\text{semi-interquartile range} = \frac{Q_3 - Q_1}{2}, \tag{5}$$

also known as the *quartile deviation*.*

If the distribution of data is symmetrical, then 50% of the measurements lie within one quartile deviation above and below the median. For Sample 1 in Example 1, $Q_1 = 1.4$ g, $Q_3 = 2.2$ g, and the interquartile range is 2.2 g $-$ 1.4 g $= 0.8$ g. And for Sample 2, $Q_1 = 1.6$ g, $Q_3 = 2.0$ g, and the interquartile range is 2.0 g $-$ 1.6 g $= 0.4$ g.

Similarly, values that partition the ordered data set into eight equal parts (or as equal as n will allow) are called *octiles*. The first octile, $\mathscr{O}_1$, is

$$\mathscr{O}_1 = X_{(n+1)/8}; \tag{6}$$

and if the subscript, $(n + 1)/8$, is not an integer or half-integer, then it is rounded up to the nearest integer or half-integer. The second, fourth, and sixth octiles are the same as quartiles; that is, $\mathscr{O}_2 = Q_1$, $\mathscr{O}_4 = Q_2 =$ median and $\mathscr{O}_6 = Q_3$. The subscript on X for the third octile, $\mathscr{O}_3$, is

$$2(\text{subscript on } X \text{ for } Q_1) - \text{subscript on } X \text{ for } \mathscr{O}_1; \tag{7}$$

the subscript on X for the fifth octile, $\mathscr{O}_5$, is

$$n + 1 - \text{subscript on } X \text{ for } \mathscr{O}_3; \tag{8}$$

and the subscript on X for the seventh octile, $\mathscr{O}_7$, is

$$n + 1 - \text{subscript on } X \text{ for } \mathscr{O}_1. \tag{9}$$

For species A, $n = 9$, $(n + 1)/8 = 1.5$ and $\mathscr{O}_1 = X_{1.5} = 35$ mo; $2(2.5) - 1.5 = 3.5$, so $\mathscr{O}_3 = X_{3.5} = 38$ mo; $n + 1 - 3.5 = 6.5$, so $\mathscr{O}_5 = X_{6.5} = 41.5$ mo; and $n + 1 - 1.5 = 8.5$, so $\mathscr{O}_7 = 61$. For species B, $n = 10$, $(n + 1)/8 = 1.25$ (which we round up to 1.5) and $\mathscr{O}_1 = X_{1.5} = 35$ mo; $2(3) - 1.5 = 4.5$, so $\mathscr{O}_3 = X_{4.5} = 39.5$ mo; $n + 1 - 4.5 = 6.5$, so $\mathscr{O}_5 = X_{6.5} = 41.5$ mo; and $n + 1 - 1.5 = 9.5$, so $\mathscr{O}_7 = 44.5$ mo.

Besides the median, quartiles, and octiles, ordered data may be divided into fifths, tenths, or hundredths by quantities that are respectively called *quintiles, deciles*, and *centiles* (the latter also called *percentiles*). Measures that divide a group of ordered data into equal parts are collectively termed *quantiles*.[†] The expression "LD$_{50}$," used in some areas of biological research, is simply the 50th percentile of the lethal doses, or the median lethal dose. That is, 50% of the experimental subjects survived this dose, whereas 50% did not. Likewise, "LC$_{50}$" is the median lethal concentration, or the 50th percentile of the lethal concentrations.

Instead of distance between the 25th and 75th percentiles, distances between other quantiles (e.g., 10th and 90th percentiles) may be used as a dispersion measure. Quantile-based measures of dispersion are valid for ordinal-, interval-, or ratio-scale data, and they do not exhibit the bias and inefficiency of the range.

---

*This measure was proposed in 1846 by L. A. J. Quetelet (1796–1874); Sir Francis Galton (1822–1911) later called it the "quartile deviation" (Walker, 1929: 84) and, in 1882, used the terms "quartile" and "interquartile range" (David, 1995).

[†]Sir Francis Galton developed the concept of percentiles, quartiles, deciles, and other quantiles in writings from 1869 to 1885 (Walker, 1929: 86–87, 177, 179). The term *quantile* was introduced in 1940 by M. G. Kendall (David, 1995).

3 THE MEAN DEVIATION

As is evident from the two samples in Example 1, the range conveys no information about how clustered about the middle of the distribution the measurements are. As the mean is so useful a measure of central tendency, one might express dispersion in terms of deviations from the mean. The sum of all deviations from the mean, that is, $\sum(X_i - \overline{X})$, will always equal zero, however, so such a summation would be useless as a measure of dispersion (as seen in Example 1).

Using the absolute values of the deviations from the mean eliminates the negative signs of the deviations, and summing those absolute values results in a quantity that is an expression of dispersion about the mean. Dividing this quantity by n yields a measure known as the *mean deviation*, or *mean absolute deviation*,* of the sample; this measure has the same units as do the data. In Example 1, Sample 1 is more variable (or more dispersed, or less concentrated) than Sample 2. Although the two samples have the same range, the mean deviations, calculated as

$$\text{sample mean deviation} = \frac{\sum |X_i - \overline{X}|}{n}, \tag{10}$$

express the differences in dispersion.[†] A different kind of mean deviation can be defined by using the sum of the absolute deviations from the median instead of from the mean.

Mean deviations are seldom encountered, because their utility is far less than that of the statistics in Sections 4 and 5.

4 THE VARIANCE

Another method of eliminating the negative signs of deviations from the mean is to square the deviations. The sum of the squares of the deviations from the mean is often simply called the *sum of squares*, abbreviated SS, and is defined as follows:[‡]

$$\text{population SS} = \sum(X_i - \mu)^2 \tag{11}$$

$$\text{sample SS} = \sum(X_i - \overline{X})^2. \tag{12}$$

It can be seen from the above two equations that as a measure of variability, or dispersion, the sum of squares considers how far the X_i's deviate from the mean. In

---

*The term *mean deviation* is apparently due to Karl Pearson (1857–1936) (Walker, 1929: 55) and *mean absolute deviation*, in 1972, to D. F. Andrews, P. J. Bickel, F. R. Hampel, P. J. Huber, W. H. Rogers, and J. W. Tukey (David, 1995).

†Karl Weierstrass, in 1841, was the first to denote the absolute value of a quantity by enclosing it within two vertical lines (Cajori, 1928/9, Vol. II: p. 123); that is, $|a| = a$ and $|-a| = a$.

‡The modern notation using raised numerals as exponents was introduced by René Descartes in 1637, and many other kinds of notation for exponents were employed before and after that (Cajori, 1928/9, Vol. I: 358; Gullberg, 1997: 134). An 1845 notation of Augustus De Morgan, $a \wedge b$ to indicate a^b (Cajori, ibid.: 358), has reemerged in modern computer use. Nicolas Chuquet (1445–1488) was the first to use negative exponents, and Nicole (also known as Nicolaus) Oresme (1323–1382) was the first to use fractional exponents, though neither of these French mathematicians employed the modern notation of Isaac Newton (1642–1727), the colossal English mathematician, physicist, and astronomer (Cajori, ibid.: 91, 102, 354–355):

$$X^{-a} = \frac{1}{X^a}; \quad X^{\frac{1}{a}} = \sqrt[a]{X}.$$

Using parentheses or brackets to group quantities dates from the mid-sixteenth century, though it was not common mathematical notation until more than two centuries later (ibid.: 392).

Sample 1 of Example 1, the sample mean is 1.8 g and it is seen (in the last column) that

$$\text{Sample SS} = (1.2 - 1.8)^2 + (1.4 - 1.8)^2 + (1.6 - 1.8)^2 + (1.8 - 1.8)^2$$
$$+ (2.0 - 1.8)^2 + (2.2 - 1.8)^2 + (2.4 - 1.8)^2$$
$$= 0.36 + 0.16 + 0.04 + 0.00 + 0.04 + 0.16 + 0.36$$
$$= 1.12$$

(where the units are grams^2).* The sum of squares may also be visualized as a measure of the average extent to which the data deviate from each other, for (using the same seven data from Sample 1 in Example 1):

$$\text{SS} = [(1.2 - 1.4)^2 + (1.2 - 1.6)^2 + (1.2 - 1.8)^2 + (1.2 - 2.0)^2$$
$$+ (1.2 - 2.2)^2 + (1.2 - 2.4)^2 + (1.4 - 1.6)^2 + (1.4 - 1.8)^2$$
$$+ (1.4 - 2.0)^2 + (1.4 - 2.2)^2 + (1.4 - 2.4)^2 + (1.6 - 1.8)^2$$
$$+ (1.6 - 2.0)^2 + (1.6 - 2.2)^2 + (1.6 - 2.4)^2 + (1.8 - 2.0)^2$$
$$+ (1.8 - 2.2)^2 + (1.8 - 2.4)^2 + (2.0 - 2.2)^2 + (2.0 - 2.4)^2$$
$$+ (2.2 - 2.4)^2]/7$$
$$= [0.04 + 0.16 + 0.36 + 0.64 + 1.00 + 1.44 + 0.04 + \cdots + 0.04 + 0.16$$
$$+ 0.04]/7$$
$$= 7.84/7 = 1.12$$

(again in grams^2).

The mean sum of squares is called the *variance* (or *mean square*,[†] the latter being short for *mean squared deviation*), and for a population is denoted by σ^2 ("sigma squared," using the lowercase Greek letter):

$$\sigma^2 = \frac{\sum(X_i - \mu)^2}{N}. \tag{14}$$

The best estimate of the population variance, σ^2, is the sample variance, s^2:

$$s^2 = \frac{\sum(X_i - \overline{X})^2}{n - 1}. \tag{15}$$

If, in Equation 14, we replace μ by $\overline{X}$ and N by n, the result is a quantity that is a biased estimate of σ^2 in that it underestimates σ^2. Dividing the sample sum of squares

---

*Owing to an important concept in statistics, known as *least squares*, the sum of squared deviations from the mean is smaller than the sum of squared deviations from any other quantity (e.g., the median). Indeed, if Equation 12 is applied using some quantity in place of the mean, the resultant "sum of squares" would be

$$SS + nd^2, \tag{13}$$

where d is the difference between the mean and the quantity used. For the population sum of squares (defined in Equation 11), the relationship would be $SS + Nd^2$.

[†]The term *mean square* dates back at least to an 1875 publication of Sir George Biddel Airy (1801–1892), Astronomer Royal of England (Walker, 1929: 54). The term *variance* was introduced in 1918 by English statistician Sir Ronald Aylmer Fisher (1890–1962) (ibid.: 189; David, 1995).

by $n - 1$ (called the *degrees of freedom*,* often abbreviated DF), rather than by n, yields an unbiased estimate, and it is Equation 15 that should be used to calculate the sample variance.

If all observations in a sample are equal, then there is no variability (that is, no dispersion) and $s^2 = 0$. And s^2 becomes increasingly large as the amount of variability, or dispersion, increases. Because s^2 is a mean sum of squares, it can never be a negative quantity.

The variance expresses the same type of information as does the mean deviation, but it has certain very important mathematical properties relative to probability and hypothesis testing that make it superior. Thus, the mean deviation is very seldom encountered in biostatistical analysis.

The calculation of s^2 can be tedious for large samples, but it can be facilitated by the use of the equality

$$\text{sample SS} = \sum X_i^2 - \frac{\left(\sum X_i\right)^2}{n}. \tag{16}$$

This formula is equivalent to Equation 12 but is much simpler to work with. Example 2 demonstrates its use to obtain a sample sum of squares.

Because the sample variance equals the sample SS divided by DF,

$$s^2 = \frac{\sum X_i^2 - \dfrac{\left(\sum X_i\right)^2}{n}}{n - 1}. \tag{17}$$

This last formula is often referred to as a "working formula," or "machine formula," because of its computational advantages. There are, in fact, two major advantages in calculating SS by Equation 16 rather than by Equation 12. First, fewer computational steps are involved, a fact that decreases chance of error. On many calculators the summed quantities, $\sum X_i$ and $\sum X_i^2$, can both be obtained with only one pass through the data, whereas Equation 12 requires one pass through the data to calculate $\overline{X}$ and at least one more pass to calculate and sum the squares of the deviations, $X_i - \overline{X}$. Second, there may be a good deal of rounding error in calculating each $X_i - \overline{X}$, a situation that leads to decreased accuracy in computation, but that is avoided by the use of Equation 16.[†]

For data recorded in frequency tables,

$$\text{sample SS} = \sum f_i X_i^2 - \frac{\left(\sum f_i X_i\right)^2}{n}, \tag{18}$$

---

*Given the sample mean ($\overline{X}$) and sample size (n) in Example 1, *degrees of freedom* means that the data could have been weights different from those shown, but when any six (i.e., $n - 1$) of the seven weights are specified, then the seventh weight is also known. The term was first used, though in a different context, by Ronald Aylmer Fisher in 1922 (David, 1955).

[†]Computational formulas advantageous on calculators may not prove accurate on computers (Wilkinson and Dallal, 1977), largely because computers may use fewer significant figures. (Also see Ling, 1974.) Good computer programs use calculation techniques designed to help avoid rounding errors.

where f_i is the frequency of observations with magnitude X_i. But with a calculator or computer it is often faster to use Equation 18 for the individual observations, disregarding the class groupings.

The variance has square units. If measurements are in grams, their variance will be in grams squared, or if the measurements are in cubic centimeters, their variance will be in terms of cubic centimeters squared, even though such squared units have no physical interpretation.

EXAMPLE 2 **"Machine Formula" Calculation of Variance, Standard Deviation, and Coefficient of Variation (These are the data of Example 1)**

Sample 1		Sample 2	
X_i (g)	X_i^2 (g^2)	X_i (g)	X_i^2 (g^2)
1.2	1.44	1.2	1.44
1.4	1.96	1.6	2.56
1.6	2.56	1.7	2.89
1.8	3.24	1.8	3.24
2.0	4.00	1.9	3.61
2.2	4.84	2.0	4.00
2.4	5.76	2.4	5.76

$\sum X_i = 12.6$ g $\sum X_i^2 = 23.80$ g^2

$n = 7$

$\overline{X} = \dfrac{12.6 \text{ g}}{7} = 1.8$ g

$\text{SS} = \sum X_i^2 - \dfrac{\left(\sum X_i\right)^2}{n}$

$\quad = 23.80 \text{ g}^2 - \dfrac{(12.6 \text{ g})^2}{7}$

$\quad = 23.80 \text{ g}^2 - 22.68 \text{ g}^2$

$\quad = 1.12 \text{ g}^2$

$s^2 = \dfrac{\text{SS}}{n-1}$

$\quad = \dfrac{1.12 \text{ g}^2}{6} = 0.1867 \text{ g}^2$

$s = \sqrt{0.1867 \text{ g}^2} = 0.43$ g

$V = \dfrac{s}{\overline{X}} = \dfrac{0.43 \text{ g}}{1.8 \text{ g}} = 0.24 = 24\%$

$\sum X_i = 12.6$ g $\sum X_i^2 = 23.50$ g^2

$n = 7$

$\overline{X} = \dfrac{12.6 \text{ g}}{7} = 1.8$ g

$\text{SS} = 23.50 \text{ g}^2 - \dfrac{(12.6 \text{ g})^2}{7}$

$\quad = 0.82 \text{ g}^2$

$s^2 = \dfrac{0.82 \text{ g}^2}{6} = 0.1367 \text{ g}^2$

$s = \sqrt{0.1367 \text{ g}^2} = 0.37$ g

$V = \dfrac{0.37 \text{ g}}{1.8 \text{ g}} = 0.21 = 21\%$

5 THE STANDARD DEVIATION

The *standard deviation*[*] is the positive square root[†] of the variance; therefore, it has the same units as the original measurements. Thus, for a population,

$$\sigma = \sqrt{\dfrac{\sum X_i^2 - \dfrac{\left(\sum X_i\right)^2}{N}}{N}}, \tag{19}$$

And for a sample,[‡]

$$s = \sqrt{\dfrac{\sum X_i^2 - \dfrac{\left(\sum X_i\right)^2}{n}}{n-1}}. \tag{20}$$

Examples 1 and 2 demonstrate the calculation of *s*. This quantity frequently is abbreviated SD, and on rare occasions is called the *root mean square deviation* or *root mean square*. Remember that the standard deviation is, by definition, always a nonnegative quantity.[§]

---

[*]It was the great English statistician Karl Pearson (1857–1936) who coined the term *standard deviation* and its symbol, σ, in 1893, prior to which this quantity was called the *mean error* (Eells, 1926; Walker, 1929: 54–55, 183, 188). In early literature (e.g., by G. U. Yule in 1919), it was termed *root mean square deviation* and acquired the symbol *s*, and (particularly in the fields of education and psychology) it was occasionally computed using deviations from the median (or even the mode) instead of from the mean (Eells, 1926).

[†]The square root sign ($\sqrt{}$) was introduced by Silesian-born Austrian mathematician Christoff Rudolff (1499–1545) in 1525; by 1637 René Descartes (1596–1650) combined this with a vinculum (a horizontal bar placed above quantities to group them as is done with parentheses or brackets) to obtain the symbol $\sqrt{}$, but Gottfried Wilhelm Leibniz (1646–1716) preferred $\sqrt{()}$, which is still occasionally seen (Cajori, 1928/9, Vol. I: 135, 208, 368, 372, 375).

[‡]The sample *s* is actually a slightly biased estimate of the population σ, in that on the average it is a slightly low estimate, especially in small samples. But this fact is generally considered to be offset by the statistic's usefulness. Correction for this bias is sometimes possible (e.g., Bliss, 1967: 131; Dixon and Massey, 1969: 136; Gurland and Tripathi, 1971; Tolman, 1971), but it is rarely employed.

[§]It can be shown that the median of a distribution is never more than one standard deviation away from the mean (μ); that is,

$$|\text{median} - \mu| \le \sigma \tag{21}$$

(Hotelling and Solomon, 1932; O'Cinneide, 1990; Page and Murty, 1982; Watson, 1994). This is a special case, where $p = 50$, of the relationship

$$\mu - \sigma\sqrt{\dfrac{1 - p/100}{p/100}} \le X_p \le \mu + \sigma\sqrt{\dfrac{p/100}{1 - p/100}}, \tag{22}$$

where X_p is the *p*th percentile of the distribution (Dharmadhikari, 1991). Also, Page and Murty (1982) have shown these population-parameter relationships between the standard deviation and the range and between the standard deviation and the mean, median, and mode:

$$\text{range}/\sqrt{2n} \le \sigma \le \text{range}/2\,; \tag{4.22a}$$

$$|\text{mode} - \mu| \le \sigma\sqrt{n/m} \quad \text{and} \quad |\text{mode} - \text{median}| \le \sigma(n/m)\,, \tag{4.22b}$$

where *m* is the number of data at the modal value.

6 THE COEFFICIENT OF VARIATION

The *coefficient of variation** or *coefficient of variability*, is defined as

$$V = \frac{s}{\overline{X}} \quad \text{or} \quad V = \frac{s}{\overline{X}} \cdot 100\%. \tag{23}$$

As $s/\overline{X}$ is generally a small quantity, it is frequently multiplied by 100% in order to express V as a percentage. (The coefficient of variation is often abbreviated as CV.)

As a measure of variability, the variance and standard deviation have magnitudes that are dependent on the magnitude of the data. Elephants have ears that are perhaps 100 times larger than those of mice. If elephant ears were no more variable, relative to their size, than mouse ears, relative to their size, the standard deviation of elephant ear lengths would be 100 times as great as the standard deviation of mouse ear lengths (and the variance of the former would be $100^2 = 10,000$ times the variance of the latter). The sample coefficient of variation expresses sample variability relative to the mean of the sample (and is on rare occasion referred to as the "relative standard deviation"). It is called a measure of *relative variability* or *relative dispersion*.

Because s and $\overline{X}$ have identical units, V has no units at all, a fact emphasizing that it is a relative measure, divorced from the actual magnitude or units of measurement of the data. Thus, had the data in Example 2 been measured in pounds, kilograms, or tons, instead of grams, the calculated V would have been the same. The coefficient of variation of a sample, namely V, is an estimate of the coefficient of variation of the population from which the sample came (i.e., an estimate of σ/μ). The coefficient of variation may be calculated only for ratio scale data; it is, for example, not valid to calculate coefficients of variation of temperature data measured on the Celsius or Fahrenheit temperature scales. Simpson, Roe, and Lewontin (1960: 89–95) present a good discussion of V and its biological application, especially with regard to zoomorphological measurements.

7 INDICES OF DIVERSITY

For nominal-scale data there is no mean or median or ordered measurements to serve as a reference for discussion of dispersion. Instead, we can invoke the concept of *diversity*, the distribution of observations among categories. Consider that sparrows are found to nest in four different types of location (vines, eaves, branches, and cavities). If, out of twenty nests observed, five are found at each of the four locations, then we would say that there was great diversity in nesting sites. If, however, seventeen nests were found in cavities and only one in each of the other three locations, then we would consider the situation to be one of very low nest-site diversity. In other words, observations distributed evenly among categories display high diversity, whereas a set of observations where most of the data occur in very few of the categories is one exhibiting low diversity.

A large number of diversity measures have been introduced, especially for ecological data (e.g., Brower, Zar, and von Ende, 1998: 177–184; Magurran, 2004), a few of which are presented here.

---

*The term *coefficient of variation* was introduced by the statistical giant Karl Pearson (1857–1936) in 1896 (David, 1995). In early literature the term was variously applied to the ratios of different measures of dispersion and different measures of central tendency (Eells, 1926).

Among the quantitative descriptions of diversity available are those based on a field known as *information theory*.* The underlying considerations of these measures can be visualized by considering *uncertainty* to be synonymous with diversity. If seventeen out of twenty nest sites were to be found in cavities, then one would be relatively certain of being able to predict the location of a randomly encountered nest site. However, if nests were found to be distributed evenly among the various locations (a situation of high nest-site diversity), then there would be a good deal of uncertainty involved in predicting the location of a nest site selected at random. If a set of nominal scale data may be considered to be a random sample, then a quantitative expression appropriate as a measure of diversity is that of Shannon (1948):

$$H' = -\sum_{i=1}^{k} p_i \log p_i \tag{24}$$

(often referred to as the Shannon-Wiener diversity index or the Shannon-Weaver index). Here, k is the number of categories and p_i is the proportion of the observations found in category i. Denoting n to be sample size and f_i to be the number of observations in category i, then $p_i = f_i/n$; and an equivalent equation for H' is

$$H' = \frac{n \log n - \sum_{i=1}^{k} f_i \log f_i}{n}, \tag{25}$$

a formula that is easier to use than Equation 24 because it eliminates the necessity of calculating the proportions (p_i). Published tables of $n \log n$ and $f_i \log f_i$ are available (e.g., Brower, Zar, and von Ende, 1998: 181; Lloyd, Zar, and Karr, 1968). Any logarithmic base may be used to compute H'; bases 10, e, and 2 (in that order of commonness) are the most frequently encountered. A value of H' (or of any other measure of this section except evenness measures) calculated using one logarithmic base may be converted to that of another base; Table 1 gives factors for doing this for bases 10, e, and 2. Unfortunately, H' is known to be an underestimate of the diversity in the sampled population (Bowman et al., 1971). However, this bias decreases with increasing sample size. Ghent (1991) demonstrated a relationship between H' and testing hypotheses for equal abundance among the k categories.

The magnitude of H' is affected not only by the distribution of the data but also by the number of categories, for, theoretically, the maximum possible diversity for a set of data consisting of k categories is

$$H'_{max} = \log k. \tag{26}$$

Therefore, some users of Shannon's index prefer to calculate

$$J' = \frac{H'}{H'_{max}} \tag{27}$$

instead of (or in addition to) H', thus expressing the observed diversity as a proportion of the maximum possible diversity. The quantity J' has been termed *evenness* (Pielou, 1966) and may also be referred to as *homogeneity* or *relative diversity*. The measure

---

*Claude Elwood Shannon (1916–2001) founded what he first called "a mathematical theory of communication" and has become known as "information theory."

TABLE 1: Multiplication Factors for Converting among Diversity Measures (H, H', H_{max}, or H'_{max}) Calculated Using Different Logarithmic Bases*

To convert to:	To convert from:		
	Base 2	**Base e**	**Base 10**
Base 2	1.0000	1.4427	3.3219
Base e	0.6931	1.0000	2.3026
Base 10	0.3010	0.4343	1.0000

For example, if $H' = 0.255$ using base 10; H' would be $(0.255)(3.3219) = 0.847$ using base 2.

*The measures J and J' are unaffected by change in logarithmic base.

$1 - J'$ may then be viewed as a measure of *heterogeneity*; it may also be considered a measure of *dominance*, for it reflects the extent to which frequencies are concentrated in a small number of categories. The number of categories in a sample (k) is typically an underestimate of the number of categories in the population from which the sample came, because some categories (especially the rarer ones) are likely to be missed in collecting the sample. Therefore, the sample evenness, J', is typically an overestimate of the population evenness. (That is, J' is a biased statistic.) Example 3 demonstrates the calculation of H' and J'.

If a set of data may not be considered a random sample, then Equation 24 (or 25) is not an appropriate diversity measure (Pielou, 1966). Examples of such

---

EXAMPLE 3 Indices of Diversity for Nominal Scale Data: The Nesting Sites of Sparrows

Category (i)	Observed Frequencies (f_i)
	Sample 1
Vines	5
Eaves	5
Branches	5
Cavities	5

$$H' = \frac{n \log n - \sum f_i \log f_i}{n} = [20 \log 20 - (5 \log 5 + 5 \log 5 + 5 \log 5 + 5 \log 5)]/20$$

$$= [26.0206 - (3.4949 + 3.4949 + 3.4949 + 3.4949)]/20$$

$$= 12.0410/20 = 0.602$$

$$H'_{max} = \log 4 = 0.602$$

$$J' = \frac{0.602}{0.602} = 1.00$$

	Sample 2
Vines	1
Eaves	1
Branches	1
Cavities	17

$$H' = \frac{n \log n - \sum f_i \log f_i}{n} = [20 \log 20 - (1 \log 1 + 1 \log 1 + 1 \log 1$$
$$+ 17 \log 17)]/20$$
$$= [26.0206 - (0 + 0 + 0 + 20.9176)]/20$$
$$= 5.1030/20 = 0.255$$
$$H'_{max} = \log 4 = 0.602$$
$$J' = \frac{0.255}{0.602} = 0.42$$

	Sample 3
Vines	2
Eaves	2
Branches	2
Cavities	34

$$H' = \frac{n \log n - \sum f_i \log f_i}{n} = [40 \log 40 - (2 \log 2 + 2 \log 2 + 2 \log 2$$
$$+ 34 \log 34)]/40$$
$$= [64.0824 - (0.6021 + 0.6021 + 0.6021$$
$$+ 52.0703)]/40$$
$$= 10.2058/40 = 0.255$$
$$H'_{max} = \log 4 = 0.602$$
$$J' = \frac{0.255}{0.602} = 0.42$$

situations may be when we have, in fact, data composing an entire population, or data that are a sample obtained nonrandomly from a population. In such a case, one may use the information-theoretic diversity measure of Brillouin (1962: 7–8):*

$$H = \frac{\log \left(\dfrac{n!}{\prod_{i=1}^{k} f_i!} \right)}{n}, \tag{28}$$

---

*The notation $n!$ is read as "n factorial" and signifies the product $(n)(n - 1)(n - 2) \cdots (2)(1)$. It was proposed by French physician and mathematician Christian Kramp (1760–1826) around 1798; he originally called this function *faculty* ("facultés" in French) but in 1808 accepted the term *factorial* ("factorielle" in French) used by Alsatian mathematician Louis François Antoine Arbogast (1759–1803) (Cajori, 1928/9, Vol. II: 72; Gullberg, 1997: 106; Miller, 2004a; O'Connor and Robertson, 1997). English mathematician Augustus De Morgan (1806–1871) decried the adoption of this symbol as a "barbarism" because it introduced into mathematics a symbol that already had an established meaning in written language, thus giving "the appearance of expressing surprise or admiration" in a mathematical result (Cajori, ibid.: 328).

where Π (capital Greek pi) means to take the product, just as Σ means to take the sum. Equation 28 may be written, equivalently, as

$$H = \frac{\log \dfrac{n!}{f_1! f_2! \ldots f_k!}}{n} \tag{29}$$

or as

$$H = \frac{(\log n! - \sum \log f_i!)}{n}. \tag{30}$$

Table 40 gives logarithms of factorials to ease this calculation. Other such tables are available, as well (e.g., Brower, Zar, and von Ende 1998: 183; Lloyd, Zar, and Karr, 1968; Pearson and Hartly, 1966: Table 51).* Ghent (1991) discussed the relationship between H and the test of hypotheses about equal abundance among k categories.

The maximum possible Brillouin diversity for a set of n observations distributed among k categories is

$$H_{max} = \frac{\log n! - (k - d) \log c! - d \log(c + 1)!}{n}, \tag{35}$$

where c is the integer portion of n/k, and d is the remainder. (For example, if $n = 17$ and $k = 4$, then $n/k = 17/4 = 4.25$ and $c = 4$ and $d = 0.25$.) The Brillouin-based evenness measure is, therefore,

$$J = \frac{H}{H_{max}}, \tag{36}$$

with $1 - J$ being a dominance measure. When we consider that we have data from an entire population, k is a population measurement, rather than an estimate of one, and J is not a biased estimate as is J'.

For further considerations of these and other diversity measures, see Brower, Zar, and von Ende (1998: Chapter 5B) and Magguran (2004: 100–121).

8 CODING DATA

Coding data may facilitate statistical computations of measures of central tendency. Such benefits are even more apparent when calculating SS, s^2, and s, because of the

---

*For moderate to large n (or f_i), "Stirling's approximation" is excellent:

$$n! = \sqrt{2\pi n}(n/e)^n = \sqrt{2\pi}\sqrt{n}e^{-n}n^n, \tag{31}$$

of which this is an easily usable derivation:

$$\log n! = (n + 0.5)\log n - 0.434294n + 0.399090. \tag{32}$$

An approximation with only half the error of the above is

$$n! = \sqrt{2\pi}\left(\frac{n + 0.5}{e}\right)^{n+0.5} \tag{33}$$

and

$$\log n! = (n + 0.5)\log(n + 0.5) - 0.434294(n + 0.5) + 0.399090. \tag{34}$$

This is named for James Stirling, who published something similar to the latter approximation formula in 1730, making an arithmetic improvement in the approximation earlier known by Abraham de Moivre (Kemp, 1989; Pearson, 1924; Walker, 1929: 16).

labor, and concomitant chances of error, involved in the unwieldy squaring of large or small numbers.

When data are coded by adding or subtracting a constant (call it A), the measures of dispersion of Sections 1 through 5 are not changed from what they were for the data before coding. This is because these measures are based upon deviations, and deviations are not changed by moving the data along the measurement scale (e.g., the deviation between 1 and 10 is the same as the deviation between 11 and 20). Sample 1 in Example 4 demonstrates this.

However, when coding by multiplying by a constant (call it M), the measures of dispersion are affected, for the magnitudes of the deviations will be changed. With such coding, the range, mean deviation, and standard deviation are changed by a factor of M, in the same manner as the arithmetic mean and the median are, whereas the sum of squares and variance are changed in accordance with the square of the coding constant (i.e., M^2), and the coefficient of variance is not affected. This is demonstrated in Sample 2 of Example 4.

A coded datum is described as

$$[X_i] = MX_i + A. \tag{37}$$

EXAMPLE 4 Coding Data to Facilitate the Calculation of Measures of Dispersion

Sample 1 (Coding by Subtraction: $A = -840$ g)

Without Coding X_i		Using Coding $[X_i]$	
X_i (g)	X_i^2 (g^2)	$[X_i]$ (g)	$[X_i]^2$ (g^2)
842	708,964	2	4
843	710,649	3	9
844	712,336	4	16
846	715,716	6	36
846	715,716	6	36
847	717,409	7	49
848	719,104	8	64
849	720,801	9	81

$\sum X_i = 6765$ g $\sum X_i^2 = 5{,}720{,}695$ g^2 $\sum [X_i] = 45$ g $\sum [X_i]^2 = 295$ g^2

$$s^2 = \frac{5720695 \text{ g}^2 - \dfrac{(6765 \text{ g})^2}{8}}{7}$$

$$= 5.98 \text{ g}^2$$
$$s = 2.45 \text{ g}$$
$$\overline{X} = 845.6 \text{ g}$$
$$V = \frac{s}{\overline{X}} = \frac{2.45 \text{ g}}{845.6 \text{ g}}$$
$$= 0.0029 = 0.29\%$$

$$[s^2] = \frac{295 \text{ g}^2 - \dfrac{(45 \text{ g})^2}{8}}{7}$$

$$= 5.98 \text{ g}^2$$
$$[s] = 2.44 \text{ g}$$
$$[\overline{X}] = 5.6 \text{ g}$$

Sample 2 (Coding by Division: $M = 0.01$)			
Without Coding X_i		*Using Coding $[X_i]$*	
X_i (sec)	X_i^2 (sec^2)	$[X_i]$ (sec)	$[X_i]^2$ (sec^2)
800	640,000	8.00	64.00
900	810,000	9.00	81.00
950	902,500	9.50	90.25
1100	1,210,000	11.00	121.00
1250	1,562,500	12.50	156.25
1300	1,690,000	13.00	169.00

$$\sum X_i = 6300 \text{ sec} \quad \sum X_i^2 = 6{,}815{,}000 \text{ sec}^2 \quad \sum [X_i] = 63.00 \text{ sec} \quad \sum [X_i]^2 = 681.50 \text{ sec}^2$$

$$s^2 = \frac{6815000 \text{ sec}^2 - \dfrac{(6300 \text{ sec})^2}{6}}{5} \qquad [s^2] = \frac{681.50 \text{ sec}^2 - \dfrac{(63.00 \text{ sec})^2}{6}}{5}$$

$$= 40{,}000 \text{ sec}^2 \qquad\qquad\qquad = 4 \text{ sec}^2$$
$$s = 200 \text{ sec} \qquad\qquad\qquad\quad [s] = 2.00 \text{ sec}$$
$$\overline{X} = 1050 \text{ sec} \qquad\qquad\qquad [\overline{X}] = 10.50 \text{ sec}$$
$$V = 0.19 = 19\% \qquad\qquad\quad [V] = 0.19 = 19\%$$

EXERCISES

1. Five body weights, in grams, collected from a population of rodent body weights are

$$66.1, \ 77.1, \ 74.6, \ 61.8, \ 71.5.$$

(a) Compute the "sum of squares" and the variance of these data using Equations 12 and 15, respectively.

(b) Compute the "sum of squares" and the variance of these data by using Equations 16 and 17, respectively.

2. Consider the following data, which are a sample of amino acid concentrations (mg/100 ml) in arthropod hemolymph:

$$240.6, \ 238.2, \ 236.4, \ 244.8, \ 240.7, \ 241.3, \ 237.9.$$

(a) Determine the range of the data.
(b) Calculate the "sum of squares" of the data.
(c) Calculate the variance of the data.
(d) Calculate the standard deviation of the data.
(e) Calculate the coefficient of variation of the data.

3. The following frequency distribution of tree species was observed in a random sample from a forest:

Species	Frequency
White oak	44
Red oak	3
Shagbark hickory	28
Black walnut	12
Basswood	2
Slippery elm	8

(a) Use the Shannon index to express the tree species diversity.
(b) Compute the maximum Shannon diversity possible for the given number of species and individuals.
(c) Calculate the Shannon evenness for these data.

4. Assume the data in Exercise 3 were an entire population (e.g., all the trees planted around a group of buildings).

(a) Use the Brillouin index to express the tree species diversity.
(b) Compute the maximum Brillouin diversity possible for the given number of species and individuals.
(c) Calculate the Brillouin evenness measure for these data.

ANSWERS TO EXERCISES

1. **(a)** SS $= 156.028 \text{ g}^2$, $s^2 = 39.007 \text{ g}^2$; **(b)** same as (a).

2. **(a)** Range $= 236.4 \text{ mg}/100 \text{ ml}$ to $244.8 \text{ mg}/100 \text{ ml} = 8.4 \text{ mg}/100 \text{ ml}$;
   (b) SS $= 46.1886 \text{ (mg}/100 \text{ ml)}^2$; **(c)** $s^2 = 7.6981$ $\text{(mg}/100 \text{ ml)}^2$; **(d)** $s = 2.77 \text{ mg}/100 \text{ ml}$;
   (e) $V = 0.0115 = 1.15\%$.

3. $k = 6, n = 97$; **(a)** $H' = 0.595$;
   (b) $H'_{max} = 0.778$; **(c)** $J' = 0.76$.

4. $k = 6, n = 97$; **(a)** $H = 0.554$; **(b)** $c = 16$, $d = 0.1667$, $H_{max} = 0.741$; **(c)** $J = 0.75$.

This page intentionally left blank

Probabilities

Everyday concepts of "likelihood," "predictability," and "chance" are formalized by that branch of mathematics called *probability*. Although earlier work on the subject was done by writers such as Giralamo Cardano (1501–1576) and Galileo Galilei (1564–1642), the investigation of probability as a branch of mathematics sprang in earnest from 1654 correspondence between two great French mathematicians, Blaise Pascal (1623–1662) and Pierre Fermat (1601–1665). These two men were stimulated by the desire to predict outcomes in the games of chance popular among the French nobility of the mid-seventeenth century; we still use the devices of such games (e.g., dice and cards) to demonstrate the basic concepts of probability.*

A thorough discourse on probability is well beyond the scope and intent of this text, but aspects of probability are of biological interest and considerations of probability theory underlie the many procedures for statistical hypothesis testing discussed. Therefore, this chapter will introduce probability concepts that bear the most pertinence to biology and biostatistical analysis.

Worthwhile presentations of probability specifically for the biologist are found in Batschelet (1976: 441–474); Eason, Coles, and Gettinby (1980: 395–414); and Mosimann (1968).

1 COUNTING POSSIBLE OUTCOMES

Suppose a phenomenon can occur in any one of k different ways, but in only one of those ways at a time. For example, a coin has two sides and when tossed will land

---

*The first published work on the subject of probability and gaming was by the Dutch astronomer, physicist, and mathematician Christiaan (also known as Christianus) Huygens (1629–1695), in 1657 (Asimov, 1982: 138; David, 1962: 113, 133). This, in turn, aroused the interest of other major minds, such as Jacob (also known as Jacques, Jakob, and James) Bernoulli (1654–1705, whose 1713 book was the first devoted entirely to probability), several other members of the remarkable Bernoulli family of Swiss mathematicians, and others such as Abraham de Moivre (1667–1754), Pierre Rémond de Montmort (1678–1719), and Pierre-Simon Laplace (1749–1827) of France. The term *probability* in its modern mathematical sense was used as early as 1718 by de Moivre (Miller, 2004a). For more detailed history of the subject, see David (1962) and Walker (1928: 5–13).

with either the "head" side (H) up or the "tail" side (T) up, but not both. Or, a die has six sides and when thrown will land with either the 1, 2, 3, 4, 5, or 6 side up.* We shall refer to each possible outcome (i.e., H or T with the coin; or 1, 2, 3, 4, 5, or 6 with the die) as an *event*.

If something can occur in any one of k_1 different ways and something else can occur in any one of k_2 different ways, then the number of possible ways for both things to occur is $k_1 \times k_2$. For example, suppose that two coins are tossed, say a silver one and a copper one. There are two possible outcomes of the toss of the silver coin (H or T) and two possible outcomes of the toss of the copper coin (H or T). Therefore, $k_1 = 2$ and $k_2 = 2$ and there are $(k_1)(k_2) = (2)(2) = 4$ possible outcomes of the toss of both coins: both heads, silver head and copper tail, silver tail and copper head, and both tails (i.e., H,H; H,T; T,H; T,T).

Or, consider tossing of a coin together with throwing a die. There are two possible coin outcomes ($k_1 = 2$) and six possible die outcomes ($k_2 = 6$), so there are $(k_1)(k_2) = (2)(6) = 12$ possible outcomes of the two events together:

$$\text{H,1; H,2; H,3; H,4; H,5; H,6; T,1; T,2; T,3; T,4; T,5; T,6.}$$

If two dice are thrown, we can count six possible outcomes for the first die and six for the second, so there are $(k_1)(k_2) = (6)(6) = 36$ possible outcomes when two dice are thrown:

$$\text{1,1; 1,2; 1,3; 1,4; 1,5; 1,6;} \quad \text{2,1; 2,2; 2,3; 2,4; 2,5; 2,6;}$$
$$\text{3,1; 3,2; 3,3; 3,4; 3,5; 3,6;} \quad \text{4,1; 4,2; 4,3; 4,4; 4,5; 4,6;}$$
$$\text{5,1; 5,2; 5,3; 5,4; 5,5; 5,6;} \quad \text{6,1; 6,2; 6,3; 6,4; 6,5; 6,6.}$$

The preceding counting rule is extended readily to determine the number of ways more than two things can occur together. If one thing can occur in any one of k_1 ways, a second thing in any one of k_2 ways, a third thing in any of k_3 ways, and so on, through an nth thing in any one of k_n ways, then the number of ways for all n things to occur together is

$$(k_1)(k_2)(k_3) \cdots (k_n).$$

Thus, if three coins are tossed, each toss resulting in one of two possible outcomes, then there is a total of

$$(k_1)(k_2)(k_3) = (2)(2)(2) = 8$$

possible outcomes for the three tosses together:

$$\text{H,H,H; H,H,T; H,T,H; H,T,T; T,H,H, T,H,T; T,T,H; T,T,T.}$$

Similarly, if three dice are thrown, there are $(k_1)(k_2)(k_3) = (6)(6)(6) = 6^3 = 216$ possible outcomes; if two dice and three coins are thrown, there are

---

*What we recognize as metallic coins originated shortly after 650 B.C.E.—perhaps in ancient Lydia (located on the Aegean Sea in what is now western Turkey). From the beginning, the obverse and reverse sides of coins have had different designs, in earliest times with the obverse commonly depicting animals and, later, deities and rulers (Sutherland, 1992). Dice have long been used for both games and religion. They date from nearly 3000 years B.C.E., with the modern conventional arrangement of dots on the six faces of a cubic die (1 opposite 6, 2 opposite 5, and 3 opposite 4) becoming dominant around the middle of the fourteenth century B.C.E. (David, 1962: 10). Of course, the arrangement of the numbers 1 through 6 on the six faces has no effect on the outcome of throwing a die.

$(k_1)(k_2)(k_3)(k_4)(k_5) = (6)(6)(2)(2)(2) = (6^2)(2^3) = 288$ outcomes; and so on. Example 1 gives two biological examples of counting possible outcomes.

EXAMPLE 1   Counting Possible Outcomes

(a) A linear arrangement of three deoxyribonucleic acid (DNA) nucleotides is called a triplet. A nucleotide may contain any one of four possible bases: adenine (A), cytosine (C), guanine (G), and thymine (T). How many different triplets are possible?

As the first nucleotide in the triplet may be any one of the four bases (A; C; G; T), the second may be any one of the four, and the third may be any one of the four, there is a total of

$$(k_1)(k_2)(k_3) = (4)(4)(4) = 64 \text{ possible outcomes;}$$

that is, there are 64 possible triplets:

$$\begin{array}{llll}
A, A, A; & A, A, C; & A, A, G; & A, A, T; \\
A, C, A; & A, C, C; & A, C, G; & A, C, T; \\
A, G, A; & A, G, C; & A, G, G; & A, G, T;
\end{array}$$

and so on.

(b) If a diploid cell contains three pairs of chromosomes, and one member of each pair is found in each gamete, how many different gametes are possible?

As the first chromosome may occur in a gamete in one of two forms, as may the second and the third chromosomes,

$$(k_1)(k_2)(k_3) = (2)(2)(2) = 2^3 = 8.$$

Let us designate one of the pairs of chromosomes as "long," with the members of the pair being L_1 and L_2; one pair as "short," indicated as S_1 and S_2; and one pair as "midsized," labeled M_1 and M_2. Then the eight possible outcomes may be represented as

$$\begin{array}{llll}
L_1, M_1, S_1; & L_1, M_1, S_2; & L_1, M_2, S_1; & L_1, M_2, S_2; \\
L_2, M_1, S_1; & L_2, M_1, S_2; & L_2, M_2, S_1; & L_2, M_2, S_2.
\end{array}$$

---

2 PERMUTATIONS

(a) Linear Arrangements. A *permutation** is an arrangement of objects in a specific sequence. For example, a horse (H), cow (C), and sheep (S) could be arranged linearly in six different ways: H,C,S; H,S,C; C,H,S; C,S,H; S,H,C; S,C,H. This set of outcomes may be examined by noting that there are three possible ways to fill the first position in the linear order; but once an animal is placed in this position, there are only two ways to fill the second position; and after animals are placed in the first two positions, there is only one possible way to fill the third position. Therefore, $k_1 = 3$, $k_2 = 2$, and $k_3 = 1$, so that by the method of counting of Section 1 there are $(k_1)(k_2)(k_3) = (3)(2)(1) = 6$ ways to align these three animals. We may say that there are six permutations of three distinguishable objects.

---

*The term *permutation* was invented by Jacob Bernoulli in his landmark posthumous 1713 book on probability (Walker, 1929: 9).

In general, if there are n linear positions to fill with n objects, the first position may be filled in any one of n ways, the second may be filled in any one of $n - 1$ ways, the third in any one of $n - 2$ ways, and so on until the last position, which may be filled in only one way. That is, the filling of n positions with n objects results in $_nP_n$ permutations, where

$$_nP_n = n(n - 1)(n - 2) \cdots (3)(2)(1). \tag{1}$$

This equation may be written more simply in *factorial* notation as

$$_nP_n = n!, \tag{2}$$

where "n factorial" is the product of n and each smaller positive integer; that is,

$$n! = n(n - 1)(n - 2) \cdots (3)(2)(1). \tag{3}$$

Example 2 demonstrates such computation of the numbers of permutations.

EXAMPLE 2   The Number of Permutations of Distinct Objects

In how many sequences can six photographs be arranged on a page?

$$_nP_n = 6! = (6)(5)(4)(3)(2)(1) = 720$$

(b) Circular Arrangements. The numbers of permutations considered previously are for objects arranged on a line. If objects are arranged on a circle, there is no "starting position" as there is on a line, and the number of permutations is

$$_nP'_n = \frac{n!}{n} = (n - 1)!. \tag{4}$$

(Observe that the notation $_nP'_n$ is used here for circular permutations to distinguish it from the symbol $_nP_n$ used for linear permutations.)

Referring again to a horse, a cow, and a sheep, there are $_nP'_n = \frac{n!}{n} = (n - 1)! = (3 - 1)! = 2! = 2$ distinct ways in which the three animals could be seated around a table, or arranged around the shore of a pond:

H   H

or

S   C   C   S

In this example, there is an assumed orientation of the observer, so clockwise and counterclockwise patterns are treated as different. That is, the animals are observed arranged around the top of the table, or observed from above the surface of the pond. But either one of these arrangements would look like the other one if observed from under the table or under the water; and if we did not wish to count the results of these two mirror-image observations as different, we would speak of there being one possible permutation, not two. For example, consider each of the preceding two diagrams to represent three beads on a circular string, one bead in the shape of a horse, one in the shape of a cow, and the other in the shape of a sheep. The two arrangements of H, C, and S shown are not really different, for there is no specific way of viewing the circle; one of the two arrangements turns into the other if the circle is turned over. If $n > 2$ and the orientation of the circle is not specified, then the number of permutations of n objects on a circle is

$$_nP''_n = \frac{n!}{2n} = \frac{(n-1)!}{2}.$$ (5)

(c) Fewer than n Positions. If one has n objects, but fewer than n positions in which to place them, then there would be considerably fewer numbers of ways to arrange the objects than in the case where there are positions for all n. For example, there are $_4P_4 = 4! = (4)(3)(2)(1) = 24$ ways of placing a horse (H), cow (C), sheep (S), and pig (P) in four positions on a line. However, there are only twelve ways of linearly arranging these four animals two at a time:

H,C; H,S; H,P; C,H; C,S; C,P; S,H; S,C; S,P; P,H; P,C; P,S.

The number of linear permutations of n objects taken X at a time is*

$$_nP_X = \frac{n!}{(n-X)!}.$$ (6)

For the preceding example,

$$_4P_2 = \frac{4!}{(4-2)!} = \frac{4!}{2!} = \frac{(4)(3)(2)(1)}{(2)(1)} = 12.$$

Equation 2 is a special case of Equation 6, where $X = n$; it is important to know that $0!$ is defined to be 1.[†]

If the arrangements are circular, instead of linear, then the number of them possible is

$$_nP'_X = \frac{n!}{(n-X)!X}.$$ (7)

So, for example, there are only $4!/[(4-2)!2] = 6$ different ways of arranging two out of our four animals around a table:

H	H	H	C	C	S
C	S	P	S	P	P

for C seated at the table opposite H is the same arrangement as H seated across from C, S seated with H is the same as H with S, and so on. Example 3 demonstrates this further. Equation 4 is a special case of Equation 7, where $X = n$; and recall that $0!$ is defined as 1.

EXAMPLE 3 The Number of Permutations of n Objects Taken X at a Time: In How Many Different Ways Can a Sequence of Four Slides Be Chosen from a Collection of Six Slides?

$$_nP_X = {_6P_4} = \frac{6!}{(6-4)!} = \frac{6!}{2!} = \frac{(6)(5)(4)(3)(2)(1)}{(2)(1)}$$
$$= (6)(5)(4)(3) = 360$$

---

*Notation in the form of $_nP_X$ to indicate permutations of n items taken X at a time was used prior to 1869 by Harvey Goodwin (Cajori, 1929: 79).

[†]Why is $0!$ defined to be 1? In general, $n! = n[(n-1)!]$; for example, $5! = 5(4!)$, $4! = 4(3!)$, $3! = 3(2!)$, and $2! = 2(1!)$. Thus, $1! = 1(0!)$, which is so only if $0! = 1$.

If $n > 2$, then for every circular permutation viewed from above there is a mirror image of that permutation, which would be observed from below. If these two mirror images are not to be counted as different (e.g., if we are dealing with beads of different shapes or colors on a string), then the number of circular permutations is

$$_nP''_X = \frac{n!}{2(n - X)!X}.\tag{8}$$

(d) If Some of the Objects Are Indistinguishable. If our group of four animals consisted of two horses (H), a cow (C), and a sheep (S), the number of permutations of the four animals would be twelve:

H,H,C,S; H,H,S,C; H,C,H,S; H,C,S,H; H,S,H,C; H,S,C,H;

C,H,H,S; C,H,S,H; C,S,H,H; S,H,H,C; S,H,C,H; S,C,H,H.

If n_i represents the number of like individuals in category i (in this case the number of animals in species i), then in this example $n_1 = 2$, $n_2 = 1$, and $n_3 = 1$, and we can write the number of permutations as

$$_nP_{n_1,n_2,n_3} = \frac{n!}{n_1!n_2!n_3!} = \frac{4!}{2!1!1!} = 12.$$

If the four animals were two horses (H) and two cows (C), then there would be only six permutations:

H,H,C,C; C,C,H,H; H,C,H,C; C,H,C,H; H,C,C,H; C,H,H,C.

In this case, $n = 4$, $n_1 = 2$, and $n_2 = 2$, and the number of permutations is calculated to be $_nP_{n_1,n_2} = n!/(n_1!n_2!) = 4!/(2!2!) = (4)(3)(2)/[(2)(2)] = 6$.

In general, if n_1 members of the first category of objects are indistinguishable, as are n_2 of the second category, n_3 of the third category, and so on through n_k members of the kth category, then the number of different permutations is

$$_nP_{n_1,n_2,\cdots,n_k} = \frac{n!}{n_1!n_2!\cdots n_k!} \text{ or } \frac{n!}{\prod\limits_{i=1}^{k} n_i!},\tag{9}$$

where the capital Greek letter pi (Π) denotes taking the product just as the capital Greek sigma (Σ) indicates taking the sum. This is shown further in Example 4.

EXAMPLE 4   Permutations with Categories Containing Indistinguishable Members

There are twelve potted plants, six of one species, four of a second species, and two of a third species. How many different linear sequences of species are possible (for example, if arranging the pots on a shelf)?

$$_nP_{n_1,n_2,n_3} = \frac{n}{\prod n_i!}$$

$$= {_{12}P_{6,4,2}} = \frac{12!}{6!4!2!}$$

$$= \frac{(12)(11)(10)(9)(8)(7)(6)(5)(4)(3)(2)(1)}{(6)(5)(4)(3)(2)(1)(4)(3)(2)(1)(2)(1)} = 13{,}860.$$

Note that the above calculation could have been simplified by writing

$$\frac{12!}{6!4!2!} = \frac{(12)(11)(10)(9)(8)(7)6!}{6!(4)(3)(2)(2)} = \frac{(12)(11)(10)(9)(8)(7)}{(4)(3)(2)(2)} = 13,860.$$

Here, "(1)" is dropped; also, "6!" appears in both the numerator and denominator, thus canceling out.

3 COMBINATIONS

In Section 2 we considered groupings of objects where the sequence within the groups was important. In many instances, however, only the components of a group, not their arrangement within the group, are important. We saw that if we select two animals from among a horse (H), cow (C), sheep (S), and pig (P), there are twelve ways of arranging the two on a line:

$$\text{H,C; H,S; H,P; C,H; C,S; C,P; S,H; S,C; S,P; P,H; P,C; P,S.}$$

However, some of these arrangements contain exactly the same kinds of animals, only in different order (e.g., H,C and C,H; H,S and S,H). If the groups of two are important to us, but not the sequence of objects within the groups, then we are speaking of *combinations*,* rather than permutations. Designating the number of combinations of n objects taken X at a time as $_nC_X$, we have[†]

$$_nC_X = \frac{_nP_X}{X!} = \frac{n!}{X!(n-X)!}. \tag{10}$$

So for the present example, $n = 4$, $\overline{X} = 2$, and

$$_4C_2 = \frac{4!}{2!(4-2)!} = \frac{4!}{2!2!} = \frac{(4)(3)(2)(1)}{(2)(1)(2)(1)} = \frac{(4)(3)}{2} = 6,$$

the six combinations of the four animals taken two at a time being

$$\text{H,C; H,S; H,P; C,S; C,P; S,P.}$$

Example 5 demonstrates the determination of numbers of combinations for another set of data.

It may be noted that

$$_nC_n = 1, \tag{11}$$

meaning that there is only one way of selecting all n items; and

$$_nC_1 = n, \tag{12}$$

indicating that there are n ways of selecting n items one at a time. Also,

$$_nC_X = {_nC_{n-X}}, \tag{13}$$

---

*The word *combination* was used in this mathematical sense by Blaise Pascal (1623–1662) in 1654 (Smith, 1953: 528).

[†]Notation in the form of $_nC_X$ to indicate combinations of n items taken X at a time was used by G. Chrystal in 1899 (Cajori, 1929: 80).

Probabilities

EXAMPLE 5   Combinations of n Objects Taken X at a Time

Of a total of ten dogs, eight are to be used in a laboratory experiment. How many different combinations of eight animals may be formed from the ten?

$$_nC_X = {}_{10}C_8 = \frac{10!}{8!(10-8)!} = \frac{10!}{8!2!} = \frac{(10)(9)(8)(7)(6)(5)(4)(3)(2)(1)}{(8)(7)(6)(5)(4)(3)(2)(1)(2)(1)}$$
$$= 45.$$

It should be noted that the above calculations with factorials could have been simplified by writing

$$_{10}C_8 = \frac{10!}{8!2!} = \frac{(10)(9)8!}{8!2!} = \frac{(10)(9)}{2} = 45,$$

so that "8!" appears in both the numerator and denominator, thus canceling each other out.

---

which means that if we select X items from a group of n, we have at the same time selected the remaining $n - X$ items; that is, an exclusion is itself a selection. For example, if we selected two out of five persons to write a report, we have simultaneously selected three of the five to refrain from writing. Thus,

$$_5C_2 = \frac{5!}{2!(5-2)!} = \frac{5!}{2!3!} = 10 \quad \text{and} \quad _5C_{5-2} = {}_5C_3 = \frac{5!}{3!(5-3)!} = \frac{5!}{3!2!} = 10,$$

meaning that there are ten ways to select two out of five persons to perform a task and ten ways to select three out of five persons to be excluded from that task. This question may be addressed by applying Equation 9, reasoning that we are asking how many distinguishable arrangements there are of two writers and three nonwriters: $_5P_{2,3} = 5!/(2!3!) = 10$.

The product of combinatorial outcomes may also be employed to address questions such as in Example 4. This is demonstrated in Example 6.

EXAMPLE 6   Products of Combinations

This example provides an alternate method of answering the question of Example 4.

There are twelve potted plants, six of one species, four of a second species, and two of a third. How many different linear sequences of species are possible?

There are twelve positions in the sequence, which may be filled by the six members of the first species in this many ways:

$$_{12}C_6 = \frac{12!}{(12-6)!6!} = 924.$$

The remaining six positions in the sequence may be filled by the four members of the second species in this many ways:

$$_6C_4 = \frac{6!}{(6-4)!4!} = 15.$$

60

And the remaining two positions may be filled by the two members of the third species in only one way:

$$_2C_2 = \frac{2!}{(2-2)!2!} = 1.$$

As each of the ways of filling positions with members of one species exists in association with each of the ways of filling positions with members of each other species, the total different sequences of species is

$$(924)(15)(1) = 13{,}860.$$

From Equation 10 it may be noted that, as $_nC_X = {}_nP_X/X!$,

$$_nP_X = X!\,{}_nC_X. \tag{14}$$

It is common mathematical convention to indicate the number of combinations of n objects taken X at a time as $\binom{n}{X}$ instead of $_nC_X$, so for the problem at the beginning of Section 3 we could have written*

$$\binom{n}{X} = \binom{4}{2} = \frac{4!}{2!(4-2)!} = 6.$$

Binomial coefficients, take this form.

4 SETS

A *set* is a defined collection of items. For example, a set may be a group of four animals, a collection of eighteen amino acids, an assemblage of twenty-five students, or a group of three genetic traits. Each item in a set is termed an *element*. If a set of animals includes these four elements: horse (H), cow (C), sheep (S), and pig (P), and a second set consists of the elements P, S, H, and C, then we say that the two sets are *equal*, as they contain exactly the same elements. The sequence of elements within sets is immaterial in defining equality or inequality of sets.

If a set consisted of animals H and P, it would be declared a *subset* of the above set (H, C, S, P). A subset is a set, all of whose elements are elements of a larger set.[†] Therefore, the determination of combinations of X items taken from a set of n items (Section 3) is really the counting of possible subsets of items from the set of n items.

In an experiment (or other phenomenon that yields results to observe), there is a set (usually very large) of possible outcomes. Let us refer to this set as the *outcome set*.[‡]

Each element of the set is one of the possible outcomes of the experiment. For example, if an experiment consists of tossing two coins, the outcome set consists of four elements: H,H; H,T; T,H; T,T, as these are all of the possible outcomes.

A subset of the outcome set is called an *event*. If the outcome set were the possible rolls of a die: 1, 2, 3, 4, 5, 6, an event might be declared to be "even-numbered rolls" (i.e., 2, 4, 6), and another event might be defined as "rolls greater than 4"

---

*This parenthetical notation for combinations was introduced by Andreas von Ettingshausen in 1826 (Miller, 2004c). Some authors have used a symbol in the form of C_X^n (or ^nC_X) instead of $_nC_X$ for combinations and P_X^n (or ^nP_X) instead of $_nP_X$ for permutations; those symbols will not be used in this text, in order to avoid confusing n with an exponent.

[†]Utilizing the terms *set* and *subset* in this fashion dates from the last half of the nineteenth century (Miller, 2004a).

[‡]Also called the *sample space*.

(i.e., 5, 6). In tossing two coins, one event could be "the two coins land differently" (i.e., T,H; H,T), and another event could be "heads do not appear" (i.e., T,T). If the two events in the same outcome set have some elements in common, the two events are said to intersect; and the *intersection* of the two events is that subset composed of those common elements. For example, the event "even-numbered rolls" of a die (2, 4, 6) and the event "rolls greater than 4" (5, 6) have an element in common (namely, the roll 6); therefore 6 is the intersection of the two events. For the events "even-numbered rolls" (2, 4, 6) and "rolls less than 5" (1, 2, 3, 4), the intersection subset consists of those elements of the events that are both even-numbered and less than 5 (namely, 2, 4).*

If two events have no elements in common, they are said to be *mutually exclusive*, and the two sets are said to be *disjoint*. The set that is the intersection of disjoint sets contains no elements and is often called the *empty set* or the *null set*. For example, the events "odd-numbered rolls" and "even-numbered rolls" are mutually exclusive and there are no elements common to both of them.

If we ask what elements are found in either one event or another, or in both of them, we are speaking of the *union* of the two events. The union of the events "even-numbered rolls" and "rolls less than 5" is that subset of the outcome set that contains elements found in either set (or both sets), namely 1, 2, 3, 4, 6.[†]

Once a subset has been defined, all other elements in the outcome set are said to be the *complement* of that subset. So, if an event is defined as "even-numbered rolls" of a die (2, 4, 6), the complementary subset consists of "odd-numbered rolls" (1, 3, 5). If subset is "rolls less than 5" (1, 2, 3, 4), the complement is the subset consisting of rolls 5 or greater (5, 6).

The above considerations may be presented by what are known as *Venn diagrams*,[‡] shown in Figure 1.

The rectangle in this diagram denotes the outcome set, the set of all possible outcomes from an experiment or other producer of observations. The circle on the

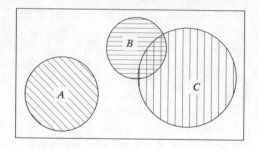

FIGURE 1: A Venn diagram showing the relationships among the outcome set represented by the rectangle and the subsets represented by circles *A*, *B*, and *C*. Subsets *B* and *C* intersect, with no intersection with *A*.

---

*The term *intersection* had been employed in this manner by 1909 (Miller, 2004a). The mathematical symbol for intersection is "∩", first used by Italian mathematician Giuseppe Peano (1858–1932) in 1888 (Miller, 2004a); so, for example, the intersection of set *A* (consisting of 2, 4, 6) and set *B* (consisting of 5, 6) is set $A \cap B$ (consisting of 6).

[†]The term *union* had been employed in this way by 1912 (Miller, 2004a). The mathematical symbol for union is "∪", first used by Giuseppe Peano in 1888 (Miller, 2004a); so, for example, if set *A* is composed of even-numbered rolls of a die (2, 4, 6), and set *B* is odd-numbered rolls (1, 3, 5), the union of the two sets, namely $A \cup B$, is 2, 4, 6, 1, 3, 5.

[‡]Named for English mathematical logician John Venn (1834–1923), who in 1880 greatly improved and popularized the diagrams (sometimes called "Euler diagrams") devised by Leonhard Euler (1707–1783) (Gullberg, 1997: 242; O'Connor and Robertson, 2003).

left represents a subset of the outcome set that we shall refer to as event A, the circle in the center signifies a second subset of the outcome set that we shall refer to as event B, and the circle on the right depicts a third subset of the outcome set that we shall call event C. If, for example, an outcome set (the rectangle) is the number of vertebrate animals in a forest, subset A might be animals without legs (namely, snakes), subset B might be mammals, and subset C might be flying animals. Figure 1 demonstrates graphically what is meant by union, intersection, mutually exclusive, and complementary sets: The union of B and C (the areas with any horizontal or vertical shading) represents all birds and mammals; the intersection of B and C (the area with both horizontal and vertical shading) represents flying mammals (i.e., bats); the portion of C with only vertical shading represents birds; A is mutually exclusive relative to the union of B and C, and the unshaded area (representing all other vertebrates—namely, amphibians and turtles) is complementary to A, B, and C (and is also mutually exclusive of A, B, and C).

5 PROBABILITY OF AN EVENT

We shall define the *relative frequency* of an event as the proportion of the total observations of outcomes that event represents. Consider an outcome set with two elements, such as the possible results from tossing a coin (H; T) or the sex of a person (male; female). If n is the total number of coin tosses and f is the total number of heads observed, then the relative frequency of heads is f/n. Thus, if heads are observed 52 times in 100 coin tosses, the relative frequency is $52/100 = 0.52$ (or 52%). If 275 males occur in 500 human births, the relative frequency of males is $f/n = 275/500 = 0.55$ (or 55%). In general, we may write

$$\text{relative frequency of an event} = \frac{\text{frequency of that event}}{\text{total number of all events}} = \frac{f}{n}. \qquad (15)$$

The value of f may, of course, range from 0 to n, and the relative frequency may, therefore, range from 0 to 1 (or 0% to 100%). A biological example is given as Example 7.

EXAMPLE 7   Relative Frequencies

A sample of 852 vertebrate animals is taken randomly from a forest. The sampling was done *with replacement*, meaning that the animals were taken one at a time, returning each one to the forest before the next one was selected. This is done to prevent the sampling procedure from altering the relative frequency in the sampled population. If the sample size is very small compared to the population size, replacement is not necessary. (Recall that random sampling assumes that each individual animal is equally likely to become a part of the sample.)

Vertebrate Subset	Number	Relative Frequency
amphibians	53	$53/852 = 0.06$
turtles	41	$41/852 = 0.05$
snakes	204	$204/852 = 0.24$
birds	418	$418/852 = 0.49$
mammals	136	$136/852 = 0.16$
total	852	1.00

The *probability* of an event is the likelihood of that event expressed either by the relative frequency observed from a large number of data or by knowledge of the system under study. In Example 7 the relative frequencies of vertebrate groups have been observed from randomly sampling forest animals. If, for the sake of the present example, we assume that each animal has the same chance of being caught as part of our sample (an unrealistic assumption in nature), we may estimate the probability, P, that the next animal captured will be a snake ($P = 0.24$). Or, using the data of the preceding paragraph, we can estimate that the probability that a human birth will be a male is 0.55, or that the probability of tossing a coin that lands head side up is 0.52. A probability may sometimes be predicted on the basis of knowledge about the system (e.g., the structure of a coin or of a die, or the Mendelian principles of heredity). If we assume that there is no reason why a tossed coin should land "heads" more or less often than "tails," we say there is an equal probability of each outcome: $P(H) = \frac{1}{2}$ and $P(T) = \frac{1}{2}$ states that "the probability of heads is 0.5 and the probability of tails is 0.5."

Probabilities, like relative frequencies, can range from 0 to 1. A probability of 0 means that the event is impossible. For example, in tossing a coin, $P(\text{neither H nor T}) = 0$, or in rolling a die, $P(\text{number} > 6) = 0$. A probability of 1 means that an event is certain. For example, in tossing a coin, $P(H \text{ or } T) = 1$; or in rolling a die, $P(1 \leq \text{number} \leq 6) = 1$.*

6 ADDING PROBABILITIES

(a) If Events Are Mutually Exclusive. If two events (call them A and B) are mutually exclusive (e.g., legless vertebrates and mammals are disjoint sets in Figure 1), then the probability of either event A or event B is the sum of the probabilities of the two events:

$$P(A \text{ or } B) = P(A) + P(B). \tag{16}$$

For example, if the probability of a tossed coin landing head up is $\frac{1}{2}$ and the probability of its landing tail up is $\frac{1}{2}$, then the probability of either head or tail up is

$$P(H \text{ or } T) = P(H) + P(T) = \frac{1}{2} + \frac{1}{2} = 1. \tag{17}$$

And, for the data in Example 7, the probability of selecting, at random, a reptile would be $P(\text{turtle or snake}) = P(\text{turtle}) + P(\text{snake}) = 0.05 + 0.24 = 0.29$.

This rule for adding probabilities may be extended for more than two mutually exclusive events. For example, the probability of rolling a 2 on a die is $\frac{1}{6}$, the probability of rolling a 4 is $\frac{1}{6}$, and the probability of rolling a 6 is $\frac{1}{6}$; so the probability of rolling an even number is

$$P(\text{even number}) = P(2 \text{ or } 4 \text{ or } 6) = P(2) + P(4) + P(6)$$
$$= \frac{1}{6} + \frac{1}{6} + \frac{1}{6} = \frac{3}{6} = \frac{1}{2}.$$

---

*A concept related to probability is the *odds* for an event, namely the ratio of the probability of the event occurring and the probability of that event not occurring. For example, if the probability of a male birth is 0.55 (and, therefore, the probability of a female birth is 0.45), then the odds in favor of male births are 0.55/0.45, expressed as "11 to 9."

Probabilities

And, for the data in Example 7, the probability of randomly selecting a reptile or amphibian would be $P(\text{turtle}) + P(\text{snake}) + P(\text{amphibian}) = 0.05 + 0.24 + 0.06 = 0.35$.

(b) If Events Are Not Mutually Exclusive. If two events are not mutually exclusive—that is, they intersect (e.g., mammals and flying vertebrates are not disjoint sets in Figure 1)—then the addition of the probabilities of the two events must be modified. For example, if we roll a die, the probability of rolling an odd number is

$$P(\text{odd number}) = P(1 \text{ or } 3 \text{ or } 5) = P(1) + P(3) + P(5)$$
$$= \frac{1}{6} + \frac{1}{6} + \frac{1}{6} = \frac{3}{6} = \frac{1}{2};$$

and the probability of rolling a number less than 4 is

$$P(\text{number} < 4) = P(1 \text{ or } 2 \text{ or } 3) = P(1) + P(2) + P(3)$$
$$= \frac{1}{6} + \frac{1}{6} + \frac{1}{6} = \frac{3}{6} = \frac{1}{2}.$$

The probability of rolling either an odd number or a number less than 4 obviously is *not* calculated by Equation 16, for that equation would yield

$$P(\text{odd number or number} < 4)$$
$$\overset{?}{=} P(\text{odd}) + P(\text{number} < 4)$$
$$= P[(1 \text{ or } 3 \text{ or } 5) \text{ or } (1 \text{ or } 2 \text{ or } 3)]$$
$$= [P(1) + P(3) + P(5)] + [P(1) + P(2) + P(3)]$$
$$= \left(\frac{1}{6} + \frac{1}{6} + \frac{1}{6}\right) + \left(\frac{1}{6} + \frac{1}{6} + \frac{1}{6}\right) = 1,$$

and that would mean that we are certain ($P = 1$) to roll either an odd number or a number less than 4, which would mean that a roll of 4 or 6 is impossible!

The invalidity of the last calculation is due to the fact that the two elements (namely 1 and 3) that lie in both events are counted twice. The subset of elements consisting of rolls 1 and 3 is the intersection of the two events and its probability needs to be subtracted from the preceding computation so that $P(1 \text{ or } 3)$ is counted once, not twice. Therefore, for two intersecting events, A and B, the probability of either A or B is

$$P(A \text{ or } B) = P(A) + P(B) - P(A \text{ and } B). \tag{18}$$

In the preceding example,

$$P(\text{odd number or number} < 4)$$
$$= P(\text{odd number}) + P(\text{number} < 4)$$
$$\quad - P(\text{odd number and number} < 4)$$
$$= P[(1 \text{ or } 3 \text{ or } 5) \text{ or } (1 \text{ or } 2 \text{ or } 3)] - P(1 \text{ or } 3)$$
$$= [P(1) + P(3) + P(5)] + [P(1) + P(2) + P(3)] - [P(1) + P(3)]$$
$$= \left(\frac{1}{6} + \frac{1}{6} + \frac{1}{6}\right) + \left(\frac{1}{6} + \frac{1}{6} + \frac{1}{6}\right) - \left(\frac{1}{6} + \frac{1}{6}\right) = \frac{4}{6} = \frac{2}{3}.$$

It may be noted that Equation 16 is a special case of Equation 18, where $P(A$ and $B) = 0$. Example 8 demonstrates these probability calculations with a different set of data.

EXAMPLE 8   **Adding Probabilities of Intersecting Events**

A deck of playing cards is composed of 52 cards, with thirteen cards in each of four suits called clubs, diamonds, hearts, and spades. In each suit there is one card each of the following thirteen denominations: ace (A), 2, 3, 4, 5, 6, 7, 8, 9, 10, jack (J), queen (Q), king (K). What is the probability of selecting at random a diamond from the deck of 52 cards?

The event in question (diamonds) is a subset with thirteen elements; therefore,

$$P(\text{diamond}) = \frac{13}{52} = \frac{1}{4} = 0.250.$$

What is the probability of selecting at random a king from the deck?
The event in question (king) has four elements; therefore,

$$P(\text{king}) = \frac{4}{52} = \frac{1}{13} = 0.077.$$

What is the probability of selecting at random a diamond or a king?
The two events (diamonds and kings) intersect, with the intersection having one element (the king of diamonds); therefore,

$$P(\text{diamond or king}) = P(\text{diamond}) + P(\text{king}) - P(\text{diamond and king})$$
$$= \frac{13}{52} + \frac{4}{52} - \frac{1}{52}$$
$$= \frac{16}{52} = \frac{4}{13} = 0.308.$$

If three events are not mutually exclusive, the situation is more complex, yet straightforward. As seen in Figure 2, there may be three two-way intersections, shown with vertical shading (A and B; A and C; and B and C), and a three-way

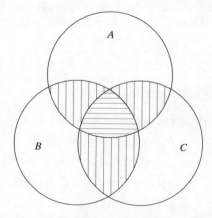

FIGURE 2: A Venn diagram showing three intersecting sets: *A*, *B*, and *C*. Here there are three two-way intersections (vertical shading) and one three-way intersection (horizontal shading).

intersection, shown with horizontal shading (*A* and *B* and *C*). If we add the probabilities of the three events, *A*, *B*, and *C*, as $P(A) + P(B) + P(C)$, we are adding the two-way intersections twice. So, we can subtract $P(A$ and $B)$, $P(A$ and $C)$, and $P(B$ and $C)$. Also, the three-way intersection is added three times in $P(A) + P(B) + P(C)$, and subtracted three times by subtracting the three two-way intersections; thus, $P(A$ and B and $C)$ must be added back into the calculation. Therefore, for three events, not mutually exclusive,

$$P(A \text{ or } B \text{ or } C) = P(A) + P(B) + P(C)$$
$$- P(A \text{ and } B) - P(A \text{ and } C) - P(B \text{ and } C) \quad (19)$$
$$+ P(A \text{ and } B \text{ and } C).$$

7 MULTIPLYING PROBABILITIES

If two or more events intersect (as *A* and *B* in Figure 1 and *A*, *B*, and *C* in Figure 2), the probability associated with the intersection is the product of the probabilities of the individual events. That is,

$$P(A \text{ and } B) = [P(A)][P(B)], \quad (20)$$
$$P(A \text{ and } B \text{ and } C) = [P(A)][P(B)][P(C)], \quad (21)$$

and so on.

For example, the probability of a tossed coin landing heads is $\frac{1}{2}$. If two coins are tossed, the probability of *both* coins landing heads is

$$P(H,H) = [P(H)][P(H)] = \left(\frac{1}{2}\right)\left(\frac{1}{2}\right) = \left(\frac{1}{4}\right) = 0.25.$$

This can be verified by examining the outcome set.

H,H; H,T; T,H; T,T,

where $P(H,H)$ is one outcome out of four equally likely outcomes. The probability that 3 tossed coins will land heads is

$$P(H,H,H) = [P(H)][P(H)][P(H)] = \left(\frac{1}{2}\right)\left(\frac{1}{2}\right)\left(\frac{1}{2}\right) = \left(\frac{1}{8}\right) = 0.125.$$

Note, however, that if one or more coins have already been tossed, the probability that the next coin toss (of the same or a different coin) will be heads is simply $\frac{1}{2}$.

8 CONDITIONAL PROBABILITIES

There are occasions when our interest will be in determining a *conditional probability*, which is the probability of one event with the stipulation that another event also occurs. An illustration of this, using a deck of 52 playing cards (as described in Example 8), would be the probability of selecting a queen, given that the card is a spade. In general, a conditional probability is

$$P(\text{event } A, \text{ given event } B) = \frac{P(A \text{ and } B \text{ jointly})}{P(B)}, \quad (22)$$

which can also be calculated as

$$P(\text{event } A, \text{ given event } B) = \frac{\text{frequency of events } A \text{ and } B \text{ jointly}}{\text{frequency of event } B}. \quad (23)$$

Probabilities

So, the probability of randomly selecting a queen, with the specification that the card is a spade, is (using Equation 22)

$$P(\text{queen, given it is a spade}) = \frac{P(\text{queen of spades})}{P(\text{spade})}$$
$$= (1/52)/(13/52) = 0.02/0.25 = 0.08,$$

which (by Equation 23) would be calculated as

$$P(\text{queen, given it is a spade}) = \frac{\text{frequency of queen of spades}}{\text{frequency of spades}}$$
$$= 1/13 = 0.8.$$

Note that this conditional probability is quite different from the probability of selecting a spade, given that the card is a queen, for that would be (by Equation 23)

$$P(\text{spade, given it is a queen}) = \frac{\text{frequency of queen of spades}}{\text{frequency of queens}}$$
$$= 1/4 = 0.25.$$

EXERCISES

1. A person may receive a grade of either high (H), medium (M), or low (L) on a hearing test, and a grade of either good (G) or poor (P) on a sight test.
    (a) How many different outcomes are there if both tests are taken?
    (b) What are these outcomes?

2. A menu lists three meats, four salads, and two desserts. In how many ways can a meal of one meat, one salad, and one dessert be selected?

3. If an organism (e.g., human) has 23 pairs of chromosomes in each diploid cell, how many different gametes are possible for the individual to produce by assortment of chromosomes?

4. In how many ways can five animal cages be arranged on a shelf?

5. In how many ways can 12 different amino acids be arranged into a polypeptide chain of five amino acids?

6. An octapeptide is known to contain four of one amino acid, two of another, and two of a third. How many different amino-acid sequences are possible?

7. Students are given a list of nine books and told that they will be examined on the contents of five of them. How many combinations of five books are possible?

8. The four human blood types below are genetic phenotypes that are mutually exclusive events. Of 5400 individuals examined, the following frequency of each blood type is observed. What is the relative frequency of each blood type?

Blood Type	Frequency
O	2672
A	2041
B	486
AB	201

9. An aquarium contains the following numbers of tropical freshwater fishes. What is the relative frequency of each species?

Species	Number
Paracheirodon innesi, neon tetra	11
Cheirodon axelrodi, cardinal tetra	6
Pterophyllum scalare, angelfish	4
Pterophyllum altum, angelfish	2
Pterophyllum dumerilii, angelfish	2
Nannostomus marginatus, one-lined pencilfish	2
Nannostomus anomalus golden pencilfish	2

10. Use the data of Exercise 8, assuming that each of the 5400 has an equal opportunity of being encountered.
  (a) Estimate the probability of encountering a person with type A blood.
  (b) Estimate the probability of encountering a person who has either type A or type AB blood.

11. Use the data of Exercise 9, assuming that each individual fish has the same probability of being encountered.
  (a) Estimate the probability of encountering an angelfish of the species *Pterophyllum scalare*.
  (b) Estimate the probability of encountering a fish belonging to the angelfish genus *Pterophyllum*.

12. Either allele A or a may occur at a particular genetic locus. An offspring receives one of its alleles from each of its parents. If one parent possesses alleles A and a and the other parent possesses a and a:
  (a) What is the probability of an offspring receiving an A and an a?
  (b) What is the probability of an offspring receiving two a alleles?
  (c) What is the probability of an offspring receiving two A alleles?

13. In a deck of playing cards (see Example 8 for a description),
  (a) What is the probability of selecting a queen of clubs?
  (b) What is the probability of selecting a black (i.e., club or spade) queen?
  (c) What is the probability of selecting a black face card (i.e., a black jack, queen, or king)?

14. A cage contains six rats, two of them white (W) and four of them black (B); a second cage contains four rats, two white and two black; and a third cage contains five rats, three white and two black. If one rat is selected randomly from each cage,
  (a) What is the probability that all three rats selected will be white?
  (b) What is the probability that exactly two of the three will be white?
  (c) What is the probability of selecting at least two white rats?

15. A group of dogs consists of three brown males, two brown females, four white males, four white females, five black males, and four black females. What is the probability of selecting at random
  (a) A brown female dog?
  (b) A female dog, if the dog is brown?
  (c) A brown dog, if the dog is a female?

ANSWERS TO EXERCISES

1. (a) $(3)(2) = 6$; **(b)** H,G H,P M,G M,P L,G L,P.

2. $(3)(4)(2) = 24$.

3. $2^{23} = 8,388,608$.

4. $_5P_5 = 5! = 120$.

5. $_{12}P_5 = 12!/7! = 95,040$.

6. $_8P_{4,2,2} = 8!/[4!2!2!] = 420$.

7. $_9C_5 = 9!/(5!4!) = 126$.

8. O: 0.49; A: 0.38; B: 0.09; AB: 0.04.

9. $n = 29$; $0.38, 0.21, 0.14, 0.07, 0.07, 0.07, 0.07$.

10. (a) $P = 0.38$; **(b)** $P = 0.38 + 0.04 = 0.42$.

11. (a) $P = 4/29 = 0.14$;
  (b) $P = 4/29 + 2/29 + 2/29 = 0.28$.

12. (a) $P = \left(\frac{1}{2}\right)(1) = \left(\frac{1}{2}\right) = 0.5$; **(b)** $P = \left(\frac{1}{2}\right)(1) = \left(\frac{1}{2}\right) = 0.5$; **(c)** $P = \left(\frac{1}{2}\right)(0) = 0$.

13. (a) $P = \left(\frac{1}{13}\right)\left(\frac{1}{4}\right) = \frac{1}{52} = 0.019$; **(b)** $P = \left(\frac{1}{4} + \frac{1}{4}\right)\left(\frac{1}{13}\right) = \frac{1}{26} = 0.038$;

  (c) $P = \left(\frac{1}{2}\right)\left(\frac{3}{13}\right) = \frac{3}{26} = 0.12$.

14. (a) $P(\text{all 3 white}) = [P(W)][P(W)][P(W)] = \left(\frac{2}{6}\right)\left(\frac{2}{4}\right)\left(\frac{3}{5}\right) = \frac{12}{120} = 0.10$; **(b)** $P(2 \text{ white}) = [P(W)][P(W)][P(B)] + [P(W)][P(B)][P(W)] + [P(B)][P(W)][P(W)] = \left(\frac{2}{6}\right)\left(\frac{2}{4}\right)\left(\frac{2}{5}\right) + \left(\frac{2}{6}\right)\left(\frac{2}{4}\right)\left(\frac{3}{5}\right) + \left(\frac{4}{6}\right)\left(\frac{2}{4}\right)\left(\frac{3}{5}\right) = \frac{8}{120} + \frac{12}{120} + \frac{24}{120} = \frac{44}{120} = 0.37$.

  (c) $P(2 \text{ or } 3 \text{ white}) = 0.10 + 0.37 = 0.47$.

15. (a) $P = 3/22 = 0.14$; **(b)** $P = 2/5 = 0.40$;
  (c) $P = 3/10 = 0.30$.

This page intentionally left blank

The Normal Distribution

The Normal Distribution

Commonly, a distribution of interval- or ratio-scale data is observed to have a preponderance of values around the mean with progressively fewer observations toward the extremes of the range of values. If n is large, the frequency polygons of many biological data distributions are "bell-shaped"* and look something like Figure 1.

Figure 1 is a frequency curve for a *normal distribution*.[†] Not all bell-shaped curves are normal; although biologists are unlikely to need to perform calculations with this equation, it can be noted that a *normal distribution* is defined as one in which height of the curve at X_i is as expressed by the relation:

$$Y_i = \frac{1}{\sigma \sqrt{2\pi}} e^{-(X_i - \mu)^2 / 2\sigma^2}.$$

(1)

The height of the curve, Y_i, is referred to as the *normal density*. It is not a frequency, for in a normally distributed population of continuous data the frequency of occurrence of a measurement *exactly* equal to X_i (e.g., exactly equal to 12.5000 cm, or exactly equal to 12.50001 cm) is zero. Equation 1 contains two mathematical constants:

---

*Comparing the curve's shape to that of a bell has been traced as far back as 1872 (Stigler, 199: 405).

[†]The normal distribution is sometimes called the *Gaussian distribution*, after [Johann] Karl Friedrich Gauss (1777–1855), a phenomenal German mathematician contributing to many fields of mathematics and for whom the unit of magnetic induction ("gauss") is named. Gauss discussed this distribution in 1809, but the influential French mathematician and astronomer Pierre-Simon Laplace (1749–1827) mentioned it in 1774, and it was first announced in 1733 by mathematician Abraham de Moivre (1667–1754; also spelled De Moivre and Demoivre), who was born in France but emigrated to England at age 21 (after three years in prison) to escape religious persecution as a Protestant (David, 1962: 161–178; Pearson, 1924; Stigler, 1980; Walker, 1934). This situation has been cited as an example of "Stigler's Law of Eponymy," which states that "no scientific discovery is named after its original discoverer" (Stigler, 1980). The distribution was first used, by de Moivre, to approximate a binomial distribution (Stigler, 1999: 407). The adjective *normal* was first used for the distribution by Charles S. Peirce in 1873, and by Wilhelm Lexis and Sir Francis Galton in 1877 (Stigler, 1999: 404–415); Karl Pearson recommended the routine use of that term to avoid "an international question of priority" although it "has the disadvantage of leading people to believe that all other distributions of frequency are in one sense or another 'abnormal' " (Pearson, 1920).

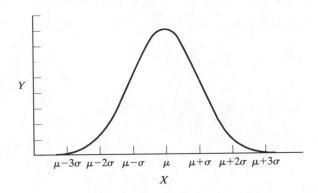

FIGURE 1: A normal distribution.

π (lowercase Greek pi),* which equals 3.14159...; and e (the base of Naperian, or natural, logarithms),† which equals 2.71828.... There are also two parameters (μ and σ^2) in the equation. Thus, for any given standard deviation, σ, there are an infinite number of normal curves possible, depending on μ. Figure 2a shows normal curves for $\sigma = 1$ and $\mu = 0, 1$, and 2. Likewise, for any given mean, μ, an infinity of normal curves is possible, each with a different value of σ. Figure 2b shows normal curves for $\mu = 0$ and $\sigma = 1, 1.5$, and 2.

A normal curve with $\mu = 0$ and $\sigma = 1$ is said to be a *standardized normal curve*. Thus, for a standardized normal distribution,

$$Y_i = 1\sqrt{2\pi}\, e^{-X_i^2/2}. \tag{2}$$

---

*The lowercase Greek letter pi, π, denotes the ratio between the circumference and the diameter of a circle. This symbol was advanced in 1706 by Wales-born William Jones (1675–1749), after it had been used for over 50 years to represent the circumference (Cajori, 1928/9, Vol. II: 9; Smith, 1953: 312); but it did not gain popularity for this purpose until Swiss Leonhard Euler (1707–1783) began using it in 1736 instead of p (Blatner, 1997: 78; Smith, 1953: 312). According to Gullberg (1997: 85), Jones probably selected this symbol because it is the first letter of the Greek word for "periphery." Pi is an "irrational number," meaning that it cannot be expressed as the ratio of two integers. To 20 decimal places its value is 3.14159 26535 89792 33846 (and it may be noted that this number rounded to 10 decimal places is sufficient to obtain, from the diameter, the circumference of a circle as large as the earth's equator to within about a centimeter of accuracy). Beckmann (1977), Blatner (1997), and Dodge (1996) present the history of π and its calculation. By 2000 B.C.E., the Babylonians knew its value to within 0.02. Archimedes of Syracuse (287–212 B.C.E.) was the first to present a procedure to calculate π to any desired accuracy, and he computed it accurate to the third decimal place. Many computational methods were subsequently developed, and π was determined to six decimal places of accuracy by around 500 C.E., to 20 decimal places by around 1600, and to 100 in 1706; 1000 decimal places were reached, using a mechanical calculating machine, before electronic computers joined the challenge in 1949. In the computer era, with advancement of machines and algorithms, one million digits were achieved in 1973, by the end of the 1980s there were calculations accurate to more than a billion digits, and more than one trillion (1,000,000,000,000) digits have now been attained.

†e is an irrational number (as is π; see the preceding footnote). To 20 decimal places e is 2.71828 18284 59045 23536. The symbol, e, for this quantity was introduced by the great Swiss mathematician Leonhard Euler (1707–1783) in 1727 or 1728 and published by him in 1736 (Cajori, 1928/9, Vol. 2: 13; Gullberg, 1997: 85). Johnson and Leeming (1990) discussed the randomness of the digits of e, and Maor (1994) presented a history of this number and its mathematical ramifications. In 2000, e was calculated to 17 billion decimal places (Adrian, 2006: 63).

The Normal Distribution

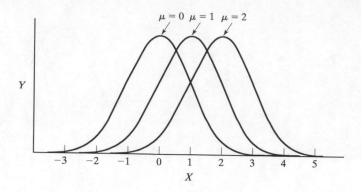

FIGURE 2a: Normal distribution with $\sigma = 1$, varying in location with different means (μ).

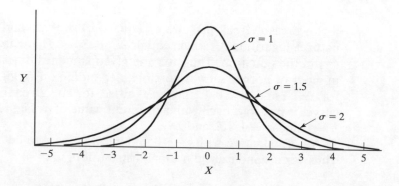

FIGURE 2b: Normal distributions with $\mu = 0$, varying in spread with different standard deviations (σ).

1 PROPORTIONS OF A NORMAL DISTRIBUTION

If a population of 1000 body weights is normally distributed and has a mean, μ, of 70 kg, one-half of the population (500 weights) is larger than 70 kg and one-half is smaller. This is true simply because the normal distribution is symmetrical. But if we desire to ask what portion of the population is larger than 80 kg, we need to know σ, the standard deviation of the population. If $\sigma = 10$ kg, then 80 kg is one standard deviation larger than the mean, and the portion of the population in question is the shaded area in Figure 3a. If, however, $\sigma = 5$ kg, then 80 kg is two standard deviations above μ, and we are referring to a relatively small portion of the population, as shown in Figure 3b.

Table 2 from *Appendix: Statistical Tables and Graphs* enables us to determine proportions of normal distributions. For any X_i value from a normal population with mean μ, and standard deviation σ, the value

$$Z = \frac{X_i - \mu}{\sigma} \tag{3}$$

tells us how many standard deviations from the mean the X_i value is located. Carrying out the calculation of Equation 3 is known as *normalizing*, or *standardizing*, X_i; and Z is known as a *normal deviate*, or a *standard score.** The mean of a set of standard scores is 0, and the variance is 1.

---

*This standard normal curve was introduced in 1899 by W. F. Sheppard (Walker, 1929: 188), and the term *normal deviate* was first used, in 1907, by F. Galton (David, 1995).

74

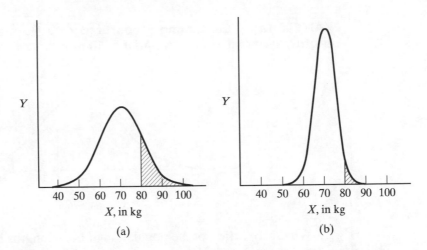

FIGURE 3: Two normal distributions with $\mu = 70$ kg. The shaded areas are the portions of the curves that lie above $X = 80$ kg. For distribution (a), $\mu = 70$ kg and $\sigma = 10$ kg; for distribution (b), $\mu = 70$ kg and $\sigma = 5$ kg.

Table 2 from *Appendix: Statistical Tables and Graphs* tells us what proportion of a normal distribution lies beyond a given value of Z.* If $\mu = 70$ kg, $\sigma = 10$ kg, and $X_i = 70$ kg, then $Z = (70 \text{ kg} - 70 \text{ kg})/10 \text{ kg} = 0$, and by consulting Table 2 from *Appendix: Statistical Tables and Graphs* we see that $P(X_i > 70 \text{ kg}) = P(Z > 0) = 0.5000$.[†] That is, 0.5000 (or 50.00%) of the distribution is larger than 70 kg. To determine the proportion of the distribution that is greater than 80 kg in weight, $Z = (80 \text{ kg} - 70 \text{ kg})/10 \text{ kg} = 1$, and $P(X_i > 80 \text{ kg}) = P(Z > 1) = 0.1587$ (or 15.87%). This could be stated as being the probability of drawing at random a measurement, X_i, greater than 80 kg from a population with a mean (μ) of 70 kg and a standard deviation (σ) of 10 kg. What, then, is the probability of obtaining, at random, a measurement, X_i, which is less than 80 kg? $P(X_i > 80 \text{ kg}) = 0.1587$, so $P(X_i < 80 \text{ kg}) = 1.0000 - 0.1587 = 0.8413$; that is, if 15.87% of the population is greater than X_i, then 100% − 15.87% (i.e., 84.13% of the population is less than X_i).[‡] Example 1a presents calculations for determining proportions of a normal distribution lying between a variety of limits.

Note that Table 2 from *Appendix: Statistical Tables and Graphs* contains no negative values of Z. However, if we are concerned with proportions in the left half of the distribution, we are simply dealing with areas of the curve that are mirror images of those present in the table. This is demonstrated in Example 1b.[§]

---

*The first tables of areas under the normal curve were published in 1799 by Christian Kramp (Walker, 1929: 58). Today, some calculators and many computer programs determine normal probabilities (e.g., see Boomsma and Molenaar, 1994).

†Read $P(X_i > 70 \text{ kg})$ as "the probability of an X_i greater than 70 kg"; $P(Z > 0)$ is read as "the probability of a Z greater than 0."

‡The statement that "$P(X_i > 80 \text{ kg}) = 0.1587$, therefore $P(X_i < 80) = 1.0000 - 0.1587$" does not take into account the case of $X_i = 80$ kg. But, as we are considering the distribution at hand to be a continuous one, the probability of X_i being *exactly* 80.000 ... kg (or being *exactly* any other stated value) is practically nil, so these types of probability statements offer no practical difficulties.

§Some old literature avoided referring to negative Z's by expressing the quantity, $Z + 5$, called a *probit*. This term was introduced in 1934 by C. I. Bliss (David, 1995).

The Normal Distribution

EXAMPLE 1a   Calculating Proportions of a Normal Distribution of Bone Lengths, Where $\mu = 60$ mm and $\sigma = 10$ mm

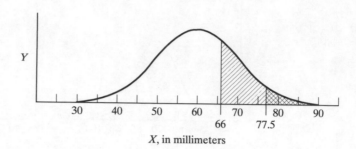

1. What proportion of the population of bone lengths is larger than 66 mm?

$$Z = \frac{X_i - \mu}{\sigma} = \frac{66 \text{ mm} - 60 \text{ mm}}{10 \text{ mm}} = 0.60$$

$$P(X_i > 66 \text{ mm}) = P(Z > 0.60) = 0.2743 \text{ or } 27.43\%$$

2. What is the probability of picking, at random from this population, a bone larger than 66 mm? This is simply another way of stating the quantity calculated in part (1). The answer is 0.2743.

3. If there are 2000 bone lengths in this population, how many of them are greater than 66 mm?

$$(0.2743)(2000) = 549$$

4. What proportion of the population is smaller than 66 mm?

$$P(X_i < 66 \text{ mm}) = 1.0000 - P(X_i > 66 \text{ mm}) = 1.0000 - 0.2743 = 0.7257$$

5. What proportion of this population lies between 60 and 66 mm? Of the total population, 0.5000 is larger than 60 mm and 0.2743 is larger than 66 mm. Therefore, $0.5000 - 0.2743 = 0.2257$ of the population lies between 60 and 66 mm. That is, $P(60 \text{ mm} < X_i < 66 \text{ mm}) = 0.5000 - 0.2743 = 0.2257$.

6. What portion of the area under the normal curve lies to the right of 77.5 mm?

$$Z = \frac{77.5 \text{ mm} - 60 \text{ mm}}{10 \text{ mm}} = 1.75$$

$$P(X_i > 77.5 \text{ mm}) = P(Z > 1.75) = 0.0401 \text{ or } 4.01\%$$

7. If there are 2000 bone lengths in the population, how many of them are larger than 77.5 mm?

$$(0.0401)(2000) = 80$$

8. What is the probability of selecting at random from this population a bone measuring between 66 and 77.5 mm in length?

$$P(66 \text{ mm} < X_i < 77.5 \text{ mm}) = P(0.60 < Z < 1.75) = 0.2743 - 0.0401$$
$$= 0.2342$$

EXAMPLE 1b   **Calculating Proportions of a Normal Distribution of Sucrose Concentrations, Where $\mu = 65$ mg/100 ml and $\sigma = 25$ mg/100 ml**

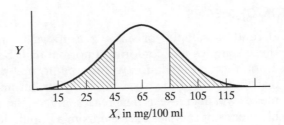

1. What proportion of the population is greater than 85 mg/100 ml?

$$Z = \frac{(X_i - \mu)}{\sigma} = \frac{85 \text{ mg/100 ml} - 65 \text{ mg/100 ml}}{25 \text{ mg/100 ml}} = 0.8$$

$$P(X_i > 85 \text{ mg/100 ml}) = P(Z > 0.8) = 0.2119 \text{ or } 21.19\%$$

2. What proportion of the population is less than 45 mg/100 ml?

$$Z = \frac{45 \text{ mg/100 ml} - 65 \text{ mg/100 ml}}{25 \text{ mg/100 ml}} = -0.80$$

$$P(X_i < 45 \text{ mg/100 ml}) = P(Z < -0.80) = P(Z > 0.80) = 0.2119$$

That is, the probability of selecting from this population an observation less than 0.80 standard deviations below the mean is equal to the probability of obtaining an observation greater than 0.80 standard deviations above the mean.

3. What proportion of the population lies between 45 and 85 mg/100 ml?

$$
\begin{aligned}
P(45 \text{ mg/100 ml} < X_i < 85 \text{ mg/100 ml}) &= P(-0.80 < Z < 0.80) \\
&= 1.0000 - P(Z < -0.80 \\
&\qquad \text{or } Z > 0.80) \\
&= 1.0000 - (0.2119 + 0.2119) \\
&= 1.0000 - 0.4238 \\
&= 0.5762
\end{aligned}
$$

Using the preceding considerations of the table of normal deviates (Table 2), we can obtain the following information for measurements in a normal population:

The interval of $\mu \pm \sigma$ will contain 68.27% of the measurements.*
The interval of $\mu \pm 2\sigma$ will contain 95.44% of the measurements.
The interval of $\mu \pm 2.5\sigma$ will contain 98.76% of the measurements.
The interval of $\mu \pm 3\sigma$ will contain 99.73% of the measurements.
50% of the measurements lie within $\mu \pm 0.67\sigma$.
95% of the measurements lie within $\mu \pm 1.96\sigma$.
97.5% of the measurements lie within $\mu \pm 2.24\sigma$.

---

*The symbol "$\pm$" indicates "plus or minus" and was first published by William Oughtred in 1631 (Cajori, 1928: 245).

99% of the measurements lie within $\mu \pm 2.58\sigma$.
99.5% of the measurements lie within $\mu \pm 2.81\sigma$.
99.9% of the measurements lie within $\mu \pm 3.29\sigma$.

2 THE DISTRIBUTION OF MEANS

If random samples of size n are drawn from a normal population, the means of these samples will conform to normal distribution. The distribution of means from a nonnormal population will not be normal but will tend to approximate a normal distribution as n increases in size.* Furthermore, the variance of the distribution of means will decrease as n increases; in fact, the variance of the population of all possible means of samples of size n from a population with variance σ^2 is

$$\sigma_{\overline{X}}^2 = \frac{\sigma^2}{n}. \tag{4}$$

The quantity $\sigma_{\overline{X}}^2$ is called the *variance of the mean*. A distribution of sample statistics is called a *sampling distribution*[†]; therefore, we are discussing the sampling distribution of means.

Since $\sigma_{\overline{X}}^2$ has square units, its square root, $\sigma_{\overline{X}}$, will have the same units as the original measurements (and, therefore, the same units as the mean, μ, and the standard deviation, σ). This value, $\sigma_{\overline{X}}$, is the *standard deviation of the mean*. The standard deviation of a statistic is referred to as a *standard error*; thus, $\sigma_{\overline{X}}$ is frequently called the *standard error of the mean* (sometimes abbreviated SEM), or simply the *standard error* (sometimes abbreviated SE)[‡]:

$$\sigma_{\overline{X}} = \sqrt{\frac{\sigma^2}{n}} \quad \text{or} \quad \sigma_{\overline{X}} = \frac{\sigma}{\sqrt{n}}. \tag{5}$$

Just as $Z = (X_i - \mu)/\sigma$ (Equation 3) is a normal deviate that refers to the normal distribution of X_i values,

$$Z = \frac{\overline{X} - \mu}{\sigma_{\overline{X}}} \tag{6}$$

is a normal deviate referring to the normal distribution of means ($\overline{X}$ values). Thus, we can ask questions such as: What is the probability of obtaining a random sample of nine measurements with a mean larger than 50.0 cm from a population having a mean of 47.0 cm and a standard deviation of 12.0 cm? This and other examples of the use of normal deviates for the sampling distribution of means are presented in Example 2.

As seen from Equation 5, to determine $\sigma_{\overline{X}}$ one must know σ^2 (or σ), which is a population parameter. Because we very seldom can calculate population parameters, we must rely on estimating them from random samples taken from the population. The best estimate of $\sigma_{\overline{X}}^2$, the population variance of the mean, is

$$s_{\overline{X}}^2 = \frac{s^2}{n}, \tag{7}$$

---

*This result is known as the *central limit theorem*.

[†]This term was apparently first used by Ronald Aylmer Fisher in 1922 (Miller, 2004a).

[‡]This relationship between the standard deviation of the mean and the standard deviation was published by Karl Friedrich Gauss in 1809 (Walker, 1929: 23). The term *standard error* was introduced in 1897 by G. U. Yule (David, 1995), though in a different context (Miller, 2004a).

EXAMPLE 2   Proportions of a Sampling Distribution of Means

1. A population of one-year-old children's chest circumferences has $\mu = 47.0$ cm and $\sigma = 12.0$ cm, what is the probability of drawing from it a random sample of nine measurements that has a mean larger than 50.0 cm?

$$\sigma_{\overline{X}} = \frac{12.0 \text{ cm}}{\sqrt{9}} = 4.0 \text{ cm}$$

$$Z = \frac{\overline{X} - \mu}{\sigma_{\overline{X}}} = \frac{50.0 \text{ cm} - 47.0 \text{ cm}}{4.0 \text{ cm}} = 0.75$$

$$P(\overline{X} > 50.0 \text{ cm}) = P(Z > 0.75) = 0.2266$$

2. What is the probability of drawing a sample of 25 measurements from the preceding population and finding that the mean of this sample is less than 40.0 cm?

$$\sigma_{\overline{X}} = \frac{12.0 \text{ cm}}{\sqrt{25}} = 2.4 \text{ cm}$$

$$Z = \frac{40.0 \text{ cm} - 47.0 \text{ cm}}{2.4 \text{ cm}} = -2.92$$

$$P(X < 40.0 \text{ cm}) = P(Z < -2.92) = P(Z > 2.92) = 0.0018$$

3. If 500 random samples of size 25 are taken from the preceding population, how many of them would have means larger than 50.0 cm?

$$\sigma_{\overline{X}} = \frac{12.0 \text{ cm}}{\sqrt{25}} = 2.4 \text{ cm}$$

$$Z = \frac{50.0 \text{ cm} - 47.0 \text{ cm}}{2.4 \text{ g}} = 1.25$$

$$P(\overline{X} > 50.0 \text{ cm}) = P(Z > 1.25) = 0.1056$$

Therefore, $(0.1056)(500) = 53$ samples would be expected to have means larger than 50.0 cm.

---

the sample variance of the mean. Thus,

$$s_{\overline{X}} = \sqrt{\frac{s^2}{n}} \text{ or } s_{\overline{X}} = \frac{s}{\sqrt{n}} \tag{8}$$

is an estimate of $\sigma_{\overline{X}}$ and is the sample standard error of the mean. Example 3 demonstrates the calculation of $s_{\overline{X}}$.

At this point, it can be noted that the magnitude of $s_{\overline{X}}$ is helpful in determining the precision to which the mean and some measures of variability may be reported. Although different practices have been followed by many, we shall employ the following (Eisenhart, 1968). We shall state the standard error to two significant figures (e.g.,

2.7 mm in Example 3;). Then the standard deviation and the mean will be reported with the same number of decimal places (e.g., $\overline{X} = 137.6$ mm in Example 3*). The variance may be reported with twice the number of decimal places as the standard deviation.

EXAMPLE 3   **The Calculation of the Standard Error of the Mean, $s_{\overline{X}}$**

The Following are Data for Systolic Blood Pressures, in mm of Mercury, of 12 Chimpanzees.

121	$n = 12$
125	
128	$\overline{X} = \dfrac{1651 \text{ mm}}{12} = 137.6$ mm
134	
136	$SS = 228{,}111 \text{ mm}^2 - \dfrac{(1651 \text{ mm})^2}{12}$
138	
139	$= 960.9167 \text{ mm}^2$
141	
144	$s^2 = \dfrac{960.9167 \text{ mm}^2}{11} = 87.3561 \text{ mm}^2$
145	
149	$s = \sqrt{87.3561 \text{ mm}^2} = 9.35$ mm
151	

$$\sum X = 1651 \text{ mm}$$
$$\sum X^2 = 228{,}111 \text{ mm}^2$$

$$s_{\overline{X}} = \frac{s}{\sqrt{n}} = \frac{9.35 \text{ mm}}{\sqrt{12}} = 2.7 \text{ mm } or$$

$$s_{\overline{X}} = \sqrt{\frac{s^2}{n}} = \sqrt{\frac{87.3561 \text{ mm}^2}{12}} = \sqrt{7.2797 \text{ mm}^2} = 2.7 \text{ mm}$$

---

3   INTRODUCTION TO STATISTICAL HYPOTHESIS TESTING

A major goal of statistical analysis is to draw inferences about a population by examining a sample from that population. A very common example of this is the desire to draw conclusions about one or more population means.

We begin by making a concise statement about the population mean, a statement called a *null hypothesis* (abbreviated H_0)[†] because it expresses the concept of "no difference." For example, a null hypothesis about a population mean (μ) might assert that μ is not different from zero (i.e., μ is equal to zero); and this would be written as

$$H_0: \mu = 0.$$

Or, we could hypothesize that the population mean is not different from (i.e., is equal to) 3.5 cm, or not different from 10.5 kg, in which case we would write $H_0: \mu = 3.5$ cm or $H_0: \mu = 10.5$ kg, respectively.

---

*In Example 3, s is written with more decimal places than the Eisenhart recommendations indicate because it is an intermediate, rather than a final, result; and rounding off intermediate computations may lead to serious rounding error. Indeed, some authors routinely report extra decimal places, even in final results, with the consideration that readers of the results may use them as intermediates in additional calculations.

[†]The term *null hypothesis* was first published by R. A. Fisher in 1935 (David, 1995; Miller, 2004a; Pearson, 1947). J. Neyman and E. S. Pearson were the first to use the symbol "H_0" and the term *alternate hypothesis*, in 1928 (Pearson, 1947; Miller, 2004a, 2004c). The concept of statistical testing of something akin to a null hypothesis was introduced 300 years ago by John Arbuthnot (1667–1725), a Scottish–English physician and mathematician (Stigler, 1986: 225–226).

The Normal Distribution

If statistical analysis concludes that it is likely that a null hypothesis is false, then an *alternate hypothesis* (abbreviated H_A or H_1) is assumed to be true (at least tentatively). One states a null hypothesis and an alternate hypothesis for each statistical test performed, and all possible outcomes are accounted for by this pair of hypotheses. So, for the preceding examples,[*]

$$H_0: \mu = 0, \quad H_A: \mu \neq 0;$$

$$H_0: \mu = 3.5 \text{ cm}, \quad H_A: \mu \neq 3.5 \text{ cm};$$

$$H_0: \mu = 10.5 \text{ kg}, \quad H_A: \mu \neq 10.5 \text{ kg}.$$

It must be emphasized that statistical hypotheses are to be stated *before* data are collected to test them. To propose hypotheses after examination of data can invalidate a statistical test. One may, however, legitimately formulate hypotheses *after* inspecting data if a new set of data is then collected with which to test the hypotheses.

(a) Statistical Testing and Probability. Statistical testing of a null hypothesis about μ, the mean of a population, involves calculating $\overline{X}$, the mean of a random sample from that population. $\overline{X}$ is the best estimate of μ; but it is only an estimate, and we can ask, What is the probability of an $\overline{X}$ at least as far from the hypothesized μ as is the $\overline{X}$ in the sample, *if* H$_0$ *is true*? Another way of visualizing this is to consider that, instead of obtaining one sample (of size n) from the population, a large number of samples (each sample of size n) could have been taken from that population. We can ask what proportion of those samples would have had means at least as far as our single sample's mean from the μ specified in the null hypothesis. This question is answered by the considerations of Section 2 and is demonstrated in Example 4.

EXAMPLE 4   **Hypothesis Testing of $H_0: \mu = 0$ and $H_A: \mu \neq 0$**

The variable, X_i, is the weight change of horses given an antibiotic for two weeks. The following measurements of X_i are those obtained from 17 horses (where a positive weight change signifies a weight gain and a negative weight change denotes a weight loss):

$$2.0, \ 1.1, \ 4.4, \ -3.1, \ -1.3, \ 3.9, \ 3.2, \ -1.6, \ 3.5$$
$$1.2, \ 2.5, \ 2.3, \ 1.9, \ 1.8, \ 2.9, \ -0.3, \text{ and } -2.4 \text{ kg}.$$

For these 17 data, the sample mean $(\overline{X})$ is 1.29 kg. Although the population variance (σ^2) is typically not known, for the demonstration purpose of this example, σ^2 is said to be 13.4621 kg^2. Then the population standard error of the mean would be

$$\sigma_{\overline{X}} = \sqrt{\frac{\sigma^2}{n}} = \sqrt{\frac{13.4621 \text{ kg}^2}{17}} = \sqrt{0.7919 \text{ kg}^2} = 0.89 \text{ kg}$$

---

[*]The symbol "$\neq$" denotes "is not equal to"; Ball (1935: 242) credits Leonhard Euler with its early, if not first, use (though it was first written with a vertical, not a diagonal, line through the equal sign).

and

$$Z = \frac{\overline{X} - \mu}{\sigma_{\overline{X}}} = \frac{1.29 \text{ kg} - 0}{0.89 \text{ kg}} = 1.45.$$

Using Table 2,

$$P(\overline{X} \geq 1.29 \text{ kg}) = P(Z \geq 1.45) = 0.0735$$

and, because the distribution of Z is symmetrical,

$$P(\overline{X} \leq -1.29 \text{ kg}) = P(Z \leq -1.45) = 0.0735.$$

Therefore,

$$P(\overline{X} \geq 1.29 \text{ kg } or \text{ } \overline{X} \leq -1.29 \text{ kg})$$
$$= P(Z \geq 1.45 \text{ } or \text{ } Z \leq -1.45)$$
$$= 0.0735 + 0.0735 = 0.1470.$$

As $0.1470 > 0.05$, do not reject H_0.

In Example 4, it is desired to ask whether treating horses with an experimental antibiotic results in a change in body weight. The data shown (X_i values) are the changes in body weight of 17 horses that received the antibiotic, and the statistical hypotheses to be tested are H_0: $\mu = 0$ kg and H_A: $\mu \neq 0$ kg. (As shown in this example, we can write "0" instead of "0 kg" in these hypotheses, because they are statements about *zero* weight change, and *zero* would have the same meaning regardless of whether the horses were weighed in kilograms, milligrams, pounds, ounces, etc.)

These 17 data have a mean of $\overline{X} = 1.29$ kg and they are considered to represent a random sample from a very large number of data, namely the body-weight changes that would result from performing this experiment with a very large number of horses. This large number of potential X_i's is the statistical population. Although one almost never knows the actual parameters of a sampled population, for this introduction to statistical testing let us suppose that the variance of the population sampled for this example is known to be $\sigma^2 = 13.4621$ kg^2. Thus, for the population of means that could be drawn from this population of measurements, the standard error of the mean is $\sigma_{\overline{X}} = \sqrt{\sigma^2/n} = \sqrt{13.4621 \text{ kg}^2/17} = \sqrt{0.7919 \text{ kg}^2} = 0.89$ kg (by Equation 5). We shall further assume that the population of possible means follows a normal distribution, which is generally a reasonable assumption even when the individual data in the population are not normally distributed.

This hypothesis test may be conceived as asking the following:

If we have a normal population with $\mu = 0$ kg, and $\sigma_{\overline{X}} = 0.89$ kg, what is the probability of obtaining a random sample of 17 data with a mean ($\overline{X}$) at least as far from 0 kg as 1.29 kg (i.e., at least 1.29 kg larger than 0 kg *or* at least 1.29 kg smaller than 0 kg)?

Section 2 showed that probabilities for a distribution of possible means may be ascertained through computations of Z (by Equation 6). The preceding null hypothesis is tested in Example 4, in which Z may be referred to as our *test statistic* (a computed quantity for which a probability will be determined). In this example, Z is calculated to be 1.45, and Table 2 from *Appendix: Statistical Tables and Graphs* informs us that the probability of a $Z \geq 1.45$ is 0.0735.* The null hypothesis asks about the deviation of the mean *in either direction* from 0 and, as the normal distribution is symmetrical, we can also say that $P(-Z \leq 1.45) = 0.0735$ and, therefore, $P(|Z| \geq 1.45) = 0.0735 + 0.0735 = 0.1470$. This tells us the probability associated with a $|Z|$ (absolute value of Z) at least as large as the $|Z|$ obtained; and this is the probability of a Z at least as extreme as that obtained, *if* the null hypothesis is true.

It should be noted that this probability,

$$P(|Z| \geq |\text{computed } Z|, \text{ if } H_0 \text{ is true}),$$

is *not* the same as

$$P(H_0 \text{ is true, if } |Z| \geq |\text{computed } Z|),$$

for these are *conditional probabilities*. In addition to the playing-card example in that section, suppose a null hypothesis was tested 2500 times, with results as in Example 5. The probability of rejecting H_0, if H_0 is true, is $P(\text{rejecting } H_0, \text{ if } H_0 \text{ is true}) = $ (number of rejections of true H_0's)/(number of true H_0's) $= 100/2000 = 0.05$. And the probability that H_0 is true, *if* H_0 is rejected, is $P(H_0 \text{ true, if } H_0 \text{ is rejected}) = $ (number of rejections of true H_0's)/(number of rejections of H_0's) $= 100/550 = 0.18$. These two probabilities (0.05 and 0.18) are decidedly not the same, for they are probabilities based on different conditions.

EXAMPLE 5　Probability of Rejecting a True Null Hypothesis

Hypothetical outcomes of testing the same null hypothesis for 2500 random samples of the same size from the same population (where the samples are taken with replacement).

	If H_0 is true	If H_0 is false	Row total
If H_0 is rejected	100	450	550
If H_0 is not rejected	1900	50	1950
Column total	2000	500	2500

Probability that H_0 is rejected if H_0 is true $= 100/2000 = 0.05$.
Probability that H_0 is true if H_0 is rejected $= 100/550 = 0.18$.

In hypothesis testing, it is correct to say that the calculated probability (for example, using Z) is

$$P(\text{the data, given } H_0 \text{ is true})$$

and it is *not* correct to say that the calculated probability is

$$P(H_0 \text{ is true, given the data}).$$

Furthermore, in reality we may not be testing $H_0: \mu = 0$ kg in order to conclude that the population mean is *exactly* zero (which it probably is *not*). Rather, we

---
*Note that "$\geq$" and "$\leq$" are symbols for "greater than or equal to" and "less than or equal to," respectively.

are interested in concluding whether there is a very small difference between the population mean and 0 kg; and what is meant by *very small* will be discussed in Section 3(d).

(b) Statistical Errors in Hypothesis Testing.   It is desirable to have an objective criterion for drawing a conclusion about the null hypothesis in a statistical test. Even if H_0 is true, random sampling might yield a sample mean ($\overline{X}$) far from the population mean (μ), and a large absolute value of Z would thereby be computed. However, such an occurrence is unlikely, and the larger the $|Z|$, the smaller the probability that the sample came from a population described by H_0. Therefore, we can ask how small a probability (which is the same as asking how large a $|Z|$) will be required to conclude that the null hypothesis is not likely to be true. The probability used as the criterion for rejection of H_0 is called the *significance level*, routinely denoted by α (the lower-case Greek letter alpha).* As indicated below, an α of 0.05 is commonly employed. The value of the test statistic (in this case, Z) corresponding to α is termed the *critical value* of the test statistic. Table 2 from *Appendix: Statistical Tables and Graphs* it is seen that $P(Z \geq 1.96) = 0.025$; and, inasmuch as the normal distribution is symmetrical, it is also the case that $P(Z \leq -1.96) = 0.025$. Therefore, the critical value for testing the above H_0 at the 0.05 level (i.e., 5% level) of significance is $Z = 1.96$ (see Figure 4). These values of Z may be denoted as $Z_{0.025(1)} = 1.96$ and $Z_{0.05(2)} = 1.96$, where the parenthetical number indicates whether one or two tails of the normal distribution are being referred to.

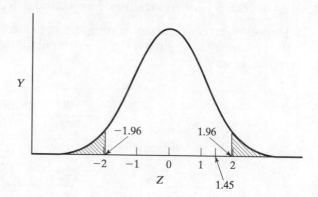

FIGURE 4: A normal curve showing (with shading) the 5% of the area under the curve that is the rejection region for the null hypothesis of Example 4. This rejection region consists of 2.5% of the curve in the right tail (demarcated by $Z_{0.05(2)} = 1.96$) and 2.5% in the left tail (delineated by $-Z_{0.05(2)} = -1.96$). The calculated test statistic in this example, $Z = 1.45$, does not lie within either tail; so H_0 is not rejected.

So, a calculated Z greater than or equal to 1.96, or less than or equal to -1.96, would be reason to reject H_0, and the shaded portion of Figure 4 is known as the "rejection region." The absolute value of the test statistic in Example 4 (namely, $|Z| = 1.45$) is not as large as the critical value (i.e., it is neither ≥ 1.9 nor ≤ -1.96), so in this example the null hypothesis is not rejected as a statement about the sampled population.

---

*David (1955) credits R. A. Fisher as the first to refer to "level of significance," in 1925. Fisher (1925b) also was the first to formally recommend use of the 5% significance level as guidance for drawing a conclusion about the propriety of a null hypothesis (Cowles and Davis, 1982), although he later argued that a fixed significance level should not be used. This use of the Greek "α" first appears in a 1936 publication of J. Neyman and E. S. Pearson (Miller, 2004c).

It is very important to realize that a true null hypothesis will sometimes be rejected, which of course means that an error has been committed in drawing a conclusion about the sampled population. Moreover, this error can be expected to be committed with a frequency of α. The rejection of a null hypothesis when it is in fact true is what is known as a *Type I error* (or "Type 1 error" or "alpha error" or "error of the first kind"). On the other hand, a statistical test will sometimes fail to detect that a H_0 is in fact false, and an erroneous conclusion will be reached by not rejecting H_0. The probability of committing this kind of error (that is, not rejecting H_0 when it is false) is represented by β (the lowercase Greek letter beta). This error is referred to as a *Type II error* (or "Type 2 error" or "beta error" or "error of the second kind"). The *power* of a statistical test is defined as $1 - \beta$: the probability of correctly rejecting the null hypothesis when it is false.* If H_0 is not rejected, some researchers refer to it as having been "accepted," but most consider it better to say "not rejected," for low statistical power often causes failure to reject, and "accept" sounds too definitive. Section 3(c) discusses how both the Type I and the Type II errors can be reduced.

Table 1 summarizes these two types of statistical errors, and Table 2 indicates their probabilities. Because, for a given n, a relatively small probability of a Type I error is associated with a relatively large probability of a Type II error, it is appropriate to ask what the acceptable combination of the two might be. By experience, and by convention, an α of 0.05 is typically considered to be a "small enough chance" of committing a Type I error while not being so small as to result in "too large a chance" of a Type II error (sometimes considered to be around 20%). But the 0.05 level of significance is not sacrosanct. It is an arbitrary, albeit customary, threshold for concluding that there is significant evidence against a null hypothesis. And caution should be exercised in emphatically rejecting a null hypothesis if $p = 0.049$ and not rejecting if $p = 0.051$, for in such borderline cases further examination—and perhaps repetition—of the experiment would be recommended.

TABLE 1: The Two Types of Errors in Hypothesis Testing

	If H_0 is true	If H_0 is false
If H_0 is rejected:	Type I error	No error
If H_0 is not rejected:	No error	Type II error

Although 0.05 has been the most widely used significance level, individual researchers may decide whether it is more important to keep one type of error

---

*The distinction between these two fundamental kinds of statistical errors, and the concept of power, date back to the pioneering work, in England, of Jerzy Neyman (1894–1981; Russian-born, of Polish roots, emigrating as an adult to Poland and then to England, and spending the last half of his life in the United States) and the English statistician Egon S. Pearson (1895–1980) (Lehmann and Reid, 1982; Neyman and Pearson, 1928a; Pearson, 1947). They conceived of the two kinds of errors in 1928 (Lehmann, 1999) and named them, and they formulated the concept of power in 1933 (David, 1995). With some influence by W. S. Gosset ("Student") (Lehmann, 1999), their modifications (e.g., Neyman and Pearson, 1933) of the ideas of the colossal British statistician (1890–1962) R. A. Fisher (1925b) provide the foundations of statistical hypothesis testing. However, from the mid-1930s until his death, Fisher disagreed intensely with the Neyman-Pearson approach, and the hypothesis testing commonly used today is a fusion of the Fisher and the Neyman-Pearson procedures (although this hybridization of philosophies has received criticism—e.g., by Hubbard and Bayarri, 2003). Over the years there has been further controversy regarding hypothesis testing, especially—but not entirely—within the social sciences (e.g., Harlow, Mulaik, and Steiger, 1997). The most extreme critics conclude that hypothesis tests should never be used, while most others advise that they may be employed but only with care to avoid abuse.

TABLE 2: The Long-Term Probabilities of Outcomes in Hypothesis Testing

	If H_0 is true	If H_0 is false
If H_0 is rejected	α	$1 - \beta$ ("power")
If H_0 is not rejected	$1 - \alpha$	β

or the other low. In some instances, we may be willing to test with an α greater than 0.05. An example of the later decision could be when there is an adverse health or safety implication if we incorrectly fail to reject a false null hypothesis. So in performing an experiment such as in Example 4, perhaps it is deemed important to the continued use of this antibiotic that it not cause a change in body weight; and we want to have a small chance of concluding that the drug causes no weight change when such a decision is incorrect. In other words, we may be especially desirous of avoiding a Type II error. In that case, an α of 0.10 (i.e., 10%) might be used, for that would decrease the probability of a Type II error, although it would concomitantly increase the likelihood of incorrectly rejecting a true H_0 (i.e., committing a Type I error). In other cases, such as indicated in Section 3(d), a 0.05 (i.e., 5%) chance of an incorrect rejection of H_0 may be felt to be unacceptably high, so a lower α would be employed in order to reduce the probability of a Type I error (even though that would increase the likelihood of a Type II error).

It is necessary, of course, to state the significance level used when communicating the results of a statistical test. Indeed, rather than simply stating whether the null hypothesis is rejected, it is good procedure to report also the sample size, the test statistic, and the best estimate of the exact probability of the statistic (and such probabilities are obtainable from many computer programs and some calculators). Note that in Example 4, it is reported that $n = 17$, $Z = 1.45$, and $P = 0.1470$, in addition to expressing the conclusion that H_0 is not rejected. In this way, readers of the research results may draw their own conclusions, even if their choice of significance level is different from the author's. It is also good practice to report results regardless of whether H_0 is rejected. Bear in mind, however, that the choice of α is to be made before seeing the data. Otherwise there is a great risk of having the choice influenced by examination of the data, introducing bias instead of objectivity into the proceedings. The best practice generally is to decide on the null and alternate hypotheses, and the significance level, before commencing with data collection and, after performing the statistical test, to express the probability that the sample came from a population for which H_0 is true. It is conventional to refer to rejection of H_0 at the 5% significance level as denoting a "statistically significant" difference between $\overline{X}$ and the μ hypothesized in H_0 (e.g., in Example 4, between $\overline{X} = 1.45$ kg and $\mu = 0$ kg).* But, in analyzing biological data, we should consider whether a statistically detected difference reflects a *biologically significant* difference, as will be discussed in Section 3(d).

(c) One-Tailed versus Two-Tailed Testing. In Section 3(a), Example 4 tests whether a population mean was significantly different from a hypothesized value, where the alternate hypothesis embodies difference in either direction (i.e., greater than or less than) from that value. This is known as *two-sided*, or *two-tailed*, testing,

*In reporting research results, some authors have attached an asterisk (*) to a test statistic if it is associated with a probability ≤ 0.05 and two asterisks (**) if the probability is ≤ 0.01, sometimes referring to results at ≤ 0.01 as "highly significant"; but the latter term is best avoided, in preference to reporting the magnitude of p.

for we reject H_0 if Z (the test statistic in this instance) is within either of the two tails of the normal distribution demarcated by the positive and negative critical values of Z (the shaded areas in Figure 4).

However, there are cases where there is good scientific justification to test for a significant difference *specifically in one direction only*. That is, on occasion there is a good reason to ask whether a population mean is significantly *larger* than μ_0, and in other situations there is a good rationale for asking whether a population mean is significantly *smaller* than μ_0. Statistical testing that examines difference in only one of the two possible directions is called *one-sided*, or *one-tailed*, testing.

Example 4 involved a hypothesis test interested in whether a drug intended to be an antibiotic caused weight change as a side effect of its use. For such a test, H_0 is rejected if Z (the test statistic in this instance) is within the rejection region in either the right-hand *or* the left-hand tail of the normal distribution (i.e., within the shaded areas of Figure 4 and Figure 5a). However, consider a similar experiment where the purpose of the drug is to cause weight loss. In that case, the statistical hypotheses would be $H_0: \mu \geq 0$ versus $H_A: \mu < 0$. That is, if the drug works as intended and there

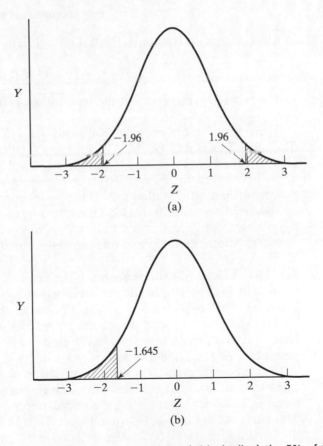

FIGURE 5: (a) As in Figure 4, a normal curve showing (with shading) the 5% of the area under the curve that is the rejection region for the two-tailed null hypotheses, $H_0: \mu = \mu_0$ versus $H_A: \mu \neq \mu_0$. This rejection region consists of 2.5% of the curve in the right tail (demarcated by $Z_{0.05(2)} = 1.96$), and 2.5% in the left tail (delineated by $-Z_{0.05(2)} = -1.96$). (b) A normal curve showing (with shading) the 5% of the area under the curve that is the rejection region for the one-tailed null hypotheses, $H_0: \mu \geq \mu_0$ vs. $H_A: \mu < \mu_0$. This rejection region consists of 5% of the curve in the left tail (demarcated by $Z_{0.05(1)} = 1.645$).

is a mean weight loss, then H_0 would be rejected; and if the drug does not work (that is, there is a mean weight gain or the mean weight did not change), H_0 would not be rejected. In such a situation, the rejection region would be entirely in one tail of the normal distribution, namely the left-hand tail. This is an example of a *one-tailed test*, whereas Example 4 represents a *two-tailed test*.

It can be seen in Table 2 from *Appendix: Statistical Tables and Graphs* that, if one employs the 5% level of significance, the one-tailed Z value is 1.645. The normal distribution's tail defined by this one-tailed Z is the shaded area of Figure 5b. If the calculated Z is within this tail, H_0 is rejected as a correct statement about the population from which this sample came.

Figure 5a shows the rejection region of a normal distribution when performing two-tailed testing of $H_0: \mu = \mu_0$ at the 5% significance level (i.e., the same shaded area as in Figure 4, namely 2.5% in each tail of the curve); and Figure 5b shows the rejection region for one-tailed testing of $H_0: \mu \geq 0$ versus $H_A: \mu < 0$ at the 5% level. (If the experimental drug were intended to result in weight gain, not weight loss, then the rejection region would be in the right-hand tail instead of in the left-hand tail.)

In general, one-tailed hypotheses about a mean are

$$H_0: \mu \geq \mu_0 \text{ and } H_A: \mu < \mu_0,$$

in which case H_0 is rejected if the test statistic is in the left-hand tail of the distribution, *or*

$$H_0: \mu \leq \mu_0 \text{ and } H_A: \mu > \mu_0,$$

in which case H_0 is rejected if the test statistic is in the right-hand tail of the distribution.*

The one-tailed critical value (let's call it $Z_{\alpha(1)}$) is found in Table 2 from *Appendix Statistical Tables and Graphs*. It is always smaller than the two-tailed critical value ($Z_{\alpha(2)}$); for example, at the 5% significance level $Z_{\alpha(1)} = 1.645$ and $Z_{\alpha(2)} = 1.96$. Thus, as will be noted in Section 3(d), for a given set of data a one-tailed test is more powerful than a two-tailed test. But it is inappropriate to employ a one-tailed test unless there is a scientific reason for expressing one-tailed, in preference to two-tailed, hypotheses. And recall that *statistical hypotheses are to be declared before examining the data.* Another example of one-tailed testing of a mean is found in Exercise 5(a).

(d) What Affects Statistical Power. The power of a statistical testing procedure was defined in Section 3(b) as the probability that a test correctly rejects the null hypothesis when that hypothesis is a false statement about the sampled population. It is useful to be aware of what affects the power of a test to estimate the power a test will have and to estimate how small a difference will be detected between a population parameter (e.g., μ) and a hypothesized value (e.g., μ_0).

Figure 6a represents a normal distribution of sample means, where each sample was the same size and each sample mean estimates the same population mean. This mean of this distribution is μ_0, the population mean specified in the null hypothesis. This curve is the same as shown in Figure 5. As in Figure 5, the shaded area in each of the two tails denotes 0.025 of the area under the curve; so both shaded areas compose an area of 0.05, the probability of a Type I error (α).

---

*Some authors write the first of these two pairs of hypotheses as $H_0: \mu = \mu_0$ and $H_A: \mu < \mu_0$, and the second pair as $H_0: \mu = \mu_0$ and $H_A: \mu > \mu_0$, ignoring mention of the tail that is not of interest.

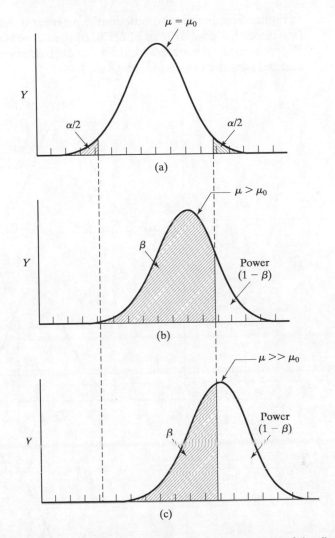

FIGURE 6: (a) A normal curve, such as that in Figure 4, where μ, the mean of the distribution, is μ_0, the value specified in the null and alternate hypotheses. The shaded area in each of the two tails is 0.025 of the area under the curve, so a total of 0.05 (i.e., 5%) of the curve is the shaded critical region, and α, the probability of a Type I error, is 0.05. (b) The same normal curve, but where μ is larger than μ_0 and the shaded area is the probability of a Type II error (β). (c) The same normal curve, but where μ is much larger than μ_0.

Figure 6b is the same normal curve, but with a population mean, μ, different from (i.e., larger than) μ_0. If H_0: $\mu = \mu_0$ is not a true statement about the population, yet we fail to reject H_0, then we have committed a Type II error, the probability of which is β, indicated by the shaded area between the vertical dashed lines in Figure 6b. The power of the hypothesis test is defined as $1 - \beta$, which is the unshaded area under this curve.

Figure 6c is the same depiction as in Figure 6b, but with a population mean, μ, even more different* from μ_0. An important result is that, the farther μ is from the μ_0 specified in H_0, the smaller β becomes and the larger the power becomes.

---

*The symbol ">" has been introduced as meaning "greater than," and "<" as meaning "less than." The symbols "W" and "≪" mean "much greater than" and "much less than," respectively.

Figure 7 indicates the outcome if a larger α is used, namely 10% instead of 5% (meaning that 5%, instead of 2.5%, of the curve is in each tail). If the probability of a Type I error (α) is increased, then the probability of a Type II error (β) is decreased, and the power of the test is increased.

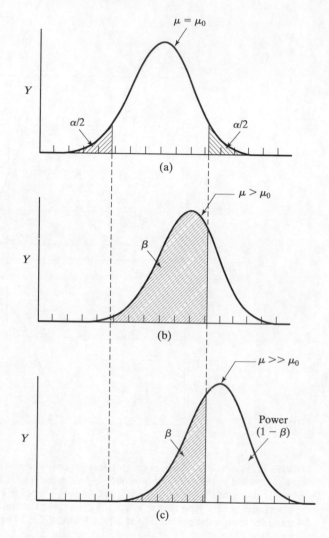

FIGURE 7: (a) A normal curve, such as that in Figure 6, where μ, the mean of the distribution, is μ_0, the value specified in the null and alternate hypotheses, but where the shaded area in each of the two tails is 0.05 of the area under the curve, so a total of 0.10 (i.e., 10%) of the curve is the shaded critical region, and α, the probability of a Type I error, is 0.10. (b) The same normal curve, but where μ is larger than μ_0 and the shaded area is the probability of a Type II error (β). (c) The same normal curve, but where μ is much larger than μ_0.

Another important outcome is seen by examining Equations 5 and 6. With larger sample size (n), or with smaller variance (σ^2), the standard error $\sigma_{\overline{X}}$ becomes smaller, which means that the shape of the normal distribution becomes narrower. Figure 3 shows an example of this narrowing as the variance decreases in a population of data, and the figures would appear similar if they were for a population of means. So, for a given value of α and of μ, either a smaller σ^2 or a larger n will result in a smaller $\sigma_{\overline{X}}$, which will result in a smaller β and greater power to reject H_0.

In some circumstances, a larger n can be used, but in other situations this would be difficult because of cost or effort. A smaller variance of the sampled population will result if the population is defined as a more homogeneous group of data. In Example 4, the experiment could have been performed using only female horses, or only horses of a specified age, or only horses of a specified breed. Then the hypothesis test would be about the specified sex, age, and/or breed, and the population variance would probably be smaller; and this would result in a greater power of the test.

To summarize what influences power,

- For given α, σ^2, and n, power is greater for larger difference between μ and μ_0.
- For given n, σ^2, and difference between μ and μ_0, power is greater for larger α.
- For given α, σ^2, and difference between μ and μ_0, power is greater for larger n.
- For given α, n, and difference between μ and μ_0, power is greater for smaller σ^2.
- For given α, n, σ^2, and difference between μ and μ_0, power is greater for one-tailed than for two-tailed tests (but one-tailed tests may be employed only when the hypotheses are appropriately one-tailed).

(e) Summary of Statistical Hypothesis Testing.   Earlier portions of Section 3 introduced the principles and practice of testing hypotheses about population parameters, using sample statistics as estimates of those parameters. It is also good practice to report an estimate of the precision with which a parameter has been estimated, by expressing what are known as "confidence limits," which will be introduced in Section 4.

To summarize the steps for testing of statistical hypotheses,

1. State H_0 and H_A, using two-tailed or one-tailed hypotheses depending upon the objective of the data analysis.
2. Declare the level of significance, α, to be employed.
3. Collect the data and calculate the test statistic (Z in this chapter).
4. Compare the test statistic to the critical value(s) of that statistic (that is, the value(s) delimiting the rejection region of the statistical distribution of the test statistic). For the testing in this chapter, the critical values are both $Z_{\alpha(2)}$ and $-Z_{\alpha(2)}$ for a two-tailed test and the critical value is $Z_{\alpha(1)}$ for a one-tailed test. If the calculated Z exceeds a critical value, H_0 is rejected.
5. State P, the probability of the test statistic if H_0 is true.
6. State confidence limits (two-tailed or one-tailed) for the population parameter, as discussed in Section 4.
7. State conclusion in terms of biological or other practical significance.

4   CONFIDENCE LIMITS

Sections 3a and 3b discussed the distribution of all possible samples of size n from a population with mean μ. It was noted that 5% of the values of Z (by Equation 6) for those sample means will be at least as large as $Z_{0.05(2)}$ *or* no larger than $-Z_{0.05(2)}$. This can be expressed as

$$P\left[-Z_{0.05(2)} \leq \frac{\overline{X} - \mu}{\sigma_{\overline{X}}} \leq Z_{0.05(2)}\right] = 95\%, \tag{9}$$

and this can be rearranged to read

$$P[\overline{X} - Z_{0.05(2)}\sigma_{\overline{X}} \leq \mu \leq \overline{X} + Z_{0.05(2)}\sigma_{\overline{X}}] = 0.95. \tag{10}$$

In general, we can say

$$P[\overline{X} - Z_{\alpha(2)}\sigma_{\overline{X}} \leq \mu \leq \overline{X} + Z_{\alpha(2)}\sigma_{\overline{X}}] = 1 - \alpha. \tag{11}$$

The *lower confidence limit* is defined as

$$L_1 = \overline{X} - Z_{\alpha(2)}\sigma_{\overline{X}}, \tag{12}$$

and the *upper confidence limit* is

$$L_2 = \overline{X} + Z_{\alpha(2)}\sigma_{\overline{X}}. \tag{13}$$

The distance between L_1 and L_2, namely

$$\overline{X} \pm Z_{\alpha(2)}\sigma_{\overline{X}} \tag{14}$$

(where "$\pm$" is read as "plus or minus"), is called a *confidence interval* (sometimes abbreviated CI).

When referring to a confidence interval, $1 - \alpha$ is known as the *confidence level* (or *confidence coefficient* or *confidence probability*).*

Although $\overline{X}$ is the best estimate of μ, it is only an estimate, and the calculation of a confidence interval for μ allows us to express the precision of this estimate. Example 6 demonstrates this for the data of Example 4, determining the confidence interval for the mean of the population from which the sample came. As the 95% confidence limits are computed to be −0.45 kg and 3.03 kg, the 95% confidence interval may be expressed as $P(-0.45 \text{ kg} \leq \mu \leq 3.03 \text{ kg}) = 95\%$. This means that, if all possible means of size n ($n = 17$ in this example) were taken from the population and a 95% confidence interval were calculated from each sample, 95% of those intervals would contain μ. (It does *not* mean that there is a 95% probability that the confidence interval computed from the one sample in Example 6 includes μ.)

EXAMPLE 6   Confidence Limits for the Mean

For the 17 data in Example 4, $\overline{X} = 1.29$ kg and $\sigma_{\overline{X}} = 0.89$ kg.
We can calculate the 95% confidence limits for μ using Equations 13 and 14 and $Z_{0.05(2)} = 1.96$:

$$L_1 = \overline{X} - Z_{\alpha(2)}\sigma_{\overline{X}}$$
$$= 1.29 \text{ kg} - (1.96)(0.89 \text{ kg})$$
$$= 1.29 \text{ kg} - 1.74 \text{ kg} = -0.45 \text{ kg}$$

---

*We owe the development of confidence intervals to Jerzy Neyman, between 1928 and 1933 (Wang, 2000), although the concept had been enunciated a hundred years before. Neyman introduced the terms *confidence interval* and *confidence coefficient* in 1934 (David, 1995). On rare occasion, biologists may see reference to "fiducial intervals," a concept developed by R. A. Fisher beginning in 1930 and identical to confidence intervals in many, but not all, situations (Pfanzagl, 1978).

$$L_2 = \overline{X} + Z_{\alpha(2)}\sigma_{\overline{X}}$$
$$= 1.29 \text{ kg} + (1.96)(0.89 \text{ kg})$$
$$= 1.29 \text{ kg} + 1.74 \text{ kg} = 3.03 \text{ kg}.$$

So, the 95% confidence interval could be stated as

$$P(-0.45 \text{ kg} \leq \mu \leq 3.03 \text{ kg}).$$

Note that the μ_0 of Example 4 (namely 0) is included between L_1 and L_2, indicating that H_0 is not rejected.

As seen in Equation 15, a small $\sigma_{\overline{X}}$ will result in a smaller confidence interval, meaning that μ is estimated more precisely when $\sigma_{\overline{X}}$ is small. And, recall from Equation 5 that $\sigma_{\overline{X}}$ becomes small as n becomes large. So, in general, a parameter estimate from a large sample is more precise than an estimate of the same parameter from a small sample.

If, instead of a 95% confidence interval, we wished to state an interval that gave us 99% confidence in estimating μ, then $Z_{0.01(2)}$ (which is 2.575) would have been employed instead of $Z_{0.05(2)}$, and we would have computed $L_1 = 1.29$ kg $-$ $(2.575)(0.89 \text{ kg}) = 1.29 \text{ kg} - 2.29 = -1.00$ and $L_2 = 1.29 \text{ kg} + (2.575)(0.89 \text{ kg}) = 1.29$ kg $+ 2.29$ kg $= 3.58$ kg. It can be seen that a larger confidence level (e.g., 99% instead of 95%) results in a larger width of the confidence interval, evincing the trade-off between confidence and utility. Indeed, if we increase the confidence to 100%, then the confidence interval would be $-\infty$ to ∞, and we would have a statement of great confidence that was useless! Note, also, that it is a two-tailed value of Z (i.e., $Z_{0.05(2)}$) that is used in the computation of a confidence interval when we set confidence limits on both sides of μ.

In summary, a narrower confidence interval will be associated with a smaller standard error ($\sigma_{\overline{X}}$), a larger sample size (n), or a smaller confidence coefficient ($1 - \alpha$).

It is recommended that a $1 - \alpha$ confidence interval be reported for μ whenever results are presented from a hypothesis test at the α significance level. If $H_0: \mu = \mu_0$ is not rejected, then the confidence interval includes μ_0 (as is seen in Example 6, where $\mu_0 = 0$ is between L_1 and L_2).

(a) One-Tailed Confidence Limits. In the case of a one-tailed hypothesis test, it is appropriate to determine a one-tailed confidence interval; and, for this, a one-tailed critical value of Z (i.e., $Z_{\alpha(1)}$) is used instead of a two-tailed critical value ($Z_{\alpha(2)}$). For $H_0: \mu \leq \mu_0$ and $H_A: \mu > \mu_0$, the confidence limits for μ are $L_1 = \overline{X} - Z_{\alpha(1)}\sigma_{\overline{X}}$ and $L_2 = \infty$. For $H_0: \mu \geq \mu_0$ and $H_A: \mu < \mu_0$, the confidence limits are $L_1 = -\infty$ and $L_2 = \overline{X} + Z_{\alpha(1)}\sigma_{\overline{X}}$. An example of a one-sided confidence interval is Exercise 6(b). If a one-tailed null hypothesis is not rejected, then the associated one-tailed confidence interval includes μ_0.

5 SYMMETRY AND KURTOSIS

Sets of data can be described by measures of central tendency and measures of variability. There are additional characteristics that help describe data sets, and they are sometimes used when we want to know whether a distribution resembles a normal distribution. Two basic features of a distribution of measurements are its *symmetry* and its *kurtosis*. A symmetric distribution (as in Figure 1) is one in which the mean and median are the same and the left half of the frequency plot is a mirror image of the right half. *Kurtosis* refers to the shape of the distribution relative to the shape of a normal distribution, as will be explained later in Section 5(b).

Ways to describe these two properties have been proposed and discussed for over a hundred years. A well-established expression of the symmetry of a population is $\sqrt{\beta_1}$ (employing the lowercase Greek beta), and it is estimated by the sample statistic $\sqrt{b_1}$. The associated kurtosis parameter is β_2, which is estimated by the statistic b_2.[*] An analogous pair of population parameters are γ_1 (using the lowercase Greek gamma) for symmetry and γ_2 for kurtosis, and they are estimated by the sample statistics g_1 and g_2, respectively.

There have been various and inconsistent ways proposed to calculate these four statistics ($\sqrt{b_1}$ and b_2 and the less commonly encountered g_1 and g_2). The computations shown in Sections 5a and 5b for $\sqrt{b_1}$ and b_2 will be used here. Although they give biased estimates of $\sqrt{\beta_1}$ and β_2, that is of no great concern because the intent in computing them will be to draw conclusions about the sampled population and not specifically to estimate the population parameters.

Statisticians refer to $\Sigma(X_i - \mu)^k/N$ as the "kth standardized moment"[†] for a population. For any population, the first standardized moment, $\Sigma(X_i - \mu)/N$, is always zero, because $\Sigma(X_i - \mu)$ is always zero. The second standardized moment for a population is $\Sigma(X_i - \mu)^2/N$, and this has already been defined as the population variance. Beyond these, the odd-numbered moments (particularly the third) describe the symmetry of the distribution, and the even-numbered moments (especially the fourth) tell us about kurtosis, another important characteristic of the shape of the distribution. Therefore, the third and fourth moments have long been used as indicators of the extent to which a distribution resembles a normal distribution.

The kth standardized moment for a sample is defined as

$$m_k = \frac{\Sigma(X_i - \overline{X})^k}{n}. \tag{15}$$

(a) Symmetry. A symmetrical set of data is one in which the left half of the frequency distribution is a mirror image of the right half, which makes the mean and median identical. The third moment, m_3, describes the *symmetry* of the set of data; but, because it has cubed units (i.e., units to the third power), the symmetry of a sample of data is expressed by a statistic called $\sqrt{b_1}$, which has no units[‡]:

$$\sqrt{b_1} = \frac{m_3}{\sqrt{m_2^3}} = \frac{\sqrt{n} \sum_{i=1}^{n}(X_i - \overline{X})^3}{\sqrt{\left[\sum_{i=1}^{n}(X_i - \overline{X})^2\right]^3}}; \tag{16}$$

[*]These beta (β) quantities were introduced by Karl Pearson in 1895 (David, 1995).

[†]This statistical concept of moments was developed by Karl Pearson around 1893, near the beginning of his long and productive publishing career (although moments themselves were described by others long before). He called them "moment coefficients" (Walker, 1929: 71, 74, 184–185).

[‡]This is related to an 1893 suggestion by Karl Pearson (Walker, 1929: 74).

$\sqrt{b_1}$ equals zero for a symmetrical distribution. If $\sqrt{b_1}$ is positive (i.e., greater than zero), the distribution is said to be positively skewed (i.e., skewed to the right), the mean is greater than the median, and the right tail is more elongated than the left tail. And if $\sqrt{b_1}$ is negative (i.e., is less than zero), the mean is less than the median, and the distribution is called negatively skewed (i.e., skewed to the left), with the left tail more elongated than the right tail.* For a sample of size n, the maximum possible $\sqrt{b_1}$ is $(n - 2)/\sqrt{n - 1}$, which will be the case if $n - 1$ of the data are the same and one of the data is any size larger than the others; the minimum possible $\sqrt{b_1}$ is $-(n - 2)/\sqrt{n - 1}$, which will be the case if $n - 1$ of the data are the same and one datum is any size smaller than the others (Kirby, 1974, 1981).

(b) Kurtosis. All symmetrical distributions are not normal distributions, so statisticians have desired an additional way to assess whether a distribution exhibits normality. This is most commonly done by employing the fourth standardized moment to express a characteristic known as *kurtosis*. This moment, m_4, has units to the fourth power. A sample kurtosis measure having no units is called b_2:

$$b_2 = \frac{m_4}{m_2^2} = \frac{n\sum_{i=1}^{n}(X_i - \overline{X})^4}{\left[\sum_{i=1}^{n}(X_i - \overline{X})^2\right]^2}; \qquad (17)$$

and b_2 equals 3 for a normal distribution.[†]

The kurtosis of a distribution of data is sometimes characterized in terms of how sharp the peak of the frequency polygon is or how elongated the distribution's tails are, but it may be best to describe kurtosis as the amount of dispersion of data around the "shoulders" of the distribution (Moors, 1986), namely around the mean minus the standard deviation and around the mean plus the standard deviation (i.e., how much the data are around $\overline{X} - s$ and $\overline{X} + s$ in a sample). A sample of data having many more values around the shoulders than a normal distribution has will exhibit $b_2 < 3$ and is said to be *platykurtic* (see Figure 8b). (Such a distribution might be the composite of two normal distributions with the same variance but different means.) In contrast, a sample having fewer values around the shoulders than does a normal distribution (i.e., more values around the mean and/or in the tails) will have $b_2 > 3$ and is called *leptokurtic* (see Figure 8c). (Such a distribution may be the composite of two normal distributions with the same mean but different standard deviations.) A normal distribution (with $b_2 = 3$), such as in Figure 8a, is said to be *mesokurtic*.[‡] The lowest possible value of b_2 is 1; the maximum possible value is n (Johnson and Lowe, 1979).

Whereas all symmetric samples have $\sqrt{b_1} = 0$ (and all symmetric populations have $\sqrt{\beta_1} = 0$), and all normally distributed samples have $b_2 = 3$ (and all normally

---

*Karl Pearson introduced the term *skewness* in 1895 (David, 1995).

†As part of the inconsistency in defining measures of skewness and kurtosis, some authors subtract 3 in computing b_2, so this statistic is zero for a normal distribution.

‡The terms *platykurtic*, *leptokurtic*, and *mesokurtic* were introduced by Karl Pearson as early as 1905 (Pearson, 1905); he also introduced the term *kurtosis* and the rarely encountered terms *isokurtic* and *allokurtic*, respectively, to refer to distributions that were symmetric and skewed (David, 1995; Walker, 1929: 182).

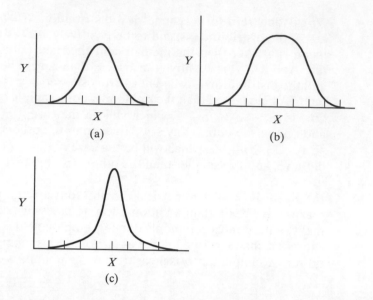

FIGURE 8: Symmetric frequency distributions. Distribution (a) is mesokurtic ("normal"), (b) is platykurtic, and (c) is leptokurtic.

distributed populations have $\beta_2 = 3$), some asymmetric distributions have a symmetry measure of 0 and some nonnormal distributions exhibit a kurtosis value of 3 (Thode, 2002: 43).

In practice, researchers seldom calculate these symmetry and kurtosis measures. When they do, however, they should be mindful that using the third and fourth powers of numbers can lead to very serious rounding errors, and they should employ computer programs that use procedures minimizing this problem.

(c) Quantile Measures of Symmetry and Kurtosis.   Denoting the ith quantile as Q_i, Q_1 is the first quartile (i.e., the 25% percentile), Q_3 is the third quartile (the 75% percentile), and Q_2 is the second quartile (the 50% percentile, namely the median). A quantile-based expression of skewness (Bowley, 1920: 116; Groeneveld and Meeden, 1984) considers the distance between Q_3 and Q_2 and that between Q_2 and Q_1:

$$\text{Quantile skewness measure} = \frac{(Q_3 - Q_2) - (Q_2 - Q_1)}{(Q_3 - Q_2) + (Q_2 - Q_1)}$$

$$= \frac{Q_3 + Q_1 - 2Q_2}{Q_3 - Q_1}, \tag{18}$$

which is a measure, without units, that may range from -1, for a distribution with extreme left skewness; to 0, for a symmetric distribution; to 1, for a distribution with extreme right skewness. Because Equation 18 measures different characteristics of a set of data than $\sqrt{b_1}$ does, these two numerical measures can be very different (and, especially if the skewness is not great, one of the measures can be positive and the other negative).

Instead of using quartiles Q_1 and Q_3, any other symmetric quantiles could be used to obtain a skewness coefficient (Groeneveld and Meeden, 1984), though the numerical value of the coefficient would not be the same as that of Equation 18. For example, the 10th and 90th percentiles could replace Q_1 and Q_3, respectively, in Equation 18, along with Q_2 (the median).

A kurtosis measure based on quantiles was proposed by Moors (1988), using octiles: $\mathcal{O}_1$, the first octile, is the 12.5th percentile; $\mathcal{O}_3$, the third octile, is the 37.5th percentile; $\mathcal{O}_5$ is the 62.5th percentile; and $\mathcal{O}_7$ is the 87.5th percentile. Also, $\mathcal{O}_2 = Q_1$, $\mathcal{O}_4 = Q_2$, and $\mathcal{O}_6 = Q_3$. The measure is

$$
\begin{aligned}
\text{Quantile kurtosis measure} &= \frac{(\mathcal{O}_7 - \mathcal{O}_5) + (\mathcal{O}_3 - \mathcal{O}_1)}{(\mathcal{O}_6 - \mathcal{O}_2)} \\
&= \frac{(\mathcal{O}_7 - \mathcal{O}_5) + (\mathcal{O}_3 - \mathcal{O}_1)}{(Q_3 - Q_1)},
\end{aligned} \tag{19}
$$

which has no units and may range from zero, for extreme platykurtosis, to 1.233, for mesokurtosis; to infinity, for extreme leptokurtosis.

Quantile-based measures of symmetry and kurtosis are rarely encountered.

6 ASSESSING DEPARTURES FROM NORMALITY

It is sometimes desired to test the hypothesis that a sample came from a population whose members follow a normal distribution. Example 7 and Figure 9 present a frequency distribution of sample data, and we may desire to know whether the data are likely to have come from a population that had a normal distribution. Comprehensive examinations of statistical methods applicable to such a question have been reported (e.g., by D'Agostino, 1986; Landry and Lepage, 1992; Shapiro, 1986; and Thode, 2002), and a brief overview of some of these techniques will be given here. The latter author discusses about 40 methods for normality testing and notes (ibid.: 143–157) that the power of a testing procedure depends upon the sample size and the nature of the nonnormality that is to be detected (e.g., asymmetry, long-tailedness, short-tailedness).

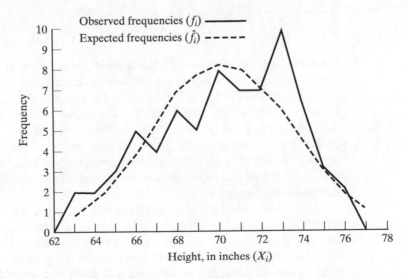

FIGURE 9: The frequency polygon for the student height data in Example 7 (solid line) with the frequency curve that would be expected if the data followed a normal distribution (broken line).

EXAMPLE 7   The Heights of the First 70 Graduate Students in My Biostatistics Course

Height (X_i) (in.)	Observed Frequency (f_i)	Cumulative Frequency (cum. f_i)	f_iX_i (in.)	$f_iX_i^2$ (in.^{2})
63	2	2	126	7,938
64	2	4	128	8,192
65	3	7	195	12,675
66	5	12	330	21,780
67	4	16	268	17,956
68	6	22	408	27,744
69	5	27	345	23,805
70	8	35	560	39,200
71	7	42	497	35,287
72	7	49	504	36,288
73	10	59	730	53,290
74	6	65	444	32,856
75	3	68	225	16,875
76	2	70	152	11,552
	$\Sigma f_i =$ $n = 70$		$\Sigma f_iX_i =$ 4,912 in.	$\Sigma f_iX_i^2 =$ 345,438 in.^{2}

$$SS = \Sigma f_iX_i^2 - \frac{\Sigma(f_iX_i)^2}{n} = 345,438 \text{ in.}^2 - \frac{(14,912 \text{ in.})^2}{70} = 755.9429 \text{ in.}^2$$

$$s^2 = \frac{SS}{n-1} = \frac{755.9429 \text{ in.}^2}{69} = 10.9557 \text{ in.}^2$$

(a) Graphical Assessment of Normality.   Many methods have been used to assess graphically the extent to which a frequency distribution of observed data resembles a normal distribution (e.g., Thode, 2002: 15–40). Recall the graphical representation of a normal distribution as a frequency curve, shown in Figure 1. A frequency polygon for the data in Example 7 is shown in Figure 9, and superimposed on that figure is a dashed curve showing what a normal distribution, with the same number of data (n) mean $(\overline{X})$, and standard deviation (s), would look like. We may wish to ask whether the observed frequencies deviate significantly from the frequencies expected from a normally distributed sample.

Figure 10 shows the data of Example 7 plotted as a cumulative frequency distribution. A cumulative frequency graph of a normal distribution will be S-shaped (called "sigmoid"). The graph in Figure 10 is somewhat sigmoid in shape, but in this visual presentation it is difficult to conclude whether that shape is pronounced enough to reflect normality. So, a different approach is desired. Note that the vertical axis on the left side of the graph expresses cumulative frequencies and the vertical axis

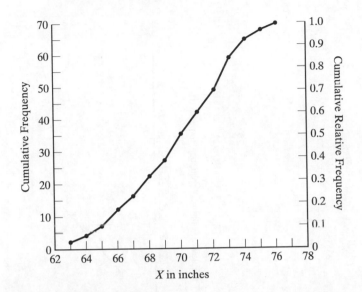

FIGURE 10: The cumulative frequency polygon of the student-height data of Example 7.

on the right side displays relative frequencies, and the latter may be thought of as percentiles. For instance, the sample of 70 measurements in Example 7 contains 22 data, where $X_i \le 68$ inches, so 68 in. on the horizontal axis is associated with a cumulative frequency of 22 on the left axis and a cumulative relative frequency of $22/70 = 0.31$ on the right axis; thus, we could say that a height of 68 in. is at the 31st percentile of this sample.

Examination of the relative cumulative frequency distribution is aided greatly by the use of the *normal probability scale*, as in Figure 11, rather than the linear scale of Figure 10. As the latter figure shows, a given increment in X_i (on the abscissa, the horizontal axis) near the median is associated with a much larger change in relative frequency (on the ordinate, the vertical axis) than is the same increment in X_i at very high or very low relative frequencies. Using the normal-probability scale on the ordinate expands the scale for high and low percentiles and compresses it for percentiles toward the median (which is the 50th percentile). The resulting cumulative frequency plot will be a straight line for a normal distribution. A leptokurtic distribution will appear as a sigmoid (S-shaped) curve on such a plot, and a platykurtic distribution will appear as a reverse S-shape. A negatively skewed distribution will show an upward curve, as the lower portion of an S, and a positively skewed distribution will manifest itself in a shape resembling the upper portion of an S. Figure 11 shows the data of Example 7 plotted as a cumulative distribution on a normal-probability scale. The curve appears to tend slightly toward leptokurtic.

Graph paper with the normal-probability scale on the ordinate is available commercially, and such graphs are produced by some computer software. One may also encounter graphs with a normal-probability scale on the abscissa and X_i on the ordinate. The shape of the plotted curves will then be converse of those described previously.

(b) Assessing Normality Using Symmetry and Kurtosis Measures. Section 5 indicated that a normally distributed population has symmetry and kurtosis parameters

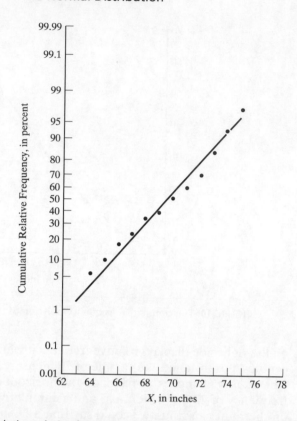

FIGURE 11: The cumulative relative frequency distribution for the data of Example 7, plotted with the normal probability scale as the ordinate. The expected frequencies (i.e., the frequencies from a normal distribution) would fall on the straight line shown.

of $\sqrt{\beta_1} = 0$ and $\beta_2 = 3$, respectively. Therefore, we can ask whether a sample of data came from a normal population by testing the null hypothesis H_0: $\sqrt{\beta_1} = 0$ (versus the alternate hypothesis, H_A: $\sqrt{\beta_1} \neq 0$) and the hypothesis H_0: $\beta_2 = 3$ (versus H_A: $\beta_2 \neq 3$). There are also procedures that employ the symmetry and kurtosis measures simultaneously, to test H_0: The sample came from a normally distributed population versus H_A: The sample came from a population that is not normally distributed (Bowman and Shenton, 1975, 1986; D'Agostino and Pearson, 1973; Pearson, D'Agostino, and Bowman, 1977; Thode, 2002: 54–55, 283).

Statistical testing using these symmetry and kurtosis measures, or the procedure of Section 6(d), is generally the best for assessing a distribution's departure from normality (Thode, 2002: 2).

(c) Goodness-of-Fit Assessment of Normality. Procedures called goodness-of-fit tests are applicable when asking whether a sample of data is likely to have come from a population with a specified distribution. Goodness-of-fit procedures known as chi-square, log-likelihood, and Kolmogorov-Smirnov, or modifications of them, have been used to test the hypothesis of normality (e.g., Zar, 1984: 88–93); and Thode (2002) notes that other goodness-of-fit tests, such as that of Kuiper (1960) may also be used. These methods perform poorly, however, in that they possess very low power; and they are not recommended for addressing hypotheses of normality (D'Agostino, 1986; D'Agostino, Belanger, and D'Agostino, 1990; Moore, 1986; Thode, 2002: 152).

(d) Other Methods of Assessing Normality. Shapiro and Wilk (1965) presented a test for normality involving the calculation of a statistic they called W. This computation requires an extensive table of constants, because a different set of $n/2$ constants is needed for each sample size, n. The authors provided a table of these constants and also of critical values of W, but only for n as large as 50. The power of this test has been shown to be excellent when testing for departures from normality (D'Agostino, 1986; Shapiro, Wilk, and Chen, 1968). Royston (1982a, 1982b) provided an approximation that extends the W test to n as large as 2000. Shapiro and Francia (1972) presented a modified procedure (employing a statistic they called W') that allows n to be as large as 99; but Pearson, D'Agostino, and Bowman (1977) noted errors in the published critical values. Among other modifications of W, Rahman and Govindarajulu (1997) offered one (with a test statistic they called $\tilde{W}$) declared to be applicable to any sample size, with critical values provided for n up to 5000. Calculation of W or its modifications is cumbersome and will most likely be done by computer; this test is unusual in that it involves rejection of the null hypothesis of normality if the test statistic is *equal to or less than* the one-tailed critical value.

The performance of the Shapiro-Wilk test is adversely affected by the common situation where there are tied data (i.e., data that are identical, as occur in Example 7, where there is more than one observation at each height) (Pearson, D'Agostino, and Bowman, 1977), but modifications of it have addressed that problem (e.g., Royston, 1986, 1989). Statistical testing using the Shapiro-Wilk test, or using symmetry and kurtosis measures (Section 6(b)), is generally the preferred method for inquiring whether an underlying population is normally distributed (Thode, 2002: 2).

EXERCISES

1. The following body weights were measured in 37 animals:

Weight (X_i) (kg)	Frequency (f_i)
4.0	2
4.3	3
4.5	5
4.6	8
4.7	6
4.8	5
4.9	4
5.0	3
5.1	1

   (a) Calculate the symmetry measure, $\sqrt{b_1}$.
   (b) Calculate the kurtosis measure, b_2.
   (c) Calculate the skewness measure based on quantiles.
   (d) Calculate the kurtosis measure based on quantiles.

2. A normally distributed population of lemming body weights has a mean of 63.5 g and a standard deviation of 12.2 g.
   (a) What proportion of this population is 78.0 g or larger?

   (b) What proportion of this population is 78.0 g or smaller?
   (c) If there are 1000 weights in the population, how many of them are 78.0 g or larger?
   (d) What is the probability of choosing at random from this population a weight smaller than 41.0 g?

3. (a) Considering the population of Exercise 2, what is the probability of selecting at random a body weight between 60.0 and 70.0 g?
   (b) What is the probability of a body weight between 50.0 and 60.0 g?

4. (a) What is the standard deviation of all possible means of samples of size 10 which could be drawn from the population in Exercise 2?
   (b) What is the probability of selecting at random from this population a sample of 10 weights that has a mean greater than 65.0 g?
   (c) What is the probability of the mean of a sample of 10 being between 60.0 and 62.0 g?

5. The following 18 measurements are obtained of a pollutant in a body of water: 10.25, 10.37, 10.66, 10.47, 10.56, 10.22, 10.44, 10.38, 10.63, 10.40, 10.39, 10.26, 10.32, 10.35, 10.54, 10.33, 10.48, 10.68 milligrams per liter. Although we would not know this in practice, for the sake of this example let us say

we know that the standard error of the mean is $\sigma_{\overline{X}} = 0.24$ mg/liter in the population from which this sample came. The legal limit of this pollutant is 10.00 milligrams per liter.

(a) Test whether the mean concentration in this body of water exceeds the legal limit (i.e., test H_0: $\mu \leq 10.00$ mg/L versus H_A: $\mu > 10.00$ mg/L), using the 5% level of significance.

(b) Calculate the 95% confidence interval for μ.

6. The incubation time was measured for 24 alligator eggs. Let's say that these 24 data came from a population with a variance of $\sigma^2 = 89.06$ days^2, and the sample mean is $\overline{X} = 61.4$ days.

(a) Calculate the 99% confidence limits for the population mean.

(b) Calculate the 95% confidence limits for the population mean.

(c) Calculate the 90% confidence limits for the population mean.

ANSWERS TO EXERCISES

1. $\sum f_i = n = 37$, $\overline{X} = 4.6514$, $\sum(X_i - \overline{X})^2 = 2.2922$, $\sum(X_i - \overline{X})^3 = -0.4049$, $\sum(X_i - \overline{X})^4 = 0.5110$, **(a)** $\sqrt{b_1} = -0.71$.
(b) $b_2 = 3.60$. **(c)** $Q_1 = X_{9.5} = 4.5$ km, $Q_2 = M = X_{19} = 4.7$ kg, $Q_3 = X_{28.5} = 4.8$ kg; skewness $= -0.100$ **(d)** $\mathcal{O}_1 = X_5 = 4.3$ kg, $\mathcal{O}_2 = Q_1 = 4.5$ kg, $\mathcal{O}_3 = X_{12} = 4.6$ kg; $\mathcal{O}_5 = X_{26} = 4.8$ kg, $\mathcal{O}_6 = Q_3 = 4.8$ kg, $\mathcal{O}_7 = X_{33} = 4.9$ kg; kurtosis $= 1.333$.

2. (a) $Z = (78.0\text{ g} - 63.5\text{ g})/12.2\text{ g} = 1.19$, $P(X \geq 78.0\text{ g}) = P(Z \geq 1.19) = 0.1170$;
(b) $P(X \leq 78.0\text{ g}) = 1.0000 - P(X \geq 78.0\text{ g}) = 1.000 - 0.1170 = 0.8830$; **(c)** $(0.1170)(1000) = 117$; **(d)** $Z = (41.0\text{ g} - 63.5\text{ g})/12.2\text{ g} = -1.84$, $P(X \leq 41.0\text{ g}) = P(Z \leq -1.84) = 0.0329$

3. (a) $P(X \leq 60.0\text{ g}) = P(Z \leq -0.29) = 0.3859$, $P(X \geq 70.0\text{ g}) = P(Z \geq 0.53) = 0.2981$, $P(60.0\text{ g} \leq X \leq 70.0\text{ g}) = 1.0000 - 0.3859 - 0.2981 = 0.3160$; **(b)** $P(X \leq 60.0\text{ g}) = P(Z \leq -0.29) = 0.3859$, $P(X \leq 50.0\text{ g}) = P(Z \leq -1.11) = 0.1335$, $P(50.0\text{ g} \leq X \leq 60.0\text{ g}) = P(-1.11 \leq Z \leq -0.29) = 0.3859 - 0.1335 = 0.2524$.

4. (a) $\sigma_{\overline{X}} = \sigma/\sqrt{n} = 12.2\text{ g}/\sqrt{10} = 3.86$ g;
(b) $Z = (65.0\text{ g} - 63.5)/3.86\text{ g} = 0.39$, $P(\overline{X} \geq 65.0\text{ g}) = P(Z \geq 0.39) = 0.3483$; **(c)** $P(\overline{X} \leq 62.0\text{ g}) = P(Z \leq -0.39) = 0.3483$, $P(\overline{X} \leq 60.0\text{ g}) = P(Z \leq -0.91) = 0.1814$, $P(60.0\text{ g} \leq \overline{X} \leq 62.0\text{ g}) = 0.3483 - 0.1814 = 0.1669$.

5. (a) $\overline{X} = 10.43$ g/L; $Z = (10.43\text{ mg/L} - 10.00\text{ mg/L})/0.24\text{ mg/L} = 1.79$; $P(Z \geq 10.00\text{ mg/L}) = 0.0367$; as $0.0367 < 0.05$, reject H_0. **(b)** $Z_{0.05(1)} = 1.645$, $L_1 = 10.43\text{ mg/L} - (1.645)(0.24\text{ mg/L}) = 10.04$ mg/L, $L_2 = \infty$; $\mu_0 = 10.00$ mg/L, $10.00\text{ mg/L} < 10.04$ mg/L so H_0 is rejected.

6. $\sigma_{\overline{X}} = \sqrt{89.06\text{ days}^2/24} = 3.71$ days; **(a)** for 99% confidence, $Z_{0.010(2)} = 2.575$, $L_1 = 61.4\text{ days} - (2.575)(1.93\text{ days}) = 61.4\text{ days} - 4.9\text{ days} = 56.5$ days, $L_2 = 61.4\text{ days} + 4.9\text{ days} = 66.3$ days; **(b)** for 95% confidence, $Z_{0.05(2)} = 1.960$, $L_1 = 57.6$ days, $L_2 = 65.2$ days; **(c)** for 90% confidence, $Z_{0.10(2)} = 1.645$, $L_1 = 58.2$ days, $L_2 = 64.6$ days.

One-Sample Hypotheses

This chapter will continue the discussion on how to draw inferences about population parameters by testing hypotheses about them using appropriate sample statistics. It will consider hypotheses about each of several population parameters, including the population mean, median, variance, standard deviation, and coefficient of variation. The chapter will also discuss procedures for expressing the confidence one can have in estimating population parameters from sample statistics.

1 TWO-TAILED HYPOTHESES CONCERNING THE MEAN

We introduced the concept of statistical testing using a pair of statistical hypotheses, the null and alternate hypotheses, as statements that a population mean (μ) is equal to some specified value (let's call it μ_0):

$$H_0: \quad \mu = \mu_0;$$
$$H_A: \quad \mu \neq \mu_0.$$

For example, let us consider the body temperatures of 25 intertidal crabs that we exposed to air at 24.3°C (Example 1). We may wish to ask whether the mean body temperature of members of this species of crab is the same as the ambient air temperature of 24.3°C. Therefore,

$$H_0: \quad \mu = 24.3°C, \text{ and}$$
$$H_A: \quad \mu \neq 24.3°C,$$

where the null hypothesis states that the mean of the population of data from which this sample of 25 came is 24.3°C (i.e., μ is "no different from 24.3°C"), and the alternate hypothesis is that the population mean is not equal to (i.e., μ is different from) 24.3°C.

EXAMPLE 1   The Two-Tailed t Test for Difference between a Population Mean and a Hypothesized Population Mean

Body temperatures (measured in °C) of 25 intertidal crabs placed in air at 24.3°C: 25.8, 24.6, 26.1, 22.9, 25.1, 27.3, 24.0, 24.5, 23.9, 26.2, 24.3, 24.6, 23.3, 25.5, 28.1, 24.8, 23.5, 26.3, 25.4, 25.5, 23.9, 27.0, 24.8, 22.9, 25.4.

$$H_0: \quad \mu = 24.3°C$$

$$H_A: \quad \mu \neq 24.3°C$$

$$\alpha = 0.05$$

$$n = 25$$

$$\overline{X} = 25.03°C$$

$$s^2 = 1.80(°C)^2$$

$$s_{\overline{X}} = \sqrt{\frac{1.80(°C)^2}{25}} = 0.27°C$$

$$t = \frac{\overline{X} - \mu}{s_{\overline{X}}} = \frac{25.03°C - 24.3°C}{0.27°C} = \frac{0.73°C}{0.27°C} = 2.704$$

$$\nu = 24$$

$$t_{0.05(2), 24} = 2.064$$

As $|t| > t_{0.05(2), 24}$, reject H_0 and conclude that the sample of 25 body temperatures came from a population whose mean is not 24.3°C.

$$0.01 < P < 0.02 \; [P = 0.012]*$$

Previously $Z = (\overline{X} - \mu)/\sigma_{\overline{X}}$ was introduced as a *normal deviate*, and it was shown how one can determine the probability of obtaining a sample with mean $\overline{X}$ from a population with a specified mean μ. The normal deviate can be used to test hypotheses about a population mean. Note, however, that the calculation of Z requires the knowledge of $\sigma_{\overline{X}}$, which we typically do not have. The best we can do is to calculate $s_{\overline{X}}$ as an estimate of $\sigma_{\overline{X}}$. If n is very, very large, then $s_{\overline{X}}$ is a good estimate of $\sigma_{\overline{X}}$, and

---

*The exact probability of a calculated test statistic (such as t), as determined by computer software, is indicated in brackets. It should not be assumed that the many decimal places given by computer programs are all accurate (McCullough, 1998, 1999); therefore, the book's examples will routinely express these probabilities to only two or three (occasionally four) decimal places. The term "software" was coined by John Wilder Tukey (Leonhardt, 2000).

we can be tempted to calculate Z using this estimate. However, for most biological situations n is insufficiently large to do this; but we can use, in place of the normal distribution (Z), a distribution known as t, the development of which was a major breakthrough in statistical methodology:*

$$t = \frac{\overline{X} - \mu}{s_{\overline{X}}}. \tag{1}$$

Because the t-testing procedure is so readily employed, we need not wonder whether n is large enough to use Z; and, in fact, Z is almost never used for hypothesis testing about means.

As do some other distributions to be encountered among statistical methods, the t distribution has different shapes for different values of what is known as *degrees of freedom* (denoted by ν, the lowercase Greek nu).[†] For hypotheses concerning a mean,

$$\nu = n - 1. \tag{2}$$

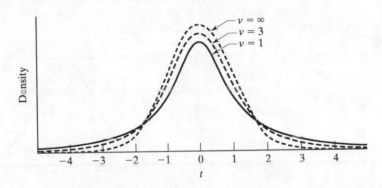

FIGURE 1: The t distribution for various degrees of freedom, ν. For $\nu = \infty$, the t distribution is identical to the normal distribution.

Recall that n is the size of the sample (i.e., the number of data from which $\overline{X}$ has been calculated). The influence of ν on the shape of the t distribution is shown in Figure 1.

---

*The t statistic is also referred to as "Student's t" because of William Sealy Gosset (1876–1937), who was an English statistician with the title "brewer" in the Guinness brewery of Dublin. He used the pen name "Student" (under his employer's policy requiring anonymity) to publish noteworthy developments in statistical theory and practice, including ("Student," 1908) the introduction of the distribution that bears his pseudonym. (See Boland, 1984, 2000; Irwin, 1978; Lehmann, 1999; Pearson, 1939; Pearson, Plackett, and Barnard, 1990; Zabell, 2008.) Gosset originally referred to his distribution as z; and, especially between 1922 and 1925, R. A. Fisher (e.g., 1925a, 1925b: 106–113, 117–125; 1928) helped develop its potential in statistical testing while modifying it; Gosset and Fisher then called the modification "t" (Eisenhart, 1979). Gosset was a modest man, but he was referred to as "one of the most original minds in contemporary science" by Fisher (1939a), himself one of the most insightful and influential statisticians of all time. From his first discussions of the t distribution, Gosset was aware that it was strictly applicable only if sampling normally distributed populations, though he surmised that only large deviations from normality would invalidate the use of t (Lehmann, 1999).

[†]In early writings of the t distribution (during the 1920s and 1930s), the symbol n or f was used for degrees of freedom. This was often confusing because these letters had commonly been used to denote other quantities in statistics so Maurice G. Kendall (1943: 292) recommended ν.

This distribution is leptokurtic, having a greater concentration of values around the mean and in the tails than does a normal distribution; but as n (and, therefore, ν) increases, the t distribution tends to resemble a normal distribution more closely, and for $\nu = \infty$ (i.e., for an infinitely large sample*), the t and normal distributions are identical; that is, $t_{\alpha,\infty} = Z_\alpha$.

The mean of the sample of 25 data (body temperatures) shown in Example 1 is 25.03°C, and the sample variance is 1.80(°C)^2. These statistics are estimates of the mean and variance of the population from which this sample came. However, this is only one of a very large number of samples of size 25 that could have been taken at random from the population. The distribution of the means of all possible samples with $n = 25$ is the t distribution for $\nu = 24$, which is represented by the curve of Figure 2. In this figure, the mean of the t distribution (i.e., $t = 0$) represents the mean hypothesized in H_0 (i.e., $\mu = \mu_0 = 24.3$°C), for, by Equation 1, $t = 0$ when $\overline{X} = \mu$. The shaded areas in this figure represent the extreme 5% of the total area under the curve (2.5% in each tail). Thus, an $\overline{X}$ so far from μ that it lies in either of the shaded areas has a probability of less than 5% of occurring by chance alone, and we assume that it occurred because H_0 is, in fact, false. Because an extreme t value in either direction from μ will cause us to reject H_0, we are said to be considering a "two-tailed" (or "two-sided") test.

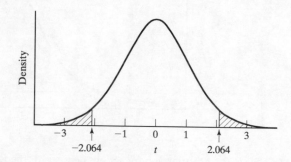

FIGURE 2: The t distribution for $\nu = 24$, showing the critical region (shaded area) for a two-tailed test using $\alpha = 0.05$. (The critical value of t is 2.064.)

For $\nu = 24$, we can consult Table 3 from *Appendix: Statistical Tables and Graphs* to find the following two-tailed probabilities (denoted as "$\alpha(2)$") of various values of t:

ν	$\alpha(2)$:	0.50	0.20	0.10	0.05	0.02	0.01
24		0.685	1.318	1.711	2.064	2.492	2.797

Thus, for example, for a two-tailed α of 0.05, the shaded areas of the curve begin at 2.064 t units on either side of μ. Therefore, we can state:

$$P(|t| \geq 2.064) = 0.05.$$

That is, 2.064 and -2.064 are the *critical values* of t; and if t (calculated from Equation 1) is equal to or greater than 2.064, or is equal to or less than -2.064, that will be considered reasonable cause to reject H_0 and consider H_A to be a true

---

*The modern symbol for infinity (∞) was introduced in 1655 by influential English mathematician John Wallis (1616–1703) (Cajori, 1928/9, Vol. 2: 44), but it did not appear again in print until a work by Jacob Bernoulli was published posthumously in 1713 by his nephew Nikolaus Bernoulli (Gullberg, 1997: 30).

statement. That portion of the t distribution beyond the critical values (i.e., the shaded areas in the figure) is called the *critical region*,* or *rejection region*. For the sample of 25 body temperatures (see Example 1), $t = 2.704$. As 2.704 lies within the critical region (i.e., $2.704 > 2.064$), H_0 is rejected, and we conclude that the mean body temperature of crabs under the conditions of our experiment is not 24.3°C.

To summarize, the hypotheses for the two-tailed test are

$$H_0: \mu = \mu_0 \quad \text{and} \quad H_A: \mu \neq \mu_0,$$

where μ_0 denotes the hypothesized value to which we are comparing the population mean. (In the above example, $\mu_0 = 24.3°C$.) The test statistic is calculated by Equation 1, and if its absolute value is larger than the two-tailed critical value of t from Table 3 from *Appendix: Statistical Tables and Graphs*, we reject H_0 and assume H_A to be true. The critical value of t can be abbreviated as $t_{\alpha(2),\nu}$, where $\alpha(2)$ refers to the two-tailed probability of α. Thus, for the preceding example, we could write $t_{0.05(2),24} = 2.064$. In general, for a two-tailed t test,

$$\text{if } |t| \geq t_{\alpha(2),\nu}, \text{ then reject } H_0.$$

Example 1 presents the computations for the analysis of the crab data. A t of 2.704 is calculated, which for 24 degrees of freedom lies between the tabled critical values of $t_{0.02(2),24} - 2.492$ and $t_{0.01(2),24} - 2.797$. Therefore, if the null hypothesis, H_0, is a true statement about the population we sampled, the probability of $\overline{X}$ being at least this far from μ is between 0.01 and 0.02; that is, $0.01 < P(|t| \geq 2.704) < 0.02$.[†] As this probability is less than 0.05, we reject H_0 and declare it is not a true statement.

Frequently, the hypothesized value in the null and alternate hypotheses is zero. For example, the weights of twelve rats might be measured before and after the animals are placed on a regimen of forced exercise for one week. The change in weight of the animals (i.e., weight after minus weight before) could be recorded, and it might have been found that the mean weight change was −0.65 g (i.e., the mean weight change is a 0.65 g weight loss). If we wished to infer whether such exercise causes any significant change in rat weight, we could state $H_0: \mu = 0$ and $H_A: \mu \neq 0$; Example 2 summarizes the t test for this H_0 and H_A. This test is two tailed, for a large $\overline{X} - \mu$ difference in either direction will constitute grounds for rejecting the veracity of H_0.

EXAMPLE 2   A Two-Tailed Test for Significant Difference between a Population Mean and a Hypothesized Population Mean of Zero

Weight change of twelve rats after being subjected to a regimen of forced exercise. Each weight change (in g) is the weight after exercise minus the weight before.

1.7	$H_0: \mu = 0$
0.7	$H_A: \mu \neq 0$
−0.4	$\alpha = 0.05$

---

*David (1995) traces the first use of this term to J. Neyman and E. S. Pearson in 1933.

[†]Some calculators and many computer programs have the capability of determining the probability of a given t (e.g., see Boomsma and Molenaar, 1994). For the present example, we would thereby find that $P(|t| \geq 2.704) = 0.012$.

$$-1.8$$
$$0.2$$
$$0.9$$
$$-1.2$$
$$-0.9$$
$$-1.8$$
$$-1.4$$
$$-1.8$$
$$-2.0$$

$$n = 12$$

$$\overline{X} = -0.65 \text{ g}$$

$$s^2 = 1.5682 \text{ g}^2$$

$$s_{\overline{X}} = \sqrt{\frac{1.5682 \text{ g}^2}{12}} = 0.36 \text{ g}$$

$$t = \frac{\overline{X} - \mu}{s_{\overline{X}}} = \frac{-0.65 \text{ g}}{0.36 \text{ g}} = -1.81$$

$$\nu = n - 1 = 11$$

$$t_{0.05(2),11} = 2.201$$

Since $|t| < t_{0.05(2),11}$, do not reject H_0.

$$0.05 < P < 0.10 \; [P = 0.098]$$

Therefore, we conclude that the exercise does not cause a weight change in the population from which this sample came.

It should be kept in mind that concluding statistical significance is not the same as determining biological significance. In Example 1, statistical significance was achieved for a difference of 0.73°C between the mean crab body temperature (25.03°C) and the air temperature (24.3°C). The statistical question posed is whether that magnitude of difference is likely to occur by chance if the null hypothesis of no difference is true. The answer is that it is unlikely (there is only a 0.012 probability) and, therefore, we conclude that H_0 is not true. Now the biological question is whether a difference of 0.73°C is of significance (with respect to the crabs' physiology, to their ecology, or otherwise). If the sample of body temperatures had a smaller standard error, $s_{\overline{X}}$, an even smaller difference would have been declared statistically significant. But is a difference of, say, 0.1°C or 0.01°C of biological importance (even if it is statistically significant)? In Example 2, the mean weight change, 0.36 g, was determined not to be significant statistically. But if the sample mean weight change had been 0.8 g (and the standard error had been the same), t would have been calculated to be 2.222 and H_0 would have been rejected. The statistical conclusion would have been that the exercise regime does result in weight change in rats, but the biological question would then be whether a weight change as small as 0.8 g has significance biologically. Thus, assertion of statistical difference should routinely be followed by an assessment of the significance of that difference to the objects of the study (in these examples, to the crabs or to the rats).

(a) Assumptions. The theoretical basis of t testing assumes that sample data came from a normal population, assuring that the mean at hand came from a normal distribution of means. Fortunately, the t test is *robust*,[*] meaning that its validity is not seriously affected by moderate deviations from this underlying assumption. The test also assumes—as other statistical tests typically do—that the data are a random sample.

The adverse effect of nonnormality is that the probability of a Type I error differs substantially from the stated α. Various studies (e.g., Cicchitelli, 1989; Pearson and Please, 1975; and Ractliffe, 1968) have shown that the detrimental effect of nonnormality is greater for smaller α but less for larger n, that there is little effect if

---

[*]The term *robustness* was introduced by G. E. P. Box in 1953 (David, 1995).

the distribution is symmetrical, and that for asymmetric distributions the effect is less with strong leptokurtosis than with platykurtosis or mesokurtosis; and the undesirable effect of nonnormality is much less for two-tailed testing than for one-tailed testing (Section 2).

It is important to appreciate that a sample used in statistical testing such as that discussed here must consist of truly replicated data, where a *replicate** is defined as the smallest experimental unit to which a treatment is independently applied. In Example 1, we desired to draw conclusions about a population of measurements representing a large number of animals (i.e., crabs). Therefore, the sample must consist of measurements (i.e., body temperatures) from n (i.e., 25) animals; it would *not* be valid to obtain 25 body temperatures from a single animal. And, in Example 2, 12 individual rats must be used; it would *not* be valid to employ data obtained from subjecting the same animal to the experiment 12 times. Such invalid attempts at replication are discussed by Hurlbert (1984), who named them *pseudoreplication*.

2 ONE-TAILED HYPOTHESES CONCERNING THE MEAN

In Section 1, we spoke of the hypotheses $H_0: \mu = \mu_0$ and $H_A: \mu \neq \mu_0$, because we were willing to consider a large deviation of $\overline{X}$ in either direction from μ_0 as grounds for rejecting H_0. However, in some instances, our interest lies only in whether $\overline{X}$ is significantly larger (or significantly smaller) than μ_0, and this is termed a "one-tailed" (or "one-sided") test situation. For example, we might be testing a drug hypothesized to cause weight reduction in humans. The investigator is interested only in whether a weight *loss* occurs after the drug is taken. (In Example 2, using a two-sided test, we were interested in determining whether either weight loss or weight gain had occurred.) It is important to appreciate that the decision whether to test one-tailed or two-tailed hypotheses must be based on the scientific question being addressed, *before* data are collected.

In the present example, if there is either weight gain or no weight change, the drug will be considered a failure. Therefore, for this one-sided test, we should state $H_0: \mu \geq 0$ and $H_A: \mu < 0$. Here, the null hypothesis states that there is no mean weight loss (i.e., the mean weight change is greater than or equal to zero), and the alternate hypothesis states that there is a mean weight loss (i.e., the mean weight change is less than zero). By examining the alternate hypothesis, H_A, we see that H_0 will be rejected if t is in the left-hand critical region of the t distribution. In general,

$$\text{for } H_A: \mu < \mu_0,$$
$$\text{if } t \leq -t_{\alpha(1),\nu}, \text{ then reject } H_0.^\dagger$$

Example 3 summarizes such a set of 12 weight change data tested against this pair of hypotheses. From Table 3 from *Appendix: Statistical Tables and Graphs* we find that $t_{0.05(1),11} = 1.796$, and the critical region for this test is shown in Figure 3. From this figure, and by examining Table 3 from *Appendix: Statistical Tables and Graphs*, we see that $t_{\alpha(1),\nu} = t_{2\alpha(2),\nu}$ or $t_{\alpha(2),\nu} = t_{\alpha/2(1),\nu}$; that is, for example, the critical value of t for a one-sided test at $\alpha = 0.05$ is the same as the critical value of t for a two-sided test at $\alpha = 0.10$.

---

*The term *replicate*, in the context of experimental design, was introduced by R. A. Fisher in 1926 (Miller, 2004a).

†For one-tailed testing of this H_0, probabilities of t up to 0.25 are indicated in Table 3 from *Appendix: Statistical Tables and Graphs*. If $t = 0$, then $P = 0.50$; so if $-t_{0.25(1),\nu} < t < 0$, then $0.25 < P < 0.50$; and if $t > 0$, then $P > 0.50$.

EXAMPLE 3   **A One-Tailed t Test for the Hypotheses $H_0: \mu \geq 0$ and $H_A: \mu < 0$**

The data are weight changes of humans, tabulated after administration of a drug proposed to result in weight loss. Each weight change (in kg) is the weight after minus the weight before drug administration.

0.2	$n = 12$
-0.5	$\overline{X} = -0.61$ kg
-1.3	$s^2 = 0.4008$ kg^2
-1.6	
-0.7	
0.4	$s_{\overline{X}} = \sqrt{\dfrac{0.4008 \text{ kg}^2}{12}} = 0.18$ kg
-0.1	
0.0	$t = \dfrac{\overline{X} - \mu}{s_{\overline{X}}} = \dfrac{-0.61 \text{ kg}}{0.18 \text{ kg}} = -3.389$
-0.6	
-1.1	$\nu = n - 1 = 11$
-1.2	
-0.8	$t_{0.05(1),11} = 1.796.$
	If $t \leq -t_{0.05(1),11}$, reject H_0.
	Conclusion: reject H_0.

$$0.0025 < P(t \leq -3.389) < 0.005 \ [P = 0.0030]$$

We conclude that the drug does cause weight loss.

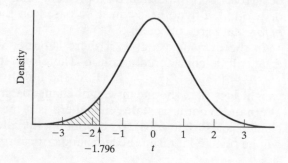

FIGURE 3: The distribution of t for $\nu = 11$, showing the critical region (shaded area) for a one-tailed test using $\alpha = 0.05$. (The critical value of t is -1.796.)

If we are interested in whether $\overline{X}$ is significantly *greater* than some value, μ_0, the hypotheses for the one-tailed test are $H_0: \mu \leq \mu_0$ and $H_A: \mu > \mu_0$. For example, a drug manufacturer might advertise that a product dissolves completely in gastric juice within 45 sec. The hypotheses appropriate for testing this claim are $H_0: \mu \leq 45$ sec and $H_A: \mu > 45$ sec, because we are not particularly interested in the possibility that the product dissolves faster than is claimed, but we wish to determine whether its dissolving time is longer than advertised. Thus, the rejection region would be in the right-hand tail, rather than in the left-hand tail (the latter being the case in Example 3). The details of such a test are shown in Example 4. In general,

$$\text{for } H_A: \mu > \mu_0,$$
$$\text{if } t \geq t_{\alpha(1),\nu}, \text{ then reject } H_0.^*$$

---

*For this H_0, if $t = 0$, then $P = 0.50$; therefore, if $0 < t < t_{0.25(1),\nu}$, then $0.25 < P < 0.50$, and if $t < 0$, then $P > 0.50$.

EXAMPLE 4 **The One-Tailed t Test for the Hypotheses H_0: $\mu \leq 45$ sec and H_A: $\mu > 45$ sec**

Dissolving times (in sec) of a drug in gastric juice: 42.7, 43.4, 44.6, 45.1, 45.6, 45.9, 46.8, 47.6.

H_0: $\mu \leq 45$ sec

H_A: $\mu > 45$ sec

$\alpha = 0.05$

$n = 8$

$\overline{X} = 45.21$ sec

$SS = 18.8288$ sec^2

$s^2 = 2.6898$ sec^2

$s_{\overline{X}} = 0.58$ sec

$t = \dfrac{45.21 \text{ sec } - 45 \text{ sec}}{0.58 \text{ sec}} = 0.36$

$\nu = 7$

$t_{0.05(1),7} = 1.895$

If $t \geq t_{0.05(1),7}$, reject H_0.

Conclusion: do not reject H_0.

$$P(t \geq 0.36) > 0.25 \; [P = 0.36]$$

We conclude that the mean dissolving time is not greater than 45 sec.

---

3 CONFIDENCE LIMITS FOR THE POPULATION MEAN

When Section 1 defined $t = (\overline{X} - \mu)/s_{\overline{X}}$, it was explained that 5% of all possible means from a normally distributed population with mean μ will yield t values that are either larger than $t_{0.05(2),\nu}$ or smaller than $-t_{0.05(2),\nu}$; that is, $|t| \geq t_{0.05(2),\nu}$ for 5% of the means. This connotes that 95% of all t values obtainable lie between the limits of $-t_{0.05(2),\nu}$ and $t_{0.05(2),\nu}$; this may be expressed as

$$P\left[-t_{0.05(2),\nu} \leq \frac{\overline{X} - \mu}{s_{\overline{X}}} \leq t_{0.05(2),\nu} \right] = 0.95. \tag{3}$$

It follows from this that

$$P[\overline{X} - t_{0.05(2),\nu}\, s_{\overline{X}} \leq \mu \leq \overline{X} + t_{0.05(2),\nu}\, s_{\overline{X}}] = 0.95. \tag{4}$$

The value of the population mean, μ, is not known, but we estimate it as $\overline{X}$, and if we apply Equation 4 to many samples from this population, for 95% of the samples the interval between $\overline{X} - t_{0.05(2),\nu}s_{\overline{X}}$ and $\overline{X} + t_{0.05(2),\nu}s_{\overline{X}}$ will include μ. This interval is called the *confidence interval* (abbreviated CI) for μ.

In general, the confidence interval for μ can be stated as

$$P[\overline{X} - t_{\alpha(2),\nu}\, s_{\overline{X}} \leq \mu \leq \overline{X} + t_{\alpha(2),\nu}\, s_{\overline{X}}] = 1 - \alpha. \tag{5}$$

$\overline{X} - t_{\alpha(2),\nu}\, s_{\overline{X}}$ is called the *lower confidence limit* (abbreviated L_1); and $\overline{X} + t_{\alpha(2),\nu}\, s_{\overline{X}}$ is the *upper confidence limit* (abbreviated L_2); and the two confidence limits can be stated as

$$\overline{X} \pm t_{\alpha(2),\nu}\, s_{\overline{X}} \tag{6}$$

(reading "$\pm$" to be "plus or minus"). In expressing a confidence interval, we call the quantity $1 - \alpha$ (namely, $1 - 0.05 = 0.95$ in the present example) the *confidence level* (or the *confidence coefficient*, or the *confidence probability*).*

Although $\overline{X}$ is the best estimate of μ, it is only an estimate (and not necessarily a very good one), and the calculation of the confidence interval for μ provides an expression of the precision of the estimate. Example 5, part (a), refers to the data of Example 1 and demonstrates the determination of the 95% confidence interval for the mean of the population from which the sample came. As the 95% confidence limits are computed to be $L_1 = 24.47°C$ and $L_2 = 25.59°C$, the 95% confidence interval may be expressed as $P(24.47°C \leq \mu \leq 25.59°C)$. The meaning of this kind of statement is commonly expressed in nonprobabilistic terms as having 95% *confidence* that the interval of 24.47°C to 25.59°C contains μ. This does *not* mean that there is a 95% *probability* that the interval constructed from this one sample contains the population mean, μ; but it does mean that 95% of the confidence limits computed for many independent random samples would bracket μ (or, this could be stated as the probability that the confidence interval from a future sample would contain μ). And, if the μ_0 in H_0 and H_A is within the confidence interval, then H_0 will not be rejected.

EXAMPLE 5   Computation of Confidence Intervals and Confidence Limits for the Mean, Using the Data of Example 1

(a) At the 95% confidence level:

$\overline{X} = 25.03°C$

$s_{\overline{X}} = 0.27°C$

$t_{0.05(2),24} = 2.064$

$\nu = 24$

95% confidence interval $= \overline{X} \pm t_{0.05(2),24}\, s_{\overline{X}}$

$= 25.03°C \pm (2.064)(0.27°C)$

$= 25.03°C \pm 0.56°C$

95% confidence limits: $L_1 = 25.03°C - 0.56°C = 24.47°C$

$L_2 = 25.03°C + 0.56°C = 25.59°C$

(b) At the 99% confidence level:

$t_{0.01(2),24} = 2.797$

99% confidence interval $= \overline{X} \pm t_{0.01(2),24}\, s_{\overline{X}}$

$= 25.03° \pm (2.797)(0.27°C)$

$= 25.03°C \pm 0.76°C$

99% confidence limits: $L_1 = 25.03°C - 0.76°C = 24.27°C$

$L_2 = 25.03°C + 0.76°C = 25.79°C$

In both parts (a) and (b), the hypothesized value, $\mu_0 = 24.3°C$ in Example 1, lies outside the confidence intervals. This indicates that H_0 would be rejected using either the 5% or the 1% level of significance.

---

*We owe the development of confidence intervals to Jerzy Neyman, between 1928 and 1933 (Wang, 2000), although the concept had been enunciated a hundred years before. Neyman introduced the terms *confidence interval* and *confidence coefficient* in 1934 (David, 1995). On rare occasion, the biologist may see reference to "fiducial intervals," a concept developed by R. A. Fisher beginning in 1930 and identical to confidence intervals in some, but not all, situations (Pfanzagl, 1978).

The smaller $s_{\overline{X}}$ is, the smaller will be the confidence interval, meaning that μ is estimated more precisely when $s_{\overline{X}}$ is small. Also, it can be observed from the calculation of $s_{\overline{X}}$ that a large n will result in a small $s_{\overline{X}}$ and, therefore, a narrower confidence interval. A parameter estimate from a large sample is generally more precise that an estimate of the same parameter from a small sample.

Setting confidence limits around μ has the same underlying assumptions as the testing of hypotheses about μ (Section 1a), and violating those assumptions can invalidate the stated level of confidence $(1 - \alpha)$.

If, instead of a 95% confidence interval, it is desired to state a higher level of confidence, say 99%, that L_1 and L_2 encompass the population mean, then $t_{0.01(1),24}$ rather than $t_{0.01(2),24}$ would be employed. From Table 3 from *Appendix: Statistical Tables and Graphs* we find that $t_{0.01(1),24} = 2.797$, so the 99% confidence interval would be calculated as shown in Example 5, part (b), where it is determined that $P(24.27°C \leq \mu \leq 25.79°C) = 0.99$.

(a) One-Tailed Confidence Limits. One-tailed confidence intervals are appropriate in situations that warrant one-tailed hypothesis tests. Such a confidence interval employs a one-tailed critical value of t (i.e., $t_{\alpha(1),\nu}$) instead of a two-tailed critical value $(t_{\alpha(2)}, \nu)$. For $H_0: \mu \leq \mu_0$ and $H_A: \mu > \mu_0$, the confidence limits for μ are $L_1 = \overline{X} - t_{\alpha(1),\nu} s_{\overline{X}}$ and $L_2 = \infty$; and for $H_0: \mu \geq \mu_0$ and $H_A: \mu < \mu_0$, the confidence limits are $L_1 = -\infty$ and $L_2 = \overline{X} + t_{\alpha(1),\nu} s_{\overline{X}}$. For the situation in Example 4, in which $H_0: \mu \leq 45$ sec and $H_A: \mu > 45$ sec, L_1 would be 45.21 sec $- (1.895)(0.58$ sec$) = 45.21$ sec $- 1.10$ sec $- 44.11$ and $L_2 - \infty$. And the hypothesized μ_0 (45 sec) lies within the confidence interval, indicating that the null hypothesis is not rejected.

(b) Prediction Limits. While confidence limits express the precision with which a population characteristic is estimated, we can also indicate the precision with which future observations from this population can be predicted.

After calculating $\overline{X}$ and s^2 from a random sample of n data from a population, we can ask what the mean would be from an additional random sample, of an additional m data, from the same population. The best estimate of the mean of those m additional data would be $\overline{X}$, and the precision of that estimate may be expressed by this two-tailed *prediction interval* (abbreviated PI):

$$\overline{X} \pm t_{\alpha(2),\nu}\sqrt{\frac{s^2}{m} + \frac{s^2}{n}}, \tag{7}$$

where $\nu = n - 1$ (Hahn and Meeker, 1991: 61–62). If the desire is to predict the value of one additional datum from that population (i.e., $m = 1$), then Equation 7 becomes

$$\overline{X} \pm t_{\alpha(2),\nu}\sqrt{s^2 + \frac{s^2}{n}}. \tag{8}$$

The prediction interval will be wider than the confidence interval and will approach the confidence interval as m becomes very large. The use of Equations 7 and 8 is demonstrated in Example 6.

One-tailed prediction intervals are not commonly obtained but are presented in Hahn and Meeker (1991: 63), who also consider another kind of interval: the *tolerance interval*, which may be calculated to contain at least a specified proportion (e.g., a specified percentile) of the sampled population; and in Patel (1989), who also discusses simultaneous prediction intervals of means from more than one future sample. The procedure is very much like that of Section 3a. If the desire is to obtain only a lower prediction limit, L_1 (while L_2 is considered to be ∞), then the first portion of Equation 7 (or 8) would be modified to be $\overline{X} - t_{\alpha(1), \nu}$ (i.e., the one-tailed t would be used); and if the intent is to express only an upper prediction limit, L_2 (while regarding L_1 to be $-\infty$), then we would use $\overline{X} + t_{\alpha(1), \nu}$. As an example, Example 6 might have asked what the highest mean body temperature is that would be predicted, with probability α, from an additional sample. This would involve calculating L_2 as indicated above, while $L_1 = -\infty$.

EXAMPLE 6 **Prediction Limits for Additional Sampling from the Population Sampled in Example 1**

From Example 1, which is a sample of 25 crab body temperatures,

$$n = 25, \quad \overline{X} = 25.03°\text{C}, \text{ and } s^2 = 1.80(°\text{C})^2.$$

(a) If we intend to collect 8 additional crab body temperatures from the same population from which the 25 data in Example 1 came, then (by Equation 7) we can be 95% confident that the mean of those 8 data will be within this prediction interval:

$$25.03°\text{C} \pm t_{0.05(2), 24}\sqrt{\frac{1.80(°\text{C})^2}{8} + \frac{1.80(°\text{C})^2}{2}}$$
$$= 25.03°\text{C} \pm 2.064(0.545°\text{C})$$
$$= 25.03°\text{C} \pm 1.12°\text{C}.$$

Therefore, the 95% prediction limits for the predicted mean of these additional data are $L_1 = 23.91°\text{C}$ and $L_2 = 26.15°\text{C}$.

(b) If we intend to collect 1 additional crab body temperature from the same population from which the 25 data in Example 1 came, then (by Equation 8) we can be 95% confident that the additional datum will be within this prediction interval:

$$25.03°\text{C} \pm t_{0.05(2), 24}\sqrt{1.80(°\text{C})^2 + \frac{1.80(°\text{C})^2}{2}}$$
$$= 25.03°\text{C} \pm 2.064(1.368°\text{C})$$
$$= 25.03°\text{C} \pm 2.82°\text{C}.$$

Therefore, the 95% prediction limits for this predicted datum are $L_1 = 22.21°\text{C}$ and $L_2 = 27.85°\text{C}$.

---

4 REPORTING VARIABILITY AROUND THE MEAN

It is very important to provide the reader of a research paper with information concerning the variability of the data reported. But authors of such papers are often unsure of appropriate ways of doing so, and not infrequently do so improperly.

If we wish to describe the population that has been sampled, then the sample mean ($\overline{X}$) and the standard deviation (s) may be reported. The range might also be reported, but in general it should not be stated without being accompanied by another measure of variability, such as s. Such statistics are frequently presented as in Table 1 or 2.

TABLE 1: Tail Lengths (in mm) of Field Mice from Different Localities

Location	n	$\overline{X} \pm \text{SD}$ (range in parentheses)
Bedford, Indiana	18	56.22 ± 1.33 (44.8 to 68.9)
Rochester, Minnesota	12	59.61 ± 0.82 (43.9 to 69.8)
Fairfield, Iowa	16	60.20 ± 0.92 (52.4 to 69.2)
Pratt, Kansas	16	53.93 ± 1.24 (46.1 to 63.6)
Mount Pleasant, Michigan	13	55.85 ± 0.90 (46.7 to 64.8)

TABLE 2: Evaporative Water Loss of a Small Mammal at Various Air Temperatures. Sample Statistics Are Mean ± Standard Deviation, with Range in Parentheses

	Air Temperature (°C)				
	16.2	24.8	30.7	36.8	40.9
Sample size	10	13	10	8	9
Evaporative water loss (mg/g/hr)	0.611 ± 0.164 (0.49 to 0.88)	0.643 ± 0.194 (0.38 to 1.13)	0.890 ± 0.212 (0.64 to 1.39)	1.981 ± 0.230 (1.50 to 2.36)	3.762 ± 0.641 (3.16 to 5.35)

If it is the author's intention to provide the reader with a statement about the precision of estimation of the population mean, the use of the standard error ($s_{\overline{X}}$) is appropriate. A typical presentation is shown in Table 3a. This table might instead be set up to show confidence intervals, rather than standard errors, as shown in Table 3b. The standard error is always smaller than the standard deviation. But this is not a reason to report the former in preference to the latter. The determination should be made on the basis of whether the desire is to describe variability within the population or precision of estimating the population mean.

There are three very important points to note about Tables 1, 2, 3a, and 3b. First, n should be stated somewhere in the table, either in the caption or in the body of the table. (Thus, the reader has the needed information to convert from SD to SE or from SE to SD, if so desired.) One should always state n when presenting sample statistics ($\overline{X}$, s, $s_{\overline{X}}$, range, etc.), and if a tabular presentation is prepared, it is very good practice to include n somewhere in the table, even if it is mentioned elsewhere in the paper.

Second, the measure of variability is clearly indicated. Not infrequently, an author will state something such as "the mean is 54.2 ± 2.7 g," with no explanation of what "± 2.7" denotes. This renders the statement worthless to the reader, because "± 2.7" will be assumed by some to indicate $\pm$ SD, by others to indicate $\pm$ SE, by others to

TABLE 3a: Enzyme Activities in the Muscle of Various Animals. Data Are $\overline{X} \pm$ SE, with n in Parentheses

Animal	Enzyme Activity (μmole/min/g of tissue)	
	Isomerase	Transketolase
Mouse	0.76 ± 0.09 (4)	0.39 ± 0.04 (4)
Frog	1.53 ± 0.08 (4)	0.18 ± 0.02 (4)
Trout	1.06 ± 0.12 (4)	0.24 ± 0.04 (4)
Crayfish	4.22 ± 0.30 (4)	0.26 ± 0.05 (4)

TABLE 3b: Enzyme Activities in the Muscle of Various Animals. Data Are $\overline{X} \pm$ 95% Confidence Limits

Animal	n	Enzyme Activity (μmole/min/g of tissue)	
		Isomerase	Transketolase
Mouse	4	0.76 ± 0.28	0.39 ± 0.13
Frog	4	1.53 ± 0.25	0.18 ± 0.05
Trout	4	1.06 ± 0.38	0.24 ± 0.11
Crayfish	4	4.22 ± 0.98	0.26 ± 0.15

indicate the 95% (or 99%, or other) confidence interval, and by others to indicate the range.* There is no widely accepted convention; one *must* state explicitly what quantity is meant by this type of statement. If such statements of "$\pm$" values appear in a table, then the explanation is best included somewhere in the table (either in the caption or in the body of the table), even if it is stated elsewhere in the paper.

Third, the units of measurement of the variable must be clear. There is little information conveyed by stating that the tail lengths of 24 birds have a mean of 8.42 and a standard error of 0.86 if the reader does not know whether the tail lengths were measured in centimeters, or inches, or some other unit. Whenever data appear in tables, the units of measurement should be stated somewhere in the table. Keep in mind that a table should be self-explanatory; one should not have to refer back and forth between the table and the text to determine what the tabled values represent.

Frequently, the types of information given in Tables 1, 2, 3a, and 3b are presented in graphs, rather than in tables. In such cases, the measurement scale is typically indicated on the vertical axis, and the mean is indicated in the body of the graph by a short horizontal line or some other symbol. The standard deviation, standard error, or a confidence interval for the mean is commonly indicated on such graphs via a vertical line or rectangle. Often the range is also included, and in such instances the SD or SE may be indicated by a vertical rectangle and the range by a vertical line. Some authors will indicate a confidence interval (generally 95%) in

---

*In older literature the $\pm$ symbol referred to yet another measure, known as the "probable error" (which fell into disuse in the early twentieth century). In a normal curve, the probable error (PE) is 0.6745 times the standard error, because $\overline{X} \pm$ PE includes 50% of the distribution. The term *probable error* was first used in 1815 by German astronomer Friedrich Wilhelm Bessel (1784–1846) (Walker, 1929: 24, 51, 186).

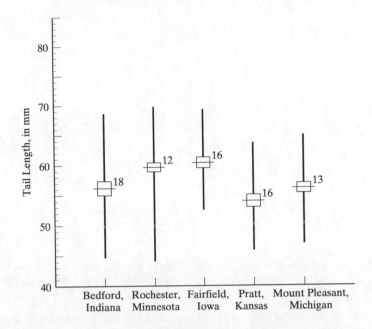

FIGURE 4: Tail lengths of male field mice from different localities, indicating the mean, the mean ± standard deviation (vertical rectangle), and the range (vertical line), with the sample size indicated for each location. The data are from Table 1.

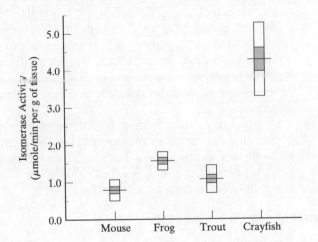

FIGURE 5: Levels of muscle isomerase in various animals. Shown is the mean ± standard error (shaded rectangle), and ± the 95% confidence interval (open rectangle). For each sample, $n = 4$. The data are from Tables 3a and 3b.

addition to the range and either SD or SE. Figures 4, 5, and 6 demonstrate how various combinations of these statistics may be presented graphically.

Instead of the mean and a measure of variability based on the variance, one may present tabular or graphical descriptions of samples using the median and quartiles (e.g., McGill, Tukey, and Larsen, 1978), or the median and its confidence interval. Thus, a graphical presentation such as in Figure 4 could have the range indicated by the vertical line, the median by the horizontal line, and the semiquartile range by the vertical rectangle. Such a graph is discussed in Section 5. Note that when the horizontal axis on the graph represents an interval or ratio scale variable (as in Figure 6),

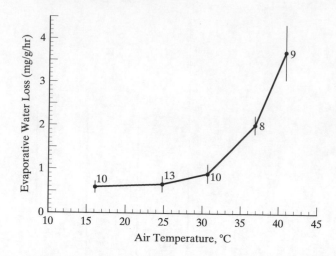

FIGURE 6: Evaporative water loss of a small mammal at various air temperatures. Shown at each temperature is the mean ± the standard deviation, and the sample size. The data are from Table 2.

adjacent means may be connected by straight lines to aid in the recognition of trends.

In graphical presentation of data, as in tabular presentation, care must be taken to indicate clearly the following either on the graph or in the caption: The sample size (n), the units of measurement, and what measures of variability (if any) are indicated (e.g., SD, SE, range, 95% confidence interval).

Some authors present $\overline{X} \pm 2s_{\overline{X}}$ in their graphs. An examination of the t table will show that, except for small samples, this expression will approximate the 95% confidence interval for the mean. But for small samples, the true confidence interval is, in fact, greater than $\overline{X} \pm 2s_{\overline{X}}$. Thus, the general use of this expression is not to be encouraged, and the calculation of the accurate confidence interval is the wiser practice.

A word of caution is in order for those who determine confidence limits, or SDs or SEs, for two or more means and, by observing whether or not the limits overlap, attempt to determine whether there are differences among the population means. Such a procedure is not generally valid; The proper methods for testing for differences between means are discussed.

5 REPORTING VARIABILITY AROUND THE MEDIAN

The median and the lower and upper quartiles (Q_1 and Q_3) form the basis of a graphical presentation that conveys a rapid sense of the middle, the spread, and the symmetry of a set of data. As shown in Figure 7, a vertical box is drawn with its bottom at Q_1 and top at Q_3, meaning that the height of the box is the semi-quartile range ($Q_3 - -Q_1$). Then, the median is indicated by a horizontal line across the box. Next, a vertical line is extended from the bottom of the box to the smallest datum that is no farther from the box than 1.5 times the interquartile range; and a vertical line is drawn from the top of the box to the largest datum that is no farther from the box than 1.5 times the interquartile range. These two vertical lines, below and above the box, are termed "whiskers," so this graphical representation is called a *box plot* or *box-and-whiskers plot.** If any data are so deviant as to lie beyond the whiskers, they

---

*The term *box plot* was introduced by John W. Tukey in 1970 (David, 1995).

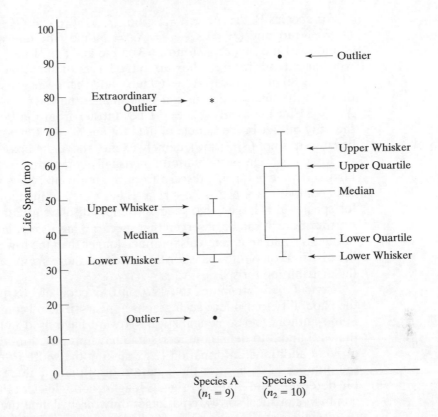

FIGURE 7: Box plots. (The wording with arrows is for instructional purposes and would not otherwise appear in such a graph.)

are termed *outliers* and are placed individually as small circles on the graph. If any are so aberrant as to lie at least 3 times the interquartile range from the box (let's call them "extraordinary outliers"), they may be placed on the graph with a distinctive symbol (such as "∗") instead of a circle.* In addition, the size of the data set (n) should be indicated, either near the box plot itself or in the caption accompanying the plot.

Figure 7 presents a box plot for the two samples in Example 3 from *Measures of Central Tendency*. For species A, the median = 40 mo, $Q_1 = Q_{10/4} = X_{2.5} = 34.5$ mo, and $Q_3 = X_{10-2.5} = X_{7.5} = 46$ mo. The interquartile range is 46 mo − 34.5 mo = 11.5 mo, so 1.5 times the interquartile range is $(1.5)(11.5$ mo$) = 17.25$ mo, and 3 times the interquartile range is $(3)(11.5) = 34.5$ mo. Therefore, the upper whisker extends from the top of the box up to the largest datum that does not exceed 46 mo + 17.25 mo = 63.25 mo (and that datum is $X_8 = 50$ mo), and the lower whisker extends from the bottom of the box down to the smallest datum that is no smaller than 34.5 mo − 17.25 mo = 17.25 mo (namely, $X_2 = 32$ mo). Two of the data, $X_1 = 16$ mo and $X_9 = 79$ mo, lie farther from the box than the whiskers; thus they are outliers and, as X_9 lies more than 3 times the interquartile range from the box (i.e., is more than 34.5 mo greater than Q_3), it is an extraordinary outlier. Therefore, X_1 is indicated with a circle below the box and X_9 is denoted with a "∗" above the box.

---

*The vertical distances above and below the box by an amount 1.5 times the interquartile range are sometimes called *inner fences*, with those using the factor of 3 being called *outer fences*. Also, the top (Q_3) and bottom (Q_1) of the box are sometimes called the *hinges* of the plot.

For species B, the median = 52 mo, $Q_1 = Q_{11/4} = Q_{2.75}$, which is rounded up to $Q_1 = 38$ mo, and $Q_3 = X_{11-3} = X_8 = 59$ mo. The interquartile range is 59 mo − 38 mo = 21 mo, so $(1.5)(21$ mo$) = 31.5$ mo and $(3)(21$ mo$) = 63$ mo. Thus, the upper whisker extends from the box up to the largest datum that does not exceed 59 mo + 21 mo = 80 mo (namely, $X_9 = 69$ mo), and the lower whisker extends from the box down to the smallest datum that is no smaller than 38 mo − 21 mo = 17 mo (namely, $X_1 = 34$ mo). As only $X_{10} = 91$ lies farther from the box than the whiskers, it is the only outlier in the sample of data for species B; it is not an extraordinary outlier because it is not (3)(21 mo), namely 63 mo, above the box.

Box plots are especially useful in visually comparing two or more sets of data. In Figure 7, we can quickly discern from the horizontal lines representing the medians that, compared to species A, species B has a greater median life span; and, as the box for species B is larger, that species' sample displays greater variability in life spans. Furthermore, it can be observed that species B has its median farther from the middle of the box, and an upper whisker much longer than the lower whisker, indicating that the distribution of life spans for this species is more skewed toward longer life than is the distribution for species A.

David (1995) attributes the 1970 introduction of box plots to J. W. Tukey, and the capability to produce such graphs appears in many computer software packages. Some authors and some statistical software have used multiplication factors other than 1.5 and 3 to define outliers, some have proposed modifications of box plots to provide additional information (e.g., making the width of each box proportional to the number of data, or to the square root of that number), and some employ quartile determination. Indeed, Frigge, Hoaglin, and Iglewicz (1989) report that, although common statistical software packages only rarely define the median (Q_2), they identified eight ways Q_1 and Q_3 are calculated in various packages. Unfortunately, the different presentations of box plots provide different impressions of the data, and some of the methods of expressing quartiles are not recommended by the latter authors.

6 SAMPLE SIZE AND ESTIMATION OF THE POPULATION MEAN

A commonly asked question is, "How large a sample must be taken to achieve a desired precision* in estimating the mean of a population?" The answer is related to the concept of a confidence interval, for a confidence interval expresses the precision of a sample statistic, and the precision increases (i.e., the confidence interval becomes narrower) as the sample size increases.

Let us write Equation 6 as $\overline{X} \pm d$, which is to say that $d = t_{\alpha(2), \nu} s_{\overline{X}}$. We shall refer to d as the half-width of the confidence interval, which means that μ is estimated to within $\pm d$. Now, the number of data we must collect to calculate a confidence interval of specified width depends upon: (1) the width desired (for a narrower confidence interval—i.e., more precision in estimating μ—requires a larger sample; (2) the variability in the population (which is estimated by s^2, and larger variability requires larger sample size); and (3) the confidence level specified (for greater confidence—e.g., 99% vs. 95%—requires a larger sample size).

---

*The precision of a sample statistic is the closeness with which it estimates the population parameter; it is not to be confused with the concept of the precision of a measurement, which is the nearness of repeated measurements to each other.

If we have a sample estimate (s^2) of the variance of a normal population, then we can estimate the required sample size for a future sample as

$$n = \frac{s^2 t^2_{\alpha(2),\nu}}{d^2}.$$ (9)

In this equation, s^2 is the sample variance, estimated with $\nu = n - 1$ degrees of freedom, d is the half-width of the desired confidence interval, and $1 - \alpha$ is the confidence level for the confidence interval. Two-tailed critical values of Student's t, with $\nu = n - 1$ degrees of freedom, are found in Table 3 from *Appendix: Statistical Tables and Graphs*.

There is a basic difficulty in solving Equation 9, however; the value of $t_{\alpha(2),(n-1)}$ depends upon n, the unknown sample size. The solution may be achieved by iteration—a process of trial and error with progressively more accurate approximations—as shown in Example 7. We begin the iterative process of estimation with an initial guess; the closer this initial guess is to the finally determined n, the faster we shall arrive at the final estimate. Fortunately, the procedure works well even if this initial guess is far from the final n (although the process is faster if it is a high, rather than a low, guess).

The reliability of this estimate of n depends upon the accuracy of s^2 as an estimate of the population variance, σ^2. As its accuracy improves with larger samples, one should use s^2 obtained from a sample with a size that is not a very small fraction of the n calculated from Equation 9.

EXAMPLE 7   Determination of Sample Size Needed to Achieve a Stated Precision in Estimating a Population Mean, Using the Data of Example 3

If we specify that we wish to estimate μ with a 95% confidence interval no wider than 0.5 kg, then $d = 0.25$ kg, $1 - \alpha = 0.95$, and $\alpha = 0.05$. From Example 3 we have an estimate of the population variance: $s^2 = 0.4008$ kg^2.

Let us guess that a sample of 40 is necessary; then,

$$t_{0.05(2),39} = 2.023.$$

So we estimate (by Equation 7):

$$n = \frac{(0.4008)(2.023)^2}{(0.25)^2} = 26.2.$$

Next, we might estimate $n = 27$, for which $t_{0.05(2),26} = 2.056$, and we calculate

$$n = \frac{(0.4008)(2.056)^2}{(0.25)^2} = 27.1.$$

Therefore, we conclude that a sample size greater than 27 is required to achieve the specified confidence interval.

---

7 SAMPLE SIZE, DETECTABLE DIFFERENCE, AND POWER IN TESTS CONCERNING THE MEAN

(a) Sample Size Required. If we are to perform a one-sample test as described in Section 1 or 2, then it is desirable to know how many data should be collected to detect a specified difference with a specified power. An estimate of the minimum sample size (n) required will depend upon σ^2, the population variance (which can be estimated by s_p^2 from previous similar studies).

We may specify that we wish to perform a t test with a probability of α of committing a Type I error and a probability of β of committing a Type II error; and we can state that we want to be able to detect a difference between μ and μ_0 as small as δ (where μ is the actual population mean and μ_0 is the mean specified in the null hypothesis).* To test at the α significance level with $1 - \beta$ power, the minimum sample size required to detect δ is

$$n = \frac{s^2}{\delta^2}(t_{\alpha, \nu} + t_{\beta(1), \nu})^2, \tag{10}$$

where α can be either $\alpha(1)$ or $\alpha(2)$, respectively, depending on whether a one-tailed or two-tailed test is to be used. However, ν depends on n, so n cannot be calculated directly but must be obtained by iteration† (i.e., by a series of estimations, each estimation coming closer to the answer than that preceding). This is demonstrated in Example 8.

Equation 10 provides better estimates of n when s^2 is a good estimate of the population variance, σ^2, and the latter estimate improves when s^2 is calculated from larger samples. Therefore, it is most desirable that s^2 be obtained from a sample with a size that is not a small fraction of the estimate of n; and it can then be estimated how large an n is needed to repeat the experiment and use the resulting data to test with the designated α, β, and δ.

EXAMPLE 8 **Estimation of Required Sample Size to Test H_0: $\mu = \mu_0$**

How large a sample is needed to reject the null hypothesis of Example 2 when sampling from the population in that example? We wish to test at the 0.05 level of significance with a 90% chance of detecting a population mean different from $\mu_0 = 0$ by as little as 1.0 g. In Example 2, $s^2 = 1.5682$ g^2.

Let us guess that a sample size of 20 would be required. Then, $\nu = 19$, $t_{0.05(2), 19} = 2.093$, $\beta = 1 - 0.90 = 0.10$, $t_{0.10(1), 19} = 1.328$, and we use Equation 8 to calculate

$$n = \frac{1.5682}{(1.0)^2}(2.093 + 1.328)^2 = 18.4.$$

We now use $n = 19$ as an estimate, in which case $\nu = 18$, $t_{0.05(2), 18} = 2.101$, $t_{0.10(1), 18} = 1.330$, and

$$n = \frac{1.5682}{(1.0)^2}(2.101 + 1.330)^2 = 18.5.$$

Thus, we conclude that a new sample of at least 19 data may be taken from this population to test the above hypotheses with the specified α, β, and δ.

---

*δ is lowercase Greek delta.

†If the population variance, σ^2, were actually known (a most unlikely situation), rather than estimated by s^2, then Z_α would be substituted for t_α in this and the other computations in this section, and n would be determined in one step instead of iteratively.

(b) Minimum Detectable Difference. By rearranging Equation 10, we can ask how small a δ (the difference between μ and μ_0) can be detected by the t test with $1 - \beta$ power, at the α level of significance, using a sample of specified size n:

$$\delta = \sqrt{\frac{s^2}{n}}\left(t_{\alpha,\nu} + t_{\beta(1),\nu}\right), \tag{11}$$

where $t_{\alpha,\nu}$, can be either $t_{\alpha(1),\nu}$ or $t_{\alpha(2),\nu}$, depending on whether a one-tailed or two-tailed test is to be performed. The estimation of δ is demonstrated in Example 9. Some literature (e.g., Cohen, 1988: 811–814) refers to the "effect size," a concept similar to minimum detectable difference.

EXAMPLE 9   Estimation of Minimum Detectable Difference in a One-Sample t Test for H_0: $\mu = \mu_0$

In the two-tailed test of Example 2, what is the smallest difference (i.e., difference between μ and μ_0) that is detectable 90% of the time using a sample of 25 data and a significance level of 0.05?

Using Equation 9:

$$\delta = \sqrt{\frac{1.5682}{25}}\left(t_{0.05(2),24} + t_{0.10(1),24}\right)$$

$$= (0.25)(2.064 + 1.318)$$

$$= 0.85 \text{ g}.$$

(c) Power of One-Sample Testing. If our desire is to express the probability of correctly rejecting a false H_0 about μ, then we seek to estimate the power of a t test. Equation 10 can be rearranged to give

$$t_{\beta(1),\nu} = \frac{\delta}{\sqrt{\frac{s^2}{n}}} - t_{\alpha,\nu}, \tag{12}$$

where α refers to either $\alpha(2)$ or $\alpha(1)$, depending upon whether the null hypothesis to be tested is two-tailed or one-tailed, respectively. As shown in Example 10, for a stipulated δ, α, s^2, and sample size, we can express $t_{\beta(1),\nu}$. Consulting Table 3 from *Appendix: Statistical Tables and Graphs* allows us to convert $t_{\beta(1),\nu}$ to β, but only roughly (e.g., $\beta > 0.25$ in Example 10). However, $t_{\beta(1),\nu}$ may be considered to be approximated by $Z_{\beta(1)}$, so Table 2 from *Appendix: Statistical Tables and Graphs* may be used to determine β.* Then, the power of the test is expected to be $1 - \beta$, as shown in Example 10. Note that this is the estimated power of a test to be run on a new sample of data from this population, *not* the power of the test performed in Example 2.

---

*Some calculators and computer programs yield β given $t_{\beta,\nu}$. Approximating $t_{\beta(1),\nu}$, by $Z_{\beta(1)}$ apparently yields a β that is an underestimate (and a power that is an overestimate) of no more than 0.01 for ν of at least 11 and no more than 0.02 for ν of at least 7.

EXAMPLE 10   Estimation of the Power of a One-Sample t Test for H_0: $\mu = \mu_0$

What is the probability of detecting a true difference (i.e., a difference between μ and μ_0) of at least 1.0 g, using $\alpha = 0.05$ for the hypotheses of Example 2, if we run the experiment again using a sample of 15 from the same population?

For $n = 15$, $\nu = 14$; $\alpha = 0.05$, $t_{0.05(2),14} = 2.145$, $s^2 = 1.5682$ g^2, and $\delta = 1.0$ g; and we use Equation 12 to find

$$t_{\beta(1),14} = \frac{1.0}{\sqrt{\dfrac{1.5682 \text{ g}^2}{15}}} - 2.145$$

$$= 0.948.$$

Consulting Table 3 from *Appendix: Statistical Tables and Graphs* tells us that, for $t_{\beta(1),14} = 0.948$, $0.10 < \beta < 0.25$, so we can say that the power would be $0.75 < 1 - \beta < 0.90$. Alternatively, by considering 0.948 to be a normal deviate and consulting Table 2 from *Appendix: Statistical Tables and Graphs*, we conclude that $\beta = 0.17$ and that the power of the test is $1 - \beta = 0.83$. (The exact probabilities, by computer, are $\beta = 0.18$ and power $= 0.82$.)

When the concept of power was introduced in the discussion "Statistical Errors in Hypothesis Testing" it was stated that, for a given sample size (n), α is inversely related to β; that is, the lower the probability of committing a Type I error, the greater the probability of committing a Type II error. It was also noted that α and β can be lowered simultaneously by increasing n. Power is also greater for one-tailed than for two-tailed tests, but recall (from Section 2) that power is *not* the criterion for performing a one-tailed instead of a two-tailed test. These relationships are shown in Table 4. Table 5 shows how power is related to n, s^2, and δ. It can be seen that, for a given s^2 and δ, an increased sample size (n) results in an increase in power. Also, for a given n and δ, power increases as s^2 decreases, so a smaller variability among the data yields greater power. And for a given n and s^2, power increases as δ increases, meaning there is greater power in detecting large differences than there is in detecting small differences.

Often a smaller s^2 is obtained by narrowing the definition of the population of interest. For example, the data of Example 2 may vary as much as they do because the sample contains animals of different ages, or of different strains, or of both sexes. It may be wiser to limit the hypothesis, and the sampling, to animals of the same sex and strain and of a narrow range of ages. And power can be increased by obtaining more precise measurements; also, greater power is associated with narrower confidence intervals. A common goal is to test with a power between 0.75 and 0.90.

8 SAMPLING FINITE POPULATIONS

In general we assume that a sample from a population is a very small portion of the totality of data in that population. Essentially, we consider that the population is infinite in size, so that the removal of a relatively small number of data from the population does not noticeably affect the probability of selecting further data.

However, if the sample size, n, is an appreciable portion of the population size (a very unusual circumstance), N (say, at least 5%), then we are said to be sampling

TABLE 4: Relationship between α, β, Power $(1 - \beta)$, and n, for the Data of Example 9
Sample Variance $(s^2 = 1.5682 \text{ g}^2)$ and True Difference $(\delta = 1.0 \text{ g})$
of Example 10, Using Equation 12

	Two-Tailed Test				One-Tailed Test		
n	α	β	$1 - \beta$	n	α	β	$1 - \beta$
10	0.10	0.25	0.75	10	0.10	0.14	0.86
10	0.05	0.40	0.60	10	0.05	0.25	0.75
10	0.01	0.76	0.24	10	0.01	0.61	0.39
12	0.10	0.18	0.82	12	0.10	0.09	0.91
12	0.05	0.29	0.71	12	0.05	0.18	0.82
12	0.01	0.73	0.27	12	0.01	0.48	0.52
15	0.10	0.10	0.90	15	0.10	0.05	0.95
15	0.05	0.18	0.82	15	0.05	0.10	0.90
15	0.01	0.45	0.55	15	0.01	0.32	0.68
20	0.10	0.04	0.96	20	0.10	0.02	0.98
20	0.05	0.08	0.92	20	0.05	0.04	0.96
20	0.01	0.24	0.76	20	0.01	0.16	0.84

TABLE 5: Relationship between n, s^2, δ, and Power (for Testing at $\alpha = 0.05$) for the Hypothesis of Example 9, Using Equation 10

n	s^2	δ	Power of Two-Tailed Test	Power of One-Tailed Test
Effect of n				
10	1.5682	1.0	0.60	0.75
12	1.5682	1.0	0.71	0.82
15	1.5682	1.0	0.82	0.90
20	1.5682	1.0	0.92	0.96
Effect of s^2				
12	2.0000	1.0	0.60	0.74
12	1.5682	1.0	0.71	0.82
12	1.0000	1.0	0.88	0.94
Effect of δ				
12	1.5682	1.0	0.71	0.82
12	1.5682	1.2	0.86	0.92
12	1.5682	1.4	0.96	0.97

a *finite population*. In such a case, $\overline{X}$ is a substantially better estimate of μ the closer n is to N; specifically,

$$s_{\overline{X}} = \sqrt{\frac{s^2}{n}} \sqrt{1 - \frac{n}{N}} = \sqrt{\left(\frac{s^2}{n}\right)\left(1 - \frac{n}{N}\right)}, \tag{13}$$

where n/N is the *sampling fraction* and $1 - n/N$ is referred to as the *finite population correction*.[*]

Obviously, from Equation 13, when n is very small compared to N, then the sampling fraction is almost zero, the finite population correction will be nearly one,

---

[*]One may also calculate $1 - n/N$ as $(N - n)/N$.

and $s_{\overline{X}}$ will be nearly $\sqrt{s^2/n}$, just as we have used ($s_{\overline{X}} = \sqrt{\frac{s^2}{n}}$ or $s_{\overline{X}} = \frac{s}{\sqrt{n}}$) when assuming the population size, N, to be infinite. As n becomes closer to N, the correction becomes smaller, and $s_{\overline{X}}$ becomes smaller, which makes sense intuitively. If $n = N$, then $1 - n/N = 0$ and $s_{\overline{X}} = 0$, meaning there is no error at all in estimating μ if the sample consists of the entire population; that is, $\overline{X} = \mu$ if $n = N$. In computing confidence intervals when sampling finite populations (i.e., when n is not a negligibly small fraction of N), Equation 13 should be used.

If we are determining the sample size required to estimate the population mean with a stated precision (Section 6), and the sample size is an appreciable fraction of the population size, then the required sample size is calculated as

$$m = \frac{n}{1 + (n - 1)/N} \tag{14}$$

(Cochran, 1977: 77–78), where n is from Equation 9.

9 HYPOTHESES CONCERNING THE MEDIAN

In Example 2 we examined a sample of weight change data in order to ask whether the mean change in the sampled population was different from zero. Analogously, we may test hypotheses about the population median, M, such as testing $H_0: M = M_0$ against $H_A: M \neq M_0$, where M_0 can be zero or any other hypothesized population median.*

A simple method for testing this two-tailed hypothesis is to determine the confidence limits for the population median, and reject H_0 (with probability $\leq \alpha$ of a Type I error) if $M_0 \leq L_1$ or $M_0 \geq L_2$. This is essentially a binomial test, where we consider the number of data $< M_0$ as being in one category and the number of data $> M_0$ being in the second category. If either of these two numbers is less than or equal to the critical value in Table 27 from *Appendix: Statistical Tables and Graphs*, then H_0 is rejected. (Data equal to M_0 are ignored in this test.)

For one-tailed hypotheses about the median, the binomial test may also be employed. For $H_0: M \geq M_0$ versus $H_A: M < M_0$, H_0 is rejected if the number of data less than M_0 is $\leq$ the one-tailed critical value, $C_{\alpha(1),n}$. For $H_0: M \leq M_0$ versus $H_A: M > M_0$, H_0 is rejected if the number of data greater than M_0 is $\geq n - C_{\alpha(1),n}$.

As an alternative to the binomial test, for either two-tailed or one-tailed hypotheses, we may use the more powerful Wilcoxon signed-rank test. The Wilcoxon procedure is applied as a one-sample median test by ranking the data and assigning a minus sign to each rank associated with a datum $< M_0$ and a plus sign to each associated with a datum $> M_0$. Any rank equal to M_0 is ignored in this procedure. The sum of the ranks with a plus sign is called T_+ and the sum of the ranks with a minus sign is T_-. The Wilcoxon test assumes that the sampled population is symmetric (in which case the median and mean are identical and this procedure becomes a hypothesis test about the mean as well as about the median, but the one-sample t test is typically a more powerful test about the mean).

10 CONFIDENCE LIMITS FOR THE POPULATION MEDIAN

The sample median is used as the best estimate of M, the population median. Confidence limits for M may be determined by considering the binomial distribution.

---

*Here M represents the Greek capital letter mu.

11 HYPOTHESES CONCERNING THE VARIANCE

The sampling distribution of means is a symmetrical distribution, approaching the normal distribution as n increases. But the sampling distribution of variances is not symmetrical, and neither the normal nor the t distribution may be employed to test hypotheses about σ^2 or to set confidence limits around σ^2. However, theory states that

$$\chi^2 = \frac{\nu s^2}{\sigma^2} \tag{15}$$

(if the sample came from a population with a normal distribution), where χ^2 represents a statistical distribution * that, like t, varies with the degrees of freedom, ν, where $\nu = n - 1$. Critical values of $\chi^2_{\alpha, \nu}$ are found in Table 1 from *Appendix: Statistical Tables and Graphs*.

Consider the pair of two-tailed hypotheses, $H_0: \sigma^2 = \sigma_0^2$ and $H_A: \sigma^2 \neq \sigma_0^2$, where σ_0^2 may be any hypothesized population variance. Then, simply calculate

$$\chi^2 = \frac{\nu s^2}{\sigma_0^2} \quad \text{or, equivalently,} \quad \chi^2 = \frac{SS}{\sigma_0^2}, \tag{16}$$

and if the calculated χ^2 is $\geq \chi^2_{\alpha/2, \nu}$ or $\leq \chi^2_{(1-\alpha/2), \nu}$, then H_0 is rejected at the α level of significance. For example, if we wished to test $H_0: \sigma^2 = 1.0(^\circ C)^2$ and $H_A: \sigma^2 \neq 1.0(^\circ C)^2$ for the data of Example 1, with $\alpha = 0.05$, we would first calculate $\chi^2 = SS/\sigma_0^2$. In this example, $\nu = 24$ and $s^2 = 1.80(^\circ C)^2$, so $SS = \nu s^2 = 43.20(^\circ C)^2$. Also, as σ^2 is hypothesized to be $1.0(^\circ C)^2$, $\chi^2 = SS/\sigma_0^2 = 43.20(^\circ C)^2/1.0(^\circ C)^2 = 43.20$. Two critical values are to be obtained from the chi-square table (Table 1 from *Appendix: Statistical Tables and Graphs*): $\chi^2_{(0.05/2),24} = \chi^2_{0.025,24} = 39.364$ and $\chi^2_{(1-0.05/2),24} = \chi^2_{0.975,24} = 12.401$. As the calculated χ^2 is more extreme than one of these critical values (i.e., the calculated χ^2 is > 39.364), H_0 is rejected, and we conclude that the sample of data was obtained from a population having a variance different from $1.0(^\circ C)^2$.

It is more common to consider one-tailed hypotheses concerning variances. For the hypotheses $H_0: \sigma^2 \leq \sigma_0^2$ and $H_A: \sigma^2 > \sigma_0^2$, H_0 is rejected if the χ^2 calculated from Equation 16 is $\geq \chi^2_{\alpha, \nu}$. For $H_0: \sigma^2 \geq \sigma_0^2$ and $H_A: \sigma^2 < \sigma_0^2$, a calculated χ^2 that is $\leq \chi^2_{(1-\alpha), \nu}$ is grounds for rejecting H_0. For the data of Example 4, a manufacturer might be interested in whether the variability in the dissolving times of the drug is greater than a certain value—say, 1.5 sec. Thus, $H_0: \sigma^2 \leq 1.5$ sec^2 and $H_A: \sigma^2 > 1.5$ sec^2 might be tested, as shown in Example 11.

EXAMPLE 11 A One-Tailed Test for the Hypotheses $H_0: \sigma^2 \leq 1.5$ sec^2 and $H_A: \sigma^2 > 1.5$ sec^2, Using the Data of Example 4

$SS = 18.8288$ sec^2

$\nu = 7$

$s^2 = 2.6898$ sec^2

$\chi^2 = \frac{SS}{\sigma_0^2} = \frac{18.8288 \text{ sec}^2}{1.5 \text{ sec}^2} = 12.553$

---

*The Greek letter "chi" (which in lowercase is χ) is pronounced as the "ky" in "sky."

$\chi^2_{0.05,7} = 14.067$

Since $12.553 < 14.067$, H_0 is not rejected.

$$0.05 < P < 0.10 \quad [P = 0.084]$$

We conclude that the variance of dissolving times is no more than 1.5 sec^2.

As is the case for testing hypotheses about a population mean, μ (Sections 1 and 2), the aforementioned testing of hypotheses about a population variance, σ, depends upon the sample's having come from a population of normally distributed data. However, the F test for variances is not as robust as the t test for means; that is, it is not as resistant to violations of this underlying assumption of normality. The probability of a Type I error will be very different from the specified α if the sampled population is nonnormal, even if it is symmetrical. And, likewise, α will be distorted if there is substantial asymmetry (say, $|\sqrt{b_1}| > 0.6$), even if the distribution is normal (Pearson and Please, 1975).

12 CONFIDENCE LIMITS FOR THE POPULATION VARIANCE

Confidence intervals may be determined for many parameters other than the population mean, in order to express the precision of estimates of those parameters.

By employing the χ^2 distribution, we can define an interval within which there is a $1 - \alpha$ chance of including σ^2 in repeated sampling. Table 1 from *Appendix: Statistical Tables and Graphs* tells us the probability of a calculated χ^2 being greater than that in the Table. If we desire to know the two χ^2 values that enclose $1 - \alpha$ of the chi-square curve, we want the portion of the curve between $\chi^2_{(1-\alpha/2),\nu}$ and $\chi^2_{\alpha/2,\nu}$ (for a 95% confidence interval, this would mean the area between $\chi^2_{0.975,\nu}$ and $\chi^2_{0.025,\nu}$). It follows from Equation 13 that

$$\chi^2_{(1-\alpha/2),\nu} \leq \frac{\nu s^2}{\sigma^2} \leq \chi^2_{\alpha/2,\nu}, \tag{17}$$

and

$$\frac{\nu s^2}{\chi^2_{\alpha/2,\nu}} \leq \sigma^2 \leq \frac{\nu s^2}{\chi^2_{(1-\alpha/2),\nu}}. \tag{18}$$

Since $\nu s^2 = \text{SS}$, we can also write Equation 16 as

$$\frac{\text{SS}}{\chi^2_{\alpha/2,\nu}} \leq \sigma^2 \leq \frac{\text{SS}}{\chi^2_{(1-\alpha/2),\nu}}. \tag{19}$$

Referring back to the data of Example 1, we would calculate the 95% confidence interval for σ^2 as follows. As $\nu = 24$ and $s^2 = 1.80(°C)^2$, $\text{SS} = \nu s^2 = 43.20(°C)^2$. From Table 1 from *Appendix: Statistical Tables and Graphs*, we find $\chi^2_{0.025,24} = 39.364$ and $\chi^2_{0.975,24} = 12.401$. Therefore, $L_1 = \text{SS}/\chi^2_{\alpha/2,\nu} = 43.20(°C)^2/39.364 = 1.10(°C)^2$, and $L_2 = \text{SS}/\chi^2_{(1-\alpha),\nu} = 43.20(°C)/12.401 = 3.48(°C)^2$. If the null hypothesis H_0: $\sigma^2 = \sigma_0^2$ would have been tested and rejected for some specified variance, σ_0, then σ_0 would be outside of the confidence interval (i.e., σ_0^2 would be either less than L_1 or greater than L_2). Note that the confidence limits, $1.10(°C)^2$ and $3.48(°C)^2$, are not symmetrical around s^2; that is, the distance from L_1 to s^2 is not the same as the distance from s^2 to L_2.

To obtain the $1 - \alpha$ confidence interval for the population standard deviation, simply use the square roots of the confidence limits for σ^2, so that

$$\sqrt{\frac{\text{SS}}{\chi^2_{\alpha/2, \nu}}} \leq \sigma \leq \sqrt{\frac{\text{SS}}{\chi^2_{(1-\alpha/2), \nu}}}. \tag{20}$$

For the preceding example, the 95% confidence interval for σ would be $\sqrt{1.10(^\circ\text{C})^2} \leq \sigma \leq \sqrt{3.48(^\circ\text{C})^2}$, or $1.0^\circ\text{C} \leq \sigma \leq 1.9^\circ\text{C}$.

The end of Section 11 cautioned that testing hypotheses about σ^2 is adversely affected if the sampled population is nonnormal (even if it is symmetrical) or if the population is not symmetrical (even if it is normal). Determination of confidence limits also suffers from this unfavorable effect.

(a) One-Tailed Confidence Limits. In a fashion analogous to estimating a population mean via a one-tailed confidence interval, a one-tailed interval for a population variance is applicable in situations where a one-tailed hypothesis test for the variance is appropriate. For $H_0: \sigma^2 \leq \sigma_0^2$ and $H_A: \sigma^2 > \sigma_0^2$, the one-tailed confidence limits for σ^2 are $L_1 = \text{SS}/\chi^2_{\alpha, \nu}$ and $L_2 = \infty$; and for $H_0: \sigma^2 \geq \sigma_0^2$ and $H_A: \sigma^2 < \sigma_0^2$, the confidence limits are $L_1 = 0$ and $L_2 = \text{SS}/\chi^2_{1-\alpha, \nu}$. Considering the data in Example 4, in which $H_0: \sigma^2 \leq 45$ sec^2 and $H_A: \sigma > 45$ sec^2, for 95% confidence, L_1 would be $\text{SS}/\chi^2_{0.05, 7} = 18.8188$ sec$^2/14.067 = 1.34$ sec^2 and $L_2 = \infty$. The hypothesized σ_0^2 (45 sec^2) lies within the confidence interval, indicating that the null hypothesis would not be rejected.

If the desire is to estimate a population's standard deviation (σ) instead of the population variance (σ^2), then simply substitute σ for σ^2 and σ_0 for σ_0^2 above and use the square root of L_1 and L_2 (bearing in mind that $\sqrt{\infty} = \infty$).

(b) Prediction Limits. We can also estimate the variance that would be obtained from an additional random sample of m data from the same population. To do so, the following two-tailed $1 - \alpha$ prediction limits may be determined:

$$L_1 = \frac{s^2}{F_{\alpha(2), n-1, m-1}} \tag{21}$$

$$L_2 = s^2 F_{\alpha(2), m-1, n-1} \tag{22}$$

(Hahn, 1972; Hahn and Meeker, 1991: 64, who also mention one-tailed prediction intervals; Patel, 1989). A prediction interval for s would be obtained by taking the square roots of the prediction limits for s^2.

The critical values of F, are given in Table 4 from *Appendix: Statistical Tables and Graphs*. These will be written in the form F_{α, ν_1, ν_2}, where ν_1 and ν_2 are termed the "numerator degrees of freedom" and "denominator degrees of freedom," respectively. So, if we wished to make a prediction about the variance (or standard deviation) that would be obtained from an additional random sample of 10 data from the population from which the sample in Example 1 came, $n = 25$, $n - 1 = 24$, $m = 10$, and $m - 1 = 9$; and to compute the 95% two-tailed prediction interval, we would consult Table 4 from *Appendix: Statistical Tables and Graphs* and obtain $F_{\alpha(2), n-1, m-1} = F_{0.05(2), 24, 9} = 3.61$ and $F_{\alpha(2), m-1, n-1} = F_{0.05(2), 9, 24} = 2.79$. Thus, the prediction limits would be $L_1 = 1.80(^\circ\text{C})^2/3.61 = 0.50(^\circ\text{C})^2$ and $L_2 = [1.80(^\circ\text{C})^2][2.79] = 5.02(^\circ\text{C})^2$.

13 POWER AND SAMPLE SIZE IN TESTS CONCERNING THE VARIANCE

(a) Sample Size Required. We may ask how large a sample would be required to perform the hypothesis tests of Section 12 at a specified power. For the hypotheses $H_0: \sigma^2 \leq \sigma_0^2$ versus $H_A: \sigma^2 > \sigma_0^2$, the minimum sample size is that for which

$$\frac{\chi^2_{1-\beta,\nu}}{\chi^2_{\alpha,\nu}} = \frac{\sigma_0^2}{s^2}, \tag{23}$$

and this sample size, n, may be found by iteration (i.e., by a directed trial and error), as shown in Example 12. The ratio on the left side of Equation 23 increases in magnitude as n increases.

EXAMPLE 12   Estimation of Required Sample Size to Test $H_0: \sigma^2 \leq \sigma_0^2$ versus $H_A: \sigma^2 > \sigma_0^2$

How large a sample is needed to reject $H_0 : \sigma^2 \leq 1.50$ sec^2, using the data of Example 11, if we test at the 0.05 level of significance and with a power of 0.90? (Therefore, $\alpha = 0.05$ and $\beta = 0.10$.)

From Example 11, $s^2 = 2.6898$ sec^2. As we have specified $\sigma_0^2 = 1.75$ sec^2, $\sigma_0^2/s^2 = 0.558$.

To begin the iterative process of estimating n, let us guess that a sample size of 30 would be required. Then,

$$\frac{\chi^2_{0.90,29}}{\chi^2_{0.05,29}} = \frac{19.768}{42.557} = 0.465.$$

Because $0.465 < 0.558$, our estimate of n is too low. So we might guess that $n = 50$ is required:

$$\frac{\chi^2_{0.90,49}}{\chi^2_{0.05,49}} = \frac{36.818}{66.339} = 0.555.$$

Because 0.555 is a little less than 0.558, $n = 50$ is a little too low and we might guess $n = 55$, for which $\chi^2_{0.90,54}/\chi^2_{0.05,54} = 41.183/70.153 = 0.571$.

Because 0.571 is greater than 0.558, our estimate of n is high, so we could try $n = 51$, for which $\chi^2_{0.90,50}/\chi^2_{0.05,50} = 37.689/67.505 = 0.558$.

Therefore, we estimate that a sample size of at least 51 is required to perform the hypothesis test with the specified characteristics.

For the hypotheses $H_0: \sigma^2 \geq \sigma_0^2$ versus $H_A: \sigma^2 < \sigma_0^2$, the minimum sample size is that for which

$$\frac{\chi^2_{\beta,\nu}}{\chi^2_{1-\alpha,\nu}} = \frac{\sigma_0^2}{s^2}. \tag{24}$$

(b) Power of the Test. If we plan to test the one-tailed hypotheses $H_0: \sigma^2 \leq \sigma_0^2$ versus $H_A: \sigma^2 > \sigma_0^2$, using the α level of significance and a sample size of n, then the power of the test would be

$$1 - \beta = P(\chi^2 \geq \chi^2_{\alpha,\nu} \sigma_0^2/s^2). \tag{25}$$

Thus, if the experiment of Example 11 were to be repeated with the same sample size, then $n = 8$, $\nu = 7$, $\alpha = 0.05$, $\chi^2_{0.05,9} = 14.067$, $s^2 = 2.6898 \text{ sec}^2$, $\sigma^2_0 = 1.5 \text{ sec}^2$, and the predicted power of the test would be

$$1 - \beta = P[\chi^2 \geq (14.067)(1.5)/2.6898] = P(\chi^2 \geq 7.845).$$

From Table 1 from *Appendix: Statistical Tables and Graphs* we see that, for $\chi^2 \geq 7.845$ with $\nu = 7$, P lies between 0.25 and 0.50 (that is, $0.25 < P < 0.50$). By linear interpolation between $\chi^2_{0.25,7}$ and $\chi^2_{0.50,7}$, we estimate $P(\chi^2 \geq 7.845)$, which is the predicted power of the test, to be 0.38.* If greater power is preferred for this test, we can determine what power would be expected if the experiment were performed with a larger sample size, say $n = 40$. In that case, $\nu = 39$, $\chi^2_{0.05,39} = 54.572$, and the estimate of the power of the test would be

$$1 - \beta = P[\chi^2 \geq (54.572)(1.5)/2.6898] = P(\chi^2 \geq 30.433).$$

Consulting Table 1 from *Appendix: Statistical Tables and Graphs* for $\nu = 39$, we see that $0.75 < P < 0.90$. By linear interpolation between $\chi^2_{0.75,39}$ and $\chi^2_{0.90,39}$, we estimate $P(\chi^2 \geq 54.572)$, the power of the test, to be 0.82.[†]

One-tailed testing of $H_0: \sigma^2 \geq \sigma^2_0$ versus $H_A: \sigma^2 \leq \sigma^2_0$ would also employ Equation 25. For two-tailed testing of $H_0: \sigma^2 = \sigma^2_0$ versus $H_A: \sigma^2 \neq \sigma^2_0$, substitute $\chi^2_{\alpha/2,\nu}$ for $\chi^2_{\alpha,\nu}$ in Equation 25.

14 HYPOTHESES CONCERNING THE COEFFICIENT OF VARIATION

Although rarely done, it is possible to ask whether a sample of data is likely to have come from a population with a specified coefficient of variation, call it $(\sigma/\mu)_0$. This amounts to testing of the following pair of two-tailed hypotheses: $H_0: \sigma/\mu = (\sigma/\mu)_0$ and $H_A: \sigma/\mu \neq (\sigma/\mu)_0$. Among the testing procedures proposed, that presented by Miller (1991) works well for a sample size of at least 10 if the sampled population is normal with a mean > 0, with a variance > 0, and with a coefficient of variation, σ/μ, no greater than 0.33. For one-tailed testing (i.e., $H_0: \sigma/\mu \leq (\sigma/\mu)_0$ vs. $H_A: \sigma/\mu > (\sigma/\mu)_0$, or $H_0: \sigma/\mu \geq (\sigma/\mu)_0$ vs. $H_A: \sigma/\mu < (\sigma/\mu)_0$), the test statistic is

$$Z = \frac{\sqrt{n-1}\,[V - (\sigma/\mu)_0]}{(\mu/\sigma)_0\sqrt{0.5 + (\sigma/\mu)^2_0}}, \tag{26}$$

the probability of which may be obtained from Table 2 from *Appendix: Statistical Tables and Graphs*; or Z may be compared to the critical values of Z_α, read from the last line of Table 3 from *Appendix: Statistical Tables and Graphs*. Miller also showed this procedure to yield results very similar to those from a χ^2 approximation by McKay (1932) that, although applicable for n as small as 5, lacks power at such small sample sizes.

Miller and Feltz (1997) present an estimate of the power of this test.

---

*In this example, linear interpolation yields $P = 0.38$, harmonic interpolation concludes $P = 0.34$, and the true probability (from appropriate computer software) is $P = 0.35$. So interpolation gave very good approximations.

[†]The actual probability (via computer) is 0.84, while linear and harmonic interpolations each produced a probability of 0.82, an excellent approximation.

15 CONFIDENCE LIMITS FOR THE POPULATION COEFFICIENT OF VARIATION

The $1 - \alpha$ confidence limits for the population coefficient of variation may be estimated as

$$V \pm \frac{V\sqrt{0.5 + V^2}Z_{\alpha/2}}{\sqrt{\nu}};\qquad(27)$$

see Miller and Feltz (1997).

16 HYPOTHESES CONCERNING SYMMETRY AND KURTOSIS

Recall the assessment of a population's departure from a normal distribution, including a consideration of a population parameter, $\sqrt{\beta_1}$, for symmetry around the mean and a parameter, β_2, for kurtosis; and their respective sample statistics are $\sqrt{b_1}$ and b_2. Methods will now be discussed for testing hypotheses about a population's symmetry and kurtosis. Such hypotheses are not often employed, but they are sometimes called upon to conclude whether a sampled population follows a normal distribution, and they do appear in some statistical computer packages.

(a) Testing Symmetry around the Mean. The two-tailed hypotheses $H_0\colon \sqrt{\beta_1} = 0$ versus $H_0\colon \sqrt{\beta_1} \neq 0$ address the question of whether a sampled population's distribution is symmetrical around its mean. The sample symmetry measure, $\sqrt{b_1}$, is an estimate of $\sqrt{\beta_1}$. Its absolute value may then be compared to critical values, $(\sqrt{b_1})_{\alpha(2),n}$, in Table 22 from *Appendix: Statistical Tables and Graphs*.

As an illustration of this, let us say that the data of Example 7 from the chapter *The Normal Distribution* yield $\sqrt{b_1} = 0.351$. To test the above H_0 at the 5% level of significance, the critical value from Table 22 from *Appendix: Statistical Tables and Graphs* is $(\sqrt{b_1})_{0.05(2),70} = 0.556$. So, H_0 is not rejected and the table indicates that $P(|\sqrt{b_1}| > 0.10)$.

One-tailed testing could be employed if the interest were solely in whether the distribution is skewed to the right ($H_0\colon \sqrt{\beta_1} \leq 0$ vs. $H_0\colon \sqrt{\beta_1} > 0$), in which case H_0 would be rejected if $\sqrt{b_1} \geq (\sqrt{b_1})_{\alpha(1),n}$. Or, a one-tailed test of $H_0\colon \sqrt{\beta_1} \geq 0$ versus $H_0\colon \sqrt{\beta_1} < 0$ could be used to test specifically whether the distribution is skewed to the left; and H_0 would be rejected if $\sqrt{b_1} \leq -(\sqrt{b_1})_{\alpha(1),n}$.

If the sample size, n, does not appear in Table 22 from *Appendix: Statistical Tables and Graphs*, a conservative approach (i.e., one with lowered power) would be to use the largest tabled n that is less than the n of our sample; for example, if n were 85, we would use critical values for $n = 80$. Alternatively, a critical value could be estimated, from the table's critical values for n's immediately above and below n under consideration, using linear or harmonic interpolation, with harmonic interpolation appearing to be a little more accurate. There is also a method (D'Agostino, 1970, 1986; D'Agostino, Belanger, and D'Agostino, 1990) by which to approximate the exact probability of H_0.

(b) Testing Kurtosis. Our estimate of a population's kurtosis (β_2) is b_2. We can ask whether the population is not mesokurtic by the two-tailed hypotheses $H_0\colon \beta_2 = 3$ versus $H_0\colon \beta_1 \neq 3$. Critical values for this test are presented in Table 23 from *Appendix: Statistical Tables and Graphs*, and H_0 is rejected if b_2 is *either* less than the lower-tail critical value for $(b_2)_{\alpha(2),n}$ *or* greater than the upper-tail critical value for $(b_2)_{\alpha(2),n}$.

For the data of Example 7 from the chapter, *The Normal Distribution*, $b_2 = 2.25$. To test the above H_0 at the 5% level of significance, we find that critical values for $n = 70$ do not appear in Table 23 from *Appendix: Statistical Tables and Graphs*. A conservative procedure (i.e., one with lowered power) is to employ the critical values for the tabled critical values for the largest n that is less than our sample's n. In our example, this is $n = 50$, and H_0 is rejected if b_2 is *either* less than the lower-tail $(b_2)_{0.05(2),50} = 2.06$ *or* greater than the upper-tail $(b_2)_{0.05(2),50} = 4.36$. In the present example, $b_2 = 2.25$ is neither less than 2.06 nor greater than 4.36, so H_0 is not rejected. And, from Table 23 from *Appendix: Statistical Tables and Graphs*, we see that $0.05 < P < 0.10$. Rather than using the nearest lower n in Table 23 from *Appendix: Statistical Tables and Graphs*, we could engage in linear or harmonic interpolation between tabled critical values, with harmonic interpolation apparently a little more accurate. There is also a method (D'Agostino, 1970, 1986; D'Agostino, Belanger, and D'Agostino, 1990) to approximate the exact probability of H_0.

One-tailed testing could be employed if the interest is solely in whether the population's distribution is leptokurtic, for which $H_0: \beta_2 \leq 3$ versus $H_0: \beta_2 > 3$ would apply; and H_0 would be rejected if $b_2 \geq$ the upper-tail $(b_2)_{\alpha(1),n}$. Or, if testing specifically whether the distribution is platykurtic, a one-tailed test of $H_0: \beta_2 \geq 3$ versus $H_0: \beta_2 < 3$ would be applicable; and H_0 would be rejected if $b_2 \leq$ the lower-tail $(b_2)_{\alpha(1),n}$.

EXAMPLE 13   Two-Tailed Nonparametric Testing of Symmetry Around the Median, Using the Data of Example 7 of *The Normal Distribution* and the Wilcoxon Test

H_0:   The population of data from which this sample came is distributed symmetrically around its median.

H_A: The population is not distributed symmetrically around its median.

$n = 70$; median $= X_{(70+1)/2} = X_{35.5} = 70.5$ in.

X (in.)	d (in.)	f	$\|d\|$ (in.)	Rank of $\|d\|$	Signed rank of $\|d\|$	(f)(Signed rank)
63	−7.5	2	7.5	69.5	−69.5	−139
64	−6.5	2	6.5	67.5	−67.5	−135
65	−5.5	3	5.5	64	−64	−192
66	−4.5	5	4.5	57.5	−57.5	−287.5
67	−3.5	4	3.5	48.5	−48.5	−194
68	−2.5	6	2.5	35.5	−35.5	−213
69	−1.5	5	1.5	21.5	−21.5	−107.5
70	−0.5	8	0.5	8	−8	−64
71	0.5	7	0.5	8	8	56
72	1.5	7	1.5	21.5	21.5	160.5
73	2.5	10	2.5	35.5	35.5	355
74	3.5	6	3.5	48.5	48.5	291
75	4.5	3	4.5	57.5	57.5	172.5
76	5.5	2	5.5	64	64	128
		70				

$T_- = 1332$

$T_+ = 1163$

$T_{0.05(2),70} = 907$ (from Table 12 from *Appendix: Statistical Tables and Graphs*)

As neither T_- nor $T_+ < T_{0.05(2),70}$, do not reject H_0. [$P > 0.50$]

(c) Testing Symmetry around the Median. Symmetry of dispersion around the median instead of the mean may be tested nonparametrically by using the Wilcoxon paired-sample test (also known as the Wilcoxon signed-rank test). For each datum (X_i) we compute the deviation from the median ($d_i = X_i - $ median) and then analyze the d_i's. For the two-tailed test (considering both T_- and T_+ in the Wilcoxon test), the null hypothesis is H_0: The underlying distribution is symmetrical around (i.e., is not skewed from) the median. For a one-tailed test, T_- is the critical value for H_0: The underlying distribution is not skewed to the right of the median; and T_+ is the critical value for H_0: The underlying distribution is not skewed to the left of the median. This test is demonstrated in Example 13.

EXERCISES

1. The following data are the lengths of the menstrual cycle in a random sample of 15 women. Test the hypothesis that the mean length of human menstrual cycle is equal to a lunar month (a lunar month is 29.5 days).

   The data are 26, 24, 29, 33, 25, 26, 23, 30, 31, 30, 28, 27, 29, 26, and 28 days.

2. A species of marine arthropod lives in seawater that contains calcium in a concentration of 32 mmole/kg of water. Thirteen of the animals are collected and the calcium concentrations in their coelomic fluid are found to be: 28, 27, 29, 29, 30, 30, 31, 30, 33, 27, 30, 32, and 31 mmole/kg. Test the appropriate hypothesis to conclude whether members of this species maintain a coelomic calcium concentration less than that of their environment.

3. Present the following data in a graph that shows the mean, standard error, 95% confidence interval, range, and number of observations for each month.

   Table of Caloric Intake (kcal/g of Body Weight) **of Squirrels**

Month	Number of Data	Mean	Standard Error	Range
January	13	0.458	0.026	0.289–0.612
February	12	0.413	0.027	0.279–0.598
March	17	0.327	0.018	0.194–0.461

4. A sample of size 18 has a mean of 13.55 cm and a variance of 6.4512 cm^2.
   (a) Calculate the 95% confidence interval for the population mean.
   (b) How large a sample would have to be taken from this population to estimate μ to within 1.00 cm, with 95% confidence?
   (c) to within 2.00 cm with 95% confidence?
   (d) to within 2.00 cm with 99% confidence?
   (e) For the data of Exercise 4, calculate the 95% prediction interval for what the mean would

be of an additional sample of 10 data from the same population.

5. We want to sample a population of lengths and to perform a test of $H_0: \mu = \mu_0$ versus $H_A: \mu \neq \mu_0$, at the 5% significance level, with a 95% probability of rejecting H_0 when $|\mu - \mu_0|$ is at least 2.0 cm. The estimate of the population variance, σ^2, is $s^2 = 8.44$ cm^2.
   (a) What minimum sample size should be used?
   (b) What minimum sample size would be required if α were 0.01?
   (c) What minimum sample size would be required if $\alpha = 0.05$ and power = 0.99?
   (d) If $n = 25$ and $\alpha = 0.05$, what is the smallest difference, $|\mu - \mu_0|$, that can be detected with 95% probability?
   (e) If $n = 25$ and $\alpha = 0.05$, what is the probability of detecting a difference, $|\mu - \mu_0|$, as small as 2.0 cm?

6. There are 200 members of a state legislature. The ages of a random sample of 50 of them are obtained, and it is found that $\overline{X} = 53.87$ yr and $s = 9.89$ yr.
   (a) Calculate the 95% confidence interval for the mean age of all members of the legislature.
   (b) If the above $\overline{X}$ and s had been obtained from a random sample of 100 from this population, what would the 95% confidence interval for the population mean have been?

7. For the data of Exercise 4:
   (a) Calculate the 95% confidence interval for the population variance.
   (b) Calculate the 95% confidence interval for the population standard deviation.
   (c) Using the 5% level of significance, test $H_0: \sigma^2 \leq 4.4000$ cm^2 versus $H_A: \sigma^2 > 4.4000$ cm^2.
   (d) Using the 5% level of significance, test $H_0: \sigma \geq 3.00$ cm versus $H_A: \sigma < 3.00$ cm.
   (e) How large a sample is needed to test $H_0: \sigma^2 \leq 5.0000$ cm^2 if it is desired to test

at the 0.05 level of significance with 75% power?

(f) For the data of Exercise 4, calculate the 95% prediction interval for what the variance and standard deviation would be of an additional sample of 20 data from the same population.

8. A sample of 100 body weights has $\sqrt{b_1} = 0.375$ and $b_2 = 4.20$.

(a) Test H_0: $\sqrt{\beta_1} = 0$ and H_A: $\sqrt{\beta_1} \neq 0$, at the 5% significance level.

(b) Test H_0: $\beta_2 = 3$ and H_A: $\beta_2 \neq 3$, at the 5% significance level.

ANSWERS TO EXERCISES

1. H_0: $\mu = 29.5$ days, H_A: $\mu \neq 29.5$ days, $\overline{X} = 27.7$ days, $s_{\overline{X}} = 0.708$ days, $n = 15$, $t = 2.542$, $v = 15 - 1 = 14$, $t_{0.05(2),14} = 2.145$, $0.02 < P(|t| \geq 2.542) < 0.05$ $[P = 0.023]$; therefore, reject H_0 and conclude that the sample came from a population with a mean that is not 29.5 days.

2. H_0: $\mu \geq 32$ mmole/kg, H_A: $\mu < 32$ mmole/kg, $\overline{X} = 29.77$ mmole/kg, $s_{\overline{X}} = 0.5$ mmole/kg, $n = 13$, $t = -4.46$, $v = 12$, $t_{0.05(1),12} = 1.782$, $P(t < -4.46) < 0.0005$ $[P = 0.00039]$; therefore, reject H_0 and conclude that the sample came from a population with a mean less than 32 mmole/kg.

3. Graph, which includes three 95% confidence intervals: 0.458 ± 0.057 kcal/g; $0.413 + 0.059$ kcal/g; 0.327 ± 0.038 kcal/g.

4. (a) $13.55 + 1.26$ cm; (b) $n = 28$; (c) $n = 9$; (d) $n = 15$. (e) For $\overline{X}$ of second sample, $m = 10$: $L_1 = 13.55$ cm $- 2.11$ cm $= 11.44$ cm, $L_2 = 13.55$ cm $+ 2.11$ cm $= 15.66$ cm.

5. (a) $n = 30$; (b) $n = 41$; (c) $n = 42$; (d) $d = 2.2$ cm; (e) $t_{\beta(1),24} = 1.378$, $0.05 < \beta < 0.10$, so $0.90 < \text{power} < 0.95$; or, by normal approximation, $\beta = 0.08$ and power $= 0.92$ $[\beta = 0.09$ and power $= 0.91]$.

6. (a) $N = 200$, $n = 50$, $s^2 = 97.8121$ yr^2, $t_{0.05(2),49} = 2.010$; $s_{\overline{X}} = 1.2113$ yr. 95% confidence interval $= 53.87 \pm 2.43$ yr; (b) $t_{0.05(2),99} = 1.984$; $s_{\overline{X}} = 0.6993$ yr, 95% confidence interval $= 53.87 \pm 1.39$ yr.

7. (a) $s^2 = 6.4512$, $n = 18$, SS $= 109.6704$ cm^2; $\chi^2_{0.025,17} = 30.191$, $\chi^2_{0.975,17} = 7.564$; $L_1 = 3.6326$ cm^2, $L_2 = 14.4990$ cm^2. (b) $s = 2.54$ cm; $L_1 = 1.91$ cm, $L_2 = 3.81$ cm. (c) $\chi^2 = 24.925$, $\chi^2_{0.05,17} = 27.587$; 24.925 is not > 27.587, so do not reject H_0; $0.05 < P < 0.10$ $[P = 0.096]$. (d) $\sigma^2 = 9.000$ cm^2, $\chi^2 = 12.186$, $\chi^2_{0.95,17} = 8.672$; 12.186 is not < 8.672, so do not reject H_0; $0.10 < P < 0.25$ $[P = 0.21]$. (e) $\sigma_0^2/s^2 = 0.682$; by trial and error: $n = 71$, $v = 70$, $\chi^2_{0.75,70}/\chi^2_{0.05,70} = 61.698/90.531 = 0.9682$. (f) For s^2 of new sample, $m = 20$: $F_{0.05(2),19,17} = 2.63$, $F_{0.05(2),17,19} = 2.57$, $L_1 = 6.4512$ cm^2/2.57 $= 2.51$ cm^2, $L_2 = (6.4512$ cm$^2)(2.63) = 16.97$ cm^2; for s: $L_1 = 1.58$ cm, $L_2 = 4.12$ cm).

8. (a) Do not reject H_0; $P > 0.10$ (b) Reject H_0; $0.02 < P < 0.05$.

This page intentionally left blank

Two-Sample Hypotheses

From Chapter 8 of *Biostatistical Analysis*, Fifth Edition, Jerrold H. Zar. Copyright © 2010 by
Pearson Education, Inc. Publishing as Pearson Prentice Hall. All rights reserved.

Two-Sample Hypotheses

Among the most commonly employed biostatistical procedures is the comparison of two samples to infer whether differences exist between the two populations sampled. This chapter will consider hypotheses comparing two population means, medians, variances (or standard deviations), coefficients of variation, and indices of diversity. In doing so, we introduce another very important sampling distribution, the *F* distribution—named for its discoverer, R. A. Fisher—and will demonstrate further use of Student's *t* distribution.

The objective of many two-sample hypotheses is to make inferences about population parameters by examining sample statistics. Other hypothesis-testing procedures, however, draw inferences about populations without referring to parameters. Such procedures are called *nonparametric* methods, and several will be discussed chapter.

1   TESTING FOR DIFFERENCE BETWEEN TWO MEANS

A very common situation for statistical testing is where a researcher desires to infer whether two population means are the same. This can be done by analyzing the difference between the means of samples taken at random from those populations.

Example 1 presents the results of an experiment in which adult male rabbits were divided at random into two groups, one group of six and one group of seven.* The members of the first group were given one kind of drug (called "B"), and the

---

*Sir Ronald Aylmer Fisher (1890–1962) is credited with the first explicit recommendation of the important concept of assigning subjects *at random* to groups for different experimental treatments (Bartlett, 1965; Fisher, 1925b; Rubin, 1990).

members of the second group were given another kind of drug (called "G"). Blood is to be taken from each rabbit and the time it takes the blood to clot is to be recorded.

EXAMPLE 1   A Two-Sample t Test for the Two-Tailed Hypotheses, H_0: $\mu_1 = \mu_2$ and H_A: $\mu_1 \neq \mu_2$ (Which Could Also Be Stated as H_0: $\mu_1 - \mu_2 = 0$ and H_A: $\mu_1 - \mu_2 \neq 0$). The Data Are Blood-Clotting Times (in Minutes) of Male Adult Rabbits Given One of Two Different Drugs

H_0: $\mu_1 = \mu_2$
H_A: $\mu_1 \neq \mu_2$

Given drug B	Given drug G
8.8	9.9
8.4	9.0
7.9	11.1
8.7	9.6
9.1	8.7
9.6	10.4
	9.5

$n_1 = 6$	$n_2 = 7$
$\nu_1 = 5$	$\nu_2 = 6$
$\overline{X}_1 = 8.75$ min	$\overline{X}_2 = 9.74$ min
$SS_1 = 1.6950$ min^2	$SS_2 = 4.0171$ min^2

$$s_p^2 = \frac{SS_1 + SS_2}{\nu_1 + \nu_2} = \frac{1.6950 + 4.0171}{5 + 6} = \frac{5.7121}{11} = 0.5193 \text{ min}^2$$

$$s_{\overline{X}_1 - \overline{X}_2} = \sqrt{\frac{s_p^2}{n_1} + \frac{s_p^2}{n_2}} = \sqrt{\frac{0.5193}{6} + \frac{0.5193}{7}} = \sqrt{0.0866 + 0.0742}$$

$$= \sqrt{0.1608} = 0.40 \text{ min}$$

$$t = \frac{\overline{X}_1 - \overline{X}_2}{s_{\overline{X}_1 - \overline{X}_2}} = \frac{8.75 - 9.74}{0.40} = \frac{-0.99}{0.40} = -2.475$$

$t_{0.05(2),\nu} = t_{0.05(2),11} = 2.201$

Therefore, reject H_0.

$$0.02 < P(|t| \geq 2.475) < 0.05 \quad [P = 0.031]$$

We conclude that mean blood-clotting time is not the same for subjects receiving drug B as it is for subjects receiving drug G.

We can ask whether the mean of the population of blood-clotting times of all adult male rabbits who might have been administered drug B (let's call that mean μ_1) is the same as the population mean for blood-clotting times of all adult male rabbits who might have been given drug G (call it μ_2). This would involve the two-tailed hypotheses H_0: $\mu_1 - \mu_2 = 0$ and H_A: $\mu_1 - \mu_2 \neq 0$; and these hypotheses are commonly expressed in their equivalent forms: H_0: $\mu_1 = \mu_2$ and H_A: $\mu_1 \neq \mu_2$. The data from this experiment are presented in Example 1.

In this example, a total of 13 members of a biological population (adult male rabbits) were divided at random into two experimental groups, each group to receive treatment with one of the drugs. Another kind of testing situation with two independent samples is where the two groups are predetermined. For example, instead of desiring to test the effect of two drugs on blood-clotting time, a researcher might want to compare the mean blood-clotting time of adult male rabbits to that of adult female rabbits, in which case one of the two samples would be composed of randomly chosen males and the other sample would comprise randomly selected females. In that situation, the researcher would not specify which rabbits will be designated as male and which as female; the sex of each animal (and, therefore, the experimental group to which each is assigned) is determined before the experiment is begun. Similarly, it might have been asked whether the mean blood-clotting time is the same in two strains (or two ages, or two colors) of rabbits. Thus, in Example 1 there is random allocation of animals to the two groups to be compared, while in the other examples in this paragraph, there is random sampling of animals within each of two groups that are already established. The statistical hypotheses and the statistical testing procedure are the same in both circumstances.

If the two samples came from two normally distributed populations, and if the two populations have equal variances, then a t value to test such hypotheses may be calculated in a manner analogous to its computation for the one-sample t test. The t for testing the preceding hypotheses concerning the difference between two population means is

$$t = \frac{\overline{X}_1 - \overline{X}_2}{s_{\overline{X}_1 - \overline{X}_2}}.$$ (1)

The quantity $\overline{X}_1 - \overline{X}_2$ is the difference between the two sample means; and $s_{\overline{X}_1 - \overline{X}_2}$ is the standard error of the difference between the sample means (explained further below), which is a measure of the variability of the data within the two samples. Therefore, Equation 1 compares the differences between two means to the differences among all the data (a concept to be enlarged upon when comparing more than two means).

The quantity $s_{\overline{X}_1 - \overline{X}_2}$, along with $s^2_{\overline{X}_1 - \overline{X}_2}$, the variance of the difference between the means, needs to be considered further. Both $s^2_{\overline{X}_1 - \overline{X}_2}$ and $s_{\overline{X}_1 - \overline{X}_2}$ are statistics that can be calculated from the sample data and are estimates of the population parameters, $\sigma^2_{\overline{X}_1 - \overline{X}_2}$ and $\sigma_{\overline{X}_1 - \overline{X}_2}$, respectively. It can be shown mathematically that the variance of the difference between two independent variables is equal to the sum of the variances of the two variables, so that $\sigma^2_{\overline{X}_1 - \overline{X}_2} = \sigma^2_{\overline{x}_1} + \sigma^2_{\overline{x}_2}$. Independence means that there is no association correlation between the data in the two populations.* As $\sigma^2_{\overline{X}} = \sigma^2/n$, we can write

$$\sigma^2_{\overline{X}_1 - \overline{X}_2} = \frac{\sigma^2_1}{n_1} + \frac{\sigma^2_2}{n_2}.$$ (2)

Because the two-sample t test requires that we assume $\sigma^2_1 = \sigma^2_2$, we can write

$$\sigma^2_{\overline{X}_1 - \overline{X}_2} = \frac{\sigma^2}{n_1} + \frac{\sigma^2}{n_2}.$$ (3)

---

*If there is a unique relationship between each datum in one sample and a specific datum in another sample, then the data are considered *paired*.

Thus, to calculate the estimate of $\sigma^2_{\overline{X}_1-\overline{X}_2}$, we must have an estimate of σ^2. Since both s_1^2 and s_2^2 are assumed to estimate σ^2, we compute the *pooled variance*, s_p^2, which is then used as the best estimate of σ^2:

$$s_p^2 = \frac{SS_1 + SS_2}{\nu_1 + \nu_2}, \tag{4}$$

and

$$s^2_{\overline{X}_1-\overline{X}_2} = \frac{s_p^2}{n_1} + \frac{s_p^2}{n_2}. \tag{5}$$

Thus,*

$$s_{\overline{X}_1-\overline{X}_2} = \sqrt{\frac{s_p^2}{n_1} + \frac{s_p^2}{n_2}}, \tag{6}$$

and Equation 1 becomes

$$t = \frac{\overline{X}_1 - \overline{X}_2}{\sqrt{\frac{s_p^2}{n_1} + \frac{s_p^2}{n_2}}}, \tag{7a}$$

which for equal sample sizes (i.e., $n_1 = n_2$, so each sample size may be referred to as n),

$$t = \frac{\overline{X}_1 - \overline{X}_2}{\sqrt{\frac{2s_p^2}{n}}}. \tag{7b}$$

Example 1 summarizes the procedure for testing the hypotheses under consideration. The critical value to be obtained from Table 3 from *Appendix: Statistical Tables and Graphs* is $t_{\alpha(2),(\nu_1+\nu_2)}$, the two-tailed t value for the α significance level, with $\nu_1 + \nu_2$ degrees of freedom. We shall also write this as $t_{\alpha(2),\nu}$, defining the pooled degrees of freedom to be

$$\nu = \nu_1 + \nu_2 \quad \text{or, equivalently,} \quad \nu = n_1 + n_2 - 2. \tag{8}$$

In the two-tailed test, H_0 will be rejected if either $t \geq t_{\alpha(2),\nu}$ or $t \leq -t_{\alpha(2),\nu}$. Another way of stating this is that H_0 will be rejected if $|t| \geq t_{\alpha(2),\nu}$.

This statistical test asks what the probability is of obtaining two independent samples with means ($\overline{X}_1$ and $\overline{X}_2$) at least this different by random sampling from populations whose means (μ_1 and μ_2) are equal. And, if that probability is α or less, then H_0: $\mu_1 = \mu_2$ is rejected and it is declared that there is good evidence that the two population means are different.[†]

H_0: $\mu_1 = \mu_2$ may be written H_0: $\mu_1 - \mu_2 = 0$ and H_A: $\mu_1 \neq \mu_2$ as H_A: $\mu_1 - \mu_2 \neq 0$; the generalized two-tailed hypotheses are H_0: $\mu_1 - \mu_2 = \mu_0$ and H_A: $\mu_1 - \mu_2 \neq \mu_0$, tested as

$$t = \frac{|\overline{X}_1 - \overline{X}_2| - \mu_0}{s_{\overline{X}_1-\overline{X}_2}}, \tag{9}$$

where μ_0 may be any hypothesized difference between population means.

---

*The standard error of the difference between means may also be calculated as $s_{\overline{X}_1-\overline{X}_2} = \sqrt{Ns_p^2/(n_1n_2)}$, where $N = n_1 + n_2$.

[†]Instead of testing this hypotheses, a hypothesis of "correlation" could be tested, which would ask whether there is a significant linear relationship between the magnitude of X and the group from which it came. This is not commonly done.

By the procedure of Section 9, one can test whether the measurements in one population are a specified amount as large as those in a second population.

(a) One-Tailed Hypotheses about the Difference between Means. One-tailed hypotheses can be tested in situations where the investigator is interested in detecting a difference in only one direction. For example, a gardener may use a particular fertilizer for a particular kind of plant, and a new fertilizer is advertised as being an improvement. Let us say that plant height at maturity is an important characteristic of this kind of plant, with taller plants being preferable. An experiment was run, raising ten plants on the present fertilizer and eight on the new one, with the resultant eighteen plant heights shown in Example 2. If the new fertilizer produces plants that are shorter than, or the same height as, plants grown with the present fertilizer, then we shall decide that the advertising claims are unfounded; therefore, the statements of $\mu_1 > \mu_2$ and $\mu_1 = \mu_2$ belong in the same hypothesis, namely the null hypothesis, H_0. If, however, mean plant height is indeed greater with the newer fertilizer, then it shall be declared to be distinctly better, with the alternate hypothesis ($H_A: \mu_1 < \mu_2$) concluded to be the true statement. The t statistic is calculated by Equation 1, just as for the two-tailed test. But this calculated t is then compared with the critical value $t_{\alpha(1),\nu}$, rather than with $t_{\alpha(2),\nu}$.

EXAMPLE 2   A Two-Sample t Test for the One-Tailed Hypotheses, H_0: $\mu_1 \geq \mu_2$ and H_A: $\mu_1 < \mu_2$ (Which Could Also Be Stated as H_0: $\mu_1 - \mu_2 \geq 0$ and H_A: $\mu_1 - \mu_2 < 0$). The Data Are Heights of Plants, Each Grown with One of Two Different Fertilizers

H_0: $\mu_1 \geq \mu_2$
H_A: $\mu_1 < \mu_2$

Present fertilizer	Newer fertilizer
48.2 cm	52.3 cm
54.6	57.4
58.3	55.6
47.8	53.2
51.4	61.3
52.0	58.0
55.2	59.8
49.1	54.8
49.9	
52.6	

$n_1 = 10$ $\qquad$ $n_2 = 8$
$\nu_1 = 9$ $\qquad$ $\nu_2 = 7$
$\overline{X}_1 = 51.91$ cm $\qquad$ $\overline{X}_2 = 56.55$ cm
$SS_1 = 102.23$ cm^2 $\qquad$ $SS_2 = 69.20$ cm^2

$$s_p^2 = \frac{102.23 + 69.20}{9 + 7} = \frac{171.43}{16} = 10.71 \text{ cm}^2$$

$$s_{\overline{X}_1 - \overline{X}_2} = \sqrt{\frac{10.71}{10} + \frac{10.71}{8}} = \sqrt{2.41} = 1.55 \text{ cm}$$

$$t = \frac{\overline{X}_1 - \overline{X}_2}{s_{\overline{X}_1 - \overline{X}_2}} = \frac{51.91 - 56.55}{1.55} = \frac{-4.64}{1.55} = -2.99$$

$t_{0.05(1),16} = 1.746$

As t of -2.99 is less than -1.746, H_0 is rejected.

$$0.0025 < P < 0.005 \quad [P = 0.0043]$$

The mean plant height is greater with the newer fertilizer.

In other cases, the one-tailed hypotheses, H_0: $\mu_1 \leq \mu_2$ and H_A: $\mu_1 > \mu_2$, may be appropriate. Just as introduced in one-sample testing, the following summary of procedures applies to two-sample t testing:

For H_A: $\mu_1 \neq \mu_2$, if $|t| \geq t_{\alpha(2),\nu}$, then reject H_0.

For H_A: $\mu_1 < \mu_2$, if $t \leq -t_{\alpha(1),\nu}$, then reject H_0.*

For H_A: $\mu_1 > \mu_2$, if $t \geq t_{\alpha(1),\nu}$, then reject H_0.†

The null and alternate hypotheses are to be decided upon *before* the data are collected.

Also, H_0: $\mu_1 \leq \mu_2$ and H_A: $\mu_1 > \mu_2$ may be written as H_0: $\mu_1 - \mu_2 \leq 0$ and H_A: $\mu_1 - \mu_2 > 0$, respectively. The generalized hypotheses for this type of one-tailed test are H_0: $\mu_1 - \mu_2 \leq \mu_0$ and H_A: $\mu_1 - \mu_2 > \mu_0$, for which the t is

$$t = \frac{\overline{X}_1 - \overline{X}_2 - \mu_0}{s_{\overline{X}_1 - \overline{X}_2}}, \tag{10}$$

and μ_0 may be any specified value of $\mu_1 - \mu_2$.

Lastly, H_0: $\mu_1 \geq \mu_2$ and H_A: $\mu_1 < \mu_2$ may be written as H_0: $\mu_1 - \mu_2 \geq 0$ and H_A: $\mu_1 - \mu_2 < 0$, and the generalized one-tailed hypotheses of this type are H_0: $\mu_1 - \mu_2 \geq \mu_0$ and H_A: $\mu_1 - \mu_2 < \mu_0$, with the appropriate t statistic being that of Equation 10. For example, the gardener collecting the data of Example 2 may have decided, because the newer fertilizer is more expensive than the other, that it should be used only if the plants grown with it averaged at least 5.0 cm taller than plants grown with the present fertilizer. Then, $\mu_0 = \mu_1 - \mu_2 = -5.0$ cm and, by Equation 10, we would calculate $t = (51.91 - 56.55 + 5.0)/1.55 = 0.36/1.55 = 0.232$, which is not $\geq$ the critical value shown in Example 2; so H_0: $\mu_1 - \mu_2 \geq -5.0$ cm is not rejected. The following summary of procedures applies to these general hypotheses:

For H_A: $\mu_1 - \mu_2 \neq \mu_0$, if $|t| \geq t_{\alpha(2),\nu}$, then reject H_0.

For H_A: $\mu_1 - \mu_2 < \mu_0$, if $t \leq -t_{\alpha(1),\nu}$, then reject H_0.

For H_A: $\mu_1 - \mu_2 > \mu_0$, if $t \geq t_{\alpha(1),\nu}$, then reject H_0.

---

*For this one-tailed hypothesis test, probabilities of t up to 0.25 are indicated in Table 3 from *Appendix: Statistical Tables and Graphs*. If $t = 0$, then $P = 0.50$; so if $-t_{0.25(1),\nu} < t < 0$, then $0.25 < P < 0.50$; and if $t > 0$ then $P > 0.50$.

†For this one-tailed hypothesis test, $t = 0$ indicates $P = 0.50$; therefore, if $0 < t < t_{0.25(1),\nu}$, then $0.25 < P < 0.50$; and if $t < 0$, then $P > 0.50$.

(b) Violations of the Two-Sample t-test Assumptions. The validity of two-sample t testing depends upon two basic assumptions: that the two samples came at random from normal populations and that the two populations had the same variance. Populations of biological data will not have distributions that are exactly normal or variances that are exactly the same. Therefore, it is fortunate that numerous studies, over 70 years, have shown that this t test is robust enough to withstand considerable nonnormality and some inequality of variances. This is especially so if the two sample sizes are equal or nearly equal, particularly when two-tailed hypotheses are tested (e.g., Boneau, 1960; Box, 1953; Cochran, 1947; Havlicek and Peterson, 1974; Posten, Yen, and Owen, 1982; Srivastava, 1958; Stonehouse and Forrester, 1998; Tan, 1982; Welch, 1938) but also in one-tailed testing (Posten, 1992).

In general, the larger and the more equal in size the samples are, the more robust the test will be; and sample sizes of at least 30 provide considerable resistance effects of violating the t-test assumptions when testing at $\alpha = 5\%$ (i.e., the 0.05 level of significance), regardless of the disparity between σ_1^2 and σ_2^2 (Donaldson, 1968; Ramsey, 1980; Stonehouse and Forrester, 1998); larger sample sizes are needed for smaller α's, smaller n's will suffice for larger significance levels, and larger samples are required for larger differences between σ_1 and σ_2.

Hsu (1938) reported remarkable robustness, even in the presence of very unequal variances and very small samples, if $n_1 = n_2 + 1$ and $\sigma_1^2 > \sigma_2^2$. So, if it is believed (by inspecting s_1^2 and s_2^2) that the population variances (σ_1^2 and σ_2^2) are dissimilar, one might plan experiments that have samples that are unequal in size by 1, where the larger sample comes from the population with the larger variance. But the procedure of Section 1c, below, has received a far greater amount of study and is much more commonly employed.

The two-sample t test is very robust to nonnormality if the population variances are the same (Kohr and Games, 1974; Posten, 1992; Posten, Yeh, and Owen, 1982; Ramsey, 1980; Stonehouse and Forrester, 1998; Tomarkin and Serlin, 1986). If the two populations have the same variance and the same shape, the test works well even if that shape is extremely nonnormal (Stonehouse and Forrester, 1998; Tan, 1982). Havlicek and Peterson (1974) specifically discuss the effect of skewness and leptokurtosis.

If the population variances are unequal but the sample sizes are the same, then the probability of a Type I error will tend to be greater than the stated α (Havlicek and Peterson, 1974; Ramsey, 1980), and the test is said to be *liberal*. As seen in Table 1a, this departure from α will be less for smaller differences between σ_1^2 and σ_2^2 and for larger sample sizes. (The situation with the most heterogeneous variances is where σ_1^2/σ_2^2 is zero (0) or infinity (∞).)

If the two variances are not equal *and* the two sample sizes are not equal, then the probability of a Type I error will differ from the stated α. If the larger σ^2 is associated with the larger sample, this probability will be less than the stated α (and the test is called *conservative*) and this probability will be greater than the stated α (and the test is called *liberal*) if the smaller sample came from the population with the larger variance (Havlicek and Peterson, 1974; Ramsey, 1980; Stonehouse and Forrester, 1998; Zimmerman, 1987).* The greater the difference between variances, the greater will be the disparity between the probability of a Type I error and the specified α, larger differences will also result in greater departure from α. Table 1b

---

*The reason for this can be seen from Equations 4–7a: If the larger s_i^2 is coupled with the larger n_i, then the numerator of s_p^2 (which is $\nu_1 s_1^2 + \nu_2 s_2^2$) is greater than if the larger variance is associated with the smaller n. This makes s_p^2 larger, which translates into a larger $s_{\overline{X}_1-\overline{X}_2}$, which produces a smaller t, resulting in a probability of a Type I error lower than the stipulated α.

TABLE 1a: Maximum Probabilities of Type I Error when Applying the Two-Tailed *(or, One-Tailed) t* Test to Two Samples of Various Equal Sizes ($n_1 = n_2 = n$), Taken from Normal Populations Having Various Variance Ratios, σ_1^2/σ_2^2

σ_1^2/σ_2^2 $n:$	3	5	10	15 [16]	20	30	∞
For $\alpha = 0.05$							
3.33 or 0.300	0.059	0.056	0.054	0.052	0.052	0.051	0.050
5.00 or 0.200	0.064	0.061	0.056	0.054	0.053	0.052	0.050
10.00 or 0.100		0.068	0.059	0.056	0.055	0.053	0.050
∞ or 0	0.109	0.082	0.065	0.060	0.057	0.055	0.050
∞ or 0	*(0.083)*	*(0.068)*	*(0.058)*	*(0.055)*	*(0.054)*	*(0.053)*	*(0.050)*
For $\alpha = 0.01$							
3.33 or 0.300	0.013	0.013	0.012	[0.011]	0.011	0.011	0.010
5.00 or 0.200	0.015	0.015	0.013	[0.012]	0.011	0.011	0.010
10.00 or 0.100	0.020	0.019	0.015	[0.013]	0.012	0.012	0.010
∞ or 0	0.044	0.028	0.018	[0.015]	0.014	0.013	0.010
∞ or 0	*(0.032)*	*(0.022)*	*(0.015)*	*(0.014)*	*(0.013)*	*(0.012)*	*(0.010)*

These probabilities are gleaned from the extensive analysis of Ramsey (1980), and from Table 1 of Posten, Yeh, and Owen (1982).

shows this for various sample sizes. For example, Table 1a indicates that if 20 data are distributed as $n_1 = n_2 = 10$ and the two-tailed *t* test is performed at the 0.05 significance level, the probability of a Type I error approaches 0.065 for greatly divergent population variances. But in Table 1b we see that if $\alpha = 0.05$ is used and 20 data are distributed as $n_1 = 9$ and $n_2 = 11$, then the probability of a Type I error can be as small as 0.042 (if the sample of 11 came from the population with the larger variance) or as large as 0.096 (if the sample of 9 came from the population with the smaller variance).

A decrease in the probability of the Type I error (α) is associated with an increase in the probability of a Type II error (β); and, because power is $1 - \beta$, an increase in β means a decrease in the power of the test ($1 - \beta$). Therefore, for situations described above as conservative—that is, P(Type I error) $<\alpha$—there will generally be less power than if the population variances were all equal; and when the test is liberal—that is, P(Type I error) $>\alpha$—there will generally be more power than if the variances were equal. (See also Zimmerman and Zumbo, 1993.)

The power of the two-tailed *t* test is affected very little by small or moderate skewness in the sampled populations, especially if the sample sizes are equal, but there can be a serious effect on one-tailed tests. As for kurtosis, the actual power of the test is less than that discussed in Section 4 when the populations are platykurtic and greater when they are leptokurtic, especially for small sample sizes (Boneau, 1960; Glass, Peckham, and Sanders, 1972). The adverse effect of nonnormality is less with large sample sizes (Srivastava, 1958).

(c) The Two-sample *t* Test with Unequal Variances. As indicated above, the *t* test for difference between two means is robust to some departure from its underlying assumptions; but it is not dependable when the two population variances are very different. The latter situation is known as the Behrens-Fisher problem, referring to the early work on it by Behrens (1929) and Fisher (e.g., 1939b), and numerous

TABLE 1b: Maximum Probabilities of Type I Error when Applying the Two-Tailed *(or One-Tailed) t* Test to Two Samples of Various Unequal Sizes, Taken from Normal Populations Having the Largest Possible Difference between Their Variances

		For $\alpha = 0.05$		For $\alpha = 0.01$	
n_1	n_2	σ_1^2 large	σ_1^2 small	σ_1^2 large	σ_1^2 small
11	9	0.042	0.096	0.0095	0.032
		(0.041)	*(0.079)*	*(0.0088)*	*(0.026)*
22	18	0.036	0.086	0.0068	0.026
		(0.038)	*(0.073)*	*(0.0068)*	*(0.021)*
33	27	0.034	0.082	0.0059	0.024
		(0.037)	*(0.072)*	*(0.0062)*	*(0.020)*
55	45	0.032	0.080	0.0053	0.022
		(0.036)	*(0.070)*	*(0.0057)*	*(0.019)*
12	8	0.025	0.13	0.0045	0.054
		(0.028)	*(0.098)*	*(0.0046)*	*(0.040)*
24	16	0.020	0.12	0.0029	0.044
		(0.025)	*(0.096)*	*(0.0033)*	*(0.034)*
36	24	0.019	0.12	0.0024	0.041
		(0.023)	*(0.094)*	*(0.0029)*	*(0.032)*
60	40	0.018	0.11	0.0021	0.039
		(0.023)	*(0.092)*	*(0.0026)*	*(0.031)*

From Posten, Yeh, and Owen (1982) and Posten (1992).

other studies of this problem have ensued (e.g., Best and Raynor, 1987; Dixon and Massey, 1969: 119; Fisher and Yates, 1963: 60–61;* Gill, 1971; Kim and Cohen, 1998; Lee and Fineberg, 1991; Lee and Gurland, 1975; Satterthwaite, 1946; Scheffé, 1970; Zimmerman and Zumbo, 1993). Several solutions have been proffered, and they give very similar results except for very small samples. One of the easiest, yet reliable, of available procedures is that attributed to Smith (1936) and is often known as the "Welch approximate t"[†] (Davenport and Webster, 1975; Mehta and Srinivasan, 1971; Wang, 1971; Welch, 1936, 1938, 1947). It has been shown to perform well with respect to Type I error, and it requires no special tables.

The test statistic is that of Equation 1 or 9, but with $s_{\overline{X}_1 - \overline{X}_2}$ (the standard error of the difference between the means) calculated with the two separate variances instead of with a pooled variance; that is,

$$s'_{\overline{X}_1 - \overline{X}_2} = \sqrt{\frac{s_1^2}{n_1} + \frac{s_2^2}{n_2}} \tag{11a}$$

instead of Equation 6. And, because $s_{\overline{X}_i} = s_i^2/n_i$, this can be written equivalently as

$$s'_{\overline{X}_1 - \overline{X}_2} = \sqrt{s_{\overline{X}_1}^2 + s_{\overline{X}_2}^2}. \tag{11b}$$

---

*In Fisher and Yates (1963), s refers to the standard error, not the standard deviation.
[†]Bernard Lewis Welch (1911–1989), English statistician. (See Mardia, 1990.)

Therefore, Equation 1 becomes

$$t' = \frac{\overline{X}_1 - \overline{X}_2}{\sqrt{s_{\overline{X}_1}^2 + s_{\overline{X}_2}^2}}, \tag{11c}$$

Equation 11 becomes

$$t' = \frac{\overline{X}_1 - \overline{X}_2 - \mu_0}{\sqrt{s_{\overline{X}_1}^2 + s_{\overline{X}_2}^2}}, \tag{11d}$$

and two-tailed and one-tailed hypotheses are tested as described earlier for t.

Tables of critical values of t' have been published, but they are not extensive. Satterthwaite (1946) and Scheffé (1970) approximated the distribution of t' well by using t with degrees of freedom of

$$\nu' = \frac{\left(s_{\overline{X}_1}^2 + s_{\overline{X}_2}^2\right)^2}{\dfrac{\left(s_{\overline{X}_1}^2\right)^2}{n_1 - 1} + \dfrac{\left(s_{\overline{X}_2}^2\right)^2}{n_2 - 1}}. \tag{12}$$

These degrees of freedom can be as small as $n_1 - 1$ or $n_2 - 1$, whichever is smaller, and as large as $n_1 + n_2 - 2$. However, ν' is typically not an integer, so the critical value of t' often will not be found in Table 3 from *Appendix: Statistical Tables and Graphs*. If ν' is not an integer, the needed critical value, $t_{\alpha,\nu}$, can be obtained via some computer software; or these values can be interpolated from the t's in Table 3 from *Appendix: Statistical Tables and Graphs* (the beginning of *Appendix: Statistical Tables and Graphs* explains interpolation, and at the end of Table 3 from *Appendix: Statistical Tables and Graphs* there is an indication of the accuracy of interpolation for t); or, less accurately, the closest integer to ν' (or, to be conservative, the nearest integer less than ν') can be used as the degrees of freedom in Table 3 from *Appendix: Statistical Tables and Graphs*. The Behrens-Fisher test is demonstrated in Example 2a.*

---

*In the highly unlikely situation where the variances (σ_1^2 and σ_2^2) of the two sampled populations are known, the test for difference between means could be effected with

$$Z = \frac{|\overline{X}_1 - \overline{X}_2| - \mu_0}{\sqrt{\sigma_1^2/n_1 + \sigma_2^2/n_2}}; \tag{12a}$$

and it can be recalled that $Z_\alpha = t_{\alpha,\infty}$. If the variance ($\sigma_1^2$) of one of the two populations is known, this test statistic and degrees of freedom may be employed (Maity and Sherman, 2006):

$$t = \frac{|\overline{X}_1 - \overline{X}_2| - \mu_0}{\sqrt{\sigma_1^2/n_1 + s_2^2/n_2}}; \tag{12b}$$

$$\nu' = \frac{\left(\sigma_1^2/n_1 + s_2^2/n_2\right)^2}{\dfrac{s_2^2/n_2}{n_2 - 1}}. \tag{12c}$$

EXAMPLE 2a   The Behrens-Fisher Test for the Two-Tailed Hypotheses, H_0: $\mu_1 = \mu_2$ and H_A: $\mu_1 \neq \mu_2$

The data are the times for seven cockroach eggs to hatch at one laboratory temperature and for eight eggs to hatch at another temperature.

H_0: $\mu_1 = \mu_2$
H_A: $\mu_1 \neq \mu_2$

At 30°C	At 10°C
40 days	36 days
38	45
32	32
37	52
39	59
41	41
35	48
	55

$n_1 = 7$ $n_2 = 8$

$\nu_1 = 6$ $\nu_2 = 7$

$\overline{X}_1 = 37.4$ days $\overline{X}_2 = 46.0$ days

$SS_1 = 57.71$ days^2 $SS_2 = 612.00$ days^2

$s_1^2 = 9.62$ days^2 $s_2^2 = 87.43$ days^2

$s_{\overline{X}_1}^2 = 1.37$ days^2 $s_{\overline{X}_2}^2 = 10.93$ days^2

$$s'_{\overline{X}_1 - \overline{X}_2} = \sqrt{s_{\overline{X}_1}^2 + s_{\overline{X}_2}^2} = \sqrt{1.37 + 10.83} = 3.51 \text{ days}$$

$$t' = \frac{\overline{X}_1 - \overline{X}_2}{s'_{\overline{X}_1 - \overline{X}_2}} = \frac{37.4 - 46.00}{3.51} = -2.450$$

$$\nu' = \frac{\left(s_{\overline{X}_1}^2 + s_{\overline{X}_2}^2 \right)^2}{\dfrac{\left(s_{\overline{X}_1}^2 \right)^2}{\nu_1} + \dfrac{\left(s_{\overline{X}_2}^2 \right)^2}{\nu_2}}$$

$$= \frac{(1.37 + 10.93)^2}{\dfrac{(1.37)^2}{6} + \dfrac{(10.93)^2}{7}}$$

$$= 8.7$$

$t_{0.05(2),8.7} = 2.274$*
Therefore, reject H_0.
[$P = 0.038$.]*

---

*These values were obtained by computer.

As with t, the robustness of t' is greater with large and with equal sample sizes. If $\sigma_1^2 = \sigma_2^2$, then either t or t' can be used, but t will be the more powerful procedure (Ramsey, 1980), but generally with only a very slight advantage over t' (Best and Rayner, 1987). If $n_1 = n_2$ and $s_1^2 = s_2^2$, then $t' = t$ and $v' = v$; but t' is not as powerful and not as robust to nonnormality as t is (Stonehouse and Forrester, 1998; Zimmerman and Zumbo, 1993). However, Best and Rayner (1987) found t' to be much better when the variances and the sample sizes are unequal. They, and Davenport and Webster (1975), reported that the probabilty of a Type I error in the t' test is related to the ratio $(n_2\sigma_1^2)/(n_1\sigma_2^2)$ (let's call this ratio r for the present): When $r > 1$ and $n_1 > n_2$, then this error is near the α specified for the significance test; when $r > 1$ and $n_1 < n_2$, then the error diverges from that α to an extent reflecting the magnitude of r and the difference betwen n_1 and n_2. And, if $r < 1$, then the error is close to the stated α if $n_1 < n_2$, and it departs from that α if $n_1 > n_2$ (differing to a greater extent as the difference between the sample sizes is larger and the size of r is greater). But, larger sample sizes result in less departure from the α used in the hypothesis test.

The effect of heterogeneous variances on the t test can be profound. For example, Best and Rayner (1987) estimated that a t test with $n_1 = 5$ and $n_2 = 15$, and $\sigma_1/\sigma_2 = 4$, has a probabilty of a Type I error using t of about 0.16; and, for those sample sizes when $\sigma_1/\sigma_2 = 0.25$, P(Type I error) is about 0.01; but the probability of that error in those cases is near 0.05 if t' is employed. When the two variances are unequal, the Brown-Forsythe test mentioned could also be employed and would be expected to perform similarly to the Behrens-Fisher test, though generally not as well.

If the Behrens-Fisher test concludes difference between the means, a confidence interval for that difference may be obtained in a manner analogous to that in Section 2: The procedure is to substitute $s'_{\overline{X}_1-\overline{X}_2}$ for $s_{\overline{X}_1-\overline{X}_2}$ and to use v' instead of v in Equation 14.

Because the t test is adversely affected by heterogeneity of variances, some authors have recommended a two-step testing process: (1) The two sample variances are compared, and (2) only if the two population variances are concluded to be similar should the t test be employed. The similarity of variances may be tested by the procedures of Section 5. However, considering that the Behrens-Fisher t' test is so robust to variance inequality (and that the most common variance-comparison test performs very poorly when the distributions are nonnormal or asymmetrical), the routine test of variances is not recommended as a precursor to the testing of means by either t or t' (even though some statistical software packages perform such a test). Gans (1991) and Markowski and Markowski (1990) enlarge upon this conclusion; Moser and Stevens (1992) explain that there is no circumstance when the testing of means using either t or t' is improved by preliminary testing of variances; and Sawilowski (2002) and Wehrhahn and Ogawa (1978) state that the t test's probability of a Type I error may differ greatly from the stated α if such two-step testing is employed.

(d) Which Two-Sample Test to Use. It is very important to inform the reader of a research report specifically what statistical procedures were used in the presentation and analysis of data. It is also generally advisable to report the size (n), the mean ($\overline{X}$), and the variability (variance, standard deviation, or standard error) of each group of data; and confidence limits for each mean and for the difference between the means (Section 2) may be expressed if the mean came from a normally distributed population. Visualization of the relative magnitudes of means and measures of variability may be aided by tables or graphs.

Major choices of statistical methods for comparing two samples are as follows:

- If the two sampled populations are normally distributed and have identical variances (or if they are only slightly to moderately nonnormal and have similar variances): The t test for difference between means is appropriate and preferable. (However, as samples nearly always come from distributions that are not exactly normal with exactly the same variances, conclusions to reject or not reject a null hypothesis should not be considered definitive when the probability associated with t is very near the specified α. For example, if testing at the 5% level of significance, it should not be emphatically declared that H_0 is false if the probability of the calculated t is 0.048. The conclusion should be expressed with caution and, if feasible, the experiment should be repeated—perhaps with more data.)

- If the two sampled populations are distributed normally (or are only slightly to moderately nonnormal), but they have very dissimilar variances: The Behrens-Fisher test of Section 1c is appropriate and preferable to compare the two means.

- If the two sampled populations are very different from normally distributed, but they have similar distribution shapes and variances: The Mann-Whitney test of Section 11 is appropriate and preferable.

- If the two sampled populations have distributions greatly different from normal and do not have similar distributions and variances: (1) Consider procedures for data that do not exhibit normality and variance equality but that can be transformed into data that are normal and homogeneous of variance; or (2) refer to the procedure mentioned at the end of Section 11, which modifies the Mann-Whitney test for Behrens-Fisher situations; or (3) report the mean and variability for each of the samples, perhaps also presenting them in tables and/or graphs, and do not perform hypothesis testing.*

(e) Replication of Data. It is important to use data that are true replicates of the variable to be tested (and recall that a replicate is the smallest experimental unit to which a treatment is independently applied). In Example 1 the purpose of the experiment was to ask whether there is a difference in blood-clotting times between persons administered two different drugs. This necessitates obtaining a blood measurement on each of n_1 individuals in the first sample (receiving one of the drugs) and n_2 individuals in the second sample (receiving the other drug). It would not be valid to use n_1 measurements from a single person and n_2 measurements from another person, and to do so would be engaging in what Hurlbert (1984), and subsequently many others, discuss as *pseudoreplication*.

2 CONFIDENCE LIMITS FOR POPULATION MEANS

We defined the confidence interval for a population mean as $\overline{X} \pm t_{\alpha(2),\nu} s_{\overline{X}}$, where $s_{\overline{X}}$ is the best estimate of $\sigma_{\overline{X}}$ and is calculated as $\sqrt{s^2/n}$. For the two-sample situation where

---

*Another procedure, seldom encountered but highly recommended by Yuen (1974), is to perform the Behrens-Fisher test on *trimmed means* (also known as "truncated means"). A trimmed mean is a sample mean calculated after deleting data from the extremes of the tails of the data distribution. There is no stipulated number of data to be deleted, but it is generally the same number for each tail. The degrees of freedom are those pertaining to the number of data remaining after the deletion.

we assume that $\sigma_1^2 = \sigma_2^2$, the confidence interval for either μ_1 or μ_2 is calculated using s_p^2 (rather than either s_1^2 or s_2^2) as the best estimate of σ^2, and we use the two-tailed tabled t value with $\nu = \nu_1 + \nu_2$ degrees of freedom. Thus, for μ_i (where i is either 1 or 2, referring to either of the two samples), the $1 - \alpha$ confidence interval is

$$\overline{X}_i \pm t_{\alpha(2),\nu}\sqrt{\frac{s_p^2}{n_i}}. \tag{13}$$

For the data of Example 1, $\sqrt{s_p^2/n_2} = \sqrt{0.5193 \text{ min}^2/7} = 0.27$ min. Thus, the 95% confidence interval for μ_2 would be 9.74 min $\pm$ (2.201)(0.27 min) = 9.74 min $\pm$ 0.59 min, so that L_1 (the lower confidence limit) = 9.15 min and L_2 (the upper confidence limit) = 10.33 min, and we can declare with 95% confidence that, for the population of blood-clotting times after treatment with drug G, the population mean, μ_2, is no smaller than 9.15 min and no larger than 10.33 min. This may be written as $P(9.15 \text{ min} \leq \mu_2 \leq 10.33 \text{ min}) = 0.95$. The confidence interval for the population mean of data after treatment with drug B would be 8.75 min $\pm$(2.201) $\sqrt{0.5193 \text{ min}^2/6} = 8.75$ min $\pm$ 0.64 min; so $L_1 = 8.11$ min and $L_2 = 9.39$ min.

Confidence limits for the difference between the two population means can also be computed. The $1 - \alpha$ confidence interval for $\mu_1 - \mu_2$ is

$$\overline{X}_1 - \overline{X}_2 \pm t_{\alpha(2),\nu}s_{\overline{X}_1-\overline{X}_2}. \tag{14}$$

Thus, for Example 1, the 95% confidence interval for $\mu_1 - \mu_2$ is (8.75 min $-$ 9.74 min) $\pm$ (2.201)(0.40 min) $-$ -0.99 min $\pm$ 0.88 min. Thus, $L_1 = -1.87$ min and $L_2 = -0.11$ min, and we can write $P(-1.87 \text{ min} \leq \mu_1 - \mu_2 \leq -0.11 \text{ min}) = 0.95$.

If $H_0: \mu_1 = \mu_2$ is not rejected, then both samples are concluded to have come from populations having identical means, the common mean being denoted as μ. The best estimate of μ is the "pooled" or "weighted" mean:

$$\overline{X}_p = \frac{n_1\overline{X}_1 + n_2\overline{X}_2}{n_1 + n_2}, \tag{15}$$

which is the mean of the combined data from the two samples. Then the $1 - \alpha$ confidence interval for μ is

$$\overline{X}_p \pm t_{\alpha(2),\nu}\sqrt{\frac{s_p^2}{n_1 + n_2}}. \tag{16}$$

If H_0 is not rejected, it is the confidence interval of Equation 16, rather than those of Equations 13 and 14, that one would calculate.

As is the case with the t test, these confidence intervals are computed with the assumption that the two samples came from normal populations with the same variance. If the sampled distributions are far from meeting these conditions, then confidence intervals should be eschewed or, if they are reported, they should be presented with the caveat that they are only approximate.

If a separate $1 - \alpha$ confidence interval is calculated for μ_1 and for μ_2, it may be tempting to draw a conclusion about $H_0: \mu_1 = \mu_2$ by observing whether the two confidence intervals overlap. Overlap is the situation where L_1 for the larger mean is less than L_2 for the smaller mean, and such conclusions are made visually enticing if the confidence intervals are presented in a graph or in a table. However, this is *not* a valid procedure for hypothesis testing (e.g., Barr, 1969; Browne, 1979; Ryan and Leadbetter, 2002; Schenker and Gentleman, 2001). If there is no overlap and the population means are consequently concluded to be different, this inference will be associated with a Type I error probability less than the specified α (very much less if the two standard errors are similar); and if there is overlap, resulting in failure to reject H_0, this conclusion will be associated with a probability of a Type II error greater than (i.e., a power less than) if the appropriate testing method were used. As an illustration if this, the data of Example 1 yield $L_1 = 8.11$ min and $L_2 = 9.39$ min for the mean of group B and $L_1 = 9.15$ min and $L_2 = 10.33$ min for the mean of group G; and the two confidence intervals overlap even though the null hypothesis is rejected.

(a) One-Tailed Confidence Limits for Difference between Means. If the two-sample t test is performed to assess one-tailed hypotheses, then it is appropriate to determine a one-tailed confidence interval (as was done following a one-tailed one-sample t test). Using one-tailed critical values of t, the following confidence limits apply:

For $H_0: \mu_1 \leq \mu_2$ versus $\mu_1 > \mu_2$, or $\mu_1 - \mu_2 \leq \mu_0$ versus $H_0: \mu_1 - \mu_2 > \mu_0$:

$$L_1 = \overline{X} - \left(t_{\alpha(1),v}\right)\left(s_{\overline{X}_1 - \overline{X}_2}\right) \text{ and } L_2 = \infty.$$

For $H_0: \mu_1 \geq \mu_2$ versus $\mu_1 < \mu_2$, or $\mu_1 - \mu_2 \geq \mu_0$ versus $\mu_1 - \mu_2 < \mu_0$:

$$L_1 = -\infty \text{ and } L_2 = \overline{X} + \left(t_{\alpha(1),v}\right)\left(s_{\overline{X}_1 - \overline{X}_2}\right).$$

In Example 2, one-tailed confidence limits would be $L_1 = -\infty$ and $L_2 = (1.746)(1.55) = 2.71$ cm.

(b) Confidence Limits for Means when Variances Are Unequal. If the population variances are judged to be different enough to warrant using the Behrens-Fisher test (Section 1c) for $H_0: \mu_1 = \mu_2$, then the computation of confidence limits is altered from that shown above. If this test rejects the null hypothesis, a confidence interval for each of the two population means (μ_1 and μ_2) and a CI for the difference between the means ($\mu_1 - \mu_2$) should be determined. The $1 - \alpha$ confidence interval for μ_i is obtained as

$$\overline{X}_i \pm t_{\alpha(2),v'}\sqrt{\frac{s_i^2}{n_i}}, \text{ which is } \overline{X}_i \pm t_{\alpha(2),v'}\sqrt{s_{\overline{X}_i}^2}, \tag{17}$$

rather than by Equation 13, where v' is from Equation 12. The confidence interval for the difference between the two population means is computed to be

$$\overline{X}_1 - \overline{X}_2 \pm \left(t_{\alpha(2),v'}\right)\left(s'_{\overline{X}_1 - \overline{X}_2}\right) \tag{18}$$

rather than by Equation 14, where $s'_{\overline{X}_1 - \overline{X}_2}$ is from Equation 11a or 11b. One-tailed confidence intervals are obtained as shown in Section 2a above, but using $s'_{\overline{X}_1 - \overline{X}_2}$ instead of $s_{\overline{X}_1 - \overline{X}_2}$. A confidence interval (two-tailed or one-tailed) for $\mu_1 - \mu_2$ includes zero when the associated H_0 is not rejected.

In Example 2a, H_0 is rejected, so it is appropriate to determine a 95% CI for μ_1, which is $37.4 \pm 2.274\sqrt{1.37} = 37.4$ days ± 2.7 days; for μ_2, which is $46.0 \pm 2.274\sqrt{10.93} = 46.0$ days ± 7.5 days; and for $\mu_1 - \mu_2$, which is $37.4 - 46.0 \pm (2.274)(3.51) = -8.6$ days ± 7.98 days.

If H_0: $\mu_1 = \mu_2$ is not rejected, then a confidence interval for the common mean, $\overline{X}_p$ (Equation 15), may be obtained by using the variance of the combined data from the two samples (call it s_t^2) and the degrees of freedom for those combined data ($\nu_t = n_1 + n_2 - 1$):

$$\overline{X}_p \pm t_{\alpha(2),\nu_t}\sqrt{\frac{s_t^2}{n_1 + n_2}}. \tag{19}$$

(c) Prediction Limits. We can predict statistical characteristics of future sampling from populations from which samples have previously been analyzed. Such a desire might arise with data from an experiment such as in Example 1. Data were obtained from six animals treated with one drug and from seven animals treated with a second drug; and the mean blood-clotting times were concluded to be different under these two treatments. Equations 14 and 18 showed how confidence intervals can be obtained for the difference between means of two samples. It could also be asked what the difference between the means would be of an additional sample of m_1 animals treated with the first drug and a sample of an additional m_2 animals treated with the second.

For those two additional samples the best prediction of the difference between the two sample means would be $X_1 - X_2$, which in Example 1 is 8.75 min $- 9.74$ min $= -0.99$ min; and there would be a $1 - \alpha$ probability that the difference between the two means would be contained in this prediction interval:

$$\overline{X}_1 - \overline{X}_2 \pm t_{\alpha(2),\nu}\sqrt{s_c^2}, \tag{19a}$$

where

$$s_c^2 = \frac{s_p^2}{m_1} + \frac{s_p^2}{n_1} + \frac{s_p^2}{m_2} + \frac{s_p^2}{n_2} \tag{19b}$$

(Hahn, 1977). For example, if an additional sample of 10 data were to be obtained for treatment with the first drug and an additional sample of 12 were to be acquired for treatment with the second drug, the 95% prediction limits for the difference between means would employ $s_p^2 = 0.5193$ min^2, $n_1 = 6$, $n_2 = 7$, $m_1 = 10$, $m_2 = 12$, $t_{0.05(2),\nu=2.201}$, and $\nu = 11$; and $s_c^2 = 0.51$ min, so the 95% prediction limits would be $L_1 = -2.11$ min and $L_2 = 0.13$ min.

As the above procedure uses the pooled variance, s_p^2, it assumes that the two sampled populations have equal variances. If the two variances are thought to be quite different (the Behrens-Fisher situation discussed in Section 1c), then it is preferable to calculate the prediction interval as

$$\overline{X}_1 - \overline{X}_2 \pm t_{\alpha(2),\nu'}\sqrt{s_c^2}, \tag{19c}$$

where

$$s_c^2 = \frac{s_1^2}{m_1} + \frac{s_1^2}{n_1} + \frac{s_2^2}{m_2} + \frac{s_2^2}{n_2} \tag{19d}$$

and

$$v' = \frac{(s_c^2)^2(n_1 - 1)(n_2 - 1)}{(n_2 - 1)(s_1^2/m_1 + s_1^2/n_1)^2 + (n_1 - 1)(s_2^2/m_2 + s_2^2/n_2)} \tag{19e}$$

(Hahn, 1977). Using the results from Example 2a, if an additional sample of 22 data were proposed to be obtained for treatment with the higher temperature and an additional sample of 26 treatment with the lower temperature, the 95% prediction limits for the difference between means would be computed with $s_1^2 = 9.62$ days^2, $s_2^2 = 87.43$ days^2, $n_1 = 7$, $n_2 = 8$, $m_1 = 22$, $m_2 = 26$, $s_c^2 = 6.1031$ days^2; $v' = 6.67$, and $t_{0.05(2),6.7} = 2.389$; and from these quantities the 95% prediction limits would be determined to be $L_1 = -23.2$ days and $L_2 = 6.98$ days.

3 SAMPLE SIZE AND ESTIMATION OF THE DIFFERENCE BETWEEN TWO POPULATION MEANS

It was shown how to determine the size of the sample that is needed to estimate a population mean, by obtaining with stated assurance a confidence interval of specified width. The same type of procedure may be employed to determine the sample size, n, required from each of two populations in order to estimate the difference between the two population means with specified precision. The estimation of this sample size is an iterative procedure, employing a series of successively improving estimates of the required n. As shown in Example 3, we use

$$n = \frac{2s_p^2 t_{\alpha,v}^2}{d^2}, \tag{20}$$

where s_p^2 is the pooled variance; and $1 - \alpha$ is the confidence level of the desired confidence interval, using $\alpha(2)$ or $\alpha(1)$, respectively, for two-tailed or one-tailed confidence limits. The critical values, $t_{\alpha,v}$, are from Table 3 from *Appendix: Statistical Tables and Graphs*, where the degrees of freedom are $v = 2(n - 1)$, and d is the half-width of the confidence interval. In general, it takes fewer iterations (i.e., it is more efficient) to employ an initial guess that is too high rather than too low.

It is best to have equal sample sizes (i.e., $n_1 = n_2$) when estimating $\mu_1 - \mu_2$, but occasionally this is impractical. If sample 1 is constrained to have a size n_1, we can, after using the above procedure to calculate n, arrive at the required n_2 by

$$n_2 = \frac{nn_1}{2n_1 - n}, \tag{21}$$

as shown in Example 3. (If $2n_1 - n \leq 0$, then n_1 must be increased, if possible, and/or s_p^2 must be decreased, if possible, and/or α must be decreased, and/or d must be increased to obtain a positive n_2.) Note the efficiency of having equal sample sizes: If $n_1 = n_2$, a total of 36 data need to be collected in Example 3, but if n_1 is limited to 14, then a total of $14 + 26 = 40$ data need to be obtained. Also, the dependability of Equations 20 and 21 depends on the accuracy of s_p^2 as an estimate of the population variance, σ^2, and this improves with increases in the sizes of the samples from which s_p^2 is computed. Therefore, the sizes of those samples should not be very small compared to those estimated by Equations 20 and 21. Sample-size estimation is most reliable if $\sigma_1^2 = \sigma_2^2$.

EXAMPLE 3   Determination of Sample Size Needed to Achieve a Stated Precision in Estimating the Difference Between Two Population Means, Using the Data of Example 1

If we specify that we wish to estimate $\mu_1 - \mu_2$ by having a 95% confidence interval no wider than 1.0 min, then $d = 0.5$ min, $1 - \alpha(2) = 0.95$, and $\alpha(2) = 0.05$. From Example 1 we have an estimate of the population variance, $s_p^2 = 0.5193$ min^2 with $\nu = 11$.

Let us guess that a sample size of 50 is necessary; then, $t_{0.05(2),98} = 1.984$, so we estimate (by Equation 20)

$$n = \frac{2(0.5193)(1.984)^2}{(0.5)^2} = 16.4.$$

Next, we might estimate $n = 17$, for which $t_{0.05(2),32} = 2.037$, and calculate

$$n = \frac{2(0.5193)(2.037)^2}{(0.5)^2} = 17.2.$$

Next, we try $n = 18$, for which $t_{0.05(2),34} = 2.032$, and calculate

$$n = \frac{2(0.5193)(2.032)^2}{(0.5)^2} = 17.2.$$

Therefore, we conclude that a sample of at least 18 (i.e., more than 17) should be taken from each of the two populations in order to achieve the specified confidence interval.

If, for some reason (say, there is a limited amount of the first drug available), n_1 is constrained to be no larger than 14, then the necessary n_2 would be determined, from Equation 21, to be

$$n_2 = \frac{(n)(n_1)}{2n_1 - n} = 25.2,$$

meaning that we should use n_2 at least 26.

4   SAMPLE SIZE, DETECTABLE DIFFERENCE, AND POWER IN TESTS FOR DIFFERENCE BETWEEN TWO MEANS

(a) Sample Size Required. Prior to performing a two-sample test for difference between means (Section 1), an investigator may ask what size samples to collect.

Lenth (2001) advises that the following be considered in planning sample sizes for data to be subjected to statistical testing such as two-sample t-testing: (1) the effectiveness of the statistical test, (2) the economy of the data collection, and (3) the ethics of the data collection:

1. Sample sizes should be large enough so a biologically significant difference will be found to be statistically significant, *and* sample sizes should not be so large that statistical significance results from a difference that is not of biological significance.

2. Sample sizes not large enough to result in detection of a difference of biological importance can expend resources without yielding useful results, *and* sample sizes larger than needed to detect a difference of biological importance can result in unnecessary expenditure of resources.

3. Sample sizes not large enough to detect a difference of biological importance can expose subjects in the study to potentially harmful factors without advancing knowledge, *and* sample sizes larger than needed to detect a difference of biological importance can expose more subjects than necessary to potentially harmful factors or deny them exposure to potentially beneficial ones.

Assuming each sample comes from a normal population and the population variances are similar, we can estimate the minimum sample size to use to achieve desired test characteristics:

$$n \geq \frac{2s_p^2}{\delta^2}(t_{\alpha,\nu} + t_{\beta(1),\nu})^2 \tag{22}$$

(Cochran and Cox, 1957: 19–21). Here, δ is the smallest population difference we wish to detect: $\delta = \mu_1 - \mu_2$ for the hypothesis test for which Equation 1 is used; $\delta = |\mu_1 - \mu_2| - \mu_0$ when Equation 9 is appropriate; $\delta = \mu_1 - \mu_2 - \mu_0$ when performing a test using Equation 10. In Equation 22, $t_{\alpha,\nu}$ may be either $t_{\alpha(1),\nu}$ or $t_{\alpha(2),\nu}$, depending, respectively, on whether a one-tailed or two-tailed test is to be performed.

Note that the required sample size depends on the following four quantities:

- δ, the minimum detectable difference between population means.* If we desire to detect a very small difference between means, then we shall need a larger sample than if we wished to detect only large differences.
- σ^2, the population variance. If the variability within samples is great, then a larger sample size is required to achieve a given ability of the test to detect differences between means. We need to know the variability to expect among the data; assuming the variance is the same in each of the two populations sampled, σ^2 is estimated by the pooled variance, s_p^2, obtained from similar studies.
- The significance level, α. If we perform the t test at a low α, then the critical value, $t_{\alpha,\nu}$, will be large and a large n is required to achieve a given ability to detect differences between means. That is, if we desire a low probability of committing a Type I error (i.e., falsely rejecting H_0), then we need large sample sizes.
- The power of the test, $1 - \beta$. If we desire a test with a high probability of detecting a difference between population means (i.e., a low probability of committing a Type II error), then $\beta(1)$ will be small, $t_{\beta(1)}$ will be large, and large sample sizes are required.

Example 4 shows how the needed sample size may be estimated. As $t_{\alpha(2),\nu}$ and $t_{\beta(1),\nu}$ depend on n, which is not yet known, Equation 22 must be solved iteratively. It matters little if the initial guess for n is inaccurate. Each iterative step will bring the estimate of n closer to the final result (which is declared when two successive

---

*δ is lowercase Greek delta. If μ_0 in the statistical hypotheses is not zero (see discussion surrounding Equations 9 and 10), then δ is the amount by which the absolute value of the difference between the population means differs from μ_0.

iterations fail to change the value of n rounded to the next highest integer). In general, however, fewer iterations are required (i.e., the process is quicker) if one guesses high instead of low.

EXAMPLE 4   Estimation of Required Sample Size for a Two-Sample t Test

We desire to test for significant difference between the mean blood-clotting times of persons using two different drugs. We wish to test at the 0.05 level of significance, with a 90% chance of detecting a true difference between population means as small as 0.5 min. The within-population variability, based on a previous study of this type (Example 1), is estimated to be 0.52 min^2.

Let us guess that sample sizes of 100 will be required. Then, $v = 2(n - 1) = 2(100 - 1) = 198, t_{0.05(2),198} \approx 1.972, \beta = 1 - 0.90 = 0.10, t_{0.10(1),198} = 1.286$, and we calculate (by Equation 22):

$$n \geq \frac{2(0.52)}{(0.5)^2}(1.972 + 1.286)^2 = 44.2.$$

Let us now use $n = 45$ to determine $v = 2(n - 1) = 88, t_{0.05(2),88} = 1.987, t_{0.10(1),88} = 1.291$, and

$$n \geq \frac{2(0.52)}{(0.5)^2}(1.987 + 1.291)^2 = 44.7.$$

Therefore, we conclude that each of the two samples should contain at least 45 data.

If n_1 were constrained to be 30, then, using Equation 21, the required n_2 would be

$$n_2 = \frac{(44.7)(30)}{2(30) - 44.7} = 88.$$

For a given total number of data $(n_1 + n_2)$, maximum test power and robustness occur when $n_1 = n_2$ (i.e., the sample sizes are equal). There are occasions, however, when equal sample sizes are impossible or impractical. If, for example, n_1 were fixed, then we would first determine n by Equation 22 and then find the required size of the second sample by Equation 21, as shown in Example 4. Note, from this example, that a total of $45 + 45 = 90$ data are required in the two equal-sized samples to achieve the desired power, whereas a total of $30 + 88 = 118$ data are needed if the two samples are as unequal as in this example. If $2n - 1 \leq 0$, then see the discussion following Equation 21.

(b) Minimum Detectable Difference. Equation 22 can be rearranged to estimate how small a population difference (δ, defined above) would be detectable with a given sample size:

$$\delta \geq \sqrt{\frac{2s_p^2}{n}}(t_{\alpha,v} + t_{\beta(1),v}). \tag{23}$$

The estimation of δ is demonstrated in Example 5.

EXAMPLE 5   Estimation of Minimum Detectable Difference in a Two-Sample t Test

In two-tailed testing for significant difference between mean blood-clotting times of persons using two different drugs, we desire to use the 0.05 level of significance and sample sizes of 20. What size difference between means do we have a 90% chance of detecting?

Using Equation 23 and the sample variance of Example 1, we calculate:

$$\delta = \sqrt{\frac{2(0.5193)}{20}}\,(t_{0.05(2),38} + t_{0.10(1),38})$$

$$= (0.2279)(2.024 + 1.304) = 0.76 \text{ min.}$$

In a Behrens-Fisher situation (i.e., if we don't assume that $\sigma_1^2 = \sigma_2^2$), Equation 23 would employ $\sqrt{s_1^2/n + s_2^2/n}$ instead of $\sqrt{2s_p^2/n}$.

(c) Power of the Test.   Further rearrangement of Equation 22 results in

$$t_{\beta(1),\nu} \leq \frac{\delta}{\sqrt{\dfrac{2s_p^2}{n}}} - t_{\alpha,\nu}, \tag{24}$$

On computing $t_{\beta(1),\nu}$, one can consult Table 3 from *Appendix: Statistical Tables and Graphs* to determine $\beta(1)$, whereupon $1 - \beta(1)$ is the power. But this generally will only result in declaring a range of power (e.g., $0.75 <$ power < 0.90). Some computer programs can provide the exact probability of $\beta(1)$, or we may, with only slight overestimation of power consider $t_{\beta(1)}$ to be approximated by a normal deviate and may thus employ Table 2 *Appendix: Statistical Tables and Graphs*.

If the two population variances are not assumed to be the same, then $\sqrt{s_1^2/n + s_2^2/n}$ would be used in place of $\sqrt{2s_p^2/n}$ in Equation 24.

The above procedure for estimating power is demonstrated in Example 6, along with the following method (which will be expanded on in the chapter on analysis of variance). We calculate

$$\phi = \sqrt{\frac{n\delta^2}{4s_p^2}} \tag{25}$$

(derived from Kirk, 1995: 182) and ϕ (lowercase Greek phi) is then located in Figure 1a from *Appendix: Statistical Tables and Graphs*, along the lower axis (taking care to distinguish between ϕ's for $\alpha = 0.01$ and $\alpha = 0.05$). Along the top margin of the graph are indicated pooled degrees of freedom, ν, for α of either 0.01 or 0.05 (although the symbol ν_2 is used on the graph for a reason that will be apparent later in your studies). By noting where ϕ vertically intersects the curve for the appropriate ν, one can read across to either the left or right axis to find the estimate of power. The calculated power is an estimate of the probability of rejecting a false null hypothesis in future statistical tests; it is not the probability of rejecting H_0 in tests performed on the present set of data.

EXAMPLE 6   Estimation of the Power of a Two-Sample t Test

What would be the probability of detecting a true difference of 1.0 min between mean blood-clotting times of persons using the two drugs of Example 1, if $n_1 = n_2 = 15$, and $\alpha(2) = 0.05$?

For $n = 15$, $\nu = 2(n - 1) = 28$ and $t_{0.05(2),28} = 2.048$. Using Equation 24:

$$t_{\beta(1),28} \leq \frac{1.0}{\sqrt{\dfrac{2(0.5193)}{15}}} - 2.048 = 1.752.$$

Consulting Table 3 from *Appendix: Statistical Tables and Graphs*, we see that, for one-tailed probabilities and $\nu = 28$: $0.025 < P(t \geq 1.752) < 0.05$, so $0.025 < \beta < 0.05$.

$$\text{Power} = 1 - \beta, \text{ so } 0.95 < \text{power} < 0.975.$$

Or, by the normal approximation, we can estimate β by $P(Z \geq 1.752) = 0.04$. So power = 0.96. [The exact figures are $\beta = 0.045$ and power = 0.955.]

To use Figure 1 from *Appendix: Statistical Tables and Graphs*, we calculate

$$\phi = \sqrt{\frac{n\delta^2}{4s_p^2}} = \sqrt{\frac{(15)(1.0)}{4(0.5193)}} = 2.69.$$

In the first page of Figure 1 from *Appendix: Statistical Tables and Graphs*, we find that $\phi = 2.69$ and $\nu(= \nu_2) = 28$ are associated with a power of about 0.96.

(d) Unequal Sample Sizes. For a given total number of data, $n_1 + n_2$, the two-sample t test has maximum power and robustness when $n_1 = n_2$. However, if $n_1 \neq n_2$, the above procedure for determining minimum detectable difference (Equation 23) and power (Equations 24 and 25) can be performed using the harmonic mean of the two sample sizes (Cohen, 1988: 42):

$$n = \frac{2n_1 n_2}{n_1 + n_2}. \tag{26}$$

Thus, for example, if $n_1 = 6$ and $n_2 = 7$, then

$$n = \frac{2(6)(7)}{6 + 7} = 6.46.$$

5   TESTING FOR DIFFERENCE BETWEEN TWO VARIANCES

If we have two samples of measurements, each sample taken at random from a normal population, we might ask if the variances of the two populations are equal. Consider the data of Example 7, where s_1^2, the estimate of σ_1^2, is 21.87 moths^2, and s_2^2, the estimate of σ_2^2, is 12.90 moths^2. The two-tailed hypotheses can be stated as $H_0: \sigma_1^2 = \sigma_2^2$ and $H_A: \sigma_1^2 \neq \sigma_2^2$, and we can ask, What is the probability of taking two samples from two populations having identical variances and having the two sample variances be as different as are s_1^2 and s_2^2? If this probability is rather low (say ≤ 0.05), then we reject the veracity of H_0 and conclude that the two samples came from populations having unequal variances. If the probability is greater than α, we conclude that there is insufficient evidence to conclude that the variances of the two populations are not the same.

(a) Variance-Ratio Test. The hypotheses may be submitted to the two-sample *variance-ratio test*, for which one calculates

$$F = \frac{s_1^2}{s_2^2} \quad \text{or} \quad F = \frac{s_2^2}{s_1^2}, \text{ whichever is larger.*} \tag{27}$$

That is, the larger variance is placed in the numerator and the smaller in the denominator. We then ask whether the calculated ratio of sample variances (i.e., F) deviates so far from 1.0 as to enable us to reject H_0 at the α level of significance. For the data in Example 7, the calculated F is 1.70. The critical value, $F_{0.05(2),10,9}$, is obtained from Table 4 from *Appendix: Statistical Tables and Graphs* and is found to be 3.59. As $1.70 < 3.59$, we do not reject H_0[†].

Note that we consider degrees of freedom associated with the variances in both the numerator and denominator of the variance ratio. Furthermore, it is important to realize that F_{α,ν_1,ν_2} and F_{α,ν_2,ν_1} are not the same (unless, of course, $\nu_1 = \nu_2$), so the numerator and denominator degrees of freedom must be referred to in the correct order.

If $H_0: \sigma_1^2 = \sigma_2^2$ is not rejected, then s_1^2 and s_2^2 are assumed to be estimates of the same population variance, σ^2. The best estimate of this σ^2 that underlies both samples is called the *pooled variance* (introduced as Equation 4):

$$s_p^2 = \frac{SS_1 + SS_2}{\nu_1 + \nu_2} = \frac{\nu_1 s_1^2 + \nu_2 s_2^2}{\nu_1 + \nu_2}. \tag{28}$$

One-tailed hypotheses may also be submitted to the variance ratio test. For $H_0: \sigma_1^2 \geq \sigma_2^2$ and $H_A: \sigma_1^2 < \sigma_2^2, s_2^2$ is always used as the numerator of the variance ratio; for $H_0: \sigma_1^2 \leq \sigma_2^2$ and $H_A: \sigma_1^2 > \sigma_2^2, s_1^2$ is always used as the numerator. (A look at the alternate hypothesis tells us which variance belongs in the numerator of F in order to make $F > 1$.)

The critical value for a one-tailed test is $F_{\alpha(1),\nu_1,\nu_2}$ from Table 4 from *Appendix: Statistical Tables and Graphs*, where ν_1 is the degrees of freedom associated with the numerator of F and ν_2 is the degrees of freedom associated with the denominator. Example 8 presents the data submitted to the hypothesis test for whether seeds planted in a greenhouse have less variability in germination time than seeds planted outside.

The variance-ratio test is not a robust test, being severely and adversely affected by sampling nonnormal populations (e.g., Box, 1953; Church and Wike, 1976; Markowski and Markowski, 1990; Pearson, 1932; Tan, 1982), with deviations from mesokurtosis somewhat more important than asymmetry; and in cases of nonnormality the probability of a Type I error can be very much greater than α.

---

*What we know as the F statistic is a ratio of the variances of two normal distributions and was first described by R. A. Fisher in 1924 (and published in 1928) (Lehmann, 1999); the statistic was named in his honor by G. W. Snedecor (1934: 15).

[†]Some calculators and many computer programs have the capability of determining the probability of a given F. For the present example, we would thereby find that $P(F \geq 1.70) = 0.44$.

EXAMPLE 7   The Two-Tailed Variance Ratio Test for the Hypothesis H_0: $\sigma_1^2 = \sigma_2^2$ and H_A: $\sigma_1^2 \neq \sigma_2^2$. The Data Are the Numbers of Moths Caught During the Night by 11 Traps of One Style and 10 Traps of a Second Style

H_0: $\sigma_1^2 = \sigma_2^2$

H_A: $\sigma_1^2 \neq \sigma_2^2$

$\alpha = 0.05$

Trap type 1	Trap type 2
41	52
35	57
33	62
36	55
40	64
46	57
31	56
37	55
34	60
30	59
38	

$n_1 = 11$	$n_2 = 10$
$\nu_1 = 10$	$\nu_2 = 9$
$SS_1 = 218.73 \text{ moths}^2$	$SS_2 = 116.10 \text{ moths}^2$
$s_1^2 = 21.87 \text{ moths}^2$	$s_2^2 = 12.90 \text{ moths}^2$

$$F = \frac{s_1^2}{s_2^2} = \frac{21.87}{12.90} = 1.70$$

$$F_{0.05(2),10,9} = 3.96$$

Therefore, do not reject H_0.

$$P(0.20 < F < 0.50)[P = 0.44]$$

$$s_p^2 = \frac{218.73 \text{ moths}^2 + 116.10 \text{ moths}^2}{10 + 9} = 17.62 \text{ moths}^2$$

The conclusion is that the variance of numbers of moths caught is the same for the two kinds of traps.

(b) Other Two-Sample Tests for Variances.   A large number of statistical procedures to test differences between variances have been proposed and evaluated (e.g., Brown and Forsythe, 1974c; Church and Wike, 1976; Draper and Hunter, 1969; Levene, 1960; Miller, 1972; O'Neill and Mathews, 2000), often with the goal of avoiding the

---

EXAMPLE 8   A One-Tailed Variance-Ratio Test for the Hypothesis That the Germination Time for Pine Seeds Planted in a Greenhouse Is Less Variable Than for Pine Seeds Planted Outside

H_0: $\sigma_1^2 \geq \sigma_2^2$

H_A: $\sigma_1^2 < \sigma_2^2$

$\alpha = 0.05$

**Germination Time (in Days)
of Pine Seeds**

Greenhouse	Outside
69.3	69.5
75.5	64.6
81.0	74.0
74.7	84.8
72.3	76.0
78.7	93.9
76.4	81.2
	73.4
	88.0

$n_1 = 7$	$n_2 = 9$
$\nu_1 = 6$	$\nu_2 = 8$
$SS_1 = 90.57$ days^2	$SS_2 = 700.98$ days^2
$s_1^2 = 15.10$ days^2	$s_2^2 = 87.62$ days^2

$F = \dfrac{87.62}{15.10} = 5.80$

$F_{0.05(1),8,6} = 4.15$

Therefore, reject H_0.

$$0.01 < P(F \geq 5.80) < 0.025 \quad [P = 0.023]$$

The conclusion is that the variance in germination time is less in plants grown in the greenhouse than in those grown outside.

---

lack of robustness of the variance-ratio test when samples come from nonnormal populations of data. A commonly encountered one is Levene's test, and its various modifications, which is typically less affected by nonnormal distributions than the variance-ratio test is.

The concept is to perform a two-sample t test (two-tailed or one-tailed, as the situation warrants; see Section 1), not on the values of X in the two samples but on values of the data after conversion to other quantities. A common conversion is to employ the deviations of each X from its group mean or median; that is, the two-sample t test is performed on $|X_{ij} - \overline{X}_i|$ or on $|X_{ij} - \text{median of group } i|$. Other data conversions, such as the square root or the logarithm of $|X_{ij} - \overline{X}_i|$, have also been examined (Brown and Forsythe, 1974c).

Levene's test is demonstrated in Example 9 for two-tailed hypotheses, and X' is used to denote $|X_i - \overline{X}|$. This procedure may also be employed to test one-tailed hypotheses about variances, either $H_0: \sigma_1^2 \geq \sigma_2^2$ vs. $H_A: \sigma_1^2 < \sigma_2^2$, or $H_0: \sigma_1^2 \leq \sigma_2^2$ vs. $H_A: \sigma_1^2 > \sigma_2^2$. This would be done by the one-tailed t-testing described in Section 1, using σ^2 in place of μ in the hypothesis statements and using $|X_i - \overline{X}|$ instead of X_i in the computations.

EXAMPLE 9 The Two-Sample Levene Test for $H_0: \sigma_1^2 = \sigma_2^2$ and $H_A: \sigma_1^2 \neq \sigma_2^2$. The Data Are Those of Example 7

$H_0: \quad \sigma_1^2 = \sigma_2^2$

$H_A: \quad \sigma_1^2 \neq \sigma_2^2$

$\alpha = 0.05$

For group 1: $\Sigma X = 401$ moths, $n = 11$, $\nu = 10$, $\overline{X} = 36.45$ moths.

For group 2: $\Sigma X = 577$ moths, $n = 10$, $\nu = 9$, $\overline{X} = 57.70$ moths.

	Trap Type 1		Trap Type 2	
X_i	$X' = $ $\lvert X_i - \overline{X}\rvert$		X_i	$X' = $ $\lvert X_i - \overline{X}\rvert$
41	4.55		52	5.70
35	1.45		57	0.70
33	3.45		62	4.30
36	0.45		55	2.70
40	3.55		64	6.30
46	9.55		57	0.70
31	5.45		56	1.70
37	0.55		55	2.70
34	2.45		60	2.30
30	6.45		59	1.30
38	1.55			

ΣX_i	$\Sigma X'_i = $ $\Sigma \lvert X_i - \overline{X}\rvert$		ΣX_i	$\Sigma X'_i = $ $\Sigma \lvert X_i - \overline{X}\rvert$
$= 401$ moths	$= 39.45$ moths		$= 577$ moths	$= 28.40$ moths

For the absolute values of the deviations from the mean:

$$X'_1 = 39.45 \text{ moths}/11 \qquad X'_2 = 28.40 \text{ moths}/10$$

$$= 3.59 \text{ moths} \qquad\qquad = 2.84 \text{ moths}$$

$$SS'_1 = 77.25 \text{ moths}^2 \qquad SS'_2 = 35.44 \text{ moths}^2$$

$$(s_p^2)' = \frac{77.25 \text{ moths}^2 + 35.44 \text{ moths}^2}{10 + 9} = 5.93 \text{ moths}^2$$

$$s'_{\overline{X}_1 - \overline{X}_2} = \sqrt{\frac{5.93 \text{ moths}^2}{11} + \frac{5.93 \text{ moths}^2}{10}} = 1.06 \text{ moths}$$

$$t' = \frac{3.59 \text{ moths} - 2.84 \text{ moths}}{1.06 \text{ moths}} = 0.71$$

$t_{0.05(2),19} = 2.093$

Therefore, do not reject H_0.

$$0.02 < P < 0.50 \ [P = 0.48]$$

We conclude that the variance of the numbers of moths caught is the same using either kind of trap.

Depending upon the data, alternate procedures, even if often more robust to non-normality than the variance-ratio test, may have drawbacks, such as the tests being adversely sensitive to unequal sample sizes or to skewed distributions, or the converted quantities involving $|X_{ij} - \overline{X}_i|$ not representing a normal distribution.

Thus, probabilities resulting from these tests, and from the variance-ratio test, are in many cases not accurate expressions of the probability of a Type I error (e.g., this probability will often greatly exceed α). Therefore, their results should be suspect or, at best, be considered only approximations.

The variance-ratio test may be employed if there is confidence that the data came from a normally distributed population. Otherwise, the procedures of the chapter *Data Transformations* can be considered, whereby data that do not exhibit normality can sometimes be transformed into data that are normal. Or, hypothesis testing might be abandoned in favor of simply reporting the variability (e.g., variance) for each of the samples, perhaps also representing them in tables and/or graphs.

Nonparametric tests have been proposed (e.g., Conover, 1999: 300–303; Hollander and Wolfe, 1999: Chapter 5; Laubscher, Steffens, and De Lange, 1968; Mood, 1954; Siegel and Tukey, 1960) to test for difference between the amounts of dispersion (i.e., variability) of two populations, but they require the use of special tables of critical values or possess other drawbacks.

(c) Which Two-Sample Test for Variances to Use. Unless the variance-ratio test is applied to samples coming from populations that are normally distributed, all of the above tests are faulty in their ability to express a probability that accurately reflects the Type I error. Therefore, the resultant probabilities should be expressed with reservation, and firm conclusions should not be drawn when P(Type I error) is close to the selected level of significance. In summary, the preferences among the tests are as follows:

- If the two sampled populations are normally distributed or very nearly normal in their data distributions: The variance-ratio test is appropriate and preferable.
- If the two sampled populations are distinctly nonnormal, but are symmetrical or nearly symmetrical and the sample sizes are equal or nearly equal: The Levene test using $|X - \overline{X}|$ is preferable.

- If the two sampled populations are distinctly nonnormal and somewhat asymmetrical, and the sample sizes are equal or nearly equal: The Levene test using $|X - \text{median}|$ is preferable.
- If the two sampled populations are very asymmetrical or the sample sizes are very different: (1) Consider the procedures of the chapter *Data Transformations* for data that do not exhibit normality and variance equality but that can be transformed into data that are normal and homogeneous of variance; or (2) report the variability for each of the samples, perhaps also representing them in tables and/or graphs, and do not perform hypothesis testing.

6 CONFIDENCE LIMITS FOR POPULATION VARIANCES AND THE POPULATION VARIANCE RATIO

If two population variances are concluded to be different, the confidence limits for each of them may be calculated via Equation 28a or Equation 28b. To obtain a CI for σ_1^2, simply substitute s_1^2, SS_1, and ν_1 for s^2, SS, and ν, respectively, in one of those equations, and the CI for σ_2^2 is obtained by substituting s_2^2, SS_2, and ν_2, respectively.

$$\frac{\nu s^2}{\chi_{\alpha/2, \nu}^2} \leq \sigma^2 \leq \frac{\nu s^2}{\chi_{(1-\alpha/2), \nu}^2}. \tag{28a}$$

$$\frac{SS}{\chi_{\alpha/2, \nu}^2} \leq \sigma^2 \leq \frac{SS}{\chi_{(1-\alpha/2), \nu}^2}. \tag{28b}$$

And a confidence limit for a population standard deviation, σ_1 or σ_2, is calculated as the square root of the corresponding confidence limit for σ_1^2 or σ_2^2.

If two population variances are not concluded to be different, then we may wish to state confidence limits of the variance common to both populations, This would be done by employing s_p^2 (defined in Equation 28) and $\nu = \nu_1 + \nu_2$; and the confidence limits for the population standard deviation, σ, common to both populations would be the square roots of the confidence limits for σ^2.

A two-tailed $1 - \alpha$ confidence interval for the variance ratio, σ_1^2/σ_2^2, is defined by its lower confidence limit,

$$L_1 = \left(\frac{s_1^2}{s_2^2}\right)\left(\frac{1}{F_{\alpha(2), \nu_1, \nu_2}}\right), \tag{29}$$

and its upper confidence limit

$$L_2 = \left(\frac{s_1^2}{s_2^2}\right) F_{\alpha(2), \nu_2, \nu_1}. \tag{30}$$

For the data of Example 8, $s_1^2/s_2^2 = 0.172$, $F_{0.05(2), 6, 8} = 4.65$, and $F_{0.05(2), 8, 6} = 5.60$. Therefore, we would calculate $L_1 = 0.037$ and $L_2 = 0.963$, and we could state

$$P\left(0.037 \leq \frac{\sigma_1^2}{\sigma_2^2} \leq 0.963\right) = 0.95. \tag{31}$$

To calculate a two-tailed confidence interval for σ_2^2/σ_1^2, simply utilize Equations 29, 30, and 31 with the two subscripts (1 and 2) reversed on s^2, ν, and σ^2.

One-tailed confidence limits for σ_1^2/σ_2^2 are also possible. If $H_A: \sigma_1^2 < \sigma_2^2$ (which could be written as $H_A: \sigma_1^2/\sigma_2^2 < 1$), then $L_1 = 0$ and L_2 is obtained from

Equation 30 using a one-tailed value of F (namely, $F_{\alpha(1),\nu_2,\nu_1}$). If $H_A: \sigma_1^2 > \sigma_2^2$ (in other words, $H_A: \sigma_1^2/\sigma_2^2 > 1$), then $L_1 =$ is obtained from Equation 29 using a one-tailed value of F (i.e., $F_{\alpha(1),\nu_1,\nu_2}$) and $L_2 = \infty$.

The above confidence limits involving variances assume that the sampled populations are normally distributed. If the populations' distributions are not normal, then the calculated confidence intervals are only approximations, with the approximation poorer the further from normality the populations are.

Meeker and Hahn (1980) discuss calculation of prediction limits for the variance ratio and provide special tables for that purpose.

7 SAMPLE SIZE AND POWER IN TESTS FOR DIFFERENCE BETWEEN TWO VARIANCES

(a) Sample Size Required. In considering the variance-ratio test of Section 5, we may ask what minimum sample sizes are required to achieve specified test characteristics. Using the normal approximation recommended by Desu and Raghavarao (1990: 35), the following number of data is needed in each sample to test at the α level of significance with power of $1 - \beta$:

$$n = \left[\frac{Z_\alpha + Z_{\beta(1)}}{\ln\left(\frac{s_1^2}{s_2^2}\right)} \right]^2 + 2. \tag{32}$$

For analysts who prefer performing calculations with "common logarithms" (those employing base 10) to using "natural logarithms" (those in base e),* Equation 32 may be written equivalently as

$$n = \left[\frac{Z_\alpha + Z_{\beta(1)}}{(2.30259) \log\left(\frac{s_1^2}{s_2^2}\right)} \right]^2 + 2. \tag{33}$$

This sample-size estimate assumes that the samples are to be equal in size, which is generally preferable. If, however, it is desired to have unequal sample sizes (which will typically require more total data to achieve a particular power), one may specify that ν_1 is to be m times the size of ν_2; then (after Desu and Raghavarao, 1990: 35):

$$m = \frac{n_1 - 1}{n_2 - 1}, \tag{34}$$

$$n_2 = \frac{(m + 1)(n - 2)}{2m} + 2, \tag{35}$$

and

$$n_1 = m(n_2 - 1) + 1. \tag{36}$$

---

*In this text, *ln* will denote the natural, or Naperian, logarithm, and *log* will denote the common, or Briggsian, logarithm. These are named for the Scottish mathematician John Napier (1550–1617), who devised and named logarithms, and the English mathematician Henry Briggs (1561–1630), who adapted this computational method to base 10; the German astronomer Johann Kepler (1550–1617) was the first to use the abbreviation "Log," in 1624, and Italian mathematician Bonaventura Cavalieri (1598–1647) was the first to use "log" in 1632 (Cajori, 1928/9, Vol. II: 105–106; Gullber, 1997: 152). Sometimes $\log_e$ and $\log_{10}$ will be seen instead of ln and log, respectively.

As in Section 5, determination of whether s_1^2 or s_2^2 is placed in the numerator of the variance ratio in Equation 32 depends upon the hypothesis test, and Z_α is either a one-tailed or two-tailed normal deviate depending upon the hypothesis to be tested; n_1 and n_2 correspond to s_1^2 and s_2^2, respectively. This procedure is applicable if the variance ratio is >1.

(b) Power of the Test. We may also estimate what the power of the variance ratio test would be if specified sample sizes were used. If the two sample sizes are the same (i.e., $n = n_1 = n_2$), then Equations 32 and 33 may be rearranged, respectively, as follows:

$$Z_{\beta(1)} = \sqrt{n - 2} \ln \left(\frac{s_1^2}{s_2^2} \right) - Z_\alpha; \tag{37}$$

$$Z_{\beta(1)} = \sqrt{n - 2}(2.30259) \log \left(\frac{s_1^2}{s_2^2} \right) - Z_\alpha. \tag{38}$$

After $Z_{\beta(1)}$ is calculated, $\beta(1)$ is determined from the last line of Table 3 from *Appendix: Statistical Tables and Graphs*, or from Table 2 from the appendix, or from a calculator or computer that gives probability of a normal deviate; and power $= 1 - \beta(1)$. If the two sample sizes are not the same, then the estimation of power may employ

$$Z_{\beta(1)} = \sqrt{\frac{2m(n_2 - 2)}{m + 1}} \ln \left(\frac{s_1^2}{s_2^2} \right) - Z_\alpha \tag{39}$$

or

$$Z_{\beta(1)} = \sqrt{\frac{2m(n_2 - 2)}{m + 1}}(2.30259) \log \left(\frac{s_1^2}{s_2^2} \right) - Z_\alpha, \tag{40}$$

where m is as in Equation 34.

8 TESTING FOR DIFFERENCE BETWEEN TWO COEFFICIENTS OF VARIATION

A very useful property of coefficients of variation is that they have no units of measurement. Thus, V's may be compared even if they are calculated from data having different units, as is the case in Example 10. And it may be desired to test the null hypothesis that two samples came from populations with the same coefficients of variation.

EXAMPLE 10   A Two-Tailed Test for Difference Between Two Coefficients of Variation

H_0: The intrinsic variability of male weights is the same as the intrinsic variability of male heights (i.e., the population coefficients of variation of weight and height are the same, namely H_0: $\sigma_1/\mu_1 = \sigma_2/\mu_2$).

H_0: The intrinsic variability of male weight is not the same as the intrinsic variability of male heights (i.e., the population coefficients of variation of weight and height are not the same, namely H_0: $\sigma_1/\mu_1 \neq \sigma_2/\mu_2$).

(a) The variance-ratio test.

Weight (kg)	Log of weight	Height (cm)	Log of height
72.5	1.86034	183.0	2.26245
71.7	1.85552	172.3	2.23629
60.8	1.78390	180.1	2.25551
63.2	1.80072	190.2	2.27921
71.4	1.85370	191.4	2.28194
73.1	1.86392	169.6	2.22943
77.9	1.89154	166.4	2.22115
75.7	1.87910	177.6	2.24944
72.0	1.85733	184.7	2.26647
69.0	1.83885	187.5	2.27300
		179.8	2.25479

$n_1 = 10$ $\qquad\qquad\qquad$ $n_2 = 11$

$\nu_1 = 9$ $\qquad\qquad\qquad\;$ $\nu_2 = 10$

$\overline{X}_1 = 70.73 \text{ kg}$ $\qquad\quad$ $\overline{X}_2 = 180.24 \text{ cm}$

$SS_1 = 246.1610 \text{ kg}^2$ $\quad$ $SS_2 = 678.9455 \text{ cm}^2$

$s_1^2 = 27.3512 \text{ kg}^2$ $\qquad$ $s_2^2 = 67.8946 \text{ cm}^2$

$s_1 = 5.23 \text{ kg}$ $\qquad\quad\;$ $s_2 = 8.24 \text{ cm}$

$V_1 = 0.0739$ $\qquad\qquad$ $V_2 = 0.0457$

$(SS_{\log})_1 = 0.00987026$ $\quad$ $(SS_{\log})_2 = 0.00400188$

$(s_{\log}^2)_1 = 0.0010967$ $\qquad$ $(s_{\log}^2)_2 = 0.00040019$

$$F = \frac{0.0010967}{0.00040019} = 2.74$$

$F_{0.05(2),9,10} = 3.78$

Therefore, do not reject H_0.

$$0.10 < P < 0.20 \quad [P = 0.13]$$

It is concluded that the coefficient of variation is the same for the population of weights as it is for the population of heights.

(b) The Z test.

$$V_p = \frac{\nu_1 V_1 + \nu_2 V_2}{\nu_1 + \nu_2} = \frac{9(0.0739) + 10(0.0457)}{9 + 10} = \frac{1.1221}{19} = 0.0591$$

$V_p^2 = 0.003493$

$$Z = \frac{V_1 - V_2}{\sqrt{\left(\dfrac{V_p^2}{\nu_1} + \dfrac{V_p^2}{\nu_2}\right)(0.5 + V_p^2)}}$$

$$= \frac{0.0739 - 0.0457}{\sqrt{\left(\dfrac{0.003493}{9} + \dfrac{0.003493}{10}\right)(0.5 + 0.003493)}}$$

$$= \frac{0.0282}{0.0193} = 1.46$$

$Z_{0.05(2)} = t_{0.05(2),\infty} = 1.960$

Do not reject H_0.

$$0.10 < P < 0.20 \quad [P = 0.14]$$

It is concluded that the coefficient of variation is the same for the population of weights as it is for the population of heights.

Lewontin (1966) showed that

$$F = \frac{\left(s_{\log}^2\right)_1}{\left(s_{\log}^2\right)_2} \quad \text{or} \quad F = \frac{\left(s_{\log}^2\right)_2}{\left(s_{\log}^2\right)_1} \tag{41}$$

may be used for a variance-ratio test, analogously to Equation 27. In Equation 41, $\left(s_{\log}^2\right)_i$ refers to the variance of the logarithms of the data in Sample i, where logarithms to any base may be employed. This procedure is applicable only if all of the data are positive (i.e., > 0), and it is demonstrated in Example 10a. Either two-tailed or one-tailed hypotheses may be tested, as shown in Section 5.

This variance-ratio test requires that the logarithms of the data in each sample come from a normal distribution. A procedure advanced by Miller (1991) allows testing when the data, not their logarithms, are from normal distributions (that have positive means and variances). The test statistic, as demonstrated in Example 10b, is

$$Z = \frac{V_1 - V_2}{\sqrt{\left(\dfrac{V_p^2}{\nu_1} + \dfrac{V_p^2}{\nu_2}\right)(0.5 + V_p^2)}}, \tag{42}$$

where

$$V_p = \frac{\nu_1 V_1 + \nu_2 V_2}{\nu_1 + \nu_2} \tag{43}$$

is referred to as the "pooled coefficient of variation," which is the best estimate of the population coefficient of variation, σ/μ, that is common to both populations if the null hypothesis of no difference is true.

This procedure is shown, as a two-tailed test, in Example 10b. Recall that critical values of Z may be read from the last line of the table of critical values of t, so $Z_{\alpha(2)} = t_{\alpha(2),\infty}$. One-tailed testing is also possible, in which case the alternate hypothesis would declare a specific direction of difference and one-tailed critical values $(t_{\infty(1),\alpha})$ would be consulted. This test works best if there are at least 10 data in each sample and each population's coefficient of variation is no larger than 0.33. An estimate of the power of the test is given by Miller and Feltz (1997).

9 CONFIDENCE LIMITS FOR THE DIFFERENCE BETWEEN TWO COEFFICIENTS OF VARIATION

Miller and Feltz (1997) have provided this $1 - \alpha$ confidence interval for $\sigma_1/\mu_1 - \sigma_2/\mu_2$, where the two sampled populations are normally distributed:

$$V_1 - V_2 \pm Z_{\alpha(2)}\sqrt{\frac{V_1^2}{\nu_1}(0.5 + V_1^2) + \frac{V_2^2}{\nu_1}(0.5 + V_2^2)}. \qquad (44)$$

10 NONPARAMETRIC STATISTICAL METHODS

There is a large body of statistical methods that do not require the estimation of population parameters (such as μ and σ) and that test hypotheses that are not statements about population parameters. These statistical procedures are termed *nonparametric tests*.* These are in contrast to procedures such as t tests, which are called *parametric tests* and which do rely upon estimates of population parameters and upon the statement of parameters in the statistical hypotheses. Although they may assume that the sampled populations have the same dispersion or shape, nonparametric methods typically do not make assumptions about the nature of the populations' distributions (e.g., there is no assumption of normality); thus they are sometimes referred to as *distribution-free tests*.[†] Both parametric and nonparametric tests require that the data have come at random from the sampled populations.

Nonparametric tests (such as the two-sample testing procedure described in Section 11) generally may be applied to any situation where we would be justified in employing a parametric test (such as the two-sample t test), as well as in some instances where the assumptions of the latter are untenable. If either the parametric or nonparametric approach is applicable, then the former will generally be more powerful than the latter (i.e., the parametric method will typically have a lower probability of committing a Type II error). However, often the difference in power is not great and can be compensated by a small increase in sample size for the nonparametric test. When the underlying assumptions of a parametric test are seriously violated, then the nonparametric counterpart may be decidedly more powerful.

Most nonparametric statistical techniques convert observed data to the ranks of the data (i.e., their numerical order). For example, measurements of 2.1, 2.3, 2.9, 3.6, and 4.0 kg would be analyzed via their ranks of 1, 2, 3, 4, and 5. A possible disadvantage of this rank transformation of data is that some information is lost (for example, the same ranks would result from measurements of 1.1, 1.3, 2.9, 4.6, and 5.0 kg). A possible advantage is that outliers will have much less influence (for example, the same ranks would result from measurements of 2.1, 2.3, 2.9, 3.6, and 25.0 kg).

It is sometimes counseled that only nonparametric testing may be employed when dealing with ordinal-scale data, but such advice is based upon what Gaito (1980) calls "an old misconception"; this issue is also discussed by Anderson (1961), Gaito (1960), Savage (1957), and Stevens (1968). Interval-scale or ratio-scale measurements are not intrinsically required for the application of parametric testing procedures. Thus parametric techniques may be considered for ordinal-scale data *if* the assumptions of such methods are met—typically, random sampling from normally distributed populations with homogeneity of variances. But ordinal data often come

---

*The term *nonparametric* was first used by Jacob Wolfowitz in 1942 (David, 1995; Noether, 1984).

[†]The terms *nonparametric* and *distribution-free* are commonly used interchangeably, but they do not both define exactly the same set of statistical techniques (Noether, 1984).

from nonnormal populations, in which case properly subjecting them to parametric analysis depends upon the robustness of the test to the extent of nonnormality present.

11 TWO-SAMPLE RANK TESTING

Several nonparametric procedures, with various characteristics and assumptions, have been proposed for testing differences between the dispersions, or variabilities, of two populations (e.g., see Hettmansperger and McKean, 1998: 118–127; Hollander and Wolfe, 1999: 141–188; Sprent and Smeeton, 2001: 175–185). A far more common desire for nonparametric testing is to compare two populations' central tendencies (i.e., locations on the measurement scale) when underlying assumptions of the t test are not met. The most frequently employed such test is that originally proposed, for equal sample sizes, by Wilcoxon (1945)* and independently presented by Mann and Whitney (1947), for equal or unequal n's. It is called the Wilcoxon-Mann-Whitney test or, more commonly, the Mann-Whitney test.

(a) The Mann-Whitney Test.   For this test, as for many other nonparametric procedures, the actual measurements are not employed, but we use instead the ranks of the measurements. The data may be ranked either from the highest to lowest or from the lowest to the highest values. Example 11 ranks the measurements from highest to lowest: The greatest height in either of the two groups is given rank 1, the second greatest height is assigned rank 2, and so on, with the shortest height being assigned rank N, where

$$N = n_1 + n_2. \tag{45}$$

A Mann-Whitney statistic is then calculated as

$$U = n_1 n_2 + \frac{n_1(n_1 + 1)}{2} - R_1, \tag{46}$$

where n_a and n_2 are the number of observations in samples 1 and 2, respectively, and R_1 is the sum of the ranks in sample 1. The Mann-Whitney statistic can also be calculated as

$$U' = n_2 n_1 + \frac{n_2(n_2 + 1)}{2} - R_2 \tag{47}$$

(where R_2 is the sum of the ranks of the observations in sample 2), because the labeling of the two samples as 1 and 2 is arbitrary.[†] If Equation 46 has been used to calculate U, then U' can be obtained quickly as

$$U' = n_1 n_2 - U; \tag{48}$$

---

*Wilcoxon may have proposed this test primarily to avoid the drudgery of performing numerous t tests in a time before ubiquitous computer availability (Noether, 1984). Kruskal (1957) gives additional history, including identification of seven independent developments of the procedure Wilcoxon introduced, two of them prior to Wilcoxon, the earliest being by the German psychologist Gustav Deuchler in 1914.

[†]The Wilcoxon two-sample test (sometimes referred to as the Wilcoxon rank-sum test) uses a test statistic commonly called W, which is R_1 or R_2; the test is equivalent to the Mann-Whitney test, for $U = R_2 - n_2(n_2 + 1)/2$ and $U' = R_1 - n_1(n_1 + 1)/2$. U (or U') is also equal to the number of data in one sample that are exceeded by each datum in the other sample. Note in Example 11: For females, ranks 7 and 8 each exceed 6 male ranks and ranks 10, 11, and 12 each exceed all 7 males ranks, for a total of $6 + 6 + 7 + 7 + 7 = 33 = U$; for males, rank 9 exceeds 2 female ranks for a total of $2 = U'$.

EXAMPLE 11   The Mann-Whitney Test for Nonparametric Testing of the Two-Tailed Null Hypothesis That There Is No Difference Between the Heights of Male and Female Students

H_0: Male and female students are the same height.

H_A: Male and female students are not the same height.

$\alpha = 0.05$

Heights of males	Heights of females	Ranks of male heights	Ranks of female heights
193 cm	178 cm	1	6
188	173	2	8
185	168	3	10
183	165	4	11
180	163	5	12
175		7	
170		9	
$n_1 = 7$	$n_2 = 5$	$R_1 = 31$	$R_2 = 47$

$$U = n_1 n_2 + \frac{n_1(n_1 + 1)}{2} - R_1$$

$$= (7)(5) + \frac{(7)(8)}{2} - 31$$

$$= 35 + 28 - 31$$

$$= 32$$

$$U' = n_1 n_2 - U$$

$$= (7)(5) - 32$$

$$= 3$$

$U_{0.05(2),7,5} = U_{0.05(2),5,7} = 30$

As $32 > 30$, H_0 is rejected.

$0.01 < P(U \geq 32 \text{ or } U' \leq 3) < 0.02 \quad [P = 0.018]*$

Therefore, we conclude that height is different for male and female students.

and if Equation 47 has been used to compute U', then U can be ascertained as

$$U = n_1 n_2 - U'. \tag{49}$$

---

*In many of the examples in this text, the exact probability of a statistic from a nonparametric test (such as U) will be given within brackets. In some cases, this probability is obtainable from published sources (e.g., Owen, 1962). It may also be given by computer software, in which case there are two cautions: The computer result may not be accurate to the number of decimal places given, and the computer may have used an approximation (such as the normal approximation in the case of U; see Section 11d), which may result in a probability departing substantially from the exact probability, especially if the sample sizes are small.

For the two-tailed hypotheses, H_0: male and female students are the same height and H_A: male and female students are not the same height, the calculated U or U' — whichever is larger — is compared with the two-tailed value of $U_{\alpha(2),n_1,n_2}$ found in Table 11 from *Appendix: Statistical Tables and Graphs*. This table is set up assuming $n_1 \leq n_2$, so if $n_1 > n_2$, simply use $U_{\alpha(2),n_2,n_1}$ as the critical value. If either U or U' is as great as or greater than the critical value, H_0 is rejected at the α level of significance. A large U or U' will result when a preponderance of the large ranks occurs in one of the samples. As shown in Example 11, neither parameters nor parameter estimates are employed in the statistical hypotheses or in the calculations of U and U'.

The values of U in the table are those for probabilities less than or equal to the column headings. Therefore, the U of 32 in Example 11 is seen to have a probability of $0.01 < P \leq 0.02$. If the calculated U would have been 31, its probability would have been expressed as $0.02 < P < 0.05$.

We may assign ranks either from large to small data (as in Example 11), or from small to large, calling the smallest datum rank 1, the next largest rank 2, and so on. The value of U obtained using one ranking procedure will be the same as the value of U' using the other procedure. In a two-tailed test both U and U' are employed, so it makes no difference from which direction the ranks are assigned.

In summary, we note that after ranking the combined data of the two samples, we calculate U and U' using either Equations 46 and 48, which requires the determination of R_1, or Equations 47 and 49, which requires R_2. That is, the sum of the ranks for only one of the samples is needed. However, we may wish to compute both R_1 and R_2 in order to perform the following check on the assignment of ranks (which is especially desirable in the somewhat more complex case of assigning ranks to tied data, as will be shown below):

$$R_1 + R_2 = \frac{N(N + 1)}{2}. \tag{50}$$

Thus, in Example 11,

$$R_1 + R_2 = 30 + 48 = 78$$

should equal

$$\frac{N(N + 1)}{2} = \frac{12(12 + 1)}{2} = 78.$$

This provides a check on (although it does not guarantee the accuracy of) the assignment of ranks.

Note that hypotheses for the Mann-Whitney test are not statements about parameters (e.g., means or medians) of the two populations. Instead, they address the more general, less specific question of whether the two population distributions of data are the same. Basically, the question asked is whether it is likely that the two samples came at random from the two populations described in the null hypothesis. If samples at least that different would occur with a probability that is small (i.e., less than the significance level, such as 0.05), then H_0 is rejected.

The Mann-Whitney procedure serves to test for difference between medians under certain circumstances (such as when the two sampled populations have symmetrical distributions), but in general it addresses the less specific hypothesis of similarity between the two populations' distributions.

(b) The Mann-Whitney Test with Tied Ranks. Example 12 demonstrates an important consideration encountered in tests requiring the ranking of observations. When two or more observations have exactly the same value, they are said to be *tied*. The rank assigned to each of the tied ranks is the mean of the ranks that would have been assigned to these ranks had they not been tied.* For example, in the present set of data, which are ranked from low to high, the third and fourth lowest values are tied at 32 words per minute, so they are each assigned the rank of $(3 + 4)/2 = 3.5$. The eighth, ninth, and tenth observations are tied at 44 words per minute, so each of them receives the rank of $(8 + 9 + 10)/3 = 9$. Once the ranks have been assigned by this procedure, U and U' are calculated as previously described.

(c) The One-Tailed Mann-Whitney Test. For one-tailed hypotheses we need to declare which tail of the Mann-Whitney distribution is of interest, as this will determine whether U or U' is the appropriate test statistic. This consideration is presented in Table 2. In Example 12 we have data that were ranked from lowest to highest and the alternate hypothesis states that the data in group 1 are greater in magnitude than those in group 2. Therefore, we need to compute U' and compare it to the one-tailed critical value, $U_{\alpha(1),n_1,n_2}$, from Table 11 from *Appendix: Statistical Tables and Graphs*.

TABLE 2: The Appropriate Test Statistic for the One-Tailed Mann-Whitney Test

	H_0: Group 1 $\geq$ Group 2 H_A: Group 1 $<$ Group 2	H_0: Group 1 $\leq$ Group 2 H_A: Group 1 $>$ Group 2
Ranking done from low to high	U	U'
Ranking done from high to low	U'	U

(d) The Normal Approximation to the Mann-Whitney Test. Note that Table 11 from *Appendix: Statistical Tables and Graphs* can be used only if the size of the smaller sample does not exceed twenty and the size of the larger sample does not exceed forty. Fortunately, the distribution of U approaches the normal distribution for larger samples. For large n_1 and n_2 we use the fact that the U distribution has a mean of

$$\mu_U = \frac{n_1 n_2}{2}, \tag{51}$$

which may be calculated, equivalently, as

$$\mu_U = \frac{U + U'}{2}, \tag{51a}$$

and a standard error of

$$\sigma_U = \sqrt{\frac{n_1 n_2 (N + 1)}{12}}, \tag{52}$$

---

*Although other procedures have been proposed to deal with ties, assigning the rank mean has predominated for a long time (e.g., Kendall, 1945).

EXAMPLE 12   The One-Tailed Mann-Whitney Test Used to Determine the Effectiveness of High School Training on the Typing Speed of College Students. This Example Also Demonstrates the Assignment of Ranks to Tied Data

H_0: Typing speed is not greater in college students having had high school typing training.

H_A: Typing speed is greater in college students having had high school typing training.

$\alpha = 0.05$

Typing Speed (words per minute)	
With training (rank in parentheses)	*Without training* (rank in parentheses)
44 (9)	32 (3.5)
48 (12)	40 (7)
36 (6)	44 (9)
32 (3.5)	44 (9)
51 (13)	34 (5)
45 (11)	30 (2)
54 (14)	26 (1)
56 (15)	
$n_1 = 8$	$n_2 = 7$
$R_1 = 83.5$	$R_2 = 36.5$

Because ranking was done from low to high and the alternate hypothesis states that the data of group one are larger than the data of group two, use U' as the test statistic (as indicated in Table 2).

$$U' = n_2 n_1 + \frac{n_2(n_2 + 1)}{2} - R_2$$

$$= (7)(8) + \frac{(7)(8)}{2} - 36.5$$

$$= 56 + 28 - 36.5$$

$$= 47.5$$

$$U_{0.05(1),8,7} = U_{0.05(1),7,8} = 43$$

As $47.5 > 43$, reject H_0.

$$0.01 < P < 0.025 \quad [P = 0.012]$$

Consequently, it is concluded that college-student typing speed is greater for students who had typing training in high school.

where $N = n_1 + n_2$, as used earlier. Thus, if a U, or a U', is calculated from data where either n_1 or n_2 is greater than that in Table 11 from *Appendix: Statistical Tables and Graphs*, its significance can be determined by computing

$$Z = \frac{U - \mu_U}{\sigma_U} \tag{53}$$

or, using a correction for continuity, by

$$Z_c = \frac{|U - \mu_U| - 0.5}{\sigma_U}. \tag{54}$$

The continuity correction is included to account for the fact that Z is a continuous distribution, but U is a discrete distribution. However, it appears to be advisable only if the two-tailed P is about 0.05 or greater (as seen from an expansion of the presentation of Lehmann, 1975: 17).

Recalling that the t distribution with $v = \infty$ is identical to the normal distribution, the critical value, Z_α, is equal to the critical value, $t_{\alpha,\infty}$. The normal approximation is demonstrated in Example 13. When using the normal approximation for two-tailed testing, only U *or* U' (not both) need be calculated. If U' is computed instead of U, then U' is simply substituted for U in Equation 53 or 54, the rest of the testing procedure remaining the same.

EXAMPLE 13   The Normal Approximation to a One-Tailed Mann-Whitney Test to Determine Whether Animals Raised on a Dietary Supplement Reach a Greater Body Weight Than Those Raised on an Unsupplemented Diet

In the experiment, 22 animals (group 1) were raised on the supplemented diet, and 46 were raised on the unsupplemented diet (group 2). The body weights were ranked from 1 (for the smallest weight) to 68 (for the largest weight), and U was calculated to be 282.

H_0: Body weight of animals on the supplemented diet are not greater than those on the unsupplemented diet.

H_A: Body weight of animals on the supplemented diet are greater than those on the unsupplemented diet.

$n_1 = 22, n_2 = 46, N = 68$

$U = 282$

$$U' = n_1 n_2 - U = (22)(46) - 282 = 1012 - 282 = 730$$

$$\mu_U = \frac{n_1 n_2}{2} = \frac{(22)(46)}{2} = 506$$

$$\sigma_U = \sqrt{\frac{n_1 n_2 (N + 1)}{12}} = \sqrt{\frac{(22)(46)(68 + 1)}{12}} = 76.28$$

$$Z = \frac{U' - \mu_U}{\sigma_U} = \frac{224}{76.28} = 2.94$$

For a one-tailed test at $\alpha = 0.05, t_{0.05(1),\infty} = Z_{0.05(1)} = 1.6449$.

As $Z = 2.94 > 1.6449$, reject H_0.   [$P = 0.0016$]

So we conclude that the supplemental diet results in greater body weight.

One-tailed testing may also be performed using the normal approximation. Here one computes either U or U', in accordance with Table 2, and uses it in either Equation 55 or 56, respectively, inserting the correction term (-0.5) if P is about

0.025 or greater:

$$Z_c = \frac{U - \mu_U - 0.5}{\sigma_U}, \text{ if } U \text{ is used, or} \tag{55}$$

$$Z_c = \frac{U' - \mu_U - 0.5}{\sigma_U}, \text{ if } U' \text{ is used.} \tag{56}$$

The resultant Z_c is then compared to the one-tailed critical value, $Z_{\alpha(1)}$, or, equivalently, $t_{\alpha(1),\infty}$; and if $Z \geq$ the critical value, then H_0 is rejected.[*]

If tied ranks exist and the normal approximation is utilized, the computations are slightly modified as follows. One should calculate the quantity

$$\sum t = \sum (t_i^3 - t_i), \tag{57}$$

where t_i is the number of ties in a group of tied values, and the summation is performed over all groups of ties. Then,

$$\sigma_U = \sqrt{\frac{n_1 n_2}{N^2 - N} \cdot \frac{N^3 - N - \sum t}{12}}, \tag{58}$$

and this value is used in place of that from Equation 52. (The computation of $\sum t$ is demonstrated, in a similar context.)

The normal approximation is best for $\alpha(2) = 0.10$ or 0.05 [or for $\alpha(1) = 0.05$ or 0.025] and is also good for $\alpha(2) = 0.20$ or 0.02 [or for $\alpha(1) = 0.10$ or 0.01], with the approximation improving as sample sizes increase; for more extreme significance levels it is not as reliable, especially if n_1 and n_2 are dissimilar. Fahoome (2002) determined that the normal approximation (Equation 53) performed well at the two-tailed 0.05 level of significance (i.e., the probability of a Type I error was between 0.045 and 0.055) for sample sizes as small as 15, and at $\alpha(2) = 0.01$ (for P(Type I error) between 0.009 and 0.011) for n_1 and n_2 of at least 29. Indeed, in many cases with even smaller sample sizes, the normal approximation also yields Type I error probabilities very close to the exact probabilities of U obtained from specialized computer software (especially if there are few or no ties).[†] Further observations on the accuracy of this approximation are given at the end of Table 11 from *Appendix: Statistical Tables and Graphs*.

Buckle, Kraft, and van Eeden (1969) propose another distribution, which they refer to as the "uniform approximation." They show it to be more accurate for $n_1 \neq n_2$, especially when the difference between n_1 and n_2 is great, and especially for small α.

Fix and Hodges (1955) describe an approximation to the Mann-Whitney distribution that is much more accurate than the normal approximation but requires very involved computation. Hodges, Ramsey, and Wechsler (1990) presented a simpler method for a modified normal approximation that provides very good results for probabilities of about 0.001 or greater. Also, the two-sample t test may be applied to the ranks of the data (what is known as using the *rank transformation* of the data), with the probability of the resultant t approaching the exact probability for very large n. But these procedures do not appear to be generally preferable to the normal approximation described above, at least for the probabilities most often of interest.

---

[*]By this procedure, Z must be positive in order to reject H_0. If it is negative, then the probability of H_0 being true is $P > 0.50$.

[†]As a demonstration of this, in Example 11 the exact probability is 0.018 and the probability by the normal approximation is 0.019; and for Example 12, the exact probability and the normal approximation are both 0.012. In Exercise 12, P for U is 0.53 and P for Z is 0.52; and in Exercise 13, P for U is 0.41 and P for Z_c is 0.41.

(e) The Mann-Whitney Test with Ordinal Data. The Mann-Whitney test may also be used for ordinal data. Example 14 demonstrates this procedure. In this example, 25 undergraduate students were enrolled in an invertebrate zoology course. Each student was guided through the course by one of two teaching assistants, but the same examinations and grading criteria were applied to all students. On the basis of the students' final grades in the course, we wish to test the null hypothesis that students (students in general, not just these 25) perform equally well under both teaching assistants. The variable measured (i.e., the final grade) results in ordinal data, and the hypothesis is amenable to examination by the Mann-Whitney test.

(f) Mann-Whitney Hypotheses Employing a Specified Difference Other Than Zero. Using the two-sample t test, one can examine hypotheses such as $H_0: \mu_1 - \mu_2 = \mu_0$, where μ_0 is not zero. Similarly, the Mann-Whitney test can be applied to hypotheses such as H_0: males are at least 5 cm taller than females (a one-tailed hypothesis with data such as those in Example 11) or H_0: the letter grades of students in one course are at least one grade higher that those of students in a second course (a one-tailed hypothesis with data such as those in Example 14). In the first hypothesis, one would list all the male heights but list all the female heights after increasing each of them by 5 cm. Then these listed heights would be ranked and the Mann-Whitney analysis would proceed as usual. For testing the second hypothesis, the letter grades for the students in the first course would be listed unchanged, with the grades for the second course increased by one letter grade before listing. Then all the listed grades would be ranked and subjected to the Mann-Whitney test.*

When dealing with ratio- or interval-scale data, it is also possible to propose hypotheses employing a multiplication, rather than an addition, constant. Consider the two-tailed hypothesis H_0: the wings of one species of insect are two times the length of the wings of a second species. We could test this by listing the wing lengths of the first species, listing the wing lengths of the second species after multiplying each length by two, and then ranking the members of the combined two lists and subjecting the ranks to the Mann-Whitney test. The parametric t testing procedure, which assumes equal population variance, ordinarily would be inapplicable for such a hypothesis, because multiplying the data by a constant changes the variance of the data by the square of the constant.

(g) Violations of the Mann-Whitney Test Assumptions. If the underlying assumptions of the parametric analog of a nonparametric test are met, then either procedure may be employed but the parametric test will be the more powerful. The Mann-Whitney test is one of the most powerful of nonparametric tests. When the t-test assumptions are met, the power of the Mann-Whitney test approaches 95.5% (i.e., $3/\pi$) of the power of the t test as sample size increases (Mood, 1954).[†] And

---

*To increase these grades by one letter each, a grade of "B" would be changed to an "A," a "C" changed to a "B," and so on; a grade of "A" would have to be increased to a grade not on the original scale (e.g., call it a "Z") and, when ranking, we simply have to keep in mind that this new grade is higher than an "A."

[†]Mood (1954) credits an earlier statement of this to 1948 lecture notes of E. J. G. Pitman and to a 1950 Dutch publication by H. R. Van der Vaart. The statement that statistical test A is 0.955 as powerful as test B means that the power of test A with sample size of n tends (as n increases) toward having the same power as test B with sample size of $0.955n$; and this is referred to as the *asymptotic relative efficiency* (ARE) of test A compared to test B. Because of its development by Australian statistician Edwin James George Pitman (1897–1993), ARE is often called *Pitman efficiency*, which distinguishes it from a less commonly encountered definition of asymptotic relative

EXAMPLE 14   The Mann-Whitney Test for Ordinal Data

H_0: The performance of students is the same under the two teaching assistants.
H_A: Students do not perform equally well under the two teaching assistants.

$\alpha = 0.05$

Teaching Assistant A		Teaching Assistant B	
Grade	Rank of grade	Grade	Rank of grade
A	3	A	3
A	3	A	3
A	3	B+	7.5
A−	6	B+	7.5
B	10	B	10
B	10	B−	12
C+	13.5	C	16.5
C+	13.5	C	16.5
C	16.5	C−	19.5
C	16.5	D	22.5
C−	19.5	D	22.5
		D	22.5
		D	22.5
		D−	25

$$n_1 = 11 \qquad\qquad n_2 = 14$$
$$R_1 = 114.5 \qquad\qquad R_2 = 210.5$$

$$U = n_1 n_2 + \frac{n_1(n_1 + 1)}{2} - R_1$$

$$= (11)(14) + \frac{(11)(12)}{2} - 114.5$$

$$= 154 + 66 - 114.5$$

$$= 105.5$$

$$U' = n_1 n_2 - U$$

$$= (11)(14) - 105.5$$

$$= 48.5$$

$$U_{0.05(2),11,14} = 114$$

As $105.5 < 114$, do not reject H_0.

$$0.10 < P(U \geq 105.5 \text{ or } U \leq 48.5) < 0.20$$

Thus, the conclusion is that student performance is the same under both teaching assistants.

---

efficiency by Bahadur (1967; Blair and Higgins, 1985). Although Pitman efficiency is defined in terms of very large n, it is generally a good expression of relative efficiency of two tests even with small n (Conover, 1999: 112).

for some extremely nonnormal distributions, the Mann-Whitney test is immensely more powerful (Blair and Higgins, 1980a, 1980b; Blair, Higgins, and Smitley, 1980; Hodges and Lehman, 1956). The power of the Mann-Whitney test will never be less than 86.4% of the power of the t test (Conover, 1997: 297; Hodges and Lehman, 1956).

The Mann-Whitney test does not assume normality of the sampled populations as the t test does, but the calculated U is affected not only by the difference between the locations of the two populations along the measurement scale but also by difference between the shapes or dispersons of the two populations' distributions (Boneau, 1962). However, the test is typically employed with the desire to conclude only whether there are differences between measurement locations, in which case it must be assumed that the two sampled populations have the same dispersion and shape, a premise that is often ignored, probably in the belief that the test is more robust to unequal dispersion than is the t test. But the Mann-Whitney test is, indeed, adversely affected by sizable differences in the variances or the shapes of the sampled populations, in that the probability of a Type I error is not the specified α (Fligner and Policello, 1981).* As with the two-sample t test, if the two sample sizes are not equal, and if the larger σ^2 is associated with the larger sample, then the probability of a Type I error will be less than α (and the test is called *conservative*); and if the smaller sample came from the population with the larger variance, then this probability will be greater than α (and the test is called *liberal*) (Zimmerman, 1987). The greater the difference between the variances, the greater the departure from α. In situations where the Mann-Whitney test is conservative, it has more power than the t test (Zimmerman, 1987). The power of the Mann-Whitney test may also be decreased, especially in the presence of outliers, to an extent to which the variances differ; but this decrease is far less than it is with t testing (Zimmerman, 1994, 1996, 1998, 2000). But in some cases unequal variances affect the probability of a Type I error using U more severely than if t or t' were employed (Zimmerman, 1998).

The Mann-Whitney test is included in the guidelines (described in Section 1d) for when various two-sample statistical procedures are appropriate.

12 TESTING FOR DIFFERENCE BETWEEN TWO MEDIANS

The null hypothesis that two samples came from populations having the same median can be tested by the *median test* described by Mood (1950: 394–395). The procedure is to determine the grand median for all the data in both samples and then to tabulate the numbers of data above and below the grand median in a 2×2 contingency table, as shown in Example 15. This contingency table can then be analyzed by the chi-square test or G test.

Example 15 demonstrates the median test for the data of Example 14. In many cases, such as this one, one or more of the data will be equal to the grand median (in this instance a grade of C+) and, therefore, the number of data above

---

*Fligner and Policello (1981; Hollander and Wolfe, 1999: 135–139) addressed situations where the sampled populations have dissimilar variances (the "Behrens-Fisher problem" discussed in Section 1c), in addition to being nonnormal. They presented a modified Mann-Whitney procedure, requiring that the underlying distributions be symmetrical, along with tables of critical values for use with sample sizes ≤ 12 and with a normal approximation good when the n's are much larger than 12.

EXAMPLE 15   The Two-Sample Median Test, Using the Data of Example 14

H_0: The two samples came from populations with identical medians (i.e., the median performance is the same under the two teaching assistants).

H_A: The medians of the two sampled populations are not equal.

$\alpha = 0.05$

The median of all 25 measurements in Example 14 is $X_{(25+1)2} = X_{13}$ = grade of C+. The following 2×2 contingency table is then produced:

Number	Sample 1	Sample 2	Total
Above median	6	6	12
Not above median	3	8	11
Total	9	14	23

Analyzing this contingency table:

$$X_c^2 = \frac{n\left(|f_{11}f_{22} - f_{12}f_{21}| - \dfrac{n}{2}\right)^2}{(C_1)(C_2)(R_1)(R_2)} \tag{59}$$

$$= 0.473.$$

$$X_{0.05,1}^2 = 3.841$$

Therefore, do not reject H_0.

$$0.25 < P < 0.50 \quad [P = 0.49]$$

So it is concluded that the two samples did not come from populations with different medians.

and below the median will be less than the number of original data. Some authors and computer programs have preferred to tabulate the row categories as "above median" and "not above median" (that is, "at or below the median") instead of "above median" and "below median." This will retain in the analysis the original number of data, but it does not test the median-comparison hypothesis as well, and it can produce conclusions very different from those resulting from analyzing the same data categorized as "below median" and "not below median" (i.e., "at or above median"). Others have suggested deleting from the analysis any data that are tied at the grand median. This, too, will give results that may be quite different from the other procedures. If there are many data at the grand median, a good option is to place all data in a contingency table with three, instead of two, rows: "above median," "at median," and "below median."

The median test is about 64% as powerful as the two-sample t test when used on data to which the latter is applicable (Mood, 1954), and about 67% as powerful as the Mann-Whitney test of the preceding section.*

If the two sampled populations have equal variances and shapes, then the Mann-Whitney test (Section 11) is a test for difference between medians (Fligner and Policello, 1981).

One can also test whether the difference between two population medians is of a specified magnitude. This would be done in a fashion similar to that indicated in Section 11f for the Mann-Whitney test. For example, to hypothesize that the median of population 1 is X units greater than the median of population 2, X would be added to each datum in sample 2 (or X would be subtracted from each datum in sample 1) prior to performing the median test.

13 TWO-SAMPLE TESTING OF NOMINAL-SCALE DATA

We may compare two samples of nominal data simply by arranging the data in a $2 \times C$ contingency table.

14 TESTING FOR DIFFERENCE BETWEEN TWO DIVERSITY INDICES

If the Shannon index of diversity, H', is obtained for each of two samples, it may be desired to test the null hypothesis that the diversities of the two sampled populations are equal. Hutcheson (1970) proposed a t test for this purpose:

$$t = \frac{H'_1 - H'_2}{s_{H'_1 - H'_2}}, \tag{60}$$

where

$$s_{H'_1 - H'_2} = \sqrt{s^2_{H'_1} + s^2_{H'_2}}. \tag{61}$$

The variance of each H' may be approximated by

$$s^2_{H'} = \frac{\sum f_i \log^2 f_i - (\sum f_i \log f_i)^2/n}{n^2} \tag{62}$$

(Basharin, 1959; Lloyd, Zar, and Karr, 1968),[†] where s, f_1, and n are as defined in Section 7 of *Measures of Variability and Dispersion*, and $\log^2 f$ signifies $(\log f)^2$. Logarithms to any base may be used for this calculation, but those to base 10 are most commonly employed. The degrees of freedom associated with the preceding t are approximated by

$$\nu = \frac{\left(s^2_{H'_1} + s^2_{H'_2}\right)^2}{\frac{\left(s^2_{H'_1}\right)^2}{n_1} + \frac{\left(s^2_{H'_2}\right)^2}{n_2}} \tag{63}$$

(Hutcheson, 1970).

---

*As the median test refers to a population parameter in hypothesis testing, it is not a nonparametric test; but it is a distribution-free procedure. Although it does not assume a specific underlying distribution (e.g., normal), it does assume that the two populations have the same shape (a characteristic that is addressed by Schlittgen, 1979).

[†]Bowman et al. (1971) give an approximation [their Equation (11b)] that is more accurate for very small n.

Example 16 demonstrates these computations. If one is faced with many calcula-tions of $s^2_{H'_2}$, the tables of $f_i \log^2 f_i$ provided by Lloyd, Zar, and Karr (1968) will be helpful. One-tailed as well as two-tailed hypotheses may be tested by this procedure. Also, the population diversity indices may be hypothesized to differ by some value, μ_0, other than zero, in which case the numerator of t would be $|H'_1 - H'_2| - \mu_0$.

EXAMPLE 16   Comparing Two Indices of Diversity

H_0: The diversity of plant food items in the diet of Michigan blue jays is the same as the diversity of plant food items in the diet of Louisiana blue jays.

H_A: The diversity of plant food items in the diet of Michigan blue jays is not the same as in the diet of Louisiana blue jays.

$\alpha = 0.05$

Michigan Blue Jays

Diet item	f_i	$f_i \log f_i$	$f_i \log^2 f_i$
Oak	47	78.5886	131.4078
Corn	35	54.0424	83.4452
Blackberry	7	5.9157	4.9994
Beech	5	3.4949	2.4429
Cherry	3	1.4314	0.6830
Other	2	0.6021	0.1812

$$s_1 = 6 \qquad n_1 = \sum f_i = 99 \qquad \sum f_i \log f_i = 144.0751 \qquad \sum f_i \log^2 f_i = 223.1595$$

$$H'_1 = \frac{n \log n - \sum f_i \log f_i}{n} = \frac{197.5679 - 144.0751}{99}$$

$$= 0.5403$$

$$s^2_{H'_1} = \frac{\sum f_i \log^2 f_i - \left(\sum f_i \log f_i\right)^2 / n}{n^2} = 0.00137602$$

Louisiana Blue Jays

Diet item	f_i	$f_i \log f_i$	$f_i \log^2 f_i$
Oak	48	80.6996	135.6755
Pine	23	31.3197	42.6489
Grape	11	11.4553	11.9294
Corn	13	14.4813	16.1313
Blueberry	8	7.2247	6.5246
Other	2	0.6021	0.1812

$$s_2 = 6 \qquad n_2 = \sum f_i = 105 \qquad \sum f_i \log f_i = 145.7827 \qquad \sum f_i \log^2 f_i = 213.0909$$

$$H_2' = \frac{n \log n - \sum f_i \log f_i}{n} = \frac{212.2249 - 145.7827}{105} = 0.6328$$

$$s_{H_2'}^2 = \frac{\sum f_i \log^2 f_i - \left(\sum f_i \log f_i\right)^2 / n}{n^2} = 0.00096918$$

$$s_{H_1' - H_2'} = \sqrt{s_{H_1'}^2 + s_{H_2'}^2} = \sqrt{0.00137602 + 0.00096918} = 0.0484$$

$$t = \frac{H_1' - H_2'}{s_{H_1' - H_2'}} = \frac{-0.0925}{0.0484} = -1.911$$

$$\nu = \frac{\left(s_{H_1'}^2 + s_{H_2'}^2\right)^2}{\dfrac{\left(s_{H_1'}^2\right)^2}{n_1} + \dfrac{\left(s_{H_2'}^2\right)^2}{n_2}} = \frac{(0.00137602 + 0.00096918)^2}{\dfrac{(0.00137602)^2}{99} + \dfrac{(0.00096918)^2}{105}}$$

$$= \frac{0.000005499963}{0.000000028071} = 196$$

$$t_{0.05(2),196} = 1.972$$

Therefore, do not reject H_0.

$$0.05 < P < 0.10 \quad [P = 0.057]$$

The conclusion is that the diversity of food items is the same in birds from Michigan and Louisiana.

15 CODING DATA

Coding raw data can sometimes simplify computations. Coding will affect the sample statistics of this chapter (i.e., measures of central tendency and of variability, and their confidence limits). The test statistics and hypothesis-test conclusions will not be altered by coding, except that coding may not be used in performing the Levene test (Section 5b). Neither may coding be used in testing for difference between two coefficients of variation (Section 8), except that it is permissible if using the F test and coding by addition (or subtraction, but not multiplication or division). There is no effect of coding on the Mann-Whitney test (Section 11) or median test (Section 12). And, for testing difference between two diversity indices (Section 14), coding by multiplication (or division, but not addition or subtraction) may be employed.

Coding affects the sample statistics (and their confidence limits). Coding may be employed for any of the hypothesis tests, except that only coding by multiplication (or division, but not addition or subtraction) may be used for testing or coefficients of variation.

EXERCISES

1. Using the following data, test the null hypothesis that male and female turtles have the same mean serum cholesterol concentrations.

Serum Cholesterol (mg/100 ml)

Male	Female
220.1	223.4
218.6	221.5
229.6	230.2
228.8	224.3
222.0	223.8
224.1	230.8
226.5	

2. It is proposed that animals with a northerly distribution have shorter appendages than animals from a southerly distribution. Test an appropriate hypothesis (by computing t), using the following wing-length data for birds (data are in millimeters).

Northern	Southern
120	116
113	117
125	121
118	114
116	116
114	118
119	123
	120

3. Two populations of animal body weights are randomly sampled, and $\overline{X}_1 = 4.6$ kg, $s_1^2 = 11.02$ kg^2, $n_1 = 18$, $\overline{X}_2 = 6.0$ kg, $s_2^2 = 4.35$ kg^2, and $n_2 = 26$. Test the hypotheses $H_0: \mu_1 \geq \mu_2$ and $H_A: \mu_1 < \mu_2$ using the Behrens-Fisher test.

4. If $\overline{X}_1 = 334.6$ g, $\overline{X}_2 = 349.8$ g, $SS_1 = 364.34$ g^2, $SS_2 = 286.78$ g^2, $n_1 = 19$, and $n_2 = 24$, test the hypothesis that the mean weight of population 2 is more than 10 g greater than the mean weight of population 1.

5. For the data of Exercise 1:

(a) If the null hypothesis is rejected, compute the 95% confidence limits for μ_1, μ_2, and $\mu_1 - \mu_2$. If H_0 is not rejected, compute the 95% confidence limits for the common population mean, μ_p.

(b) Calculate the 95% prediction interval for the difference between the mean of an additional 25 data from the male population and an additional 20 data from the female population.

6. A sample is to be taken from each of two populations from which previous samples of size 14 have had $SS_1 = 244.66$ (km/hr)^2 and $SS_2 = 289.18$ (km/hr)^2. What size sample should be taken from each population in order to estimate $\mu_1 - \mu_2$ to within 2.0 km/hr, with 95% confidence?

7. Consider the populations described in Exercise 6.

(a) How large a sample should we take from each population if we wish to detect a difference between μ_1 and μ_2 of at least 5.0 km/hr, using a 5% significance level and a t test with 90% power?

(b) If we take a sample of 20 from one population and 22 from the other, what is the smallest difference between μ_1 and μ_2 that we have a 90% probability of detecting with a t test using $\alpha = 0.05$?

(c) If $n_1 = n_2 = 50$, and $\alpha = 0.05$, what is the probability of rejecting $H_0: \mu_1 = \mu_2$ when $\mu_1 - \mu_2$ is as small as 2.0 km/hr?

8. The experimental data of Exercise 1 might have been collected to determine whether serum cholesterol concentrations varied as much in male turtles as in female turtles. With those data, use the variance-ratio test to assess $H_0: \sigma_1^2 = \sigma_2^2$ versus $\sigma_1^2 \neq \sigma_2^2$.

9. Let us propose that wings of a particular bird species vary in length more in the northern part of the species' range than in the southern portion. Use the variance ratio test for $H_0: \sigma_1 \leq \sigma_2$ versus $H_A: \sigma_1 > \sigma_2$ with the data of Exercise 2.

10. A sample of 21 data from one population has a variance of 38.71 g^2, and a sample of 20 data from a second population has a variance of 21.35 g^2.

(a) Calculate the 95% two-tailed confidence interval for the ratio of σ_1^2/σ_2^2.

(b) How large a sample must be taken from each population if we wish to have a 90% chance of rejecting $H_0: \sigma_1^2 \leq \sigma_2^2$ when $H_A: \sigma_1^2 > \sigma_2^2$ is true and we apply the variance-ratio test at the 5% level of significance?

(c) What would be the power of a variance-ratio test of this H_0, with $\alpha = 0.05$, if sample sizes of 20 were used?

11. A sample of twenty-nine plant heights of members of a certain species had $\overline{X}_1 = 10.74$ cm and $s^2 = 14.62$ cm^2, and the heights of a sample of twenty-five from a second species had $\overline{X}_2 = 14.32$ cm and $s^2 = 8.45$ cm^2. Test the null hypothesis that the coefficients of variation of the two sampled populations are the same.

12. Using the Mann-Whitney test, test the appropriate hypotheses for the data in Exercise 1.

13. Using the Mann-Whitney procedure, test the appropriate hypotheses for the data in Exercise 2.

14. The following data are volumes (in cubic microns) of avian erythrocytes taken from normal (diploid) and intersex (triploid) individuals. Test the hypothesis (using the Mann-Whitney test) that the volume of intersex cells is 1.5 times the volume of normal cells.

Normal	Intersex
248	380
236	391
269	377
254	392
249	398
251	374
260	
245	
239	
255	

ANSWERS TO EXERCISES

1. $H_0: \mu_1 = \mu_2$, $H_A: \mu_1 \neq \mu_2$, $n_1 = 7$, $SS_1 = 108.6171 \ (mg/100 \ ml)^2$, $\overline{X}_1 = 224.24 \ mg/100 \ ml$, $v_1 = 6$, $n_2 = 6$, $SS_2 = 74.7533 \ (mg/100 \ ml)^2$, $\overline{X}_2 = 225.67 \ mg/100 \ ml$, $v_2 = 5$, $s_p^2 = 16.6700 \ (mg/100 \ ml)^2$, $s_{\overline{X}_1 - \overline{X}_2} = 2.27 \ mg/100 \ ml$, $t = -0.630$, $t_{0.05(2), 11} = 2.201$; therefore, do not reject H_0; $P > 0.50 \ [P = 0.54]$.

2. $H_0: \mu_1 \geq \mu_2$, $H_A: \mu_1 < \mu_2$, $n_1 = 7$, $SS_1 = 98.86 \ mm^2$, $v_1 = 6$, $\overline{X}_1 = 117.9 \ mm$, $n_2 = 8$, $SS_2 = 62.88 \ mm^2$, $v_2 = 7$, $\overline{X}_2 = 118.1 \ mm$, $s_p^2 = 12.44 \ mm^2$, $s_{\overline{X}_1 - \overline{X}_2} = 1.82 \ mm$, $t = -0.11$, $t_{0.05(1), 13} = 1.771$; therefore, do not reject H_0; $P > 0.25 \ [P = 0.46]$.

3. $H_0: \mu_1 \geq \mu_2$, $H_0: \mu_1 < \mu_2$, $\overline{X}_1 = 4.6 \ kg$, $s_1^2 = 11.02 \ kg^2$, $n_1 = 18$; $v_1 = 17$, $\overline{X}_2 = 6.0 \ kg^2$, $s_2^2 = 4.35 \ kg^2$, $n_2 = 26$, $v_2 = 25$, $s_{\overline{X}_1 - \overline{X}_2} = 0.88 \ kg$, $t = -1.59$, $t_{0.05(1), 42} = 1.682$; therefore, do not reject H_0, $0.05 < P < 0.10 \ [P = 0.060]$.

4. $H_0: \mu_2 - \mu_1 \leq 10 \ g$, $H_A: \mu_2 - \mu_1 > 10 \ g$, $\overline{X}_1 = 334.6 \ g$, $SS_1 = 364.34 \ g^2$, $n_1 = 19$, $v_1 = 18$, $\overline{X}_2 = 349.8 \ g$, $SS_2 = 286.78 \ g^2$, $n_2 = 24$, $v_2 = 23$, $s_p^2 = 15.88 \ g^2$, $s_{\overline{X}_1 - \overline{X}_2} = 1.22 \ g$, $t = 4.26$, $t_{0.05(1), 41} = 1.683$ therefore, reject H_0 and conclude that μ_2 is at least 10 g greater than μ_1; $P < 0.0005 \ [P = 0.00015]$.

5. **(a)** H_0 is not rejected; $\overline{X}_p = 224.90 \ mg/100 \ ml$; $t_{0.05(2), 11} = 2.201$; $s_p^2 = 16.6700 \ (mg/100 \ ml)^2$; 95% confidence interval $= 22.490 \pm \sqrt{16.5700/13} = 224.90 \pm 1.13$; $L_1 = 223.77 \ mg/100 \ ml$, $L_2 = 226.03 \ mg/100 \ ml$. **(b)** $\overline{X}_1 - \overline{X}_2 = -3.43 \ mg/100 \ ml$; $s_c^2 = 6.6601 \ (mg/100 \ ml)^2$; 95% prediction interval $= -3.43 \pm 2.201\sqrt{6.6601} = -3.43 \pm 5.68$; $L_1 = -9.29 \ mg/100 \ ml$, $L_2 = 2.25 \ mg/100 \ ml$.

6. $s_p^2 = (244.66 + 289.18)/(13 + 13) = 20.53 (km/hr)^2$; $d = 2.0 \ km/hr$. If we guess $n = 50$,

then $v = 2(50 - 1) = 98$, $t_{0.05(2), 98} = 1.984$, and $n = 40.4$. Then, guess $n = 41$; $v = 80$, $t_{0.05(2), 80} = 1.990$, and $n = 40.6$. So, the desired $n = 41$.

7. **(a)** If we guess $n = 25$, then $v = 2(24) = 48$. $t_{0.05(2), 48} = 2.011$, $t_{0.10(1), 48} = 1.299$, and $n = 18.0$. Then, guess $n = 18$; $v = 34$, $t_{0.05(2), 34} = 2.032$, $t_{0.10(1), 34} = 1.307$, and $n = 18.3$. So, the desired sample size is $n = 19$. **(b)** $n = 20.95$, $v = 40$, $t_{0.05(2), 40} = 2.021$, $t_{0.10(1), 40} = 1.303$, and $\delta = 4.65 \ km/hr$. **(c)** $n = 50$, $v = 98$, $t_{0.05(2), 98} = 1.984$, and $t_{\beta(1), 98} = 0.223$; $\beta > 0.25$, so power < 0.75 (or, by the normal approximation, $\beta = 0.41$, so power $= 0.59$).

8. $n_1 = 7$, $v_1 = 6$, $s_1^2 = 18.1029 \ (mg/100 \ ml)^2$, $n_2 = 6$, $v_2 = 5$, $s_2^2 = 14.9507 \ (mg/100 \ ml)^2$; $F = 1.21$, $F_{0.05(2), 6, 5} = 6.98$; do not reject H_0; $P > 0.50 \ [P = 0.85]$.

9. $n_1 = 7$, $v_1 = 6$, $s_1^2 = 16.48 \ mm^2$, $n_2 = 8$, $v_2 = 7$, $s_2^2 = 8.98 \ mm^2$; $F = 1.83$, $F_{0.05(2), 6, 7} = 3.87$; do not reject H_0; $0.10 < P < 0.25 \ [P = 0.22]$.

10. $n_1 = 21$, $v_1 = 20$, $s_1^2 = 38.71 \ g^2$, $n_2 = 20$, $v_2 = 19$, $s_2^2 = 21.35 \ g^2$, $s_1^2/s_2^2 = 1.81$. **(a)** $F_{0.05(2), 20, 19} = 2.51$, $F_{0.05(2), 19, 20} = 2.48$; $L_1 = 0.72 \ g^2$, $L_2 = 4.49 \ g^2$. **(b)** $Z_{0.05(1)} = 1.6449$, $Z_{0.10(1)} = 1.2816$, $n = 26.3$ (so a sample of at least 27 should be taken from each population). **(c)** $Z_{\beta(1)} = 0.87$, $\beta(1) = 0.19$, power $= 0.81$.

11. $H_0: \sigma_1/\mu_1 = \sigma_2/\mu_2$, $H_A: \sigma_1/\mu_1 \neq \sigma_2/\mu_2$; $s_1 = 3.82 \ cm$, $V_1 = 0.356$, $s_2 = 2.91 \ cm$, $V_2 = 0.203$; $V_p = 0.285$; $Z = 2.533$, $Z_{0.05(2)} = 1.960$; reject H_0; $0.01 < P < 0.02 \ [P = 0.013]$.

12. H_0: Male and female turtles have the same serum cholesterol concentrations; H_A: Male and female turtles do not have the same serum cholesterol concentrations.

Male ranks	Female ranks
2	5
1	3
11	12
10	8
4	6
7	13
9	

$R_1 = 44$, $n_1 = 7$, $R_2 = 47$, $n_2 = 6$; $U = 26$; $U' = (7)(6) - 26 = 16$; $U_{0.05(2),7,6} = U_{0.05(2),6,7} = 36$; therefore, do not reject H_0; $P > 0.20$ $[P = 0.53]$.

13. H_0: Northern birds do not have shorter wings than southern birds; H_A: Northern birds have shorter wings than southern birds.

Northern ranks	Southern ranks
11.5	5
1	7
15	13
8.5	2.5
5	5
2.5	8.5
10	14
	11.5

$R_1 = 53.5$, $n_1 = 7$, $n_2 = 8$; $U = 30.5$; $U' = 25.5$; $U_{0.05(1),7,8} = 43$; therefore, do not reject H_0; $P > 0.10$ $[P \approx 0.41]$.

14. H_0: Intersex cells have 1.5 times the volume of normal cells; H_A: Intersex cells do not have 1.5 times the volume of normal cells.

Normal $\times$ 1.5	Rank	Intersex	Rank
372	4	380	9
354	1	391	13
403.5	16	377	8
381	10	392	14
373.5	5	398	15
376.5	7	374	6
390	12		
367.5	3		
358.5	2		
382.5	11		

$R_1 = 71$, $n_1 = 10$, $n_2 = 6$; $U = 44$, $U' = 16$; $U_{0.05(2),10,6} = 49$; therefore, do not reject H_0; $0.10 < P < 0.20$.

This page intentionally left blank

Paired-Sample Hypotheses

Two-sample testing procedures apply when the two samples are independent, independence implying that each datum in one sample is in no way associated with any specific datum in the other sample. However, there are instances when each observation in Sample 1 is in some way physically associated with an observation in Sample 2, so that the data may be said to occur in pairs.

For example, we might wish to test the null hypothesis that the left foreleg and left hindleg lengths of deer are equal. We could make these two measurements on a number of deer, but we would have to remember that the variation among the data might be owing to two possible factors. First, the null hypothesis might be false, there being, in fact, a difference between foreleg and hindleg length. Second, deer are of different sizes, and for each deer the hindleg length is correlated with the foreleg length (i.e., a deer with a large front leg is likely to have a large hind leg). Thus, as Example 1 shows, the data can be tabulated in pairs, one pair (i.e., one hindleg measurement and one foreleg measurement) per animal.

1   TESTING MEAN DIFFERENCE BETWEEN PAIRED SAMPLES

The two-tailed hypotheses implied by Example 1 are $H_0: \mu_1 - \mu_2 = 0$ and $H_A: \mu_1 - \mu_2 \neq 0$ (which could also be stated $H_0: \mu_1 = \mu_2$ and $H_A: \mu_1 \neq \mu_2$). However, we can define a mean population difference, μ_d, as $\mu_1 - \mu_2$, and write the hypotheses as $H_0: \mu_d = 0$ and $H_A: \mu_d \neq 0$. Although the use of either μ_d or $\mu_1 - \mu_2$ is correct, the former will be used here when it implies the paired-sample situation.

The test statistic for the null hypothesis is

$$t = \frac{\overline{d}}{s_{\overline{d}}}.$$

(1)

Therefore, we do not use the original measurements for the two samples, but only the difference within each pair of measurements. One deals, then, with a sample of d_j values, whose mean is $\overline{d}$ and whose variance, standard deviation, and standard error are denoted as s_d^2, s_d, and $s_{\overline{d}}$ respectively. Thus, the *paired-sample t test*, as this procedure may be called, is essentially a one-sample t test. In the paired-sample

EXAMPLE 1   The Two-Tailed Paired-Sample t Test

H_0: $\mu_d = 0$

H_A: $\mu_d \neq 0$

$\alpha = 0.05$

Deer (j)	Hindleg length (cm) (X_{1j})	Foreleg length (cm) (X_{2j})	Difference (cm) ($d_j = X_{1j} - X_{2j}$)
1	142	138	4
2	140	136	4
3	144	147	−3
4	144	139	5
5	142	143	−1
6	146	141	5
7	149	143	6
8	150	145	5
9	142	136	6
10	148	146	2

$n = 10$ $\overline{d} = 3.3$ cm

$s_d^2 = 9.3444$ cm^2 $s_{\overline{d}} = 0.97$ cm

$\nu = n - 1 = 9$ $t = \dfrac{\overline{d}}{s_{\overline{d}}} = \dfrac{3.3}{0.97} = 3.402$

$t_{0.05(2),9} = 2.262$ Therefore, reject H_0.

$$0.005 < P(|t| \geq 3.402) < 0.01 \quad [P = 0.008]$$

t test, n is the number of differences (i.e., the number of pairs of data), and the degrees of freedom are $\nu = n - 1$. Note that the hypotheses used in Example 1 are special cases of the general hypotheses H_0: $\mu_d = \mu_0$ and H_A: $\mu_d \neq \mu_0$, where μ_0 is usually, but not always, zero.

For one-tailed hypotheses with paired samples, one can test either H_0: $\mu_d \geq \mu_0$ and H_A: $\mu_d < \mu_0$, or H_0: $\mu_d \leq \mu_0$ and H_A: $\mu_d > \mu_0$, depending on the question to be asked. Example 2 presents data from an experiment designed to test whether a new fertilizer results in an increase of more than 250 kg/ha in crop yield over the old fertilizer. For testing this hypothesis, 18 test plots of the crop were set up. It is probably unlikely to find 18 field plots having exactly the same conditions of soil, moisture, wind, and so on, but it should be possible to set up two plots with similar environmental conditions. If so, then the experimenter would be wise to set up nine pairs of plots, applying the new fertilizer randomly to one plot of each pair and the old fertilizer to the other plot of that pair. As Example 2 shows, the statistical hypotheses to be tested are H_0: $\mu_d \leq 250$ kg/ha and H_A: $\mu_d > 250$ kg/ha.

Paired-sample t-testing assumes that each datum in one sample is associated with one, *but only one*, datum in the other sample. So, in the last example, each yield using

---

EXAMPLE 2   A One-Tailed Paired-Sample t Test

H_0: $\mu_d \leq 250$ kg/ha

H_A: $\mu_d > 250$ kg/ha

$\alpha = 0.05$

Plot (j)	Crop Yield (kg/ha) With new fertilizer (X_{1j})	With old fertilizer (X_{2j})	d_j
1	2250	1920	330
2	2410	2020	390
3	2260	2060	200
4	2200	1960	240
5	2360	1960	400
6	2320	2140	180
7	2240	1980	260
8	2300	1940	360
9	2090	1790	300

$n = 9$ $\overline{d} = 295.6$ kg/ha

$s_d^2 = 6502.78$ (kg/ha)^2 $s_{\overline{d}} = 26.9$ kg/ha

$v = n - 1 = 8$ $t = \dfrac{\overline{d} - 250}{s_{\overline{d}}} = 1.695$

$t_{0.05(1),8} = 1.860$ Therefore, do not reject H_0.

$0.05 < P < 0.10$ $[P = 0.064]$

---

new fertilizer is paired with only one yield using old fertilizer; and it would have been inappropriate to have some tracts of land large enough to collect two or more crop yields using each of the fertilizers.

The paired-sample t test does not have the normality and equality of variances assumptions of the two-sample t test, but it does assume that the differences, d_i, come from a normally distributed population of differences. If a nonnormal distribution of differences is doubted, the nonparametric test of Section 5 should be considered.

If there is, in fact, pairwise association of data from the two samples, then analysis by the two-sample t test will often be less powerful than if the paired-sample t test was employed, and the two-sample test will not have a probability of a Type I error equal to the specified significance level, α. It appears that the latter probability will be increasingly less than α for increasingly large correlation between the pairwise data (and, in the less common situation where there is a negative correlation between the data, the probability will be greater than α); and only a small relationship is needed to make the paired-sample test advantageous (Hines, 1996; Pollak and Cohen, 1981; Zimmerman, 1997). If the data from Example 1 were subjected (inappropriately) to the two-sample t test, rather than to the paired-sample t test, a difference would not have been concluded, and a Type II error would have been committed.

2 CONFIDENCE LIMITS FOR THE POPULATION MEAN DIFFERENCE

In paired-sample testing we deal with a sample of differences, d_j. The $1 - \alpha$ confidence interval for μ_d is

$$\overline{d} \pm t_{\alpha(2),\nu} s_{\overline{d}}. \tag{2}$$

For example, for the data in Example 1, we can compute the 95% confidence interval for μ_d to be 3.3 cm $\pm$ (2.262)(0.97 cm) = 3.3 cm $\pm$ 2.2 cm; the 95% confidence limits are $L_1 = 1.1$ cm and $L_2 = 5.5$ cm.

Furthermore, we may ask how large a sample is required to be $1 - \alpha$ confident in estimating μ_d to within $\pm d$ (using the equation $\overline{X} \pm t_{\alpha(2),\nu}\sqrt{\frac{s^2}{m} + \frac{s^2}{n}}$,).

3 POWER, DETECTABLE DIFFERENCE AND SAMPLE SIZE IN PAIRED-SAMPLE TESTING OF MEANS

By considering the paired-sample test to be a one-sample t test for a sample of differences, we acquire estimates of required sample size (n), minimum detectable difference (δ), and power ($1 - \beta$), using Equations A, B, and C, respectively.

$$n = \frac{s^2}{\delta^2}(t_{\alpha,\nu} + t_{\beta(1),\nu})^2, \tag{A}$$

$$\delta = \sqrt{\frac{s^2}{n}}(t_{\alpha,\nu} + t_{\beta(1),\nu}), \tag{B}$$

$$t_{\beta(1),\nu} = \frac{\delta}{\sqrt{\frac{s^2}{n}}} - t_{\alpha,\nu}, \tag{C}$$

4 TESTING FOR DIFFERENCE BETWEEN VARIANCES FROM TWO CORRELATED POPULATIONS

If we wanted to compare the variance of the lengths of deer forelegs with the variance of deer hindlegs, we could measure a sample of foreleg lengths of several deer and a sample of hindleg lengths from a different group of deer. As these are independent samples, the variance of the foreleg sample could be compared to the variance of the hindleg sample. However, just as the paired-sample comparison of means is more powerful than independent-sample comparison of means when the data are paired (i.e., when there is an association between each member of one sample and a member of the other sample), there is a variance-comparison test more powerful if the data are paired (as they are in Example 1). This test takes into account the amount of association between the members of the pairs of data, as presented by Snedecor and Cochran (1989: 192–193) based upon a procedure of Pitman (1939). We compute:

$$t = \frac{(F - 1)\sqrt{n - 2}}{2\sqrt{F(1 - r^2)}}. \tag{3}$$

Here, n is the sample size common to both samples, r is the correlation coefficient, and the degrees of freedom associated with t are $\nu = n - 2$. For a two-tailed test

($H_0: \sigma_1^2 = \sigma_2^2$ vs. $H_A: \sigma_1^2 \neq \sigma_2^2$), either $F = s_1^2/s_2^2$ or $F = s_2^2/s_1^2$ may be used, and H_0 is rejected if $|t| \geq t_{\alpha(2),\nu}$. This is demonstrated in Example 3. For the one-tailed hypotheses, $H_0: \sigma_1^2 \leq \sigma_2^2$ versus $H_A: \sigma_1^2 > \sigma_2^2$, use $F = s_1^2/s_2^2$; for $H_0: \sigma_1^2 \geq \sigma_2^2$ versus $H_A: \sigma_1^2 < \sigma_2^2$, use $F = s_2^2/s_1^2$; and a one-tailed test rejects H_0 if $t \geq t_{\alpha(1),\nu}$.

McCulloch (1987) showed that this t test is adversely affected if the two sampled populations do not have a normal distribution, in that the probability of a Type I error can depart greatly from the stated α. He demonstrated a testing procedure that is very little affected by nonnormality and is only slightly less powerful than t when the underlying populations are normal. It utilizes the differences and the sums of the members of the pairs, just as described in the preceding paragraph; but, instead of the parametric correlation coefficient (r) referred to above, it employs the nonparametric correlation coefficient (r_s). This technique may be used for two-tailed or one-tailed testing.

EXAMPLE 3   Testing for Difference Between the Variances of Two Paired Samples

H_0: $\sigma_1^2 = \sigma_2^2$

H_A: $\sigma_1^2 \neq \sigma_2^2$

$\alpha = 0.05$

Using the paired-sample data of Example 1:

$n = 10$; $\nu = 8$

$\sum x^2 = 104.10$; $\sum y^2 = 146.40$

$\sum xy = 83.20$

$s_1^2 = 11.57 \text{ cm}^2$; $s_2^2 = 16.27 \text{ cm}^2$

$F = 11.57 \text{ cm}^2/16.27 \text{ cm}^2 = 0.7111$

Using Equation 19.1, $r = 0.6739$.

Using Equation 3:

$t = -0.656$ and $t_{0.05(2),8} = 2.306$, so H_0 is not rejected.

$$P > 0.50 \quad [P = 0.54]$$

---

5 PAIRED-SAMPLE TESTING BY RANKS

The *Wilcoxon paired-sample test* (Wilcoxon, 1945; Wilcoxon and Wilcox, 1964: 9) is a nonparametric analogue to the paired-sample t test, just as the Mann-Whitney test is a nonparametric procedure analogous to the two-sample t test. The literature refers to the test by a variety of names, but usually in conjunction with Wilcoxon's name* and some wording such as "paired sample" or "matched pairs," sometimes together with a phrase like "rank sum" or "signed rank."

Whenever the paired-sample t test is applicable, the Wilcoxon paired-sample test is also applicable. The Wilcoxon procedure is a nonparametric one-sample test, but it

---

*Frank Wilcoxon (1892–1965), American (born in Ireland) chemist and statistician, a major developer of statistical methods based on ranks (Bradley and Hollander, 1978).

is also very useful for paired-sample testing, just as the one-sample t and the paired-sample t test are basically the same. If the d_j values are from a normal distribution, then the Wilcoxon test has $3/\pi$ (i.e., 95.5%) of the power in detecting differences as the t test has (Conover, 1999: 363; Mood, 1954). But when the d_j's cannot be assumed to be from a normal distribution, the parametric paired-sample t test should be avoided, for with nonnormality, the Wilcoxon paired-sample test will be more powerful, sometimes much more powerful (Blair and Higgins, 1985). However, the Wilcoxon test assumes the population of differences is symmetrical (which the t test also does, for the normal distribution is symmetrical). The sign test could also be used for one-sample testing of the d_j's. It has only $2/\pi$ (64%) of the power of the t test, and only 67% of the power of the Wilcoxon test, when the normality assumption of the t test is met (Conover, 1999: 164). But the sign test does not assume symmetry and is therefore preferable to the Wilcoxon test when the differences come from a very asymmetric population.

Example 4 demonstrates the use of the Wilcoxon paired-sample test with the ratio-scale data of Example 1, and it is best applied to ratio- or interval-scale data. The testing procedure involves the calculation of differences, as does the paired-sample t test. Then one ranks the absolute values of those differences, from low to high, and affixes the sign of each difference to the corresponding rank. The rank assigned to tied observations is the mean of the ranks that would have been assigned to the observations had they not been tied. Differences of zero are ignored in this test.

Then we sum the ranks having a plus sign (calling this sum T_+) and the ranks with a minus sign (labeling this sum T_-). For a two-tailed test (as in Example 4), we reject H_0 if either T_+ or T_- is *less than or equal to* the critical value, $T_{\alpha(2),n}$, from Table 12 from *Appendix: Statistical Tables and Graphs*. In doing so, n is the number of differences that are not zero.

Having calculated either T_+ or T_-, the other can be determined as

$$T_- = \frac{n(n+1)}{2} - T_+ \tag{4}$$

or

$$T_+ = \frac{n(n+1)}{2} - T_- . \tag{5}$$

A different value of T_+ (call it T'_+) or T_- (call it T'_-) will be obtained if rank 1 is assigned to the largest, rather than the smallest, d_i (i.e., the absolute values of the d_j's are ranked from high to low). If this is done, the test statistics are obtainable as

$$T_+ = m(n+1) - T'_+ \tag{6}$$

and

$$T_- = m(n+1) - T'_- , \tag{7}$$

where m is the number of ranks with the sign being considered.

Pratt (1959) recommended maintaining differences of zero until after ranking, and thereafter ignoring the ranks assigned to the zeros. This procedure may yield slightly better results in some circumstances, though worse results in others (Conover, 1973). If used, then the critical values of Rahe (1974) should be consulted or the normal approximation employed (see the following section) instead of using critical values of T from Table 12 from *Appendix: Statistical Tables and Graphs*.

If data are paired, the undesirable use of the Mann-Whitney test, instead of the Wilcoxon paired-sample test, may lead to a greater Type II error, with the concomitant inability to detect actual population differences.

EXAMPLE 4   The Wilcoxon Paired-Sample Test Applied to the Data of Example 1

H_0:   Deer hindleg length is the same as foreleg length.

H_A: Deer hindleg length is not the same as foreleg length.

$\alpha = 0.05$

Deer (j)	Hindleg length (cm) (X_{1j})	Foreleg length (cm) (X_{2j})	Difference $(d_j = X_{1j} - X_{2j})$	Rank of $\lvert d_j \rvert$	Signed rank of $\lvert d_j \rvert$
1	142	138	4	4.5	4.5
2	140	136	4	4.5	4.5
3	144	147	−3	3	−3
4	144	139	5	7	7
5	142	143	−1	1	−1
6	146	141	5	7	7
7	149	143	6	9.5	9.5
8	150	145	5	7	7
9	142	136	6	9.5	9.5
10	148	146	2	2	2

$n = 10$

$T_+ = 4.5 + 4.5 + 7 + 7 + 9.5 + 7 + 9.5 + 2 = 51$

$T_- = 3 + 1 = 4$

$T_{0.05(2),10} = 8$

Since $T_- < T_{0.05(2),10},$ H_0 is rejected.

$$0.01 < P(T_- \text{ or } T_+ \leq 4) < 0.02 \quad [P = 0.014]$$

The Wilcoxon paired-sample test has an underlying assumption that the sampled population of d_j's is symmetrical about the median. Another nonparametric test for paired samples is the sign test, which does not have this assumption but is less powerful if the assumption is met.

Recall the Mann-Whitney test for hypotheses dealing with differences of specified magnitude. The Wilcoxon paired-sample test can be used in a similar fashion. For instance, it can be asked whether the hindlegs in the population sampled in Example 4 are 3 cm longer than the lengths of the forelegs. This can be done by applying the Wilcoxon paired-sample test after subtracting 3 cm from each hindleg length in the sample (or adding 3 cm to each foreleg length).

(a) The One-Tailed Wilcoxon Paired-Sample Test. For one-tailed testing we use one-tailed critical values from Table 12 from *Appendix: Statistical Tables and Graphs* and either T_+ or T_- as follows. For the hypotheses

H_0: Measurements in population 1 $\leq$ measurements in population 2

and H_A: Measurements in population 1 $>$ measurements in population 2,

H_0 is rejected if $T_- \leq T_{\alpha(1),n}$. For the opposite hypotheses:

H_0: Measurements in population 1 $\geq$ measurements in population 2

and H_A: Measurements in population 1 $<$ measurements in population 2,

reject H_0 if $T_+ \leq T_{\alpha(1),n}$.

(b) The Normal Approximation to the Wilcoxon Paired-Sample Test. For data consisting of more than 100 pairs* (the limit of Table 12 from *Appendix: Statistical Tables and Graphs*), the significance of T (where either T_+ or T_- may be used for T) may be determined by considering that for such large samples the distribution of T is closely approximated by a normal distribution with a mean of

$$\mu_T = \frac{n(n + 1)}{4} \tag{8}$$

and a standard error of

$$\sigma_T = \sqrt{\frac{n(n + 1)(2n + 1)}{24}}. \tag{9}$$

Thus, we can calculate

$$Z = \frac{|T - \mu_T|}{\sigma_T}, \tag{10}$$

where for T we may use, with identical results, either T_+ or T_-. Then, for a two-tailed test, Z is compared to the critical value, $Z_{\alpha(2)}$, or, equivalently, $t_{\alpha(2),\infty}$ (which for $\alpha = 0.05$ is 1.9600); if Z is greater than or equal to $Z_{\alpha(2)}$, then H_0 is rejected.

A normal approximation with a correction for continuity employs

$$Z_c = \frac{|T - \mu_T| - 0.5}{\sigma_T}. \tag{11}$$

As shown at the end of Table 12 from *Appendix: Statistical Tables and Graphs*, the normal approximation is better using Z for $\alpha(2)$ from 0.001 to 0.05 and is better using Z_c for $\alpha(2)$ from 0.10 to 0.50.

If there are tied ranks, then use

$$\sigma_T = \sqrt{\frac{n(n + 1)(2n + 1) - \dfrac{\sum t}{2}}{24}}, \tag{12}$$

where

$$\sum t = \sum (t_i^3 - t_i) \tag{13}$$

is the correction for ties introduced in using the normal approximation to the Mann-Whitney test, applied here to ties of nonzero differences.

---

*Fahoome (2002) concluded that the normal approximation also works well for sample sizes smaller than 100. She found that the probability of a Type I error is between 0.045 and 0.055 for two-tailed testing at the 0.05 level of significance with n as small as 10 and is between 0.009 and 0.011 when testing at $\alpha(2) = 0.01$ with n as small as 22. Additional information regarding the accuracy of this approximation is given at the end of Table 12 from *Appendix: Statistical Tables and Graphs*.

If we employ the Pratt procedure for handling differences of zero (described above), then the normal approximation is

$$Z = \frac{\left| T - \dfrac{n(n+1) - m'(m'+1)}{4} \right| - 0.5}{\sqrt{\dfrac{n(n+1)(2n+1) - m'(m'+1)(2m+1) - \dfrac{\sum t}{2}}{24}}} \tag{14}$$

(Cureton, 1967), where n is the total number of differences (including zero differences), and m' is the number of zero differences; $\sum t$ is as in Equation 13, applied to ties other than those of zero differences. We calculate T_+ or T_- by including the zero differences in the ranking and then deleting from considerations both the zero d_j's and the ranks assigned to them. For T in Equation 14, either T_+ or T_- may be used. If neither tied ranks nor zero d_j's are present, then Equation 14 becomes Equation 11.

One-tailed testing may also be performed using the normal approximation (Equation 10 or 11) or Cureton's procedure (Equation 14). The calculated Z is compared to $Z_{\alpha(1)}$ (which is the same as $t_{\alpha(1),\infty}$), and the direction of the arrow in the alternate hypothesis must be examined. If the arrow points to the left ("<"), then H_0 is rejected if $Z \geq Z_{\alpha(1)}$ and $T_+ < T_-$; if it points to the right (">"), then reject H_0 if $Z \geq Z_{\alpha(1)}$ and $T_+ > T_-$.

Iman (1974a) presents an approximation based on Student's t:

$$t = \frac{T - \mu_T}{\sqrt{\dfrac{n^2(n+1)(2n+1)}{2(n-1)} - \dfrac{(T - \mu_T)^2}{n-1}}}, \tag{15}$$

with $n - 1$ degrees of freedom. As shown at the end of Table 12 from *Appendix: Statistical Tables and Graphs*, this performs slightly better than the normal approximation (Equation 10). The test with a correction for continuity is performed by subtracting 0.5 from $|T - \mu_T|$ in both the numerator and denominator of Equation 15. This improves the test for $\alpha(2)$ from 0.001 to 0.10, but the uncorrected t is better for $\alpha(2)$ from 0.20 to 0.50. One-tailed t-testing is effected in a fashion similar to that described for Z in the preceding paragraph.*

Fellingham and Stoker (1964) discuss a more accurate approximation, but it requires more computation, and for sample sizes beyond those in Table 12 from *Appendix: Statistical Tables and Graphs* the increased accuracy is of no great consequence.

(c) The Wilcoxon Paired-Sample Test for Ordinal Data. The Wilcoxon test nonparametrically examines differences between paired samples when the samples consist of interval-scale or ratio-scale data (such as in Example 4), which is legitimate because the paired differences can be meaningfully ordered. However, it may not work well with samples comprising ordinal-scale data because the differences between ordinal scores may not have a meaningful ordinal relationship to each other. For example, each of several frogs could have the intensity of its green skin color recorded on a scale of 1 (very pale green) to 10 (very deep green). Those data would represent an ordinal scale of measurement because a score of 10 indicates a more intense green than a score of 9, a 9 represents an intensity greater than an 8, and so on. Then the

---

*When Table 12 from *Appendix: Statistical Tables and Graphs* cannot be used, a slightly improved approximation is effected by comparing the mean of t and Z to the mean of the critical values of t and Z (Iman, 1974a).

skin-color intensity could be recorded for these frogs after they were administered hormones for a period of time, and those data would also be ordinal. However, the *differences* between skin-color intensities before and after the hormonal treatment would not necessarily be ordinal data because, for example, the difference between a score of 5 and a score of 2 (a difference of 3) cannot be said to represent a difference in skin color that is greater than the difference between a score of 10 and a score of 8 (a difference of 2).

To deal with such a situation, Kornbrot (1990) presented a modification of the Wilcoxon paired-sample test (which she called the "rank difference test"), along with tables to determine statistical significance of its results.

(d) Wilcoxon Paired-Sample Test Hypotheses about a Specified Difference Other Than Zero. As indicated in Section 1, the paired-sample t test can be used for hypotheses proposing that the mean difference is something other than zero. Similarly, the Wilcoxon paired-sample test can examine whether paired differences are centered around a quantity other than zero. Thus, for data such as in Example 2, the nonparametric hypotheses could be stated as H_0 : crop yield does not increase more than 250 kg/ha with the new fertilizer, versus H_A : crop yield increases more than 250 kg/ha with the new fertilizer. In that case, each datum for the old-fertilizer treatment would be increased by 250 kg/ha (resulting in nine data of 1920, 2020, 2060 kg/ha, etc.) to be paired with the nine new-fertilizer data of 2250, 2410, 2260 kg/ha, and so on. Then, the Wilcoxon paired-sample test would be performed on those nine pairs of data.

With ratio- or interval-scale data, it is also possible to propose hypotheses considering a multiplication, rather than an addition, constant.

6 CONFIDENCE LIMITS FOR THE POPULATION MEDIAN DIFFERENCE

In Section 2, confidence limits were obtained for the mean of a population of differences. Given a population of differences, one can also determine confidence limits for the population median. Simply consider the observed differences between members of pairs (d_j) as a sample from a population of such differences.

EXERCISES

1. Concentrations of nitrogen oxides and of hydrocarbons (recorded in $\mu g/m^3$) were determined in a certain urban area.
   (a) Test the hypothesis that both classes of air pollutants were present in the same concentration.

(b) Calculate the 95% confidence interval for μ_d.
2. Using the data of Exercise 1, test the appropriate hypotheses with Wilcoxon's paired-sample test.
3. Using the data of Exercise 1, test for equality of the variances of the two kinds of air pollutants.

Day	Nitrogen oxides	Hydrocarbons
1	104	108
2	116	118
3	84	89
4	77	71
5	61	66
6	84	83
7	81	88
8	72	76
9	61	68
10	97	96
11	84	81

ANSWERS TO EXERCISES

1. $H_0: \mu_d = 0$, $H_A: \mu_d \neq 0$; $\bar{d} = -2.09 \ \mu\mathrm{g/m^3}$, $s_{\bar{d}} = 1.29 \ \mu\mathrm{g/m^3}$. **(a)** $t = -1.62$, $n = 11$, $\nu = 10$, $t_{0.05(2),10} = 2.228$; therefore, do not reject H_0; $0.10 < P < 0.20$ $[P = 0.14]$. **(b)** 95% confidence interval for $\mu_d = -2.09 \pm (2.228)(1.29) = -2.09 \pm 2.87$; $L_1 = -4.96 \ \mu\mathrm{g/m^3}$, $L_2 = 0.78 \ \mu\mathrm{g/m^3}$.

2.

d_i	Signed rank
−4	−5.5
−2	−3
−5	−7.5
6	9
−5	−7.5
1	1.5
−7	−10.5
−4	−5.5
−7	−10.5
1	1.5
3	4

$T = 9 + 1.5 + 1.5 + 4 = 16$; $T_{0.05(2),11} = 10$; since T is not ≤ 10, do not reject H_0; $0.10 < P < 0.20$.

3. $s_1^2 = 285.21 \ (\mu\mathrm{g/m^3})^2$, $s_2^2 = 270.36 \ (\mu\mathrm{g/m^3})^2$; $F = 1.055$; $r = 0.9674$; $t = 0.317$; $t_{0.05(2),9} = 2.262$; do not reject $H_0: \sigma_1^2 = \sigma_2^2$; $P > 0.50$ $[P = 0.76]$.

This page intentionally left blank

Multisample Hypotheses and the Analysis of Variance

Biologists often obtain data in the form of three or more samples, which are from three or more populations, a situation calling for multisample analyses, as introduced in this chapter.

It is tempting to some to test multisample hypotheses by applying two-sample tests to all possible pairs of samples. In this manner, for example, one might proceed to test the null hypothesis H_0: $\mu_1 = \mu_2 = \mu_3$ by testing each of the following hypotheses by the two-sample t test: H_0: $\mu_1 = \mu_2$, H_0: $\mu_1 = \mu_3$, H_0: $\mu_2 = \mu_3$. But such a procedure, employing a series of two-sample tests to address a multisample hypothesis, is invalid.

The calculated test statistic, t, and the critical values we find in the t table are designed to test whether the two sample statistics, $\overline{X_1}$ and $\overline{X_2}$, are likely to have come from the same population (or from two populations with identical means). In properly employing the two-sample test, we could randomly draw two sample means from the same population and wrongly conclude that they are estimates of two different populations' means; but we know that the probability of this error (the Type I error) will be no greater than α. However, consider that three random samples were taken from a single population. In performing the three possible two-sample t tests indicated above, with $\alpha = 0.05$, the probability of wrongly concluding that two of the means estimate different parameters is 14%, considerably greater than α. Similarly, if α is set at 5% and four means are tested, two at a time, by the two-sample t test, there are six pairwise H_0's to be tested in this fashion, and there is a 26% chance of wrongly concluding a difference between one or more of the means. Why is this?

For each two-sample t test performed at the 5% level of significance, there is a 95% probability that we shall correctly conclude not to reject H_0 when the two population means are equal. For the set of three hypotheses, the probability of *correctly* declining to reject all of them is only $0.95^3 = 0.86$. This means that the probability of *incorrectly* rejecting at least one of the H_0's is $1 - (1 - \alpha)^C = 1 - (0.95)^3 = 0.14$, where C

is the number of possible different pairwise combinations of k samples (see footnote to Table 1). As the number of means increases, it becomes almost certain that performing all possible two-sample t tests will conclude that some of the sample means estimate different values of μ, even if all of the samples came from the same population or from populations with identical means. Table 1 shows the probability of committing a Type I error if multiple t tests are employed to assess differences among more than two means. If, for example, there are 10 sample means, then $k = 10$, $C = 45$, and $1 - 0.95^{45} = 1 - 0.10 = 0.90$ is the probability of at least one Type I error when testing at the 0.05 level of significance. Two-sample tests, it must be emphasized, should *not* be applied to multisample hypotheses. The appropriate procedures are introduced in the following sections.

TABLE 1: Probability of Committing at Least One Type I Error by Using Two-Sample t Tests for All C Pairwise Comparisons of k Means*

		Level of Significance, α, Used in the t Tests				
k	C	0.10	0.05	0.01	0.005	0.001
2	1	0.10	0.05	0.01	0.005	0.001
3	3	0.27	0.14	0.03	0.015	0.003
4	6	0.47	0.26	0.06	0.030	0.006
5	10	0.65	0.40	0.10	0.049	0.010
6	15	0.79	0.54	0.14	0.072	0.015
10	45	0.99	0.90	0.36	0.202	0.044
∞		1.00	1.00	1.00	1.000	1.000

*There are $C = k(k - 1)/2$ pairwise comparisons of k means. This is the number of combinations of k items taken two at a time.

1 SINGLE-FACTOR ANALYSIS OF VARIANCE

To test the null hypothesis H_0: $\mu_1 = \mu_2 = \cdots = \mu_k$, where k is the number of experimental groups, or samples, we need to become familiar with the topic of *analysis of variance*, often abbreviated ANOVA (or less commonly, ANOV or AOV). Analysis of variance is a large area of statistical methods, owing its name and much of its early development to R. A. Fisher;* in fact, the F statistic was named in his honor by G. W. Snedecor[†] (1934: 15). There are many ramifications of analysis of variance considerations, the most common of which will be discussed in this chapter. More complex applications and greater theoretical coverage are found in the many books devoted specifically to analysis of variance and experimental design. At this point, it may appear strange that a procedure used for testing the equality of *means* should be named analysis of *variance*, but the reason for this terminology soon will become apparent.

---

*Sir Ronald Aylmer Fisher (1890–1962), British statistician and geneticist, who introduced the name and basic concept of the technique in 1918 (David, 1995; Street, 1990) and stressed the importance of randomness as discussed in this section. When he introduced analysis of variance, he did so by way of intraclass correlation (Box, 1978: 101), to which it is related.

[†]George W. Snedecor (1881–1974), American statistician.

Let us assume that we wish to test whether four different feeds result in differ-ent body weights in pigs. Since we are to test for the effect of only one *factor* (feed type) on the variable in question (body weight), the appropriate analysis is termed a single-factor (or "single-criterion" or "single-classification" or "one-way") analysis of variance.* Furthermore, each type of feed is said to be a *level* of the factor. The design of this experiment should have each experimental animal being assigned at random to receive one of the four feeds, with approximately equal numbers of pigs receiving each feed.

As with other statistical testing, it is of fundamental importance that each sample is composed of a random set of data from a population of interest. In Example 1, each of four populations consists of body weights of pigs on one of four experimental diets, and 19 pigs were assigned, at random, to the four diets. In other instances, researchers do not actually perform an experiment but, instead, collect data from populations defined other than by the investigator. For example, the interest might be in compar-ing body weights of four strains of pigs. If that were the case, the strain to which each animal belonged would not have been under the control of the researcher. Instead, he or she would measure a sample of weights for each strain, and the important con-sideration would be having each of the four samples consist of data assumed to have come at random from one of the four populations of data being studied.

EXAMPLE 1   A Single-Factor Analysis of Variance (Model I)

Nineteen pigs are assigned at random among four experimental groups. Each group is fed a different diet. The data are pig body weights, in kilograms, after being raised on these diets. We wish to ask whether pig weights are the same for all four diets.

H_0: $\mu_1 = \mu_2 = \mu_3 = \mu_4$.

H_A: The mean weights of pigs on the four diets are not all equal.

$\alpha = 0.05$

	Feed 1	Feed 2	Feed 3	Feed 4
	60.8	68.7	69.6	61.9
	67.0	67.7	77.1	64.2
	65.0	75.0	75.2	63.1
	68.6	73.3	71.5	66.7
	61.7	71.8		60.3
i	1	2	3	4
n_i	5	5	4	5
$\sum_{j=1}^{n_i} X_{ij}$	323.1	356.5	293.4	316.2
$\overline{X_i}$	64.62	71.30	73.35	63.24

Because the pigs are assigned to the feed groups at random (as with the aid of a random-number table, the single factor ANOVA is said to represent a *completely*

---

*Some authors would here refer to the feed as the "independent variable" and to the weight as the "dependent variable."

randomized experimental design, or "completely randomized design" (sometimes abbreviated "CRD"). In general, statistical comparison of groups of data works best if each group has the same number of data (a situation referred to as being a *balanced,* or *orthogonal, experimental design*), and the power of the test is heightened by having sample sizes as nearly equal as possible. The present hypothetical data might represent a situation where there were, in fact, five experimental animals in each of four groups, but the body weight of one of the animals (in group 3) was not used in the analysis for some appropriate reason. (Perhaps the animal died, or perhaps it became ill or was discovered to be pregnant, thus introducing a factor other than feed into the experiment.) The performance of the test is also enhanced by having all pigs as similar as possible in all respects except for the experimental factor, diet (i.e., the animals should be of the same breed, sex, and age, should be kept at the same temperature, etc.).

Example 1 shows the weights of 19 pigs subjected to this feed experiment, and the null hypothesis to be tested would be $H_0: \mu_1 = \mu_2 = \mu_3 = \mu_4$. Each datum in the experiment may be uniquely represented by the double subscript notation, where X_{ij} denotes datum j in experimental group i. For example, X_{23} denotes the third pig weight in feed group 2, that is, $X_{23} = 74.0$ kg. Similarly, $X_{34} = 96.5$ kg, $X_{41} = 87.9$ kg, and so on. We shall let the mean of group i be denoted by $\overline{X}_i$, and the grand mean of all observations will be designated by $\overline{X}$. Furthermore, n_i will represent the size of sample i, and $N = \sum_{i=1}^{k} n_i$ will be the total number of data in the experiment. The alternate hypothesis for this experiment is H_A: The mean weight of pigs is not the same on these four diets. Note that H_A is *not* $\mu_1 \neq \mu_2 \neq \mu_3 \neq \mu_4$ nor $\mu_1 \neq \mu_2 = \mu_3 = \mu_4$ nor any other specification of which means are different from which; we can only say that, if H_0 is rejected, then there is at least one difference among the four means.*

The four groups in this example (namely, types of feed) represent a nominal-scale variable in that the groups could be arranged in any sequence. However, in some situations the groups represent a measurement made on a ratio or interval scale. For example, the animal weights of Example 1 might have been measured at each of four environmental temperatures or, instead of the groups being different types of feed, the groups might have been different daily amounts of the same kind of feed. In other situations the groups might be expressed on an ordinal scale; for example, body weights could be measured at four environmental temperatures defined as cold, medium, warm, and hot, or as quantities of feed defined as very low, low, medium, and high. The analysis of variance of this section is appropriate when the groups are defined on a nominal or ordinal scale. If they represent a ratio or interval scale, regression procedures may be more appropriate, but the latter methods require more levels of the factor than are generally present in an analysis-of-variance experimental design.

(a) Sources of Variation. The statistical technique widely known as *analysis of variance* (ANOVA) examines the several sources of variation among all of the data

---

*There may be situations where the desired hypothesis is not whether k means are equal to each other, but whether they are all equal to some particular value. Mee, Shah, and Lefante (1987) proposed a procedure for testing $H_0: \mu_1 = \mu_2 = \cdots = \mu_k = \mu_0$, where μ_0 is the specified mean to which all of the other means are to be compared. The alternate hypothesis would be that at least one of the means is different from μ_0 (i.e., $H_A: \mu_i \neq \mu_0$ for at least one i).

in an experiment, by determining a *sum of squares* (meaning "sum of squares of deviations from the mean" for each source. Those sums of squares are as shown below.

In an experimental design with k groups, there are n_i data in group i; that is, n_1 designates the number of data in group 1, n_2 the number in group 2, and so on. The total number of data in all k groups will be called N; that is,

$$N = \sum_{i=1}^{k} n_i, \tag{1}$$

which in Example 1 is $N = 5 + 5 + 4 + 5 = 19$. The sum of squares for all N data is

$$\text{total SS} = \sum_{i=1}^{k} \sum_{j=1}^{n_i} (X_{ij} - \overline{X})^2, \tag{2}$$

where X_{ij} is datum j in group i and $\overline{X}$ is the mean of all N data:

$$\overline{X} = \frac{\sum_{i=1}^{k} \sum_{j=1}^{n_i} X_{ij}}{N}. \tag{2a}$$

This is the same as considering the N data in all the groups to compose a single group for which the sum of squares is as shown in Equation 2b. For the data in Example 1, these calculations are demonstrated in Example 1a.

$$\text{sample SS} = \sum (X_i - \overline{X})^2. \tag{2b}$$

EXAMPLE 1a   Sums of Squares and Degrees of Freedom for the Data of Example 1.

	Feed 1	Feed 2	Feed 3	Feed 4
$\sum_{j=1}^{n_i} X_{ij}$	323.1	356.5	293.4	316.2
$\overline{X}_i$	64.62	71.30	73.35	63.24

$$\sum_{i=1}^{k} \sum_{j=1}^{n_i} X_{ij} = 60.8 + 67.0 + 65.0 + \cdots + 63.1 + 66.7 + 60.3 = 1289.2$$

$$\overline{X} = \frac{1289.2}{19} = 67.8526$$

$$\text{Total SS} = \sum_{i=1}^{k} \sum_{j=1}^{n_i} (X_{ij} - \overline{X})^2$$

$$= (60.8 - 67.8526)^2 + (67.0 - 67.8526)^2$$

$$+ \cdots + (66.7 - 67.8526)^2 + (60.3 - 6.8526)^2$$

$$= 49.7372 + 0.7269 + \cdots + 1.3285 + 57.0418 = 479.6874.$$

$$\text{total DF} = N - 1 = 19 - 1 = 18$$

$$\text{groups SS} = \sum_{i=1}^{k} n_i (\overline{X}_i - \overline{X})^2$$

$$= 5(64.62 - 67.8526)^2 + 5(71.30 - 67.8526)^2$$
$$+ 4(73.35 - 67.8526)^2 + 5(63.24 - 67.8526)^2$$
$$= 52.2485 + 59.4228 + 120.8856 + 106.3804 = 338.9372$$

$$\text{groups DF} = k - 1$$

$$\text{within-groups (error) SS} = \sum_{i=1}^{k} \left[\sum_{j=1}^{n_i} (X_{ij} - \overline{X}_i)^2 \right]$$

$$= (60.8 - 64.62)^2 + (67.0 - 64.62)^2$$
$$+ \cdots + (66.7 - 63.24)^2 + (60.3 - 63.24)^2$$
$$= 14.5924 + 5.6644 + \cdots + 11.9716 + 8.6436$$
$$= 140.7500$$

or, alternatively,

$$\text{within-groups (error) SS} = \text{Total SS} - \text{Groups SS}$$
$$= 479.6874 - 338.9373 = 140.7501.$$

$$\text{within-groups (error) DF} = \sum_{i=1}^{k} (n_i - 1)$$

$$= (5 - 1) + (5 - 1) + (4 - 1) + (5 - 1) = 15$$

or within-groups (error) DF $= N - k = 19 - 4 = 15$

or within-groups (error) DF $=$ Total DF $-$ Groups DF $= 18 - 3 = 15$.

Note: The quantities involved in the sum-of-squares calculations are carried to several decimal places (as computers typically do) to avoid rounding errors. All of these sums of squares (and the subsequent mean squares) have $(\text{kg})^2$ as units. However, for typographic convenience and ease in reading, the units for ANOVA computations are ordinarily not printed.

The degrees of freedom associated with the total sum of squares are

$$\text{total DF} = N - 1, \tag{3}$$

which for the data in Example 1 are $19 - 1 = 18$.

A portion of this total amount of variability of the N data is attributable to differences among the means of the k groups; this is referred to as the *among-groups sum of squares* or, simply, as the *groups sum of squares*:

$$\text{groups SS} = \sum_{i=1}^{k} n_i (\overline{X}_i - \overline{X})^2, \tag{4}$$

where $\overline{X}_i$ is the mean of the n_i data in sample i and $\overline{X}$ is the mean of all N data. Example 1a shows this computation for the data in Example 1. Associated with this sum of squares are these degrees of freedom:

$$\text{groups DF} = k - 1, \tag{5}$$

which for Example 1 are $4 - 1 = 3$.

Furthermore, the portion of the total sum of squares that is not explainable by differences *among* the group means is the variability *within* the groups:

$$\text{within-groups SS} = \sum_{i=1}^{k}\left[\sum_{j=1}^{n_i}(X_{ij} - \overline{X}_i)^2\right] \tag{6}$$

and is commonly called the *error sum of squares*. Within the brackets Equation 2b is applied to the data in each one of the k groups, and the within-groups sum of squares is the sum of all k of these applications of Equation 2b. For the data of Example 1, this is shown in Example 1a.

The degrees of freedom associated with the within-groups sum of squares are

$$\text{within-groups DF} = \sum_{i=1}^{k}(n_i - 1) = N - k, \tag{7}$$

also called the *error DF*, which for Example 1 are $4 + 4 + 3 + 4 = 15$ or, equivalently, $19 - 4 = 15$.

The within-groups SS and DF may also be obtained by realizing that they represent the difference between the total variability among the data and the variability among groups:

$$\text{within-groups SS} = \text{total SS} - \text{groups SS} \tag{8}$$

and

$$\text{within-groups DF} = \text{total DF} - \text{groups DF}. \tag{8a}$$

In summary, each deviation of an observed datum from the grand mean of all data is attributable to a deviation of that datum from its group mean plus the deviation of that group mean from the grand mean; that is,

$$(X_{ij} - \overline{X}) = (X_{ij} - \overline{X}_i) + (\overline{X}_i - \overline{X}). \tag{9}$$

Furthermore, sums of squares and degrees of freedom are additive, so

$$\text{total SS} = \text{groups SS} + \text{error SS} \tag{10}$$

and

$$\text{total DF} = \text{groups DF} + \text{error DF}. \tag{11}$$

(b) "Machine Formulas." The total sum of squares (Equation 2) may be calculated readily by a "machine formula"

$$\text{total SS} = \sum_{i=1}^{k}\sum_{j=1}^{n_i} X_{ij}^2 - C, \tag{12}$$

where*

$$C = \frac{\left(\sum\sum X_{ij}\right)^2}{N}.$$

(13)

Example 1b demonstrates these calculations.

EXAMPLE 1b   Sums of Squares and Degrees of Freedom for the Data of Example 1, Using Machine Formulas

	Feed 1	Feed 2	Feed 3	Feed 4
i	1	2	3	4
n_i	5	5	4	5
$\sum\limits_{j=1}^{n_i} X_{ij}$	323.1	356.5	293.4	316.2

$$\frac{\left(\sum\limits_{j=1}^{n_i} X_{ij}\right)^2}{n_i} \qquad 20878.7220 \quad 25418.4500 \quad 21520.8900 \quad 19996.4480 \qquad \sum_{i=1}^{k} \frac{\left(\sum\limits_{j=1}^{n_i} X_{ij}\right)}{n_i} = 87814.5500$$

$$\sum_i\sum_j X_{ij} = 1289.2 \qquad \text{total DF} = N - 1 = 19 - 1 = 18$$

$$\sum_i\sum_j X_{ij}^2 = 87955.30 \qquad \text{groups DF} = k - 1 = 4 - 1 = 3$$

$$\text{error DF} = N - k = 19 - 4 = 15$$

$$C = \frac{\left(\sum\limits_i\sum\limits_j X_{ij}\right)^2}{N} = \frac{(1289.2)^2}{19} = 87475.6126$$

$$\text{total SS} = \sum_i\sum_j X_{ij}^2 - C = 87955.3000 - 87475.6126 = 479.6874$$

$$\text{groups SS} = \sum_{i=1}^{k} \frac{\left(\sum\limits_{j=1}^{n_i} X_{ij}\right)^2}{n_i} - C = 87814.5500 - 87475.6126 = 338.9374$$

$$\text{error SS} = \text{total SS} - \text{groups SS} = 479.68747 - 338.9374 = 140.7500$$

A machine formula for the groups sum of squares (Equation 4) is

$$\text{groups SS} = \sum_{i=1}^{k} \frac{\left(\sum\limits_{j=1}^{n_i} X_{ij}\right)^2}{n_i} - C,$$

(14)

where $\sum_{j=1}^{n_i} X_{ij}$ is the sum of the n_i data from group i.

---

*The term "machine formula" derives from the formula's utility when using calculating machines. The quantity C is often referred to as a "correction term"—an unfortunate expression, for it implies that some miscalculation needs to be rectified.

The error SS may be calculated as

$$\text{error SS} = \sum_{i=1}^{k} \sum_{j=1}^{n_l} X_{ij}^2 - \sum_{i=1}^{k} \frac{\left(\sum_{j=1}^{n_i} X_{ij} \right)^2}{n_i}, \tag{15}$$

which is the machine formula for Equation 6.

As shown in Example 1b, machine formulas such as these can be very convenient when using simple calculators; but they are of less importance if the statistical computations are performed by computer. As demonstrated in Examples 1a and 1b, the sums of squares are the same using the two computational formulas.

(c) Testing the Null Hypothesis. Dividing the groups SS or the error SS by the respective degrees of freedom results in a variance, referred to in ANOVA terminology as a *mean square* (abbreviated MS and short for *mean squared deviation from the mean*). Thus,

$$\text{groups MS} = \frac{\text{groups SS}}{\text{groups DF}} \tag{16}$$

and

$$\text{error MS} = \frac{\text{error SS}}{\text{error DF}}, \tag{17}$$

and the latter quantity, which may also be called the *within-groups mean square*, is occasionally abbreviated as MSE (for "mean square error"). As will be seen below, testing the null hypothesis of equality among population means involves the examination of the groups mean square and the error mean square. Because a mean square is a kind of variance, this procedure is named *analysis of variance*. A total mean square could also be calculated, as (total SS)/(total DF), but it is not used in the ANOVA.

Statistical theory informs us that if the null hypothesis is a true statement about the populations, then the groups MS and the error MS will each be an estimate of σ^2, the variance common to all k populations. But if the k population means are not equal, then the groups MS in the population will be greater than the population's error MS.[*] Therefore, the test for the equality of means is a one-tailed variance ratio test, where the groups MS is always placed in the numerator so as to ask whether it is significantly larger than the error MS:[†]

$$F = \frac{\text{groups MS}}{\text{error MS}}. \tag{18}$$

---

[*]Two decades before R. A. Fisher developed analysis of variance techniques, the Danish applied mathematician, Thorvald Nicolai Thiele (1838–1910) presented the concept of comparing the variance among groups to the variance within groups (Thiele, 1897: 41–44). Stigler (1986: 244) reported that an 1860 book by Gustav Theodor Fechner included the most extensive discussion of the concepts of experimental design prior to R. A. Fisher.

[†]An equivalent computation of F is

$$F = \left(\frac{\text{error DF}}{\text{groups DF}} \right) \left(\frac{(\text{groups SS})/(\text{total SS})}{1 - (\text{groups SS})/(\text{total SS})} \right), \tag{18a}$$

and Levin, Serlin, and Webne-Behrman (1989) show how ANOVA can be performed by considering the correlation between observations and their group means.

This quantity expresses how the variability of data among groups compares to the variability of data within groups.

The critical value for this test is $F_{\alpha(1),(k-1),(N-k)}$, which is the value of F at the one-tailed significance level α and with numerator degrees of freedom ($\nu_1 =$ groups DF) of $k - 1$ and denominator degrees of freedom ($\nu_2 =$ error DF) of $N - k$. If the calculated F is as large as, or larger than, the critical value then we reject H_0; and a rejection indicates that the probability is $\leq \alpha$ that the observed data came from populations described by H_0. But remember that all we conclude in such a case is that all the k population means are not equal.

Example 1c shows the conclusion of the analysis of variance performed on the data and hypotheses of Example 1. Table 2 summarizes the single-factor ANOVA calculations.*

EXAMPLE 1c   The Conclusion of the ANOVA of Example 1, Using the Results of Either Example 1a or 1b

Summary of the Analysis of Variance

Source of variation	SS	DF	MS
Total	479.6874	18	
Groups	338.9374	3	112.9791
Error	140.7500	15	9.3833

$$F = \frac{\text{groups MS}}{\text{error MS}} = \frac{112.9791}{9.3833} = 12.04$$

$F_{0.05(1),3,15} = 3.29$, so reject H_0.

$P < 0.0005 \quad [P = 0.00029]$

(d) The Case where $k = 2$.   If $k = 2$, then H_0: $\mu_1 = \mu_2$, and either the two-sample t test or the single-factor ANOVA may be applied; the conclusions obtained from these two procedures will be identical. The error MS will, in fact, be identical to the pooled variance, s_p^2, in the t test; the groups DF will be $k - 1 = 1$; the F value determined by the analysis of variance will be the square of the t value from the t test; and $F_{\alpha(1),1,(N-2)} = (t_{\alpha(2),(N-2)})^2$. If a one-tailed test between means is required, or if the hypothesis H_0: $\mu_1 - \mu_2 = \mu_0$ is desired for a μ_0 not equal to zero, then the t test is applicable, whereas the ANOVA is not.

---

*Occasionally the following quantity (or its square root) is called the *correlation ratio*:

$$\eta^2 = \frac{\text{groups SS}}{\text{total SS}}. \tag{18b}$$

This is also called *eta squared*, for it is represented using the lowercase Greek letter eta. It is always between 0 and 1, it has no units of measurement, and it expresses the proportion of the total variability of X that is accounted for by the effect of differences among the groups (and is, therefore, reminiscent of the coefficient of determination). For Example 1, $\eta^2 = 338.9374/479.6874 = 0.71$, or 71%.

TABLE 2: Summary of the Calculations for a Single-Factor Analysis of Variance

Source of variation	Sum of squares (SS)	Degrees of freedom (DF)	Mean square (MS)
Total $[X_{ij} - \overline{X}]$	Equation 2 or 12	$N - 1$	
Groups (i.e., among groups) $[\overline{X}_i - \overline{X}]$	Equation 4 or 14	$k - 1$	$\dfrac{\text{groups SS}}{\text{groups DF}}$
Error (i.e., within groups) $[X_{ij} - \overline{X}_i]$	Equation 6 or 8	$N - k$ or Equation 8a	$\dfrac{\text{error SS}}{\text{error DF}}$

Note: For each source of variation, the bracketed quantity indicates the variation being assessed; k is the number of groups; X_{ij} is datum j in group i; $\overline{X}_i$ is the mean of the data in group i; $\overline{X}$ is the mean of all N data.

(e) ANOVA Using Means and Variances. The above discussion assumes that all the data from the experiment to be analyzed are in hand. It may occur, however, that all we have for each of the k groups is the mean and some measure of variability based on the variances of each group. That is, we may have $\overline{X}_i$ and either SS_i, s_i^2, s_i, or $s_{\overline{X}_i}$ for each group, rather than all the individual values of X_{ij}. For example, we might encounter presentations such as Tables 4, 5, or 6. If the sample sizes, n_i, are also known, then the single-factor analysis of variance may still be performed, in the following manner.

First, determine the sum of squares or sample variance for each group; recall that

$$SS_i = (n_i - 1)s_i^2 \quad \text{and} \quad s_i^2 = (s_i)^2 = n_i(s_{\overline{X}_i})^2. \tag{19}$$

Then calculate

$$\text{error SS} = \sum_{i=1}^{k} SS_i = \sum_{i=1}^{k} (n_i - 1)s_i^2 \tag{20}$$

and

$$\text{groups SS} = \sum_{i=1}^{k} n_i \overline{X}_i^2 - \frac{\left(\sum_{i=1}^{k} n_i \overline{X}_i \right)^2}{\sum_{i=1}^{k} n_i}. \tag{21}$$

Knowing the groups SS and error SS, the ANOVA can proceed in the usual fashion.

(f) Fixed-Effects amd Random-Effects ANOVA. In Example 1, the biologist designing the experiment was interested in whether all of these particular four feeds have the same effect on pig weight. That is, these four feeds were not randomly selected from a feed catalog but were specifically chosen. When the levels of a factor are specifically chosen one is said to have designed a *fixed-effects model*, or a *Model I*, ANOVA. In such a case, the null hypothesis H_0: $\mu_1 = \mu_2 = \mu_3 = \cdots = \mu_k$ is appropriate.

However, there are instances where the levels of a factor to be tested are indeed chosen at random. For example, we might have been interested in the effect of geographic location of the pigs, rather than the effect of their feed. It is possible that our concern might be with certain specific locations, in which case we would be employing a fixed-effects mode ANOVA. But we might, instead, be interested in testing the statement that in general there is a difference in pig weights in animals from different locations. That is, instead of being concerned with only the particular locations used in the study, the intent might be to generalize, considering the locations in our study to be a random sample from all possible locations. In this *random-effects model*, or *Model II*, ANOVA,* all the calculations are identical to those for the fixed-effects model, but the null hypothesis is better stated as H_0: there is no difference in pig weight among geographic locations (or H_0: there is no variability in weights among locations). Examination of Equation 18 shows that what the analysis asks is whether the variability among locations is greater than the variability within locations. Example 2 demonstrates the ANOVA for a random-effects model. The relevant sums of squares could be computed as in Section 1a or 1b; the machine formulas of Section 1b are used in this example. Most biologists will encounter Model I analyses more commonly than Model II situations. When dealing with more than one experimental factor, the distinction between the two models becomes essential, as it will determine the calculation of F.

(g) Violation of Underlying Assumptions. To test H_0: $\mu_1 = \mu_2$ by the two-sample t test, we assume that $\sigma_1^2 = \sigma_2^2$ and that each of the two samples came at random from a normal population. Similarly, in order to apply the analysis of variance to $\mu_1 = \mu_2 = \cdots = \mu_k$, we assume that $\sigma_1^2 = \sigma_2^2 = \cdots = \sigma_k^2$ and that each of the k samples came at random from a normal population.

However, these conditions are never *exactly* met, so the question becomes how serious the consequences are when there are departures from these underlying assumptions. Fortunately, under many circumstances the analysis of variance is a robust test, meaning that its Type I and Type II error probabilities are not always seriously altered by violation of the test's assumptions. Reports over several decades of research have not agreed on every aspect of this issue, but the following general statements can be made about fixed-effects (i.e., Model I) ANOVA:

As with the two-sample t test, the adverse effect of nonnormality is greater with greater departures from normality, but the effect is relatively small if samples sizes are equal, or if the n_i's are unequal but large (with the test less affected by nonnormality as the n_i's increase), or if the variances are equal; and asymmetric distributions have a greater adverse effect than do symmetric distributions (Box and Anderson, 1955; Büning, 1997; Donaldson, 1968; Glass, Peckham, and Sanders, 1972; Harwell et al., 1992; Lix, Keselman, and Keselman, 1996; Srivastava, 1959; Tiku, 1971).

If the variances of the k populations are not equal, the analysis of variance is generally liberal for equal sample sizes, and the extent to which the test is liberal (i.e., the probability of a Type I error exceeds α) increases with greater variance heterogeneity (Büning, 1997; Clinch and Keselman, 1982; Rogan and Keselman, 1977) and decreases with increased sample size (Rogan and Keselman, 1977). Myers and Well (2003: 221) report that this inflation of P(Type I error) is usually less than 0.02 at the 0.05 significance level and less than 0.005 when using $\alpha = 0.01$, when n is at least 5 and the largest variance is no more than four times the smallest variance.

If the group variances are not equal and the n_i's also are unequal, then there can be very serious effects on P(Type I error). The effect will be greater for greater variance heterogeneity (Box, 1954), and if the larger variances are associated with the larger sample sizes (what we shall call a "direct" relationship), the test is conservative

---

*Also referred to as a components of variance model. The terms *components of variance, fixed effects*, *random effects*, *Class I*, and *Class II* for analysis of variance were introduced by Eisenhart (1947).

EXAMPLE 2   A Single-Factor Analysis of Variance for a Random-Effects Model (i.e., Model II) Experimental Design

A laboratory employs a technique for determining the phosphorus content of hay. The question arises: "Do phosphorus determinations differ among the technicians performing the analysis?" To answer this question, each of four randomly selected technicians was given five samples from the same batch of hay. The results of the 20 phosphorus determinations (in mg phosphorus/g of hay) are shown.

H_0:   Determinations of phosphorus content do not differ among technicians.

H_A: Determinations of phosphorus content do differ among technicians.

$\alpha = 0.05$

	Technician			
	1	2	3	4
	34	37	34	36
	36	36	37	34
	34	35	35	37
	35	37	37	34
	34	37	36	35
Group sums:	173	182	179	176

$$\sum_i \sum_j X_{ij} = 710$$

$$\sum_i \sum_j X_{ij}^2 = 25234$$

$N = 20$

$C = \dfrac{(710)^2}{20} = 25205.00$

total SS $= 25234 - 25205.00 = 29.00$

$$\text{groups (i.e., technicians) SS} = \frac{(173)^2}{5} + \frac{a(182)^2}{5}$$
$$+ \frac{(179)^2}{5} + \frac{(176)^2}{5} - 25205.00$$
$$= 25214.00 - 25205.00 = 9.00$$

error SS $= 29.00 - 9.00 = 20.00$

Source of variation	SS	DF	MS
Total	29.00	19	
Groups (technicians)	9.00	3	3.00
Error	20.00	16	1.25

$F = \dfrac{3.00}{1.25} = 2.40$

$F_{0.05(1),3,16} = 3.24$

Do not reject H_0.

$$0.10 < P < 0.25 \quad [P = 0.11]$$

(i.e., the probability of a Type I error is less than α), while larger variances affiliated with smaller samples (what we'll call an "inverse" relationship) cause the test to be liberal—that is, P(Type I error) $> \alpha$ (Brown and Forsythe, 1974a; Büning, 1997; Clinch and Keselman, 1982; Donaldson, 1968; Glass, Peckham, and Sanders, 1972; Harwell et al., 1992; Kohr and Games, 1974; Maxwell and Delaney, 2004: 131; Stonehouse and Forrester, 1998; Tomarkin and Serlin, 1986). For example, if testing at $\alpha = 0.05$ and the largest n is twice the size of the smallest, the probability of a Type I error can be as small as 0.006 for a direct relationship between variances and sample sizes and as large as 0.17 for an inverse relationship; if the ratio between the largest and smallest variances is 5, P(Type I error) can be as small as 0.00001 or as large as 0.38, depending upon whether the relationship is direct or inverse, respectively (Scheffé, 1959: 340). This huge distortion of P(Type I error) may be reason to avoid employing the analysis of variance when there is an inverse relationship (if the researcher is primarily concerned about avoiding a Type I error) or when there is a direct relationship (if the principal concern is to evade a Type II error). The adverse effect of heterogeneous variances appears to increase as k increases (Tomarkin and Serlin, 1986).

Recall that a decrease in the probability of the Type I error (α) is associated with an increase in the Type II error (β), and an increase in β means a decrease in the power of the test ($1 - \beta$). Therefore, for situations described above as conservative [i.e., P(Type I error) $< \alpha$], there will generally be less power than if the population variances were all equal; and when the test is liberal [i.e., P(Type I error) $> \alpha$], there will generally be more power than if the variances were equal.

If the sample sizes are all equal, nonnormality generally affects the power of the analysis of variance to only a small extent (Clinch and Keselman, 1982; Glass, Peckham, and Sanders, 1972; Harwell et al., 1992; Tan, 1982), and the effect decreases with increased n (Donaldson, 1968). However, extreme skewness or kurtosis can severely alter (and reduce) the power (Games and Lucas, 1966), and nonnormal kurtosis generally has a more adverse effect than skewness (Sahai and Ageel, 2000: 85). With small samples, for example, very pronounced platykurtosis in the sampled populations will decrease the test's power, and strong leptokurtosis will increase it (Glass, Peckham, and Sanders, 1972). When sample sizes are not equal, the power is much reduced, especially when the large samples have small means (Boehnke, 1984).

The robustness of random-effects (i.e., Model II) analysis of variance (Section 1f) has not been studied as much as that of the fixed-effects (Model I) ANOVA. However, the test appears to be robust to departures of normality within the k populations, though not as robust as Model I ANOVA (Sahai, 2000: 86), provided the k groups (levels) of data can be considered to have been selected at random from all possible groups and that the effect of each group on the variable can be considered to be from a normally distributed set of group effects. When the procedure is nonrobust, power appears to be affected more than the probability of a Type I error; and the lack of robustness is not very different if sample sizes are equal or unequal (Tan, 1982; Tan and Wong, 1980). The Model II analysis of variance (as is the case with the Model I ANOVA) also assumes that the k sampled populations have equal variances.

(h) Testing of Multiple Means when Variances Are Unequal. Although testing hypotheses about means via analysis of variance is tolerant to small departures from the assumption of variance homogeneity when the sample sizes are equal, it can yield very misleading results in the presence of more serious heterogeneity of variances and/or unequal sample sizes. Having unequal variances represents a multisample Behrens-Fisher problem (i.e., an extension of the two-sample Behrens-Fisher situation).

Several approaches to this analysis have been proposed (e.g., see Keselman et al., 2000; Lix, Keselman, and Keselman, 1996). A very good one is that described by Welch (1951), which employs

$$F' = \frac{\displaystyle\sum_{i=1}^{k} c_i(\overline{X}_i - \overline{X}_w)^2}{(k-1)\left[1 + \dfrac{2A(k-2)}{k^2-1}\right]}, \tag{22}$$

where

$$c_i = \frac{n_i}{s_i^2} \tag{23}$$

$$C = \sum_{i=1}^{k} c_i \tag{24}$$

$$\overline{X}_w = \frac{\displaystyle\sum_{i=1}^{k} c_i \overline{X}_i}{C} \tag{25}$$

$$A = \sum_{i=1}^{k} \frac{(1 - c_i/C)^2}{\nu_i}, \text{ where } \nu_i = n_i - 1 \tag{26}$$

and F' is associated with degrees of freedom of $\nu_1 = k - 1$ and

$$\nu_2 = \frac{k^2 - 1}{3A}, \tag{27}$$

which should be rounded to the next lower integer when using Table 4 from *Appendix: Statistical Tables and Graphs*. This procedure is demonstrated in Example 3.

A modified ANOVA advanced by Brown and Forsythe (1974a, b) also works well:

$$F'' = \frac{\text{groups SS}}{B}, \tag{28}$$

EXAMPLE 3   Welch's Test for an Analysis-of-Variance Experimental Design with Dissimilar Group Variances

The potassium content (mg of potassium per 100 mg of plant tissue) was measured in five seedlings of each of three varieties of wheat.

H_0: $\mu_1 = \mu_2 = \mu_3$.

H_A: The mean potassium content is not the same for seedlings of all three wheat varieties.

$\alpha = 0.05$

	Variety G	Variety A	Variety L
	27.9	24.2	29.1
	27.0	24.7	27.7
	26.0	25.6	29.9
	26.5	26.0	30.7
	27.0	27.4	28.8
	27.5	26.1	31.1

i	1	2	3
n_i	6	6	6
ν_i	5	5	5
$\overline{X}_i$	26.98	25.67	29.55
s_i^2	0.4617	1.2787	1.6070

$$c_i = n_i/s_i^2 \qquad 12.9955 \qquad 4.6923 \qquad 3.7337 \qquad C = \sum_i c_i = 21.4215$$

$$c_i\overline{X}_i \qquad 350.6186 \qquad 120.4513 \qquad 110.3308 \qquad \sum_i c_i\overline{X}_i = 581.4007$$

$$\frac{\left(1 - \dfrac{c_i}{C}\right)^2}{\nu_i} \qquad 0.0309 \qquad 0.1220 \qquad 0.1364 \qquad A = \sum_i \frac{\left(1 - \dfrac{c_i}{C}\right)^2}{\nu_i} = 0.2893$$

$$\overline{X}_w = \frac{\sum\limits_i c_i\overline{X}_i}{C} = \frac{581.4007}{21.4215} = 27.14$$

$$F' = \frac{\sum c_i(\overline{X}_i - \overline{X}_w)^2}{(k - 1)\left[1 + \dfrac{2A(k - 2)}{k^2 - 1}\right]}$$

$$= \frac{12.9955(26.98 - 27.14)^2 + 4.6923(25.67 - 27.14)^2}{+ 3.7337(29.55 - 27.14)}{(3 - 1)\left[1 + \dfrac{2(0.2893)(3 - 2)}{3^2 - 1}\right]}$$

$$= \frac{0.3327 + 10.1396 + 21.6857}{2(0.9268)} = \frac{32.4144}{1.8536} = 17.5$$

For critical value of F:

$$\nu_1 = k - 1 = 3 - 1 = 2$$

$$\nu_2 = \frac{k^2 - 1}{3A} = \frac{3^2 - 1}{3(0.2893)} = \frac{8}{0.8679} = 9.22$$

By harmonic interpolation in Table 4 from *Appendix: Statistical Tables and Graphs* or by computer program:

$F_{0.05(1),2,9.22} = 4.22$. So, reject H_0.

$$0.0005 < P < 0.001 \quad [P = 0.0073]$$

where

$$b_i = \left(1 - \frac{n_i}{N}\right) s_i^2, \tag{28a}$$

and

$$B = \sum_{i=1}^{k} b_i. \tag{29}$$

F'' has degrees of freedom of $v_1 = k - 1$ and

$$v_2 = \frac{B^2}{\sum\limits_{i=1}^{k} \dfrac{b_i^2}{v_i}}. \tag{30}$$

If $k = 2$, both the F' and F'' procedures are equivalent to the t' test. The Welch method (F') has been shown (by Brown and Forsythe, 1974a; Büning, 1997; Dijkstra and Werter, 1981; Harwell et al., 1992; Kohr and Games, 1974; Levy, 1978a; Lix, Keselman, and Keselman, 1996) to generally perform better than F or F'' when population variances are unequal, especially when n_i's are equal. However, the Welch test is liberal if the data come from highly skewed distributions (Clinch and Keselman, 1992; Lix, Keselman, and Keselman, 1996).

Browne and Forsythe (1974a) reported that when variances are equal, the power of F is a little greater than the power of F', and that of F' is a little less than that of F''. But if variances are not equal, F' has greater power than F'' in cases where extremely low and high means are associated with low variances, and the power of F'' is greater than that of F' when extreme means are associated with large variances. Also, in general, F' and F'' are good if all $n_i \geq 10$ and F' is reasonably good if all $n_i \geq 5$.

(i) Which Multisample Test to Use. As with all research reports, the reader should be informed of explicitly what procedures were used for any statistical analysis. And, when results involve the examination of means, that reporting should include the size (n), mean ($\overline{X}$), and variability (e.g., standard deviation or standard error) of each sample. If the samples came from populations having close to normal distributions, then presentation of each sample's confidence limits (Section 2) might also be included. Additional interpretation of the results could include displaying the means and measures of variability via tables or graphs.

Although it not possible to generalize to all possible situations that might be encountered, the major approaches to comparing the means of k samples, where k is more than two, are as follows:

- If the k sampled populations are normally distributed and have identical variances (or if they are only slightly to moderately nonnormal and have similar variances): The analysis of variance, using F, is appropriate and preferable to test for difference among the means. (However, samples nearly always come from distributions that are not *exactly* normal with *exactly* the same variances, so conclusions to reject or not reject a null hypothesis should not be considered definitive when the probability associated with F is very near the α specified for the hypothesis test; in such a situation the statistical conclusion should be expressed with some caution and, if feasible, the experiment should be repeated (perhaps with more data).

- If the k sampled populations are distributed normally (or are only slightly to moderately nonnormal), but they have very dissimilar variances: The Behrens-Fisher testing of Section 1h is appropriate and preferable to compare the k means. If extremely high and low means are associated with small variances, F' is preferable; but if extreme means are associated with large variances, then F'' works better.

- If the k sampled populations are very different from normally distributed, but they have similar distributions and variances: The Kruskal-Wallis test of Section 4 is appropriate and preferable.

- If the k sampled populations have distributions greatly different from normal and do not have similar distributions and variances: (1) Consider the procedures for data that do not exhibit normality and variance equality but that can be transformed into data that are normal and homogeneous of variance; or (2) report the mean and variability for each of the k samples, perhaps also presenting them in tables and/or graphs, but do not perform hypothesis testing.

(j) Outliers. A small number of data that are much more extreme than the rest of the measurements are called *outliers*, and they may cause a sample to depart seriously from the assumptions of normality and variance equality. If, in the experiment of Example 1, a pig weight of 652 kg, or 7.12 kg, or 149 kg was reported, the researcher would likely suspect an error. Perhaps the first two of these measurements were the result of the careless reporting of weights of 65.2 kg and 71.2 kg, respectively; and perhaps the third was a weight measured in pounds and incorrectly reported as kilograms. If there is a convincingly explained error such as this, then an offending datum might be readily corrected. Or, if it is believed that a greatly disparate datum is the result of erroneous data collection (e.g., an errant technician, a contaminated reagent, or an instrumentation malfunction), then it might be discarded or replaced. In other cases outliers might be valid data, and their presence may indicate that one should not employ statistical analyses that require population normality and variance equality. There are statistical methods that are sometimes used to detect outliers, some of which are discussed by Barnett and Lewis (1994: Chapter 6), Snedecor and Cochran (1989: 280–281), and Thode (2002: Chapter 6).

True outliers typically will have little or no influence on analyses employing nonparametric two-sample tests or multisample tests (Section 4).

2 CONFIDENCE LIMITS FOR POPULATION MEANS

When $k > 2$, confidence limits for each of the k population means may be computed in a fashion analogous to that for the case where $k = 2$, under the same assumptions of normality and homogeneity of variances applicable to the ANOVA. The $1 - \alpha$ confidence interval for μ_i is

$$\overline{X}_i \pm t_{\alpha(2),\nu}\sqrt{\frac{s^2}{n_i}}, \tag{31}$$

where s^2 is the error mean square and ν is the error degrees of freedom from the analysis of variance. For example, let us consider the 95% confidence interval for μ_4

in Example 1. Here, $\overline{X}_4 = 63.24$ kg, $s^2 = 9.383$ kg^2, $n_4 = 5$, and $t_{0.05(2),15} = 2.131$. Therefore, the lower 95% confidence limit, L_1, is 63.24 kg $- 2.131\sqrt{9.383 \text{ kg}^2/5} =$ 63.24 kg $- 3.999$ kg $= 59.24$ kg, and L_2 is 63.24 kg $+ 2.131\sqrt{9.383 \text{ kg}^2/5} =$ 63.24 kg $+ 3.999$ kg $= 67.24$ kg.

Computing a confidence interval for μ_i would only be warranted if that population mean was concluded to be different from each other population mean. And calculation of a confidence interval for each of the k μ's may be performed only if it is concluded that $\mu_1 \neq \mu_2 \neq \cdots \neq \mu_k$. However, the analysis of variance does not enable conclusions as to which population means are different from which. Therefore, we must first perform multiple comparison testing, after which confidence intervals may be determined for each different population mean.

3 SAMPLE SIZE, DETECTABLE DIFFERENCE, AND POWER IN ANALYSIS OF VARIANCE

When dealing with the difference between two means, we saw how to estimate the sample size required to predict a population difference with a specified level of confidence.

Methods were presented for estimating the power of the two-sample t test, the minimum sample size required for such a test, and the minimum difference between population means that is detectable by such a test. There are also procedures for analysis-of-variance situations, namely for dealing with more than two means. (The following discussion begins with consideration of Model I—fixed-effects model—analyses of variance.)

If H_0 is true for an analysis of variance, then the variance ratio of Equation 18 follows the F distribution, this distribution being characterized by the numerator and denominator degrees of freedom (ν_1 and ν_2, respectively). If, however, H_0 is false, then the ratio of Groups MS to error MS follows instead what is known as the *noncentral F distribution*, which is defined by ν_1, ν_2, and a third quantity known as the *noncentrality parameter*. As power refers to probabilities of detecting a false null hypothesis, statistical discussions of the power of ANOVA testing depend upon the noncentral F distribution.

A number of authors have described procedures for estimating the power of an ANOVA, or the required sample size, or the detectable difference among means (e.g., Bausell and Li, 2002; Cohen, 1988: Ch. 8; Tiku, 1967, 1972), but the charts prepared by Pearson and Hartley (1951) provide one of the best of the methods and will be described below.

(a) Power of the Test. Prior to performing an experiment and collecting data from it, it is appropriate and desirable to estimate the power of the proposed test. (Indeed, it is possible that on doing so one would conclude that the power likely will be so low that the experiment needs to be run with many more data or with fewer groups or, perhaps, not run at all.)

Let us specify that an ANOVA involving k groups will be performed at the α significance level, with n data (i.e., replications) per group. We can then estimate the power of the test if we have an estimate of σ^2, the variability within the k populations (e.g., this estimate typically is s^2 from similar experiments, where s^2 is the error MS), and an estimate of the variability among the populations. From this information we may calculate a quantity called ϕ (lowercase Greek phi), which is related to the non-centrality parameter.

The variability among populations might be expressed in terms of deviations of the k population means, μ_i, from the overall mean of all populations, μ, in which case

$$\phi = \sqrt{\frac{n \sum_{i=1}^{k} (\mu_i - \mu)^2}{k\sigma^2}} \tag{32}$$

(e.g., Guenther, 1964: 47; Kirk, 1995: 182). The grand population mean is

$$\mu = \frac{\sum_{i=1}^{k} \mu_i}{k} \tag{33}$$

if all the samples are the same size. In practice, we employ the best available estimates of these population means.

Once ϕ has been obtained, we consult Figure 1 from *Appendix: Statistical Tables and Graphs*. This figure consists of several pages, each with a different ν_1 (i.e., groups DF) indicated at the upper left of the graph. Values of ϕ are indicated on the lower axis of the graph for both $\alpha = 0.01$ and $\alpha = 0.05$. Each of the curves on a graph is for a different ν_2 (i.e., error DF), for $\alpha = 0.01$ or 0.05, identified on the top margin of a graph. After turning to the graph for the ν_1 at hand, one locates the point at which the calculated ϕ intersects the curve for the given ν_2 and reads horizontally to either the right or left axis to determine the power of the test. This procedure is demonstrated in Example 4.

EXAMPLE 4   Estimating the Power of an Analysis of Variance When Variability among Population Means Is Specified

A proposed analysis of variance of plant root elongations is to comprise ten roots at each of four chemical treatments. From previous experiments, we estimate σ^2 to be 7.5888 mm^2 and estimate that two of the population means are 8.0 mm, one is 9.0 mm, and one is 12.0 mm. What will be the power of the ANOVA if we test at the 0.05 level of significance?

$k = 4$

$n = 10$

$\nu_1 = k - 1 = 3$

$\nu_2 = k(n - 1) = 4(9) = 36$

$\mu = \dfrac{8.0 + 8.0 + 9.0 + 12.0}{4} = 9.25$

$\phi = \sqrt{\dfrac{n \sum (\mu_i - \mu)^2}{ks^2}}$

$= \sqrt{\dfrac{10[(8.0 - 9.25)^2 + (8.0 - 9.25)^2 + (9.0 - 9.25)^2 + (12.0 - 9.25)^2]}{4(7.5888)}}$

$$= \sqrt{\frac{10(10.75)}{4(7.5888)}}$$

$$= \sqrt{3.5414}$$

$$= 1.88$$

In Figure 1c from *Appendix: Statistical Tables and Graphs*, we enter the graph for $\nu_1 = 3$ with $\phi = 1.88, \alpha = 0.05$, and $\nu_2 = 36$ and read a power of about 0.88. Thus, there will be a 12% chance of committing a Type II error in the proposed analysis.

An alternative, and common, way to estimate power is to specify the smallest difference we wish to detect between the two most different population means. Calling this minimum detectable difference δ, we compute

$$\phi = \sqrt{\frac{n\delta^2}{2ks^2}} \tag{34}$$

and proceed to consult Figure 1 from the appendix as above, and as demonstrated in Example 5. This procedure leads us to the statement that the power will be at least that determined from Figure 1 from the appendix (and, indeed, it typically is greater).

EXAMPLE 5   Estimating the Power of an Analysis of Variance When Minimum Detectable Difference Is Specified

For the ANOVA proposed in Example 3, we do not estimate the population means, but rather specify that, using ten data per sample, we wish to detect a difference between population means of at least 4.0 mm.

$$k = 4 \qquad\qquad \phi = \sqrt{\frac{n\delta^2}{2ks^2}}$$

$$\nu_1 = 3 \qquad\qquad =$$

$$n = 10 \qquad\qquad \sqrt{\frac{10(4.0)^2}{2(4)(7.5888)}}$$

$$\nu_2 = 36 \qquad\qquad$$

$$\delta = 4.0 \text{ mm} \qquad = \sqrt{2.6355}$$

$$s^2 = 7.5888 \text{ mm}^2 \qquad = 1.62$$

In Appendix Figure 1 from the appendix, we enter the graph for $\nu_1 = 3$ with $\phi = 1.62, \alpha = 0.05$, and $\nu_2 = 36$ and read a power of about 0.72. That is, there will be a 28% chance of committing a Type II error in the proposed analysis.

It can be seen in Figure 1 from *Appendix: Statistical Tables and Graphs* that power increases rapidly as ϕ increases, and Equations 32 and 34 show that the power is affected in the following ways:

- Power is greater for greater differences among group means (as expressed by $\Sigma(\mu_i - \mu)^2$ or by the minimum detectable difference, δ).
- Power is greater for larger sample sizes, n_i (and it is greater when the sample sizes are equal).

Multisample Hypotheses and the Analysis of Variance

- Power is greater for fewer groups, k.
- Power is greater for smaller within-group variability, σ^2 (as estimated by s^2, which is the error mean square).
- Power is greater for larger significance levels, α.

These relationships are further demonstrated in Table 3a (which shows that for a given total number of data, N, power increases with increased δ and decreases with increased k) and Table 3b (in which, for a given sample size, n_i, power is greater for larger δ's and is less for larger k's).

The desirable power in performing a hypothesis test is arbitrary, just as the significance level (α) is arbitrary. A goal of power between 0.75 and 0.90 is often used, with power of 0.80 being common.

TABLE 3a: Estimated Power of Analysis of Variance Comparison of Means, with k Samples, with Each Sample of Size $n_i = 20$, with $N = \sum_{i=1}^{k} n_i$ Total Data, and with a Pooled Variance (s^2) of 2.00, for Several Different Minimum Detectable Differences (δ)

δ	k : N :	2 40	3 60	4 80	5 100	6 120
1.0		0.59	0.48	0.42	0.38	0.35
1.2		0.74	0.64	0.58	0.53	0.49
1.4		0.86	0.78	0.73	0.68	0.64
1.6		0.94	0.89	0.84	0.81	0.78
1.8		0.97	0.95	0.92	0.90	0.88
2.0		0.99	0.98	0.97	0.95	0.94
2.2		>0.99	0.99	0.99	0.98	0.98
2.4		>0.99	>0.99	>0.99	0.99	0.99

The values of power were obtained from UNISTAT (2003: 473–474).

TABLE 3b: Estimated Power of Analysis of Variance Comparison of Means, with k Samples, with the k Sample Sizes (n_i) Totaling $N = \sum_{i=1}^{k} n_i = 60$ Data, and with a Pooled Variance (s^2) of 2.00, for Several Different Minimum Detectable Differences (δ)

δ	k : n_i :	2 30	3 20	4 15	5 12	6 10
1.0		0.77	0.48	0.32	0.23	0.17
1.2		0.90	0.64	0.44	0.32	0.24
1.4		0.96	0.78	0.58	0.43	0.32
1.6		0.99	0.89	0.71	0.54	0.42
1.8		>0.99	0.95	0.82	0.66	0.52
2.0		>0.99	0.98	0.90	0.76	0.63
2.2		>0.99	0.99	0.95	0.85	0.72
2.4		>0.99	>0.99	0.98	0.91	0.81

The values of power were obtained from UNISTAT (2003: 473–474).

Estimating the power of a proposed ANOVA may effect considerable savings in time, effort, and expense. For example, such an estimation might conclude that the power is so very low that the experiment, as planned, ought not to be performed. The proposed experimental design might be revised, perhaps by increasing n, or decreasing k, so as to render the results more likely to be conclusive. One may also strive to increase power by decreasing s^2, which may be possible by using experimental subjects that are more homogeneous. For instance, if the 19 pigs in Example 1 were not all of the same age and breed and not all maintained at the same temperature, there might well be more weight variability within the four dietary groups than if all 19 were the same in all respects except diet.

As noted for one-sample and two-sample testing, calculations of power (and of minimum required sample size and minimum detectable difference) and estimates apply to future samples, not to the samples already subjected to the ANOVA. There are both theoretical and practical reasons for this (Hoenig and Heisey, 2001).

(b) Sample Size Required. Prior to performing an analysis of variance, we might ask how many data need to be obtained in order to achieve a desired power. We can specify the power with which we wish to detect a particular difference (say, a difference of biological significance) among the population means and then ask how large the sample from each population must be. This is done, with Equation 34, by iteration (i.e., by making an initial guess and repeatedly refining that estimate), as shown in Example 6.

How well Equation 34 performs depends upon how good an estimate s^2 is of the population variance common to all groups. As the excellence of s^2 as an estimate improves with increased sample size, one should strive to calculate this statistic from a sample with a size that is not a very small fraction of the n estimated from Equation 34.

EXAMPLE 6   Estimation of Required Sample Size for a One-Way Analysis of Variance

Let us propose an experiment such as that described in Example 1. How many replicate data should be collected in each of the four samples so as to have an 80% probability of detecting a difference between population means as small as 3.5 kg, testing at the 0.05 level of significance?

In this situation, $k = 4$, $v_1 = k - 1 = 3$, $\delta = 3.5$ kg, and we shall assume (from the previous experiment in Example 1) that $s^2 = 9.383$ kg^2 is a good estimate of σ^2.

We could begin by guessing that $n = 15$ is required. Then, $v_2 = 4(15 - 1) = 56$, and by Equation 34,

$$\phi = \sqrt{\frac{n\delta^2}{2ks^2}} = \sqrt{\frac{15(3.5)^2}{2(4)(9.383)}} = 1.56.$$

Consulting Figure 1 from *Appendix: Statistical Tables and Graphs*, the power for the above v_1, v_2, α, and ϕ is approximately 0.73. This is a lower power than we desire, so we guess again with a larger n, say $n = 20$:

$$\phi = \sqrt{\frac{20(3.5)^2}{2(4)(9.383)}} = 1.81.$$

Figure 1 from the appendix indicates that this ϕ, for $\nu_2 = 4(20 - 1) = 76$, is associated with a power of about 0.84. This power is somewhat higher than we specified, so we could recalculate power using $n = 18$:

$$\phi = \sqrt{\frac{18(3.5)^2}{2(4)(9.383)}} = 1.71$$

and, for $\nu_2 = 4(18 - 1) = 68$, Figure 1 from the appendix indicates a power slightly above 0.80.

Thus, we have estimated that using sample sizes of at least 18 will result in an ANOVA of about 80% for the described experiment. (It will be seen that the use of Figure 1 from the appendix allows only approximate determinations of power; therefore, we may feel more comfortable in specifying that n should be at least 19 for each of the four samples.)

(c) Minimum Detectable Difference.   If we specify the significance level and sample size for an ANOVA and the power that we desire the test to have, and if we have an estimate of σ^2, then we can ask what the smallest detectable difference between population means will be. This is sometimes called the "effect size." By entering on Figure 1 from the appendix the specified α, ν_1, and power, we can read a value of ϕ on the bottom axis. Then, by rearrangement of Equation 34, the minimum detectable difference is

$$\delta = \sqrt{\frac{2ks^2\phi^2}{n}}. \tag{35}$$

Example 7 demonstrates this estimation procedure.

EXAMPLE 7   Estimation of Minimum Detectable Difference in a One-Way Analysis of Variance

In an experiment similar to that in Example 1, assuming that $s^2 = 9.3833$ $(kg)^2$ is a good estimate of σ^2, how small a difference between μ's can we have 90% confidence of detecting if $n = 10$ and $\alpha = 0.05$ are used?

As $k = 4$ and $n = 10$, $\nu_2 = 4(10 - 1) = 36$. For $\nu_1 = 3$, $\nu_2 = 36$, $1 - \beta = 0.90$, and $\alpha = 0.05$, Figure 1c from *Appendix: Statistical Tables and Graphs* gives a ϕ of about 2.0, from which we compute an estimate of

$$\delta = \sqrt{\frac{2ks^2\phi^2}{n}} = \sqrt{\frac{2(4)(9.3833)(2.0)^2}{10}} = 5.5 \text{ kg.}$$

(d) Maximum Number of Groups Testable.   For a given α, n, δ, and σ^2, power will decrease as k increases. It may occur that the total number of observations, N, will be limited, and for given ANOVA specifications the number of experimental groups, k, may have to be limited. As Example 8 illustrates, the maximum k can be determined by trial-and-error estimation of power, using Equation 34.

EXAMPLE 8   Determination of Maximum Number of Groups to be Used in a One-Way Analysis of Variance

Consider an experiment such as that in Example 1. Perhaps we have six feeds that might be tested, but we have only space and equipment to examine a total of 50 pigs. Let us specify that we wish to test, with $\alpha = 0.05$ and $\beta \leq 0.20$ (i.e., power of at least 80%), and to detect a difference as small as 4.5 kg between population means.

If $k = 6$ were used, then $n = 50/6 = 8.3$ (call it 8), $v_1 = 5$, $v_2 = 6(8 - 1) = 42$, and (by Equation 34)

$$\phi = \sqrt{\frac{(8)(4.5)^2}{2(6)(9.3833)}} = 1.20,$$

for which Figure 1e from the appendix indicates a power of about 0.55.

If $k = 5$ were used, $n = 50/5 = 10$, $v_1 = 4$, $v_2 = 5(10 - 1) = 45$, and

$$\phi = \sqrt{\frac{(10)(4.5)^2}{2(5)(9.3833)}} = 1.47,$$

for which Figure 1d from the appendix indicates a power of about 0.70.

If $k = 4$ were used, $n = 50/4 = 12.5$ (call it 12), $v_1 = 3$, $v_2 = 4(12 - 1) = 44$, and

$$\phi = \sqrt{\frac{(12)(4.5)^2}{2(4)(9.3833)}} = 1.80,$$

for which Figure 1c from *Appendix: Statistical Tables and Graphs* indicates a power of about 0.84.

Therefore, we conclude that no more than four of the feeds should be tested in an analysis of variance if we are limited to a total of 50 experimental pigs.

(e) Random-Effects Analysis of Variance.   If the analysis of variance is a random-effects model (described in Section 1f), the power, $1 - \beta$, may be determined from

$$F_{(1-\beta),v_1,v_2} = \frac{v_2 s^2 F_{\alpha(1),v_1,v_2}}{(v_2 - 2)(\text{groups MS})} \tag{36}$$

(after Scheffé, 1959: 227; Winer, Brown, and Michels, 1979: 246). This is shown in Example 9. As with the fixed-effects ANOVA, power is greater with larger n, larger differences among groups, larger α, and smaller s^2.

EXAMPLE 9   Estimating the Power of the Random-Effects Analysis of Variance of Example 2

$$\text{Groups MS} = 3.00; \ s^2 = 1.25; \ v_1 = 3, v_2 = 16$$

$$F_{\alpha(1),v_1,v_2} = F_{0.05(1),3,16} = 3.24$$

$$F_{(1-\beta),\nu_1,\nu_2} = \frac{\nu_2 s^2 F_{\alpha(1),\nu_1,\nu_2}}{(\nu_2 - 2)(\text{groups MS})}$$

$$= \frac{(16)(1.25)(3.24)}{(14)(3.00)} = 1.54$$

By consulting Table 4 from *Appendix: Statistical Tables and Graphs*, it is seen that an F of 1.54, with degrees of freedom of 3 and 16, is associated with a one-tailed probability between 0.10 and 0.25. (The exact probability is 0.24.) This probability is the power.

To determine required sample size in a random-effects analysis, one can specify values of α, groups MS, s^2, and k. Then, $\nu_1 = k - 1$ and $\nu_2 = k(n - 1)$; and, by iterative trial and error, one can apply Equation 36 until the desired power (namely, $1 - \beta$) is obtained.

4 NONPARAMETRIC ANALYSIS OF VARIANCE

If a set of data is collected according to a completely randomized design where $k > 2$, it is possible to test nonparametrically for difference among groups. This may be done by the *Kruskal-Wallis test** (Kruskal and Wallis, 1952), often called an "analysis of variance by ranks."[†] This test may be used in any situation where the parametric single-factor ANOVA (using F) of Section 1 is applicable, and it will be $3/\pi$ (i.e., 95.5%) as powerful as the latter; and in other situations its power, relative to F, is never less than 86.4% (Andrews, 1954; Conover 1999: 297). It may also be employed in instances where the latter is not applicable, in which case it may in fact be the more powerful test. The nonparametric analysis is especially desirable when the k samples do not come from normal populations (Keselman, Rogan, and Feir-Walsh, 1977; Krutchkoff, 1998). It also performs acceptably if the populations have no more than slightly different dispersions and shapes; but if the k variances are not the same, then (as with the Mann-Whitney test) the probability of a Type I error departs from the specified α in accordance with the magnitude of those differences (Zimmerman, 2000).[‡]

As with the parametric analysis of variance (Section 1), the Kruskal-Wallis test tends to be more powerful with larger sample sizes, and the power is less when the n_i's are not equal, especially if the large means are associated with the small n_i's (Boehnke, 1984); and it tends to be conservative if the groups with large n_i's have high within-groups variability and liberal if the large samples have low variability (Keselman, Rogan, and Feir-Walsh, 1997). Boehnke (1984) advises against using the Kruskal-Wallis test unless $N > 20$.

If $k = 2$, then the Kruskal-Wallis test is equivalent to the Mann-Whitney test. Like the Mann-Whitney test, the Kruskal-Wallis procedure does not test whether means

---

*William Henry Kruskal (b. 1919), American statistician, and Wilson Allen Wallis (b. 1912), American statistician and econometrician.

[†]As will be seen, this procedure does not involve variances, but the term *nonparametric analysis of variance* is commonly applied to it in recognition that the test is a nonparametric analog to the parametric ANOVA.

[‡]Modifications of the Kruskal-Wallis test have been proposed for nonparametric situations where the k variances are not equal (the "Behrens-Fisher problem" addressed parametrically in Section 1h) but the k populations are symmetrical (Rust and Fligner, 1984; Conover 1999: 223–224).

(or medians or other parameters) may be concluded to be different from each other, but instead addresses the more general question of whether the sampled populations have different distrubutions. However, if the shapes of the distributions are very similar, then the test does become a test for central tendency (and is a test for means if the distributions are symmetric).

The Type I error rate with heterogeneous variances is affected less with the Kruskal-Wallis test than with the parametric analysis of variance if the groups with large variances have small sample sizes (Keselman, Rogan, and Feir-Walsh, 1977; Tomarkin and Serlin, 1986).

Example 10 demonstrates the Kruskal-Wallis test procedure. As in other nonparametric tests, we do not use population parameters in statements of hypotheses, and neither parameters nor sample statistics are used in the test calculations. The Kruskal-Wallis test statistic, H, is calculated as

$$H = \frac{12}{N(N + 1)} \sum_{i=1}^{k} \frac{R_i^2}{n_i} - 3(N + 1), \tag{37}$$

where n_i is the number of observations in group i, $N = \sum_{i=1}^{k} n_i$ (the total number of observations in all k groups), and R_i is the sum of the ranks of the n_i observations in group i.* A good check (but not a guarantee) of whether ranks have been assigned correctly is to see whether the sum of all the ranks equals $N(N + 1)/2$.

Critical values of H for small sample sizes where $k \leq 5$ are given in Table 13 from *Appendix: Statistical Tables and Graphs*. For larger samples and/or for $k > 5$, H may be considered to be approximated by χ^2 with $k - 1$ degrees of freedom. Chi-square, χ^2, is a statistical distribution that is shown in Table 1 from *Appendix: Statistical Tables and Graphs*, where probabilities are indicated as column headings and degrees of freedom (ν) designate the rows.

If there are tied ranks, as in Example 11, H is a little lower than it should be, and a correction factor may be computed as

$$C = 1 - \frac{\sum t}{N^3 - N}, \tag{40}$$

and the corrected value of H is

$$H_c = \frac{H}{C}. \tag{41}$$

---

*Interestingly, H (or H_c of Equation 41) could also be computed as

$$H = \frac{\text{groups SS}}{\text{total MS}}, \tag{38}$$

applying the procedures of Section 1 to the ranks of the data in order to obtain the Groups SS and Total MS. And, because the Total MS is the variance of all N ranks, if there are no ties the Total MS is the variance of the integers from 1 to N, which is

$$\frac{N(N + 1)(2N + 1)/6 - N^2(N + 1)^2/4N}{N - 1} \tag{38a}$$

The following alternate formula (Pearson and Hartley, 1976: 49) shows that H is expressing the differences among the κ groups' mean ranks ($\overline{R}_i = R_i/n_i$) and the mean of all N ranks, which is $\overline{R} = N(N - 1)/2$:

$$H = \frac{12 \sum_{i=1}^{\kappa} n_i(\overline{R}_i - \overline{R})^2}{N(N - 1)}. \tag{39}$$

EXAMPLE 10   The Kruskal-Wallis Single-Factor Analysis of Variance by Ranks

An entomologist is studying the vertical distribution of a fly species in a deciduous forest and obtains five collections of the flies from each of three different vegetation layers: herb, shrub, and tree.

H_0:   The abundance of the flies is the same in all three vegetation layers.

H_A: The abundance of the flies is not the same in all three vegetation layers.

$\alpha = 0.05$

The data are as follows (with ranks of the data in parentheses):*

\multicolumn{3}{c}{**Numbers of Flies/m³ of Foliage**}		
Herbs	*Shrubs*	*Trees*
14.0 (15)	8.4 (11)	6.9 (8)
12.1 (14)	5.1 (2)	7.3 (9)
9.6 (12)	5.5 (4)	5.8 (5)
8.2 (10)	6.6 (7)	4.1 (1)
10.2 (13)	6.3 (6)	5.4 (3)
$n_1 = 5$	$n_2 = 5$	$n_3 = 5$
$R_1 = 64$	$R_2 = 30$	$R_3 = 26$

$$N = 5 + 5 + 5 = 15$$

$$H = \frac{12}{N(N+1)} \sum_{i=1}^{k} \frac{R_i^2}{n_1} - 3(N+1)$$

$$= \frac{12}{15(16)} \left[\frac{64^2}{5} + \frac{30^2}{5} + \frac{26^2}{5} \right] - 3(16)$$

$$= \frac{12}{240}[1134.400] - 48$$

$$= 56.720 - 48$$

$$= 8.720$$

$H_{0.05,5,5,5} = 5.780$

Reject H_0.

$$0.005 < P < 0.01$$

*To check whether ranks were assigned correctly, the sum of the ranks (or sum of the rank sums: $64 + 30 + 26 = 120$) is compared to $N(N+1)/2 = 15(16)/2 = 120$. This check will not guarantee that the ranks were assigned properly, but it will often catch errors of doing so.

EXAMPLE 11   The Kruskal-Wallis Test with Tied Ranks

A limnologist obtained eight containers of water from each of four ponds. The pH of each water sample was measured. The data are arranged in ascending order within each pond. (One of the containers from pond 3 was lost, so $n_3 = 7$, instead of 8; but the test procedure does not require equal numbers of data in each group.) The rank of each datum is shown parenthetically.

H_0: pH is the same in all four ponds.

H_A: pH is not the same in all four ponds.

$\alpha = 0.05$

Pond 1	Pond 2	Pond 3	Pond 4
7.68 (1)	7.71 (6*)	7.74 (13.5*)	7.71 (6*)
7.69 (2)	7.73 (10*)	7.75 (16)	7.71 (6*)
7.70 (3.5*)	7.74 (13.5*)	7.77 (18)	7.74 (13.5*)
7.70 (3.5*)	7.74 (13.5*)	7.78 (20*)	7.79 (22)
7.72 (8)	7.78 (20*)	7.80 (23.5*)	7.81 (26*)
7.73 (10*)	7.78 (20*)	7.81 (26*)	7.85 (29)
7.73 (10*)	7.80 (23.5*)	7.84 (28)	7.87 (30)
7.76 (17)	7.81 (26*)		7.91 (31)

*Tied ranks.

$n_1 = 8$	$n_2 = 8$	$n_3 = 7$	$n_4 = 8$
$R_1 = 55$	$R_2 = 132.5$	$R_3 = 145$	$R_4 = 163.5$

$N = 8 + 8 + 7 + 8 = 31$

$$H = \frac{12}{N(N+1)} \sum_{i=1}^{k} \frac{R_i^2}{n_i} - 3(N+1)$$

$$= \frac{12}{31(32)} \left[\frac{55^2}{8} + \frac{132.5^2}{8} + \frac{145^2}{7} + \frac{163.5^2}{8} \right] - 3(32)$$

$$= 11.876$$

Number of groups of tied ranks $= m = 7$.

$$\sum t = \sum (t_i^3 - t_i)$$

$$= (2^3 - 2) + (3^3 - 3) + (3^3 - 3) + (4^3 - 4)$$
$$+ (3^3 - 3) + (2^3 - 2) + (3^3 - 3)$$

$$= 168$$

$$C = 1 - \frac{\sum t}{N^3 - N} = 1 - \frac{168}{31^3 - 31} = 1 - \frac{168}{29760} = 0.9944$$

$$H_c = \frac{H}{C} = \frac{11.876}{0.9944} = 11.943$$

$$\nu = k - 1 = 3$$

$$\chi^2_{0.05,3} = 7.815$$

Reject H_0.

$$0.005 < P < 0.01 \quad [P = 0.0076]$$

or, by Equation 43,

$$F = \frac{(N - k)H_c}{(k - 1)(N - 1 - H_c)} = \frac{(31 - 4)(11.943)}{(4 - 1)(31 - 1 - 11.943)} = 5.95$$

$$F_{0.05(1),3,26} = 2.98$$

Reject H_0.

$$0.0025 < P < 0.005 \quad [P = 0.0031]$$

Here,

$$\sum t = \sum_{i=1}^{m} (t_i^3 - t_i), \tag{42}$$

where t_i is the number of ties in the ith group of ties, and m is the number of groups of tied ranks. H_c will differ little from H when the t_i's are very small compared to N.

Kruskal and Wallis (1952) give two approximations that are better than chi-square when the n_i's are small or when significance levels less than 1% are desired; but they are relatively complicated to use. The chi-square approximation is slightly conservative for $\alpha = 0.05$ or 0.10 (i.e., the true Type I probability is a little less than α) and more conservative for $\alpha = 0.01$ (Gabriel and Lachenbruch, 1969): it performs better with larger n_i's. Fahoome (2002) found the probability of a Type I error to be between 0.045 and 0.055 when employing this approximation at the 0.05 significance level if each sample size is at least 11, and between 0.009 and 0.011 when testing at $\alpha = 0.01$ when each $n_i \geq 22$.

Because the χ^2 approximation tends to be conservative, other approximations have been proposed that are better in having Type I error probabilities closer to α. A good alternative is to calculate

$$F = \frac{(N - k)H}{(k - 1)(N - 1 - H)}, \tag{43}$$

which is also the test statistic that would be obtained by applying the parametric ANOVA of Section 1 to the ranks of the data (Iman, Quade and Alexander, 1975). For the Kruskal-Wallis test, this F gives very good results, being only slightly liberal (with the probability of a Type I error only a little larger than the specified α), and the preferred critical values are F for the given α and degrees of freedom of $\nu_1 = k - 1$ and $\nu_2 = N - k - 1$.* This is demonstrated at the end of Example 11.

---

*A slightly better approximation in some, but not all, cases is to compare

$$\frac{H}{2}\left[1 + \frac{N - k}{N - 1 - H}\right] \quad \text{to} \quad \frac{(k - 1)F_{\alpha(1),k-1,N-k} + \chi^2_{\alpha,k-1}}{2}. \tag{43a}$$

5 TESTING FOR DIFFERENCE AMONG SEVERAL MEDIANS

The *median test* for the two-sample case may be expanded to multisample considerations (Mood, 1950: 398–399). The method requires the determination of the grand median of all observations in all k samples considered together. The numbers of data in each sample that are above and below this median are tabulated, and the significance of the resultant $2 \times k$ contingency table is then analyzed, generally by chi-square, alternatively by the G test. For example, if there were four populations being compared, the statistical hypotheses would be H_0: all four populations have the same median, and H_A: all four populations do not have the same median. The median test would be the testing of the following contingency table:

	Sample 1	Sample 2	Sample 3	Sample 4	Total
Above median	f_{11}	f_{12}	f_{13}	f_{14}	R_1
Below median	f_{21}	f_{22}	f_{23}	f_{24}	R_2
Total	C_1	C_2	C_3	C_4	n

This multisample median test is demonstrated in Example 12.

EXAMPLE 12   The Multisample Median Test

H_0: Median elm tree height is the same on all four sides of a building.

H_A: Median elm tree height is not the same on all four sides of a building.

A total of 48 seedlings of the same size were planted at the same time, 12 on each of a building's four sides. The heights, after several years of growth, were as follows:

	North	East	South	West
	7.1 m	6.9 m	7.8 m	6.4 m
	7.2	7.0	7.9	6.6
	7.4	7.1	8.1	6.7
	7.6	7.2	8.3	7.1
	7.6	7.3	8.3	7.6
	7.7	7.3	8.4	7.8
	7.7	7.4	8.4	8.2
	7.9	7.6	8.4	8.4
	8.1	7.8	8.6	8.6
	8.4	8.1	8.9	8.7
	8.5	8.3	9.2	8.8
	8.8	8.5	9.4	8.9

medians: 7.7 m   7.35 m   8.4 m   8.0 m

grand median $= 7.9$ m

The 2×4 contingency table is as follows, with expected frequencies in parentheses:

	North	*East*	*South*	*West*	
Above median	4 (5.5000)	3 (6.0000)	10 (5.5000)	6 (6.0000)	23
Below median	7 (5.5000)	9 (6.0000)	1 (5.5000)	6 (6.0000)	23
Total	11	12	11	12	46

$\chi^2 = 11.182$

$\chi^2_{0.05,3} = 7.815$

Reject H_0.

$$0.0005 < P < 0.001 \ [P = 0.00083]$$

If the k samples came from populations having the same variance and shape, then the Kruskal-Wallis test may be used as a test for difference among the k population medians.

6 HOMOGENEITY OF VARIANCES

You may recall a discussion on testing the null hypothess $H_0: \sigma_1^2 = \sigma_2^2$ against the alternate, $H_A: \sigma_1^2 \neq \sigma_2^2$. This pair of two-sample hypotheses can be extended to more than two samples (i.e., $k > 2$) to ask whether all k sample variances estimate the same population variance. The null and alternate hypotheses would then be $H_0: \sigma_1^2 = \sigma_2^2 = \cdots = \sigma_k^2$ and H_0: the k population variances are not all the same. The equality of variances is called homogeneity of variances, or *homoscedasticity*; variance heterogeneity is called *heteroscedasticity*.[*]

(a) Bartlett's Test. A commonly encountered method employed to test for homogeneity of variances is *Bartlett's test*[†] (Bartlett, 1937a, 1937b; based on a principle of Neyman and Pearson, 1931). In this procedure, the test statistic is

$$B = (\ln s_p^2) \left(\sum_{i=1}^{k} \nu_i \right) - \sum_{i=1}^{k} \nu_i \ln s_i^2, \tag{44}$$

where $\nu_i = n_i - 1$ and n_i is the size of sample i. The pooled variance, s_p^2, is calculated as before as $\sum_{i=1}^{k} SS_i / \sum_{i=1}^{k} \nu_i$. Many researchers prefer to operate with common logarithms (base 10), rather than with natural logarithms (base e);[‡] so Equation 44 may be written as

$$B = 2.30259[(\log s_p^2) \left(\sum_{i=1}^{k} \nu_i \right) - \sum_{i=1}^{k} \nu_i \log s_i^2]. \tag{45}$$

The distribution of B is approximated by the chi-square distribution,[§] with $k - 1$ degrees of freedom, but a more accurate chi-square approximation is obtained by

---

[*]The two terms were introduced by K. Pearson in 1905 (Walker, 1929: 181); since then they have occasionally been spelled *homoskedasticity* and *heteroskedasticity*, respectively.

[†]Maurice Stevenson Bartlett (1910–2002), English statistician.

[‡]See footnote in Section 8.7.

[§]A summary of approximations is given by Nagasenker (1984).

computing a correction factor,

$$C = 1 + \frac{1}{3(k-1)} \left(\sum_{i=1}^{k} \frac{1}{v_i} - \frac{1}{\sum_{i=1}^{k} v_i} \right), \tag{46}$$

with the corrected test statistic being

$$B_c = \frac{B}{C}. \tag{47}$$

Example 13 demonstrates these calculations. The null hypothesis for testing the homogeneity of the variances of four populations may be written symbolically as $H_0: \sigma_1^2 = \sigma_2^2 = \sigma_3^2 = \sigma_4^2$, or, in words, as "the four population variances are homogeneous (i.e., are equal)." The alternate hypothesis can be stated as "The four population variances are not homogeneous (i.e., they are not all equal)," or "There is difference (or heterogeneity) among the four population variances." If H_0 is rejected, further testing will allow us to ask which population variances are different from which.

Bartlett's test is powerful if the sampled populations are normal, but it is very badly affected by nonnormal populations (Box, 1953; Box and Anderson, 1955; Gartside, 1972). If the population distribution is platykurtic, the true α is less than the stated α (i.e., the test is conservative and the probability of a Type II error is increased); if it is leptokurtic, the true α is greater than the stated α (i.e., the probability of a Type I error is increased).

When $k = 2$ and $n_1 = n_2$, Bartlett's test is equivalent to the variance-ratio test. However, with two samples of unequal size, the two procedures may yield different results; one will be more powerful in some cases, and the other more powerful in others (Maurais and Ouimet, 1986).

(b) Other Multisample Tests for Variances. There are other tests for heterogeneity (Levene's test and others) but that all are undesirable in many situations. The Bartlett test remains commendable when the sampled populations are normal, and no procedure is especially good when they are not.

Because of the poor performance of tests for variance homogeneity and the robustness of analysis of variance for multisample testing among means (Section 1), it is not recommended that the former be performed as tests of the underlying assumptions of the latter.

7 HOMOGENEITY OF COEFFICIENTS OF VARIATION

The two-sample procedure has been extended by Feltz and Miller (1996) for hypotheses where $k \geq 3$ and each coefficient of variation (V_i) is positive:

$$\chi^2 = \frac{\sum_{i=1}^{k} v_i V_i^2 - \frac{\left(\sum_{i=1}^{k} v_i V_i \right)^2}{\sum_{i=1}^{k} v_i}}{V_p^2(0.5 + V_p^2)}, \tag{48}$$

EXAMPLE 13   Bartlett's Test for Homogeneity of Variances

Nineteen pigs were divided into four groups, and each group was raised on a different food. The data, which are those of Example 1, are weights, in kilograms, and we wish to test whether the variance of weights is the same for pigs fed on all four feeds.

H_0: $\sigma_1^2 = \sigma_2^2 = \sigma_3^2 = \sigma_4^2$

H_A: The four population variances are not all equal (i.e., are heterogeneous).

$\alpha = 0.05$

	Feed 1	Feed 2	Feed 3	Feed 4
	60.8	68.7	69.6	61.9
	67.0	67.7	77.1	64.2
	65.0	75.0	75.2	63.1
	68.6	73.3	71.5	66.7
	61.7	71.8		60.3

i	1	2	3	4	
n_i	5	5	4	5	
v_i	4	4	3	4	$\sum\limits_{i=1}^{k} v_i = 15$
SS_i	44.768	37.660	34.970	23.352	$\sum\limits_{i=1}^{k} SS_i = 140.750$

s_i^2	11.192	9.415	11.657	5.838	
$\log s_i^2$	1.0489	0.9738	1.0666	0.7663	
$v_i \log s_i^2$	4.1956	3.8952	3.1998	3.0652	$\sum\limits_{i=1}^{k} v_i \log s_i^2 = 14.3558$
$1/v_i$	0.250	0.250	0.333	0.250	$\sum\limits_{i=1}^{k} 1/v_i = 1.083$

$$s_p^2 = \frac{\sum SS_i}{\sum v_i} = \frac{140.750}{15} = 9.3833 \qquad C = 1 + \frac{1}{3(k-1)}$$

$$\times \left(\sum \frac{1}{v_i} - \frac{1}{\sum v_i} \right)$$

$$\log s_p^2 = 0.9724$$

$$= 1 + \frac{1}{3(3)} \left(1.083 - \frac{1}{15} \right)$$

$$B = 2.30259 \left[(\log s_p^2) \left(\sum v_i \right) \right. \qquad = 1.113$$

$$\left. - \sum v_i \log s_i^2 \right]$$

$$= 2.30259[(0.9724)(15) - 14.3558] \qquad B_c = \frac{B}{C} = \frac{0.530}{1.113} = 0.476$$

$$= 2.30259(0.2302) \qquad \chi_{0.05,3}^2 = 7.815$$

$$= 0.530 \qquad \text{Do not reject } H_0.$$

$$0.90 < P < 0.95 \quad [P = 0.92]$$

where the common coefficient of variation is

$$V_p = \frac{\sum\limits_{i=1}^{k} \nu_i V_i}{\sum\limits_{i=1}^{k} \nu_i}. \tag{49}$$

This test statistic approximates the chi-square distribution with $k - 1$ degrees of freedom and its computation is shown in Example 14. When $k = 2$, the test yields results identical to the two-sample test using the equation,

$$Z = \frac{V_1 - V_2}{\sqrt{\left(\dfrac{V_p^2}{\nu_1} + \dfrac{V_p^2}{\nu_2}\right)(0.5 + V_p^2)}},$$

(and $\chi^2 = Z^2$). As with other tests, the power is greater with larger sample size; for a given sample size, the power is greater for smaller coefficients of variation and for greater differences among coefficients of variation. If the null hypothesis of equal population coefficients of variation is not rejected, then V_p is the best estimate of the coefficient of variation common to all k populations.

EXAMPLE 14   Testing for Homogeneity of Coefficients of Variation

For the data of Example 1:

H_0:   The coefficients of the four sampled populations are the same; i.e., $\sigma_1^2/\mu_1 = \sigma_2^2/\mu_2 = \sigma_3^2/\mu_3 = \sigma_4^2/\mu_4$.

H_A: The coefficients of variation of the four populations are not all the same.

	Feed 1	Feed 2	Feed 3	Feed 4
n_i	5	5	4	5
ν_i	4	4	3	4
$\overline{X}_i$ (kg)	64.62	68.30	73.35	66.64
s_i^2 (kg^2)	11.192	16.665	11.657	9.248
s_i (kg)	3.35	4.08	3.41	3.04
V_i	0.0518	0.0597	0.0465	0.0456

$$\sum_{i=1}^{n} \nu_i = 4 + 4 + 3 + 4 = 15$$

$$\sum_{i=1}^{n} \nu_i V_i = (4)(0.0518) + (4)(0.0597) + (3)(0.0465) + (4)(0.0456) = 0.7679$$

$$V_p = \frac{\sum \nu_i V_i}{\sum \nu_i} = \frac{0.7679}{15} = 0.05119 \qquad V_p^2 = (0.7679)^2 = 0.002620$$

$$\sum_{i=1}^{n} \nu V_i^2 = (4)(0.0518)^2 + (4)(0.0597)^2 + (3)(0.0465)^2 + (4)(0.0456)^2$$

$$= 0.03979$$

$$\chi^2 = \frac{\sum \nu_i V_i^2 - \dfrac{\left(\sum \nu_i V_i\right)^2}{\sum \nu_i}}{V_p^2(0.5 + V_p^2)} = \frac{0.03979 - \dfrac{(0.7679)^2}{15}}{0.002620(0.5 + 0.002620)} = \frac{0.0004786}{0.001317} = 0.363$$

For chi-square: $\nu = 4 - 1 = 3$; $\chi^2_{0.05,3} = 7.815$. Do not reject H_0.

$$0.90 < P < 0.95 \ [P = 0.948]$$

Miller and Feltz (1997) reported that this test works best if each sample size (n_i) is at least 10 and each coefficient of variation (V_i) is no greater than 0.33; and they describe how the power of the test (and, from such a calculation, the minimum detectable difference and the required sample size) may be estimated.

8 CODING DATA

In the parametric ANOVA, coding the data by addition or subtraction of a constant causes no change in any of the sums of squares or mean squares, so the resultant F and the ensuing conclusions are not affected at all. If the coding is performed by multiplying or dividing all the data by a constant, the sums of squares and the mean squares in the ANOVA each will be altered by an amount equal to the square of that constant, but the F value and the associated conclusions will remain unchanged.

A test utilizing ranks (such as the Kruskal-Wallis procedure) will not be affected at all by coding of the raw data. Thus, the coding of data for analysis of variance, either parametric or nonparametric, may be employed with impunity, and coding frequently renders data easier to manipulate. Bartlett's test is also unaffected by coding. The testing of coefficients of variation is unaffected by coding by multiplication or division, but coding by addition or subtraction may not be used.

9 MULTISAMPLE TESTING FOR NOMINAL-SCALE DATA

A $2 \times c$ contingency table may be analyzed to compare frequency distributions of nominal data for two samples. In a like fashion, an $r \times c$ contingency table may be set up to compare frequency distributions of nominal-scale data from r samples.

Other procedures have been proposed for multisample analysis of nominal-scale data (e.g., Light and Margolin, 1971; Windsor, 1948).

EXERCISES

1. The following data are weights of food (in kilograms) consumed per day by adult deer collected at different times of the year. Test the null hypothesis that food consumption is the same for all the months tested.

Feb.	May	Aug.	Nov.
4.7	4.6	4.8	4.9
4.9	4.4	4.7	5.2
5.0	4.3	4.6	5.4
4.8	4.4	4.4	5.1
4.7	4.1	4.7	5.6
	4.2	4.8	

2. An experiment is to have its results examined by analysis of variance. The variable is temperature (in degrees Celsius), with 12 measurements to be taken in each of five experimental groups. From previous experiments, we estimate the within-groups variability, σ^2, to be $1.54(°C)^2$. If the 5% level of significance is employed, what is the probability of the ANOVA detecting a difference as small as $2.0°C$ between population means?

3. For the experiment of Exercise 2, how many replicates are needed in each of the five groups to detect a difference as small as $2.0°C$ between population means, with 95% power?

4. For the experiment of Exercise 2, what is the smallest difference between population means that we are 95% likely to detect with an ANOVA using 10 replicates per group?

5. Using the Kruskal-Wallis test, test nonparametrically the appropriate hypotheses for the data of Exercise 1.

6. Three different methods were used to determine the dissolved-oxygen content of lake water. Each of the three methods was applied to a sample of water six times, with the following results. Test the null hypothesis that the three methods yield equally variable results ($\sigma_1^2 = \sigma_2^2 = \sigma_3^2$).

Method 1 (mg/kg)	Method 2 (mg/kg)	Method 3 (mg/kg)
10.96	10.88	10.73
10.77	10.75	10.79
10.90	10.80	10.78
10.69	10.81	10.82
10.87	10.70	10.88
10.60	10.82	10.81

7. The following statistics were obtained from measurements of the circumferences of trees of four species. Test whether the coefficients of variation of circumferences are the same among the four species.

	Species B	Species A	Species Q	Species H
n.	40	54	58	32
$\overline{X}$ (m):	2.126	1.748	1.350	1.392
s^2 (m^2):	0.488219	0.279173	0.142456	0.203208

TABLE 4: Tail Lengths (in mm) of Field Mice from Different Localities

Location	n	$\overline{X} \pm SD$ (range in parentheses)
Bedford, Indiana	18	56.22 ± 1.33 (44.8 to 68.9)
Rochester, Minnesota	12	59.61 ± 0.82 (43.9 to 69.8)
Fairfield, Iowa	16	60.20 ± 0.92 (52.4 to 69.2)
Pratt, Kansas	16	53.93 ± 1.24 (46.1 to 63.6)
Mount Pleasant, Michigan	13	55.85 ± 0.90 (46.7 to 64.8)

TABLE 5: Evaporative Water Loss of a Small Mammal at Various Air Temperatures. Sample Statistics Are Mean ± Standard Deviation, with Range in Parentheses

	Air Temperature (°C)				
	16.2	24.8	30.7	36.8	40.9
Sample size	10	13	10	8	9
Evaporative water loss (mg/g/hr)	0.611 ± 0.164 (0.49 to 0.88)	0.643 ± 0.194 (0.38 to 1.13)	0.890 ± 0.212 (0.64 to 1.39)	1.981 ± 0.230 (1.50 to 2.36)	3.762 ± 0.641 (3.16 to 5.35)

TABLE 6: Enzyme Activities in the Muscle of Various Animals. Data Are $\overline{X}$ ± SE, with n in Parentheses

Animal	Enzyme Activity (μmole/min/g of tissue)	
	Isomerase	Transketolase
Mouse	0.76 ± 0.09 (4)	0.39 ± 0.04 (4)
Frog	1.53 ± 0.08 (4)	0.18 ± 0.02 (4)
Trout	1.06 ± 0.12 (4)	0.24 ± 0.04 (4)
Crayfish	4.22 ± 0.30 (4)	0.26 ± 0.05 (4)

ANSWERS TO EXERCISES

1. $H_0: \mu_1 = \mu_2 = \mu_3 = \mu_4$; H_A: The mean food consumption is not the same for all four months; $F = 0.7688/0.0348 = 22.1$; $F_{0.05(1),3,18} = 3.16$; reject H_0; $P < 0.0005$ [$P = 0.0000029$].

2. $k = 5$, $\nu_1 = 4$, $n = 12$, $\nu_2 = 55$, $\sigma^2 = 1.54(°C)^2$, $\delta = 2.0°C$; $\phi = 1.77$; from Appendix Figure 1d we find that the power is about 0.88.

3. $n = 16$, for which $\nu_2 = 75$ and $\phi = 2.04$. (The power is a little greater than 0.95; for $n = 15$ the power is about 0.94.)

4. $\nu_2 = 45$, power $= 0.95$, $\phi = 2.05$; minimum detectable difference is about 2.5°C.

5. H_0: The amount of food consumed is the same during all four months; H_A: The amount of food consumed is not the same during all four months; $n_1 = 5, n_2 = 6, n_3 = 6, n_4 = 5$; $R_1 = 69.5$,

$R_2 = 23.5$, $R_3 = 61.5$, $R_4 = 98.5$; $N = 22$; $H = 17.08$; $\chi^2_{0.05,3} = 7.815$; reject H_0; $P \ll 0.001$. H_c (i.e., H corrected for ties) would be obtained as $\sum t = 120$, $C = 0.9887$, $H_c = 17.28$. $F = 27.9$, $F_{0.05(1),3,17} = 3.20$; reject H_0; $P \ll 0.0005$ [$P = 0.00000086$].

6. $H_0: \sigma_1^2 = \sigma_2^2 = \sigma_3^2$; H_A: The three population variances are not all equal; $B = 5.94517$, $C = 1.0889$, $B_c = B/C = 5.460$; $\chi^2_{0.05,2} = 5.991$; do not reject H_0; $0.05 < P < 0.10$ [$P = 0.065$].

7. $H_0: \mu_1/\sigma_1 = \mu_2/\sigma_2 = \mu_3/\sigma_3 = \mu_4/\sigma_4$; $s_1 = 0.699$, $V_1 = 0.329$, $s_2 = 0.528$, $V_2 = 0.302$, $s_3 = 0.377$, $V_3 = 0.279$, $s_4 = 0.451$, $V_4 = 0.324$; $V_p = 0.304$; $\chi^2 = 1.320$, $\chi^2_{0.05,3} = 7.815$; do not reject H_0; $0.50 < P < 0.75$ [$P = 0.72$].

Multiple Comparisons

From Chapter 11 of *Biostatistical Analysis*, Fifth Edition, Jerrold H. Zar. Copyright © 2010 by
Pearson Education, Inc. Publishing as Pearson Prentice Hall. All rights reserved.

Multiple Comparisons

The Model I single-factor analysis of variance (ANOVA) tests the null hypothesis H_0: $\mu_1 = \mu_2 = \cdots = \mu_k$. However, the rejection of H_0 does not imply that all k population means are different from one another, and we don't know how many differences there are or where differences lie among the k means. For example, if $k = 3$ and H_0: $\mu_1 = \nu_2 = \mu_3$ is rejected, we are not able to conclude whether there is evidence of $\mu_1 \neq \nu_2 = \mu_3$ or of $\mu_1 = \nu_2 \neq \mu_3$ or of $\mu_1 \neq \nu_2 \neq \mu_3$.

It is invalid to employ multiple two-sample t tests to examine the difference among more than two means, for to do so would increase the probability of a Type I error. This chapter presents statistical procedures that may be used to compare k means with each other; they are called *multiple-comparison procedures** (MCPs). Except for the procedure known as the least significance difference test, all of the tests referred to in this chapter may be performed even without a preliminary analysis of variance. Indeed, power may be lost if a multiple-comparison test is performed only if the ANOVA concludes a significant difference among means (Hsu, 1996: 177–178; Myers and Well, 2003: 261). And all except the Scheffé test of Section 4 are for a set of comparisons to be specified before the collection of data.

The most common principle for multiple-comparison testing is that the significance level, α, is the probability of committing at least one Type I error when making all of the intended comparisons for a set of data. These are said to be a *family* of comparisons, and this error is referred to as *familywise error* (FWE) or, sometimes, *experimentwise error*. Much less common are tests designed to express *comparison-wise error*, the probability of a Type I error in a single comparison.

A great deal has been written about numerous multiple-comparison tests with various objectives, and the output of many statistical computer packages enhances misuse of them (Hsu, 1996: xi). Although there is not unanimity regarding what the "best" procedure is for a given situation, this chapter will present some frequently encountered highly regarded tests for a variety of purposes.

If the desire is to test for differences between members of all possible pairs of means, then the procedures of Section 1 would be appropriate, using Section 1a

---

*The term *multiple comparisons* was introduced by D. E. Duncan in 1951 (David, 1995).

if sample sizes are unequal and Section 1b if variances are not the same. If the data are to be analyzed to compare the mean of one group (typically called the control) to each of the other group means, then Section 3 would be applicable. And if the researcher wishes to examine sample means after the data are collected and compare specific means, or groups of means, of interest, then the testing in Section 4 is called for.

Just as with the parametric analysis of variance, the testing procedures of Sections 1–4 are premised upon there being a normal distribution of the population from which each of the k samples came; but, like the ANOVA, these tests are somewhat robust to deviations from that assumption. However, if it is suspected that the underlying distributions are far from normal, then the analyses of Section 5, or data transformations, should be considered. Multiple-comparison tests are adversely affected by heterogeneous variances among the sampled populations, in the same manner as in ANOVA (Keselman and Toothaker, 1974; Petrinovich and Hardyck, 1969), though to a greater extent (Tukey, 1993).

In multiple-comparison testing—except when comparing means to a control—equal sample sizes are desirable for maximum power and robustness, but the procedures presented can accommodate unequal n's. Petrinovich and Hardyck (1969) caution that the power of the tests is low when sample sizes are less than 10.

This chapter discusses multiple comparisons for the single-factor ANOVA experimental design.*

1 TESTING ALL PAIRS OF MEANS

There are $k(k - 1)/2$ different ways to obtain pairs of means from a total of k means.[†] For example, if $k = 3$, the $k(k - 1)/2 = 3(2)/2 = 3$ pairs are μ_1 and μ_2, μ_1 and μ_3, and μ_2 and μ_3; and for $k = 4$, the $k(k - 1)/2 = 4(3)/2 = 6$ pairs are μ_1 and μ_2, μ_1 and μ_3, μ_1 and μ_4, μ_2 and μ_3, μ_2 and μ_4, and μ_3 and μ_4. So each of $k(k - 1)/2$ null hypotheses may be tested, referring to them as H_0: $\mu_B = \mu_A$, where the subscripts A and B represent each pair of subscripts; each corresponding alternate hypothesis is H_0: $\mu_B \neq \mu_A$.

An excellent way to address these hypotheses is with the *Tukey test* (Tukey, 1953), also known as the honestly significant difference test (HSD test) or wholly significant difference test (WSD test). Example 1 demonstrates the Tukey test, utilizing an ANOVA experimental design, where all groups have equal numbers of data (i.e., all of the n_i's are equal). The first step in examining these multiple-comparison hypotheses is to arrange and number all five sample means in order of increasing magnitude. Then pairwise differences between the means, $\overline{X}_A - \overline{X}_B$, are tabulated. Just as a difference between means, divided by the appropriate standard error, yields

---

*For nonparametric testing, Conover and Iman (1981) recommend applying methods as those in Sections 1–4 on the ranks of the data. However Hsu (1996: 177); Sawilowsky, Blair, and Higgins (1999); and Toothaker (1991: 109) caution against doing so.

[†]The number of combinations of k groups taken 2 at a time is:

$$_kC_2 = \frac{k!}{2!(k - 2)!} = \frac{k(k - 1)(k - 2)!}{2!(k - 2)!} = \frac{k(k - 1)}{2}. \tag{1}$$

EXAMPLE 1   Tukey Multiple Comparison Test with Equal Sample Sizes.

The data are strontium concentrations (mg/ml) in five different bodies of water. First an analysis of variance is performed.

H_0: $\mu_1 = \mu_2 = \mu_3 = \mu_4 = \mu_5$.

H_A: Mean strontium concentrations are not the same in all five bodies of water.

$\alpha = 0.05$

Grayson's Pond	Beaver Lake	Angler's Cove	Appletree Lake	Rock River
28.2	39.6	46.3	41.0	56.3
33.2	40.8	42.1	44.1	54.1
36.4	37.9	43.5	46.4	59.4
34.6	37.1	48.8	40.2	62.7
29.1	43.6	43.7	38.6	60.0
31.0	42.4	40.1	36.3	57.3

$\overline{X}_1 = 32.1$ mg/ml $\overline{X}_2 = 40.2$ mg/ml $\overline{X}_3 = 44.1$ mg/ml $\overline{X}_4 = 41.1$ mg/ml $\overline{X}_5 = 58.3$ mg/ml

$n_1 = 6$ $n_2 = 6$ $n_3 = 6$ $n_4 = 6$ $n_5 = 6$

Source of variation	SS	DF	MS
Total	2437.5720	29	
Groups	2193.4420	4	548.3605
Error	244.1300	25	9.7652

$k = 5, n = 6$

Samples number (i) of ranked means: 1   2   4   3   5

Ranked sample mean ($\overline{X}_i$):   32.1   40.2   41.1   44.1   58.3

To test each H_0: $\mu_B = \mu_A$,

$$SE = \sqrt{\frac{9.7652}{6}} = \sqrt{1.6275} = 1.28.$$

As $q_{0.05,25,k}$ does not appear in Table 5 from *Appendix: Statistical Tables and Graphs*, the critical value with the next lower DF is used: $q_{0.05,24,5} = 4.166$.

Comparison B vs. A	Difference $(\overline{X}_B - \overline{X}_A)$	SE	q	Conclusion
5 vs. 1	$58.3 - 32.1 = 26.2$	1.28	20.47	Reject H_0: $\mu_5 = \mu_1$
5 vs. 2	$58.3 - 40.2 = 18.1$	1.28	14.14	Reject H_0: $\mu_5 = \mu_2$
5 vs. 4	$58.3 - 41.1 = 17.2$	1.28	13.44	Reject H_0: $\mu_5 = \mu_4$
5 vs. 3	$58.3 - 44.1 = 14.2$	1.28	11.09	Reject H_0: $\mu_5 = \mu_3$
3 vs. 1	$44.1 - 32.1 = 12.0$	1.28	9.38	Reject H_0: $\mu_3 = \mu_1$
3 vs. 2	$44.1 - 40.2 = 3.9$	1.28	3.05	Do not reject H_0: $\mu_3 = \mu_2$
3 vs. 4	Do not test			
4 vs. 1	$44.1 - 32.1 = 9.0$	1.28	7.03	Reject H_0: $\mu_4 = \mu_1$
4 vs. 2	Do not test			
2 vs. 1	$40.2 - 32.1 = 8.1$	1.28	6.33	Reject H_0: $\mu_2 = \mu_1$

Thus, we conclude that μ_1 is different from the other means, that μ_5 is different from the other means, and that μ_2, μ_4, and μ_3 are indistinguishable from each other: $\mu_1 \neq \mu_2 = \mu_4 = \mu_3 \neq \mu_5$.

a t value, the Tukey test statistic, q, is calculated by dividing a difference between two means by

$$SE = \sqrt{\frac{s^2}{n}}, \tag{2}$$

where n is the number of data in each of groups B and A, and s^2 is the error mean square by ANOVA computation. Thus

$$q = \frac{\overline{X}_B - \overline{X}_A}{SE}, \tag{3}$$

which is known as the *studentized range** (and is sometimes designated as T). The null hypothesis H_0: $\overline{X}_B = \overline{X}_A$ is rejected if q is equal to or greater than the critical value, $q_{\alpha,\nu,k}$, from Table 5 from *Appendix: Statistical Tables and Graphs*, where ν is the error degrees of freedom.

The significance level, α, is the probability of committing at least one Type I error (i.e., the probability of incorrectly rejecting at least one H_0) during the course of comparing all pairs of means. And the Tukey test has good power and maintains the probability of the familywise Type I error at or below the stated α.

The conclusions reached by this multiple-comparison testing may depend upon the order in which the pairs of means are compared. The proper procedure is to compare first the largest mean against the smallest, then the largest against the next smallest, and so on, until the largest has been compared with the second largest. Then one compares the second largest with the smallest, the second largest with the next smallest, and so on. Another important procedural rule is that if no significant difference is found

*E. S. Pearson and H. O. Hartley first used this term in 1953 (David, 1995).

between two means, then it is concluded that no significant difference exists between any means enclosed by those two, and no differences between enclosed means are tested for. Thus, in Example 1, because we conclude no difference between population means 3 and 2, no testing is performed to judge the difference between means 3 and 4, or between means 4 and 2. The conclusions in Example 1 are that Sample 1 came from a population having a mean different from that of any of the other four sampled populations; likewise, it is concluded that the population mean from which Sample 5 came is different from any of the other population means, and that samples 2, 4, and 3 came from populations having the same means. Therefore, the overall conclusion is that $\mu_1 \neq \mu_2 = \mu_4 = \mu_3 \neq \mu_5$. As a visual aid in Example 1, each time a null hypothesis was not rejected, a line was drawn beneath means to connect the two means tested and to encompass any means between them.

The null hypothesis $H_0: \mu_B = \mu_A$ may also be written as $\mu_B - \mu_A = 0$. The hypothesis $\mu_B - \mu_A = \mu_0$, where $\mu_0 \neq 0$, may also be tested; this is done by replacing $\overline{X}_B - \overline{X}_A$ with $| \overline{X}_B - \overline{X}_A | - \mu_0$ in the numerator of Equation 3.

Occasionally, a multiple-comparison test, especially if $n_B \neq n_A$, will yield ambiguous results in the form of conclusions of overlapping spans of nonsignificance. For example, one might arrive at the following:

$$\overline{X}_1 \quad \overline{X}_2 \quad \overline{X}_3 \quad \overline{X}_4$$

for an experimental design consisting of four groups of data. Here the four samples seem to have come from populations among which there were two different population means: Samples 1 and 2 appear to have been taken from one population, and Samples 2, 3, and 4 from a different population. But this is clearly impossible, for Sample 2 has been concluded to have come from both populations. Because the statistical testing was not able to conclude decisively from which population Sample 2 came, at least one Type II error has been committed. Therefore, it can be stated that $\mu_1 \neq \mu_3 \neq \mu_4$, but it cannot be concluded from which of the two populations Sample 2 came (or if it came from a third population). Repeating the data collection and analysis with a larger number of data might yield more conclusive results.

(a) Multiple Comparisons with Unequal Sample Sizes. If the sizes of the k samples are not equal, the Tukey-Kramer procedure (Kramer, 1956; supported by Dunnett, 1980a; Stoline, 1981; Jaccard, Becker, and Wood, 1984)* is desirable to maintain the probability of a Type I error near α and to operate with good power. For each comparison involving unequal n's, the standard error for use in Equation 3 is calculated as

$$SE = \sqrt{\frac{s^2}{2} \left(\frac{1}{n_B} + \frac{1}{n_A} \right)}, \tag{4}$$

which is inserting the harmonic mean of n_B and n_A in place of n in Equation 2;[†] and Equation 4 is equivalent to 2 when $n_B = n_A$. This test is shown in Example 2.

---

*This procedure has been shown to be excellent (e.g., Dunnett, 1980a; Hayter, 1984; Keselman, Murray, and Rogan, 1976; Smith, 1971; Somerville, 1993; Stoline, 1981), with the probability of a familywise Type I error no greater than the stated α.

[†]Some researchers have replaced n in Equation 2 with the harmonic mean of all k samples or with the median or arithmetic mean of the pair of means examined. Dunnett (1980a); Keselman, Murray, and Rogan (1976); Keselman and Rogan (1977); and Smith (1971) concluded the Kramer approach to be superior to those methods.

EXAMPLE 2   The Tukey-Kramer Test with Unequal Sample Sizes

The data (in kg).

$k = 4$

$s^2 = \text{Error MS} = 9.383$

Error DF $= 15$

$q_{0.05,15,4} = 4.076$

Sample number (i) of ranked means:	4	1	2	3
Ranked sample mean ($\overline{X}_i$):	63.24	64.62	71.30	73.35
Sample sizes (n_i):	4	5	5	5

If $n_B = n_A$ (call it n), then $\text{SE} = \sqrt{\dfrac{s^2}{n}} = \sqrt{\dfrac{9.383}{5}} = \sqrt{1.877} = 1.370.$

If $n_B \neq n_A$, then $\text{SE} = \sqrt{\dfrac{s^2}{2}\left(\dfrac{1}{n_B} + \dfrac{1}{n_A}\right)} = \sqrt{\dfrac{0.383}{2}\left(\dfrac{1}{5} + \dfrac{1}{4}\right)}$

$$= \sqrt{2.111} = 1.453.$$

Comparison B vs. A	Difference $(\overline{X}_B - \overline{X}_A)$	SE	q	Conclusion
3 vs. 4	$73.35 - 63.24 = 10.11$	1.453	6.958	Reject H_0: $\mu_3 = \mu_4$
3 vs. 1	$73.35 - 64.62 = 8.73$	1.370	6.371	Reject H_0: $\mu_3 = \mu_1$
3 vs. 2	$73.35 - 71.30 = 2.05$	1.370	1.496	Do not reject H_0: $\mu_3 = \mu_2$
2 vs. 4	$71.30 - 63.24 = 8.06$	1.453	5.547	Reject H_0: $\mu_2 = \mu_4$
2 vs. 1	$71.30 - 64.62 = 6.68$	1.370	4.876	Reject H_0: $\mu_2 = \mu_1$
1 vs. 4	$64.62 - 63.24 = 1.38$	1.453	0.950	Do not reject H_0: $\mu_1 = \mu_4$

Thus, we conclude that μ_4 and μ_1 are indistinguishable, that μ_2 and μ_3 are indistinguishable, and that μ_4 and μ_1 are different from μ_2 and μ_3: $\mu_4 = \mu_1 \neq \mu_2 = \mu_3$.

(b) Multiple Comparisons with Unequal Variances. Although the Tukey test can withstand some deviation from normality (e.g., Jaccard, Becker, and Wood, 1984), it is less resistant to heterogeneous variances, especially if the sample sizes are not equal. The test is conservative if small n's are associated with small variances and undesirably liberal if small samples come from populations with large variances, and in the presence of both nonnormality and heteroscedasticity the test is very liberal. Many investigations* have determined that the Tukey-Kramer test is also adversely affected by heterogeneous variances.

---

*These include those of Dunnett (1980b, 1982); Games and Howell (1976); Jaccard, Becker, and Wood (1984); Keselman, Games, and Rogan (1979); Keselman and Rogan (1978); Keselman and Toothaker (1974); Keselman, Toothaker, and Shooter (1975); Ramseyer and Tcheng (1973); Jenkdon and Tamhane (1979).

As a solution to this problem, Games and Howell (1976) proposed the use of the Welch approximation to modify Equation 4 to be appropriate when the k population variances are not assumed to be the same or similar:

$$\text{SE} = \sqrt{\frac{1}{2}\left(\frac{s_B^2}{n_B} + \frac{s_A^2}{n_A}\right)}, \tag{5}$$

and q will be associated with the degrees of freedom of Equation 5b;

$$\nu' = \frac{\left(s_{\bar{X}_1}^2 + s_{\bar{X}_2}^2\right)^2}{\dfrac{\left(s_{\bar{X}_1}^2\right)^2}{n_1 - 1} + \dfrac{\left(s_{\bar{X}_2}^2\right)^2}{n_2 - 1}}. \tag{5b}$$

but each sample size should be at least 6. This test maintains the probability of a familywise Type I error around α (though it is sometimes slightly liberal) and it has good power (Games, Keselman, and Rogan, 1981; Keselman, Games, and Rogan, 1979; Keselman and Rogan, 1978; Tamhane, 1979). If the population variances are the same, then the Tukey or Tukey-Kramer test is preferable (Kirk, 1995: 147–148). If there is doubt about whether there is substantial heteroscedacity, it is safer to use the Games and Howell procedure, for if the underlying populations do not have similar variances, that test will be far superior to the Tukey-Kramer test; and if the population variances are similar, the former will have only a little less power than the latter (Levy, 1978c).

(c) Other Multiple-Comparison Methods. Methods other than the Tukey and Tukey-Kramer tests have been employed by statisticians to examine pairwise differences for more than two means. The *Newman-Keuls test* (Newman, 1939; Keuls, 1952), also referred to as the *Student-Newman-Keuls test*, is employed as is the Tukey test, except that the critical values from Table 5 from *Appendix: Statistical Tables and Graphs* are those for $q_{\alpha,\nu,p}$ instead of $q_{\alpha,\nu,k}$, where p is the range of means for a given H_0. So, in Example 2, comparing means 3 and 4 would use $p = 4$, comparing means 3 and 1 would call for $p = 3$, and so on (with p ranging from 2 to k). This type of multiple-comparison test is called a *multiple-range test*. There is considerable opinion against using this procedure (e.g., by Einot and Gabriel, 1975; Ramsey, 1978) because it may falsely declare differences with a probability undesirably greater than α.

The *Duncan test* (Duncan, 1955) is also known as the *Duncan new multiple range test* because it succeeds an earlier procedure (Duncan, 1951). It has a different theoretical basis, one that is not as widely accepted as that of Tukey's test, and it has been declared (e.g., by Carmer and Swanson, 1973; Day and Quinn, 1899) to perform poorly. This procedure is executed as is the Student-Newman-Keuls test, except that different critical-value tables are required.

Among other tests, there is also a procedure called the *least significant difference test* (LSD), and there are other tests, such as with *Dunn* or *Bonferroni* in their names (e.g., Howell, 2007: 356–363). The name *wholly significant difference test* (WSD test) is sometimes applied to the Tukey test (Section 1) and sometimes as a compromise between the Tukey and Student-Newman-Keuls procedures by employing a critical value midway between $q_{\alpha,\nu,k}$ and $q_{\alpha,\nu,p}$. The Tukey test is preferred here because of its simplicity and generally good performance with regard to Type I and Type II errors.

2 CONFIDENCE INTERVALS FOR MULTIPLE COMPARISONS

Expressing a $1 - \alpha$ confidence interval using a sample mean denotes that there is a probability of $1 - \alpha$ that the interval encloses its respective population mean. Once multiple-comparison testing has concluded which of three or more sample means are significantly different, confidence intervals may be calculated for each different population mean. If one sample mean ($\overline{X}_i$) is concluded to be significantly different from all others, then Equation 5b is used:

$$\overline{X}_i \pm t_{\alpha(2),v} \sqrt{\frac{s^2}{n_i}}. \tag{5b}$$

In these calculations, s^2 (essentially a pooled variance) is the same as the error mean square would be for an analysis of variance for these groups of data. If two or more sample means are not concluded to be significantly different, then a pooled mean of those samples is the best estimate of the mean of the population from which those samples came:

$$\overline{X}_p = \frac{\sum n_i \overline{X}_i}{\sum n_i}, \tag{6}$$

where the summation is over all samples concluded to have come from the same population. Then the confidence interval is

$$\overline{X}_p \pm t_{\alpha(2),v} \sqrt{\frac{s^2}{\sum n_i}}, \tag{6a}$$

again summing over all samples whose means are concluded to be indistinguishable. This is analogous to the two-sample situation, and it is demonstrated in Example 3.

If a pair of population means, μ_B and μ_A, are concluded to be different, the $1 - \alpha$ confidence interval for the difference ($\mu_B - \mu_A$) may be computed as

$$(\overline{X}_B - \overline{X}_A) \pm (q_{\alpha,v,k})(\text{SE}). \tag{7}$$

Here, as in Section 1, v is the error degrees of freedom appropriate to an ANOVA, k is the total number of means, and SE is obtained from either Equation 2 or Equation 4, depending upon whether n_B and n_A are equal, or Equation 5 if the underlying population variances are not assumed to be equal. This calculation is demonstrated in Example 3 for the data in Example 1.

(a) Sample Size and Estimation of the Difference between Two Population Means. Recall how to estimate the sample size required to obtain a confidence interval of specified width for a difference between the two population means associated with the two-sample t test. In a multisample situation, a similar procedure may be used with the difference between population means, employing q instead of the t statistic. Iteration is necessary, whereby n is determined such that

$$n = \frac{s^2(q_{\alpha,v,k})^2}{d^2}. \tag{8}$$

Here, d is the half-width of the $1 - \alpha$ confidence interval, s^2 is the estimate of error variance, and k is the total number of means; v is the error degrees of freedom with the estimated n, namely $v = k(n - 1)$.

EXAMPLE 3 Confidence Intervals (CI) for the Population Means from Example 1

It was concluded in Example 1 that $\mu_1 \neq \mu_2 = \mu_4 = \mu_3 \neq \mu_5$. Therefore, we may calculate confidence intervals for μ_1 for $\mu_{2,4,3}$ and for μ_5 (where $\mu_{2,4,3}$ indicates the mean of the common population from which Samples 2, 4, and 3 came).

Using Equation 5b:

$$95\% \text{ CI for } \mu_1 = \overline{X}_1 \pm t_{0.05(2),25}\sqrt{\frac{s^2}{n_1}} = 32.1 \pm (2.060)\sqrt{\frac{9.7652}{6}}$$
$$= 32.1 \text{ mg/ml} \pm 2.6 \text{ mg/ml.}$$

Again using Equation 5b:

$$95\% \text{ CI for } \mu_5 = \overline{X}_5 \pm t_{0.05(2),25}\sqrt{\frac{s^2}{n_5}} = 58.3 \text{ mg/ml} \pm 2.6 \text{ mg/ml.}$$

Using Equation 6:

$$\overline{X}_p = \overline{X}_{2,4,3} = \frac{n_2\overline{X}_2 + n_4\overline{X}_4 + n_3\overline{X}_3}{n_2 + n_4 + n_3}$$
$$= \frac{(6)(40.2) + (6)(41.1) + (6)(44.1)}{6 + 6 + 6} = 41.8 \text{ mg/ml.}$$

Using Equation 6a:

$$95\% \text{ CI for } \mu_{2,4,3} = \overline{X}_{2,4,3} \pm t_{0.05(2),25}\sqrt{\frac{s^2}{6 + 6 + 6}} = 41.8 \text{ mg/ml}$$
$$\pm 1.5 \text{ mg/ml.}$$

Using Equation 7:

$$95\% \text{ CI for } \mu_5 - \mu_{2,4,3} = \overline{X}_5 - \overline{X}_{2,4,3} \pm q_{0.05,25,5}\sqrt{\frac{s^2}{2}\left(\frac{1}{n_5} + \frac{1}{n_2 + n_4 + n_3}\right)}$$
$$= 58.3 - 41.8 \pm (4.166)(1.04)$$
$$= 16.5 \text{ mg/ml} \pm 4.3 \text{ mg/ml.}$$

Using Equation 7:

$$95\% \text{ CI for } \mu_{2,4,3} - \mu_1 = \overline{X}_{2,4,3} - \overline{X}_1 \pm q_{0.05,25,5}\sqrt{\frac{s^2}{2}\left(\frac{1}{n_2 + n_4 + n_3} + \frac{1}{n_1}\right)}$$
$$= 41.8 - 32.1 \pm (4.166)(1.04)$$
$$= 9.7 \text{ mg/ml} \pm 4.3 \text{ mg/ml.}$$

3 TESTING A CONTROL MEAN AGAINST EACH OTHER MEAN

Sometimes means are obtained from k groups with the a priori objective of concluding whether the mean of one group, commonly designated as a control, differs significantly from each of the means of the other $k - 1$ groups. Dunnett (1955) provided an excellent procedure for such testing. Thus, whereas the data described in Section 1 were collected with the intent of comparing each sample mean with each other sample mean, the Dunnett test is for multisample data where the objective of the analysis was stated as comparing the control group's mean to the mean of each other group. Tukey's test could be used for this purpose, but it would be less powerful (Myers and Well, 2003: 255). If $k = 2$, Dunnett's test is equivalent to the two-sample t test.

As in the previous section, s^2 denotes the error mean square, which is an estimate of the common population variance underlying each of the k samples. The Dunnett's test statistic (analogous to that of Equation 3) is

$$q' = \frac{\overline{X}_{\text{control}} - \overline{X}_A}{\text{SE}}, \tag{9}$$

where the standard error, when the sample sizes are equal, is

$$\text{SE} = \sqrt{\frac{2s^2}{n}}, \tag{10}$$

and when the sample sizes are not equal, it is

$$\text{SE} = \sqrt{s^2 \left(\frac{1}{n_A} + \frac{1}{n_{\text{control}}} \right)}, \tag{11}$$

and when the variances are not equal:

$$\text{SE} = \sqrt{\frac{s_A^2}{n_A} + \frac{s_{\text{control}}^2}{n_{\text{control}}}}. \tag{11.11a}$$

For a two-tailed test, critical values, $q'_{\alpha(2),v,k}$, are given in Table 7 from *Appendix: Statistical Tables and Graphs*. If $|q'| \geq q'_{\alpha(2),v,k}$, then H_0: $\mu_{\text{control}} = \mu_A$ is rejected. Critical values for a one-sample test, $q'_{\alpha(1),v,k}$, are given in Table 6 from *Appendix: Statistical Tables and Graphs*. In a one-tailed test, H_0: $\mu_{\text{control}} \leq \mu_A$ is rejected if $q' \geq q'_{\alpha(1),v,k}$; and H_0: $\mu_{\text{control}} \geq \mu_A$ is rejected if $|q'| \geq q'_{\alpha(1),v,k}$ and $\overline{X}_{\text{control}} < \mu_A$ (i.e., if $q \leq -q'_{\alpha(1),v,k}$). This is demonstrated in Example 4. These critical values ensure that the familywise Type I error $= \alpha$.

The null hypothesis H_0: $\mu_{\text{control}} = \mu_A$ is a special case of H_0: $\mu_{\text{control}} - 0 = \mu_0$ where $\mu_0 = 0$. However, other values of μ_0 may be placed in the hypothesis, and Dunnett's test would proceed by placing $|\overline{X}_{\text{control}} - \overline{X}_A - \mu_0|$ in the numerator of the q' calculation. In an analogous manner, H_0: $\mu_{\text{control}} - \mu_0 \leq \mu$ (or H_0: $\mu_{\text{control}} - \mu_0 \geq \mu$) may be tested.

When comparison of group means to a control mean is the researcher's stated desire, the sample from the group designated as the control ought to contain more observations than the samples representing the other groups. Dunnett (1955) showed that the optimal size of the control sample typically should be a little less than $\sqrt{k-1}$ times the size of each other sample.

(a) Sample Size and Estimation of the Difference between One Population Mean and the Mean of a Control Population. This situation is similar to that discussed in Section 2a, but it pertains specifically to one of the k means being designated as

EXAMPLE 4   Dunnett's Test for Comparing the Mean of a Control Group to the Mean of Each Other Group

The yield (in metric tons per hectare) of each of several plots (24 plots, as explained below) of potatoes has been determined after a season's application of a standard fertilizer. Likewise, the potato yields from several plots (14 of them) were determined for each of four new fertilizers. A manufacturer wishes to promote at least one of these four fertilizers by claiming a resultant increase in crop yield. A total of 80 plots is available for use in this experiment.

Optimum allocation of plots among the five fertilizer groups will be such that the control group (let us say that it is group 2) has a little less than $\sqrt{k-1} = \sqrt{4} = 2$ times as many data as each of the other groups. Therefore, it was decided to use $n_2 = 24$ and $n_1 = n_3 = n_4 = m_5 = 14$, for a total N of 80.

Using analysis-of-variance calculations, the error MS (s^2) was found to be 10.42 (metric tons/ha)^2 and the error DF = 75.

$$SE = \sqrt{10.42\left(\frac{1}{14} + \frac{1}{24}\right)} = 1.1 \text{ metric tons/acre}$$

Group number (i) of ranked means:	1	2	3	4	5
Ranked group mean ($\overline{X}_i$):	17.3	21.7	22.1	23.6	27.8

As the control group (i.e., the group with the standard fertilizer) is group 2, each H_0: $\mu_2 \geq \mu_A$ will be tested against H_A: $\mu_2 < \mu_A$. And for each hypothesis test, $q'_{\alpha,v,k} = q'_{0.05(1),75,5}$.

Comparison B vs. A	Difference $(\overline{X}_2 - \overline{X}_A)$	SE	$|q'|$	Conclusion
2 vs. 1	21.7 − 17.3 = 4.4			Because $\overline{X}_2 > \overline{X}_1$, do not reject H_0: $\mu_2 \geq \mu_1$
2 vs. 5	21.7 − 27.8 = −6.1	1.1	5.55	Reject H_0: $\mu_2 \geq \mu_5$
2 vs. 4	21.7 − 23.6 = −1.9	1.1	1.73	Reject H_0: $\mu_2 \geq \mu_4$
2 vs. 3	Do not test			

We conclude that only fertilizer 5 produces a yield greater than the yield from the control fertilizer (fertilizer 2).

from a control group. The procedure uses this modification of Equation 8:

$$n = \frac{2s^2(q'_{\alpha,v,k})^2}{d^2}. \tag{12}$$

(b) Confidence Intervals for Differences between Control and Other Group Means. Using Dunnett's q' statistic and the SE of Equation 10, 11, or 11a, two-tailed confidence limits can be calculated for the difference between the control mean and each of the other group means:

$$1 - \alpha \text{ CI for } \mu_{control} - \mu_A = (\overline{X}_{control} - \overline{X}_A) \pm (q'_{\alpha(2),v,k})(SE). \tag{13}$$

One-tailed confidence limits are also possible. The $1 - \alpha$ confidence can be expressed that a difference, $\mu_{\text{control}} - \mu_A$, is not less than (i.e., is at least as large as)

$$(\overline{X}_{\text{control}} - \overline{X}_A) - (q'_{\alpha(1),v,k})(\text{SE}),\tag{14}$$

or it might be desired to state that the difference is no greater than

$$(\overline{X}_{\text{control}} - \overline{X}_A) + (q'_{\alpha(1),v,k})(\text{SE}).\tag{15}$$

4 MULTIPLE CONTRASTS

Inspecting the sample means after performing an analysis of variance can lead to a desire to compare combinations of samples to each other, by what are called *multiple contrasts*. The method of Scheffé* (1953; 1959: Sections 3.4, 3.5) is an excellent way to do this while ensuring a familywise Type I error rate no greater than α.

The data in Example 1 resulted in ANOVA rejection of the null hypothesis H_0: $\mu_1 = \mu_2 = \mu_3 = \mu_4 = \mu_5$; and, upon examining the five sample means, perhaps by arranging them in order of magnitude $(\overline{X}_1 < \overline{X}_2 < \overline{X}_4 < \overline{X}_3 < \overline{X}_5)$, the researcher might then want to compare the mean strontium concentration in the river (group 5) with that of the bodies of water represented by groups 2, 4, and 3. The relevant null hypothesis would be H_0: $(\mu_2 + \mu_4 + \mu_3)/3 = \mu_5$, which can also be expressed as H_0: $\mu_2/3 + \mu_4/3 + \mu_3/3 - \mu_5 = 0$. The Scheffé test considers that each of the four μ's under consideration is associated with a coefficient, c_i: $c_2 = \frac{1}{3}$, $c_4 = \frac{1}{3}$, $c_3 = \frac{1}{3}$, and $c_5 = -1$ (and the sum of these coefficients is always zero). The test statistic, S, is calculated as

$$S = \frac{\left|\sum c_i \overline{X}_i\right|}{\text{SE}},\tag{16}$$

where

$$\text{SE} = \sqrt{s^2 \left(\sum \frac{c_i^2}{n_i}\right)},\tag{17}$$

and the critical value of the test is

$$S_\alpha = \sqrt{(k-1)F_{\alpha(1),k-1,N-k}}\ .\tag{18}$$

Also, with these five groups of data, there might be an interest in testing H_0: $\mu_1 - (\mu_2 + \mu_4 + \mu_3)/3 = 0$ or H_0: $(\mu_1 + \mu_5)/2 - (\mu_2 + \mu_4 + \mu_3)/3 = 0$ or H_0: $(\mu_1 + \mu_4)/2 - (\mu_2 + \mu_3)/2 = 0$, or other contrasts. Any number of such hypotheses may be tested, and the familywise Type I error rate is the probability of falsely rejecting at least one of the possible hypotheses. A significant F from an ANOVA of the k groups indicates that there is at least one significant contrast among the groups, although the contrasts that are chosen may not include one to be rejected by the Scheffé test. And if F is not significant, testing multiple contrasts need not be done, for the probability of a Type I error in that testing is not necessarily at α (Hays, 1994: 458). The testing of several of these hypotheses is shown in Example 5. In employing the Scheffé test, the decision of which means to compare with which others occurs after inspecting the data, so this is referred to as an a posteriori, or post hoc test.

---

*Henry Scheffé (1907–1977), American statistician.

EXAMPLE 5   Scheffé's Test for Multiple Contrasts, Using the Data of Example 1

For $\alpha = 0.05$, the critical value, S_α, for each contrast is (via Equation 18) $\sqrt{(k-1)F_{0.05(1),k-1,N-k}}$

$$= \sqrt{(5-1)F_{0.05(1),4,25}}$$

$$= \sqrt{4(2.76)}$$

$$= 3.32.$$

Example 1 showed $s^2 = 9.7652$ and $n = 6$.

Contrast	SE		S	Conclusion
$\dfrac{\overline{X}_2 + \overline{X}_3 + \overline{X}_4}{3} - \overline{X}_5$ $= 41.8 - 58.3$ $= -16.5$	$\sqrt{9.7652\left[\dfrac{\left(\frac{1}{3}\right)^2}{6} + \dfrac{\left(\frac{1}{3}\right)^2}{6} + \dfrac{\left(\frac{1}{3}\right)^2}{6} + \dfrac{(1)^2}{6}\right]}$	$= 1.47$	11.22	Reject H_0: $\dfrac{\mu_2 + \mu_3 + \mu_4}{3}$ $-\mu_5 = 0$
$\overline{X}_1 - \dfrac{\overline{X}_2 + \overline{X}_3 + \overline{X}_4}{3}$ $= 32.1 - 41.8$ $= -9.7$	$\sqrt{9.7652\left[\dfrac{(1)^2}{6} + \dfrac{\left(\frac{1}{3}\right)^2}{6} + \dfrac{\left(\frac{1}{3}\right)^2}{6} + \dfrac{\left(\frac{1}{3}\right)^2}{6}\right]}$	$= 1.47$	6.60	Reject H_0: $\mu_1 - \dfrac{\mu_2 + \mu_3 + \mu_4}{3}$ $= 0$
$\dfrac{\overline{X}_1 + \overline{X}_5}{2}$ $-\dfrac{\overline{X}_2 + \overline{X}_3 + \overline{X}_4}{3}$ $= 45.2 - 41.8$ $= 3.4$	$\sqrt{9.7652\left[\dfrac{\left(\frac{1}{2}\right)^2}{6} + \dfrac{\left(\frac{1}{2}\right)^2}{6} + \dfrac{\left(\frac{1}{3}\right)^2}{6} + \dfrac{\left(\frac{1}{3}\right)^2}{6} + \dfrac{\left(\frac{1}{3}\right)^2}{6}\right]}$	$= 1.16$	2.93	Accept H_0: $\dfrac{\mu_1 + \mu_5}{2}$ $-\dfrac{\mu_2 + \mu_3 + \mu_4}{3}$ $= 0$
$\dfrac{\overline{X}_1 + \overline{X}_4}{2} - \dfrac{\overline{X}_2 + \overline{X}_3}{2}$ $= 36.6 - 42.15$ $= -5.55$	$\sqrt{9.7652\left[\dfrac{\left(\frac{1}{2}\right)^2}{6} + \dfrac{\left(\frac{1}{2}\right)^2}{6} + \dfrac{\left(\frac{1}{2}\right)^2}{6} + \dfrac{\left(\frac{1}{2}\right)^2}{6}\right]}$	$= 1.28$	4.34	Reject H_0: $\dfrac{\mu_1 + \mu_4}{2}$ $-\dfrac{\mu_2 + \mu_3}{2} = 0$

The Scheffé test may also be used to compare one mean with one other. It is then testing the same hypotheses as is the Tukey test. It is less sensitive than the Tukey test to nonnormality and heterogeneity of variances (Hays, 1994: 458, 458; Sahai and Ageel, 2000: 77); but is less powerful and it is recommended that it not be used for pairwise comparisons (e.g., Carmer and Swanson, 1973; Kirk, 1995: 154; Toothaker, 1991: 51, 77, 89–90). Shaffer (1977) described a procedure, more powerful than Scheffé's, specifically for comparing a combination of groups to a group specified as a control.

(a) Multiple Contrasts with Unequal Variances. The Scheffé test is suitable when the samples in the contrast each came from populations with the same variance.

When the population variances differ but the sample sizes are equal, the probability of a Type I error can be different from 0.05 when α is set at 0.05. If the variances *and* the sample sizes are unequal, then (as with the ANOVA) the test will be very conservative if the large variances are associated with large sample sizes and very liberal if the small samples come from the populations with the large variances (Keselman and Toothaker, 1974). If the σ_i^2's cannot be assumed to be the same or similar, the procedure of Brown and Forsythe (1974b) may be employed. This is done in a fashion analogous to the two-sample Welch modification of the t test, using

$$t' = \frac{\left| \sum c_i \overline{X}_i \right|}{\sqrt{\sum \dfrac{c_i^2 s_i^2}{n_i}}} \tag{19}$$

with degrees of freedom of

$$\nu' = \frac{\left(\sum \dfrac{c_i^2 s_i^2}{n_i} \right)^2}{\sum \dfrac{(c_i^2 s_i^2)^2}{n_i^2 (n_i - 1)}}. \tag{20}$$

(b) Confidence Intervals for Contrasts. The Scheffé procedure enables the establishment of $1 - \alpha$ confidence limits for a contrast:

$$\sum c_i \overline{X}_i \pm S_\alpha \text{SE} \tag{21}$$

(with SE from Equation 17). Shaffer's (1977) method produces confidence intervals for a different kind of contrast, that of a group of means with the mean of a control group.

Example 6 demonstrates the determination of confidence intervals for two of the statistically significant contrasts of Example 5.

5 NONPARAMETRIC MULTIPLE COMPARISONS

In the multisample situation where the nonparametric Kruskal-Wallis test is appropriate, the researcher usually will desire to conclude which of the samples are significantly different from which others, and the experiment will be run with that goal. This may be done in a fashion paralleling the Tukey test of Section 1, by using rank sums instead of means, as demonstrated in Example 7. The rank sums, determined as in the Kruskal-Wallis test, are arranged in increasing order of magnitude. Pairwise differences between rank sums are then tabulated, starting with the difference between the largest and smallest rank sums, and proceeding in the same sequence as described in Section 1. The standard error is calculated as

$$\text{SE} = \sqrt{\frac{n(nk)(nk + 1)}{12}} \tag{22}$$

Multiple Comparisons

(Nemenyi, 1963; Wilcoxon and Wilcox, 1964: 10),* and the Studentized range (Table 5 from *Appendix: Statistical Tables and Graphs* to be used is $q_{\alpha,\infty,\kappa}$.)

EXAMPLE 6 Confidence Intervals for Multiple Contrasts

The critical value, S_α, for each confidence interval is that of Equation 8: $\sqrt{(k-1)F_{\alpha(1),k-1,N-k}}$, and for $\alpha = 0.05$, $S_\alpha = 3.32$ and $s^2 = 9.7652$ as in Example 5.

(a) A confidence interval for $\dfrac{\mu_2 + \mu_3 + \mu_4}{3} - \mu_5$ would employ SE = 1.47 from Example 5, and the 95% confidence interval is

$$\left(\frac{\overline{X}_2 + \overline{X}_3 + \overline{X}_4}{3} - \overline{X}_5\right) \pm S_\alpha\text{SE} = -16.5 \pm (3.32)(1.47)$$

$$= -16.5 \text{ mg/ml} \pm 4.9 \text{ mg/ml}$$

$$L_1 = -21.4 \text{ mg/ml}$$

$$L_2 = -11.6 \text{ mg/ml}.$$

(b) A confidence interval for $\mu_1 - \dfrac{\mu_2 + \mu_3 + \mu_4}{3}$ would employ SE = 1.47 from Example 5, and the 95% confidence interval is

$$\left(\overline{X}_1 - \frac{\overline{X}_2 + \overline{X}_3 + \overline{X}_4}{3}\right) \pm S_\alpha\text{SE} = -9.7 \pm (3.32)(1.47)$$

$$= -9.7 \text{ mg/ml} \pm 4.9 \text{ mg/ml}$$

$$L_1 = -14.6 \text{ mg/ml}$$

$$L_2 = -4.8 \text{ mg/ml}.$$

(a) Nonparametric Multiple Comparisons with Unequal Sample Sizes. Multiple-comparison testing such as in Example 7 requires that there be equal numbers of data in each of the k groups. If such is not the case, then we may use the procedure of Section 7, but a more powerful test is that proposed by Dunn (1964), using a standard error of

$$\text{SE} = \sqrt{\frac{N(N+1)}{12}\left(\frac{1}{n_A} + \frac{1}{n_B}\right)} \tag{24}$$

for a test statistic we shall call

$$Q = \frac{\overline{R}_B - \overline{R}_A}{\text{SE}}, \tag{25}$$

---

*Some authors (e.g., Miller 1981: 166) perform this test in an equivalent fashion by considering the difference between mean ranks ($\overline{R}_A$ and $\overline{R}_B$) rather than rank sums (R_A and R_B), in which case the appropriate standard error would be

$$\text{SE} = \sqrt{\frac{k(nk+1)}{12}}. \tag{23}$$

EXAMPLE 7   Nonparametric Tukey-Type Multiple Comparisons, Using the Nemenyi Test

By Equation 11.22:

$$SE = \sqrt{\frac{n(nk)(nk + 1)}{12}} = \sqrt{\frac{5(15)(16)}{12}} = \sqrt{100} = 10.00$$

Sample number (i) of ranked rank sums:	3	2	1
Rank sum (R_i):	26	30	64

Comparison (B vs. A)	Difference ($R_B - R_A$)	SE	q	$q_{0.05,\infty,3}$	Conclusion
1 vs. 3	$64 - 26 = 38$	10.00	3.80	3.314	Reject H_0: Fly abundance is the same at vegetation heights 3 and 1.
1 vs. 2	$64 - 30 = 34$	10.00	3.40	3.314	Reject H_0: Fly abundance is the same at vegetation heights 2 and 1.
2 vs. 3	$30 - 26 = 4$	10.00	0.40	3.314	Do not reject H_0: Fly abundance is the same at vegetation heights 3 and 2.

Overall conclusion: Fly abundance is the same at vegetation heights 3 and 2 but is different at height 1.

where $\overline{R}$ indicates a mean rank (i.e., $\overline{R}_A = R_A/n_A$ and $\overline{R}_B = R_B/n_B$). Critical values for this test, $Q_{\alpha,k}$, are given in Table 15 from *Appendix: Statistical Tables and Graphs*. Applying this procedure to the situation of Example 7 yields the same conclusions, but this will not always be the case as this is only an approximate method and conclusions based upon a test statistic very near the critical value should be expressed with reservation. It is advisable to conduct studies that have equal sample sizes so Equation 22 or 23 may be employed.

If tied ranks are present, then the following is an improvement over Equation 24 (Dunn, 1964):

$$SE = \sqrt{\left(\frac{N(N + 1)}{12} - \frac{\sum t}{12(N - 1)}\right)\left(\frac{1}{n_A} + \frac{1}{n_B}\right)}. \tag{26}$$

In the latter equation, $\sum t$ is used in the Kruskal-Wallis test when ties are present. The testing procedure is demonstrated in Example 8; note that it is the mean ranks ($\overline{R}_i$), rather than the ranks sums (R_i), that are arranged in order of magnitude.

A procedure developed independently by Steel (1960, 1961b) and Dwass (1960) is somewhat more advantageous than the tests of Nemenyi and Dunn (Critchlow and Fligner, 1991; Miller, 1981: 168–169), but it is less convenient to use and it tends to be very conservative and less powerful (Gabriel and Lachenbruch, 1969). And

EXAMPLE 8   Nonparametric Multiple Comparisons with Unequal Sample Sizes

The data are those from Example 11 from *Multisample Hypotheses and the Analysis of Variance*, where the Kruskal-Wallis test rejected the null hypothesis That water pH was the same in all four ponds examined

$$\sum t = 168.$$

For $n_A = 8$ and $n_B = 8$,

$$SE = \sqrt{\left(\frac{N(N+1)}{12} - \frac{\sum t}{12(N-1)}\right)\left(\frac{1}{n_A} + \frac{1}{n_B}\right)}$$

$$= \sqrt{\left(\frac{31(32)}{12} - \frac{168}{12(30)}\right)\left(\frac{1}{8} + \frac{1}{8}\right)}$$

$$= \sqrt{20.5500} = 4.53$$

For $n_A = 7$ and $n_B = 8$,

$$SE = \sqrt{\left(\frac{31(32)}{12} - \frac{168}{12(30)}\right)\left(\frac{1}{7} + \frac{1}{8}\right)} = \sqrt{22.0179} = 4.69.$$

Sample number (i) of ranked means:	1	2	4	3
Rank sum (R_i):	55	132.5	163.5	145
Sample size (n_i):	8	8	8	7
Mean rank ($\overline{R}_i$)	6.88	16.56	20.44	20.71

To test at the 0.05 significance level, the critical value is $Q_{0.05,4} = 2.639$.

Comparison B vs. A	Difference $(\overline{R}_B - \overline{R}_A)$	SE	Q	Conclusion
3 vs. 1	$20.71 - 6.88 = 13.831$	4.69	2.95	Reject H_0: Water pH is the same in ponds 3 and 1.
3 vs. 2	$20.71 - 16.56 = 4.15$	4.69	0.88	Do not reject H_0: Water pH is the same in ponds 3 and 2.
3 vs. 4	Do not test			
4 vs. 1	$20.44 - 6.88 = 13.56$	4.53	2.99	Reject H_0: Water pH is the same in ponds 4 and 1.
4 vs. 2	Do not test			
2 vs. 1	$16.56 - 6.88 = 9.68$	4.53	2.14	Do not reject H_0: Water pH is the same in ponds 2 and 1.

Overall conclusion: Water pH is the same in ponds 4 and 3 but is different in pond 1, and the relationship of pond 2 to the others is unclear.

this test can lose control of Type I error if the data come from skewed populations (Toothaker, 1991: 108).

(b) Nonparametric Comparisons of a Control to Other Groups. Subsequent to a Kruskal-Wallis test in which H_0 is rejected, a nonparametric analysis may be performed to seek either one-tailed or two-tailed significant differences between one group (designated as the "control") and each of the other groups of data. This is done in a manner paralleling that of the procedure of Section 4, but using group rank sums instead of group means. The standard error to be calculated is

$$SE = \sqrt{\frac{n(nk)(nk + 1)}{6}} \qquad (27)$$

(Wilcoxon and Wilcox, 1964: 11), and one uses as critical values either $q'_{\alpha(1),\infty,k}$ or $q'_{\alpha(2),\infty,k}$ (from Table 6 or Table 7, respectively from *Appendix: Statistical Tables and Graphs*) for one-tailed or two-tailed hypotheses, respectively.*

The preceding nonparametric test requires equal sample sizes. If the n's are not all equal, then the procedure suggested by Dunn (1964) may be employed. By this method, group B is considered to be the control and uses Equation 27, where the appropriate standard error is that of Equation 26 or 28, depending on whether there are ties or no ties, respectively. We shall refer to critical values for this test, which may be two tailed or one tailed, as $Q'_{\alpha,k}$; and they are given in Table 16 from *Appendix: Statistical Tables and Graphs*. The test presented by Steel (1959) has drawbacks compared to the procedures above (Miller, 1981: 133).

6   NONPARAMETRIC MULTIPLE CONTRASTS

Multiple contrasts, introduced in Section 4, can be tested nonparametrically using the Kruskal-Wallis H statistic instead of the F statistic. As an analog of Equation 16, we compute

$$S = \frac{\left| \sum c_i \overline{R}_i \right|}{SE}, \qquad (29)$$

where c_i is as in Section 4, and

$$SE = \sqrt{\left(\frac{N(N + 1)}{12} \right) \left(\sum \frac{c_i^2}{n_i} \right)}, \qquad (30)$$

unless there are tied ranks, in which cases we use

$$SE = \sqrt{\left(\frac{N(N + 1)}{12} - \frac{\sum t}{12(N - 1)} \right) \left(\sum \frac{c_i^2}{n_i} \right)}, \qquad (31)$$

---

*If mean ranks, instead of rank sums, are used, then

$$SE = \sqrt{\frac{k(nk + 1)}{6}}. \qquad (28)$$

The critical value for these multiple contrasts is $\sqrt{H_{\alpha,n_1,n_2,\ldots}}$, using Table 13 from *Appendix: Statistical Tables and Graphs* to obtain the critical value of H. If the needed critical value of H is not on that table, then $\chi^2_{\alpha,(k-1)}$ may be used.

7 MULTIPLE COMPARISONS AMONG MEDIANS

If the null hypothesis is rejected in a multisample median test, then it is usually desirable to ascertain among which groups significant differences exist. A Tukey-type multiple comparison test has been provided by Levy (1979), using

$$q = \frac{f_{1B} - f_{1A}}{SE}. \tag{32}$$

As shown in Example 19, we employ the values of f_{1j} for each group, where f_{1j} is the number of data in group j that are greater than the grand median. The values of f_{1j} are ranked, and pairwise differences among the ranks are examined as in other Tukey-type tests. The appropriate standard error, when N (the total number of data in all groups) is an even number, is

$$SE = \sqrt{\frac{n(N+1)}{4N}}, \tag{33}$$

and, when N is an odd number, the standard error is

$$SE = \sqrt{\frac{nN}{4(N-1)}}. \tag{34}$$

The critical values to be used are $q_{\alpha,\infty,k}$. This multiple-comparison test appears to possess low statistical power. If the sample sizes are slightly unequal, as in Example 9, the test can be used by employing the harmonic mean of the sample sizes,

$$n = \frac{k}{\sum_{j=1}^{k} \frac{1}{n_j}}, \tag{35}$$

for an approximate result.

8 MULTIPLE COMPARISONS AMONG VARIANCES

If the null hypothesis that k population variances are all equal is rejected, then we may wish to determine which of the variances differ from which others. Levy (1975a, 1975c) suggests multiple-comparison procedures for this purpose based on a logarithmic transformation of sample variances.

A test analogous to the Tukey test of Section 1 is performed by calculating

$$q = \frac{\ln s_B^2 - \ln s_A^2}{SE}, \tag{36}$$

EXAMPLE 9 **Tukey-Type Multiple Comparison for Differences among Medians, Using the Data of Example 12 from** *Multisample Hypotheses and the Analysis of Variance*

Sample number (j) of samples ranked by f_{1j} : 2 1 4 3

Ranked f_{1j} : 2 3 6 9
Sample size (n_j) : 12 11 12 11

$k = 4$

$N = 11 + 12 + 11 + 12 = 46$

$n = 12$

By Equation 35,

$$n = \frac{4}{\dfrac{1}{12} + \dfrac{1}{11} + \dfrac{1}{12} + \dfrac{1}{11}} = 11.48$$

By Equation 36,

$$SE = \sqrt{\frac{(11.48)(46)}{4(46 - 1)}} = 1.713.$$

H_0: Median of population B = Median of population A.
H_A: Median of population B ≠ Median of population A.

Comparison	$f_{1B} - f_{1A}$	SE	q	$q_{0.05,4,\infty}$	Conclusion
3 vs. 2	$9 - 2 = 7$	1.713	4.086	3.633	Reject H_0.
3 vs. 1	$9 - 3 = 6$	1.713	3.503	3.633	Do not reject H_0.
3 vs. 4	Do not test				
4 vs. 2	$6 - 2 = 4$	1.713	2.335	3.633	Do not reject H_0.
4 vs. 1	Do not test				
1 vs. 2	Do not test				

Overall conclusion: The medians of populations 3 and 2 (i.e., south and east) are not the same; but the test lacks the power to allow clear conclusions about the medians of populations 4 and 1.

where

$$SE = \sqrt{\frac{2}{\nu}}, \tag{37}$$

if both samples being compared are of equal size. If $\nu_A \neq \nu_B$, we can employ

$$SE = \sqrt{\frac{1}{\nu_B} + \frac{1}{\nu_A}}. \tag{38}$$

EXAMPLE 10   Tukey-Type Multiple Comparison Test for Differences among Four Variances (i.e., $k = 4$)

i	s_i^2	n_i	ν_i	$\ln s_i^2$
1	2.74 g^2	50	49	1.0080
2	2.83 g^2	48	47	1.0403
3	2.20 g^2	50	49	0.7885
4	6.42 g^2	50	49	1.8594

Sample ranked by variances (i):	3	1	2	4
Logarithm of ranked sample variance ($\ln s_i^2$):	0.7885	1.0080	1.0403	1.8594
Sample degrees of freedom (ν_i):	49	49	47	49

Comparison (B vs. A)	Difference ($\ln s_B^2 - \ln s_A^2$)	SE	q	$q_{0.05,\infty,4}$	Conclusions
4 vs. 3	$1.8594 - 0.7885 = 1.0709$	0.202*	5.301	3.633	Reject H_0: $\sigma_4^2 = \sigma_3^2$
4 vs. 1	$1.8594 - 1.0080 = 0.8514$	0.202	4.215	3.633	Reject H_0: $\sigma_4^2 = \sigma_1^2$
4 vs. 2	$1.8594 - 1.0403 = 0.8191$	0.204†	4.015	3.633	Reject H_0: $\sigma_4^2 = \sigma_2^2$
2 vs. 3	$1.0403 - 0.7885 = 0.2518$	0.204	1.234	3.633	Do not reject H_0: $\sigma_2^2 = \sigma_3^2$
2 vs. 1	Do not test				
1 vs. 3	Do not test				

*As $\nu_4 = \nu_3$: $\text{SE} = \sqrt{\dfrac{2}{\nu}} = \sqrt{\dfrac{2}{49}} = 0.202.$

†As $\nu_4 \neq \nu_2$: $\text{SE} = \sqrt{\dfrac{1}{\nu_4} + \dfrac{1}{\nu_2}} = \sqrt{\dfrac{1}{49} + \dfrac{1}{47}} = 0.204.$

Overall conclusion: $\sigma_3^2 = \sigma_1^2 = \sigma_2^2 \neq \sigma_4^2.$

Just as in Sections 1 and 2, the subscripts A and B refer to the pair of groups being compared; and the sequence of pairwise comparisons must follow that given in those sections. This is demonstrated in Example 10.* The critical value for this test is $q_{\alpha,\infty,k}$.

A Newman-Keuls-type test can also be performed using the logarithmic transformation. For this test, we calculate q using Equation 36; but the critical value,

---

*Recall that "ln" refers to natural logarithms (i.e., logarithms using base e). If one prefers using common logarithms ("log"; logarithms in base 10), then

$$q = \frac{2.30259(\log s_B^2 - \log s_A^2)}{\text{SE}}. \tag{39}$$

$q_{\alpha,\infty,p}$, depends on p, the range of variances being tested (just as p is the range of means being tested in Section 1c).

It must be pointed out that the methods of this section, are valid only if the sampled populations are normal or very close to normal and are severely affected if this assumption is not satisfied. Stevens (1989) discussed nonparametric multiple comparison testing for dispersion.

(a) Comparing a Control-Group Variance to Each Other Group Variance. If the investigator's intent in multiple-comparison testing is to compare each possible pair of variances, then the procedures above are applicable. If, however, it is desired to stipulate that one of the variances (call it the "control," or Sample B, variance) is to be compared with each other variance (but the others are not to be compared with each other), then the Dunnett-type test of Levy (1975b) may be employed for more powerful testing.

Here, in a fashion analogous to that of Section 3 for means, we calculate

$$q = \frac{\ln s^2_{control} - \ln s^2_A}{SE}, \tag{40}$$

where

$$SE = \sqrt{\frac{4}{\nu}} \tag{41}$$

if the control sample and Sample A are of equal size. If $\nu_A \neq \nu_{control}$, then use

$$SE = \sqrt{\frac{2}{\nu_A} + \frac{2}{\nu_{control}}}. \tag{42}$$

The appropriate critical value for the two-tailed test (i.e., $H_0: \sigma^2_{control} = \sigma^2_A$) is $q'_{\alpha(2),\infty,p}$ and for the one-tailed test (i.e., either $H_0: \sigma^2_{control} \leq \sigma^2_A$ or $H_0: \sigma^2_{control} \geq \sigma^2_A$) (the critical value is $q'_{\alpha(1),\infty,p}$ from Table 6 from *Appendix: Statistical Tables and Graphs*). This testing should be used only if it can be assumed that the underlying populations are normally distributed (or very nearly so).

EXERCISES

1. (a) Apply the Tukey test procedure to the following results: $k = 3$ and the three sample means are 14.8, 20.2, and 16.2; the error mean square and degrees of freedom are 8.46 and 21, respectively; there are eight data in each of the three groups.

   (b) Calculate the 95% confidence interval for each different population mean and for each difference between means.

2. Apply the Tukey-Kramer test to the data below.

Feb.	May	Aug.	Nov.
4.7	4.6	4.8	4.9
4.9	4.4	4.7	5.2
5.0	4.3	4.6	5.4
4.8	4.4	4.4	5.1
4.7	4.1	4.7	5.6
	4.2	4.8	

3. Apply the Games and Howell test to the data from Exercise 2 so we do not have to assume that the variances of all four sampled populations are the same.

4. Assume that in the experiment described, group 1 was set up to be a control. Use a two-tailed Dunnett's test to compare the control mean with each other mean.

5. Use Scheffé's S test on the data.

   (a) Test the hypothesis that the means of the populations represented by groups 1 and 4 are the same as the means of groups 2 and 3.

   (b) Test the hypothesis that the means of groups 2 and 4 are the same as the mean of group 3.

6. The following ranks result in a significant Kruskal-Wallis test. Employ nonparametric multiple-range testing to conclude between which of the three groups population differences exist.

Group 1	Group 2	Group 3
8	10	14
4	6	13
3	9	7
5	11	12
1	2	15

ANSWERS TO EXERCISES

1. **(a, b)** Ranked sample means: <u>14.8   16.2</u>   20.2; $k = 3, n = 8, \alpha = 0.05, s^2 = 8.46, \nu = 21$ (which is not in Table 5 from *Appendix: Statistical Tables and Graphs*, so use $\nu = 20$, which is in the table); reject $H_0: \mu_2 = \mu_1$; reject $H_0: \mu_2 = \mu_3$; do not reject $H_0: \mu_3 = \mu_1$. Therefore, the overall conclusion is $\mu_1 = \mu_3 \neq \mu_2$.
   (c) $\overline{X}_p = \overline{X}_{1,3} = 15.5, t_{0.05(2),21} = 2.080,$ $n_1 + n_2 = 16, 95\%$ CI for $\mu_{1,3} = 15.5 \pm 1.5$; 95% CI for $\mu_2 = 20.2 \pm 2.1; \overline{X}_{1,3} - \overline{X}_2 = -4.7,$ SE $= 1.03, 95\%$ CI for $\mu_{1,3} - \mu_2 = -4.7 \pm 3.2.$

2. $\overline{X}_1 = 4.82, n_1 = 5; \overline{X}_2 = 4.33, n_2 = 6; \overline{X}_3 = 4.67,$ $n_3 = 6; \overline{X}_4 = 5.24, n_4 = 5; s^2 = 0.0348; \nu = 18;$ $q_{0.05,18,4} = 3.997$; conclusion: $\mu_2 \neq \mu_3 = \mu_1 \neq \mu_4.$

3. Means, sample sizes, and $q_{0.05,18,14}$ as in Exercise 2; $s_1^2 = 0.0170, s_2^2 = 0.0307, s_3^2 = 0.0227,$ $s_4^2 = 0.0730$, conclusion: $\mu_2 \neq \mu_3 = \mu_1 \neq \mu_4.$

4. Ranked sample means: 60.62, 69.30, 86.24, 100.35; sample sizes of 5, 5, 5, and 4, respectively; $k = 4,$ $\nu = 15, \alpha = 0.05, s^2 = 8.557$; control group is group 1; $q'_{0.05(2),15,4} = 2.61$; reject $H_0: \mu_4 = \mu_1,$ reject $H_0: \mu_3 = \mu_1$, reject $H_0: \mu_2 = \mu_1$. Overall conclusion: The mean of the control population is different from the mean of each other population.

5. Ranked sample means: 60.62, 69.30, 86.24, 100.35; sample sizes of 5, 5, 5, and 4, respectively; $k = 4,$ $\nu = 15, \alpha = 0.05, s^2 = 8.557$; critical value of S is 3.14; for $H_0: (\mu_1 + \mu_4)/2 - (\mu_2 + \mu_3)/2 = 0,$ $S = 8.4$, reject H_0; for $H_0: (\mu_2 + \mu_4)/2 - \mu_3 = 0, S = 13.05$, reject $H_0.$

6. $R_1 = 21, R_2 = 38, R_3 = 61.$ Overall conclusion: The variable being measured is the same magnitude in populations 1 and 2. The variable is of different magnitude in population 3.

Two-Factor Analysis of Variance

The present chapter will discuss how the effects of two factors can be assessed using a single statistical procedure.

The simultaneous analysis to be considered, of the effect of more than one factor on population means, is termed a *factorial analysis of variance*,* and there can be important advantages to such an experimental design. Among them is the fact that a single set of data can suffice for the analysis and it is not necessary to perform a one-way ANOVA for each factor. This may be economical with respect to time, effort, and money; and factorial analysis of variance also can test for the interactive effect of factors. The two-factor analysis of variance is introduced in this chapter.

There have been attempts to devise dependable nonparametric statistical tests for experimental designs with two or more factors. For a one-factor ANOVA, the Mann-Whitney test or an ANOVA on the ranks of the data may be employed for nonparametric testing. But, except for the situation in Section 7, nonparametric procedures with more than one factor have not been generally acceptable. For multifactor analyses, it has been proposed that a parametric ANOVA may be performed on the ranks of the data (and this rank transformation is employed by some computer packages) or that the Kruskal-Wallis test may be expanded. However, Akritas (1990); Blair, Sawilowsky, and Higgins (1987); Brunner and Neumann (1986); McKean and Vidmar (1994); Sawilowsky, Blair, and Higgins (1989); Seaman et al. (1994); Toothaker

---

*Some concepts on two-factor analysis of variance were discussed as early as 1899 (Thiele, 1899). In 1926, R. A. Fisher was the first to present compelling arguments for factorial analysis (Box, 1978: 158; Street, 1990).

and Chang (1980); and Toothaker and Newman (1994) found that these procedures perform poorly and should not be employed.

1 TWO-FACTOR ANALYSIS OF VARIANCE WITH EQUAL REPLICATION

Example 1 presents data from an experiment suited to a two-way analysis of variance. The two factors are fixed (discussed further in Section 1d). The variable under consideration is blood calcium concentration in birds, and the two factors being simultaneously tested are hormone treatment and sex. Because there are two levels in the first factor (hormone-treated and nontreated) and two levels in the second factor (female and male), this experimental design* is termed a 2×2 (or 2^2) factorial. The two factors are said to be "crossed" because each level of one factor is found in combination with each level of the second factor.[†] There are $n = 5$ replicate observations (i.e., calcium determinations on each of five birds) for each of the $2 \times 2 = 4$ combinations of the two factors; therefore, there are a total of $N = 2 \times 2 \times 5 = 20$ data in this experiment. In general, it is advantageous to have equal replication (what is sometimes called a "balanced" or "orthogonal" experimental design), but Section 2 will consider cases with unequal numbers of data per cell, and Section 3 will discuss analyses with only one datum per combination of factors.

For the general case of the two-way factorial analysis of variance, we can refer to one factor as A and to the other as B. Furthermore, let us have a represent the number of levels in factor A, b the number of levels in factor B, and n the number of replicates. A triple subscript on the variable, as X_{ijl}, will enable us to identify uniquely the value that is replicate l of the combination of level i of factor A and level j of factor B. In Example 1, $X_{213} = 32.3$ mg/100 ml, $X_{115} = 9.5$ mg/100 ml, and so on. Each combination of a level of factor A with a level of factor B is called a *cell*. The cells may be visualized as the "groups" in a one-factor ANOVA. There are four cells in Example 1: females without hormone treatment, males without hormone treatment, females with hormone treatment, and males with hormone treatment. And there are n replicate data in each cell. For the cell formed by the combination of level i of factor A and level j of factor B, $\overline{X}_{ij}$ denotes the cell mean; for the data in Example 1, the mean of a cell is the cell total divided by 5, so $\overline{X}_{11} = 14.88$, $\overline{X}_{12} = 12.12$, $\overline{X}_{21} = 32.52$, and $\overline{X}_{22} = 27.78$ (with the units for each mean being mg/100 ml). The mean of all bn data in level i of factor A is $\overline{X}_{i \cdot}$, and the mean of all an data in level j of factor B is $\overline{X}_{j}$. That is, the mean for the 10 non-hormone-treated birds is $\overline{X}_{1 \cdot}$, which is an estimate of the population mean, $\mu_{1 \cdot}$; the mean for the hormone-treated birds is $\overline{X}_{2 \cdot}$, which estimates $\mu_{2 \cdot}$; the mean of the female birds is $\overline{X}_{\cdot 1}$; which estimates $\mu_{\cdot 1}$; and the mean of the male birds is $\overline{X}_{\cdot 2}$, which estimates $\mu_{\cdot 2}$. There are a total of $abn = 20$ data in the experiment, and (just as in the single-factor ANOVA) the mean of all N data (the "grand mean")

---

*R. A. Fisher (1890–1962) is credited with creating and promoting the concept of *experimental design* (Savage, 1976), by which is meant the use of statistical considerations in the planning and executing of experiments.

[†]Two (or more) factors can exist in an ANOVA without being crossed.

EXAMPLE 1   Hypotheses and Data for a Two-Factor Analysis of Variance with Fixed-Effects Factors and Equal Replication

The data are plasma calcium concentrations (in mg/100 ml) of birds of both sexes, half of the birds of each sex being treated with a hormone and half not treated with the hormone.

H_0: There is no effect of hormone treatment on the mean plasma calcium concentration of birds (i.e., $\mu_{\text{no hormone}} = \mu_{\text{hormone}}$ or $\mu_{1.} = \mu_{2.}$).

H_A: There is an effect of hormone treatment on the mean plasma calcium concentration of birds (i.e., $\mu_{\text{no hormone}} \neq \mu_{\text{hormone}}$ or $\mu_{1.} \neq \mu_{2.}$).

H_0: There is no difference in mean plasma calcium concentration between female and male birds (i.e., $\mu_{\text{female}} = \mu_{\text{male}}$ or $\mu_{.1} = \mu_{.2}$).

H_A: There is a difference in mean plasma calcium concentration between female and male birds (i.e., $\mu_{\text{female}} \neq \mu_{\text{male}}$ or $\mu_{.1} \neq \mu_{.2}$).

H_0: There is no interaction of sex and hormone treatment on the mean plasma calcium concentration of birds.

H_A: There is interaction of sex and hormone treatment on the mean plasma calcium concentration of birds.

$\alpha = 0.05$

No Hormone Treatment		Hormone Treatment	
Female	*Male*	*Female*	*Male*
16.3	15.3	38.1	34.0
20.4	17.4	26.2	22.8
12.4	10.9	32.3	27.8
15.8	10.3	33.8	25.0
9.5	6.7	30.2	29.3

Cell totals : $\displaystyle\sum_{l=1}^{5} X_{11l}=74.4 \quad \sum_{l=1}^{5} X_{12l}=60.6 \quad \sum_{l=1}^{5} X_{21l}=162.6 \quad \sum_{l=1}^{5} X_{22l}=138.9$

Cell means : $\overline{X}_{11} = 14.88 \quad \overline{X}_{12} = 12.12 \quad \overline{X}_{21} = 32.52 \quad \overline{X}_{22} = 27.78$

---

is the sum of all the data divided by the total number of data. That is, the grand mean is

$$\overline{X} = \sum_{i=1}^{a}\sum_{j=1}^{b}\sum_{l=1}^{n} X_{ijl}/N. \tag{1}$$

(a) Sources of Variation.   Recall that the total sum of squares is a measure of variability among all the data in a sample. For the two-factor analysis of variance this is conceptually the same as for the single-factor ANOVA:

$$\text{total SS} = \sum_{i=1}^{a}\sum_{j=1}^{b}\sum_{l=1}^{n}(X_{ijl} - \overline{X})^2, \tag{2}$$

Two-Factor Analysis of Variance

with

$$\text{total DF} = N - 1. \tag{3}$$

Next we may consider the variability among cells (each cell being a combination of a level of factor A and a level of factor B), handling cells as we did treated "groups" in the single-factor ANOVA:

$$\text{cells SS} = n \sum_{i=1}^{a} \sum_{j=1}^{b} (\overline{X}_{ij} - \overline{X})^2; \tag{4}$$

and, as the number of cells is ab,

$$\text{cells DF} = ab - 1. \tag{5}$$

Furthermore, the quantity analogous to the within-groups SS in the single-factor ANOVA is

$$\text{within-cells SS} = \sum_{i=1}^{a} \sum_{j=1}^{b} \left[\sum_{l=1}^{n} (X_{ijl} - \overline{X}_{ij})^2 \right], \tag{6}$$

which may also be calculated as

$$\text{within-cells SS} = \text{total SS} - \text{cells SS} \tag{7}$$

and has degrees of freedom of

$$\text{within-cells DF} = ab(n - 1), \tag{8}$$

which is also

$$\text{within-cells DF} = \text{total DF} - \text{cells DF}. \tag{9}$$

The terms *Error SS* and *Error DF* are very commonly used for within-cells SS and within-cells DF, respectively.

The calculations indicated above are analogous to those for the one-way analysis of variance. But a major desire in the two-factor ANOVA is not to consider differences among the cells, but to assess the effects of each of the two factors independently of the other. This is done by considering factor A to be the sole factor in a single-factor ANOVA and then by considering factor B to be the single factor. For factor A this is done as follows:

$$\text{factor } A \text{ SS} = bn \sum_{i=1}^{a} (\overline{X}_{i \cdot} - \overline{X})^2, \tag{10}$$

which is associated with degrees of freedom of

$$\text{factor } A \text{ DF} = a - 1. \tag{11}$$

Similarly, for factor B,

$$\text{factor } B \text{ SS} = an \sum_{j=1}^{b} (\overline{X}_{\cdot j} - \overline{X})^2, \tag{12}$$

for which the degrees of freedom are

$$\text{factor } B \text{ DF} = b - 1. \tag{13}$$

In general, the variability among cells is not equal to the variability among levels of factor A plus the variability among levels of factor B (i.e., the result of Equation 4 is not equal to the sum of the results from Equations 10 and 12). The amount of variability not accounted for is that due to the effect of *interaction** between factors A and B. This is designated as the $A \times B$ interaction, and its sum of squares and degrees of freedom are readily calculated as representing the difference between the variability within cells and the variability due to the two factors:

$$A \times B \text{ interaction SS} = \text{cells SS} - \text{factor} A \text{ SS} - \text{factor } B \text{ SS}, \qquad (14)$$

and

$$A \times B \text{ interaction DF} = \text{cells DF} - \text{factor } A \text{ DF} - \text{factor } B \text{ DF}, \qquad (15)$$

or, equivalently,

$$A \times B \text{ interaction DF} = (\text{factor } A \text{ DF})(\text{factor } B \text{ DF})$$
$$= (a - 1)(b - 1). \qquad (16)$$

Example 1a shows the above calculations of sums of squares and degrees of freedom for the data of Example 1, and Example 2 shows the ANOVA results.

EXAMPLE 1a   Sums of Squares and Degrees of Freedom for the Data of Example 1

grand mean: $\overline{X} = (74.4 + 60.6 + 162.6 + 138.9)/20 = 21.825$

treatment means:

  no hormone: $\overline{X}_1. = (74.4 + 60.6)/10 = 13.50$

  hormone: $\overline{X}_2. = (162.6 + 138.9)/10 = 30.15$

sex means:

  female: $\overline{X}._1 = (74.4 + 162.6)/10 = 23.70$

  male: $\overline{X}._2 = (60.6 + 138.9)/10 = 19.95$

total SS $= \displaystyle\sum_{i=1}^{a}\sum_{j=1}^{b}\sum_{k=1}^{n}(X_{ijk} - \overline{X})^2$

$= (16.3 - 21.825)^2 + (20.4 - 21.825)^2 + \cdots + (29.3 - 21.825)^2$

$= 1762.7175$

total DF $= N - 1 = 20 - 1 = 19$

cells SS $= n\displaystyle\sum_{i=1}^{a}\sum_{j=1}^{b}(\overline{X}_{ij} - \overline{X})^2$

$= 5\Big[(14.88 - 21.825)^2 + (12.12 - 21.825)^2$

$+ (32.52 - 21.825)^2 + (27.78 - 21.825)^2\Big]$

$= 1461.3255$

cells DF $= ab - 1 = (2)(2) - 1 = 4 - 1 = 3$

---

*The term *interaction* was introduced for ANOVA by R. A. Fisher (David, 1995).

$$\text{within-cells (error) SS} = \sum_{i=1}^{a} \sum_{j=1}^{b} \left[\sum_{l=1}^{n} (X_{ijk} - \overline{X}_{ij})^2 \right]$$

$$= (16.3 - 14.88)^2 + (20.4 - 14.88)^2$$
$$+ \cdots + (29.3 - 27.78)^2$$
$$= 301.3920$$

or, equivalently,

within-cells (error) SS = Total SS − Cells SS

$$= 1762.7115 - 1461.3255 = 301.3920$$

within-cells (error) DF $= ab(n - 1) = (2)(2)(5 - 1) = (4)(4) = 16$

or, equivalently,

within-cells (error) DF = total DF − cells DF $= 19 - 3 = 16$

$$\text{Factor } A \text{ SS} = bn \sum_{i=1}^{a} (\overline{X}_{i\cdot} - \overline{X})^2$$

$$= (2)(5) \left[(13.50 - 21.825)^2 + (30.15 - 21.825)^2 \right]$$
$$= 1386.1125$$

factor A DF $= a - 1 = 2 - 1 = 1$

$$\text{factor } B \text{ SS} = an \sum_{j=1}^{b} (\overline{X}_{\cdot j} - \overline{X})^2$$

$$= (2)(5) \left[(23.70 - 21.825)^2 + (19.95 - 21.825)^2 \right]$$
$$= 70.3125$$

factor B DF $= b - 1 = 2 - 1 = 1$

$A \times B$ interaction SS = cells SS − factor A SS − factor B SS

$$= 1461.3255 - 1386.1125 - 70.3125 = 4.9005$$

$A \times B$ interaction DF = cells DF − factor A DF − factor B DF

$$= 3 - 1 - 1 = 1$$

or, equivalently,

$A \times B$ interaction DF = (factor A DF)(factor B DF)

$$= (2 - 1)(2 - 1) = 1$$

An interaction between two factors means that the effect of one factor is not independent of the presence of a particular level of the other factor. In Example 1, no interaction would imply that the difference in the effect of hormone treatment on plasma calcium between males and females is the same under both hormone treatments.* Therefore, interaction among factors is an effect on the variable (e.g.,

---

*Symbolically, the null hypothesis for interaction effect could be stated as $H_0: \mu_{11} - \mu_{12} = \mu_{21} - \mu_{22}$ or $H_0 : \mu_{11} - \mu_{21} = \mu_{12} - \mu_{22}$, where μ_{ij} is the population mean of the variable in the presence of level i of factor A and level j of factor B.

EXAMPLE 2
of Example 1

Two-Factor ANOVA Summary for the Data and Hypotheses

Analysis of Variance Summary Table

Source of variation	SS	DF	MS
Total	1762.2175	19	
Cells	1461.3255	3	
Factor A (hormone)	1386.1125	1	1386.1125
Factor B (sex)	70.3125	1	70.3125
$A \times B$	4.9005	1	4.9005
Within-Cells (Error)	301.3920	16	18.8370

For H_0: There is no effect of hormone treatment on the mean plasma calcium concentration of birds in the population sampled.

$$F = \frac{\text{hormone MS}}{\text{within-cells MS}} = \frac{1386.1125}{18.8370} = 73.6$$

$F_{0.05(1),1,16} = 4.49$
Therefore, reject H_0.

$$P < 0.0005 \quad [P = 0.00000022]$$

For H_0: There is no difference in mean plasma calcium concentration between male and female birds in the population sampled.

$$F = \frac{\text{sex MS}}{\text{within-cells MS}} = \frac{70.3125}{18.8370} = 3.73$$

$F_{0.05(1),1,16} = 4.49$
Therefore, do not reject H_0.

$$0.05 < P < 0.10 \quad [P = 0.071]$$

For H_0: There is no interaction of sex and hormone treatment affecting the mean plasma calcium concentration of birds in the population sampled.

$$F = \frac{\text{hormone} \times \text{sex interaction MS}}{\text{within-cells MS}} = \frac{4.9005}{22.8370} = 0.260$$

$F_{0.05(1)1,16} = 4.49$
Therefore, do not reject H_0.

$$P > 0.25 \quad [P = 0.62]$$

plasma calcium) that is in addition to the sum of the effects of each factor considered separately.

For the one-factor ANOVA it was shown how alternative formulas, referred to as "machine formulas," make the sum-of-squares calculations easier because they do not require computing deviations from means, squaring those deviations, and summing the squared deviations. There are also machine formulas for two-factor analyses of variance that avoid the need to calculate grand, cell, and factor means and the several squared deviations associated with them. Familiarity with these formulas is not necessary if the ANOVA calculations are done by an established computer program, but they can be very useful if a calculator is used. These are shown in Section 1b.

Table 1 summarizes the sums of squares, degrees of freedom, and mean squares for the two-factor analysis of variance.

(b) Machine Formulas. Just as with one-factor analysis of variance, there are so-called machine formulas for two-factor ANOVA that allow the computation of sums of squares without first calculating overall, cell, within-cell, and factor means. These calculations are shown in Example 2a, and they yield the same sums of squares as shown in Example 1a.

TABLE 1: Summary of the Calculations for a Two-Factor Analysis of Variance with Fixed Effects and Equal Replication

Source of variation	Sum of squares (SS)	Degrees of freedom (DF)	Mean square (MS)
Total $[X_{ijl} - \overline{X}]$	Equation 2 or 17	$N - 1$	
Cells $[\overline{X}_{ij} - \overline{X}]$	Equation 4 or 19	$ab - 1$	$\dfrac{\text{cells SS}}{\text{cells DF}}$
Factor A $[\overline{X}_{i.} - \overline{X}]$	Equation 10 or 20	$a - 1$	$\dfrac{\text{factor } A \text{ SS}}{\text{factor } A \text{ DF}}$
Factor B $[\overline{X}_{.j} - \overline{X}]$	Equation 12 or 21	$b - 1$	$\dfrac{\text{factor } B \text{ SS}}{\text{factor } B \text{ DF}}$
$A \times B$ interaction	cells SS − factor A SS − factor B SS	$(a - 1)(b - 1)$	$\dfrac{A \times B \text{ SS}}{A \times B \text{ DF}}$
Within cells (Error) $[X_{ijl} - \overline{X}_{ij}]$	Equation 6 or total SS − cells SS	$ab(n - 1)$ or total DF − cells DF	$\dfrac{\text{error SS}}{\text{error DF}}$

Note: For each source of variation, the bracketed quantity indicates the variation being assessed; a is the number of levels in factor A; b is the number of factors in factor B; n is the number of replicate data in each cell; N is the total number of data (which is abn); X_{ijl} is datum l in the cell formed by level i of factor A and level j of factor B; $\overline{X}_{i.}$ is the mean of the data in level i of factor A; $\overline{X}_{.j}$ is the mean of the data in level j of factor B; $\overline{X}_{ij}$ is the mean of the data in the cell formed by level i of factor A and level j of factor B; and $\overline{X}$ is the mean of all N data.

EXAMPLE 2a **Using Machine Formulas for the Sums of Squares in Example 2**

$$\sum_{i=1}^{a}\sum_{j=1}^{b}\sum_{l=1}^{l}X_{ijl} = 436.5$$

$$\sum_{i=1}^{a}\sum_{j=1}^{b}\sum_{l=1}^{l}X_{ijl}^{2} = 11354.31$$

total for no hormone $= \sum_{j=1}^{b}\sum_{l=1}^{l}X_{1jl} = 74.4 + 60.6 = 135.0$

total for hormone $= \sum_{j=1}^{2}\sum_{l=1}^{l}X_{2jl} = 162.6 + 138.9 = 301.5$

total for females $= \sum_{i=1}^{a}\sum_{l=1}^{l}X_{i1l} = 74.4 + 162.6 = 237.0$

total for males $= \sum_{i=1}^{a}\sum_{l=1}^{l}X_{i2l} = 60.6 + 138.9 = 199.5$

$$C = \frac{\left(\sum_{i=1}^{a}\sum_{j=1}^{b}\sum_{l=1}^{n}X_{ijl}\right)^{2}}{N} = \frac{(436.5)^{2}}{20} = 9526.6125$$

total SS $= \sum_{i=1}^{a}\sum_{j=1}^{b}\sum_{l=1}^{n}X_{ijl}^{2} - C = 11354.31 - 9526.6125 = 1827.6975$

cells SS7 $= \sum_{i=1}^{a}\sum_{j=1}^{b}\frac{\left(\sum_{l=1}^{n}X_{ijl}\right)^{2}}{n} - C$

$$= \frac{(74.4)^{2} + (60.6)^{2} + (162.6)^{2} + (138.9)^{2}}{5} - 9526.6125$$

$$= 1461.3255$$

within-cells (i.e., error) SS $=$ total SS $-$ cells SS

$$= 1827.6975 - 1461.3255 = 366.3720$$

factor A (hormone group) SS $= \dfrac{\sum_{i=1}^{a}\left(\sum_{j=1}^{b}\sum_{l=1}^{n}X_{ijl}\right)^{2}}{bn} - C$

$$= \frac{(\text{sum without hormone})^{2} + (\text{sum with hormone})^{2}}{\text{number of data per hormone group}} - C$$

$$= \frac{(135.0)^2 + (301.5)^2}{(2)(5)} - 9526.6125$$

$$= 1386.1125$$

$$\text{factor } B \text{ (sex) SS} = \frac{\sum\limits_{j=1}^{b}\left(\sum\limits_{i=1}^{a}\sum\limits_{l=1}^{n} X_{ijl}\right)^2}{an} - C$$

$$= \frac{(\text{sum for females})^2 + (\text{sum for males})^2}{\text{number of data per sex}} - C$$

$$= \frac{(237.0)^2 + (199.5)^2}{(2)(5)} - 9526.6125 = 70.3125$$

$$A \times B \text{ interaction SS} = \text{cells SS} - \text{factor } A \text{ SS} - \text{factor } B \text{ SS}$$

$$= 1461.3255 - 1386.1125 - 70.3125 = 4.9005$$

The total variability is expressed by

$$\text{total SS} = \sum_{i=1}^{a}\sum_{j=1}^{b}\sum_{l=1}^{n} X_{ijl}^2 - C, \tag{17}$$

where

$$C = \frac{\left(\sum\limits_{i=1}^{a}\sum\limits_{j=1}^{b}\sum\limits_{l=1}^{n} X_{ijl}\right)^2}{N}. \tag{18}$$

The variability among cells is

$$\text{cells SS} = \frac{\sum\limits_{i=1}^{a}\sum\limits_{j=1}^{b}\left(\sum\limits_{l=1}^{n} X_{ijl}\right)^2}{n} - C. \tag{19}$$

And the variability among levels of factor A is

$$\text{factor } A \text{ SS} = \frac{\sum\limits_{i=1}^{a}\left(\sum\limits_{j=1}^{b}\sum\limits_{l=1}^{n} X_{ijl}\right)^2}{bn} - C. \tag{20}$$

Simply put, the factor A SS is calculated by considering factor A to be the sole factor in a single-factor analysis of variance of the data. That is, we obtain the sum for each level of factor A (ignoring the fact that the data are also categorized into levels of factor B); the sum of a level is what is in parentheses in Equation 20. Then we square each of these level sums and divide the sum of these squares by the number of data per level (i.e., bn). On subtracting the "correction term," C, we arrive at the factor A SS. If the data were in fact analyzed by a single-factor ANOVA, then the groups SS would indeed be the same as the factor A SS just described, and the groups DF would be what the two-foctor ANOVA considers as factor A DF; but the error SS in the one-way ANOVA would be what the two-factor ANOVA considers as the within-cells SS plus the factor B SS and the interaction sum of squares, and the error DF would be the sum of the within-cells, factor B, and interaction degrees of freedom.

For factor B computations, we simply ignore the division of the data into levels of factor A and proceed as if factor B were the single factor in a one-way ANOVA:

$$\text{factor } B \text{ SS} = \frac{\sum\limits_{j=1}^{b}\left(\sum\limits_{i=1}^{a}\sum\limits_{l=1}^{n}X_{ijl}\right)^2}{an} - C. \tag{21}$$

(c) Graphical Display. The cell, column, and row means of Example 1 are summarized in Table 2. Using these means, the effects of each of the two factors, and the presence of interaction, may be visualized by a graph such as Figure 1. We shall refer to the two levels of factor A as A_1 and A_2, and the two levels of

TABLE 2: Cell, Row, and Column Means of the Data of Example 2 (in mg/100 ml)

	Female (B_1)	Male (B_2)	
No hormone (A_1)	14.9	12.1	13.5
Hormone (A_2)	32.5	27.8	30.2
	23.7	20.0	

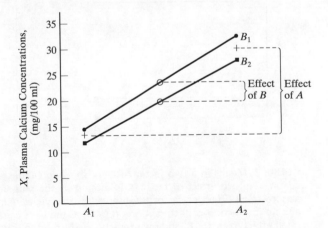

FIGURE 1: The means of the two-factor ANOVA data of Example 1, as given in Table 1. The A_i are the levels of factor A, the B_j are the levels of factor B. A plus sign indicates the mean of an A_i over all (i.e., both) levels of factor B, and an open circle indicates the mean of a B_j over all (i.e., both) levels of factor A.

factor B as B_1 and B_2. The variable, X, is situated on the vertical axis of the figure and on the horizontal axis we indicate A_1 and A_2. The two cell means for B_1 (14.9 and 32.5 mg/100 ml, which are indicated by black circles) are plotted and connected by a line; and the two cell means for B_2 (12.1 and 27.8 mg/100 ml, which are indicated by black squares) are plotted and connected by a second line. The mean of all the data in each level of factor A is indicated with a plus sign, and the mean of all the data in each level of factor B is denoted by an open circle. Then the effect of factor A is observed as the vertical distance between the plus signs; the effect of factor B is expressed as the vertical distance between the open circles; and nonparallelism of the lines indicates interaction between factors A and B. Thus, the ANOVA results of Example 1 are readily seen in this plot: there is a large effect of factor A (which is found to be significant by the F statistic) and a small effect of factor B (which is found to be nonsignificant). There is a small interaction effect, indicated in the figure by the two lines departing a little from being parallel (and this effect is also concluded to be nonsignificant). Various possible patterns of such plots are shown in Figure 2. Such figures may be drawn for situations with more than two levels within factors. And one may place either factor A or factor B on the horizontal axis; usually the factor with the larger number of levels is placed on this axis, so there are fewer lines to examine.

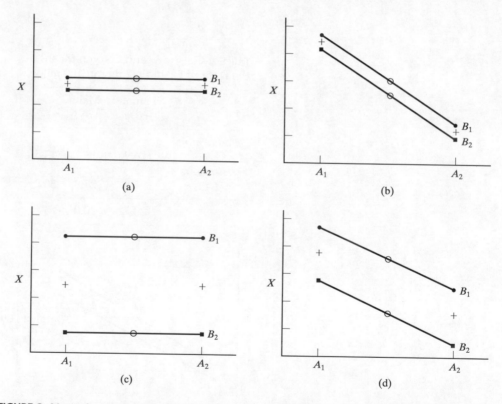

FIGURE 2: Means in a two-factor ANOVA, showing various effects of the two factors and their interaction. (a) No effect of factor A (indicated by the plus signs at the same vertical height on the X axis), a small effect of factor B (observed as the circles being only a small distance apart vertically), and no interaction of factors A and B (seen as the lines being parallel). (b) Large effect of factor A, small effect of factor B, and no interaction (which is the situation in Figure 1). (c) No effect of A, large effect of B, and no interaction. (d) Large effect of A, large effect of B, and no interaction. (e) No effect of A, no effect of B, but interaction between A and B. (f) Large effect of A, no effect of B, with slight interaction. (g) No effect of A, large effect of B, with large interaction. (h) Effect of A, large effect of B, with large interaction.

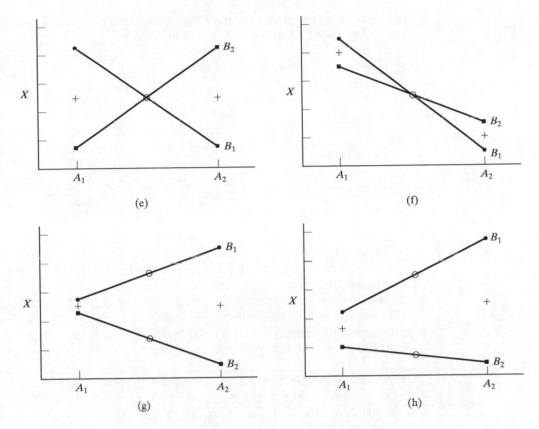

FIGURE 2: *(continued)*

(d) Model I ANOVA. The distinction between fixed and random factors. Example 1 is an ANOVA where the levels of both factors are fixed; we did not simply pick these levels at random. A factorial analysis of variance in which all (in this case both) factors are fixed effects is termed a *Model I ANOVA*. In such a model, the null hypothesis of no difference among the levels of a factor is tested using F = factor MS/error MS. In Example 2, the appropriate F tests conclude that there is a highly significant effect of the hormone treatment on the mean plasma calcium content, and that there is not a significantly different mean plasma calcium concentration between males and females.

In addition, we can test for significant interaction in a Model I ANOVA by F = interaction MS/error MS and find, in our present example, that there is no significant interaction between the sex of the bird and whether it had the hormone treatment. This is interpreted to mean that the effect of the hormone treatment on calcium is not different in males and females (i.e., the effect of the hormone is not dependent on the sex of the bird). This concept of interaction (or its converse, independence) is analogous to that employed in the analysis of contingency tables.

If, in a two-factor analysis of variance, the effects of one or both factors are significant, the interaction effect may or may not be significant. In fact, it is possible to encounter situations where there is a significant interaction even though each of the individual factor effects is judged to be insignificant. A significant interaction implies that the difference among levels of one factor is not constant at all levels of the second factor. Thus, it is generally not useful to speak of a factor effect—even if its F is significant—if there is a significant interaction effect.

TABLE 3: Computation of the F Statistic for Tests of Significance in a Two-Factor ANOVA with Replication

Hypothesized effect	Model I (factors A and B both fixed)	Model II (factors A and B both random)	Model III (factor A fixed: factor B random)
Factor A	$\dfrac{\text{factor } A \text{ MS}}{\text{error MS}}$	$\dfrac{\text{factor } A \text{ MS}}{A \times B \text{ MS}}$	$\dfrac{\text{factor } A \text{ MS}}{A \times B \text{ MS}}$
Factor B	$\dfrac{\text{factor } B \text{ MS}}{\text{error MS}}$	$\dfrac{\text{factor } B \text{ MS}}{A \times B \text{ MS}}$	$\dfrac{\text{factor } B \text{ MS}}{\text{error MS}}$
$A \times B$ interaction	$\dfrac{A \times B \text{ MS}}{\text{error MS}}$	$\dfrac{A \times B \text{ MS}}{\text{error MS}}$	$\dfrac{A \times B \text{ MS}}{\text{error MS}}$

(e) Model II ANOVA. If a factorial design is composed only of factors with random levels, then we are said to be employing a *Model II ANOVA* (a relatively uncommon situation). In such a case, where two factors are involved, the appropriate hypothesis testing for significant factor effects is accomplished by calculating $F =$ factor MS/interaction MS (see Table 3). We test for the interaction effect, as before, by $F =$ interaction MS/error MS, and it is generally not useful to declare factor effects significant if there is a significant interaction effect.

(f) Model III ANOVA. If a factorial design has both fixed-effect and random-effect factors, then it is said to be a *mixed-model*,* or a *Model III ANOVA*. The appropriate F statistics are calculated as shown in Table 3. Special cases of this will be discussed in Section 4. This text observes Voss's (1999) "resolution" of a controversy over the appropriate F for testing a factor effect in mixed models.

(g) Underlying Assumptions. The assumptions underlying the appropriate application of the two-factor analysis of variance are basically those for the single-factor ANOVA: The data in each cell came at random from a normally distributed population of measurements and the variance is the same in all of the populations represented by the cells. This population variance is estimated by the within-cells mean square (i.e., the error mean square).

Although these hypothesis tests are robust enough that minor deviations from these assumptions will not appreciably affect them, the probabilities associated with the calculated F values lose dependability as the sampled populations deviate from normality and homoscedasticity, especially if the populations have skewed distributions. If there is doubt about whether the data satisfy these assumptions, then conclusions regarding the rejection of a null hypothesis should not be made if the associated P is near the α specified for the test.

Few alternatives to the ANOVA exist when the underlying assumptions are seriously violated. In a procedure analogous to that described for single-factor analysis of variance with heterogeneous group variances, Brown and Forsythe (1974b) present a two-factor ANOVA procedure applicable when the cell variances are not assumed to have come from populations with similar variances. In some cases an appropriate data transformation can convert a set of data so the extent of nonnormality and heteroscedasticity is small. As indicated at the end of the introduction to this chapter,

---

*The term *mixed model* was introduced by A. M. Mood in 1950 (David, 1995).

there appears to be no nonparametric procedures to be strongly recommended for factorial ANOVA, with the exception of that of Section 7.

(h) Pooling Mean Squares. If it is not concluded that there is a significant interaction effect, then the interaction MS and the within-cells (i.e., the error) MS are theoretically estimates of the same population variance. Because of this, some authors suggest the pooling of the interaction and within-cells sums of squares and degrees of freedom in such cases. From these pooled SS and DF values, one can obtain a pooled mean square, which then should be a better estimate of the population random error (i.e., within-cell variability) than either the error MS or the interaction MS alone; and the pooled MS will always be a quantity between the interaction MS and the error MS.

The conservative researcher who does not engage in such pooling can be assured that the probability of a Type I error is at the stated α level. But the probability of a Type II error may be greater than is acceptable to some. The chance of the latter type of error is reduced by the pooling described, but confidence in stating the probability of committing a Type I error may be reduced (Brownlee, 1965: 509). Rules of thumb for deciding when to pool have been proposed (e.g., Paull, 1950; Bozivich, Bancroft, and Hartley, 1956), but statistical advice beyond this text should be obtained if such pooling is contemplated. The analyses in this text will proceed according to the conservative nonpooling approach, which Hines (1996), Mead, Bancroft, and Han (1995), and Myers and Well (2003: 333) conclude is generally advisable.

(i) Multiple Comparisons. If significant differences are concluded among the levels of a factor, then multiple comparison procedures may be employed. For such purposes, s^2 is the within-cells MS, ν is the within-cells DF, and the n is replaced in the present situation with the total number of data per level of the factor being tested (i.e., what we have noted in this section as bn data per level of factor A and an data per level of factor B). If there is significant interaction between the two factors, then the means of levels should not be compared. Instead, multiple comparison testing may be performed among cell means.

(j) Confidence Limits for Means. We may compute confidence intervals for population means of levels of a fixed factor by the methods in Section 2 from *Multisample Hypotheses and the Analysis of Variance*. The error mean square, s^2, is the within-cells MS of the present discussion; the error degrees of freedom, ν, is the within-cells DF; and n is replaced in the present context by the total number of data in the level being examined. Confidence intervals for differences between population means are obtained by the procedures of Section 2 from *Multiple Comparisons*. This is demonstrated in Example 3.

EXAMPLE 3   Confidence Limits for the Results of Example 2

We concluded that mean plasma calcium concentration is different between birds with the hormone treatment and those without.

$$\overline{X}_1 = \frac{\text{total for nonhormone group}}{\text{number in nonhormone group}} = \frac{135.0 \text{ mg}/100 \text{ ml}}{10} = 13.50 \text{ mg}/100 \text{ ml}$$

$$\overline{X}_2 = \frac{\text{total for hormone group}}{\text{number in hormone group}} = \frac{301.5 \text{ mg}/100 \text{ ml}}{10} = 30.15 \text{ mg}/100 \text{ ml}$$

$$95\% \text{ CI for } \mu_1 = \overline{X}_1 \pm t_{0.05(2),\,\nu}\sqrt{\frac{s^2}{bn}}$$

$$= 13.50 \pm t_{0.05(2),16}\sqrt{\frac{18.8370}{(2)(5)}}$$

$$= 13.50 \pm (2.120)(0.940)$$

$$= 13.50 \text{ mg}/100 \text{ ml} \pm 1.99 \text{ mg}/100 \text{ ml}$$

$$L_1 = 11.51 \text{ mg}/100 \text{ ml}; \; L_2 = 15.49 \text{ mg}/100 \text{ ml}$$

$$95\% \text{ CI for } \mu_2 = \overline{X}_2 \pm t_{0.05(2),\,\nu}\sqrt{\frac{s^2}{bn}}$$

$$= 30.15 \pm 1.99 \text{ mg}/100 \text{ ml}$$

$$L_1 = 28.16 \text{ mg}/100 \text{ ml}; \; L_2 = 32.14 \text{ mg}/100 \text{ ml}$$

$$95\% \text{ CI for } \mu_1 - \mu_2 = \overline{X}_1 - \overline{X}_2 \pm q_{0.05,\,\nu,k}\sqrt{\frac{s^2}{bn}}$$

$$= 13.50 - 30.15 \pm q_{0.05,16,2}\sqrt{\frac{18.8370}{(2)(5)}}$$

$$= -16.65 \pm (2.998)(0.940)$$

$$= -16.65 \text{ mg}/100 \text{ ml} \pm 2.82 \text{ mg}/100 \text{ ml}$$

$$L_1 = -19.47 \text{ mg}/100 \text{ ml}; \; L_2 = -13.82 \text{ mg}/\text{ ml}$$

We concluded that mean calcium concentration is not different in males and females. That is, the conclusion is that there is no significant difference between $\mu_{\male}$ and $\mu_{\female}$,* and we would not speak of separate confidence intervals for each of these two means or for the difference between the means. If we desired, we could pool the means and speak of a confidence interval for the pooled population mean, μ_p:

$$\text{pooled } \overline{X} = \overline{X}_p = \frac{\text{total for females} + \text{total for males}}{\text{number of females} + \text{number of males}}$$

$$= \frac{237.0 + 199.5}{10 + 10} = 21.82 \text{ mg}/100 \text{ ml}$$

$$95\% \text{ CI for } \mu_p = \overline{X}_p \pm t_{0.05(2),16}\sqrt{\frac{s^2}{20}}$$

$$= 21.82 \pm (2.120)(0.970)$$

$$= 21.82 \text{ mg}/100 \text{ ml} \pm 2.06 \text{ mg}/100 \text{ ml}$$

$$L_1 = 19.76 \text{ mg}/100 \text{ ml}; \; L_2 = 23.88 \text{ mg}/100 \text{ ml}$$

---

*The symbols $\male$ and $\female$ are standard biological designators of male and female, respectively. The former is the same as the astronomical symbol for the planet Mars, named for the Roman god of war (Ares in Greek mythology) and depicting a shield and spear; the latter is the astronomical symbol for the planet Venus, named for the Roman goddess of love and beauty (Aphrodite to the ancient Greeks) and representing a hand mirror. The male symbol has also been used in chemistry and alchemy to denote iron, and the female symbol has indicated copper.

2 TWO-FACTOR ANALYSIS OF VARIANCE WITH UNEQUAL REPLICATION

The procedures outlined in Section 1 for two-factor analysis of variance require that n, the number of replicates per cell, be the same in all cells (a condition sometimes called *orthogonality*). In general, it is desirable, for optimum power, to design experiments with equal cell sizes, but occasionally this is impossible or impractical. Figure 3 shows two-factor experimental designs with various kinds of replication.

(a) Proportional Replication. A two-factor experimental design exhibits proportional replication if the number of data in the cell in row i and column j is

$$n_{ij} = \frac{(\text{number of data in row } i)(\text{number of data in column } j)}{N}, \tag{22}$$

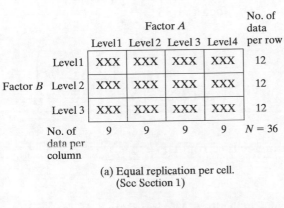

(a) Equal replication per cell. (See Section 1)

(b) Equal replication within rows; proportional replication within columns. (See Section 2a)

(c) Proportional replication within rows and within columns (See Section 2a)

(d) Disproportional replication (See Section 2b)

(e) No replication (See Section 3)

FIGURE 3: Various kinds of replication in a two-factor analysis of variance. In all cases shown, there are four levels of factor A and three levels of factor B.

where N is the total number of data in all cells.* (For example, in Figure 3c, there are two data on row 3, column 1; and $(16)(9)/72 = 2$. The appropriate hypothesis tests are the same as those in Section 1. The sums of squares, degrees of freedom, and mean squares may be calculated by some factorial ANOVA computer programs. Or, the machine formulas referred to in Table 1 may be applied with the following modifications: For sums of squares, substitute n_{ij} for n in Equations 17 and 18, and use

$$\text{cells SS} = \sum_{i=1}^{a} \sum_{j=1}^{b} \frac{\left(\sum_{l=1}^{n_{ij}} X_{ijl} \right)^2}{n_{ij}} : - C, \tag{23}$$

$$\text{factor } A \text{ SS} = \sum_{i=1}^{a} \frac{\left(\sum_{j=1}^{b} \sum_{l=1}^{n_{ij}} X_{ijl} \right)^2}{n_{ij}} : - C, \tag{24}$$

$$\text{factor } B \text{ SS} = \sum_{j=1}^{b} \frac{\left(\sum_{i=1}^{a} \sum_{l=1}^{n_{ij}} X_{ijl} \right)^2}{n_{ij}} : - C, \tag{25}$$

$$\text{within-cells (error) DF} = \sum_{i=1}^{a} \sum_{j=1}^{b} (n_{ij} - 1) : . \tag{26}$$

(b) Disproportional Replication; Missing Data. In factorial analysis of variance, it is generally advisable to have data with equal replication in the cells (Section 1), or at least to have proportional replication (Section 2a). If equality or proportionality is not the case, we may employ computer software capable of performing such analyses of variance with disproportional replication. Alternatively, if only a very few cells have numbers of data in excess of those representing equal or proportional replications, then data may be deleted, at random, within such cells, so that equality or proportionality is achieved. Then the ANOVA can proceed as usual, as described in Section 1 or 2a.

If one cell is one datum short of the number required for equal or proportional replication, a value may be estimated[†] for inclusion in place of the missing datum, as follows (Shearer, 1973):

$$\hat{X}_{ijl} = \frac{aA_i + bB_j - \sum_{i=1}^{a} \sum_{j=1}^{b} \sum_{l=1}^{n_{ij}} X_{ijl}}{N + 1 - a - b}, \tag{27}$$

---

*The number of replicates in each of the ab cells need not be checked against Equation 22 to determine whether proportional replication is present. One need check only one cell in each of $a - 1$ levels of factor A and one in each of $b - 1$ levels of factor B (Huck and Layne, 1974).

[†]The estimation of missing values is often referred to as *imputation* and is performed by some computer routines. However, there are many different methods for imputing missing values, especially when more than one datum is missing, and these methods do not all yield the same results.

where $\hat{X}_{ijl}$ is the estimated value for replicate l in level i of factor A and level j of factor B; A_i is the sum of the other data in level i of factor A; B_j is the sum of the other data in level j of factor B; $\sum\sum\sum X_{ijl}$ is the sum of all the known data, and N is the total number of data (including the missing datum) in the experimental design. For example, if datum X_{124} had been missing in Example 1, it could have had a quantity inserted in its place, estimated by Equation 27, where $a = 2$, $b = 2$, $N = 20$, $A_i = A_1$ = the sum of all known data from animals receiving no hormone treatment; $B_j = B_2$ = the sum of all known data from males; and $\sum\sum\sum X_{ijl}$ = the sum of all 19 known data from both hormone treatments and both sexes. After the missing datum has been estimated, it is inserted into the data set and the ANOVA computations may proceed, with the provision that a missing datum is not counted in determining total and within-cells degrees of freedom. (Therefore, if a datum were missing in Example 1, the total DF would have been 18 and the within-cells DF would have been 15.)

If more than one datum is missing (but neither more than 10% of the total number of data nor more data than the number of levels of any factor), then Equation 27 could be used iteratively to derive estimates of the missing data (e.g., using cell means as initial estimates). The number of such estimates would not enter into the total or within-cells degrees-of-freedom determinations.

If only a few cells (say, no more than the number of levels in either factor) are each one datum short of the numbers required for equal or proportional replication, then the mean of the data in each such cell may be inserted as an additional datum in that cell. In the latter situation, the analysis proceeds as usual but with the total DF and the within-cells DF each being determined without counting such additional inserted data. Instead of employing these cell means themselves, however, they could be used as starting values for employing Equation 27 in iterative fashion. Another procedure for dealing with unequal, and nonproportional, replication is by so-called unweighted means analysis, which employs the harmonic mean of the n_{ij}'s. This will not be discussed here.

None of these procedures is as desirable as when the data are equally or proportionally distributed among the cells.

3 TWO-FACTOR ANALYSIS OF VARIANCE WITHOUT REPLICATION

It is generally advisable that a two-factor experimental design have more than one datum in each cell, but situations are encountered in which there is only one datum for each combination of factors (i.e., $n = 1$ for all cells). It is sometimes feasible to collect additional data, to allow the use of the procedures of Section 1 or 2, but it is also possible to perform a two-factorial ANOVA with nonreplicated data. In a situation of no replication, each datum may be denoted by a double subscript, as X_{ij}, where i denotes a level of factor A and j indicates a level of factor B.

For a levels of factor A and b levels of factor B, the appropriate computations of sums of squares, degrees of freedom, and mean squares are shown directly below. These are analogous to equations in Section 1, modified by eliminating n and any summation within cells.

$$\text{total SS} = \sum_{i=1}^{a}\sum_{j=1}^{b}\left(X_{ij} - \overline{X}\right)^2 , \tag{28}$$

where the mean of all N data (and $N = ab$) is

$$\overline{X} = \frac{\sum\limits_{i=1}^{a}\sum\limits_{j=1}^{b}X_{ij}}{N}.$$ (29)

Further,

$$\text{factor } A \text{ SS} = b\sum\limits_{i=1}^{a}\left(\overline{X}_{i\cdot} - \overline{X}\right)^2;$$ (30)

$$\text{factor } B \text{ SS} = a\sum\limits_{j=1}^{a}\left(\overline{X}_{\cdot j} - \overline{X}\right)^2.$$ (31)

When there is no replication within cells (i.e., $n = 1$), the cells SS of Section 1 is identical to the total SS, and the cells DF is the same as the total DF. Consequently, the within-cells sum of squares and degrees of freedom are both zero; that is, with only one datum per cell, there is no variability within cells. The variability among the N data that is not accounted for by the effects of the two factors is the *remainder** variability:

$$\text{remainder SS} = \text{total SS} - \text{factor } A \text{ SS} - \text{factor } B \text{ SS}$$ (32)

$$\text{remainder DF} = \text{total DF} - \text{factor } A \text{ DF} - \text{factor } B \text{ DF}$$ (33)

These sums of squares and degrees of freedom, and the relevant mean squares, are summarized in Table 4. Note that Equations 32 and 33 are what are referred to as "interaction" quantities when replication is present; with no replicates it is not possible to assess interaction in the population that was sampled. Table 5

TABLE 4: Summary of the Calculations for a Two-Factor Analysis of Variance with No Replication

Source of variation	Sum of squares (SS)	Degrees of freedom (DF)	Mean square (MS)
Total $[X_{ij} - \overline{X}]$	Equation 27 or 34	$N - 1$	
Factor A $[\overline{X}_{i\cdot} - \overline{X}]$	Equation 29 or 35	$a - 1$	$\dfrac{\text{factor } A \text{ SS}}{\text{factor } A \text{ DF}}$
Factor B $[\overline{X}_{\cdot j} - \overline{X}]$	Equation 30 or 26	$b - 1$	$\dfrac{\text{factor } B \text{ SS}}{\text{factor } B \text{ DF}}$
Remainder	Equation 31	$(a - 1)(b - 1)$ or total DF $-$ factor A DF $-$ factor B DF	$\dfrac{\text{remainder SS}}{\text{remainder DF}}$

Note: For each source of variation, the bracketed quantity indicates the variation being assessed; a is the number of levels in factor A; b is the number of factors in factor B; N is the total number of data (which is ab); X_{ij} is the datum in level i of factor A and level j of factor B; $\overline{X}_{i\cdot}$ is the mean of the data in level i of factor A; $\overline{X}_{\cdot j}$ is the mean of the data in level j of factor B; and $\overline{X}$ is the mean of all N data.

*Some authors refer to "remainder" as "error" or "residual."

TABLE 5: Computation of the F Statistic for Tests of Significance in a Two-Factor ANOVA without Replication

(a) If It Is Assumed That There May Be a Significant Interaction Effect

Hypothesized effect	Model I (factors A and B both fixed)	Model II (factors A and B both random)	Model III (factor A fixed; factor B random)
Factor A	Test with caution*	$\dfrac{\text{factor } A \text{ MS}}{\text{remainder MS}}$	$\dfrac{\text{factor } A \text{ MS}}{\text{remainder MS}}$
Factor B	Test with caution*	$\dfrac{\text{factor } B \text{ MS}}{\text{remainder MS}}$	Test with caution*
$A \times B$ interaction	No test possible	No test possible	No test possible

*Analysis can be performed as in Model II, but with increased chance of Type II error.

(b) If It Is Correctly Assumed That There Is No Significant Interaction Effect

Hypothesized effect	Model I	Model II	Model III
Factor A	$\dfrac{\text{factor } A \text{ MS}}{\text{remainder MS}}$	$\dfrac{\text{factor } A \text{ MS}}{\text{remainder MS}}$	$\dfrac{\text{factor } A \text{ MS}}{\text{remainder MS}}$
Factor B	$\dfrac{\text{factor } B \text{ MS}}{\text{remainder MS}}$	$\dfrac{\text{factor } B \text{ MS}}{\text{remainder MS}}$	$\dfrac{\text{factor } B \text{ MS}}{\text{remainder MS}}$
$A \times B$ interaction	No test possible	No test possible	No test possible

summarizes the significance tests that may be performed to test hypotheses about each of the factors. Testing for the effect of each of the two factors in a Model I analysis (or testing for the effect of the random factor in a Model III design) is not advisable if there may, in fact, be interaction between the two factors (and there will be decreased test power); but if a significant difference is concluded, then that conclusion may be accepted. The presence of interaction, also called nonadditivity, may be detectable by the testing procedure of Tukey (1949).

(a) "Machine Formulas." If there is no replication in a two-factor ANOVA, the machine formulas for sums of squares are simplifications of those in Section 1b:

$$C = \frac{\left(\sum\limits_{i=1}^{a} \sum\limits_{j=1}^{b} X_{ij} \right)^2}{N}, \tag{34}$$

$$\text{total SS} = \sum\limits_{i=1}^{a} \sum\limits_{j=1}^{b} X_{ij}^2 - C, \tag{35}$$

$$\text{factor : } A \text{ SS} = \frac{\sum\limits_{i=1}^{a} \left(\sum\limits_{j=1}^{b} X_{ij}^2 \right)}{b} - C, \tag{36}$$

$$\text{factor : } B \text{ SS} = \frac{\sum\limits_{j=1}^{b} \left(\sum\limits_{i=1}^{a} X_{ij}^2 \right)}{a} - C, \tag{37}$$

and the remainder sum of squares is as in Equation 32.

(b) Multiple Comparisons and Confidence Limits.   In a two-factor ANOVA with no replication, the multiple-comparison and confidence-limit considerations of Sections 1i and 1j may be applied. However, there is no within-cells (error) mean square or degrees of freedom in the absence of replication. If it can be assumed that there is no interaction between the two factors, then the remainder MS may be employed where s^2 is specified in those sections, the remainder DF is used in place of v, a is the same as an, and b is the same as bn. If, however, there may be interaction, then multiple comparisons and confidence limits should be avoided.

4 TWO-FACTOR ANALYSIS OF VARIANCE WITH RANDOMIZED BLOCKS OR REPEATED MEASURES

The analysis of variance procedure (the *completely randomized design*) is for situations where the data are all independent of each other and the experimental units are assigned to the k treatments in a random fashion (that is, random except for striving for equal numbers in each treatment). The following discussion is of two types of ANOVA, in which each datum in one of the k groups *is* related to one datum in each of the other groups.

(a) Randomized Blocks.   Each of four animals from the same litter could be assigned to be raised on each of four diets. The body weights of each set of four animals (i.e., the data for each litter) would be said to constitute a *block*, for the data in a litter are related to each other (namely by having the same mother). With a experimental groups (denoted as k) and b blocks, there would be $N = ab$ data in the analysis. The concept of blocks is an extension, for more than two groups, the concept of pairs. This experimental plan is called a *randomized-complete-block design*, and each block contains a measurement for each of the a treatments. Only complete blocks will be considered here, so this will simply be called a *randomized-block design*.* When the analysis employs blocks of data within which the data are related, the hypothesis testing of differences among groups can be more powerful than in the completely randomized design.

An illustration of a randomized-block ANOVA is in Example 4. The intent of the experiment shown is to determine whether there is a difference among three anesthetic drugs in the time it takes for the anesthetic to take effect when injected intramuscularly into cats of a specified breed. Three cats are obtained from each of five laboratories; because the laboratories may differ in factors such as the food and exercise the animals have had, the three from each laboratory are considered to be a block. Thus, the experiment has $a = 3$ treatment groups and $b = 5$ blocks, and the variable in anesthetic group i and block j is indicated as X_{ij}. The sum of the data in group i can be denoted as $\sum_{j=1}^{b} X_{ij}$, the total of the measurements in block j as $\sum_{i=1}^{a} X_{ij}$, and the sum of all ab data as $N = \sum_{i=1}^{a} \sum_{j=1}^{b} X_{ij}$. In this example, there is only one datum in each of the ab cells (i.e., one per combination of treatment and block), a very common situation when working with randomized blocks.

In Example 4, the interest is whether there is any difference among the effects of the different anesthetic drugs, not whether there is any difference due to laboratory source of the animals. (Indeed, the factor defining the blocks is sometimes referred to

---

*The randomized-block experimental design was developed and so named by R. A. Fisher (1926; David, 1995).

EXAMPLE 4   A Randomized Complete Block Analysis of Variance (Model III Two-Factor Analysis of Variance) without Within-Cell Replication

H_0: The mean time for effectiveness is the same for all three anesthetics (i.e., $\mu_1 = \mu_2 = \mu_3$).

H_A: The mean time for effectiveness is not the same for all three anesthetics.

$\alpha = 0.05$

Each block consists of three cats from a single source, and each block is from a different source. Within a block, the cats are assigned one of the anesthetics at random, by numbering the cats 1, 2, and 3 and assigning each of them treatment 1, 2, or 3 at random. For this experiment the randomly designated treatments, from 1 to 3, for each block were as follows, with the anesthetic's time for effect (in minutes) given in parentheses:

	Animal 1	Animal 2	Animal 3
Block 1:	Treatment 3 (10.75)	Treatment 1 (8.25)	Treatment 2 (11.25)
Block 2:	Treatment 1 (10.00)	Treatment 3 (11.75)	Treatment 2 (12.50)
Block 3:	Treatment 3 (11.25)	Treatment 1 (10.25)	Treatment 2 (12.00)
Block 4:	Treatment 1 (9.50)	Treatment 2 (9.75)	Treatment 3 (9.00)
Block 5:	Treatment 2 (11.00)	Treatment 1 (8.75)	Treatment 3 (10.00)

These data are rearranged as follows in order to tabulate the treatment, block, and grand totals (and, if not using the machine formulas, the treatment, block, and grand means).

Block (j)	Treatment (i) 1	2	3	Block Total $\left(\sum_{i=1}^{a} X_{ij}\right)$	Block Mean $\left(\overline{X}_{\cdot j}\right)$
1	8.25	11.25	10.75	30.25	10.08
2	11.00	12.50	11.75	35.25	11.75
3	10.25	12.00	11.25	33.50	11.17
4	9.50	9.75	9.00	28.25	9.42
5	8.75	11.00	10.00	29.75	9.92

Treatment total: $\sum_{j=1}^{b} X_{ij}$ 47.75   56.50   52.75

Treatment mean: $\overline{X}_{i\cdot}$ 9.55   11.30   10.55

Grand total $= \sum_{i=1}^{a} \sum_{j=1}^{b} X_{ij} = 157.00$ Grand mean $= \overline{X} = 10.47$

The sums of squares required in the following table may be obtained using the equations in Section 3, as referenced in Table 4.

Source of variation	SS	DF	MS
Total	21.7333	14	
Treatments	7.7083	2	3.8542
Blocks	11.0667	4	
Remainder	2.9583	8	0.3698

$$F = \frac{\text{treatments MS}}{\text{remainder MS}} = \frac{3.8542}{0.3698} = 10.4$$

$F_{0.05(1),2,8} = 4.46$, so reject H_0.

$0.005 < P < 0.01$ $[P = 0.0060]$

as a "nuisance factor" or "nuisance variable.") The anesthetics, therefore, are three levels of a fixed-effects factor, and the laboratories are five levels of a random-effects factor. So the completely randomized experimental design calls for a mixed-model (i.e., Model III) analysis of variance (Section 1f). Blocking by laboratory is done to account for more of the total variability among all N data than would be accounted for by considering the fixed-factor effects alone. This will decrease the mean square in the denominator of the F that assesses the difference among the treatments, with the intent of making the ANOVA more powerful than if the data were not collected in blocks.

The assignment of an experimental unit to each of the animals in a block should be done at random. For this purpose, Table 41 from *Appendix: Statistical Tables and Graphs* or other source of random numbers may be consulted. In the present example, the experimenter could arbitrarily assign numbers 1, 2, and 3 to each cat from each laboratory. The random-number table should then be entered at a random place, and for each block a random sequence of the numerals 1, 2, and 3 (ignoring all other numbers and any repetition of a 1, 2, or 3) will indicate which treatments should be applied to the animals numbered 1, 2, and 3 in that block. So, in Example 4, a sequence of animals numbered 3, 1, 2 was obtained for the first block; 1, 3, 2 for the second; 3, 2, 1 for the third; and so on.

The randomized-block experimental design has found much use in agricultural research, where b plots of ground are designated as blocks and where the environmental (for example, soil and water) conditions are very similar within each block (though not necessarily among blocks). Then a experimental treatments (e.g., fertilizer of pesticide treatment) are applied to random portions of each of the b blocks.

(b) Randomized Blocks with Replication. It is common for the randomized-complete-block experimental design to contain only one datum per cell. That is what is demonstrated in Example 4, with the calculation of F executed as shown in the last column of Tables 5a and 5b. If there are multiple data per cell (what is known as the generalized randomized-block design), this would be handled as a mixed-model two-factor ANOVA with replication. Doing so would obtain the mean squares and degrees of freedom as indicated earlier in this chapter; and the appropriate F's would be those indicated in the last column of Table 3. A case where this would be applicable is where the experiment of Example 4 employed a total of six cats—instead of three—from each laboratory, assigning two of the six at random to each of the three treatments.

(c) Repeated Measures. The hypotheses of Example 4 could also be tested using an experimental design differing from, but related to, that of randomized blocks. In an experimental procedure using what are called *repeated measures*, each of b experimental animals would be tested with one of the a anesthetic drugs; then, after the effects of the drug had worn off, one of the other anesthetics would be applied to the same animal; and after the effects of that drug were gone, the third drug would be administered to that animal. Thus, b experimental animals would be needed, far fewer than the ab animals required in the randomized-block experiment of Example 4, for each block of data contains a successive measurements from the same experimental animal (often referred to as an experimental "subject"). If possible, the application of the a treatments to each of the b subjects should be done in a random sequence, comparable to the randomization within blocks in Section 4a. Also, when collecting data for a repeated-measures analysis of variance, sufficient time should be allowed between successive treatments so the effect of a treatment is not contaminated with the effect of the previous treatment (i.e., so there is no "carryover effect" from treatment to treatment).*

Thus, the arrangement of data from a repeated-measures experiment for the hypotheses of Example 4 would look exactly like that in that example, except that each of the five blocks of data would be measurements from a single animal, instead of from a animals, from a specified laboratory.

There are some repeated-measures studies where the treatments are not administered in a random sequence to each subject. For example, we might wish to test the effect of a drug on the blood sugar of horses at different times (perhaps at 1, 2, and 5 hours) after the drug's administration. Each of b horses ("subjects") could be given the drug and its blood sugar measured before administering the drug and subsequently at each of the three specified times (so a would be 4). In such a situation, the levels of factor A (the four times) are fixed and are the same for all blocks, and carryover effects are a desired part of the study.[†]

The repeated-measures experimental design is commonly used by psychological researchers, where the behavioral response of each of several subjects is recorded for each of several experimental circumstances.

(d) Randomized-Block and Repeated-Measures Assumptions. In a randomized-block or repeated-measures experiment, we assume that there are correlations among the measurements within a block or among measurements repeated on a subject. For the randomized-block data in Example 4, it may be reasonable to suppose that if an animal is quickly affected by one anesthetic, it will be quickly affected by each

---

*Although the experiment should be conducted to avoid carryover effects, the times of administering the drug (first, second, or third) could be considered the levels of a third factor, and a three-factor ANOVA could be performed in what is referred to as a "crossover experimental design".

[†]Kirk (1995: 255) calls randomized levels of factor A within blocks a "subjects-by-treatment" experimental design and uses the term *subjects-by-trials* to describe a design where the sequence of application of the levels of factor A to is the same in each block.

of the others. And for the repeated-measures situation described in Section 4c, it might well be assumed that the effect of a drug at a given time will be related to the effect at a previous time. However, for the probability of a calculated F to be compared dependably to tabled values of F, there should be equal correlations among all pairs of groups of data. So, for the experiment in Example 4, the correlation between the data in groups 1 and 2 is assumed to be the same as the correlation between the data in groups 1 and 3, and the same as that between data in groups 2 and 3. This characteristic, referred to as *compound symmetry*, is related to what statisticians call *sphericity* (e.g., Huynh and Feldt, 1970), or *circularity* (e.g., Rouanet and Lépine, 1970), and it—along with the usual ANOVA assumptions (Section 1g)—is an underlying assumption of randomized-block and repeated-measures analyses of variance. Violation of this assumption is, unfortunately, common but difficult to test for, and the investigator should be aware that the Type I error in such tests may be greater than the specified α. An alternative procedure for analyzing data from repeated-measures experiments, one that does not depend upon the sphericity assumption, is *multivariate analysis of variance*, which has gained in popularity with the increased availability of computer packages to handle the relatively complex computations. This assumption and this alternative are discussed in major works on analysis of variance and multivariate analysis (e.g., Girden, 1992; Kirk, 1995; Maxwell and Delaney, 2004; O'Brien and Kaiser, 1985; and Stevens, 2002).

If there are missing data, the considerations of Section 2b apply. If the experimental design has only one datum for each combination of the factors, and one of the data is missing, then the estimation of Equation 26 becomes

$$\hat{X}_{ij} = \frac{aA_i + bB_j - \sum_{i=1}^{a}\sum_{j=1}^{b}X_{ij}}{(a-1)(b-1)}. \tag{38}$$

If more than one datum is missing in a block, the entire block can be deleted from the analysis.

(e) More Than One Fixed-Effects Factor. There are many possible experimental designs when the effects of more than one factor are being assessed. One other situation would be where the experiment of Example 4 employed blocks as a random-effects factor along with *two* fixed-effects factors, perhaps the drug and the animal's sex. The needed computations of sums of squares, degrees of freedom, and mean squares would likely be performed by computer.

5 MULTIPLE COMPARISONS AND CONFIDENCE INTERVALS IN TWO-FACTOR ANALYSIS OF VARIANCE

If a two-factor analysis of variance reveals a significant effect among levels of a fixed-effects factor having more than two levels, then we can determine between which levels the significant difference(s) occur(s). If the desire is to compare all pairs of means for levels in a factor, this may be done using the Tukey test. The appropriate SE is calculated by $\text{SE} = \sqrt{\frac{s^2}{n}}$,, substituting for n the number of data in each level (i.e., there are bn data in each level of factor A and an data in levels of factor B); s^2 is the within-cells MS and ν is the within-cells degrees of freedom. If there is no replication in the experiment, then we are obliged to use the remainder MS in place of the within-cells MS and to use the remainder DF as ν.

The calculation of confidence limits for the population mean estimated by each significantly different level mean can be performed by the procedures, as can the computation of confidence limits for differences between members of pairs of significantly different level means.

If it is desired to compare a control mean to each of the other level means, Dunnett's test, may be used; and that section also shows how to calculate confidence limits for the differences between such means. Scheffé's procedure for multiple contrasts may also be applied to the levels of a factor, where the critical value in Equation 38a employs either a or b in place of k (depending, respectively, on whether the levels of factor A or B are being examined), and the within-cells DF is used in place of $N - k$. n in the standard-error computation is to be replaced by the number of data per level, and s^2 and v are the within-cells MS and DF, respectively.

Multiple-comparison testing and confidence-interval determination are appropriate for levels of a fixed-effects factor but are not used with random-effects factors.

$$S_\alpha = \sqrt{(k - 1)F_{\alpha(1),k-1,N-k}}. \tag{38a}$$

(a) If Interaction Is Significant. On concluding that there is a significant interaction between factors A and B, it is generally not meaningful to test for differences among levels of either of the factors. However, it may be desired to perform multiple comparison testing to seek significant differences among cell means. This can be done with any of the above-mentioned procedures, where n (the number of data per cell) is appropriate instead of the number of data per level. For the Scheffé test critical value (Equation 38a), k is the number of cells (i.e., $k - ab$) and $N - k$ is the within-cells DF.

(b) Randomized Blocks and Repeated Measures. In randomized-block and repeated-measures experimental designs, the sphericity problem mentioned in Section 4d is reason to recommend that multiple-comparison testing not use a pooled variance but, instead, employ the Games and Howell procedure. In doing so, the two sample sizes (n_B and n_A) for calculating SE will each be b.

6 SAMPLE SIZE, DETECTABLE DIFFERENCE, AND POWER IN TWO-FACTOR ANALYSIS OF VARIANCE

The concepts and procedures of estimating power, sample size, and minimum detectable difference for a single-factor ANOVA can be applied to fixed-effects factors in a two-factor analysis of variance. (The handling of the fixed factor in a mixed-model ANOVA will be explained in Section 6e.)

We can consider either factor A or factor B (or both, but one at a time). Let us say k' is the number of levels of the factor being examined. (That is, $k' = a$ for factor A; $k' = b$ for factor B.) Let us define n' as the number of data in each level. (That is, $n' = bn$ for factor A; $n' = an$ for factor B.) We shall also have s^2 refer to the within-cells MS. The mean of the population from which level m came is denoted as μ_m.

(a) Power of the Test.

$$\phi = \sqrt{\frac{n' \sum_{m=1}^{k'} (\mu_m - \mu)^2}{k's^2}}, \tag{39}$$

$$\mu = \frac{\sum_{m=1}^{k'} \mu_m}{k'}, \tag{40}$$

$$\phi = \sqrt{\frac{n'\delta^2}{2k's^2}}, \tag{41}$$

in order to estimate the power of the analysis of variance in detecting differences among the population means of the levels of the factor under consideration.

After any of the computations of ϕ have taken place, either as above or as below, then we proceed to employ Figure 1 from *Appendix: Statistical Tables and Graphs*, with ν_1 being the factor DF (i.e., $k' - 1$), and ν_2 referring to the within-cells (i.e., error) DF.

Later in this text there are examples of ANOVAs where the appropriate denominator for F is some mean square other than the within-cells MS. In such a case, s^2 and ν_2 will refer to the relevant MS and DF.

(b) Sample Size Required. By using Equation 41 with a specified significance level, and detectable difference between means, we can determine the necessary minimum number of data per level, n', needed to perform the experiment with a desired power. This is done iteratively.

(c) Minimum Detectable Difference. Recall estimating the smallest detectable difference between population means, given the significance level, sample size, and power of a one-way ANOVA. We can pose the same question in the two-factor experiment

$$\delta = \sqrt{\frac{2k's^2\phi^2}{n'}}. \tag{42}$$

(d) Maximum Number of Levels Testable. The considerations of Example 8 from *Multisample Hypotheses and the Analysis of Variance* can be applied to the two-factor case by using Equation 41.

(e) Mixed-Model ANOVA. All the preceding considerations of this section can be applied to the fixed factor in a mixed-model (Model III) two-factor analysis of variance with the following modifications.

For factor A fixed, with replication within cells, substitute the interaction MS for the within-cells MS, and use the interaction DF for ν_2.

For factor A fixed, with no replication (i.e., a randomized block experimental design), substitute the remainder MS for the within-cells MS, and use the remainder DF for v_2. If there is no replication, then $n = 1$, and $n' = b$. (Recall that if there is no replication, we do not test for interaction effect.)

7 NONPARAMETRIC RANDOMIZED-BLOCK OR REPEATED-MEASURES ANALYSIS OF VARIANCE

Friedman's* test (1937, 1940) is a nonparametric analysis that may be performed on a randomized-block experimental design, and it is especially useful with data that do not meet the parametric analysis of variance assumptions of normality and homoscedasticity, namely that the k samples (i.e., the k levels of the fixed-effect factor) come from populations that are each normally distributed and have the same variance. Kepner and Robinson (1988) showed that the Friedman test compares favorably with other nonparametric procedures. If the assumptions of the parametric ANOVA are met, the Friedman test will be $3k/[\pi(k + 1)]$ as powerful as the parametric method (van Elteren and Noether, 1959). (For example, the power of the nonparametric test ranges from 64% of the power of the parametric test when $k = 2$, to 72% when $k = 3$, to 87% when $k = 10$, to 95% when k approaches ∞.) If the assumptions of the parametric test are seriously violated, it should not be used and the Friedman test is typically advisable. Where $k = 2$, the Friedman test is equivalent to the sign test.

In Example 5, Friedman's test is applied to the data of Example 4. The data within each of the b blocks are assigned ranks. The ranks are then summed for each of the a groups, each rank sum being denoted as R_i. The test statistic, χ_r^2, is calculated as[†]

$$\chi_r^2 = \frac{12}{ba(a + 1)} \sum_{i=1}^{a} R_i^2 - 3b(a + 1).$$ (44)

Critical values of χ_r^2, for many values of a and b, are given in Table 14 from *Appendix: Statistical Tables and Graphs*.

When $a = 2$, the Wilcoxon paired-sample test should be used; if $b = 2$, then the Spearman rank correlation should be employed. Table 14 from *Appendix: Statistical Tables and Graphs* should be used when the a and b of an experimental design are contained therein. For a and b beyond this table, the distribution of χ_r^2 may be considered to be approximated by the χ^2 distribution, with $a - 1$ degrees of freedom. Fahoome (2002) advised that the chi-square approximation is acceptable when b is at least 13 when testing at the 0.05 level of significance and at least 23 when $\alpha = 0.01$ is specified. However, Iman and Davenport (1980) showed that this commonly used approximation tends to be conservative (i.e., it may have a high likelihood

---

*Milton Friedman (1912–2006), American economist and winner of the 1976 Nobel Memorial Prize in Economic Science. He is often credited with popularizing the statement, "There's no such thing as a free lunch," and in 1975 he published a book with that title; F. Shapiro reported that Friedman's statement had been in use by others more than 20 years before (Hafner, 2001).

[†]An equivalent formula is

$$\chi_r^2 = \frac{12 \sum_{i=1}^{a} (R_i - \overline{R})^2}{ba(a + 1)}$$ (43)

(Pearson and Hartley, 1976: 52), showing that we are assessing the difference between the rank sums (R_i) and the mean of the rank sums ($\overline{R}$).

EXAMPLE 5   Friedman's Analysis of Variance by Ranks Applied to the Randomized Block Data of Example 4

H_0: The time for effectiveness is the same for all three anesthetics.

H_A: The time for effectiveness is not the same for all three anesthetics.

$\alpha = 0.05$

The data from Example 4, for the three treatments and five blocks, are shown here, with ranks (1, 2, and 3) within each block shown in parentheses.

	Treatment (i)		
Block (j)	1	2	3
1	8.25	11.25	10.75
	(1)	(3)	(2)
2	11.00	12.50	11.75
	(1)	(3)	(2)
3	10.25	12.00	11.25
	(1)	(3)	(2)
4	9.50	9.75	9.00
	(2)	(3)	(1)
5	8.75	11.00	10.00
	(1)	(3)	(2)
Rank sum (R_i)	6	15	9
Mean rank ($\overline{R}_i$)	1.2	3.0	1.8

$a = 3, \quad b = 5$

$$\chi_r^2 = \frac{12}{ba(a+1)} \sum R_i^2 (3b(a+1))$$

$$= \frac{12}{(5)(3)(3+1)} \left(6^2 + 15^2 + 9^2\right) - 3(5)(3+1)$$

$$= 0.200(342) - 60 = 8.400$$

$\left(\chi_r^2\right)_{0.05,3,5} = 6.400$

Reject H_0.

$$P < 0.01 \quad [P = 0.0085]$$

$$F_F = \frac{(b-1)\chi_r^2}{b(a-1) - \chi_r^2} = \frac{(5-1)(8.4)}{5(3-1) - 8.4} = \frac{33.6}{1.6} = 21.0$$

$F_{0.05(1),2,4} = 6.94$

Reject H_0.

$$0.005 < P < 0.01 \quad [P = 0.0076]$$

of a Type II error—and, therefore, low power) and that

$$F_F = \frac{(b-1)\chi_r^2}{b(a-1) - \chi_r^2}$$ (45)

is generally superior. To test, H_0, F_F is compared to F with degrees of freedom of $a - 1$ and $(a - 1)(b - 1)$.[*]

Because this nonparametric test employs ranks, it could have been used even if the measurements (in minutes) were not known for the times for the anesthetics to take effect. All that would be needed would be the ranks in each block. In Example 4, this would be the knowledge that, for litter 1 (i.e., block 1) drug 1 was effective in a shorter time than was drug 3; and drug 3 acted faster than drug 2; and so on for all b blocks.

Another approach to testing of this experimental design is that of *rank transformation*, by which one ranks all ab data and performs the analysis of variance of Section 4 on those ranks (Conover 1974b; Iman and Iman, 1976, 1981; and Iman, Hora, and Conover, 1984). Quade (1979) presented a test that is an extension of the Wilcoxon paired-sample test that may be preferable in some circumstances (Iman, Hora, and Conover, 1984). The rank-transformation procedure, however, often gives results better than those from the Friedman or Quade tests. But its proponents do not recommend that it be routinely employed as an alternative to the parametric ANOVA when it is suspected that the underlying assumptions of the latter do not apply. Instead, they propose that it be employed along with the usual ANOVA and, if both yield the same conclusion, one can feel comfortable with that conclusion.

If tied ranks are present, they may be taken into consideration by computing

$$\left(\chi_r^2\right)_c = \frac{\chi_r^2}{C},$$ (46)

(Marascuilo and McSweeney, 1967),[†] (Kendall, 1962: Chapter 6), where

$$C = 1 - \frac{\sum t}{b(a^3 - a)}$$ (48)

and $\sum t$ are as defined in Equation 48a.

$$\sum t = \sum_{i=1}^{m}(t_i^3 - t_i),$$ (48a)

The *Kendall coefficient of concordance (W)* is another form of Friedman's χ_r^2:

$$W = \frac{\chi_r^2}{b(a-1)}$$ (49)

---

[*]Iman and Davenport (1980) also show that comparing the mean of χ_r^2 and F_F to the mean of the critical values of χ^2 and F provides an improved approximation cannot be used.

[†]Equivalently,

$$\left(\chi_r^2\right)_c = \frac{\displaystyle\sum_{i=1}^{a} R_i^2 - \frac{\left(\displaystyle\sum_{i=1}^{a} R_i\right)^2}{a}}{\dfrac{ba(a+1)}{12} - \dfrac{\sum t}{a-1}}.$$ (47)

(Kendall and Babington Smith, 1939). It is used as a measure of the agreement of rankings within blocks.

(a) Multiple Observations per Cell.   In the experiment of Example 5, there is one datum for each combination of treatment and block. Although this is the typical situation, one might also encounter an experimental design in which there are multiple observations recorded for each combination of block and treatment group. As in Section 1, each combination of level of factor A (group) and level of factor B is called a cell; for n replicate data per cell,

$$\chi_r^2 = \frac{12}{ban^2(na + 1)} \sum_{i=1}^{a} R_i^2 - 3b(na + 1) \tag{50}$$

(Marascuilo and McSweeney, 1977: 376–377), with a critical value of $\chi_{\alpha,a-1}^2$. Note that if $n = 1$, Equation 50 reduces to Equation 44. Benard and van Eltern (1953) and Skillings and Mack (1981) present procedures applicable when there are unequal numbers of data per cell.

(b) Multiple Comparisons.   A multiple-comparison analysis applicable to ranked data in a randomized block is similar to the Tukey procedure for ranked data in a one-way ANOVA design. In this case, Equation 50a is used with the difference between rank sums; that is, $R_B - R_A$ in the numerator and

$$q = \frac{\overline{X}_B - \overline{X}_A}{SE}, \tag{50a}$$

$$SE = \sqrt{\frac{ba(a + 1)}{12}} \tag{51}$$

in the denominator.* (Nemenyi, 1963; Wilcoxon and Wilcox, 1964); and this is used in conjunction with the critical value of $q_{\alpha,\infty,k}$.

If the various groups are to be compared one at a time with a control group, then

$$SE = \sqrt{\frac{ba(a + 1)}{6}} \tag{53}$$

may be used in Dunnett's procedure.

The preceding multiple comparisons are applicable to the levels of the fixed-effect factor, not to the blocks (levels of the random-effect factor).

---

*If desired, mean ranks ($\overline{R}_A = R_A/b$ and $\overline{R}_B = R_B/a$) can be used in the numerator of Equation 50a, in which case the denominator will be

$$SE = \sqrt{\frac{a(a + 1)}{12b}}. \tag{52}$$

Multiple contrasts, may be performed using rank sums.[*]

$$SE = \sqrt{\frac{ba(a + 1)}{12} \left(\sum_i c_i^2 \right)},$$ (55)

unless there are tied ranks, in which case[†]

$$SE = \sqrt{\left(\frac{\dfrac{a(a + 1)}{b} - \dfrac{\sum t}{b^2(a - 1)}}{12} \right) \left(\sum_i c_i^2 \right)}$$ (57)

(Marascuilo and McSweeney, 1967). The critical value for the multiple contrasts is $\sqrt{(\chi_r^2)_{\alpha,a,b}}$, using Appendix Table 14 from *Appendix: Statistical Tables and Graphs* to obtain $(\chi_r^2)_{\alpha,a,b}$. If the needed critical value is not on that table, then $\sqrt{\chi_{\alpha,a-1}^2}$ may be used as an approximation to it.

If there is replication per cell (as in Equation 50), the standard errors of this section are modified by replacing a with an and b with bn wherever they appear. (See Marascuilo and McSweeney, 1977: 378.) Norwood et al. (1989) and Skillings and Mack (1981) present multiple-comparison methods applicable when there are unequal numbers of data per cell.

8 DICHOTOMOUS NOMINAL-SCALE DATA IN RANDOMIZED BLOCKS OR FROM REPEATED MEASURES

The data for a randomized-block or repeated-measures experimental design may be for a dichotomous variable (i.e., a variable with two possible values: e.g., "present" or "absent," "dead" or "alive," "true" or "false," "left" or "right," "male" or "female," etc.), in which case Cochran's Q test[‡] (Cochran, 1950) may be applied. For such an analysis, one value of the attribute is recorded with a "1," and the other with a "0." In Example 8, the data are the occurrence or absence of mosquito attacks on humans wearing one of several types of clothing. The null hypothesis is that the proportion of people attacked is the same for each type of clothing worn.

---

[*]If mean ranks are used,

$$SE = \sqrt{\frac{a(a + 1)}{12b} \left(\sum_i c_i^2 \right)}.$$ (54)

[†]If mean ranks are used,

$$SE = \sqrt{\left(\frac{\dfrac{a(a + 1)}{b^2} - \dfrac{\sum t}{b^3(a - 1)}}{12} \right) \left(\sum_i c_i^2 \right)}.$$ (56)

[‡]William Gemmell Cochran (1909–1980), born in Scotland and influential in the United States after some early important work in England (Dempster, 1983; Watson, 1982).

EXAMPLE 6   Cochran's Q Test

H_0: The proportion of humans attacked by mosquitoes is the same for all five clothing types.

H_A: The proportion of humans attacked by mosquitoes is not the same for all five clothing types.

$\alpha = 0.05$

A person attacked is scored as a "1"; a person not attacked is scored as a "0."

	Clothing Type					
Person (block)	Light, loose	Light, tight	Dark, long	Dark, short	None	Totals (B_j)
1	0	0	0	1	0	1
2*	1	1	1	1	1	*
3	0	0	0	1	1	2
4	1	1	0	1	0	3
5	0	1	1	1	1	4
6	0	1	0	0	1	2
7	0	0	1	1	1	3
8	0	0	1	1	0	2
Totals* (G_i)	1	3	3	6	4	$\sum_{i=1}^{a} G_i = \sum_{j=1}^{b} B_j = 17$

$a = 5; \qquad b = 7^*$

$$Q = \frac{(a-1)\left[\sum_{i=1}^{a} G_i^2 - \dfrac{\left(\sum_{i=1}^{a} G_i\right)^2}{a}\right]}{\sum_{j=1}^{b} B_j - \dfrac{\sum_{j=1}^{b} B_j^2}{a}}$$

$$= \frac{(5-1)\left[1 + 9 + 9 + 36 + 16 - \dfrac{17^2}{5}\right]}{17 - \dfrac{(1 + 4 + 9 + 16 + 4 + 9 + 4)}{5}} = \frac{52.8}{7.6} = 6.947$$

$\nu = a - 1 = 4$

$\chi^2_{0.05,4} = 9.488$

Therefore, do not reject H_0.

$$0.10 < P < 0.25 \quad [P = 0.14]$$

---

*The data for block 2 are deleted from the analysis, because 1's occur for all clothing. (See test discussion in Section 8.)

For a groups and b blocks, where G_i is the sum of the l's in group i and B_j is the sum of the l's in block j,

$$Q = \frac{(a-1)\left[\sum_{i=1}^{a} G_i^2 - \frac{\left(\sum_{i=1}^{a} G_i\right)^2}{a}\right]}{\sum_{j=1}^{b} B_j - \frac{\sum_{j=1}^{b} B_j^2}{a}}. \tag{58}$$

Note, as shown in Example 8, that $\sum B = \sum G$, which is the total number of l's in the set of data. This test statistic, Q, is distributed approximately as chi-square with $a - 1$ degree of freedom. Tate and Brown (1970) explain that the value of Q is unaffected by having blocks containing either all 0's or all 1's. Thus, any such block may be disregarded in the calculations. They further point out that the approximation of Q to χ^2 is a satisfactory one only if the number of data is large. These authors suggest as a rule of thumb that a should be at least 4 and ba should be at least 24, where b is the number of blocks remaining after all those containing either all 0's or all 1's are disregarded. For sets of data smaller than these suggestions allow, the analysis may proceed but with caution exercised if Q is near a borderline of significance. In these cases it would be better to use the tables of Tate and Brown (1964) or Patil (1975).

If $a = 2$, then Cochran's test is identical to McNemar's test, except that the latter employs a correction for continuity.

9 MULTIPLE COMPARISONS WITH DICHOTOMOUS RANDOMIZED-BLOCK OR REPEATED-MEASURES DATA

Marascuilo and McSweeney (1967) present a multiple-comparison procedure that may be used for multiple contrasts as well as for pairwise comparisons for data subjected to the Cochran Q test of Section 8. It may be performed using group means, $\overline{R}_i = G_i/b$.

For pairwise comparisons, the test statistic is

$$S = \frac{\overline{R}_B - \overline{R}_A}{\text{SE}} \tag{59}$$

where

$$\text{SE} = \sqrt{2\left(\frac{a\sum_j B_j - \sum_j B_j^2}{ab^2(a-1)}\right)}. \tag{60}$$

$\overline{R}_i$ replaces $\overline{X}_i$ and

$$\text{SE} = \sqrt{\left(\frac{a\sum_j B_j - \sum_j B_j^2}{ab^2(a-1)}\right)\sum_i c_i^2}. \tag{61}$$

The critical value for such multiple comparisons is $S_\alpha = \sqrt{\chi^2_{\alpha,a-1}}$.

10 INTRODUCTION TO ANALYSIS OF COVARIANCE

Each of the two factors in a two-way ANOVA generally consists of levels that are nominal-scale categories. For instance, the variable of interest was the body weight of pigs, and the one factor tested was diet. In a two-factor ANOVA, we might ask about the effect of diet and also introduce the sex of the animal (or the breed) as a second factor, with the levels of sex (or breed) being on a nominal scale.

We would attempt to employ animals of the same age (and weight), so differences in the measured variable could be attributed to the effect of the diets. However, if the beginning ages (or weights) were markedly not alike, then we might wish to introduce age (or weight) as a second factor. The relationship between ending weight and age (or ending weight and beginning weight) may be thought of as a regression, while the relationship between ending weight and diet is a one-way analysis of variance. The concepts of these two kinds of analyses, and their statistical assumptions, are combined in what is known as *analysis of covariance* (abbreviated ANCOVA),* and the factor that acts as an independent variable in regression is called a *concomitant variable*. This is a large area of statistical methodology beyond the scope of this text but found in many references, including several dealing with experimental design.

EXERCISES

1. A study is made of amino acids in the hemolymph of millipedes. For a sample of four males and four females of each of three species, the following concentrations of the amino acid alanine (in mg/100 ml) are determined:

	Species 1	Species 2	Species 3
Male	21.5	14.5	16.0
	19.6	17.4	20.3
	20.9	15.0	18.5
	22.8	17.8	19.3
Female	14.8	12.1	14.4
	15.6	11.4	14.7
	13.5	12.7	13.8
	16.4	14.5	12.0

(a) Test the hypothesis that there is no difference in mean hemolymph alanine concentration among the three species.

(b) Test the hypothesis that there is no difference between males and females in mean hemolymph alanine concentration.

(c) Test the hypothesis that there is no interaction between sex and species in the mean concentration of alanine in hemolymph.

(d) Prepare a graph of the row, column, and cell means, as done in Figure 1, and interpret it in terms of the results of the above hypothesis tests.

(e) If the null hypothesis of part a, above, is rejected, then perform a Tukey test to assess the mean differences among the species.

2. Six greenhouse benches were set up as blocks. Within each block, one of each of four varieties of house plants was planted. The plant heights (in centimeters) attained are tabulated as follows. Test the hypothesis that all four varieties of plants reach the same maximum height.

*The first use (and the name) of this statistical technique is attributed to R. A. Fisher prior to 1930 (e.g., Fisher, 1932: 249–262; Yates, 1964).

Block	Variety 1	Variety 2	Variety 3	Variety 4
1	19.8	21.9	16.4	14.7
2	16.7	19.8	15.4	13.5
3	17.7	21.0	14.8	12.8
4	18.2	21.4	15.6	13.7
5	20.3	22.1	16.4	14.6
6	15.5	20.8	14.6	12.9

3. Consider the data of Exercise 2. Nonparametrically test the hypothesis that all four varieties of plants reach the same maximum height.

4. A textbook distributor wishes to assess potential acceptance of four general biology textbooks. He asks 15 biology professors to examine the books and to respond as to which ones they would seriously consider for their courses. In the table, a positive response is recorded as 1 and a negative response as a 0. Test the hypothesis that there is no difference in potential acceptance among the four textbooks.

Professor	Textbook 1	Textbook 2	Textbook 3	Textbook 4
1	1	1	0	0
2	1	1	0	1
3	1	0	0	0
4	1	1	1	1
5	1	1	0	1
6	0	1	0	0
7	0	1	1	0
8	1	1	1	0
9	0	0	1	0
10	1	0	1	0
11	0	0	0	0
12	1	1	0	1
13	1	0	0	1
14	0	1	1	0
15	1	1	0	0

ANSWERS TO EXERCISES

1. (a) H_0: There is no difference in mean hemolymph alanine among the three species; H_A: There is difference in mean hemolymph alanine among the three species; $F = 27.6304/2.1121 = 13.08$; $F_{0.05(1),2,18} = 3.55$; reject H_0; $P < 0.0005$ $[P = 0.00031]$. **(b)** H_0: There is no difference in mean hemolymph alanine between males and females; H_A: There is difference in mean hemolymph alanine between males and females; $F = 138.7204/2.1121 = 65.68$; $F_{0.05(1),1,18} = 4.41$ reject H_0; $P \ll 0.0005$ $[P = 0.00000020]$. **(c)** H_0: There is no species × sex interaction in mean hemolymph alanine; H_A: There is species × sex interaction in mean hemolymph alanine; $F = 3.4454/2.1121 = 1.63$; $F_{0.05(1),2,18} = 3.55$; do not reject H_0; $0.10 < P < 0.25$ $[P = 0.22]$. **(d)** See graph below; the wide vertical distance between the open circles indicates difference between sexes; the vertical distances among the plus signs indicates difference among the three species; the parallelism of the male and female lines indicates no interaction effect. **(e)** Ranked sample means: <u>14.43</u> 16.13 <u>18.14</u> (means 2, 3, and 1, respectively); $k = 3$; $n = 8$; $\alpha = 0.05$; $s^2 = 2.1121$; $\nu = 18$; reject H_0: $\mu_1 = \mu_2$, reject H_0: $\mu_1 = \mu_3$, do not reject H_0: $\mu_3 = \mu_2$.

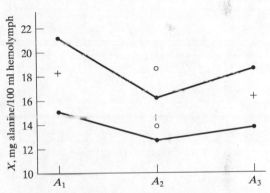

2. H_0: All four plant varieties reach the same mean height (i.e., H_0: $\mu_1 = \mu_2 = \mu_3 = \mu_4$); H_A: All four plant varieties do not reach the same mean height; $F = 62.8461/0.4351 = 144$; $F_{0.05(1),3,15} = 3.29$; reject H_0; $P \ll 0.0005$ $[P < 10^{-10}]$.

3. H_0: All four plant varieties reach the same height; H_A: All four plant varieties do not reach the same height; $R_1 = 18$, $R_2 = 24$, $R_3 = 12$, $R_4 = 6$; $\chi^2_{0.05,3} = 7.815$; reject H_0; $P < 0.001$.

4. H_0: There is no difference in potential acceptance among the three textbooks; H_A: The three textbooks do not have the same potential acceptance; $a = 4$, $b = 13$ (blocks 4 and 11 are deleted from the analysis); $Q = 5.53$; $\nu = 3$; $\chi^2_{0.05,3} = 7.815$; do not reject H_0; $0.10 < P < 0.25$ $[P = 0.14]$.

This page intentionally left blank

Data Transformations

Data Transformations

We have previously discussed underlying assumptions of several statistical procedures, such as t-testing, analysis of variance, and parametric multiple comparisons. Three assumptions were noted: (1) Each sample of data was obtained randomly from the sampled population, (2) each sampled population was normally distributed, and (3) all of the sampled populations had the same variance. These three assumptions will also apply to several statistical procedures discussed. If the assumptions are not well satisfied, then the probabilities associated with the test statistics may be incorrect, and conclusions whether to reject the null hypothesis may not be warranted.

If an analysis of variance is to be performed for the effects of two or more factors (i.e., a factorial ANOVA), then it is important to consider the effects of interactions among the factors. For example, in a two-factor analysis of variance, the effect of factor A on the variable, X, can be assessed, as can the effect of factor B on that variable. It is also important to examine the interaction effect of the two factors. If the effect of one of the factors on X is the same at all levels of the other factor, then there is no interaction effect, and the effects of the two variables are said to be *additive*.*

If there is replication within cells in a factorial ANOVA, then the hypothesis of no interaction can be tested, as well as a hypothesis about each of the factors. If, however, there is no replication, then hypothesis testing is problematic and limited, especially in Model I ANOVA. Thus, in the absence of replication, a fourth assumption—that of no interaction (i.e., of additivity)—can be added to the three assumptions mentioned previously.

There are data sets that violate one or more of underlying premises 2, 3, and 4 for which a *transformation* of the data from their original values (X) to values (call them X') that constitute a data set more closely satisfying the assumptions. Also, a reduction in the interaction effect can enhance the power of hypothesis testing for factor effects. For some kinds of data it is known, on theoretical grounds, that a transformation will result in data more amenable to the intended statistical analysis.

---

*The term *additivity* in this context was introduced by C. Eisenhart in 1947 (David, 1995).

Data transformation will *not* compensate for the absence of random sampling (violation of assumption 1). As with untransformed data, analysis of transformed data can be adversely affected by the presence of outliers.

Many authors provided early recommendations on the use of data transformations that have become commonly used (e.g., Bartlett, 1947; Box and Cox, 1964; Kendall and Stuart, 1966: 87–96; Thöni, 1967). This chapter will concentrate on three of those transformations: The logarithmic transformation (Section 1) is applicable when there is heterogeneity of variances among groups and the group standard deviations are directly proportional to the means, and in cases where two or more factors have a multiplicative (instead of an additive) effect. The square-root transformation (Section 2) applies to heteroscedastic data where the group variances are directly proportional to the means, a situation often displayed when the data come from a population of randomly distributed counts. The arcsine transformation (Section 3) is germane when the data come from binomial distributions, such as when the data consist of percentages within the limits of 0 and 100% (or, equivalently, proportions within the range of 0 to 1).

1 THE LOGARITHMIC TRANSFORMATION

If the factor effects in an analysis of variance are, in fact, multiplicative instead of additive, then the data will not exhibit additivity, but logarithms of the data will. This is demonstrated in Example 1: Example 1a shows data for which, for each level of factor B, each datum in factor-A level 2 differs from the corresponding datum in factor-A level 1 by the addition by the same weight (namely, 20 g − 10 g = 10 g and 30 g − 20 g = 10 g), and each datum in level 3 differs from its corresponding level-2 datum by the same amount (i.e., 25 g − 20 g = 5 g and 35 g − 30 g = 5 g). And for factor B, there is a constant difference (20 g − 10 g = 30 g − 20 g = 35 g − 25 g = 10 g) between the two levels, at all three levels of factor A. Thus, in Example 1a, no data transformation is needed to achieve additivity. However, in the data of Example 1b, the effect of each factor is multiplicative instead of additive. For each level of factor B, the datum in level 2 differs from its corresponding datum in level 1 by a factor of 3 (i.e., 30 g = 3 × 10 g and 60 g = 3 × 20 g); each X in level 3 differs from its level-2 neighbor by a factor of 2 (60 g = 2 × 30 g and 120 g = 2 × 60 g); and there is a multiplicative difference of 2 between the data, at each level of factor A (20 g = 2 × 10 g; 60 g = 2 × 30 g; 120 g = 2 × 60 g) for each of the two levels of factor B. In such a situation, the logarithms of the data will exhibit additivity. This is shown by the logarithmically transformed data of Example 1c. In Example 1, the six quantities may be data in a 3 × 2 ANOVA without replication or they may represent the six cell means if there is replication.

Figure 1 graphs the data in Example 1. Figure 1a shows the values of X in Example 1a. For the two levels of factor B (B_1 and B_2), the line segments are parallel between factor-A levels 1 and 2 (A_1 and A_2) and between levels 2 and 3 (A_2 and A_3), indicating the additive effect of the two factors (i.e., no interaction between the two factors). Figure 1b graphs the values of X found in Example 1b. In comparing the plots for the two levels of factor B (B_1 and B_2), it is seen that the line segments between the first two levels of factor A (A_1 and A_2) are not parallel, nor are the line segments between the second and third levels of that factor (A_2 and A_3), indicating that the effects of the two factors are not additive (i.e., that there is an interactive effect between factors A and B). Figure 1c shows the graph for the

Data Transformations

data transformed into their logarithms (Example 1c). Here it is seen that the two line segments representing the two levels of factor B are parallel for the comparison of levels 1 and 2 of factor A and for the comparison of levels 2 and 3. Thus, Example 1c achieved additivity by using the logarithmic transformation of the data in Example 1b.

EXAMPLE 1 Additive and Multiplicative Effects

(a) A hypothetical two-way analysis-of-variance design, where the effects of the factors are additive. (Data are in grams.)

Factor B	Factor A Level 1	Level 2	Level 3
Level 1	10	20	25
Level 2	20	30	35

(b) A hypothetical two-way analysis-of-variance design, where the effects of the factors are multiplicative. (Data are in grams.)

Factor B	Factor A Level 1	Level 2	Level 3
Level 1	10	30	60
Level 2	20	60	120

(c) The two-way analysis-of-variance design of Example 1b, showing the logarithms (rounded to two decimal places) of the data.

Factor B	Factor A Level 1	Level 2	Level 3
Level 1	1.00	1.48	1.78
Level 2	1.30	1.78	2.08

The logarithmic transformation is also applicable when there is heteroscedasticity and the groups' standard deviations are directly proportional to their means (i.e., there is a constant coefficient of variation among the groups). Such a situation is shown in Example 2. This transformation may also convert a positively skewed distribution into a symmetrical one.

Instead of the transformation $X' = \log(X)$, however,

$$X' = \log(X + 1) \tag{1}$$

is preferred as the logarithmic transformation on theoretical grounds and is especially preferable when some of the data are small numbers (particularly zero) (Bartlett, 1947). Logarithms in base 10 are generally utilized, but any logarithmic base may be employed. Equation 1 is what is used in Example 2.

Data Transformations

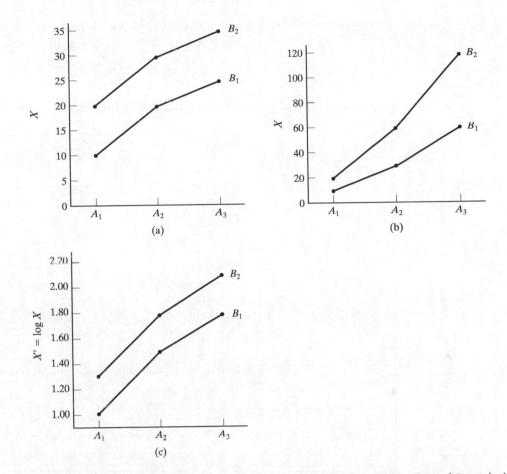

(a)

(b)

(c)

FIGURE 1: The effects of the two factors in a 3 × 2 analysis of variance. (a) The data of Example 1a, where the parallel line segments reflect lack of interaction between the two factors (i.e., additivity). (b) The data of Example 1b, where the nonparallel line segments indicate interaction (i.e., nonadditivity) of the two factors. (c) The data of Example 1b transformed to their logarithms and shown as Example 1c; the parallelism of these line segments shows that the transformation has resulted in the absence of interaction (i.e., the result is factor additivity).

EXAMPLE 2   The Logarithmic Transformation for Data in Which There Is Heterogeneity of Variance and the Standard Deviations Are Directly Proportional to the Means (i.e., the Coefficients of Variation Are the Same)

A prime symbol on a statistic denotes a quantity obtained using the transformed data (e.g., X', s', L').

The original data (leaf lengths, in centimeters):

Group 1	Group 2
3.1	7.6
2.9	6.4
3.3	7.5
3.6	6.9
3.5	6.3

$$\overline{X}_1 = 3.28 \text{ cm} \qquad \overline{X}_2 = 6.94 \text{ cm}$$
$$s_1^2 = 0.0820 \text{ cm}^2 \qquad s_2^2 = 0.3630 \text{ cm}^2$$
$$s_1 = 0.29 \text{ cm} \qquad s_2 = 0.60 \text{ cm}$$
$$V_1 = 0.09 \qquad V_2 = 0.09$$

The logarithmically transformed data, using Equation (1):

Group 1	Group 2
0.61278	0.93450
0.59106	0.86923
0.63347	0.92942
0.66276	0.89763
0.65321	0.86332

$$\overline{X}_1' = 0.63066 \qquad \overline{X}_2' = 0.89882$$
$$(s_1^2)' = 0.0008586657 \qquad (s_2^2)' = 0.0010866641$$
$$s_1' = 0.02930 \qquad s_2' = 0.03296$$
$$V_1' = 0.04646 \qquad V_2' = 0.03667$$
$$s_{\overline{X}_1}' = 0.01310 \qquad s_{\overline{X}_2}' = 0.01474$$

Calculating confidence limits for the mean, using the transformed data from Group 1:

$$95\% \text{ confidence interval for } \mu_1' = \overline{X}_1' \pm (t_{0.05(2),4})(0.01310)$$
$$= 0.63066 \pm (2.776)(0.01310)$$
$$= 0.63066 \pm 0.03637$$

$$L_1' = 0.59429 \text{ and } L_2' = 0.66703$$

95% confidence limits for μ_1, in the original units:

$$L_1 = \text{antilog } 0.59429 - 1 = 3.93 - 1 = 2.93 \text{ cm}$$
$$L_2 = \text{antilog } 0.66703 - 1 = 4.65 - 1 = 3.65 \text{ cm}$$

The 95% confidence intervals for μ_2' and for μ_2 may be calculated in the same manner.

After data transformation, hypothesis testing and expression of confidence intervals may be done on the transformed data. Subtracting 1 from the antilogarithm of the mean of the logarithmically transformed data expresses the sample mean in the units of the original data,* and subtracting 1 from the antilogarithm of each confidence limit for the mean of the transformed data gives the confidence limits for the mean in terms of the nontransformed data. This is demonstrated in Example 2. Note that, when performing these calculations on the transformed data, the confidence interval is not symmetrical around the mean in the original units.

---

*Thöni (1967: 16) has shown that an unbiased estimate of μ would be obtained by adding $(1 - 1/n)s^2$ to the $\overline{X}$ derived by untransforming $\overline{X}'$, where s^2 is the variance of the transformed data. Bias is less for large samples. The antilogarithm of the mean of the transformed data (i.e., the antilogarithm of $(\overline{X}')$ is the geometric mean of the untransformed data.

If the distribution of X' is normal, the distribution of X is said to be *lognormal*.[*]

2 THE SQUARE-ROOT TRANSFORMATION

The square-root transformation is applicable when the group variances are directly proportional to the means; that is, when the variances increase as the means increase. This most often occurs in biological data when samples are taken from a Poisson distribution (i.e., when the data consist of counts of randomly occurring objects or events). Transforming such data by utilizing their square roots results in a sample whose underlying distribution is normal. However, Bartlett (1936) proposed that

$$X' = \sqrt{X + 0.5} \tag{2}$$

is preferable to $X' = \sqrt{X}$, especially when there are very small data and/or when some of the observations are zero (see Example 3). Actually,

$$X' = \sqrt{X + \frac{3}{8}} \tag{3}$$

has even better variance-stabilizing qualities than Equation 2 (Kihlberg, Herson, and Schutz, 1972), and Freeman and Tukey (1950) show

$$X' = \sqrt{X} + \sqrt{X + 1} \tag{4}$$

to yield similar results but to be preferable for $X \leq 2$.

Equation 2 is most commonly employed. Statistical computation may then be performed on the transformed data. The mean of those data can be expressed in terms of the original data by squaring it and then subtracting 0.5, although the resultant statistic is slightly biased.[†] Budescu and Appelbaum (1981) examined ANOVA for Poisson data and concluded that data transformation is not desirable unless the largest variances are found in the largest samples and the largest sample is more than five times the size of the smallest.

3 THE ARCSINE TRANSFORMATION

It is known from statistical theory that percentages from 0 to 100% or proportions from 0 to 1 form a binomial, rather than a normal, distribution, the deviation from normality being great for small or large percentages (0 to 30% and 70 to 100%).[‡] If the square root of each proportion, p, in a binomial distribution is transformed to its arcsine (i.e., the angle whose sine is $\sqrt{p}$), then the resultant data will have an underlying distribution that is nearly normal. This transformation,

$$p' = \arcsin \sqrt{p}, \tag{5}$$

---

[*]The term *lognormal* was introduced by J. H. Gaddam in 1945 (David, 1995).

[†]Also, an antilogarithmic transformation to obtain $\overline{X}$ in terms of the original units is known to result in a somewhat biased estimator of μ, the estimator being less biased for larger variances of $\overline{X}'$ values.

[‡]The symbol for percent, "%," appeared around 1650 (Cajori, 1928/1929, Vol. I: 312).

EXAMPLE 3   The Square Root Transformation for Poisson Data

Original data (number of parasites in the lungs of 20 frogs allocated to four experimental groups):

	Group 1	Group 2	Group 3	Group 4
	2	6	9	2
	0	4	5	4
	2	8	6	1
	3	2	5	0
	0	4	11	2
$\overline{X}_1$	1.4	4.8	7.2	1.8
s_i^2	1.8	5.2	7.2	2.2

Transformed data; by Equation 2:

	Group 1	Group 2	Group 3	Group 4
	1.581	2.550	3.082	1.581
	0.707	2.121	2.345	2.121
	1.581	2.915	2.550	1.225
	1.871	1.581	2.345	0.707
	0.707	2.121	3.391	1.581
$\overline{X}_i'$	1.289	2.258	2.743	1.443
$(s_i^2)'$	0.297	0.253	0.222	0.272
$s'_{\overline{X}_i}$	0.244	0.225	0.211	0.233
$(L_1')_i$	0.612	1.633	2.157	0.796
$(L_2')_i$	1.966	2.883	3.329	2.090

On transforming back to original units [e.g., $\overline{X} = (\overline{X}')^2 - 0.5$]:

	Group 1	Group 2	Group 3	Group 4
$\overline{X}_i$	1.2	4.6	7.0	1.6
$(L_1)_i$	−0.1	2.2	4.2	0.1
$(L_2)_i$	3.4	7.8	10.6	3.9

is performed easily with the aid of Table 24 from *Appendix: Statistical Tables and Graphs*. For proportions of 0 to 1.00 (i.e., percentages of 0 to 100%), the transformed values will range between 0 and 90 degrees (although some authors' tables present the transformation in terms of radians*).[†]

The arcsine transformation ("arcsine" is abbreviated "arcsin") frequently is referred to as the "angular transformation," and "inverse sine" or "sin^{-1}" is sometimes written to denote "arcsine."[‡]

Example 4 demonstrates calculations using data submitted to the arcsine transformation. Transformed values (such as means or confidence limits) may be transformed back to proportions, as

$$p = (\sin p')^2; \tag{6}$$

and Table 25 from *Appendix: Statistical Tables and Graphs* is useful for this purpose.[§] Confidence limits for proportions will not generally be symmetrical around the mean. This transformation is not as good at the extreme ends of the range of possible values (i.e., near 0 and 100%) as it is elsewhere. If, instead of simply having data consisting of percentages, the researcher knows the count (X) and sample size (n) composing each percentage ($p = X/n$), then the arcsine transformation is improved by replacing $0/n$ with $1/4n$ and n/n with $1 - 1/4n$ (Bartlett, 1937a). Anscombe (1948) proposed an even better transformation:

$$p' = \arcsin \sqrt{\frac{X + \dfrac{3}{8}}{n + \dfrac{3}{4}}}. \tag{7}$$

And a slight modification of the Freeman and Tukey (1950) transformation, namely,

$$p' = \frac{1}{2}\left[\arcsin \sqrt{\frac{X}{n + 1}} + \arcsin \sqrt{\frac{X + 1}{n + 1}}\right], \tag{8}$$

yields very similar results, except for small and large proportions where it appears to be preferable. Determination of transformed proportions, p', by either Equation 7 or 8 is facilitated by using Table 24 from *Appendix: Statistical Tables and Graphs*.

---

*A radian is $180°/\pi = 57.29577951308232\ldots$ degrees. Expressing angles in radians, instead of degrees, would have Equation 5 yield arcsines (p') of 0 to 1.5708 for proportions (p) of 0 to 1; sometimes the use of radians is associated with substituting

$$p' = 2 \arcsin \sqrt{p} \tag{5a}$$

for Equation 5, resulting in values of p' that can range from 0 to 3.1416 (that is, a range of zero to pi). The choice between degrees and radians will not affect the conclusions of statistical procedures employing the arcsine transformation.

[†]The arcsine transformation is applicable only if the data came from a distribution of data that can lie between 0 and 100% (e.g., not if data are percent increases, which can be greater than 100%).

[‡]The arcsine of a number is the angle whose sine is that number. The term was initiated in the latter part of the eighteenth century; and the abbreviation "sin^{-1}" was introduced in 1813 by the English astronomer Sir John Frederick William Herschel (1792–1871) (Cajori, 1928/1929, Vol. II: 175–176).

[§]The mean, $\bar{p}$, that is obtained from $\bar{p}'$ by consulting Table 25 from *Appendix: Statistical Tables and Graphs* is, however, slightly biased. Quenouille (1950) suggests correcting for the bias by adding to $\bar{p}$ the quantity $0.5 \cos(2\bar{p}')(1 - e^{-2s^2})$, where s^2 is the variance of the p' values.

EXAMPLE 4   The Arcsine Transformation for Percentage Data

Original data (p, the percentage of insects killed in each of seven groups of insects subjected to one of two insecticides):

Insecticide 1 (%)	Insecticide 2 (%)
84.2	92.3
88.9	95.1
89.2	90.3
83.4	88.6
80.1	92.6
81.3	96.0
85.8	93.7
$\overline{p}_2 = 84.7\%$	$\overline{p}_2 = 92.7\%$
$s_1^2 = 12.29(\%)^2$	$s_2^2 = 6.73(\%)^2$
$s_1 = 3.5\%$	$s_2 = 2.6\%$

Transformed data (by using Equation 5 or Table 24 from *Appendix: Statistical Tables and Graphs*) (p'):

Insecticide 1 (°)	Insecticide 2 (°)
66.58	73.89
70.54	77.21
70.81	71.85
65.96	70.27
63.51	74.21
64.38	78.46
67.86	75.46
$\overline{p}_1' = 67.09$	$\overline{p}_2' = 74.48$
$(s_1^2)' = 8.0052$	$(s_2^2)' = 8.2193$
$s_1' = 2.83$	$s_2' = 2.87$
$s_{\overline{X}_1}' = 1.07$	$s_{\overline{X}_2}' = 1.08$

Calculating confidence limits:

95% confidence interval for $\mu_1' : \overline{p}_1' \pm (t_{0.05(2),6})(1.07) = 67.09 \pm 2.62$

$$L_1' = 64.47° \text{ and } L_2' = 69.71°$$

By using Table 25 from *Appendix: Statistical Tables and Graphs* to transform backward from $L_1', L_2',$ and p_1':

95% confidence limits for $\mu_1 : L_1 = 81.5\%$ and $L_2 = 88.0\%$.

$$\overline{p}_1 = 84.9\%$$

4 OTHER TRANSFORMATIONS

The logarithmic, arcsine, and square-root transformations are those most commonly required to handle nonnormal, heteroscedastic, or nonadditive data. Other transformations are only rarely called for.

If the standard deviations of groups of data are proportional to the square of the means of the groups, then the *reciprocal transformation*,

$$X' = \frac{1}{X},\qquad\qquad (9)$$

may be employed. (If counts are being transformed, then

$$X' = \frac{1}{X+1}\qquad\qquad (10)$$

may be used to allow for observations of zero.) See Thöni (1967: 32) for further discussion of the use of this transformation.

If the standard deviations decrease as the group means increase, and/or if the distribution is skewed to the left, then

$$X' = X^2\qquad\qquad (11)$$

might prove useful.

If the data come from a population with what is termed a "negative binomial distribution," then the use of inverse hyperbolic sines may be called for (see Anscombe, 1948; Bartlett, 1947; Beall, 1940, 1942; Thöni, 1967: 20–24).

Thöni (1967) mentions other, infrequently employed, transformations.

EXERCISES

1. Perform the logarithmic transformation on the following data (using Equation 1) and calculate the 95% confidence interval for μ. Express the confidence limits in terms of the original units (i.e., ml). The data are 3.67, 4.01, 3.85, 3.92, 3.71, 3.88, 3.74, and 3.82 ml.

2. Transform the following proportions by the arcsine transformation (using Table 24 from *Appendix: Statistical Tables and Graphs*) and calculate the 95% confidence interval for μ. Express the confidence limits in terms of proportions (using Table 25 from *Appendix: Statistical Tables and Graphs*).

   0.733, 0.804, 0.746, 0.781, 0.772, and 0.793

3. Apply the square-root transformation to the following data (using Equation 2) and calculate the 95% confidence interval for μ. Transform the confidence limits back to the units of the original data. The data are 4, 6, 3, 8, 10, 3.

ANSWERS TO EXERCISES

1. $\overline{X}' = 0.68339$, $s'_{\overline{X}} = 0.00363$; $L'_1 = 0.67481$, $L'_2 = 0.69197$; $L_1 = 3.73$ ml, $L_2 = 3.92$ ml.

2. $\overline{X}' = 61.48$, $s'_{\overline{X}} = 0.76$; $L'_1 = 59.53$, $L'_2 = 63.43$; $L_1 = 0.742$, $L_2 = 0.800$.

3. $\overline{X}' = 2.4280$, $s'_{\overline{X}} = 0.2329$; $L'_1 = 1.8292$, $L'_2 = 3.0268$; $L_1 = 2.85$, $L_2 = 8.66$.

This page intentionally left blank

Multiway Factorial Analysis of Variance

Multiway Factorial Analysis of Variance

In analyzing the effects on a variable of two factors acting simultaneously. In such a procedure—a two-way, or two-factor, analysis of variance—we can conclude whether either of the factors has a significant effect on the magnitude of the variable and also whether the interaction of the two factors significantly affects the variable. By expanding the considerations of the two-way analysis of variance, we can assess the effects on a variable of the simultaneous application of three or more factors, this being done by what is referred to as multiway factorial analysis of variance.*

It is not unreasonable for a researcher to perform a one-way or two-way analysis of variance by hand (i.e., using a calculator), although computer programs are routinely employed, especially when the experiment consists of a large number of data. However, it has become rare for analyses of variance with more than two factors to be analyzed other than via statistical software, for considerations of time, ease, and accuracy. Therefore, this chapter will presume that established computer programs will be used to perform the necessary calculations, but it will consider the subsequent examination and interpretation of the numerical results of the computer's labor.

1 THREE-FACTOR ANALYSIS OF VARIANCE

For a particular variable, we may wish to assess the effects of three factors; let us refer to them as factors A, B, and C. For example, we might desire to determine what effect the following three factors have on the rate of oxygen consumption of crabs: species, temperature, and sex. Example 1a shows experimental data collected for crabs of both sexes, representing three species, and measured at three temperatures. For each cell (i.e., each combination of species, temperature, and sex) there was an oxygen consumption datum for each of four crabs (i.e., there were four replicates); therefore, 72 animals were used in the experiment ($N = 2 \times 3 \times 3 \times 4 = 72$).

---

*The concept of the factorial analysis of variance was introduced by the developer of ANOVA, R. A. Fisher (Bartlett, 1965), and Fisher's first use of the term *factorial* was in 1935 (David, 1995).

EXAMPLE 1a   A Three-Factor Analysis of Variance (Model I), Where the Variable Is Respiratory Rate of Crabs (in ml O_2/hr)

1 $\begin{cases} H_0: & \text{Mean respiratory rate is the same in all three crab species (i.e., } \mu_1 = \mu_2 = \mu_3\text{).} \\ H_A: & \text{Mean respiratory rate is not the same in all three crab species.} \end{cases}$

2 $\begin{cases} H_0: & \text{Mean respiratory rate is the same at all three experimental temperatures (i.e., } \mu_{low} = \mu_{med} = \mu_{high}\text{).} \\ H_A: & \text{Mean respiratory rate is not the same at all three experimental temperatures.} \end{cases}$

3 $\begin{cases} H_0: & \text{Mean respiratory rate is the same for males and females (i.e., } \mu_\male = \mu_\female) \\ H_A: & \text{Mean respiratory rate is not the same for males and females (i.e., } \mu_\male \neq \mu_\female) \end{cases}$

4 $\begin{cases} H_0: & \text{Differences in mean respiratory rate among the three species are independent of (i.e., the population means are the same at) the three experimental temperatures; or, differences in mean respiratory rate among the three temperatures are independent of (i.e., are the same in) the three species. (Testing for } A \times B \text{ interaction.)} \\ H_A: & \text{Differences in mean respiratory rate among the species are not independent of the experimental temperatures.} \end{cases}$

5 $\begin{cases} H_0: & \text{Differences in mean respiratory rate among the three species are independent of sex (i.e., the population means are the same for both sexes); or, differences in mean respiratory rate between males and females are independent of (i.e., are the same in) the three species. (Testing for } A \times C \text{ interaction.)} \\ H_A: & \text{Differences in mean respiratory rate among the species are not independent of sex.} \end{cases}$

6 $\begin{cases} H_0: & \text{Differences in mean respiratory rate among the three experimental temperatures are independent of (i.e., the population means are the same in) the two sexes; or, differences in mean respiration rate between the sexes are independent of (i.e., are the same at) the three temperatures. (Testing for } B \times C \text{ interaction.)} \\ H_A: & \text{Differences in mean respiratory rate among the three temperatures are not independent of sex.} \end{cases}$

7 $\begin{cases} H_0: & \text{Differences in mean respiratory rate among the species (or temperatures, or sexes) are independent of the other two factors. (Testing for } A \times B \times C \text{ interaction.)} \\ H_A: & \text{Differences in mean respiratory rate among the species (or temperature, or sexes) are not independent of the other two factors.} \end{cases}$

| **Species 1** | | | | | |
| Low temp. | | Med. temp. | | High temp. | |
♂	♀	♂	♀	♂	♀
1.9	1.8	2.3	2.4	2.9	3.0
1.8	1.7	2.1	2.7	2.8	3.1
1.6	1.4	2.0	2.4	3.4	3.0
1.4	1.5	2.6	2.6	3.2	2.7

| **Species 2** | | | | | |
| Low temp. | | Med. temp. | | High temp. | |
♂	♀	♂	♀	♂	♀
2.1	2.3	2.4	2.0	3.6	3.1
2.0	2.0	2.6	2.3	3.1	3.0
1.8	1.9	2.7	2.1	3.4	2.8
2.2	1.7	2.3	2.4	3.2	3.2

| **Species 3** | | | | | |
| Low temp. | | Med. temp. | | High temp. | |
♂	♀	♂	♀	♂	♀
1.1	1.4	2.0	2.4	2.9	3.2
1.2	1.0	2.1	2.6	2.8	2.9
1.0	1.3	1.9	2.3	3.0	2.8
1.4	1.2	2.2	2.2	3.1	2.9

Table 1 presents the computer output for the analysis of these experimental results, such output typically giving the sums of squares, degrees of freedom, and mean squares pertaining to the hypotheses to be tested. Some computer programs also give the F values calculated, assuming that the experiment calls for a Model I analysis, which is the case with most biological data, and some present the probability of each F. The major work that the computer software has performed for us is the calculation of the sums of squares. We could easily have arrived at the degrees of freedom for

TABLE 1: Computer Output from a Three-Factor Analysis of Variance of the Data Presented in Example 1

Source of variation	Sum of squares	DF	Mean square
Factor A	1.81750	2	0.90875
Factor B	24.65583	2	12.32791
Factor C	0.00889	1	0.00889
A × B	1.10167	4	0.27542
A × C	0.37028	2	0.18514
B × C	0.17528	2	0.08764
A × B × C	0.22056	4	0.05514
Error	2.00500	54	0.03713

each factor as the number of levels -1 (so, for factor A, DF $= 3 - 1 = 2$; for factor B, DF $= 3 - 1 = 2$; and for factor C, DF $= 2 - 1 = 1$). The degrees of freedom for each interaction are $A \times B$ DF $=$ factor A DF $\times$ factor B DF $= 2 \times 2 = 4$; $A \times C$ DF $=$ factor A DF $\times$ factor C DF $= 2 \times 1 = 2$; $B \times C$ DF $=$ factor B DF $\times$ factor C DF $= 2 \times 1 = 2$; and $A \times B \times C$ DF $=$ factor A DF $\times$ factor B DF $\times$ factor C DF $= 2 \times 2 \times 1 = 4$. The error DF is, then, the total DF (i.e., $N - 1$) minus all other degrees of freedom. Each needed mean square is then obtained by dividing the appropriate sum of squares by its associated DF. As we are dealing with a Model I (fixed-effects model) ANOVA, the computation of each F value consists of dividing a factor or interaction mean square by the error MS.

Example 1b demonstrates testing of the hypotheses stated in Example 1a. To test whether oxygen consumption is the same among all three species, the species F is compared to the critical value, $F_{0.05(1),2,54} \approx 3.17$; because the former exceeds the later, the null hypothesis is rejected.* In a similar fashion, we test the hypothesis concerning each of the other two factors, as well as each of the four hypotheses regarding the interactions of factors, by comparing the calculated F values with the critical values from Table 4 from *Appendix: Statistical Tables and Graphs*.

The test for a two-way interaction asks whether differences in the variable among levels of one factor are the same at all levels of the second factor. A test for a three-factor interaction may be thought of as asking if the interaction between any two of the factors is the same at all levels of the third factor. As shown at the conclusion of Example 1b, statistical differences among levels of a factor must be expressed with caution if that factor has a significant interaction with another factor.

It is only for a factorial ANOVA with all factors fixed that we compute all F values utilizing the error MS. If any of the factors are random effects, then the analysis becomes more complicated.

If there are not equal numbers of replicates in each cell of a factorial analysis of variance design, then the usual ANOVA computations are not valid (see Section 5).

A factorial ANOVA experimental design may also include nesting. For example, in Example 1 we might have performed two or more respiratory-rate determinations on each of the four animals per cell. Some of the available computer programs for factorial analysis of variance also provide for nested (also called hierarchical) experimental designs.

If one or two of the three factors are measured on an interval or ratio scale, this is an analysis of covariance situation, and computer programs are available for such analyses. For instance, the variable is plasma calcium concentration and two factors are hormone treatment and sex. A third factor might be age, or weight, or hemoglobin concentration, or temperature.

2 THE LATIN-SQUARE EXPERIMENTAL DESIGN

A special case of a three-factor analysis of variance is an extension of the randomized-complete-block ANOVA or of the repeated-measures ANOVA. The two-factor experimental design is composed of a fixed-effects factor (factor A) about which

---

*There are no critical values in Appendix Table B.4 for $v_2 = 54$, so the values for the next lower degrees of freedom ($v_2 = 50$) were utilized. The symbol "$\approx$" indicates "approximately equal to." Alternatively, harmonic interpolation could have been employed; for this example, the interpolation would calculate critical values different from those above by only 0.01. Also, some computer routines can produce critical values.

EXAMPLE 1b   **The Analysis of Variance Summary for the Experiment in Example 1a**

The following results are obtained for the information in Table 1:

For Factor A: $F = \dfrac{0.90875}{0.03713} = 24.45$ For $A \times B$ interaction: $F = \dfrac{0.27542}{0.03713} = 7.42$

For Factor B: $F = \dfrac{12.32791}{0.03713} = 332.02$ For $A \times C$ interaction: $F = \dfrac{0.18514}{0.03713} = 4.99$

For Factor C: $F = \dfrac{0.00889}{0.03713} = 0.24$ For $B \times C$ interaction: $F = \dfrac{0.08764}{0.03713} = 2.36$

For $A \times B \times C$ interaction: $F = \dfrac{0.05514}{0.03713} = 1.49$

Effect in hypothesis	Calculated F	Critical F (see footnote 1)	Conclusion	P (see footnote 2)
1. Species (Factor A)	24.45	$F_{0.05(1),2,54} \approx 3.17$	Reject H_0	$P \ll 0.00001$
2. Temperature (Factor B)	332.02	$F_{0.05(1),2,54} \approx 3.17$	Reject H_0	$P \ll 0.00001$
3. Sex (Factor C)	0.24	$F_{0.05(1),1,54} \approx 4.03$	Do not reject H_0	$P = 0.63$
4. $A \times B$	7.42	$F_{0.05(1),4,54} \approx 2.56$	Reject H_0	$P = 0.000077$
5. $A \times C$	4.99	$F_{0.05(1),2,54} \approx 3.17$	Reject H_0	$P = 0.010$
6. $B \times C$	2.36	$F_{0.05(1),2,54} \approx 3.17$	Do not reject H_0	$P = 0.10$
7. $A \times B \times C$	1.49	$F_{0.05(1),4,54} \approx 2.56$	Do not reject H_0	$P = 0.22$

[1]There are no critical values in Table 4 from *Appendix: Statistical Tables and Graphs* for $v_2 = 54$, so the values for the next lower DF ($v_2 = 50$) were used.
[2]These probabilities were obtained from a computer program.

Thus, the hypothesis of equal effects of species and the hypothesis of equal effects of the temperatures are both rejected. However, there is also concluded to be significant interaction between species and temperature, and significant interaction between species and sex. Therefore, it must be realized that, although mean respiratory rates in the sampled populations are concluded to be different for the three species and different at the three temperatures, the differences among species are dependent on both temperature and sex.

there is a null hypothesis of interest, and a random-effects factor (factor B) whose levels are termed "blocks," and blocking is intended to reduce the unexplained variability among the data. The *Latin-square** experimental design typically consists of a fixed-effects factor of interest in hypothesis testing (let us call it factor A) and *two*

---

*The term *Latin square* derives from an ancient game of arranging Latin letters in cells within a square. Latin squares were first studied by Swiss mathematician Leonhard Euler (1707–1783) late his very productive life (Norton, 1939), long before they were employed in analysis of variance. He used the French term *quarré latin*, and A. Cayley may have been the first to use the English term, in 1890 (David, 1995); in English it is sometimes written without capitalization.

blocking factors (sometimes referred to as "nuisance factors"), which we shall call factor B and factor C. Having two blocking factors may reduce the remainder MS even further, thus increasing the power of the test for difference among levels of factor A. However, a disadvantage may be that the remainder degrees of freedom are so small that the power of the test is low.

The data in a Latin-square experiment may be displayed conveniently in a tabulation that has the levels of one blocking factor (factor B) as rows and the levels of the other blocking factor (factor C) shown as columns. For example, for three levels of factors A, of factor B, and of factor C, we have the following table:

	Factor C		
Factor B	Level 1	Level 2	Level 3
Level 1	X	X	X
Level 2	X	X	X
Level 3	X	X	X

This table has $3 \times 3 = 9$ cells, and each cell contains one datum: "X." The data to which each of the three levels of factor A are applied can be denoted as A_1, A_2, and A_3; and a Latin square must always contain the same number of levels of each of the three factors. There are 12 possible arrangements for a 3×3 Latin square,* one of which is this:

$$A_2 \quad A_1 \quad A_3$$
$$A_3 \quad A_2 \quad A_1$$
$$A_1 \quad A_3 \quad A_2$$

Consider an experiment designed to test the null hypothesis that the mean time to take effect is the same for three different anesthetics. In a Latin-square arrangement for testing this H_0, one blocking factor could be the source of the animals (factor B, a random-effects factor), and the other could be the source of the drugs (factor C, also a random-effects factor). One of the advantages of a Latin-square design is that it requires fewer data than if a crossed-factor ANOVA were used. For example, for three factors, each consisting of three levels, the Latin square employs nine data; but for a crossed three-factor ANOVA with three levels per factor (see Section 1), there would be 27 (i.e., $3^3 = 27$) cells, and 27 data would be required if there were one datum per cell. Therefore, the Latin-square procedure demands far less experimental resources than does a crossed-factor analysis.

In other situations, a block could represent repeated measures on a subject. So the experiment, which consisted of measurements of effect time for three drugs (factor A, a fixed-effects factor) using animals from three sources (factor B, a random-effects factor), could have been expanded into a Latin-square experiment for each animal tested on three different days (factor C, a repeated-measure factor, for which the experimenter needs to be cautious about avoiding carryover effects.

The Latin-square arrangement of treatments (levels of factor A) should be selected at random from all possible arrangements. The experimenter can arrange the rows randomly and the columns randomly, and then assign each level of factor A randomly within each row with the stipulation that each level of factor A appears only once in

---

*There are 12 configurations possible for a 3×3 Latin square, 576 possible for a 4×4 square, 161,280 for a 5×5 square, 812,851,299 for a 6×6 square, and 61,479,419,904,000 for a 7×7 square.

each column and only once in each row. Several sets of such configurations are listed by Cochran and Cox (1957: 145–146) and Fisher and Yates (1963: 86–89) to facilitate setting up Latin squares. Or, the designation of one of the possible Latin-square configurations may be done by an appropriate computer routine.

A level of factor A is to be assigned to each cell randomly, with the stipulation that each level must appear in each row only once and in each column only once. Also, it must be assumed that there is no interaction between any of the factors (i.e., there must be additivity), and there must be a reasonable expectation that this is so, for there is no good test for interaction in this experimental design.

The only hypothesis generally of interest in a Latin-square analysis is that of equality among the levels of the fixed-effects factor. The total factor-A, factor-B, and factor-C sums of squares are obtained as in other three-factor analyses of variance without replication; because there is only one datum per cell, there are no interactions to be examined. Using a to denote the number of levels in factor A (which is the same as the number of levels in factor B and in factor C), the degrees of freedom for the Latin-square ANOVA are as follows:

$$\text{total DF} = a^2 - 1; \tag{1}$$
$$\text{factor } A \text{ DF} = a - 1; \tag{2}$$
$$\text{factor } B \text{ DF} = a - 1; \tag{3}$$
$$\text{factor } C \text{ DF} = a - 1; \tag{4}$$
$$\text{remainder DF} = a^2 - 1 - (a - 1) - (a - 1) - (a - 1)$$
$$= (a - 1)(a - 2). \tag{5}$$

The H_0 of no difference among the population means for the a levels of factor A is tested by

$$F = \frac{\text{factor } A \text{ MS}}{\text{remainder MS}}, \tag{6}$$

with factor A and remainder DF. Because of the small number of data typically in a Latin-square analysis, it is generally not advisable to proceed if there are missing data. However, a missing datum can be estimated as in Myers and Well (2003: 465) or by some computer software.

The Latin-square design, including situations with data replication, is discussed elsewhere (e.g., Maxwell and Delaney, 2004: 557–561, 611–615; Montgomery, 2005: 136–145; Myers and Well, 2003: 469–477; Snedecor and Cochran, 1989: Section 10; Steel, Torrie, and Dickey, 1997: 227–237; and Winer, Brown, and Michels, 1991: Chapter 9).

(a) Crossover Design. Consider the two-factor ANOVA experimental design where one of the factors is subjects (B, a random-effects factor) upon which repeated measurements are taken, one measurement for each level of the fixed-effects factor (A). It was noted there that enough time should be allowed between successive measurements so there is no carryover effect on X from one measurement to the next on the same subject. As a precaution against there being carryover effects of levels of the fixed-effects factor, the time at which measurements are made on a subject may be considered a third ANOVA factor. Considering subjects and times to be random-effects factors, with no interaction among the three factors, the so-called crossover experimental design takes the form of a Latin square. The crossover design is discussed, for example, in Kirk (1995: 349ff) and Montgomery (2005: 141ff).

(b) Greco-Latin-Square Design. The Latin-square experimental design comprises a fixed-effects factor of interest and two random-effects blocking factors. This concept can be expanded to a design having a fixed-effects factor and *three* blocking factors ("nuisance factors"). This design is rarely encountered. Its reduction of the remainder mean square by using three blocking factors can result in increased power; but the remainder degrees of freedom, $(a - 1)(a - 3)$, are so small that power is decreased. Also, as with Latin squares, this design assumes that there is no interaction among the factors (in this case, among four factors), an assumption that may not be warranted. Greco-Latin-square experiments are described by Cochran and Cox (1957: 132–133); Montgomery (2005: 142–145); Myers and Well (2003: 476–477); and Winer, Brown, and Michels (1991: 680–681, 699–702, 733–734).

3   HIGHER-ORDER FACTORIAL ANALYSIS OF VARIANCE

More than three factors may be analyzed simultaneously, but the number of possible interactions to be dealt with will soon become unwieldy as larger analyses are considered (see Table 2). For more than three or four factors, prohibitively large amounts of data are needed and interpretations of factor and interaction effects become very difficult.

The major effort in performing an ANOVA is the calculation of the several factor and interaction sums of squares. The factor and interaction degrees of freedom may be obtained as indicated in Section 1, and each needed mean square is the relevant sum of squares divided by the respective degrees of freedom. Available computer software provides the needed sums of squares, degrees of freedom, and mean squares. If all factors to be examined are for fixed effects, the F required to test each null hypothesis is obtained by dividing the appropriate factor or interaction mean square by the error MS. If, however, any of the factors represents random effects, then the analysis is more complex and, in some cases, impossible. Presents the procedures applicable to hypothesis testing in several such cases. See Section 5 for consideration of analyses with unequal replication.

If any (but not all) of the factors in a multiway ANOVA are measured on an interval or ratio scale, then we have an analysis of covariance situation. If all of the factors are on an interval or ratio scale, then a multiple regression may be called for.

TABLE 2: Number of Hypotheses Potentially Testable in Factorial Analyses of Variance

	Number of factors			
	2	3	4	5
Main factor	2	3	4	5
2-way interactions	1	3	6	10
3-way interactions		1	4	10
4-way interactions			1	5
5-way interactions				1

Note: The number of mth-order interactions in a k-factor ANOVA is the number of ways k factors can be combined m at a time:

$$_kC_m = \frac{k!}{m!(k - m)!}.$$

4 MULTIWAY ANALYSIS OF VARIANCE WITH BLOCKS OR REPEATED MEASURES

Experimental designs can be devised having three or more factors where one or more factors are blocks or are subjects upon which repeated measures are taken. In such a situation, the analysis may proceed as a factorial ANOVA with the blocking factor or the subjects considered as a random-effects factor. After the sums of squares, degrees of freedom, and mean squares are calculated, an appropriate computer program, can assist in deriving the appropriate F's to test the hypotheses of interest.

There are also designs in which the same block is applied to some—*but not all*—of the combinations of other factors, and such cases are known as *split-plot* experimental designs. If the same subject is exposed to some—*but not all*—combinations of the other factors, this is one of many kinds of *repeated-measures* designs. Discussions of these topics are found in texts on experimental design such as those of Maxwell and Delaney (2004: Chapters 12–15); Mickey, Dunn, and Clark (2004: Chapter 11); Montgomery (2005: Section 5); Myers and Well (2003: Chapter 14); Quinn and Keough (2002: Chapter 11); Snedecor and Cochran (1989, Sections 16.15 and 16.16); Steel, Torrie, and Dickey (1997: Chapter 16); and Winer, Brown, and Michels (1991: Section 5.15, Chapters 7 and 8).

5 FACTORIAL ANALYSIS OF VARIANCE WITH UNEQUAL REPLICATION

Although equal replication is always desirable for optimum power and ease of computation in analysis of variance, it is not essential for the performance of the computations in a single-factor ANOVA. However, all the techniques thus far discussed for ANOVA designs consisting of two or more factors require equal numbers of data per cell (with the exception of the case of proportional replication). For example, the data in Example 1 are composed of four replicates in each combination of species, temperature, and sex. If there were five or more replicates in a very small number of cells, then it is not highly criticizable to discard (at random within a cell) those few data necessary to arrive at equal numbers of replicate data. However, a more general approach is available, a procedure by which data suffering from replication inequality can be analyzed and interpreted by analysis-of-variance considerations. The mathematical manipulations involved are sufficiently complex as to be attempted reasonably only by a computer, but it is worthwhile to be aware of the fact that programs for such an analysis are available. (These procedures may employ a type of multiple linear regression and may be referred to as "general linear models.") An introduction to regression methods for ANOVA experimental designs is given by Glantz and Slinker (2001).*

If inequality is due to one or a few cells containing one fewer datum than the others, then a factorial analysis of variance may be performed after inserting an estimate of each missing datum. If one datum is missing, an estimate of its value may be found as follows (Shearer, 1973):

$$\hat{X} = \frac{aA_i + bB_j + cC_l + \cdots - (k-1)\sum X}{N + k - 1 - a - b - c - \cdots},$$ (7)

---

*R. A. Fisher described the relationship between regression and analysis of variance in 1923 (Peters, 1987: 136).

where $\hat{X}$ is the estimated value for a missing datum in level i of factor A, level j of factor B, level l of factor C, and so on; a, b, c, and so on are the numbers of levels in factors A, B, C, and so on, respectively; A is the sum of all the other data in level i of factor A, B is the sum of the other data in level j of factor B, and so on; the summation of $aA_i + bB_j + cC_l + \ldots$ is over all factors; k is the number of factors; $\sum X$ is the sum of all the other data in all levels of all factors; and N is the total number of data (including the missing one) in the experimental design. This estimated value may then be inserted with the other data in the analysis-of-variance computations.

An alternative method of handling experimental designs with one or a few cells containing one fewer datum than the other is much simpler than, but not as desirable as, the aforementioned procedure. For each small cell the mean of the cell's observed data can be inserted as an additional datum. The analysis of variance is then performed as usual, but with the total DF and within-cells DF calculated without including the number of such additional data. (That is, the total and within-cells DF are those appropriate to the set of original observations.) A better estimation procedure for missing data is to use the cell means as starting values for employing Equation 7 iteratively.

6 MULTIPLE COMPARISONS AND CONFIDENCE INTERVALS IN MULTIWAY ANALYSIS OF VARIANCE

As we have seen, for each factor a hypothesis may be tested concerning the equality of the population means of levels of that factor. If the null hypothesis of equality is rejected for a fixed-effects factor, then it may be desirable to ascertain between which levels the difference(s) lie(s). This can be done by the multiple-comparison procedures prescribed for two-way analyses of variance. Also mentioned in that section is the calculation of confidence intervals with respect to level means in a two-factor analysis of variance; those considerations also apply to an ANOVA with more than two factors. It should be remembered that the sample size, n, is replaced in the present context by the total number of data per level (i.e., the number of data used to calculate the level mean); k is replaced by the number of levels of the factor being tested; s^2 will be replaced by the MS appropriate in the denominator of the F ratio used to test for significance of the factor being examined; the degrees of freedom, v (in q, q', and t) is the DF associated with this MS; and F in Scheffé's test is the same as in the ANOVA.

7 POWER, DETECTABLE DIFFERENCE, AND SAMPLE SIZE IN MULTIWAY ANALYSIS OF VARIANCE

The principles and procedures for two-way ANOVA may be readily expanded to multifactor analysis of variance. k' is the number of levels of the factor under consideration, n' is the total number of data in each level of that factor, s^2 is the appropriate MS in the denominator of the F used for the desired hypothesis test, v_2 is the DF associated with that MS, and $v_1 = k' - 1$. Then, the power of the ANOVA in detecting differences among level means may be estimated.

The equation $\delta = \sqrt{\frac{2k's^2\phi^2}{n'}}$, may be used in the fashion shown in Example 6 from *Two-Factor Analysis of Variance*, to estimate the minimum number of data per level that would be needed to achieve a specified power, given the significance level and detectable difference desired among means.

The equation

$$\chi_r^2 = \frac{12\sum_{i=1}^{a}(R_i - \overline{R})^2}{ba(a + 1)}$$

enables us to estimate the smallest difference among level means detectable with the ANOVA. We can also use equation $\delta = \sqrt{\frac{2k's^2\phi^2}{n'}}$ to estimate the maximum number of levels testable.

(a) The Mixed-Model ANOVA. The aforementioned procedures are applicable when all the factors are fixed effects (i.e., we have a Model I ANOVA). They may also be applied to any fixed-effects factor in a mixed-model ANOVA, but in such cases we must modify our method as follows.

Consider the appropriate denominator for the F calculated to test for the significance of the factor in question. Then, substitute this denominator for the within-cells MS (s^2); and substitute this denominator DF for ν_2.

EXERCISES

1. Use an appropriate computer program to test for all factor and interaction effects in the following $4 \times 3 \times 2$ Model I analysis of variance, where a_i is a level of factor A, b_i is a level of factor B, and c_i is a level of factor C.

	a_1			a_2			a_3			a_4		
	b_1	b_2	b_3	b_1	b_2	b_3	b_1	b_2	b_3	b_1	b_2	b_3
	4.1	4.6	3.7	4.9	5.2	4.7	5.0	6.1	5.5	3.9	4.4	3.7
c_1	4.3	4.9	3.9	4.6	5.6	4.7	5.4	6.2	5.9	3.3	4.3	3.9
	4.5	4.2	4.1	5.3	5.8	5.0	5.7	6.5	5.6	3.4	4.7	4.0
	3.8	4.5	4.5	5.0	5.4	4.5	5.3	5.7	5.0	3.7	4.1	4.4
	4.8	5.6	5.0	4.9	5.9	5.0	6.0	6.0	6.1	4.1	4.9	4.3
c_2	4.5	5.8	5.2	5.5	5.3	5.4	5.7	6.3	5.3	3.9	4.7	4.1
	5.0	5.4	4.6	5.5	5.5	4.7	5.5	5.7	5.5	4.3	4.9	3.8
	4.6	6.1	4.9	5.3	5.7	5.1	5.7	5.9	5.8	4.0	5.3	4.7

2. Use an appropriate computer program to test for the effects of all factors and interactions in the following $2 \times 2 \times 2 \times 3$ Model I analysis of variance design, where a_i is a level of factor A, b_i is a level of factor B, c_i is a level of factor C, and d_i is a factor of level D.

		a_1				a_2			
		b_1		b_2		b_1		b_2	
		c_1	c_2	c_1	c_2	c_1	c_2	c_1	c_2
		12.2	13.4	12.2	13.1	10.9	12.1	10.1	11.2
d_1		12.6	13.1	12.4	13.0	11.3	12.0	10.2	10.8
		12.5	13.5	12.3	13.4	11.2	11.7	9.8	10.7
		11.9	12.8	11.8	12.7	10.6	11.3	10.0	10.9
d_2		11.8	12.6	11.9	12.5	10.4	11.1	9.8	10.6
		12.1	12.4	11.6	12.3	10.3	11.2	9.8	10.7
		12.6	13.0	12.5	13.0	11.1	11.9	10.0	10.9
d_3		12.8	12.9	12.7	12.7	11.1	11.8	10.4	10.5
		12.9	13.1	12.4	13.2	11.4	11.7	10.1	10.8

3. Using an appropriate computer program, test for all factor and interaction effects in the following Model I 3×2 analysis of variance with unequal replication.

	a_1		a_2		a_3	
	b_1	b_2	b_1	b_2	b_1	b_2
	34.1	35.6	38.6	40.3	41.0	42.1
	36.9	36.3	39.1	41.3	41.4	42.7
	33.2	34.7	41.3	42.7	43.0	43.1
	35.1	35.8	41.4	41.9	43.4	44.8
				40.8		44.5

4. A Latin-square experimental design was used to test for the effect of four hormone treatments (A_1, A_2, A_3, and A_4) on the blood calcium levels (measured in mg Ca per 100 ml of blood) of adult farm-raised male ostriches. The two blocking factors are farms (factor B) and analytical methods (factor C). Test the null hypothesis H_0: The mean blood-calcium level is the same with all four hormone treatments.

	Analytical methods			
Farms	C_1	C_2	C_3	C_4
	A_3	A_4	A_2	A_1
B_1	12.5	9.7	12.0	9.4
	A_2	A_3	A_1	A_4
B_2	10.3	13.1	9.5	13.0
	A_1	A_2	A_4	A_3
B_3	8.8	11.7	12.4	14.3
	A_4	A_1	A_3	A_2
B_4	10.6	7.1	12.0	10.1

ANSWERS TO EXERCISES

1. H_0: No effect of factor A; H_A: Factor A has an effect; $F = 10.2901/0.0805 = 127.8$; as $F_{0.05(1),3,72} \cong 2.74$, H_0 is rejected; $P \ll 0.0005$ $[P < 10^{-12}]$. H_0: No effect of factor B; $F = 3.8295/0.0805 = 47.6$; as $F_{0.05(1),2,72} \cong 3.13$, H_0 is rejected; $P \ll 0.0005$ $[P < 10^{-13}]$. H_0: No effect of factor C; $F = 4.2926/0.0805 = 53.3$; as $F_{0.05(1),1,72} \cong 3.98$, H_0 is rejected; $P \ll 0.0005$ $[P < 10^{-9}]$. H_0: No interaction between factors A and B; H_A: There is $A \times B$ interaction; $F = 0.1182/0.0805 = 1.47$; as $F_{0.05(1),6,72} \cong 2.23$, H_0 is not rejected; $P > 0.25$ $[P = 0.20]$. H_0: No interaction between factors A and C; $F = 0.6159/0.0805 = 7.65$; as $F_{0.05(1),3,72} \cong 2.74$, H_0 is rejected; $P < 0.0005$ $[P = 0.00017]$. H_0: No interaction between factors B and C; $F = 0.0039/0.0805 = 0.048$; as $F_{0.05(1),2,72} \cong 1.41$, H_0 is not rejected; $P > 0.25$ $[P = 0.95]$. H_0: No interaction between factors A, B, and C; $F = 0.1459/0.0805 = 1.81$; as $F_{0.05(1),6,72} \cong 2.23$, H_0 is not rejected; $0.10 < P < 0.25$ $[P = 0.11]$.

2. H_0: No effect of factor A; H_A: Factor A has an effect; $F = 56.00347/0.03198 = 1751$; reject H_0; $P \ll 0.0005$ $[P < 10^{-14}]$. H_0: No effect of factor B; $F = 4.65125/0.03198 = 145.4$; reject H_0; $P \ll 0.0005$ $[P < 10^{-13}]$. H_0: No effect of factor C; $F = 8.6125/0.03198 = 269.3$; reject H_0: $P \ll 0.0005$ $[P < 10^{-13}]$. H_0: No effect of factor D; $F = 2.17056/0.03198 = 67.9$; reject H_0; $P \ll 0.0005$ $[P < 10^{-14}]$. H_0: No interaction between factors A and B; $F = 2.45681/0.03198 = 76.8$; reject H_0; $P \ll 0.0005$ $[P < 10^{-11}]$. H_0: No interaction between factors A and C; $F = 0.05014/0.03198 = 1.57$; do not reject H_0;

$0.10 < P < 0.25$ $[P = 0.22]$. H_0: No interaction between factors A and D; $F = 0.06889/0.03198 = 2.15$; do not reject H_0; $0.10 < P < 0.25$ $[P = 0.13]$. H_0: No interaction between factors B and C; $F = 0.01681/0.03198 = 0.53$; do not reject H_0; $P > 0.25$ $[P = 0.47]$. H_0: No interaction between factors B and D; $F = 0.15167/0.03198 = 4.74$; reject H_0; $0.01 < P < 0.025$. H_0: No interaction between factor C and D; $F = 0.26000/0.03198 = 8.13$; reject H_0; $0.0005 < P < 0.001$ $[P = 0.00091]$. H_0: No interaction among factors A, B, and C; $F = 0.00125/0.03198 = 0.039$; do not reject H_0; $P > 0.25$ $[P = 0.84]$. H_0: No interaction among factors A, B, and D; $F = 0.14222/0.03198 = 2.11$; do not reject H_0; $0.10 < P < 0.25$ $[P = 0.13]$. H_0: No interaction among factors B, C, and D; $F = 0.00222/0.03198 = 0.069$; do not reject H_0; $P > 0.25$ $[P = 0.093]$. H_0: No interaction among factors A, B, C, and D; $F = 0.01167/0.03198 = 0.36$; do not reject H_0; $P > 0.25$ $[P = 0.70]$.

3. H_0: No effect of factor A; H_A: There is an effect of factor A; $F = 239.39048/2.10954 = 113.5$; as $F_{0.05(1),1,22} = 4.30$, reject H_0; $P \ll 0.0005$ $[P < 10^{-9}]$. H_0: No effect of factor B; $F = 8.59013/2.10954 = 4.07$; as $F_{0.05(1),1,22} = 4.30$; do not reject H_0; $0.05 < P < 0.10$ $[P = 0.056]$. H_0: No interaction between factors A and B; $F = 0.10440/2.10954 = 0.05$; as $F_{0.05(1),1,22} = 4.30$, do not reject H_0; $P > 0.25$ $[P = 0.83]$.

4. For Factor A: SS $= 37.4719$, DF $= 3$, MS $= 12.4906$; for remainder: SS $= 3.5287$, MS $= 0.5881$, DF $= 6$; $F = 21.2$; $F_{0.05(1),3,6} = 4.76$; $0.01 < P < 0.0025$; $P = 0.0014$; reject H_0.

This page intentionally left blank

Nested (Hierarchical) Analysis of Variance

A crossed experiment is one where all possible combinations of levels of the factors exist; the cells of data are formed by each level of one factor being in combination with each level of every other factor.

In some experimental designs, however, we may have some levels of one factor occurring in combination with the levels of one or more other factors, and other distinctly different levels occurring in combination with others. In Example 1a, where blood-cholesterol concentration is the variable, there are two factors: drug type and drug source. Each drug was obtained from two sources, but the two sources are not the same for all the drugs. Thus, the experimental design is not crossed; rather, we say it is *nested* (or *hierarchical*). One factor (drug source) is nested within another factor (drug type). A nested factor, as in the present example, is typically a random-effects factor, and the experiment may be viewed as a modified one-way ANOVA where the levels of this factor (drug source) are samples and the cholesterol measurements within a drug source are called a "subsample."

Sometimes experiments are designed with nesting in order to test a hypothesis about difference among the samples. More typical, however, is the inclusion of a random-effects nested factor in order to account for some within-groups variability and thus make the hypothesis testing for the other factor (usually a fixed-effects factor) more powerful.

1 NESTING WITHIN ONE FACTOR

In the experimental design such as in Example 1a, the primary concern is to detect population differences among levels of the fixed-effects factor (drug type). We can often employ a more powerful test by nesting a random-effects factor that can account for some of the variability within the groups of interest. The partitioning of the variability in a nested ANOVA may be observed in this example.

(a) Calculations for the Nested ANOVA. Testing the hypotheses in Example 1a involves calculating relevant sums of squares and mean squares. This is often done by computer; it can also be accomplished with a calculator as follows

EXAMPLE 1a   A Nested (Hierarchical) Analysis of Variance

The variable is blood cholesterol concentration in women (in mg/100 ml of plasma). This variable was measured after the administration of one of three different drugs to each of 12 women, and each administered drug was obtained from one of two sources.

	Drug 1		Drug 2		Drug 3		
	Source A	Source Q	Source D	Source B	Source L	Source S	
	102	103	108	109	104	105	
	104	104	110	108	106	107	
n_{ij}	2	2	2	2	2	2	$N = 12$
$\sum_{l=1}^{n_{ij}} X_{ijl}$	206	207	218	217	210	212	
$\overline{X}_i$	103	103.5	109	108.5	105	106	$\overline{X} = 105.8333$
n_i	4		4		4		$N = 12$
$\sum_{j=1}^{b}\sum_{l=1}^{n_{ij}} X_{ijl}$	413		435		422		$\sum_{i=1}^{a}\sum_{j=1}^{b}\sum_{l=1}^{n_{ij}} X_{ijl}$ $= 1270$
$\overline{X}_i$	103.25		108.75		105.5		

(see Example 1b). In the hierarchical design described, we can uniquely designate each datum by using a triple-subscript notation, where X_{ijl} indicates the lth datum in subgroup j of group i. Thus, in Example 1a, $X_{222} = 108$ mg/100 ml, $X_{311} = 104$ mg/100 ml, and so on. For the general case, there are a groups, numbered 1 through a, and b is the number of subgroups in each group. For Example 1a, there are three levels of factor A (drug type), and b (the number of levels of factor B, i.e., sources for each drug) is 2. The number of data in subgroup j of group i may be denoted by n_{ij} (2 in this experiment), and the total number of data in group i is n_i (in this example, 4). The total number of observations in the entire experiment is $N = \sum_{i=1}^{k} n_i$ (which could also be computed as $N = \sum_{i=1}^{a}\sum_{j=1}^{b} n_{ij}$). The sum of the data in subgroup j of group i is calculated as $\sum_{l=1}^{n_{ij}} X_{ijl}$; the sum of the data in group i is $\sum_{j=1}^{b}\sum_{l=1}^{n_{ij}} X_{ijl}$; and the mean of group i is

$$\overline{X}_i = \frac{\sum_{j=1}^{b}\sum_{l=1}^{n_{ij}} X_{ijl}}{n_i}. \tag{1}$$

The grand mean of all the data is

$$\overline{X} = \frac{\sum_{i=1}^{a}\sum_{j=1}^{b}\sum_{l=1}^{n_{ij}} X_{ijl}}{N}. \tag{2}$$

EXAMPLE 1b   Computations for the Nested ANOVA of Example 1a

	Drug 1	**Drug 2**	**Drug 3**
	Source A Source Q	*Source D Source B*	*Source L Source S*

$$\frac{\left(\sum\limits_{l=1}^{n_{ij}} X_{ijl}\right)^2}{n_{ij}} \quad 21218.0 \quad 21424.5 \quad 23762.0 \quad 23544.5 \quad 22050.0 \quad 22472.0 \quad \sum_{i=1}^{a}\sum_{j=1}^{b} \frac{\left(\sum\limits_{l=1}^{n_{ij}} X_{ijl}\right)^2}{n_{ij}} = 134471.0$$

$$\frac{\left(\sum\limits_{j=1}^{b}\sum\limits_{l=1}^{n_{ij}} X_{ijl}\right)^2}{n_i} \quad 42642.25 \quad 47306.25 \quad 44521.00 \quad \sum_{i=1}^{a} \frac{\left(\sum\limits_{j=1}^{b}\sum\limits_{l=1}^{n_{ij}} X_{ijl}\right)^2}{n_i} = 134469.50$$

$$\sum_{i=1}^{a}\sum_{j=1}^{b}\sum_{l=1}^{n_{ij}} x_{ijl}^2 = 134480.00 \qquad C = \frac{\left(\sum\limits_{i=1}^{a}\sum\limits_{j=1}^{b}\sum\limits_{l=1}^{n_{ij}} X_{ijl}\right)^2}{N} = \frac{(1270)^2}{12} = 134408.33$$

$$\text{total SS} = \sum_{i=1}^{a}\sum_{j=1}^{b}\sum_{l=1}^{n_{ij}} X_{ijl}^2 - C = 134480.00 - 134408.33 = 71.67$$

$$\text{among all subgroups SS} = \sum_{i=1}^{a}\sum_{j=1}^{b} \frac{\left(\sum\limits_{l=1}^{n_{ij}} X_{ijl}\right)^2}{n_{ij}} - C = 134471.00 - 134408.33 = 62.67$$

$$\text{error SS} = \text{total SS} - \text{among all subgroups SS} = 71.67 - 62.67 = 9.00$$

$$\text{groups SS} = \sum_{i=1}^{a} \frac{\left(\sum\limits_{j=1}^{b}\sum\limits_{l=1}^{n_{ij}} X_{ijl}\right)^2}{n_i} - C = 134469.50 - 134408.33 = 61.17$$

$$\text{subgroups SS} = \text{among all subgroups SS} - \text{groups SS} = 62.67 - 61.17 = 1.50$$

Source of variation	SS	DF	MS
Total	71.67	11	
Among all subgroups (Sources)	62.67	5	
Groups (Drugs)	61.17	2	30.58
Subgroups	1.50	3	0.50
Error	9.00	6	1.50

H_0: There is no difference among the drug sources in affecting mean blood cholesterol concentration.

H_A: There is difference among the drug sources in affecting mean blood cholesterol concentration.

$$F = \frac{0.50}{1.50} = 0.33. \quad F_{0.05(1),3,6} = 4.76. \quad \text{Do not reject } H_0.$$
$$P > 0.50 \quad [P = 0.80]$$

H_0: There is no difference in mean cholesterol concentrations owing to the three drugs (i.e., $\mu_1 = \mu_2 = \mu_3$).

H_A: There is difference in mean cholesterol concentrations owing to the three drugs.

$$F = \frac{30.58}{0.50} = 61.16. \quad F_{0.05(1),2,3} = 9.55. \quad \text{Reject } H_0.$$

$$0.0025 < P < 0.005 \quad [P = 0.0037]$$

The total sum of squares for this ANOVA design considers the deviations of all the X_{ijl} from $\overline{X}$ and may be calculated as

$$\text{total SS} = \sum_{i=1}^{a} \sum_{j=1}^{b} \sum_{l=1}^{n_{ij}} \left(X_{ijl} - \overline{X} \right)^2 \tag{3}$$

or by this "machine formula":

$$\text{total SS} = \sum_{i=1}^{a} \sum_{j=1}^{b} \sum_{l=1}^{n_{ij}} X_{ijl}^2 - C, \tag{3a}$$

where

$$C = \frac{\left(\displaystyle\sum_{i=1}^{a} \sum_{j=1}^{b} \sum_{l=1}^{n_{ij}} X_{ijl} \right)^2}{N}. \tag{4}$$

For the total variability,

$$\text{total DF} = N - 1. \tag{5}$$

The variability among groups (i.e., the deviations $\overline{X}_i - \overline{X}$) is expressed as the "among groups SS" or, simply,

$$\text{groups SS} = \sum_{i=1}^{a} n_i \left(\overline{X}_i - \overline{X} \right)^2 \tag{6}$$

or

$$\text{groups SS} = \sum_{i=1}^{a} \frac{\left(\displaystyle\sum_{j=1}^{b} \sum_{l=1}^{n_{ij}} X_{ijl} \right)^2}{n_i} - C; \tag{6a}$$

and

$$\text{groups DF} = a - 1. \tag{7}$$

There is a total of ab subgroups in the design, and, considering them as if they were groups in a one-way ANOVA, we can calculate a measure of the deviations $\overline{X}_{ij} - \overline{X}$ as

$$\text{among all subgroups SS} = \sum_{i=1}^{a} \sum_{j=1}^{b} n_{ij} \left(\overline{X}_{ij} - \overline{X} \right)^2 \tag{8}$$

or

$$\text{among all subgroups SS} = \sum_{i=1}^{a} \sum_{j=1}^{b} \frac{\left(\sum_{l=1}^{n_{ij}} X_{ijl}\right)^2}{n_{ij}} - C; \tag{8a}$$

and

$$\text{among all subgroups DF} = ab - 1. \tag{9}$$

The variability due to the subgrouping within groups is evidenced by the deviations of subgroup means from their group means, $\overline{X}_{ij} - \overline{X}_i$, and the appropriate sum of squares is the "among subgroups within groups" SS, which will be referred to as

$$\text{subgroups SS} = \text{among all subgroups SS} - \text{groups SS}; \tag{10}$$

and

$$\text{subgroups DF} = \text{among all subgroups DF} - \text{groups DF} = a(b-1). \tag{11}$$

The within-subgroups, or "error," variability expresses the deviations $X_{ijl} - \overline{X}_{ij}$, namely the deviations of data from their subgroup means. The appropriate sum of squares is obtained by difference:

$$\text{error SS} = \text{total SS} - \text{among all subgroups SS}, \tag{12}$$

with

$$\text{error DF} = \text{total DF} - \text{among all subgroups DF} = N - ab. \tag{13}$$

The summary of this hierarchical analysis of variance is presented in Table 1. Recall that MS = SS/DF. In the nested ANOVA of Table 1 we cannot speak of interaction between the two factors. Calculations for the data and hypotheses of Example 1a are shown in Example 1b.

(b) Hypothesis Testing in the Nested ANOVA. For the data in Example 1a, we can test the null hypothesis that no difference in cholesterol occurs among subgroups (i.e., the source of the drugs has no effect on the mean concentration of blood cholesterol). We do this by examining

$$F = \frac{\text{subgroups MS}}{\text{error MS}}. \tag{14}$$

For Example 1b, this is $F = 0.50/1.50 = 0.33$; since $F_{0.05(1),3,6} = 4.76$, H_0 is not rejected. (The exact probability of an F at least this large if H_0 is true is 0.80.)

The null hypothesis that there is no difference in cholesterol with the administration of the three different drugs can be tested by

$$F = \frac{\text{groups MS}}{\text{subgroups MS}}, \tag{15}$$

which in the present example is $F = 30.58/0.50 = 61.16$. As $F_{0.05(1),2,3} = 9.55$, H_0 is rejected. (The exact probability is 0.0037.) In an experimental design having subgroups nested within groups, as shown here, the groups most often represent a

Nested (Hierarchical) Analysis of Variance

TABLE 1: Summary of Hierarchical (Nested) Single-Factor Analysis of Variance Calculations

Source of variation	SS	DF
Total $[X_{ijl} - \overline{X}]$	$\displaystyle\sum_{i=1}^{a}\sum_{j=1}^{b}\sum_{l=1}^{n_{ij}} X_{ijl}^2 - C$	$N-1$
Among all subgroups $[\overline{X}_{ij} - \overline{X}]$	$\displaystyle\sum_{i=1}^{a}\sum_{j=1}^{b} \frac{\left(\displaystyle\sum_{l=1}^{n_{ij}} X_{ijl}\right)^2}{n_{ij}} - C$	$ab-1$
Groups (i.e., Among groups) $[\overline{X}_i - \overline{X}]$	$\displaystyle\sum_{i=1}^{a} \frac{\left(\displaystyle\sum_{j=1}^{b}\sum_{l=1}^{n_{ij}} X_{ijl}\right)^2}{n_i} - C$	$a-1$
Subgroups (i.e., Among subgroups within groups) $[\overline{X}_{ij} - \overline{X}_i]$	among all subgroups SS− groups SS	$a(b-1)$
Error (i.e., Within subgroups) $[X_{ijl} - \overline{X}_{ij}]$	total SS− among all subgroups SS	$N-ab$

Note: $C = \dfrac{\left(\displaystyle\sum_{i=1}^{a}\sum_{j=1}^{b}\sum_{l=1}^{n_{ij}} X_{ijl}\right)^2}{N}$; a = number of groups; b = number of subgroups within each group; n_i = number of data in group i; n_{ij} = number of data in subgroup j for group i; N = total number of data in entire experiment.

fixed-effects factor. But the hypotheses testing is the same if, instead, the groups are a random-effects factor.

If we do not reject the null hypothesis of no difference among subgroups within groups, then the subgroups MS might be considered to estimate the same population variance as does the error MS. Thus, some statisticians suggest that in such cases a pooled mean square can be calculated by pooling the sums of squares and pooling the degrees of freedom for the subgroups variability and the error variability, for this will theoretically provide the ability to perform a more powerful test for differences among groups. However, there is not widespread agreement on this matter, so the suggested procedure is to be conservative and not engage in pooling, at least not without consulting a statistician.

If there are unequal numbers of subgroups in each group, then the analysis becomes more complex, and the preceding calculations are not applicable. This situation is generally submitted to analysis by computer, perhaps by a procedure referred to in Glantz and Slinker (2001).

A hierarchical experimental design might have two (or more) layers of nesting with each subgroup composed of sub-subgroups, thus involving an additional step in the hierarchy. For instance, for the data of Example 1a, the different drugs define the groups, the different sources define the subgroups, and if different technicians or different instruments were used to perform the cholesterol analyses within each subgroup, then these technicians or instruments would define the sub-subgroups. Sokal and Rohlf (1995: 288–292) describe the calculations for a design with sub-subgroups, although one generally resorts to computer calculation for hierarchical designs with more than the two steps in the hierarchy discussed in the preceding paragraphs.

Brits and Lemmer (1990) discuss nonparametric ANOVA with nesting within a single factor.

2 NESTING IN FACTORIAL EXPERIMENTAL DESIGNS

Experimental designs are encountered where there are two or more crossed factors as well as one or more nested factors. For example, imagine that the two crossed factors are sex and hormone treatment, and five birds of each sex were given each hormone treatment. In addition, the experimenter might have obtained three syringes of blood (that is, three subsamples) from each bird, so that individual birds would be samples and the triplicate blood collections would be subsamples. The birds represent a nested, rather than a crossed, factor because the same animal is not found at every combination of the other two factors. The analysis-of-variance table would then look like that in Example 2. The computation of sums of squares could be obtained by computer. Some available computer programs can operate with data where there is not equal replication. In a factorial ANOVA with nesting, the determination of F's for hypothesis testing depends upon whether the crossed factors are fixed effects or random effects.

The concept of hierarchical experimental designs could be extended further in this example by considering that each subsample (i.e., each syringe of blood) in Example 2 was subjected to two or more (i.e., replicate) chemical analyses. Then chemical analysis would be a factor nested within the syringe factor, syringe nested within animal, and animal nested within the two crossed factors.

EXAMPLE 2   An Analysis of Variance with a Random-effects Factor (Animal) Nested within the Two-factor Crossed Experimental Design of Example 1 from *Two-Factor Analysis of Variance*

For each of the four combinations of two sexes and two hormone treatments ($a = 2$ and $b = 2$), there are five animals ($c = 5$), from each of which three blood collections are taken ($n = 3$). Therefore, the total number of data collected is $N = abcn = 60$.

Source of variation	SS	DF	MS
Total		$N - 1 = 59$	
Cells		$ab - 1 = 3$	
Hormone treatment (Factor A)	*	$a - 1 = 1$	†
Sex (Factor B)	*	$b - 1 = 1$	†
$A \times B$	*	$(a - 1)(b - 1) = 1$	†
Among all animals		$abc - 1 = 19$	
Cells		$ab - 1 = 3$	
Animals (Within cells) (Factor C)	*	$ab(c - 1) = 16$	†
Error (Within animals)	*	$abc(n - 1) = 40$	†

* These sums of squares can be obtained from appropriate computer software; the other sums of squares in the table might not be given by such a program, or MS might be given but not SS.

† The mean squares can be obtained from an appropriate computer program. Or, they may be obtained from the sums of squares and degrees of freedom (as MS = SS/DF). The degrees of freedom might appear in the computer output, or they may have to be determined by hand.

H_0: There is no difference in mean blood calcium concentration between males and females.

H_A: There is a difference in mean blood calcium concentration between males and females.

$$F = \frac{\text{factor } A \text{ MS}}{\text{factor } C \text{ MS}} \qquad F_{0.05(1),1,16} = 4.49$$

H_0: The mean blood calcium concentration is the same in birds receiving and not receiving the hormone treatment.

H_A: The mean blood calcium concentration is not the same in birds receiving and not receiving the hormone treatment.

$$F = \frac{\text{factor } B \text{ MS}}{\text{factor } C \text{ MS}} \qquad F_{0.05(1),1,16} = 4.49$$

H_0: There is no interactive effect of sex and hormone treatment on mean blood calcium concentration.

H_A: There is interaction between sex and hormone treatment in affecting mean blood calcium concentration.

$$F = \frac{A \times B \text{ MS}}{\text{factor } C \text{ MS}} \qquad F_{0.05(1),1,16} = 4.49$$

H_0: There is no difference in blood calcium concentration among animals within combinations of sex and hormone treatment.

H_A: There is difference in blood calcium concentration among animals within combinations of sex and hormone treatment.

$$F = \frac{\text{factor } C \text{ MS}}{\text{error MS}} \qquad F_{0.05(1),16,40} = 1.90$$

3 MULTIPLE COMPARISONS AND CONFIDENCE INTERVALS

Whenever a fixed-effects factor is concluded by an ANOVA to have a significant effect on the variable, we may turn to the question of which of the factor's levels are different from which others. If there are only two levels of the factor, then of course we have concluded that their population means are different by the ANOVA. But if there are more than two levels, then a multiple-comparison test must be employed.

Keep the following in mind

1. k refers to the number of levels being compared. (In Example 1a, $k = a$, the number of levels in factor A. In Example 2, $k = a$ when comparing levels of factor A, and $k = b$ when testing levels of factor B.)

2. The sample size, n, refers to the total number of data from which a level mean is calculated. (In Example 1a, the sample size $bn = 4$ would be used in place of n. In Example 2, we would use $bcn = 30$ to compare level means for factor A and $acn = 30$ for factor B.)

3. The mean square, s^2, refers to the MS in the denominator of the F ratio appropriate to testing the effect in question in the ANOVA. (In Example 1a, the subgroups [sources] MS would be used. In Example 2, the factor C MS would be used.)

4. The degrees of freedom, v, for the critical value of q or q' are the degrees of freedom associated with the mean square indicated in item 3. (In Examples 1a and 2, these would be 3 and 16, respectively.)

5. The critical value of F in the Scheffé test has the same degrees of freedom as it does in the ANOVA for the factor under consideration. (In Example 1a, these are 2 and 3. In Example 2, they are 1 and 16.)

Once a multiple-comparison test has determined where differences lie among level means, we can express a confidence interval for each different mean, keeping in mind the sample sizes, mean squares, and degrees of freedom defined in the preceding list.

4 POWER, DETECTABLE DIFFERENCE, AND SAMPLE SIZE IN NESTED ANALYSIS OF VARIANCE

Previously, power and sample size for factorial analyses of variance were discussed. The same types of procedures may be employed for a fixed-effects factor within which nesting occurs. As previously used, k' is the number of levels of the factor, n' is the total number of data in each level, and $v = k' - 1$. The appropriate mean square, s^2, is that appearing in the denominator of the F ratio used to test that factor in the ANOVA, and v_2 is the degrees of freedom associated with s^2.

The power of a nested ANOVA to detect differences among level means may be estimated using Equations 16–18. Equation 17 may be used to estimate the minimum number of data per level that would be needed to achieve a specified power, and Equation 18 allows estimation of the smallest detectable difference among level means.

$$\mu = \frac{\sum\limits_{m-1}^{k'} \mu_m}{k'}, \tag{16}$$

$$\psi = \sqrt{\frac{n'\delta^2}{2k's^2}}, \tag{17}$$

$$\delta = \sqrt{\frac{2k's^2\phi^2}{n'}}. \tag{18}$$

EXERCISES

1. Using the data and conclusions of Example 1, perform the following:
   (a) Use the Tukey test to conclude which of the three drug means are statistically different from which.
   (b) Determine the 95% confidence interval for each significantly different drug mean.

2. Three water samples were taken from each of three locations. Two determinations of fluoride content were performed on each of the nine samples. The data are as follows, in milligrams of fluoride per liter of water:

(a) Test the hypothesis that there is no difference in mean fluoride content among the samples within locations.
(b) Test the hypothesis that there is no difference in mean fluoride content among the locations.
(c) If the null hypothesis in part (b) is rejected, use the Tukey test to conclude which of the three populations means differ from which.
(d) If the null hypothesis in part (b) is rejected, determine the 95% confidence interval for each different population mean.

Locations	1			2			3		
Samples	1	2	3	1	2	3	1	2	3
	1.1	1.3	1.2	1.3	1.3	1.4	1.8	2.1	2.2
	1.2	1.1	1.0	1.4	1.5	1.2	2.0	2.0	1.9

ANSWERS TO EXERCISES

1. **(a)** $q_{\alpha,\nu,k} = q_{0.05,3,3} = 5.910$; $s^2 = 0.50$; SE $= 0.3536$; reject H_0: $\mu_2 = \mu_1$, $q = 15.56$ [$0.001 < P < 0.005$]; reject $H_0 : \mu_2 = \mu_3$, $q = 9.19$ [$0.01 < P < 0.025$], reject H_0: $\mu_3 = \mu_1$, $q = 6.35$ [$0.025 < P < 0.05$]. **(b)** $t_{0.05(2),3} = 3.182$; 95% CI $= \overline{X}_i \pm 1.13$ mg/100 ml.

2. **(a)** H_0: The mean fluoride concentrations are the same for all three samples at a given location; H_A: The mean fluoride concentrations are not the same for all three samples at a given location; $F = 0.008333/0.01778 = 0.469$; as $F < 1.0$, so do not reject H_0; $P > 0.25$ [$P = 0.82$]. **(b)** H_0: The mean fluoride concentration is the same at all three locations; H_A: The mean fluoride concentration is not the same at all three locations: $F = 1.1850/0.008333 = 142$; $F_{0.05(1),2,6} = 5.14$; reject H_0; $P \ll 0.0005$ [$P = 0.0000086$]. **(c)** $q_{\alpha,\nu,k} = q_{0.05,6,3} = 4.339$; $s^2 = 0.08333$; SE $= 0.0373$; $\overline{X}_1 = 1.15$, $\overline{X}_2 = 1.35$, $\overline{X}_3 = 2.00$; reject H_0: $\mu_3 = \mu_1$, $q = 22.79$ [$P < 0.001$]; reject H_0: $\mu_3 = \mu_2$, $q = 17.43$ [$P < 0.001$], reject H_0: $\mu_2 = \mu_1$, $q = 5.36$ [$0.01 < P < 0.025$]. **(d)** $t_{0.05(2),6} = 2.447$; 95% confidence interval (mg/L): $\overline{X}_i \pm 0.09$.

Multivariate Analysis of Variance

Multivariate Analysis of Variance

We have previously discussed various experimental designs categorized as analysis of variance (ANOVA), wherein a variable is measured in each of several categories, or levels, of one or more factors. The hypothesis testing asked whether the population mean of the variable differed among the levels of each factor. These are examples of what may be termed *univariate analyses of variance*, because they examine the effect of the factor(s) on only one variable.

An expansion of this concept is an experimental design where more than one variable is measured on each experimental subject. 19 animals were allocated at random to four experimental groups, and each group was fed a different diet. Thus, diet was the experimental factor, and there were four levels of the factor. In that example, only one variable was measured on each animal: the body weight. But other measurements might have been made on each animal, such as blood cholesterol, blood pressure, or body fat. If two or more variables are measured on each subject in an ANOVA design, we have a *multivariate analysis of variance* (abbreviated MANOVA).*

There are several uses to which multivariate analysis of variance may be put (e.g., Hair et al., 2006: 399–402). This chapter presents a brief introduction to this type of analysis, a multifaceted topic often warranting consultation with knowledgeable practitioners. More extensive coverage is found in many texts on the subject (e.g., Bray and Maxwell, 1985; Hair et al., 2006: Chapter 6; Hand and Taylor, 1987: Chapter 4; Johnson and Wichern, 2002: Chapter 6; Marcoulides and Hershberger, 1997: Chapters 3–4; Sharma, 1996: Chapters 11–12; Srivastava, 2002: Chapter 6; Stevens, 2002: Chapters 4–6; and Tabachnik and Fidell, 2001: Chapter 9).

The multivariate analysis-of-variance experimental design discussed here deals with a single factor. There are also MANOVA procedures for blocked, repeated-measures, and factorial experimental designs, as presented in the references just cited.

1   THE MULTIVARIATE NORMAL DISTRIBUTION

Recall that univariate analysis of variance assumes that the sample of data for each group of data came from a population of data that were normally distributed, and univariate normal distributions may be shown graphically. In such two-dimensional

---

*Sometimes the variables are referred to as *dependent variables* and the factors as *independent variables* or *criterion variables*.

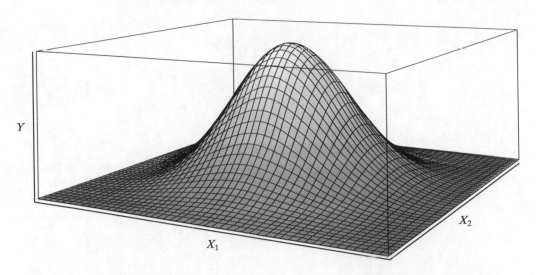

FIGURE 1: A bivariate normal distribution, where X_1 and X_2 have identical standard deviations.

figures, the height of the curve, Y, representing the frequency of observations at a given magnitude of the variable X, is plotted against that value of X; and the highest point of the normal curve is at the mean of X.

The simplest multivariate case is where there are two variables (call them X_1 and X_2) and they can be plotted on a graph with three axes representing Y, X_1, and X_2. The two-variable extension of the single-variable normal curve is a surface representing a *bivariate normal distribution*, such as that shown in Figure 1.* The three-dimensional normal surface rises like a hill above a flat floor (where the floor is the plane formed by the two variables, X_1 and X_2), and the highest point of the curved surface is at the means of X_1 and X_2. A plot of more than three dimensions would be required to depict multivariate distributions with more than two measured variables.

Multivariate normality requires, among other characteristics (e.g., Stevens, 2002: 262), that for *each* X_i (in this example, X_i and X_2) there is a normal distribution of Y values. Univariate normal distributions with smaller standard deviations form narrower curves than those with larger standard deviations. Similarly, the hill-shaped bivariate graph of Figure 1 will be narrow when the standard deviations of X_1 and X_2 are small and broad when they are large.

Rather than drawing bivariate normal graphs such as Figure 1, we may prefer to depict these three-dimensional plots using two dimensions, just as mapmakers represent an elevated or depressed landscape using contour lines. Figure 2a shows the distribution of Figure 1 with a small plane passing through it parallel to the X_1 and X_2 plane at the base of the graph. A circle is delineated where the small plane intersects the normal-distribution surface. Figure 2b shows two planes passing through the normal surface of Figure 1 parallel to its base, and their intersections form two concentric circles. If Figures 2a and 2b are viewed from above the surface, looking straight down toward the plane of X_1 and X_2, those circles would appear as in Figures 3a and 3b, respectively. Three such intersecting planes would result in three circles, and so on. If the standard deviations

---

*This three-dimensional surface is what gave rise to the term *bell curve*, named by E. Jeuffret in 1872 (Stigler, 1999: 404).

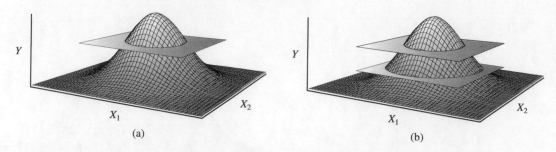

(a)　　　　　　　　　　　　　　　　　(b)

FIGURE 2: The bivariate normal distribution of Figure 1, with **(a)** an intersecting plane parallel to the X_1-X_2 plane, and **(b)** two intersection planes parallel to the X_1-X_2 plane.

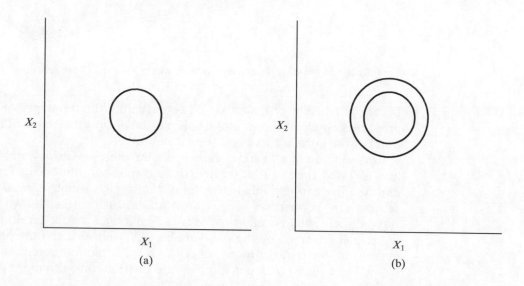

(a)　　　　　　　　　　　　　　　　　(b)

FIGURE 3: Representations of Figures 2a and 2b, showing the circles defined by the intersecting planes.

of X_1 and X_2 are not the same, then parallel planes will intersect the bivariate-normal surface to form ellipses instead of circles. This is shown, for three planes, in Figure 4. In such plots, the largest ellipses (or circles, if X_1 and X_2 have equal standard deviations) will be formed nearest the tails of the distribution. Only ellipses (and not circles) will be discussed hereafter, for it is unlikely that the two variables will have exactly the same variances (that is, the same standard deviations).

If an increase in magnitude of one of the two variables is not associated with a change in magnitude of the other, it is said that there is no correlation between X_1 and X_2. If an increase in magnitude of one of the variables *is* associated with either an increase or a decrease in the other, then the two variables are said to be correlated. If X_1 and X_2 are not correlated, graphs such as Figure 4 will show the long axis of all the ellipses parallel to either the X_1 or X_2 axis of the graph (Figures 4a and 4b, respectively). If, however, X_1 and X_2 are positively correlated, the ellipses appear as running from the lower left to the upper right of the graph (Figure 4c); and if the two variables are negatively correlated, the ellipses run from the lower right to the upper left (Figure 4d).

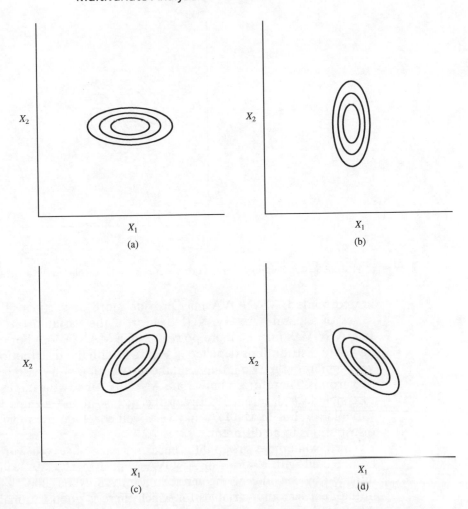

FIGURE 4: Representations of bivariate normal distributions where the standard deviations of X_1 and X_2 are not the same. **(a)** X_1 and X_2 are not correlated. **(b)** X_1 and X_2 are not correlated. **(c)** X_1 and X_2 are positively correlated. **(d)** X_1 and X_2 are negatively correlated.

2 MULTIVARIATE ANALYSIS OF VARIANCE HYPOTHESIS TESTING

At the beginning of the discussion of univariate analysis of variance, it was explained that, when comparing a variable's mean among more than two groups, to employ multiple t-tests would cause a substantial inflation of α, the probability of a Type I error. In multivariate situations, we desire to compare two or more variables' means among two or more groups, and to do so with multiple ANOVAs would also result in an inflated chance of a Type I error*. Multivariate analysis of variance is a method of comparing the population means for each of the multiple variables of interest at the same time while maintaining the chosen magnitude of Type I error.

A second desirable trait of MANOVA is that it considers the correlation among multiple variables, which separate ANOVAs cannot do. Indeed, if the variables

---

*For m variables, the probability of a Type I error will range from α, if all of the variables are perfectly correlated, to $1 - (1 - \alpha)^m$, if there is no correlation among them. So, for example, if testing with two variables at the 5% significance level, $P(\text{Type I error}) = 0.10$ (Hair et al., 2006: 400).

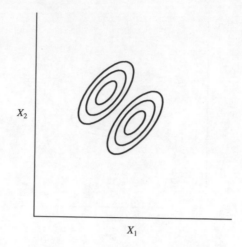

FIGURE 5: Two bivariate normal groups of positively correlated data differing in both dimensions.

are correlated, MANOVA may provide more powerful testing than performing a series of separate ANOVAs. (However, if the variables are not correlated, separate ANOVAs may be more powerful than MANOVA.) For example, there are two bivariate distributions depicted in Figure 5, with the distribution of variable X_1 being very similar in the two groups and the distribution of X_2 also being very similar in the two groups. Therefore, a univariate ANOVA (or a two-sample t-test) will be unlikely to conclude a difference between the means of the groups for variable X_1 or for variable X_2, but MANOVA may very well conclude the means of the two bivariate distributions to be different.

Third, sometimes group differences for each of several variables are too small to be detected with a series of ANOVAs, but a MANOVA will conclude the groups different by considering the variables jointly. Stevens (2002: 245) cautions to include in the analysis only variables for which there is good rationale, because very small differences between means for most of them may obscure substantial differences for some of them.

In univariate ANOVA with k groups, a typical null hypothesis is

$$H_0: \mu_1 = \mu_2 = \cdots = \mu_k,$$

which says that all k population means are the same. And the corresponding alternate hypothesis is

$$H_A: \text{The } k \text{ population means are not all equal.}$$

Recall that H_A does not say that all means are different, only that at least one is different from the others.

Thus, for example, consider the experiment to ask whether the mean body weight of pigs is the same when the animals are raised on four different feeds. And

$$H_0: \quad \mu_1 = \mu_2 = \mu_3 = \mu_4$$
$$H_A: \quad \text{All four population means are not equal.}$$

In a MANOVA with two variables (X_1 and X_2) and k groups, the null hypothesis may be stated as

$$H_0: \quad \mu_{11} = \mu_{12} = \cdots = \mu_{1k} \quad and \quad \mu_{21} = \mu_{22} = \cdots = \mu_{2k},$$

Multivariate Analysis of Variance

where μ_{ij} denotes the population mean of variable i in group j. This H_0 says that the means of variable 1 are the same for all k groups *and* the means of variable 2 are the same for all k groups. The corresponding MANOVA alternate hypothesis is

H_A: The k populations do not have the same group means for variable 1 *and* the same group means for variable 2.

Thus, H_0 is rejected if any of the μ_{1j}'s are concluded to differ from each other *or* if any of the μ_{2j}'s are concluded to differ from each other.

Example 1 is similar to univariate analysis, except it comprises two variables (X_1 and X_2): the weight of each animal's body fat and the dry weight of each animal without its body fat. The null hypothesis is that mean weight of body fat is the same on all four diets *and* the mean fat-free dry body weight is the same on all of these diets. (It is *not* being hypothesized that mean body-fat weight is the same as mean fat-free dry body weight!) In this example,

H_0: $\mu_{11} = \mu_{12} = \mu_{13} = \mu_{14}$ *and* $\mu_{21} = \mu_{22} = \mu_{23} = \mu_{24}$

and

H_A: The four feeds do not result in the same mean weight of body fat and the same mean fat-free dry body weight.

If, in the sampled populations, one or more of the six equals signs in H_0 is untrue, then H_0 should be rejected.

There are several methods for comparing means to test MANOVA hypotheses. This chapter will refer to four test statistics employed for this purpose and encountered in MANOVA computer programs. None of these four is considered "best" in all situations. Each captures different characteristics of the differences among means; thus, the four have somewhat different abilities to detect differences in various circumstances. The computations of these test statistics are far from simple and—especially for more than two variables—cannot readily be expressed in algebraic equations. (They are represented with much less effort as matrix calculations, which are beyond the scope of this text.) Therefore, we shall depend upon computer programs to calculate these statistics and shall not attempt to demonstrate the numerical manipulations. It can be noted, however, that the necessary calculations involve total, groups, and error sums of squares (SS) and sums of cross products. And, just as mean squares are derived from sums of squares, quantities known as "covariances" are derived from sums of crossproducts.

The four common MANOVA test statistics are the following.* They are all given by most MANOVA computer software; they often result in the same or very similar conclusions regarding H_0 and operate with similar power (especially with large samples); and they yield identical results when only one variable is being analyzed (a univariate ANOVA) or when $k = 2$.

- *Wilks' lambda.* Wilks' Λ (capital Greek lambda), also called Wilks' likelihood ratio (or Wilks' U),[†] is the oldest and most commonly encountered multivariate analysis-of-variance statistic, dating from the original formulation of the

---

*Each of the four MANOVA statistics is a function of what are called *eigenvalues*, or *roots*, of matrices. Matrix algebra is explained in many tests on multivariate statistics.

[†]Named for American statistician Samuel Stanley Wilks (1906–1964), who made numerous contributions to theoretical and applied statistics, including to MANOVA (David and Morrison, 2006).

EXAMPLE 1   A Bivariate Analysis of Variance

Several members of a species of sparrow were collected at the same location at four different times of the year. Two variables were measured for each bird: the fat content (in grams) and the fat-free dry weight (in grams). For the statement of the null hypothesis, μ_{ij} denotes the population mean for variable i (where i is 1 or 2) and month j (i.e., j is 1, 2, 3, or 4).

H_0: $\mu_{11} = \mu_{12} = \mu_{13} = \mu_{14}$ and $\mu_{21} = \mu_{22} = \mu_{23} = \mu_{24}$

H_A: Sparrows do not have the same weight of fat *and* the same weight of fat-free dry body tissue at these four times of the year.

$\alpha = 0.05$

	December		January		February		March	
	Fat weight	*Lean dry weight*	*Fat weight*	*Lean dry weight*	*Fat weight*	*Lean dry weight*	*Fat weight*	*Lean dry weight*
	2.41	4.57	4.35	5.30	3.98	5.05	1.98	4.19
	2.52	4.11	4.41	5.61	3.48	5.09	2.05	3.81
	2.61	4.79	4.38	5.83	3.36	4.95	2.17	4.33
	2.42	4.35	4.51	5.75	3.52	4.90	2.00	3.70
	2.51	4.36			3.41	5.38	2.02	4.06
$\bar{X}$:	2.49	4.44	4.41	5.62	3.55	5.07	2.04	4.02

Computer computation yields the following output:

Wilks' $\Lambda = 0.0178$, $F = 30.3$,   DF $= 6, 28$, $P \ll 0.0001$.

  Reject H_0.

Pillai's trace $= 1.0223$, $F = 5.23$,   DF $= 6, 30$, $P = 0.0009$.

  Reject H_0.

Lawley-Hotelling trace $= 52.9847$, $F = 115$,   DF $= 6, 26$, $P \ll 0.0001$.

  Reject H_0.

Roy's maximum root $= 52.9421$, $F = 265$,   DF $= 3, 15$, $P \ll 0.0001$.

  Reject H_0.

MANOVA procedure (Wilks, 1932). Wilks' Λ is a quantity ranging from 0 to 1; that is, a measure of the amount of variability among the data that *is not* explained by the effect of the levels of the factor.* So, unlike typical test statistics

---

*Thus, a measure of the proportion of the variability that *is* explained by the levels of the factor is

$$\eta^2 = 1 - \Lambda, \tag{1}$$

and η^2 has a meaning like that of R^2 in another kind of multivariate analysis, that of multiple regression or multiple correlation.

(e.g., F or t or χ^2), H_0 is rejected for *small*, instead of large, values of Λ. Tables of critical values have been published (e.g., Rencher, 2002: 161, 566–573), but computer programs may present Λ transformed into a value of F or χ^2 (tables of which are far more available) with the associated probability (P); as elsewhere, large values of F or χ^2 yield small P's. In Example 1, Λ is an expression of the amount of variability among fat weights and among lean dry body weights that is *not* accounted for by the effect of the four times of year.

- *Pillai's trace.* This statistic, based on Pillai (1955), is also called the Pillai-Bartlett trace (or V).* Many authors recommend this statistic as the best test for general use (see Section 2a). Large values of V result in rejection of H_0. There are tables of critical values of this statistic (e.g., Rencher, 2002: 166–167, 578–581), but it is often transformed to a value of F.

- *Lawley-Hotelling trace.* This statistic (sometimes symbolized by U) was developed by Lawley (1938) and modified by Hotelling (1951). Tables of critical values exist for this (e.g., Rencher, 2002: 167, 582–686), but it is commonly transformed into values of F.

- *Roy's maximum root.* This is also known by similar names, such as Roy's largest, or greatest, root and sometimes denoted by θ (lowercase Greek theta). This statistic (Roy, 1945) may be compared to critical values (e.g., Rencher, 2002: 165, 574–577), or it may be converted to an F.

Wilks' Λ is a very widely used test statistic for MANOVA, but Olson (1974, 1976, 1979) concluded that Pillai's trace is usually more powerful (see Section 2b), and others have found that the Pillai statistic appears to be the most robust and most desirable for general use. If the four MANOVA test statistics do not result in the same conclusion about H_0, further scrutiny of the data may include examination of scatter plots, with axes as in Figures 5 and 6, for correlation among variables. Correlation will suggest favoring the conclusion reached by Pillai's trace, and noncorrelation will suggest relying on Roy's statistic for superior power. In Example 1, all four test statistics conclude that H_0 is to be rejected.

(a) Assumptions. As in univariate ANOVA, the underlying mathematical foundations of MANOVA depend upon certain assumptions. Although it is unlikely that all of the assumptions will be exactly met for a given set of data, it is important to be cognizant of them and of whether the statistical procedures employed are robust to departures from them.

A very important underlying assumption in MANOVA is that the data represent random samples from the populations of interest and that the observations on each subject are independent. In Example 1, the body-fat weights of sparrows in each month are assumed to have come at random from the population of body-fat weights of all sparrows from that month from that location, and the lean dry weights at each month are assumed to have come randomly from a population of such weights. Also, the body-fat weight of each subject (i.e., each sparrow) must be independent of the body-fat weight of each other subject, and the lean dry body weights of all the subjects must be independent of each other. MANOVA is invalidated by departure from the assumption of random and independent data.

---

*A *trace* is the result of a specific mathematical operation on a matrix.

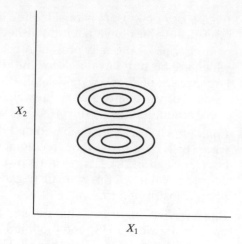

FIGURE 6: Two bivariate normal groups of positively correlated data differing in one dimension.

There is also an assumption of multivariate normality (see Section 1) for each group of data. However, departures from normality may have only a slight affect on the Type I error rate,* especially if the sample sizes are large.

MANOVA appears to be resistant to nonnormality due to skewness, but platykurtosis can severely reduce power. There are procedures for assessing multivariate normality (e.g., Johnson and Wichern, 2002: Section 4.6; Sharma, 1996: 380–382; Srivastava, 2002: 71–88; Stevens, 2002: 263), although they are seldom applied. And there are nonparametric tests (e.g., multivariate analogs of the Kruskal-Wallis and Friedman tests) that may be used in some cases where the underlying distributions are far from multivariate normal.

Just as ANOVA assumes that there is equality of variances over all k groups, MANOVA assumes equality of variances across all k groups. Furthermore, MANOVA requires that the correlation between any two variables must be the same in all k groups; this is known as the assumption of equal *covariance*. If the group sample sizes are equal, or nearly equal (defined by Stevens, 2002: 328, as the largest n no greater than 1.5 times the smallest n; see also Hair et al., 2006: 409; Rencher, 2002: 177), then departures from this assumption will have little effect on the Type I error using the Pillai trace, somewhat more effect using the Wilks or Lawley-Hotelling statistics, and considerable adverse effect using Roy's criterion. However, with large sample sizes, the first three test statistics are nearly equivalent in performance (Olson, 1974, 1976, 1979).[†] With very disparate sample sizes, the adverse effect may be serious using any of the four statistics, causing the hypothesis testing to reject H_0 either more often or less often than it should, depending upon the magnitudes of their variances and correlations and their relationships to the sample sizes. If the sample sizes are not equal and there is variance and covariance heterogeneity, then, as in ANOVA, the testing is conservative if the larger n's are associated with the larger variances and covariances, and liberal if the larger n's are associated with the smaller variances and covariances.

---

*The *central limit theorem* also applies to multivariate distributions.

[†]Olson declares that the three may be considered equivalent if the denominator degrees of freedom of F are at least $10m$ times the numerator DF, where m is the number of variables.

There are statistical tests for this assumption, such as that which is analogous to the Bartlett test of the homogeneity of variances in the univariate ANOVA (Box, 1949, 1950) but which is seriously affected by nonnormality (e.g., Olson, 1974; Stevens, 2002: 271) and thus is not generally recommended. Data transformations may be useful to reduce nonnormality or to reduce heterogeneity of variances and covariances. Although MANOVA is typically robust to departures from the variability and variance-correlations assumptions, the Pillai trace (according to Olson, 1974, 1976) is generally the most robust of the four methods.

(b) Power. The power of a MANOVA depends upon a complex set of characteristics, including the extent to which the underlying assumptions (see Section 2a) are met. In general, increased sample size is associated with increased power, but power decreases with increase in the number of variables. Thus, if it is desired to employ an experimental design with several variables, larger samples will be needed than would be the case if there were only two variables. Also, as with ANOVA, the power of MANOVA is greater when the population differences between means are larger and when the variability within groups is small. The magnitude of correlations between variables can cause MANOVA to be either more or less powerful than separate ANOVAs. Some MANOVA computer programs calculate power, and Rencher (2002: Section 4.4) and Stevens (2002: 192–202) discuss the estimation of power and of the sample size required in MANOVA. Many computer programs provide a calculation of power, but recall that power estimated from a set of data should be considered as applying to future data sets.

Differences in power among the four test statistics are often not great, but there can be differences. If the group means differ in only one direction (i.e., they are uncorrelated, as in Figure 6), a relatively uncommon situation, Roy's statistic is the most powerful of the four, followed—in order of power—by the Lawley-Hotelling trace, Wilks' Λ, and Pillai's trace. However, in the more common situation where the group means differ among more than one dimension (i.e., the variables are correlated, as in Figure 5), then the relative powers of these statistics are in the reverse order: The Pillai trace is the most powerful, followed by Wilks' Λ, the Lawley-Hotelling trace, and then Roy's statistic. In intermediate situations, the four statistics tend more toward the latter ordering than the former.

(c) Two-Sample Hypotheses. The preceding four procedures may be used in the case of only two groups, for when $k = 2$ all four will yield the same results. But another test encountered for two-group multivariate analysis is *Hotelling's T^2* (Hotelling, 1931). This is analogous to the univariate two-group situation, where either ANOVA or Student's t may be employed.

T^2 is related to the MANOVA statistics of Section 2 as follows (Rencher, 2002: 130)*:

$$T^2 = (n_1 + n_2 - 2)\left(\frac{1 - \Lambda}{\Lambda}\right), \tag{3}$$

$$T^2 = (n_1 + n_2 - 2)\left(\frac{V}{1 - V}\right), \tag{4}$$

---

*T^2 is also related to the multiple-regression coefficient of determination, R^2, as

$$T^2 = (n_1 + n_2 - 2)\left(\frac{R^2}{1 - R^2}\right). \tag{2}$$

$$T^2 = (n_1 + n_2 - 2)\, U, \tag{5}$$

$$T^2 = (n_1 + n_2 - 2)\left(\frac{\theta}{1 - \theta}\right), \tag{6}$$

where m is the number of variables.

Upon calculation of T^2 (usually by computer), tables of critical values of T^2 (e.g., Rencher, 2002: 558–561) may be consulted, or

$$F = \frac{n_1 + n_2 - m - 1}{(n_1 + n_2 - 2)m}\, T^2 \tag{7}$$

may be used, with m and $n_1 + n_2 - m - 1$ degrees of freedom or, equivalently,

$$F = \frac{n_1 + n_2 - m}{(n_1 + n_2 - 1)m}\, T^2 \tag{7a}$$

with m and $n_1 + n_2$ degrees of freedom.

Nonparametric testing is also available for two-sample multivariate tests (i.e., as analogs to the univariate Mann-Whitney test) and for paired-sample tests (analogous to the univariate Wilcoxon or sign tests).

3 FURTHER ANALYSIS

When a MANOVA rejects H_0, there are procedures that might be employed to expand the analysis of difference among groups (e.g., Bray and Maxwell, 1985: 40–45; Hand and Taylor, 1987: Chapter 5; Hair et al., 2006: 422–426); Hummel and Sligo, 1971; Stevens, 2002: 217–225; Weinfurt, 1995). One approach is to perform a univariate ANOVA on each of the variables (Rencher, 2002: 162–164, followed perhaps by multiple comparisons) to test the difference among means for each variable separately. However, this procedure will ignore relationships among the variables, and other criticisms have been raised (e.g., Weinfurt, 1995). Multiple-comparison tests are described in some of the references cited in this discussion.

In univariate ANOVA, one can reject H_0 and have none of the μ's declared different by further analysis. Similarly, a MANOVA may reject H_0 with subsequent ANOVAs detecting no differences (either because of lack of power or because the interrelations among variables are important in rejection of the multivariate H_0). Some computer programs perform ANOVAs along with a MANOVA.

If $k = 2$ and H_0 is rejected by MANOVA or Hotelling's T^2 test, then two-sample t tests and univariate ANOVAs will yield identical results.

4 OTHER EXPERIMENTAL DESIGNS

The data of Example 1 are subjected to a multivariate analysis of variance composed of one factor (time of year) and two variables (weight of fat and fat-free dry body weight). This is the same experimental design as the ANOVA except that two variables, instead of one, are measured on each animal. A MANOVA may also involve more than two variables. For example, the blood-cholesterol concentration might have been a third variable measured for each of the animals, and the null hypothesis regarding the four factor levels (months) would be

$$H_0: \quad \mu_{11} = \mu_{12} = \mu_{13} = \mu_{14} \quad \text{and} \quad \mu_{21} = \mu_{22} = \mu_{23} = \mu_{24}$$

$$\text{and} \quad \mu_{31} = \mu_{32} = \mu_{33} = \mu_{34}.$$

Other sets of data may be the result of consideration of more than one variable *and* more than one factor. For example, measurements of two or more variables, such as those in Example 1, might have been collected for sparrows collected at more than one time of year *and* at more than one geographic location. This would be a multivariate factorial experimental design (which could include an examination of factor interactions), and multivariate versions of repeated-measures and hierarchical analyses of variance are also possible, as is multivariate analysis of covariance (MANCOVA).

Multivariate one-sample testing and paired-sample testing are possible (e.g., Rencher, 2002: Sections 5.3.2 and 5.7.1).

Analysis of covariance (ANCOVA), also may be extended to experimental designs with multiple dependent variables. This is done via MANCOVA, for which computer routines are available.

EXERCISES

1. Using multivariate analysis of variance, analyze the following data for the concentration of three amino acids in centipede hemolymph (mg/100 ml), asking whether the mean concentration of these amino acids is the same in males and females:

Male			Female		
Alanine	Aspartic Acid	Tyrosine	Alanine	Aspartic Acid	Tyrosine
7.0	17.0	19.7	7.3	17.4	22.5
7.3	17.2	20.3	7.7	19.8	24.9
8.0	19.3	22.6	8.2	20.2	26.1
8.1	19.8	23.7	8.3	22.6	27.5
7.9	18.4	22.0	6.4	23.4	28.1
6.4	15.1	18.1	7.1	21.3	25.8
6.6	15.9	18.7	6.4	22.1	26.9
8.0	18.2	21.5	8.6	18.8	25.5

2. The following data for deer are for two factors (species and sex), where for each combination of factors there is a measurement of two variables (rate of oxygen consumption, in ml O_2/g/hr, and rate of evaporative water loss, in mg/min). Perform a multivariate analysis of variance to test for equality of the population means of these two variables for each of the two factors and the factor interaction.

Species 1				Species 2			
Female		Male		Female		Male	
0.165	76	0.145	80	0.391	71	0.320	65
0.184	71	0.110	72	0.262	70	0.238	69
0.127	64	0.108	77	0.213	63	0.288	67
0.140	66	0.143	69	0.358	59	0.250	56
0.128	69	0.100	74	0.402	60	0.293	52

ANSWERS TO EXERCISES

1. H_0: $\mu_{11} = \mu_{12} = \mu_{13}$ and $\mu_{21} = \mu_{22} = \mu_{23}$; Wilks' $\Lambda = 0.0872$, Pillai's trace $= 0.9128$, Lawley-Hotelling trace $= 10.4691$, Roy's maximum root $= 10.4681$; for each, $F = 41.8723$, $P \ll 0.0001$; reject H_0.

2. For species, H_0: $\mu_{11} = \mu_{12}$ and $\mu_{21} = \mu_{22}$; Wilks' $\Lambda = 0.1820$, Pillai's trace $= 0.8180$, Lawley-Hotelling trace $= 4.4956$, Roy's maximum root $= 4.4956$; for each, $F = 33.7167$, $P \ll 0.0001$; reject H_0. For sex, H_0: $\mu_{11} = \mu_{12}$ and $\mu_{21} = \mu_{22}$; Wilks' $\Lambda = 0.8295$, Pillai's trace $= 0.1705$, Lawley-Hotelling trace $= 0.2055$, Roy's maximum root $= 0.2055$; for each, $F = 1.5415$, $P = 0.2461$; do not reject H_0. For species $\times$ sex interaction, H_0: There is no interaction; Wilks' $\Lambda = 0.8527$, Pillai's trace $= 0.1473$, Lawley-Hotelling trace $= 0.1727$, Roy's maximum root $= 0.1727$; for each, $F = 1.2954$, $P = 0.3027$; do not reject H_0.

This page intentionally left blank

Simple Linear Regression

From Chapter 17 of *Biostatistical Analysis*, Fifth Edition, Jerrold H. Zar. Copyright © 2010 by Pearson Education, Inc. Publishing as Pearson Prentice Hall. All rights reserved.

Simple Linear Regression

Techniques that consider relationships between two variables are described in this chapter.

1   REGRESSION VERSUS CORRELATION*

The relationship between two variables may be one of functional dependence of one on the other. That is, the magnitude of one of the variables (the *dependent variable*) is assumed to be determined by—that is, is a function of—the magnitude of the second variable (the *independent variable*), whereas the reverse is not true. For example, in the relationship between blood pressure and age in humans, blood pressure may be considered the dependent variable and age the independent variable; we may reasonably assume that although the magnitude of a person's blood pressure might be a function of age, age is not determined by blood pressure. This is not to say that age is the only biological determinant of blood pressure, but we do consider it to be one determining factor.[†] The term *dependent* does not necessarily imply a cause-and-effect relationship between the two variables. (See Section 4.)

Such a dependence relationship is called a *regression*. The term *simple regression* refers to the simplest kind of regression, one in which only two variables are considered.[‡]

---

[*]The historical developments of regression and correlation are strongly related, owing their discovery—the latter following the former—to Sir Francis Galton, who first developed these procedures during 1875–1885 (Walker, 1929: 103–104, 187); see also the first footnote in Section 19.1. He first used the term *regression* in 1885 (Desmond, 2000).

[†]Some authors refer to the independent variable as the predictor, regressor, explanatory, or exogenous variable and the dependent variable as the response, criterion, or endogenous variable.

[‡]In the case of simple regression, the adjective *linear* may be used to refer to the relationship between the two variables being a straight line, but to a statistician it describes the relationship of the parameters discussed in Section 2.

Simple Linear Regression

Data amenable to simple regression analysis consist of pairs of data measured on a ratio or interval scale. These data are composed of measurements of a dependent variable (Y) that is a random effect and an independent variable (X) that is either a fixed effect or a random effect.*

It is convenient and informative to graph simple regression data using the ordinate $(Y$ axis) for the dependent variable and the abscissa $(X$ axis) for the independent variable. Such a graph is shown in Figure 1 for the $n = 13$ data of Example 1, where the data appear as a scatter of 13 points, each point representing a pair of X and Y values.[†] One pair of X and Y data may be designated as (X_1, Y_1), another as (X_2, Y_2), another as (X_3, Y_3), and so on, resulting in what is called a *scatter plot* of all n of the (X_i, Y_i) data. (The line passing through the data in this figure will be explained in Section 2.)

EXAMPLE 1   Wing Lengths of 13 Sparrows of Various Ages. The Data Are Plotted in Figure 1.

Age (days) (X)	*Wing length* (cm) (Y)
3.0	1.4
4.0	1.5
5.0	2.2
6.0	2.4
8.0	3.1
9.0	3.2
10.0	3.2
11.0	3.9
12.0	4.1
14.0	4.7
15.0	4.5
16.0	5.2
17.0	5.0

$n = 13$

---

*On rare occasions, we want to describe a regression relationship where the dependent variable (Y) is recorded on a nominal scale. This requires *logistic regression*, a procedure discussed.

[†]Royston (1956) observed that "the basic idea of using co-ordinates to determine the location of a point in space dates back to the Greeks at least, although it was not until the time of Descartes that mathematicians systematically developed the idea." The familiar system of specifying the location of a point by its distance from each of two perpendicular axes (now commonly called the X and Y axes) is referred to as Cartesian coordinates, after the French mathematician and philosopher René Descartes (1596–1650), who wrote under the Latinized version of his name, Renatus Cartesius. His other enduring mathematical introductions included (in 1637) the use of numerals as exponents, the square root sign with a vinculum (i.e., with a horizontal line: $\sqrt{}$), and the use of letters at the end of the alphabet (e.g., X, Y, Z) to denote variables and those near the beginning (e.g., a, b, c) to represent constants (Asimov, 1982: 117; Cajori, 1928: 205, 208, 375).

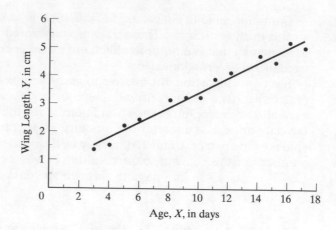

FIGURE 1: Sparrow wing length as a function of age. The data are from Example 1.

In many kinds of biological data, however, the relationship between two variables is not one of dependence. In such cases, the magnitude of one of the variables changes as the magnitude of the second variable changes, but it is not reasonable to consider there to be an independent and a dependent variable. In such situations, *correlation*, rather than regression, analyses are called for, and both variables are theoretically to be random-effects factors. An example of data suitable for correlation analysis would be measurements of human arm and leg lengths. It might be found that an individual with long arms will in general possess long legs, so a relationship may be describable; but there is no justification in stating that the length of one limb is dependent upon the length of the other.

2 THE SIMPLE LINEAR REGRESSION EQUATION

The simplest functional relationship of one variable to another in a population is the *simple linear regression*

$$Y_i = \alpha + \beta X_i. \tag{1}$$

Here, α and β are population parameters (and, therefore, constants), and this expression will be recognized as the general equation for a straight line.* However, in a population the data are unlikely to be exactly on a straight line, so Y may be said to be related to X by

$$Y_i = \alpha + \beta X_i + \epsilon_i, \tag{1a}$$

where ϵ_i (lowercase Greek epsilon) is referred to as an "error," or "residual," which is a departure of an actual Y_i from what Equation 1 predicts Y_i to be; and the sum of the ϵ_i's is zero.

---

*α and β are commonly used for these population parameters, and as such should not be confused with the standard use of the same Greek letters to denote the probabilities of a Type I and Type II error, respectively. Sometimes α and β in regression are designated as β_0 and β_1, respectively. The additive (linear) relationship of the two parameters in Equation 1 leads to the term *linear regression equation*.

Consider the data in Example 1, where wing length is the dependent variable and age is the independent variable. From a scatter plot of these data (Figure 1), it appears that our sample of measurements from 13 birds represents a population of data in which wing length is linearly related to age. Thus, we would like to estimate the values of α and β that would uniquely describe the functional relationship existing in the population.

If all the data in a scatter diagram such as Figure 1 occurred in a straight line, it would be an unusual situation. Generally, as is shown in this figure, there is considerable variability of data around any straight line we might draw through them. What we seek to define is what is commonly termed the "best-fit" line through the data. The criterion for "best fit" that is generally employed utilizes the concept of *least squares*.* Figure 2 is an enlarged portion of Figure 1. Each value of X will have a corresponding value of Y lying on the line that we might draw through the scatter of data points. This value of Y is represented as $\hat{Y}$ to distinguish it from the Y value actually observed in our sample.[†] Thus, as Figure 2 illustrates, an observed data point is denoted as (X_i, Y_i), and a point on the regression line is $(X_i, \hat{Y}_i)$.

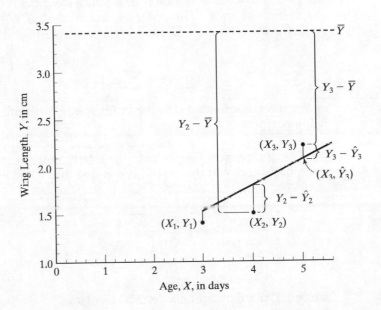

FIGURE 2: An enlarged portion of Figure 1, showing the partitioning of Y deviations.

The criterion of least squares considers the vertical deviation of each point from the line (i.e., the deviation describable as $Y_i - \hat{Y}_i$) and defines the best-fit line as that which results in the smallest value for the sum of the squares of these deviations for all values of Y_i and $\hat{Y}_i$. That is, $\sum_{i=1}^{n}(Y_i - \hat{Y}_i)^2$ is to be a minimum, where n is the number of data points composing the sample. The sum of squares of these deviations

---

*The French mathematician Adrien Marie Legendre (1752–1833) published the method now known as least squares (also called ordinary least squares or OLS) in 1805, but the brilliant German mathematician and physicist Karl Friedrich Gauss (1777–1855) claimed—probably truthfully—that he had used it at least 10 years prior to that. (See also Eisenhart, 1978; Seal, 1967.) David (1995) asserts that the term least squares (published in French as moindres quarrés) is properly attributed to Legendre's 1805 publication.

[†] Statisticians refer to $\hat{Y}$ as "Y hat."

is called the *residual sum of squares* (or, sometimes, the *error sum of squares*) and will be discussed in Section 3.*

The only way to determine the population parameters α and β would be to possess all the data for the entire population. Since this is nearly always impossible, we have to estimate these parameters from a sample of n data, where n is the number of pairs of X and Y values. The calculations required to arrive at such estimates, as well as to execute the testing of a variety of important hypotheses, involve the computation of sums of squared deviations from the mean. Recall that the "sum of squares" of X_i values is defined as $\sum(X_i - \overline{X})^2$, which is more easily obtained on a calculator as $\sum X_i^2 - (\sum X_i)^2/n$. It will be convenient to define $x_i = X_i - \overline{X}$, so that this sum of squares can be abbreviated as $\sum x_i^2$, or, more simply, as $\sum x^2$.

Another quantity needed for regression analysis is referred to as the *sum of the cross products* of deviations from the mean:

$$\sum xy = \sum (X_i - \overline{X})(Y_i - \overline{Y}), \tag{2}$$

where y denotes a deviation of a Y value from the mean of all Y's just as x denotes a deviation of an X value from the mean of all X's. The sum of the cross products, analogously to the sum of squares, has a simple-to-use "machine formula":

$$\sum xy = \sum X_i Y_i - \frac{\left(\sum X_i\right)\left(\sum Y_i\right)}{n}, \tag{3}$$

and it is recommended that the latter formula be employed if the calculation is not being performed by computer.

(a) The Regression Coefficient. The parameter β is termed the *regression coefficient*, or the *slope* of the best-fit regression line. The best sample estimate of β is

$$b = \frac{\sum xy}{\sum x^2} = \frac{\sum (X_i - \overline{X})(Y_i - \overline{Y})}{\sum (X_i - \overline{X})^2} = \frac{\sum X_i Y_i - \dfrac{\left(\sum X_i\right)\left(\sum Y_i\right)}{n}}{\sum X_i^2 - \dfrac{\left(\sum X_i\right)^2}{n}}. \tag{4}$$

Although the denominator in this calculation is always positive, the numerator may be positive, negative, or zero, and the value of b theoretically can range from $-\infty$ to $+\infty$, including zero (see Figure 3).

---

*Another method of regression was proposed in 1757 by Roger Joseph Boscovich (1711–1787, born in what is now Croatia and known also by the Italian name Ruggiero Giuseppe Boscovich). This defined the "best-fit" line as that which minimizes the sum of the absolute values of deviations (that is, $\sum_{n=1}^{n} |Y_i - \hat{Y}_i|$) instead of the sum of squared deviations (Heyde and Seneta, 2001: 82–85). This is referred to as *least absolute deviations* (or LAD). It is rarely seen and employs different (and computationally more difficult) statistical procedures than least-squares regression but may be preferable if there are major outliers or substantial departures from some least-squares assumptions (Section 2). A regression method differing from that of least-squares regression and least-absolute-deviations regression is M-*regression* (employing what statisticians call "maximum-likelihood estimation"), described by Birkes and Dodge (1993: Chapter 5), Draper and Smith (1998: Section 25.2), and Huber (2004: Section 7.8) and based upon the concept of Huber (1964). There also exist nonparametric regression methods (Birkes and Dodge, 1993: Chapter 6). These procedures are more robust than least-squares regression and may be preferable when there are prominent outliers or other serious departures from least-squares assumptions.

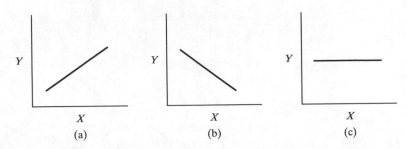

FIGURE 3: The slope of a linear regression line may be (a) positive, (b) negative, or (c) zero.

Example 2 demonstrates the calculation of b for the data of Example 1. Note that the units of b are the units of Y divided by the units of X. The regression coefficient expresses what change in Y is associated, on the average, with a unit change in X. In the present example, $b = 0.270$ cm/day indicates that, for this sample, there is a mean wing growth of 0.270 cm per day for ages 3.0 to 17.0 days. Section 5 discusses how to express the precision of b.

EXAMPLE 2   The Simple Linear Regression Equation Calculated (Using the "Machine Formula") by the Method of Least Squares, for the Data from the 13 Birds of Example 1

$n = 13$

$$\sum X = 3.0 + 4.0 + \cdots + 17.0 \qquad\qquad \sum Y = 1.4 + 1.5 + \cdots + 5.0$$
$$= 130.0 \qquad\qquad\qquad\qquad\qquad = 44.4$$

$$\overline{X} = 130.0/13 = 10.0 \qquad\qquad \overline{Y} = 44.4/13 = 3.415$$

$$\sum X^2 = 3.0^2 + \cdots + 17.0^2 \qquad \sum XY = (3.0)(1.4) + \cdots$$
$$= 1562.00 \qquad\qquad\qquad\qquad + (17.0)(5.0) = 514.80$$

$$\sum x^2 = 1562.00 - \frac{(130.0)^2}{13} \qquad \sum xy = 514.80 - \frac{(130.0)(44.4)}{13}$$
$$= 1562.00 - 1300.00 = 262.00 \qquad = 514.80 - 444.00 = 70.80$$

$$b = \frac{\sum xy}{\sum x^2} = \frac{70.80}{262.00} = 0.270 \text{ cm/day}$$

$$a = \overline{Y} - b\overline{X} = 3.415 \text{ cm} - (0.270 \text{ cm/day})(10.0 \text{ days})$$

$$= 3.415 \text{ cm} - 2.700 \text{ cm} = 0.715 \text{ cm}$$

So the simple linear regression equation is $\hat{Y} = 0.715 + 0.270X$.

(b) The Y Intercept. An infinite number of lines possess any stated slope, all of them parallel (see Figure 4). However, each such line can be defined uniquely by stating, in addition to β, any one point on the line—that is, any pair of coordinates, $(X_i, \hat{Y}_i)$. The point conventionally chosen is the point on the line where $X = 0$. The

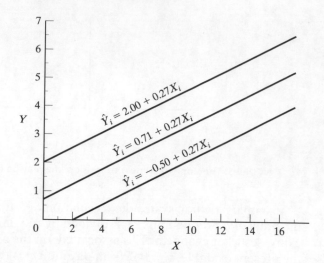

FIGURE 4: For any given slope, there exists an infinite number of possible regression lines, each with a different Y intercept. Three of this infinite number are shown here.

value of Y in the population at this point is the parameter α, which is called the Y *intercept*.

It can be shown mathematically that the point $(\overline{X}, \overline{Y})$ always lies on the best-fit regression line. Thus, substituting $\overline{X}$ and $\overline{Y}$ in Equation (1), we find that

$$\overline{Y} = \alpha + \beta\overline{X} \tag{5}$$

and

$$\alpha = \overline{Y} - \beta\overline{X}. \tag{6}$$

The best estimate of the population parameter α is the sample statistic

$$a = \overline{Y} - b\overline{X}. \tag{7}$$

The calculation of a is shown in Example 2. Note that the Y intercept has the same units as any other Y value. (The precision of the statistic a is considered in Section 5.) The sample regression equation (which estimates the population relationship between Y and X stated in Equation 1) may be written as

$$\hat{Y}_i = a + bX_i, \tag{8}$$

although some authors write

$$\hat{Y}_i = \overline{Y} + b(X_i - \overline{X}), \tag{9}$$

which is equivalent.

Figures 4 and 5 demonstrate that the knowledge of either a or b allows only an incomplete description of a regression function. But by specifying both a and b, a line is uniquely defined. Also, because a and b were calculated using the criterion of least squares, the residual sum of squares from this line is smaller than the residual sum of squares that would result from any other line (i.e., a line with any other a or b) that could be drawn through the data points. This regression line (i.e., the line with this a and b) is not the same line that would result if Y were the independent variable and X the dependent variable.

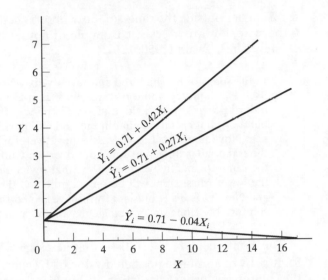

FIGURE 5: For any given Y intercept, there exist an infinite number of possible regression lines, each with a different slope. Three of this infinite number are shown here.

(c) Predicting Values of Y. Knowing the parameter estimates a and b for the linear regression equation, we can predict the value of the dependent variable expected in the population at a stated value of X_i. For the regression in Example 2, the wing length of a sparrow at 13.0 days of age would be predicted to be

$$\hat{Y} = a + bX_i$$
$$= 0.715 \text{ cm} + (0.270 \text{ cm/day})(13.0 \text{ day}) = 4.225 \text{ cm}.$$

The wing length in the population at 7.0 days of age would be estimated to be $\hat{Y} = 0.715 \text{ cm} + (0.270 \text{ cm/day})(7.0 \text{ day}) = 2.605 \text{ cm}$, and so on.

To plot a linear regression line graphically, we need to know only two points that lie on the line. We already know two points, namely $(\overline{X}, \overline{Y})$ and $(0, a)$; however, for ease and accuracy in drawing the line by hand, two points that lie near extreme ends of the observed range of X are most useful. For drawing the line in Figure 1, the values of $\hat{Y}_i$ for $X_i = 3.0$ days and $X_i = 17.0$ days were used. These were found to be $\hat{Y} = 1.525$ and 5.305 cm, respectively. A regression line should always be drawn using predicted points, and never drawn "by eye."

A word of caution is in order concerning predicting $\hat{Y}_i$ values from a regression equation. Generally, it is an unsafe procedure to extrapolate from a regression equation—that is, to predict $\hat{Y}_i$ values for X_i values outside the observed range of X_i. It would, for example, be unjustifiable to attempt to predict the wing length of a 20-day-old sparrow, or a 1-day-old sparrow, using the regression calculated for birds ranging from 3.0 to 17.0 days in age. Indeed, applying the equation of Example 2 to a one-year-old sparrow would predict a wing nearly one meter long! What the linear regression describes is Y as a function of X *within the range of observed values of X.* Thus, a regression equation is often used to interpolate; that is, to estimate a value of Y for an X lying between X's in the sample. But for values of X above or below this range, the function may not be the same (i.e., α and/or β may be different); indeed, the relationship may not even be linear in such ranges, even though it is linear within the observed range. If there is good reason to believe that the described function holds for X values outside the range of those observed, then we may cautiously extrapolate. Otherwise, beware. A classic example of nonsensical extrapolation was provided in 1874 by Mark Twain (1950: 156):

> In the space of one hundred and seventy-six years the Lower Mississippi has shortened itself two hundred and forty-two miles. That is an average of a trifle over one mile and a third per year [i.e., a slope of -1.375 mi/yr]. Therefore, any calm person, who is not blind or idiotic, can see that in the Old Oölitic Silurian period, just a million years ago next November, the Lower Mississippi River was upward of one million three hundred thousand miles long, and stuck out over the Gulf of Mexico like a fishing rod. And by the same token any person can see that seven hundred and forty-two years from now, the lower Mississippi will be only a mile and three-quarters long, and Cairo [Illinois] and New Orleans [Louisiana] will have joined their streets together, and be plodding comfortably along under a single mayor and a mutual board of aldermen.*

The Y intercept, a, is a statistic that helps specify a regression equation (Equation 8). It is, by definition, a predicted value of Y (namely, the $\hat{Y}$ at $X = 0$), but because of the caution against extrapolation, it should not necessarily be considered to represent the magnitude of Y in the population at an X of zero if $X = 0$ is outside the range of X's in the sample. Thus, in Examples 1 and 2 it should not be proposed that a newly hatched bird in the population (i.e., one 0 days old) has a mean wing length of 0.715 cm.

Section 4 discusses the estimation of the error and confidence intervals associated with predicting $\hat{Y}_i$ values.

(d) Assumptions of Regression Analysis. Certain basic assumptions must be met to validly test hypotheses about regressions or to set confidence intervals for regression parameters, although these assumptions are not necessary to compute the regression coefficient, b, the Y intercept, a, and the coefficient of determination, r^2:

1. For each value of X, the values of Y are to have come at random from the sampled population and are to be independent of one another. That is, obtaining a particular Y from the population is in no way dependent upon the obtaining of any other Y.

2. For any value of X in the population there exists a normal distribution of Y values. (This also means that for each value of X there exists in the population a normal distribution of ϵ's.)

3. There is homogeneity of variances in the population; that is, the variances of the distributions of Y values must all be equal to each other. (Indeed, the residual mean square—to be described in Section 3—estimates the common variance assumed in the analysis of variance.)

4. In the population, the mean of the Y's at a given X lies on a straight line with the mean of all other Y's at all other X's. That is, the actual relationship between Y and X is linear.

5. The measurements of X were obtained without error. This, of course, is typically impossible; so what we do in practice is assume that the errors in measuring X are negligible, or at least small, compared with errors in measuring Y. If that

---

*Author Twain concludes by noting that "There is something fascinating about science. One gets such a wholesale return of conjecture out of a trifling investment of fact." It can also be noted that the Silurian period is now considered to have occurred well over 400 million years ago. Even at the time Mark Twain wrote this, scientific opinion placed it at least 20 million years ago.

assumption is not reasonable, other, more complex methods may be considered (Montgomery, Peck, and Vining, 2001: 502).

Violations of assumptions 2, 3, or 4 can sometimes be countered by transformation of data (presented in Section 10). Data in violation of assumption 3 will underestimate the residual mean square (Section 3) and result in an inflation of the test statistic (F or t), thus increasing the probability of a Type I error (Caudill, 1988). Heteroscedastic data may sometimes be analyzed advantageously by a procedure known as *weighted regression*, which will not be discussed here.

Regression statistics are known to be robust with respect to at least some of these underlying assumptions (e.g., Jacques and Norusis, 1973), so violations of them are not usually of concern unless they are severe. One kind of datum that causes violation of the assumption of normality and homogeneity of variance is the *outlier*, which in regression is a recorded measurement that lies very much apart from the trend in the bulk of the data. (For example, in Figure 1 a data point at $X = 4$ days and $Y = 4$ cm would have been an outlier.) Procedures known as nonparametric (or distribution-free) regression analyses make no assumptions about underlying statistical distributions. Several versions exist (including regression using ranks) and are discussed by several authors, including Birkes and Dodge (1993: Chapter 6); Cleveland, Mallows, and McRae (1993); Daniel (1990: Chapter 10); Härdle (1990); Hollander and Wolfe (1999: Chapter 9); Montgomery, Peck, and Vining, 2001: Section 7.3); Neave and Worthington (1988: Chapter 10); and Wang and Scott (1994).

(e) Two Kinds of Independent Variables. In regression, measurements of the dependent variable, Y, are considered to be data that have come at random from a population of such data. However, the independent variable, X, may be one of two types.

There are two kinds of factors in analysis of variance: A fixed-effect factor has its levels specifically selected by the experimenter, whereas the levels of a random-effect factor are obtained at random from all possible levels of the factor. Analogously, the values of X in regression may be fixed or random. In Example 1, X is a variable with fixed values if the X's were selected by the experimenter (i.e., 13 specific ages of birds were obtained at which to measure wing length). Alternatively, the values of X may have come at random from the sampled population (meaning that the ages were recorded for 13 birds that were selected at random).

Whether the independent variable is random or fixed has no effect on the calculations and hypothesis testing for regression analysis, so the distinction is seldom noted.

3 TESTING THE SIGNIFICANCE OF A REGRESSION

The slope, b, of the regression line computed from the sample data expresses quantitatively the straight-line dependence of Y on X in the sample. But what is really desired is information about the functional relationship (if any) in the population from which the sample came. Indeed, the finding of a dependence of Y on X in the sample (i.e., $b \neq 0$) does not necessarily mean that there is a dependence in the population (i.e., $\beta \neq 0$). Consider Figure 6, a scatter plot representing a population of data points with no dependence of Y on X; the best-fit regression line for this population would be parallel to the X axis (i.e., the slope, β, would be zero). However, it is possible, by random sampling, to obtain a sample of data points having

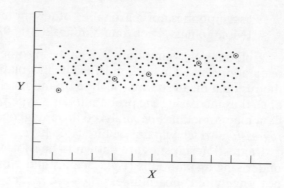

FIGURE 6: A hypothetical population of data points, having a regression coefficient, β, of zero. The circled points are a possible sample of five.

the five values circled in the figure. By calculating b for this sample of five, we would estimate that β was positive, even though it is, in fact, zero.

We are not likely to obtain five such points out of this population, but we desire to assess just how likely it is; therefore, we can set up a null hypothesis, $H_0: \beta = 0$, and the alternate hypothesis, $H_A: \beta \neq 0$, appropriate to that assessment. If we conclude that there is a reasonable probability (i.e., a probability greater than the chosen level of significance—say, 5%) that the calculated b could have come from sampling a population with a $\beta = 0$, the H_0 is not rejected. If the probability of obtaining the calculated b is small (say, 5% or less), then H_0 is rejected, and H_A is assumed to be true.

(a) Analysis-of-Variance Testing. The preceding H_0 may be tested by an analysis-of-variance (ANOVA) procedure. First, the overall variability of the dependent variable is calculated by computing the sum of squares of deviations of Y_i values from $\overline{Y}$, a quantity termed the *total sum of squares*:

$$\text{total SS} = \sum (Y_i - \overline{Y})^2 = \sum y^2 = \sum Y_i^2 - \frac{\left(\sum Y_i\right)^2}{n}. \tag{10}$$

Then we determine the amount of variability among the Y_i values that is attributable to there being a linear regression; this is termed the *linear regression sum of squares*:

$$\text{regression SS} = \sum (\hat{Y}_i - \overline{Y})^2 = \frac{\left(\sum xy\right)^2}{\sum x^2} = \frac{\left(\sum X_i Y_i - \frac{\sum X_i \sum Y_i}{n}\right)^2}{\sum X_i^2 - \frac{\left(\sum X_i\right)^2}{n}}; \tag{11}$$

because $b = \sum xy / \sum x^2$ (Equation 4), this can also be calculated as

$$\text{regression SS} = b \sum xy. \tag{12}$$

The value of the regression SS will be equal to that of the total SS only if each data point falls exactly on the regression line, a very unlikely situation. The scatter of data points around the regression line has been alluded to, and the residual, or error, sum of squares is obtained as

$$\text{residual SS} = \sum (Y_i - \hat{Y}_i)^2 = \text{total SS} - \text{regression SS}. \qquad (13)$$

Table 1 presents the analysis-of-variance summary for testing the hypothesis $H_0: \beta = 0$ against $H_A: \beta \neq 0$. Example 3 performs such an analysis for the data from Examples 1 and 2. The degrees of freedom associated with the total variability of Y_i values are $n - 1$. The degrees of freedom associated with the variability among Y_i's due to regression are always 1 in a simple linear regression. The residual degrees of freedom are calculable as residual DF = total DF − regression DF = $n - 2$. Once the regression and residual mean squares are calculated (MS = SS/DF, as usual), H_0 may be tested by determining

$$F = \frac{\text{regression MS}}{\text{residual MS}}, \qquad (14)$$

TABLE 1: Summary of the Calculations for Testing $H_0: \beta = 0$ against $H_A: \beta \neq 0$ by an Analysis of Variance

Source of variation	Sum of squares (SS)	DF	Mean square (MS)
Total $[Y_i - \overline{Y}]$	$\sum y^2$	$n - 1$	
Linear regression $[\hat{Y}_i - \overline{Y}]$	$\dfrac{\left(\sum xy\right)^2}{\sum x^2}$	1	$\dfrac{\text{regression SS}}{\text{regression DF}}$
Residual $[Y_i - \hat{Y}_i]$	total SS − regression SS	$n - 2$	$\dfrac{\text{residual SS}}{\text{residual DF}}$

Note: To test the null hypothesis, we compute $F = $ regression MS/residual MS. The critical value for the test is $F_{\alpha(1),1,(n-2)}$.

EXAMPLE 3   Analysis of Variance Testing of $H_0: \beta = 0$ Against $H_A: \beta \neq 0$, Using the Data of Examples 1 and 2

$n = 13$ $\sum xy = 70.80$ (from Example 2)

$\sum Y = 44.4$ $\sum x^2 = 262.00$ (from Example 2)

$\sum Y^2 = 171.30$

$$\text{total SS} = \sum y^2 = 171.30 - \frac{(44.4)^2}{13} \qquad \text{regression SS} = \frac{\left(\sum xy\right)^2}{\sum x^2} = \frac{(70.80)^2}{262.00}$$

$$= 171.30 - 151.6431 \qquad\qquad\qquad = \frac{5012.64}{262.00}$$

$$= 19.656923 \qquad\qquad\qquad\qquad = 19.132214$$

$$\text{total DF} = n - 1 = 12$$

Source of variation	SS	DF	MS
Total	19.656923	12	
Linear regression	19.132214	1	19.132214
Residual	0.524709	11	0.047701

$$F = \frac{19.132214}{0.047701} = 401.1$$

$$F_{0.05(1),1,11} = 4.84$$

Therefore, reject H_0.

$$P \ll 0.0005 \quad [P = 0.00000000053]$$

$$r^2 = \frac{19.132214}{19.656923} = 0.97$$

$$s_{Y \cdot X} = \sqrt{0.047701} = 0.218 \text{ cm}$$

which is then compared to the critical value, $F_{\alpha(1),\nu_1,\nu_2}$, where $\nu_1 =$ regression DF $= 1$ and $\nu_2 =$ residual DF $= n - 2$.

The residual mean square is often written as $s_{Y \cdot X}^2$, a representation denoting that it is the variance of Y after taking into account the dependence of Y on X. The square root of this quantity (i.e., $s_{Y \cdot X}$) is called the *standard error of estimate* (occasionally termed the "standard error of the regression"). In Example 3, $s_{Y \cdot X} = \sqrt{0.047701} \text{ cm}^2$ $= 0.218$ cm. The standard error of estimate is an overall indication of the accuracy with which the fitted regression function predicts the dependence of Y on X. The magnitude of $s_{Y \cdot X}$ is proportional to the magnitude of the dependent variable, Y, making examination of $s_{Y \cdot X}$ a poor method for comparing regressions. Thus, Dapson (1980) recommends using $s_{Y \cdot X}/\overline{Y}$ (a unitless measure) to examine similarities among two or more regression fits.

The proportion (or percentage) of the total variation in Y that is explained or accounted for by the fitted regression is termed the *coefficient of determination*, r^2, which is often used as a measure of the strength of the straight-line relationship:*

$$r^2 = \frac{\text{regression SS}}{\text{total SS}}; \tag{15}$$

r^2 is sometimes referred to as expressing the goodness of fit of the line to the data or as the precision of the regression.

For Example 3, $r^2 = 0.97$, or 97%. That portion the total variation not explained by the regression is, of course, $1 - r^2$, or residual SS/total SS, and this is called the *coefficient of nondetermination*, a quantity seldom referred to.[†] In Example 3, $1 - r^2 = 1.00 - 0.97 = 0.03$, or 3%. (The quantity r is the correlation coefficient.)[‡]

---

*However, Ranney and Thigpen (1981) and others caution against declaring r^2 to be a measure of strength of the relationship in cases where X is a fixed-effect variable (see Section 2e).

[†]The standard error of estimate is directly related to the coefficient of nondetermination and to the variability of Y as

$$s_{Y \cdot X} = s_Y \sqrt{(1 - r^2)(n - 1)/(n - 2)}. \tag{16}$$

[‡]Sutton (1990) showed that

$$r^2 = \frac{F}{F + \nu_2}. \tag{17}$$

Another good way to express the accuracy of a regression, or to compare accuracies of several regressions, is to compute confidence intervals for predicted values of Y, as described in Section 5.

(b) t Testing. The preceding null hypothesis concerning β can also be tested by using Student's t statistic. Indeed, the more general two-tailed hypotheses, H_0: $\beta = \beta_0$ and H_A: $\beta \neq \beta_0$, can be tested in this fashion.* Most frequently, β_0 is zero in these hypotheses, in which case either the analysis of variance or the t test may be employed and the conclusion will be the same. But if any other value of β_0 is hypothesized, then the following procedure is applicable, whereas the analysis of variance is not. Also, the t-testing procedure allows for the testing of one-tailed hypotheses: either H_0: $\beta \leq \beta_0$ and H_A: $\beta > \beta_0$, or H_0: $\beta \geq \beta_0$ and H_A: $\beta < \beta_0$.

Since the t statistic is in general calculated as

$$t = \frac{\text{(parameter estimate)} - \text{(parameter value hypothesized)}}{\text{standard error of parameter estimate}}, \tag{18}$$

we need to compute s_b, the standard error of the regression coefficient.

The variance of b is calculated as

$$s_b^2 = \frac{s_{Y \cdot X}^2}{\sum x^2}. \tag{19}$$

Therefore,

$$s_b = \sqrt{\frac{s_{Y \cdot X}^2}{\sum x^2}}, \tag{20}$$

and

$$t = \frac{b - \beta_0}{s_b}. \tag{21}$$

To test H_0: $\beta = 0$ against H_A: $\beta \neq 0$ in Example 4, $s_b = 0.0135$ cm/day and $t = 20.000$. The degrees of freedom for this testing procedure are $n - 2$; thus, the critical value in this example, at the 5% significance level, is $t_{0.05(1),11} = 2.201$, and H_0 is rejected. For this two-tailed hypothesis, H_0: $\beta = 0$, either t or F may be employed, with the same result; and $F = t^2$ and $F_{\alpha,1,(n-2)} = t_{\alpha(2),(n-2)}^2$. The relevant concepts and procedures involved in one-tailed hypothesis testing.

4 INTERPRETATIONS OF REGRESSION FUNCTIONS

Potential misinterpretation of regression relationships has been alluded to earlier in this chapter and warrants further discussion. If we calculate the two constants, a and b, that define a linear regression question, then we have quantitatively described the average rate of change of Y with a change in X. However, although a mathematical dependence between Y and X has been determined, it must not automatically be assumed that there is a biological cause-and-effect relationship. Causation should be suggested only with insight into the phenomenon being investigated and should not be declared by statistical testing alone. Indeed, it is often necessary to determine the interrelationships among variables beyond the two variables under study, for an observed dependence may, in fact, be due to the influence of one or more variables not yet analyzed.

---

*The use of t for testing regression coefficients emanates from Fisher (1922a).

EXAMPLE 4   **Use of Student's t to Test $H_0: \beta = 0$ Against $H_A: \beta \neq 0$, Employing the Data of Examples 1 and 2**

$n = 13$

$b = 0.270$ cm/day

$$s_b = \sqrt{\frac{s_{Y \cdot X}^2}{\sum x^2}} = \sqrt{\frac{0.047701}{262.00}} = \sqrt{0.00018206} = 0.0135 \text{ cm/day}$$

$$t = \frac{b - 0}{s_b} = \frac{0.270}{0.0135} = 20.000$$

$t_{0.05(2),11} = 2.201$

Therefore, reject H_0.

$$P \ll 0.01 \quad [P = 0.00000000027]$$

We must also remember that a linear regression function is mathematically nothing more than a straight line forced to fit through a set of data points, and it may not at all describe a natural phenomenon. The biologist may be chagrined when attempting to explain why the observed relationship is well described by a linear function or what biological insights are to be unfolded by the consideration of a particular slope or a particular magnitude of a Y intercept. That is, although a derived regression function often provides a satisfactory and satisfying description of a natural phenomenon, sometimes it does not.

Even if a regression function does not help us to explain the functional anatomy of a natural system, it may still be useful in its ability to predict Y, given X. In the sciences, equations may inaccurately represent natural processes yet may be employed advantageously to predict the magnitude of one variable given the magnitude of an associated variable. Thus, predicting $\hat{Y}$ values (or $\hat{X}$ values; see Section 6) and their standard errors is frequently a useful end in itself. But, as stressed in Section 2, great caution should be exercised in predicting a $\hat{Y}$ for an X outside the range of the X's used to obtain the regression equation. In addition, while a (the Y intercept) has utility in expressing a regression relationship, expressing a as the predicted value of Y when $X = 0$ may not have biological significance—and may even be meaningless if $X = 0$ lies outside of the range of the observed X's.

If the relationship between two variables is not that of an independent variable and a dependent variable, then correlation analysis, instead of regression analysis, should be considered.

5   CONFIDENCE INTERVALS IN REGRESSION

In many (though not all) cases, knowing the standard error of a statistic allows us to calculate a confidence interval for the parameter being estimated, as

$$\text{confidence interval} = \text{statistic} \pm (t)(\text{SE of statistic}). \tag{22}$$

This was first demonstrated for the confidence interval for a mean. In addition, the second significant figure of the standard error of a statistic may be used as an indicator of the precision to which that statistic should be reported. The standard error of b has been given by Equation 20. For the data in Example 4, the second significant figure of $s_b = 0.0135$ cm/day enables us to express b to the third decimal place (i.e., $b = 0.270$ cm/day).

(a) Confidence Interval for the Regression Coefficient. For the $(1 - \alpha)$ confidence limits of β,

$$b \pm t_{\alpha(2),(n-2)}s_b. \tag{23}$$

Therefore,

$$L_1 = b - t_{\alpha(2),(n-2)}s_b \tag{24}$$

and

$$L_2 = b + t_{\alpha(2),(n-2)}s_b. \tag{25}$$

For Example 2, the 95% confidence interval for β would be $b \pm t_{0.05(2),11}s_b = 0.270 \pm (2.201)(0.0135) = 0.20 \pm 0.030$ cm/day. Thus, the 95% confidence limits are $L_1 = 0.270 - 0.030 = 0.240$ cm/day and $L_2 = 0.270 + 0.030 = 0.300$ cm/day; and we can state, with 95% confidence (i.e., we state that there is no greater than a 5% chance that we are wrong), that 0.240 cm/day and 0.300 cm/day form an interval that includes the population regression coefficient, β. Figure 7 shows, by the broken lines, these confidence limits for the slope of the regression line. Within these limits, the various possible b values rotate the line about the point $(\overline{X}, \overline{Y})$.

(b) Confidence Interval for an Estimated Y. As shown in Section 1, a regression equation allows the estimate of the value of Y (namely, $\hat{Y}$) existing in the population

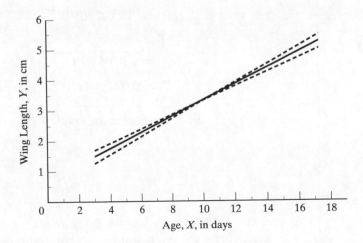

FIGURE 7: The regression line from Figure 1, showing, by broken lines, the lines with slopes equal to the upper and lower 95% confidence limits for β.

Simple Linear Regression

at a given value of X. The standard error of such a population estimate is

$$s_{\hat{Y}_i} = \sqrt{s_{Y \cdot X}^2 \left[\frac{1}{n} + \frac{(X_i - \overline{X})^2}{\sum x^2} \right]}. \tag{26}$$

Example 5a shows how $s_{\hat{Y}_i}$ can be used in Equation 22 to calculate confidence intervals. It is apparent from Equation 26 that the standard error is a minimum for $X_i = \overline{X}$, and that it increases as estimates are made at values of X_i farther from the mean. If confidence limits were calculated for all points on the regression line, the result would be the curved *confidence bands* shown in Figure 8.

EXAMPLE 5   **Standard Errors of Predicted Values of Y**

The regression equation derived in Example 2 is used for the following considerations. For this regression, $a = 0.72$ cm, $b = 0.270$ cm/day, $\overline{X} = 10.0$ days, $\sum x^2 = 262.00$ days^2, $n = 13, s_{Y \cdot X}^2 = 0.047701$ cm^2, and $t_{0.05(2),11} = 2.201$.

a. Equation 26 is used when we wish to predict the mean value of $\hat{Y}_i$, given X_i, in the entire population. For example, we could ask, "What is the mean wing length of all 13.0-day-old birds in the population under study?"

$$\hat{Y}_i = a + bX_i$$
$$= 0.715 + (0.270)(13.0)$$
$$= 0.715 + 3.510$$
$$= 4.225 \text{ cm}$$

$$s_{\hat{Y}_i} = \sqrt{s_{Y \cdot X}^2 \left[\frac{1}{n} + \frac{(X_i - \overline{X})^2}{\sum x^2} \right]}$$

$$= \sqrt{0.047701 \left[\frac{1}{13} + \frac{(13.0 - 10.0)^2}{262.00} \right]}$$

$$= \sqrt{(0.047701)(0.111274)}$$

$$= 0.073 \text{ cm}$$

$$95\% \text{ confidence interval} = \hat{Y}_i \pm t_{0.05(2),11} s_{\hat{Y}_i}$$
$$= 4.225 \pm (2.201)(0.073)$$
$$= 4.225 \pm 0.161 \text{ cm}$$
$$L_1 = 4.064 \text{ cm}$$
$$L_2 = 4.386 \text{ cm}$$

b. Equation 28 is used when we propose taking an additional sample of m individuals from the population and wish to predict the mean Y value, at a given X, for these m new data. For example, we might ask, "If ten 13.0-day-old birds were taken from the population, what would be their mean wing length?"

$$\hat{Y}_i = 0.715 + (0.270)(13.0) = 4.225 \text{ cm}$$

$$(s_{\hat{Y}_i})_{10} = \sqrt{0.047701\left[\frac{1}{10} + \frac{1}{13} + \frac{(13.0 - 10.0)^2}{262.00}\right]}$$

$$= \sqrt{(0.047701)(0.211274)}$$

$$= 0.100 \text{ cm}$$

$$95\% \text{ prediction interval} = \hat{Y}_i \pm t_{0.05(2),11}(s_{\hat{Y}_i})_{10}$$

$$= 4.225 \pm (2.201)(0.100)$$

$$= 4.225 \pm 0.220 \text{ cm}$$

$$L_1 = 4.005 \text{ cm}$$

$$L_2 = 4.445 \text{ cm}$$

c. Equation 29 is used when we wish to predict the Y value of a single observation taken from the population as a specified X. For example, we could ask, "If one 13.0-day-old bird were taken from the population, what would be its wing length?"

$$\hat{Y}_i = 0.715 + (0.270)(13.0) = 4.225 \text{ cm}$$

$$(s_{\hat{Y}_i})_1 = \sqrt{0.047701\left[1 + \frac{1}{13} + \frac{(13.0 - 10.0)^2}{262.00}\right]}$$

$$= \sqrt{(0.047701)(1.111274)}$$

$$= 0.230 \text{ cm}$$

$$95\% \text{ prediction interval} = \hat{Y}_i \pm t_{0.05(2),11}(s_{\hat{Y}_i})_1$$

$$= 4.225 \pm (2.201)(0.230)$$

$$= 4.225 \pm 0.506 \text{ cm}$$

$$L_1 = 3.719 \text{ cm}$$

$$L_2 = 4.731 \text{ cm}$$

Note from these three examples that the accuracy of prediction increases as does the number of data upon which the prediction is based. For example, predictions about a mean for the entire population will be more accurate than a prediction about a mean from 10 members of the population, which is more accurate than a prediction about a single member of the population.

If $X_i = 0$, then $\hat{Y} = a$ (the Y intercept). Therefore,

$$s_a = \sqrt{s_{Y \cdot X}^2\left[\frac{1}{n} + \frac{\overline{X}^2}{\sum x^2}\right]}. \tag{27}$$

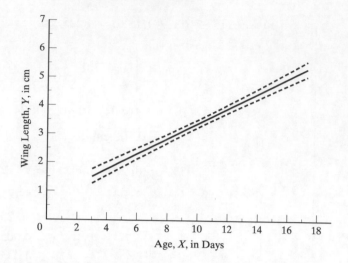

FIGURE 8: The 95% confidence bands (broken lines) for the regression line from Figure 1 (the regression of Example 2).

(c) Prediction Interval for an Estimated Y. If we predict a value of $\hat{Y}$ that is the mean of m additional measurements at a given X (Example 5b), its standard error would be

$$(s_{\hat{Y}_i})_m = \sqrt{s^2_{Y \cdot X}\left[\frac{1}{m} + \frac{1}{n} + \frac{(X_i - \overline{X})^2}{\sum x^2}\right]}. \tag{28}$$

A special case of Equation 28, shown as Example 5c, exists when it is desired to know the standard error associated with estimating $\hat{Y}_i$ for a single additional measurement at Y_i:

$$(s_{\hat{Y}_i})_1 = \sqrt{s^2_{Y \cdot X}\left[1 + \frac{1}{n} + \frac{(X_i - \overline{X})^2}{\sum x^2}\right]}. \tag{29}$$

Equation 26 is equal to Equation 28 when m approaches infinity. Examples 5b and 5c demonstrate the use of these standard errors of prediction.

(d) Testing Hypotheses about Estimated Y Values. Once we have computed the standard error of a predicted Y, we can test hypotheses about that prediction. For example, we might ask whether the mean population wing length of 13.0-day-old sparrows, call it $\mu_{\hat{Y}_{13.0}}$, is equal to some specified value (two-tailed test) or is greater than (or less than) some specified value (one-tailed test). We simply refer to Equation 18, as Example 6 demonstrates.

(e) Confidence Interval and Hypothesis Testing for the Residual Mean Square. The sample residual mean square, $s^2_{Y \cdot X}$, is an estimate of the residual mean square in the population, $\sigma^2_{Y \cdot X}$. Confidence limits may be calculated for $\sigma^2_{Y \cdot X}$ as they are for the population variance, σ^2. Simply use $\nu = n - 2$, instead of $\nu = n - 1$, and replace σ^2 with $\sigma^2_{Y \cdot X}$ and SS with residual SS in Equation 29a or 29b. Also, a confidence interval for the population standard error of estimate, $\sigma_{Y \cdot X}$, may be obtained by analogy to Equation 29c.

$$\frac{\nu s^2}{\chi^2_{\alpha/2, \nu}} \leq \sigma^2 \leq \frac{\nu s^2}{\chi^2_{(1-\alpha/2), \nu}}. \tag{29a}$$

$$\frac{SS}{\chi^2_{\alpha/2,\,\nu}} \leq \sigma^2 \leq \frac{SS}{\chi^2_{(1-\alpha/2),\,\nu}}. \tag{29b}$$

$$\sqrt{\frac{SS}{\chi^2_{\alpha/2,\,\nu}}} \leq \sigma \leq \sqrt{\frac{SS}{\chi^2_{(1-\alpha/2),\,\nu}}}. \tag{29c}$$

EXAMPLE 6   Hypothesis Testing with an Estimated Y Value

H_0: The mean population wing length of 13.0-day-old birds is not greater than 4 cm (i.e., H_0: $\mu_{\hat{Y}_{13.0}} \leq 4$ cm).

H_A: The mean population wing length of 13.0-day-old birds is greater than 4 cm (i.e., H_A: $\mu_{\hat{Y}_{13.0}} > 4$ cm).

From Example 5b, $Y_{13.0} = 4.225$ cm and $s_{\hat{Y}_{13.0}} = 0.073$ cm,

$$t = \frac{4.225 - 4}{0.073} = \frac{0.225}{0.073} = 3.082$$

$$t_{0.05(1),11} = 1.796$$

Therefore, reject H_0.

$$0.005 < P < 0.01 \quad [P = 0.0052]$$

---

6 INVERSE PREDICTION

Situations exist where we desire to predict the value of the independent variable (X_i) that is to be expected in the population at a specified value of the dependent variable (Y_i), a procedure known as *inverse prediction*. In Example 1, for instance, we might ask, "How old is a bird that has a wing 4.5 cm long?" By simple algebraic rearrangement of the linear regression relationship of Equation 8, we obtain

$$\hat{X}_i = \frac{Y_i - a}{b}. \tag{30}$$

From Figure 8, it is clear that, although confidence limits calculated around the predicted $\hat{Y}_i$ are symmetrical above and below $\hat{Y}_i$, confidence limits associated with the predicted $\hat{X}_i$ are not symmetrical to the left and to the right of $\hat{X}_i$. The $1 - \alpha$ confidence limits for the X predicted at a given Y may be calculated as follows, which is demonstrated in Example 7:

$$\bar{X} + \frac{b(Y_i - \bar{Y})}{K} \pm \frac{t}{K}\sqrt{s^2_{Y \cdot X}\left[\frac{(Y_i - \bar{Y})^2}{\sum x^2} + K\left(1 + \frac{1}{n}\right)\right]}, \tag{31}$$

where* $K = b^2 - t^2 s_b^2$. This computation is a special case of the prediction of the $\hat{X}$ associated with multiple values of Y at that X. For the study of Example 1, age can be predicted of m birds to be taken from the population and having a mean body weight of $\bar{Y}_i$:

$$\hat{X}_i = \frac{\bar{Y}_i - a}{b}, \tag{32}$$

---

*Recall that $F_{\alpha(1),1,\nu} = t^2_{\alpha(2),\nu}$. Therefore, we could compute $K = b^2 - Fs_b^2$, where $F = t^2_{\alpha(2),(n-2)} = F_{\alpha(1),1,(n-2)}$. Snedecor and Cochran (1989: 171) presented an alternative, yet equivalent, computation of these confidence limits.

EXAMPLE 7   Inverse Prediction

We wish to estimate, with 95% confidence, the age of a bird with a wing length of 4.5 cm.

Predicted age:

$$\hat{X} = \frac{Y_i - a}{b}$$

$$= \frac{4.5 - 0.715}{0.270}$$

$$= 14.019 \text{ days}$$

To compute 95% confidence interval:

$$t = t_{0.05(2),11} = 2.201$$

$$K = b^2 - t^2 s_b^2$$

$$= 0.270^2 - (2.201)^2 (0.0135)^2$$

$$= 0.0720$$

95% confidence interval:

$$\overline{X} + \frac{b(Y_i - \overline{Y})}{K} \pm \frac{t}{K} \sqrt{s_{Y \cdot X}^2 \left[\frac{(Y_i - \overline{Y})^2}{\sum x^2} + K \left(1 + \frac{1}{n} \right) \right]}$$

$$= 10.0 + \frac{0.270(4.5 - 3.415)}{0.0720}$$

$$\pm \frac{2.201}{0.0720} \sqrt{0.047701 \left[\frac{(4.5 - 3.415)^2}{262.00} + 0.0720 \left(1 + \frac{1}{13} \right) \right]}$$

$$= 10.0 + 4.069 \pm 30.569 \sqrt{0.003913}$$

$$= 14.069 \pm 1.912 \text{ days}$$

$$L_1 = 12.157 \text{ days}$$
$$L_2 = 15.981 \text{ days}$$

---

where $\overline{Y}_i$ is the mean of the m values of Y_i; and the confidence limits would be calculated as

$$\overline{X} + \frac{b(\overline{Y}_i - \overline{Y})}{K} \pm \frac{t}{K} \sqrt{(s_{Y \cdot X}^2)' \left[\frac{(\overline{Y}_i - \overline{Y})^2}{\sum x^2} + K \left(\frac{1}{m} + \frac{1}{n} \right) \right]}, \qquad (33)$$

where* $t = t_{\alpha(2),(n+m-3)}, K = b^2 - t^2 (s_b^2)'$,

$$(s_b^2)' = \frac{(s_{Y \cdot X}^2)'}{\sum x^2}, \qquad (34)$$

---

*Alternatively, we may compute $K = b^2 - F(s_b^2)_*$, where $F = t_{\alpha(2),(n+m-3)}^2 = F_{\alpha(1),1,(n+m+3)}$.

Simple Linear Regression

and

$$(s_{Y \cdot X}^2)' = \text{residual SS} + \sum_{j=1}^{m}(Y_{ij} - \overline{Y}_i)^2/(n + m - 3) \tag{35}$$

(Ostle and Malone, 1988: 241; Seber and Lee, 2003: 147–148).

7 REGRESSION WITH REPLICATION AND TESTING FOR LINEARITY

If, in Example 1, we had wing measurements for more than one bird for at least some of the recorded ages, then we could test the null hypothesis that the population regression is linear.* (Note that true replication requires that there are multiple birds at a given age, not that there are multiple wing measurements on the same bird.) Figure 9 presents the data of Example 8a. A least-squares, best-fit, linear regression equation can be calculated for any set of at least two data, but neither the equation itself nor the testing for a significant slope (which requires at least three data) indicates whether Y is, in fact, a straight-line function of X in the population sampled.

EXAMPLE 8a   Regression Data Where There Are Multiple Values of Y for Each Value of X

	Age (yr)	Systolic blood pressure (mm Hg)		
i	X_i	Y_{ij}	n_i	$\overline{Y}_i$
1	30	108, 110, 106	3	108.0
2	40	125, 120, 118, 119	4	120.5
3	50	132, 137, 134	3	134.3
4	60	148, 151, 146, 147, 144	5	147.2
5	70	162, 156, 164, 158, 159	5	159.8

$$k = 5; \quad i = 1 \text{ to } 5; \quad j = 1 \text{ to } n_i; \quad N = 20$$

$$\sum\sum X_{ij} = 1050 \qquad \sum\sum Y_{ij} = 2744$$

$$\sum\sum X_{ij}^2 = 59{,}100 \qquad \sum\sum Y_{ij}^2 = 383{,}346 \qquad \sum\sum X_{ij}Y_{ij} = 149{,}240$$

$$\sum x^2 = 3975.00 \qquad \sum y^2 = 6869.20 \qquad \sum xy = 5180.00$$

$$\overline{X} = 52.5 \qquad \overline{Y} = 137.2$$

$$b = \frac{\sum xy}{\sum x^2} = \frac{5180.00}{3975.00} = 1.303 \text{ mm Hg/yr}$$

$$a = \overline{Y} - b\overline{X} = 137.2 - (1.303)(52.5) = 68.79 \text{ mm Hg}$$

Therefore, the least-squares regression line is $\hat{Y}_{ij} = 68.79 + 1.303X_{ij}$.

We occasionally encounter the suggestion that for data such as those in Figure 9 the mean Y at each X be utilized for a regression analysis. However, to do so would be to discard information, and such a procedure is not recommended (Freund, 1971).

*Thornby (1972) presents a procedure to test the hypothesis of linearity even when there are not multiple observations of Y. But the computation is rather tedious.

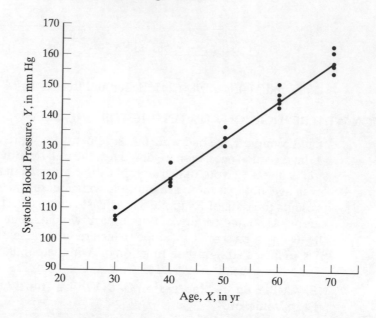

FIGURE 9: A regression where there are multiple values of Y for each value of X.

EXAMPLE 8b   **Statistical Analysis of the Regression Data of Example 8a**

H_0:   The population regression is linear.

H_A: The population regression is not linear.

$$\text{total SS} = \sum y^2 = 6869.20 \quad \text{total DF} = N - 1 = 19$$

$$\text{among-groups SS} = \sum_{i=1}^{k} \frac{\left(\sum_{j=1}^{n_i} Y_{ij} \right)^2}{n_i} - \frac{\left(\sum_{i=1}^{k} \sum_{j=1}^{n_i} Y_{ij} \right)^2}{N}$$

$$= 383{,}228.73 - 376{,}476.80 = 6751.93$$

$\text{among-groups DF} = k - 1 = 4$

$\text{within-groups SS} \qquad = \text{total SS} - \text{among-groups SS}$

$$= 6869.20 - 6751.93 = 117.27$$

$\text{within-groups DF} \qquad = \text{total DF} - \text{among-groups DF}$

$$= 19 - 4 = 15$$

$\text{deviations-from-linearity SS} = \text{among-groups SS} - \text{regression SS}$

$$= 6751.93 - 6750.29 = 1.64$$

$\text{deviations-from-linearity DF} = \text{among-groups DF} - \text{regression DF}$

$$= 4 - 1 = 3$$

Source of variation	SS	DF	MS
Total	6869.20	19	
Among groups	6751.93	4	
Linear regression	6750.29	1	
Deviations from linearity	1.64	3	0.55
Within groups	117.27	15	7.82

$$F = \frac{0.55}{7.82} = 0.070$$

Since $F < 1.00$, do not reject H_0.

$$P > 0.25 \quad [P = 0.975]$$

H_0: $\beta = 0$.

H_A: $\beta \neq 0$.

$$\text{regression SS} = \frac{\left(\sum xy\right)^2}{\sum x^2} = \frac{(5180.00)^2}{3975.00} = 6750.29$$

Source of variation	SS	DF	MS
Total	6869.20	19	
Linear regression	6750.29	1	6750.29
Residual	118.91	18	6.61

$$F = \frac{6750.29}{6.61} = 1021.2$$

$$F_{0.05(1),1,18} = 4.41$$

Therefore, reject H_0.

$$P \ll 0.0005 \quad [P < 0.00000000001]$$

$$r^2 = \frac{6750.29}{6869.20} = 0.98$$

$$s_{Y \cdot X} = \sqrt{6.61} = 2.57 \text{ mm Hg}$$

Example 8b appropriately analyzes data consisting of multiple Y values at each X value, and Figure 9 presents the data graphically. For each of the k unique X_i values, we can speak of each of n_i values of Y (denoted by Y_{ij}) using the double subscript on i exactly as in the one-way analysis of variance. In Example 8a, $n_1 = 3, n_2 = 4, n_3 = 3$, and so on; and $X_{11} = 50$ cm, $Y_{ii} = 108$ mm; $X_{12} = 50$ cm, $Y_{12} = 110$ mm; $X_{13} = 50$ cm, $Y_{13} = 106$ mm; and so on through $X_{55} = 70$ cm, $Y_{55} = 159$ mm. Therefore,

$$\sum xy = \sum_{i=1}^{k} \sum_{j=1}^{n_i} (X_{ij} - \overline{X})(Y_{ij} - \overline{Y}) = \sum \sum X_{ij} Y_{ij} - \frac{\sum X_{ij} \sum Y_{ij}}{N}, \quad (36)$$

where $N = \sum_{i=1}^{k} n_i$ the total number of pairs of data. Also,

$$\sum x^2 = \sum_{i=1}^{k}\sum_{j=1}^{n_i}(X_{ij} - \overline{X})^2 = \sum\sum X_{ij}^2 - \frac{(\sum X_{ij})^2}{N} \qquad (37)$$

and

$$\text{total SS} = \sum_{i=1}^{k}\sum_{j=1}^{n_i}(Y_{ij} - \overline{Y})^2 = \sum\sum Y_{ij}^2 - C, \qquad (38)$$

where

$$C = \frac{(\sum\sum Y_{ij})^2}{N} \quad \text{and} \quad N = \sum_{i=1}^{k} n_i. \qquad (39)$$

Examples 8a and 8b show the calculations of the regression coefficient, b, the Y intercept a, and the regression and residual sums of squares, using Equations 4, 7, 11, and 13, respectively. The total, regression, and residual degrees of freedom are $N - 1, 1$, and $N - 2$, respectively.

As shown in Section 3, the analysis of variance for significant slope involves the partitioning of the total variability of Y (i.e., $Y_{ij} - \overline{Y}$) into that variability due to regression ($\hat{Y}_i - \overline{Y}$) and that variability remaining (i.e., residual) after the regression line is fitted ($Y_{ij} - \hat{Y}_i$). However, by considering the k groups of Y values, we can also partition the total variability exactly as we did in the one-way analysis of variance, by describing variability among groups ($\overline{Y}_i - \overline{Y}$) and within groups ($Y_{ij} - \overline{Y}_i$):

$$\text{among-groups SS} = \sum_{i=1}^{k} n_i\left(\overline{Y}_i - \overline{Y}\right)^2 = \sum_{i=1}^{k} \frac{\left(\sum_{j=1}^{n_i} Y_{ij}\right)^2}{n_i} - C, \qquad (40)$$

$$\text{among-groups DF} = k - 1, \qquad (41)$$

$$\text{within-groups SS} = \text{total SS} - \text{among-groups SS}, \qquad (42)$$

$$\text{within-groups DF} = \text{total DF} - \text{among-groups DF} = N - k. \qquad (43)$$

The variability among groups ($\overline{Y}_i - \overline{Y}$) can also be partitioned. Part of this variability ($\hat{Y}_i - \overline{Y}$) results from the linear regression fit to the data, and the rest ($\overline{Y}_i - \hat{Y}_i$) is due to the deviation of each group of data from the regression line, as shown in Figure 10. Therefore,

$$\text{deviations-from-linearity SS} = \text{among-groups SS} - \text{regression SS} \qquad (44)$$

and

$$\text{deviations-from-linearity DF} = \text{among-groups DF} - \text{regression DF}$$
$$= k - 2. \qquad (45)$$

Table 2 summarizes this partitioning of sums of squares.

Alternatively, and with identical results, we may consider the residual variability ($Y_{ij} - \hat{Y}_i$) to be divisible into two components: within-groups variability ($Y_{ij} - \overline{Y}_i$)

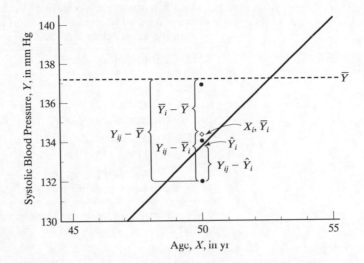

FIGURE 10: An enlarged portion of Figure 9, showing the partitioning of Y deviations. The mean Y at $X = 50$ yr is 134.3 mm Hg, shown by the symbol "⋄"; the mean of all Y's is $\overline{Y}$, shown by a dashed line; Y_{ij} is the jth Y at the ith X; and $\overline{Y}_i$ is the mean of the n_i Y's at X_i.

TABLE 2: Summary of the Analyses of Variance Calculations for Testing H_0: the Population Regression Is Linear, and for Testing H_0: $\beta = 0$

Source of variation	Sum of squares (SS)	DF	Mean Square (MS)
Total $[Y_{ij} - \overline{Y}]$	$\sum y^2$	$N - 1$	
Linear regression $[\hat{Y}_i - \overline{Y}]$	$\dfrac{\left(\sum xy\right)^2}{\sum x^2}$	1	$\dfrac{\text{regression SS}}{\text{regression DF}}$
Residual $[Y_{ij} - \hat{Y}_i]$	total SS − regression SS	$N - 2$	$\dfrac{\text{residual SS}}{\text{residual DF}}$
Among groups $[\overline{Y}_i - \overline{Y}]$	$\displaystyle\sum_{i=1}^{k} \dfrac{\left(\sum_{j=1}^{n_i} Y_{ij}\right)^2}{n_i} - \dfrac{\left(\sum_{i=1}^{k}\sum_{j=1}^{n_i} Y_{ij}\right)^2}{N}$	$k - 1$	
Linear regression $[\hat{Y}_i - \overline{Y}]$	$\dfrac{\left(\sum xy\right)^2}{\sum x^2}$	1	
Deviations from linearity $[\overline{Y}_i - \hat{Y}_i]$	among-groups SS − regression SS	$k - 2$	$\dfrac{\text{deviations SS}}{\text{deviations DF}}$
Within groups $[Y_{ij} - \overline{Y}_i]$	total SS − among-groups SS	$N - k$	$\dfrac{\text{within-groups SS}}{\text{within-groups DF}}$

Note: To test H_0: the population regression is linear, we use $F = $ deviations MS/within-groups MS, with a critical value of $F_{\alpha(1),(k-2),(N-k)}$. If the null hypothesis of linearity is rejected, then H_0: $\beta = 0$ is tested using $F = $ regression MS/within-groups MS, with a critical value of $F_{\alpha(1),1,(N-k)}$.

Simple Linear Regression

TABLE 3: Summary of Analysis of Variance Partitioning of Sources of Variation for Testing Linearity, as an Alternative to That in Table 2

Source of variation	DF
Total $[Y_{ij} - \overline{Y}]$	$N - 1$
Among groups $[\overline{Y}_i - \overline{Y}]$	$k - 1$
Within groups $[Y_{ij} - \overline{Y}_i]$	$N - k$
Linear regression $[\hat{Y}_i - \overline{Y}]$	1
Residual $[Y_{ij} - \hat{Y}_i]$	$N - 2$
Within groups $[Y_{ij} - \overline{Y}_i]$	$N - k$
Deviations from linearity $[\overline{Y}_i - \hat{Y}_i]$	$k - 2$

Note: Sums of squares and mean squares are as in Table 2.

and deviations-from-linearity $(\overline{Y}_i - \hat{Y}_i)$. This partitioning of sums of squares and degrees of freedom is summarized in Table 3.*

If the population relationship between Y and X is a straight line (i.e., "H_0: The population regression is linear" is a true statement), then the deviations-from-linearity MS and the within-groups MS will be estimates of the same variance; if the relationship is not a straight line (H_0 is false), then the deviations-from-linearity MS will be significantly greater than the within-groups MS. Thus, as demonstrated in Example 8b,

$$F = \frac{\text{deviations-from-linearity MS}}{\text{within-groups MS}} \tag{46}$$

provides a one-tailed test of the null hypothesis of linearity. (If all n_i's are equal, then performing a regression using the k $\overline{Y}$'s will result in the same b and a as will the calculations using all N Y_i's but the significance test for β will be much less powerful and the preceding test for linearity will not be possible.) The power for the test for linearity will be greater for larger numbers of replicate Y's at each X.

If the null hypothesis of linearity is not rejected, then the deviations-from-linearity MS and the within-groups MS may be considered to be estimates of the same population variance. The latter will be the better estimate, as it is based on more degrees of freedom; but an even better estimate is the residual MS, which is $s_{Y \cdot X}^2$, for it constitutes a pooling of the deviations MS and the within-groups MS. Therefore, if a regression is assumed to be linear, $s_{Y \cdot X}^2$ is the appropriate variance to use in the computation of standard errors (e.g., by Equations 20, 26–29) and confidence intervals resulting from them, and this residual mean square ($s_{Y \cdot X}^2$) is also appropriate in testing the hypothesis H_0: $\beta = 0$ (either by Equation 14 or by Equations 20 and 21), as demonstrated in Example 8b.

If the population regression is concluded not to be linear, then the investigator can consider the procedures of Section 10. If, however, it is desired to test H_0: $\beta = 0$, then the within-groups MS should be substituted for the residual MS ($s_{Y \cdot X}^2$); but it would not be advisable to engage in predictions with the linear-regression equation.

---

*Some authors refer to deviations from linearity as "lack of fit" and to within-groups variability as "error" or "pure error."

Simple Linear Regression

(a) Regression versus Analysis of Variance. Data consisting of replicate values of Y at each of several values of X (such as in Example 8) could also be submitted to a single-factor analysis of variance. This would be done by considering the X's as levels of the factor and the Y's as the data whose means are to be compared. This would test H_0: $\mu_1 = \mu_2 = \cdots = \mu_k$ instead of H_0: $\beta = 0$, where k is the number of different values of X (e.g., $k = 5$ in Example 8). When there are only two levels of the ANOVA factor (i.e., two different X's in regression), the power of testing these two hypotheses is the same. Otherwise, the regression analysis will be more powerful than the ANOVA (Cottingham, Lennon, and Brown, 2005).

8 POWER AND SAMPLE SIZE IN REGRESSION

Although there are basic differences between regression and correlation (see Section 1), a set of data for which there is a statistically significant regression coefficient (i.e., H_0: $\beta = 0$ is rejected, as explained in Section 3) would also yield a statistically significant correlation coefficient (i.e., we would reject H_0: $\rho = 0$). In addition, conclusions about the power of a significance test for a regression coefficient can be obtained by estimating power associated with the significance test for the correlation coefficient that would have been obtained from the same set of data.

After performing a regression analysis for a set of data, we may obtain the sample correlation coefficient, r,

$$r = b \sqrt{\frac{\sum x^2}{\sum y^2}}, \tag{47}$$

or we may take the square root of the coefficient of determination r^2 (Equation 15), assigning to it the sign of b. Then, with r in hand, estimate power and minimum required sample size for the hypothesis test for the regression coefficient, H_0: $\beta = 0$.

9 REGRESSION THROUGH THE ORIGIN

Although not of common biological importance, a special type of regression procedure is called for when we are faced with sets of data for which we know, a priori, that in the population Y will be zero when X is zero (i.e., the population Y intercept is known to be zero). Since the point on the graph with coordinates $(0, 0)$ is termed the *origin* of the graph, this regression situation is known as regression through the origin. In this type of regression analysis, both variables must be measured on a ratio scale, for only such a scale has a true zero.

For regression through the origin, the linear regression equation is

$$\hat{Y}_i = bX_i, \tag{48}$$

and some of the calculations pertinent to such a regression are as follows:

$$b = \frac{\sum X_i Y_i}{\sum X_i^2}, \tag{49}$$

$$\text{total SS} = \sum Y_i^2, \quad \text{with total DF} = n, \tag{50}$$

$$\text{regression SS} = \frac{\left(\sum X_i Y_i\right)^2}{\sum X_i^2}, \quad \text{with regression DF} = 1, \tag{51}$$

$$\text{residual SS} = \text{total SS} - \text{regression SS}, \text{with residual DF} = n - 1, \tag{52}$$

$$s_b^2 = \frac{s_{Y \cdot X}^2}{\sum X_i^2}, \tag{53}$$

where $s_{Y \cdot X}^2$ is residual mean square (residual SS/residual DF). Tests of hypotheses about the slope of the line are performed, as explained earlier in this chapter, with the exception that the preceding values are used; $n - 1$ is used as degrees of freedom whenever $n - 2$ is used for regressions not assumed to pass through the origin. Some statisticians (e.g., Kvålseth, 1985) caution against expressing a coefficient of determination, r^2, for this kind of regression. Bissell (1992) discusses potential difficulties with, and alternatives to, this regression model. A regression line forced through the origin does not necessarily pass through point $(\overline{X}, \overline{Y})$.

(a) Confidence Intervals. For regressions passing through the origin, confidence intervals may be obtained in ways analogous to the procedures in Section 5. That is, a confidence interval for the population regression coefficient, β, is calculated using Equation 53 for s_b^2 and $n - 1$ degree of freedom in place of $n - 2$. A confidence interval for an estimated $\hat{Y}$ is

$$s_{\hat{Y}} = \sqrt{s_{Y \cdot X}^2 \left(\frac{X_i^2}{\sum X^2}\right)}, \tag{54}$$

using the $s_{Y \cdot X}^2$ as in Equation 53; a confidence interval for $\hat{Y}_i$ predicted as the mean of m additional measurements at X_i is

$$(s_{\hat{Y}})_m = \sqrt{s_{Y \cdot X}^2 \left(\frac{1}{m} + \frac{X_i^2}{\sum X^2}\right)}; \tag{55}$$

and a confidence interval for the $\hat{Y}_i$ predicted for one additional measurement of X_i is

$$(s_{\hat{Y}})_1 = \sqrt{s_{X \cdot Y}^2 \left(1 + \frac{X_i^2}{\sum X^2}\right)} \tag{56}$$

(Seber and Lee, 2003: 149).

(b) Inverse Prediction. For inverse prediction (see Section 6) with a regression passing through the origin,

$$\hat{X}_i = \frac{Y_i}{b}, \tag{57}$$

and the confidence interval for the X_i predicted at a given Y is

$$\overline{X} + \frac{bY_i}{K} \pm \frac{t}{K} \sqrt{s_{Y \cdot X}^2 \left(\frac{Y_i^2}{\sum X_i^2} + K\right)}, \tag{58}$$

where $t = t_{\alpha(2),(n-1)}$ and* $K = b^2 - t^2 s_b^2$ (Seber and Lee, 2003: 149).

---

*Alternatively, $K = b^2 - F s_b^2$, where $F = t_{\alpha(2),(n-1)}^2 = F_{\alpha(1),1,(n-1)}$.

If X is to be predicted for multiple values of Y at that X, then

$$\hat{X}_i = \frac{\overline{Y}_i}{b}, \tag{59}$$

where $\overline{Y}_i$ is the mean of m values of Y; and the confidence limits would be calculated as

$$\overline{X} + \frac{b\overline{Y}_i}{K} \pm \frac{t}{K}\sqrt{(s^2_{Y\cdot X})' \left(\frac{\overline{Y}_i^2}{\sum X^2} + \frac{K}{m} \right)}, \tag{60}$$

where $t = t_{\alpha(2),(n+m-2)}$ and* $K = b^2 - t^2(s^2_b)'$;

$$(s^2_b)' = \frac{(s^2_{Y\cdot X})'}{\sum X^2}; \tag{61}$$

and

$$(s^2_{Y\cdot X})' = \frac{\text{residual SS} + \sum_{j=1}^{m}(Y_{ij} - \overline{Y}_i)^2}{n + m - 2} \tag{62}$$

(Seber and Lee, 2003: 149).

10 DATA TRANSFORMATIONS IN REGRESSION

As noted in Section 2d, the testing of regression hypotheses and the computation of confidence intervals—though not the calculation of a and b—depend upon the assumptions of normality and homoscedasticity, with regard to the values of Y, the dependent variable. Consciously striving to satisfy the assumptions often (but without guaranty) appeases the others. The same considerations are applicable to regression data.

Transformation of the independent variable will not affect the distribution of Y, so transformations of X generally may be made with impunity, and sometimes they conveniently convert a curved line into a straight line. However, transformations of Y do affect least-squares considerations and will therefore be discussed. Acton (1966: Chapter 8); Glantz and Slinker (2001: 150–154); Montgomery, Peck, and Vining (2001: 173–193); and Weisberg (2005: Chapter 7) present further discussions of transformations in regression.

If the values of Y are from a Poisson distribution (i.e., the data are counts, especially small counts), then the square-root transformation is usually desirable:

$$Y' = \sqrt{Y + 0.5}, \tag{63}$$

where the values of the variable after transformation (Y') are then submitted to regression analysis.

If the Y values are from a binomial distribution (e.g., they are proportions or percentages), then the arcsine transformation is appropriate:

$$Y' = \arcsin \sqrt{Y}. \tag{64}$$

Table 24 from *Appendix: Statistical Tables and Graphs* allows for ready use of this transformation.

---

*Alternatively, $K = b^2 - F(s^2_b)'$, where $F = t^2_{\alpha(2),(n+m-2)} = F_{\alpha(1),1,(n+m-2)}$.

The most commonly used transformation in regression is the logarithmic transformation, although it is sometimes employed for the wrong reasons. This transformation,

$$Y' = \log Y, \tag{65}$$

or

$$Y' = \log(Y + 1), \tag{66}$$

is appropriate when there is heteroscedasticity owing to the standard deviation of Y at any X increasing in proportion to the value of X. When this situation exists, it implies that values of Y can be measured more accurately at low than at high values of X. Figure 11 shows such data (from Example 9) before and after the transformation.

EXAMPLE 9   Regression Data Before and After Logarithmic Transformation of Y

Original data (as plotted in Figure 11a), indicating the variance of Y (namely, s_Y^2) at each X:

X	Y	s_Y^2
5	10.72, 11.22, 11.75, 12.31	0.4685
10	14.13, 14.79, 15.49, 16.22	0.8101
15	18.61, 19.50, 20.40, 21.37	1.4051
20	24.55, 25.70, 26.92, 28.18	2.4452
25	32.36, 33.88, 35.48, 37.15	4.2526

Transformed data (as plotted in Figure 11b), indicating the variance of $\log Y$ (namely, $s_{\log Y}^2$) at each X:

X	$\log Y$	$s_{\log Y}^2$
5	1.03019, 1.04999, 1.07004, 1.09026	0.000668
10	1.15014, 1.16997, 1.19005, 1.21005	0.000665
15	1.26975, 1.29003, 1.30963, 1.32980	0.000665
20	1.39005, 1.40993, 1.43008, 1.44994	0.000665
25	1.51001, 1.52994, 1.54998, 1.56996	0.000666

Many scatter plots of data imply a curved, rather than a straight-line, dependence of Y on X (e.g., Figure 11a). Often, logarithmic or other transformations of the values of Y and/or X will result in a straight-line relationship (as Figure 11b) amenable to linear regression techniques. *However*, if original, nontransformed values of Y agree with our assumptions of normality and homoscedasticity, then the data resulting from any of the preceding transformations will not abide by these assumptions. This is often not considered, and many biologists employing transformations do so simply to straighten out a curved line and neglect to consider whether the transformed data might indeed be analyzed legitimately by least-squares regression methods.

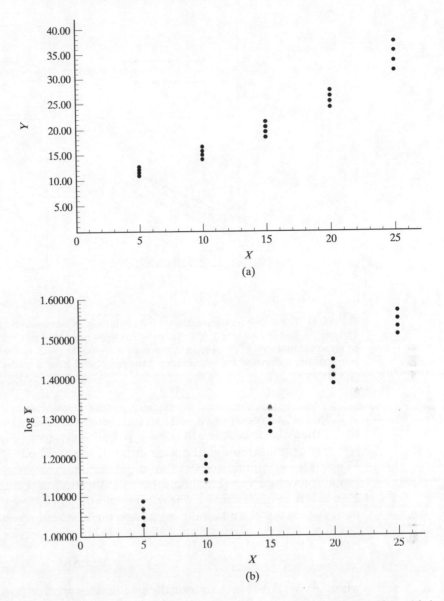

FIGURE 11: Regression data (of Example 9) exhibiting an increasing variability of Y with increasing magnitude of X. (a) The original data. (b) The data after logarithmic transformation of Y.

There exist some other, less commonly employed, data transformations. Iman and Conover (1979) discuss rank transformation (i.e., performing a regression of the ranks of Y on the ranks of X).

(a) Examination of Residuals. Since the logarithmic transformation is frequently proposed and employed to try to achieve homoscedasticity, we should consider how a justification for such a transformation might be obtained. If a regression is fitted by least squares, then the sample residuals (i.e., the values of $Y_i - \hat{Y}_i$) may be plotted against their corresponding X's, as in Figure 12 (see Draper and Smith, 1998: 62–64). If homoscedasticity exists, then the residuals should be distributed evenly above and below zero (i.e., within the shaded area in Figure 12a).

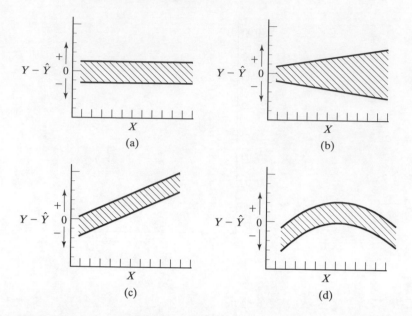

FIGURE 12: The plotting of residuals. (a) Data exhibiting homoscedasticity. (b) Data with heteroscedasticity of the sort in Example 9. (c) Data for which there was likely an error in the regression calculations, or an additional variable is needed in the regression model. (d) Data for which a linear regression does not accurately describe the relationship between Y and X, and a curvilinear relationship should be considered.

If there is heteroscedasticity due to increasing variability in Y with increasing values of X, then the residuals will form a pattern such as in Figure 12b, and a logarithmic transformation might be warranted. If the residuals form a pattern such as in Figure 12c, we should suspect that a calculation error has occured or, that an additional important variable should be added to the regression model. The pattern in Figure 12d indicates that a *linear* regression is an improper model to describe the data; for example, a quadratic regression might be employed.

Glejser (1969) suggests fitting the simple linear regression

$$E_i = a + bX_i, \tag{67}$$

where $E_i = |Y_i - \hat{Y}_i|$. A statistically significant b greater than zero indicates Figure 12b to be the case, and the logarithmic transformation may be attempted. Then, after the application of the transformation, a plot of the new residuals (i.e., $\log Y_i - \widehat{\log Y_i}$) should be examined and Equation 67 fitted, where $E_i = \left| \log Y_i - \widehat{\log Y_i} \right|$. If this regression has a b not significantly different from zero, then we may assume that the transformation was justified. An outlier (see Section 2d) will appear on plots such as Figure 12 as a point very far outside the pattern indicated by the shaded area.

Tests for normality in the distribution of residuals may be made by assessing departures from normality (employing $Y_i - \hat{Y}_i$ in place of X_i); graphical examination of normality is often convenient.

11 THE EFFECT OF CODING DATA

Either X or Y data, or both, may be coded prior to the application of regression analysis, and coding may facilitate computations, especially when the data are very large or very small in magnitude. Coding may consist of adding a constant to (or subtracting it from) X, or multiplying (or dividing) X by a constant; or both addition (or subtraction) and multiplication (or division) may be applied simultaneously. Values of Y may be coded in the same fashion; this may even be done simultaneously with the coding of X values, using either the same or different coding constants. If we let M_X and M_Y represent constants by which X and Y, respectively, are to be multiplied, and let A_X and A_Y be constants then to be added to $M_X X$ and $M_Y Y$, respectively, then the transformed variables, $[X]$ and $[Y]$, are

$$[X] = M_X X + A_X \tag{68}$$

and

$$[Y] = M_Y Y + A_Y. \tag{69}$$

The slope, b, will not be changed by adding constants to X and/or Y, for such transformations have the effect of simply sliding the scale of one or both axes. But if multiplication factors are used in coding, then the resultant slope, $[b]$, will be equal to $(b)(M_Y/M_X)$. Note that coding in no way alters the value of r^2 or the t or F statistics calculated for hypothesis testing.

A common situation involving multiplicative coding factors is one where the variables were recorded using certain units of measurement, and we want to determine what regression statistics would have resulted if other units of measurement had been used.

For the data in Examples 1, 2, and 3, $a = 0.715$ cm, and $b = 0.270$ cm/day, and $s_{Y \cdot X} = 0.218$ cm. If the wing length data were measured in inches, instead of in centimeters, there would have to be a coding by multiplying by 0.3937 in./cm (for there are 0.3937 inches in one centimeter). With $M_Y = 0.3937$ in./cm, $A_Y = 0$, $M_X = 1$, and $A_X = 0$, we can calculate that if a regression analysis were run on these data, where X was recorded in inches, the slope would be $[b] = (0.270$ cm/day$)(0.3937$ in./cm$) = 0.106$ in./day; the Y intercept would be $[a] = (0.715$ cm$)(0.3937$ in./cm$) = 0.281$ in.; and the standard error of estimate would be $s_{Y \cdot X} = (0.3937$ in./cm$)(0.218$ cm$) = 0.086$.

A situation employing coding by both adding a constant and multiplying a constant is when we have temperature measurements in degrees Celsius (or Fahrenheit) and wish to determine the regression equation that would have resulted had the data been recorded in degrees Fahrenheit (or Celsius). The appropriate coding constants for use in Appendix are determined by knowing that Celsius and Fahrenheit temperatures are related as follows:

$$\text{degrees Celsius} = \left(\frac{5}{9}\right)(\text{degrees Fahrenheit}) - \left(\frac{5}{9}\right)(32)$$

$$\text{degrees Fahrenheit} = \left(\frac{9}{5}\right)(\text{degrees Celsius}) + 32.$$

This is summarized elsewhere (Zar, 1968), as are the effects of multiplicative coding on logarithmically transformed data (Zar, 1967).

Simple Linear Regression

EXERCISES

1. The following data are the rates of oxygen consumption of birds, measured at different environmental temperatures:

Temperature (°C)	Oxygen consumption (ml/g/hr)
−18.	5.2
−15.	4.7
−10.	4.5
− 5.	3.6
0.	3.4
5.	3.1
10.	2.7
19.	1.8

(a) Calculate a and b for the regression of oxygen consumption rate on temperature.

(b) Test, by analysis of variance, the hypothesis $H_0: \beta = 0$.

(c) Test, by the t test, the hypothesis $H_0: \beta = 0$.

(d) Calculate the standard error of estimate of the regression.

(e) Calculate the coefficient of determination of the regression.

(f) Calculate the 95% confidence limits for β.

2. Utilize the regression equation computed for the data of Exercise 1.

(a) What is the mean rate of oxygen consumption in the population for birds at 15°C?

(b) What is the 95% confidence interval for this mean rate?

(c) If we randomly chose one additional bird at 15°C from the population, what would its rate of oxygen consumption be estimated to be?

(d) We can be 95% confident of this value lying between what limits?

3. The frequency of electrical impulses emitted from electric fish is measured from three fish at each of several temperatures. The resultant data are as follows:

Temperature (°C)	Impulse frequency (number/sec)
20	225, 230, 239
22	251, 259, 265
23	266, 273, 280
25	287, 295, 302
27	301, 310, 317
28	307, 313, 325
30	324, 330, 338

(a) Compute a and b for the linear regression equation relating impulse frequency to temperature.

(b) Test, by analysis of variance $H_0: \beta = 0$.

(c) Calculate the standard error of estimate of the regression.

(d) Calculate the coefficient of determination of the regression.

(e) Test H_0: The population regression is linear.

ANSWERS TO EXERCISES

1. (a) $b = -0.0878$ ml/g/hr/°C, $a = 3.78$ ml/g/hr.
(b) $H_0: \beta = 0$, $H_A: \beta \neq 0$; $F = 309$; reject H_0; $P \ll 0.0005$ [$P = 0.0000022$]. **(c)** $H_0: \beta = 0$, $H_A: \beta \neq 0$; $t = -17.6$; reject $H_0: P \ll 0.001$ [$P = 0.0000022$]. **(d)** $s_{Y \cdot X} = 0.17$ ml/g/hr; **(e)** $r^2 = 0.98$; **(f)** 95% confidence interval for $\beta = -0.0878 \pm 0.0122$; $L_1 = -0.1000$ ml/g/hr/°C, $L_2 = -0.0756$ ml/g/hr/°C.

2. (a) $\hat{Y} = 3.47 - (0.0878)(15) = 2.15$ ml/g/hr.
(b) $s_{\hat{Y}} = 0.1021$ ml/g/hr; $L_1 = 1.90$ ml/g/hr, $L_2 = 2.40$ ml/g/hr. **(c)** $\hat{Y} = 2.15$ ml/g/hr.

(d) $s_{\hat{Y}} = 0.1960$ ml/g/hr; $L_1 = 1.67$ ml/g/hr, $L_2 = 2.63$ ml/g/hr.

3. (a) $b = 9.73$ impulses/sec/°C, $a = 44.2$ impulses/sec. **(b)** $H_0: \beta = 0$, $H_A: \beta \neq 0$; $F = 311$; reject H_0; $P \ll 0.0005$ [$P < 10^{-13}$]. **(c)** $s_{Y \cdot X} = 8.33$ impulses/sec. **(d)** $r^2 = 0.94$. **(e)** H_0: The population regression is linear; H_A: The population regression is not linear; $F = 1.78$, do not reject H_0; $0.10 < P < 0.25$ [$P = 0.18$].

Comparing Simple Linear Regression Equations

A regression equation may be calculated for each of two or more samples of data to compare the regression relationships in the populations from which the samples came. We may ask whether the slopes of the regression lines are significantly different (as opposed to whether they may be estimating the same population slope, β). Then, if it is concluded that the slopes of the lines are not significantly different, we may want to test whether the several sets of data are from populations in which the population Y intercepts, as well as the slopes, are the same. In this chapter, procedures for testing differences among regression lines will be presented, as summarized in Figure 1.

1 COMPARING TWO SLOPES

The comparison of the slopes of two regression lines is demonstrated in Example 1. The regression relationship to be studied is the amount of water lost by salamanders maintained at various environmental temperatures. A regression line is determined using data from each of two species of salamanders. The regression line for 26 animals of species 1 is $10.57 + 2.97X$, and that for 30 animals of species 2 is $24.91 + 2.17X$; these two regression lines are shown in Figure 2. Temperature, the independent variable (X), is measured in degrees Celsius, and the dependent variable (Y) is measured in microliters (μl) of water per gram of body weight per hour. Example 1 shows the calculations of the slope of each of the two regression lines. In this example, the slope of the line expresses water loss, in μl/g/hr, for each temperature increase of 1°C. The raw data (the 26 X and Y data for species 1 and the 30 pairs of data for species 2) are not shown, but the sums of squares (Σx^2 and Σy^2) and sum of crossproducts (Σxy) for each line are given in this example. (The calculation of the Y intercepts is not shown.)

As shown is Example 1, a simple method for testing hypotheses about equality of two population regression coefficients involves the use of Student's t in a fashion analogous to that of testing for differences between two population means. The test statistic is

$$t = \frac{b_1 - b_2}{s_{b_1 - b_2}}, \tag{1}$$

Comparing Simple Linear Regression Equations

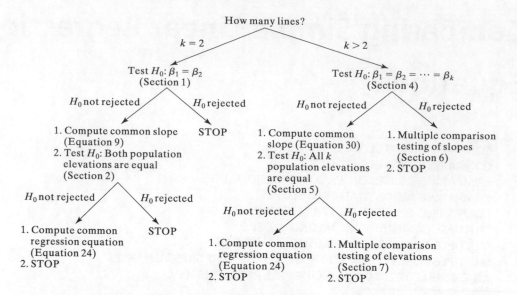

FIGURE 1: Flow chart for the comparison of regression lines.

EXAMPLE 1   Testing for Difference Between Two Population Regression Coefficients

For each of two species of salamanders, the data are for water loss (Y, measured as μl/g/hr) and environmental temperature (X, in °C).

$$H_0: \quad \beta_1 = \beta_2$$
$$H_A: \quad \beta_1 \neq \beta_2$$

For Species 1:

$n = 26$

$\sum x^2 = 1470.8712$

$\sum xy = 4363.1627$

$\sum y^2 = 13299.5296$

$b = \dfrac{4363.1627}{1470.8712} = 2.97$

residual SS = 13299.5296

$$-\frac{(4363.1627)^2}{1470.8712}$$

$$= 356.7317$$

residual DF $= 26 - 2 = 24$

For Species 2:

$n = 30$

$\sum x^2 = 2272.4750$

$\sum xy = 4928.8100$

$\sum y^2 = 10964.0947$

$b = \dfrac{4928.8100}{2272.4750} = 2.17$

residual SS = 10964.0947

$$-\frac{(4928.8100)^2}{2272.4750}$$

$$= 273.9142$$

residual DF $= 30 - 2 = 28$

$$(s_{Y \cdot X}^2)_p = \frac{356.7317 + 273.9142}{24 + 28} = 12.1278$$

$$s_{b_1 - b_2} = \sqrt{\frac{12.1278}{1470.8712} + \frac{12.1278}{2272.4750}} = 0.1165$$

388

$$t = \frac{2.97 - 2.17}{0.1165} = 6.867$$

$$\nu = 24 + 28 = 52$$

Reject H_0 if $|t| \geq t_{\alpha(2),\nu}$

$t_{0.05(2),52} = 2.007$; Reject H_0.

$$P < 0.001 \quad [P = 0.0000000081]$$

Calculation not shown:

$$a_1 = 10.57 \quad a_2 = 24.91$$

where the standard error of the difference between regression coefficients is

$$s_{b_1-b_2} = \sqrt{\frac{(s_{Y.X}^2)_p}{\left(\sum x^2\right)_1} + \frac{(s_{Y.X}^2)_p}{\left(\sum x^2\right)_2}}, \tag{2}$$

and the pooled residual mean square is calculated as

$$(s_{Y.X}^2)_p = \frac{(\text{residual SS})_1 + (\text{residual SS})_2}{(\text{residual DF})_1 + (\text{residual DF})_2}, \tag{3}$$

the subscripts 1 and 2 referring to the two regression lines being compared. The critical value of t for this test has $(n_1 - 2) + (n_2 - 2)$ degrees of freedom (i.e., the sum of the two residual degrees of freedom), namely

$$\nu = n_1 + n_2 - 4. \tag{4}$$

Just as the t test for difference between means assumes that $\sigma_1^2 = \sigma_2^2$, the preceding t test assumes that $(\sigma_{Y.X}^2)_1 = (\sigma_{Y.X}^2)_2$. The presence of the latter condition can be tested by the variance ratio test, $F = (s_{Y.X}^2)_{\text{larger}}/(s_{Y.X}^2)_{\text{smaller}}$; but this is usually not done due to the limitations of that test.

The $1 - \alpha$ confidence interval for the difference between two slopes, β_1 and β_2, is

$$(b_1 - b_2) \pm t_{\alpha(2),\nu} s_{b_1-b_2}, \tag{5}$$

where ν is as in Equation 4. Thus, for Example 1,

$$95\% \text{ confidence interval for } \beta_1 - \beta_2 = (2.97 - 2.17) \pm (t_{0.05(2),52})(0.1165)$$
$$= 0.80 \pm (2.007)(0.1165)$$
$$= 0.80 \ \mu\text{l/g/hr/}^\circ\text{C} \pm 0.23 \ \mu\text{l/g/hr/}^\circ\text{C};$$

and the upper and lower 95% confidence limits for $\beta_1 - \beta_2$ are $L_1 = 0.57 \ \mu\text{l/g/hr/}^\circ\text{C}$ and $L_2 = 1.03 \ \mu\text{l/g/hr/}^\circ\text{C}$.

If $H_0: \beta_1 = \beta_2$ is rejected (as in Example 1), we may wish to calculate the point where the two lines intersect. The intersection is at

$$X_I = \frac{a_2 - a_1}{b_1 - b_2}, \tag{6}$$

at which the value of $\hat{Y}$ may be computed either as

$$\hat{Y}_I = a_1 + b_1 X_I \tag{7}$$

or

$$\hat{Y}_I = a_2 + b_2 X_I. \tag{8}$$

The point of intersection of the two lines in Example 1 is at

$$X_I = \frac{24.91 - 10.57}{2.97 - 2.17} = 17.92°C$$

and

$$\hat{Y}_I = 1.057 + (2.97)(17.92) = 63.79 \ \mu l/g/hr/°C.$$

Figure 2 illustrates this intersection.

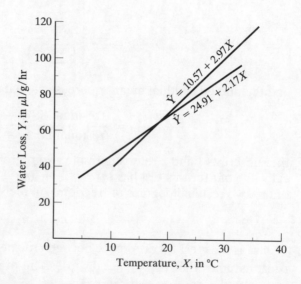

FIGURE 2: The two regression lines of Example 1. The two slopes are concluded to be significantly different and the two lines are found to intersect at $X_I = 17.92°C$ and $\hat{Y}_I = 63.79 \ \mu l/g/hr$.

If $H_0: \beta_1 = \beta_2$ is not rejected (as will be shown in Example 2), then an estimate of the population regression coefficient, β, underlying both b_1 and b_2 is called the *common* (or *weighted*) *regression coefficient:*

$$b_c = \frac{\left(\sum xy\right)_1 + \left(\sum xy\right)_2}{\left(\sum x^2\right)_1 + \left(\sum x^2\right)_2} \tag{9}$$

or, equivalently (but with more chance of rounding error),

$$b_c = \frac{\left(\sum x^2\right)_1 b_1 + \left(\sum x^2\right)_2 b_2}{\left(\sum x^2\right)_1 + \left(\sum x^2\right)_2}. \tag{10}$$

Equation 1 is a special case of

$$t = \frac{|b_1 - b_2| - \beta_0}{s_{b_1 - b_2}},\tag{11}$$

namely when $\beta_0 = 0$. By using Equation 11, we may test the hypothesis that the difference between two population regression coefficients is a specified magnitude; that is, $H_0: \beta_1 - \beta_2 = \beta_0$ may be tested against $H_A: \beta_1 - \beta_2 \neq \beta_0$.

One-tailed testing is also possible, asking whether one population regression coefficient is greater than the other. If we test $H_0: \beta_1 \geq \beta_2$ and $H_A: \beta_1 < \beta_2$, or $H_0: \beta_1 - \beta_2 \geq \beta_0$ versus $H_A: \beta_1 - \beta_2 < \beta_0$, then H_0 is rejected if $t \leq -t_{\alpha(1),\nu}$; if we test $H_0: \beta_1 \leq \beta_2$ and $H_A: \beta_1 > \beta_2$, or $H_0: \beta_1 - \beta_2 \leq \beta_0$ versus $H_A: \beta_1 - \beta_2 > \beta_0$, then we reject H_0 if $t \geq t_{\alpha(1),\nu}$. In either case, t is computed by Equation 1, or by Equation 11 if $\beta_0 \neq 0$.

An alternative method of testing $H_0: \beta_1 = \beta_2$ is by the analysis of covariance procedure of Section 4. However, if a computer program is not used, the preceding t test generally involves less computational effort.

(a) Power and Sample Size in Comparing Regressions. The procedure for consideration of power in correlation analysis can be used to estimate power and sample size in a regression analysis. Power and sample-size estimation when testing for difference between two correlation coefficients. Unfortunately, utilization of that procedure for comparing two regression coefficients is not valid—unless one has the rare case of $\left(\sum x^2\right)_1 = \left(\sum x^2\right)_2$ and $\left(\sum y^2\right)_1 = \left(\sum y^2\right)_2$ (Cohen, 1988: 110).

2 COMPARING TWO ELEVATIONS

If $H_0: \beta_1 = \beta_2$ is rejected, we conclude that two different populations of data have been sampled. However, if two population regression lines are not concluded to have different slopes (i.e., $H_0: \beta_1 = \beta_2$ is not rejected), then the two lines are assumed to be parallel. In the latter case, we often wish to determine whether the two population regressions have the same elevation (i.e., the same vertical position on a graph) and thus coincide.

To test the null hypothesis that the elevations of the two population regression lines are the same, the following quantities may be used in a t test:

sum of squares of X for common regression

$$= A_c = \left(\sum x^2\right)_1 + \left(\sum x^2\right)_2,\tag{12}$$

sum of crossproducts for common regression

$$= B_c = \left(\sum xy\right)_1 + \left(\sum xy\right)_2,\tag{13}$$

sum of squares of Y for common regression

$$= C_c = \left(\sum y^2\right)_1 + \left(\sum y^2\right)_2,\tag{14}$$

residual SS for common regression

$$= \mathrm{SS}_c = C_c - \frac{B_c^2}{A_c},\tag{15}$$

$$\text{residual DF for common regression} = \text{DF}_c = n_1 + n_2 - 3, \tag{16}$$

and

$$\text{residual MS for common regression} = (s^2_{Y \cdot X})_c = \frac{\text{SS}_c}{\text{DF}_c} \tag{17}$$

Then, the appropriate test statistic is

$$t = \frac{(\overline{Y}_1 - \overline{Y}_2) - b_c(\overline{X}_1 - \overline{X}_2)}{\sqrt{(s^2_{Y \cdot X})_c \left[\dfrac{1}{n_1} + \dfrac{1}{n_2} + \dfrac{(\overline{X}_1 - \overline{X}_2)^2}{A_c}\right]}}, \tag{18}$$

and the relevant critical value of t is that for $\nu = \text{DF}_c$. Example 2 and Figure 3 consider the regression of human systolic blood pressure on age for men over 40 years old. A regression equation was fitted for data for men in each of two different occupations. The two-tailed null hypothesis is that in the two sampled populations the regression elevations are the same. This also says that blood pressure is the same in both groups, after accounting for the effect of age. In the example, the H_0 of equal elevations is rejected, so we conclude that men in these two occupations do not have the same blood pressure. As an alternative to this t-testing procedure, the analysis of covariance of Section 4 may be used to test this hypothesis, but it generally requires more computational effort unless a computer package is used.

EXAMPLE 2   Testing for Difference Between Two Population Regression Coefficients and Elevations

The data are for systolic blood pressure (the dependent variable, Y, in millimeters of mercury [i.e., mm Hg]) and age (the independent variable, X, in years) for men over 40 years of age; the two samples are from different occupations.

For Sample 1:	For Sample 2:
$n = 13$	$n = 15$
$\overline{X} = 54.65$ yr	$\overline{X} = 56.93$ yr
$\overline{Y} = 170.23$ mm Hg	$\overline{Y} = 162.93$ mm Hg
$\sum x^2 = 1012.1923$	$\sum x^2 = 1659.4333$
$\sum xy = 1585.3385$	$\sum xy = 2475.4333$
$\sum y^2 = 2618.3077$	$\sum y^2 = 3848.9333$
$b = 1.57$ mm Hg/yr	$b = 1.49$ mm Hg/yr
$a = 84.6$ mm Hg	$a = 78.0$ mm Hg
residual SS = 135.2833	residual SS = 156.2449
residual DF = 11	residual DF = 13

$H_0: \quad \beta_1 = \beta_2$
$H_A: \quad \beta_1 \neq \beta_2$

$$(s_{Y \cdot X}^2)_p = \frac{135.2833 + 156.2449}{11 + 13} = 12.1470$$

$$\nu = 11 + 13 = 24$$

$$s_{b_1 - b_2} = 0.1392$$

$$t = \frac{1.57 - 1.49}{0.1392} = 0.575$$

$$t_{0.05(2),24} = 2.064; \text{ do not reject } H_0.$$

$$P > 0.50 \quad [P = 0.57]$$

H_0: The two population regression lines have the same elevation.
H_A: The two population regression lines do not have the same elevation.

$$A_c = 1012.1923 + 1659.4333 = 2671.6256$$

$$B_c = 1585.3385 + 2475.4333 = 4060.7718$$

$$C_c = 2618.3077 + 3848.9333 = 6467.2410$$

$$b_c = \frac{4060.7718}{2671.6256} = 1.520 \text{ mm Hg/yr}$$

$$SS_c = 6467.2410 - \frac{(4060.7718)^2}{2671.6256} = 295.0185$$

$$DF_c = 13 + 15 - 3 = 25$$

$$(s_{Y \cdot X}^2)_c = \frac{295.0185}{25} = 11.8007$$

$$t = \frac{(170.23 - 162.93) - 1.520(54.65 - 56.93)}{\sqrt{11.8007\left[\frac{1}{13} + \frac{1}{15} + \frac{(54.65 - 56.93)^2}{2671.6256}\right]}} = \frac{10.77}{1.3105} = 8.218$$

$$t_{0.05(2),25} = 2.060; \text{ reject } H_0.$$

$$P < 0.001 \quad [P = 0.0000000072]$$

If it is concluded that two population regressions do not have different slopes but do have different elevations, then the slopes computed from the two samples are both estimates of the common population regression coefficient, and the Y intercepts of the two samples are

$$a_1 = \overline{Y}_1 - b_c \overline{X}_1 \tag{19}$$

and

$$a_2 = \overline{Y}_2 - b_c \overline{X}_2, \tag{19a}$$

and the two regression equations may be written as

$$\hat{Y}_i = a_1 + b_c X_i \tag{20}$$

and

$$\hat{Y}_i = a_2 + b_c X_i \tag{20a}$$

393

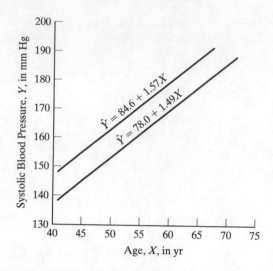

FIGURE 3: The two regression lines of Example 2.

(though this can be misleading if $X = 0$ is far from the range of X's in the sample). For the two lines in Example 2 and Figure 3,

$$\hat{Y}_i = 84.6 + 1.52X_i$$

and

$$\hat{Y}_i = 78.0 + 1.52X_i.$$

If it is concluded that two population regressions have neither different slopes nor different elevations, then both sample regressions estimate the same population regression, and this estimate may be expressed using the common regression coefficient, b_c, as well as a common Y intercept:

$$a_c = \overline{Y}_p - b_c\overline{X}_p, \tag{21}$$

where the pooled sample means of the two variables may be obtained as

$$\overline{X}_p = \frac{n_1\overline{X}_1 + n_2\overline{X}_2}{n_1 + n_2} \tag{22}$$

and

$$\overline{Y}_p = \frac{n_1\overline{Y}_1 + n_2\overline{Y}_2}{n_1 + n_2}. \tag{23}$$

Thus, when two samples have been concluded to estimate the same population regression, a single regression equation representing the regression in the sampled population would be

$$\hat{Y}_i = a_c + b_cX_i. \tag{24}$$

We may also use t to test one-tailed hypotheses about elevations. For data such as those in Example 2 and Figure 3, it might have been the case that one occupation was considered to be more stressful, and we may want to determine whether men in that occupation had higher blood pressure than men in the other occupation.

This t test of elevations is preferable to testing for difference between the two population Y intercepts. Difference between Y intercepts would be tested with the null hypothesis $H_0 : \alpha_1 = \alpha_2$, using the sample statistics a_1 and a_2, and could proceed with

$$t = \frac{a_1 - a_2}{s_{a_1 - a_2}},$$ (25)

where

$$s_{a_1 - a_2} = \sqrt{(s_{Y \cdot X}^2)_p \left[\frac{1}{n_1} + \frac{1}{n_2} + \frac{\overline{X}_1^2}{\left(\sum x^2\right)_1} + \frac{\overline{X}_2^2}{\left(\sum x^2\right)_2} \right]}$$ (26)

(the latter two equations are a special case of Equations 27 and 28). However, a test for difference between Y intercepts is generally not as advisable as a test for difference between elevations because it uses a point on each line that may lie far from the observed range of X's. There are many regressions for which the Y intercept has no importance beyond helping to define the line and in fact may be a sample statistic prone to misleading interpretation. In Figure 3, for example, discussion of the Y intercepts (and testing hypotheses about them) would require a risky extrapolation of the regression lines far below the range of X for which data were obtained. This would assume that the linear relationship that was determined for ages above 40 years also holds between $X = 0$ and $X = 40$ years, a seriously incorrect assumption in the present case dealing with blood pressures. Also, because the Y intercepts are so far from the mean values of X, their standard errors would be very large, and a test of H_0: $\alpha_1 = \alpha_2$ would lack statistical power.

3 COMPARING POINTS ON TWO REGRESSION LINES

If the slopes of two regression lines and the elevations of the two lines have not been concluded to be different, then the two lines are estimates of the same population regression line. If the slopes of two lines are not concluded to be different, but their elevations are declared different, then the population lines are assumed to be parallel, and for a given X_i, the corresponding $\hat{Y}_i$ on one line is different from that on the other line.

If the slopes of two population regression lines are concluded different, then the lines are intersecting rather than parallel. In such cases we may wish to test whether a $\hat{Y}$ on one line is the same as the $\hat{Y}$ on the second line at a particular X. For a two-tailed test, we can state the null hypothesis as H_0: $\mu_{\hat{Y}_1} = \mu_{\hat{Y}_2}$ and the alternate as H_A: $\mu_{\hat{Y}_1} \neq \mu_{\hat{Y}_2}$. The test statistic is

$$t = \frac{\hat{Y}_1 - \hat{Y}_2}{s_{\hat{Y}_1 - \hat{Y}_2}},$$ (27)

where

$$s_{\hat{Y}_1 - \hat{Y}_2} = \sqrt{(s_{Y \cdot X}^2)_p \left[\frac{1}{n_1} + \frac{1}{n_2} + \frac{(X - \overline{X}_1)^2}{\left(\sum x^2\right)_1} + \frac{(X - \overline{X}_2)^2}{\left(\sum x^2\right)_2} \right]},$$ (28)

and the degrees of freedom are the pooled degrees of freedom of Equation 4. Such a test is demonstrated in Example 3.

EXAMPLE 3   Testing for Difference Between Points on the Two Nonparallel Regression Lines of Example 1 and Figure 2. We Are Testing Whether the Volumes (Y) Are Different in the Two Groups at $X = 12°C$

H_0: $\mu_{\hat{Y}_1} = \mu_{\hat{Y}_2}$

H_A: $\mu_{\hat{Y}_1} \neq \mu_{\hat{Y}_2}$

Beyond the statistics given in Example 1, we need to know the following:

$$\overline{X}_1 = 22.93°C \quad \text{and} \quad \overline{X}_2 = 18.95°C.$$

We then compute:

$$\hat{Y}_1 = 10.57 + (2.97)(12) = 46.21 \ \mu l/g/hr$$
$$\hat{Y}_2 = 24.91 + (2.17)(12) = 50.95 \ \mu l/g/hr$$

$$s_{\hat{Y}_1 - \hat{Y}_2} = \sqrt{12.1278 \left[\frac{1}{26} + \frac{1}{30} + \frac{(12 - 22.93)^2}{1470.8712} + \frac{(12 - 18.95)^2}{2272.4750} \right]}$$

$$= \sqrt{2.1135} = 1.45 \ \mu l/g/hr$$

$$t = \frac{46.21 - 50.95}{1.45} = -3.269$$

$$\nu = 26 + 30 - 4 = 52$$

$$t_{0.05(2),52} = 2.007$$

As $|t| > t_{0.05(2),52}$, reject H_0.

$$0.001 < P < 0.002 \quad [P = 0.0019]$$

One-tailed testing is also possible. However, it should be applied with caution, as it assumes that each of the two predicted $\hat{Y}$'s has associated with it the same variance. Therefore, the test works best when the two lines have the same $\overline{X}$, the same $\sum x^2$, and the same n.

4   COMPARING MORE THAN TWO SLOPES

If the slopes of more than two regression equations are to be compared, the null hypothesis $H_0: \beta_1 = \beta_2 = \cdots = \beta_k$ may be tested, where k is the number of regressions. The alternate hypothesis would be that, in the k sampled populations, all k slopes are not the same. These hypotheses are analogous to those used in testing whether the means are the same in k samples. The hypothesis about equality of regression slopes may be tested by a procedure known as *analysis of covariance*.

Analysis of covariance (ANCOVA) encompasses a large body of statistical methods, and various kinds of ANCOVA are presented in many comprehensive texts. The following version of analysis of covariance suffices to test for the equality (sometimes called homogeneity) of regression coefficients (i.e., slopes). Just as an analysis of variance for $H_0: \mu_1 = \mu_2 = \cdots = \mu_k$ assumes that all k population variances are equal (i.e., $\sigma_1^2 = \sigma_2^2 = \cdots = \sigma_k^2$), the testing of $\beta_1 = \beta_2 = \cdots = \beta_k$ assumes that the residual mean squares in the k populations are all the same (i.e., $(\sigma^2_{Y \cdot X})_1 = (\sigma^2_{Y \cdot X})_2 = \cdots = (\sigma^2_{Y \cdot X})_k$). Heterogeneity of the k residual mean squares can be tested by Bartlett's test, but this generally is not done for the same reasons that the test is not often employed as a prelude to analysis-of-variance procedures.

The basic calculations necessary to compare k regression lines require quantities already computed: $\sum x^2$, $\sum xy$, $\sum y^2$ (i.e., total SS), and the residual SS and DF for each computed line (Table 1). The values of the k residual sums of squares may then be summed, yielding what we shall call the *pooled residual sum of squares*, SS_p; and the sum of the k residual degrees of freedom is the *pooled residual degrees of freedom*, DF_p. The values of $\sum x^2$, $\sum xy$, and $\sum y^2$ for the k regressions may each be summed, and from these sums a residual sum of squares may be calculated. The latter quantity will be termed the *common residual sum of squares*, SS_c.

TABLE 1: Calculations for Testing for Significant Differences Among Slopes and Elevations of k Simple Linear Regression Lines

	$\sum x^2$	$\sum xy$	$\sum y^2$	Residual SS	Residual DF
Regression 1	A_1	B_1	C_1	$SS_1 = C_1 - \dfrac{B_1^2}{A_1}$	$DF_1 = n_1 - 2$
Regression 2	A_2	B_2	C_2	$SS_2 = C_2 - \dfrac{B_2^2}{A_2}$	$DF_2 = n_2 - 2$
$\vdots$	$\vdots$	$\vdots$	$\vdots$	$\vdots$	$\vdots$
Regression k	A_k	B_k	C_k	$SS_k = C_k - \dfrac{B_k^2}{A_k}$	$DF_k = n_k - 2$
Pooled regression				$SS_p = \displaystyle\sum_{i=1}^{k} SS_i$	$DF_p = \displaystyle\sum_{i=1}^{k}(n_i - 2)$ $= \displaystyle\sum_{i=1}^{k} n - 2k$
Common regression	$A_c = \displaystyle\sum_{i=1}^{k} A_i$	$B_c = \displaystyle\sum_{i=1}^{k} B_i$	$C_c = \displaystyle\sum_{i=1}^{k} C_i$	$SS_c = C_c - \dfrac{B_c^2}{A_c}$	$DF_c = \displaystyle\sum_{i=1}^{k} n_i - k - 1$
Total regression*	A_t	B_t	C_t	$SS_t = C_t - \dfrac{B_t^2}{A_t}$	$DF_t = \displaystyle\sum_{i=1}^{k} n_i - 2$

* See Section 5 for explanation.

To test H_0: $\beta_1 = \beta_2 = \cdots = \beta_k$, we may calculate

$$F = \frac{\left(\dfrac{SS_c - SS_p}{k - 1}\right)}{\dfrac{SS_p}{DF_p}}, \tag{29}$$

a statistic with numerator and denominator degrees of freedom of $k - 1$ and DF_p, respectively.* Example 4 demonstrates this testing procedure for three regression lines calculated from three sets of data (i.e., $k = 3$).

---

*The quantity $SS_c - SS_p$ is an expression of variability among the k regression coefficients; hence, it is associated with $k - 1$ degrees of freedom.

If H_0: $\beta_1 = \beta_2 = \cdots = \beta_k$ is rejected, then we may wish to employ a multiple comparison test to determine which of the k population slopes differ from which others. This is analogous to the multiple-comparison testing employed after rejecting H_0: $\mu_1 = \mu_2 = \cdots = \mu_k$, and it is presented in Section 6.

If H_0: $\beta_1 = \beta_2 = \cdots = \beta_k$ is not rejected, then the common regression coefficient, b_c, may be used as an estimate of the β underlying all k samples:

$$b_c = \frac{\sum_{i=1}^{k} \left(\sum xy \right)_i}{\sum_{i=1}^{k} \left(\sum x^2 \right)_i}. \tag{30}$$

For Example 4, this is $b_c = 2057.66/1381.10 = 1.49$.

EXAMPLE 4 Testing for Difference Among Three Regression Functions*

	$\sum x^2$	$\sum xy$	$\sum y^2$	n	b	Residual SS	Residual DF
Regression 1	430.14	648.97	1065.34	24	1.51	86.21	22
Regression 2	448.65	694.36	1184.12	29	1.55	109.48	27
Regression 3	502.31	714.33	1186.52	30	1.42	170.68	28
Pooled regression						366.37	77
Common regression	1381.10	2057.66	3435.98		1.49	370.33	79
Total regression	2144.06	3196.78	5193.48	83		427.10	81

* The italicized values are those computed from the raw data; all other values are derived from them.

To test for differences among slopes: H_0, $\beta_1 = \beta_2 = \beta_3$; H_A: All three β's are not equal.

$$F = \frac{\dfrac{370.33 - 366.37}{3 - 1}}{\dfrac{366.37}{77}} = 0.42$$

As $F_{0.05(1),2,77} \cong 3.13$, do not reject H_0.

$$P > 0.25 \quad [P = 0.66]$$

$$b_c = \frac{2057.66}{1381.10} = 1.49$$

To test for differences among elevations,

H_0: The three population regression lines have the same elevation.

H_A: The three lines do not have the same elevation.

$$F = \frac{\dfrac{427.10 - 370.33}{3 - 1}}{\dfrac{370.33}{79}} = 6.06$$

As $F_{0.05(1),2,79} \cong 3.13$, reject H_0.

$$0.0025 < P < 0.005 \quad [P = 0.0036]$$

5 COMPARING MORE THAN TWO ELEVATIONS

Consider the case where it has been concluded that all k population slopes underlying our k samples of data are equal (i.e., H_0: $\beta_1 = \beta_2 = \cdots = \beta_k$ is not rejected). In this situation, it is reasonable to ask whether all k population regressions are, in fact, identical; that is, whether they have equal elevations as well as slopes, and thus the lines all coincide.

The null hypothesis of equality of elevations may be tested by a continuation of the analysis of covariance considerations outlined in Section 4. We can combine the data from all k samples and from the summed data compute $\sum x^2$, $\sum xy$, $\sum y^2$, a residual sum of squares, and residual degrees of freedom; the latter will be called the total residual sum of squares (SS_t) and total residual degrees of freedom (DF_t). (See Table 1.) The null hypothesis of equal elevations is tested with

$$F = \frac{\dfrac{SS_t - SS_c}{k - 1}}{\dfrac{SS_c}{DF_c}} \tag{31}$$

with $k - 1$ and DF_c degrees of freedom. An example of this procedure is offered in Example 4.

If the null hypothesis is rejected, we can then employ multiple comparisons to determine the location of significant differences among the elevations, as described in Section 6. If it is not rejected, then all k sample regressions are estimates of the same population regression, and the best estimate of that underlying population regression is given by Equation 24 using Equations 9 and 21.

6 MULTIPLE COMPARISONS AMONG SLOPES

If an analysis of covariance concludes that k population slopes are not all equal, we may employ a multiple-comparison procedure to determine which β's are different from which others. For example, the Tukey test may be employed to test for differences between each pair of β values, by H_0: $\beta_B = \beta_A$ and H_A: $\beta_B \neq \beta_A$, where A and B represent two of the k regression lines.

The test statistic is

$$q = \frac{b_B - b_A}{SE}. \tag{32}$$

If $\sum x^2$ is the same for lines A and B, use the standard error

$$SE = \sqrt{\frac{(s_{Y \cdot X}^2)_p}{\sum x^2}}. \tag{33}$$

If $\sum x^2$ is different for lines A and B, then use

$$\text{SE} = \sqrt{\frac{(s^2_{Y \cdot X})_p}{2}\left[\frac{1}{(\sum x^2)_B} + \frac{1}{(\sum x^2)_A}\right]}. \tag{34}$$

The degrees of freedom for determining the critical value of q are the pooled residual DF (i.e., DF_p in Table 1). Although it is not mandatory to have first performed the analysis of covariance before applying the multiple-comparison test, such a procedure is commonly followed.

The confidence interval for the difference between the slopes of population regressions A and B is

$$(b_B - b_A) \pm (t_{\alpha,\nu,k})(\text{SE}), \tag{35}$$

where $q_{\alpha,\nu,k}$ is from Table 5 from *Appendix: Statistical Tables and Graphs* and ν is the pooled residual DF (i.e., DF_p in Table 1).

If one of several regression lines is considered to be a control to which each of the other lines is to be compared, then the procedures of Dunnett's test are appropriate. Here,

$$\text{SE} = \sqrt{\frac{2(s^2_{Y \cdot X})_p}{\sum x^2}} \tag{36}$$

if $\sum x^2$ is the same for the control line and the line that is compared to the control line (line A), and

$$\text{SE} = \sqrt{(s^2_{Y \cdot X})_p\left[\frac{1}{(\sum x^2)_A} + \frac{1}{(\sum x^2)_{\text{control}}}\right]} \tag{37}$$

if it is not. Either two-tailed or one-tailed hypotheses may be thus tested.

The $1 - \alpha$ confidence interval for the difference between the slopes of the control line and the line that is compared to it (line A) is

$$(b_A - b_{\text{control}}) \pm \left(q'_{\alpha(2),\nu,k}\right)(\text{SE}), \tag{38}$$

where $q'_{\alpha(2),\nu,k}$ is from Table 6 from *Appendix: Statistical Tables and Graphs*.

To apply Scheffé's procedure, calculate SE as Equation 36 or 37, depending on whether $\sum x^2$ is the same for both lines.

7 MULTIPLE COMPARISONS AMONG ELEVATIONS

If the null hypothesis $H_0 : \beta_1 = \beta_2 = \cdots = \beta_k$ has *not* been rejected and the null hypothesis of all k elevations being equal *has* been rejected, then multiple-comparison procedures may be applied to conclude between which elevations there are differences in the populations sampled. The test statistic for the Tukey test is

$$q = \frac{|(\overline{Y}_A - \overline{Y}_B) - b_c(\overline{X}_A - \overline{X}_B)|}{\text{SE}}, \tag{39}$$

with DF_c degrees of freedom (see Table 1), where the subscripts A and B refer to the two lines the elevations of which are being compared, b_c is from Equation 30,

and

$$SE = \sqrt{\frac{(s_{Y \cdot X}^2)_c}{2}\left[\frac{1}{n_B} + \frac{1}{n_A} + \frac{(\overline{X}_A - \overline{X}_B)^2}{(\sum x^2)_B + (\sum x^2)_A}\right]}. \qquad (40)$$

If Dunnett's test is used to compare the elevation of a regression line (call it line A) and another line considered to be for a control set of data,

$$SE = \sqrt{(s_{Y \cdot X}^2)_c\left[\frac{1}{n_A} + \frac{1}{n_{\text{control}}} + \frac{(\overline{X}_A - \overline{X}_B)^2}{(\sum x^2)_A + (\sum x^2)_{\text{control}}}\right]}. \qquad (41)$$

Equation 41 would also be employed if Scheffé's test were being performed on elevations.

8 MULTIPLE COMPARISONS OF POINTS AMONG REGRESSION LINES

If it is concluded that there is no significant difference among the slopes of three or more regression lines (i.e., H_0: $\beta_1 = \beta_2 = \cdots = \beta_k$ is not rejected; see Section 4), then it would be appropriate to test for differences among elevations (see Sections 5 and 7). Occasionally, when the above null hypothesis is rejected it is desired to ask whether points on the several regression lines differ at a specific value of X. This can be done, as a multisample extension of Section 3, by modifying Equations 27 and 28. For each line the value of $\hat{Y}$ is computed at the specified X, as

$$\hat{Y}_i = a_i + b_c X \qquad (42)$$

and a Tukey test is performed for H_0: $\mu_{\hat{Y}_B} = \mu_{\hat{Y}_A}$ as

$$q = \frac{\hat{Y}_B - \hat{Y}_A}{SE}, \qquad (43)$$

where

$$SE = \sqrt{\frac{(s_{Y \cdot X}^2)_p}{2}\left[\frac{1}{n_B} + \frac{1}{n_A} + \frac{(X - \overline{X}_A)^2}{(\sum x^2)_B} + \frac{(X - \overline{X}_B)^2}{(\sum x^2)_A}\right]}, \qquad (44)$$

with DF_p degrees of freedom. An analogous Dunnett or Scheffé test would employ

$$SE = \sqrt{(s_{Y \cdot X}^2)_p\left[\frac{1}{n_B} + \frac{1}{n_A} + \frac{(X - \overline{X}_A)^2}{(\sum x^2)_B} + \frac{(X - \overline{X}_B)^2}{(\sum x^2)_A}\right]}. \qquad (45)$$

A special case of this testing is where we wish to test for differences among the Y intercepts (i.e., the values of $\hat{Y}$ when $X = 0$), although such a test is rarely appropriate. Equations 43 and 44 for the Tukey test would become

$$q = \frac{a_B - a_A}{SE}, \qquad (46)$$

and

$$SE = \sqrt{\frac{(s^2_{Y \cdot X})_p}{2} \left[\frac{1}{n_B} + \frac{1}{n_A} + \frac{(\overline{X}_B)^2}{(\sum x^2)_B} + \frac{(\overline{X}_A)^2}{(\sum x^2)_A} \right]},$$ (47)

respectively. The analogous Dunnett or Scheffé test for Y intercepts would employ

$$SE = \sqrt{(s^2_{Y \cdot X})_p \left[\frac{1}{n_B} + \frac{1}{n_A} + \frac{(\overline{X}_B)^2}{(\sum x^2)_B} + \frac{(\overline{X}_A)^2}{(\sum x^2)_A} \right]}.$$ (48)

9 AN OVERALL TEST FOR COINCIDENTAL REGRESSIONS

It is also possible to perform a single test for the null hypothesis that all k regression lines are coincident; that is, that the β's are all the same *and* that all of the α's are identical. This test would employ

$$F = \frac{\dfrac{SS_t - SS_p}{2(k-1)}}{\dfrac{SS_p}{DF_p}}$$ (49)

with $2(k-1)$ and DF_p degrees of freedom. If this F is not significant, then all k sample regressions are concluded to estimate the same population regression, and the best estimate of that population regression is that given by Equation 24.

Some statistical workers prefer this test to those of the preceding sections in this chapter. However, if the null hypothesis is rejected, it is still necessary to employ the procedures of the previous sections if we wish to determine whether the differences among the regressions are due to differences among slopes or among elevations.

EXERCISES

1. Given:
   For Sample 1; $n = 28$, $\sum x^2 = 142.35$, $\sum xy = 69.47$, $\sum y^2 = 108.77$, $\overline{X} = 14.7$, $\overline{Y} = 32.0$.
   For Sample 2: $n = 30$, $\sum x^2 = 181.32$, $\sum xy = 97.40$, $\sum y^2 = 153.59$, $\overline{X} = 15.8$, $\overline{Y} = 27.4$.

   (a) Test H_0: $\beta_1 = \beta_2$ vs. H_A: $\beta_1 \neq \beta_2$.

   (b) If H_0 in part (a) is not rejected, test H_0: The elevations of the two population regressions are the same, versus H_A: The two elevations are not the same.

2. Given:
   For Sample 1: $n = 33$, $\sum x^2 = 744.32$, $\sum xy = 2341.37$, $\sum y^2 = 7498.91$.

   For Sample 2: $n = 34$, $\sum x^2 = 973.14$, $\sum xy = 3147.68$, $\sum y^2 = 10366.97$.
   For Sample 3: $n = 29$, $\sum x^2 = 664.42$, $\sum xy = 2047.73$, $\sum y^2 = 6503.32$.
   For the total of all 3 samples: $n = 96$, $\sum x^2 = 3146.72$, $\sum xy = 7938.25$, $\sum y^2 = 20599.33$.

   (a) Test H_0: $\beta_1 = \beta_2 = \beta_3$, vs. H_A: All three β's are not equal.

   (b) If H_0: in part (a) is not rejected, test H_0: The three population regression lines have the same elevation, versus H_A: The lines do not have the same elevation.

ANSWERS TO EXERCISES

1. **(a)** H_0: $\beta_1 = \beta_2$; $b_1 = 0.488$, $b_2 = 0.537$; $s_{b_1 - b_2} = 0.202$; $t = -0.243$; as $t_{0.05(2),54} = 2.005$, do not reject H_0; $P > 0.50$ $[P = 0.81]$. **(b)** H_0: The elevations of the two population regressions are the same; H_A: The elevations of the two population regressions are not the same; $b_c = 0.516$; $t = 10.7$; as $t_{0.05(2),55} \cong 2.004$, reject H_0; $P \ll 0.001$ $[P = 2 \times 10^{-14}]$.

2. **(a)** H_0: $\beta_1 = \beta_2 = \beta_3$; H_A: All three β's are not equal; $F = 0.84$; as $F_{0.05(1),2,90} = 3.10$, do not reject H_0; $P > 0.25$ $[P = 0.44]$; $b_c = 3.16$. **(b)** H_0: The three population regression lines have the same elevation; H_A: The three lines do not all have the same elevation; $F = 4.61$; as $F_{0.05(1),2,90} = 3.10$, reject H_0; $0.01 < P < 0.025$ $[P = 0.012]$.

This page intentionally left blank

Simple Linear Correlation

Simple linear regression is the linear dependence of one variable (termed the dependent variable, Y) on a second variable (called the independent variable, X). In simple linear correlation, we also consider the linear relationship between two variables, but neither is assumed to be functionally dependent upon the other. An example of a correlation situation is the relationship between the wing length and tail length of a particular species of bird. The adjective *simple* refers to there being only two variables considered simultaneously. Coefficients of correlation are sometimes referred to as *coefficients of association*.

1 THE CORRELATION COEFFICIENT

Some authors refer to the two variables in a simple correlation analysis as X_1 and X_2. Here we employ the more common designation of X and Y, which does not, however, imply dependence of Y on X as it does in regression; nor does it imply a cause-and-effect relationship between the two variables. Indeed, correlation analysis yields the same results regardless of which variable is labeled X and which is Y.

The *correlation coefficient* (sometimes called the *simple* correlation coefficient,* indicating that the relationship of only two variables is being examined) is

---

*It is also called the Pearson product-moment correlation coefficient because of the algebraic expression of the coefficient, and the pioneering work on it, by Karl Pearson (1857–1936), who in 1896 was the first to refer to this measure as a correlation coefficient (David, 1995; Seal, 1967). This followed the major elucidation of the concept of correlation by Sir Francis Galton (1822–1911,

calculated as[*]

$$r = \frac{\sum xy}{\sqrt{\sum x^2 \sum y^2}}$$ (1)

Among other methods (e.g., Symonds, 1926), Equation 1 may be computed by this "machine formula":

$$r = \frac{\sum XY - \dfrac{\sum X \sum Y}{n}}{\sqrt{\left(\sum X^2 - \dfrac{(\sum X)^2}{n}\right)\left(\sum Y^2 - \dfrac{(\sum Y)^2}{n}\right)}}.$$ (2)

Although the denominator of Equations 1 and 2 is always positive, the numerator may be positive, zero, or negative, thus enabling r to be either positive, zero, or negative, respectively. A positive correlation implies that for an increase in the value of one of the variables, the other variable also increases in value; a negative correlation indicates that an increase in value of one of the variables is accompanied by a decrease in value of the other variable.[†] If $\sum xy = 0$, then $r = 0$, and there is zero correlation, denoting that there is no linear association between the magnitudes of the two variables; that is, a change in magnitude of one does not imply a change in magnitude of the other. Figure 1 presents these considerations graphically.[‡]

Also important is the fact that the absolute value of the numerator of Equation 1 can never be larger than the denominator. Thus, r can never be greater than 1.0 nor

---

cousin of Charles Darwin and proponent of human eugenics) in 1888 (who published it first with the terms *co-relation* and *reversion*). The symbol r can be traced to Galton's 1877–1888 discussion of regression in heredity studies (he later used r to indicate the slope of a regression line), and Galton developed correlation from regression. Indeed, in the early history of correlation, correlation coefficients were called Galton functions. The basic concepts of correlation, however, predated Galton's and Pearson's work by several decades (Pearson, 1920; Rodgers and Nicewander, 1988; Stigler, 1989; Walker, 1929: 92–102, 106, 109–110, 187). The term *coefficient of correlation* was used as early as 1892 by Francis Ysidro Edgeworth (1845–1926; Irish statistician and economist, whose uncle and grand-uncle [sic] was Sir Francis Beaufort, 1774–1857; Beaufort conceived the Beaufort Wind Scale) (Desmond, 2000; Pearson, 1920).

[*]The computation depicted in Equation 2 was first published by Harris (1910). The correlation coefficient may also be calculated as $r = \sum xy/[(n-1)s_X s_Y]$ (Walker, 1929: 111). It is also the case that $|r| = \sqrt{b_Y b_X}$, where b_Y is the regression coefficient if Y is treated as the dependent variable and b_X is the regression coefficient if X is treated as the dependent variable; that is, r is the geometric mean of b_Y and b_X; also, $|r| = \sqrt{(\text{Regression SS})/(\text{Total SS})}$; see also Rodgers and Nicewander (1988). In literature appearing within a couple of decades of Pearson's work, it was sometimes suggested that a correlation coefficient be computed using deviations from the median instead of from the mean (Eells, 1926; Pearson, 1920), which would result in a quantity not only different from r but without the latter's theoretical and practical advantages.

[†]The first explanation of negative correlation was in an 1892 paper (on shrimp anatomy) by English marine biologist Walter Frank Raphael Weldon (1860–1906) (Pearson, 1920).

[‡]Galton published the first two-variable scatter plot of data in 1885 (Rodgers and Nicewander, 1988).

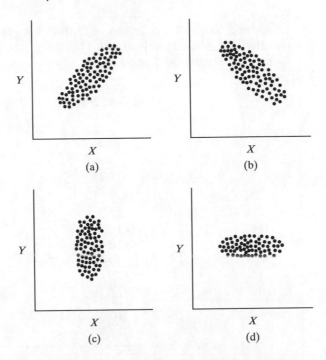

FIGURE 1: Simple linear correlation. (a) Positive correlation. (b) Negative correlation. (c) No correlation. (d) No correlation.

less than -1.0. Inspection of this equation further will reveal also that r has no units of measurement, for the units of both X and Y appear in both the numerator and denominator and thus cancel out arithmetically. A regression coefficient, b, may lie in the range of $-\infty \le b \le \infty$, and it expresses the magnitude of a change in Y associated with a unit change in X. But a correlation coefficient is unitless and $-1 \le r \le 1$. The correlation coefficient is not a measure of quantitative change of one variable with respect to the other, but it is a measure of strength of association between the two variables. That is, a large value of $|r|$ indicates a strong association between X and Y.

The coefficient of determination, r^2, was introduced as a measure of how much of the total variability in Y is accounted for by regressing Y on X. In a correlation analysis, r^2 (occasionally called the correlation index) may be calculated simply by squaring the correlation coefficient, r. It may be described as the amount of variability in one of the variables (either Y or X) accounted for by correlating that variable with the second variable.* As in regression analysis, r^2 may be considered to be a measure of the strength of the straight-line relationship.[†] The calculation of r and r^2 is demonstrated in Example 1a. Either r or r^2 can be used to express the strength of the relationship between the two variables.

---

*$1 - r^2$ may be referred to as the *coefficient of nondetermination*. A term found in older literature is *coefficient of alienation*: $\sqrt{1 - r^2}$, given by Galton in 1889 and named by T. L. Kelley in 1919 (Walker, 1929: 175).

[†]Ozer (1985) argued that there are circumstances where $|r|$ is a better coefficient of determination than r^2.

EXAMPLE 1a   Calculation of the Simple Correlation Coefficient and Coefficient of Determination. The Data Are Wing and Tail Lengths Among Birds of a Particular Species

Wing length (cm) (X)	Tail length (cm) (Y)
10.4	7.4
10.8	7.6
11.1	7.9
10.2	7.2
10.3	7.4
10.2	7.1
10.7	7.4
10.5	7.2
10.8	7.8
11.2	7.7
10.6	7.8
11.4	8.3

$n = 12$

$$\sum X = 128.2 \text{ cm} \qquad \sum Y = 90.8 \text{ cm}$$

$$\sum X^2 = 1371.32 \text{ cm}^2 \qquad \sum Y^2 = 688.40 \text{ cm}^2 \qquad \sum XY = 971.37 \text{ cm}^2$$

$$\sum x^2 = 1.7167 \text{ cm}^2 \qquad \sum y^2 = 1.3467 \text{ cm}^2 \qquad \sum xy = 1.3233 \text{ cm}^2$$

$$\text{correlation coefficient} = r = \frac{1.3233 \text{ cm}^2}{\sqrt{(1.7167 \text{ cm}^2)(1.3467 \text{ cm}^2)}} = 0.870$$

$$\text{coefficient of determination} = r^2 = 0.757$$

The standard error of the correlation coefficient may be computed as

$$s_r = \sqrt{\frac{1 - r^2}{n - 2}}. \tag{3}$$

The location of the decimal place at which the second significant digit of s_r is located may be noted and r may be expressed rounded off to that decimal place.

(a) Assumptions of Correlation Analysis.   Although no statistical assumptions need be satisfied in order to compute a correlation coefficient, there are assumptions underlying the testing of hypotheses about, and the determination of confidence intervals for, correlation coefficients.

In regression we assume that for each X the Y values have come at random from a normal population. However, in correlation, not only are the Y's at each X assumed to be normal, but also the X values at each Y are assumed to have come at random from a normal population. This situation is referred to as sampling from a bivariate normal distribution. The effect of deviations from the assumption of bivariate normality appears unimportant when there is, in fact, only slight correlation in the population; but if there is substantial population correlation, then there may be a marked adverse effect of such nonnormality, this effect not being diminished by increasing sample size (Kowalkski, 1972; Norris and Hjelm, 1961). Data transformation may be considered for X, for Y, or for both variables, in striving to achieve bivariate normality.

2 HYPOTHESES ABOUT THE CORRELATION COEFFICIENT

The correlation coefficient, r, that we calculate from a sample is an estimate of a population parameter, namely the correlation coefficient in the population that was sampled. This parameter is denoted by ρ, lowercase Greek rho. If we wish to ask whether there is, in fact, a correlation between Y and X in the population, we can test $H_0: \rho = 0$. We do this, as in Example 1b, by the familiar Student's t considerations, using

$$t = \frac{r}{s_r}, \tag{4}$$

EXAMPLE 1b   **Testing $H_0: \rho = 0$ versus $H_A: \rho \neq 0$. The Data Are Those of Example 1a**

From Example 1a: $r = 0.870$.

To test $H_0: \rho = 0$ versus $H_A: \rho \neq 0$:

$$\text{standard error of } r = s_r = \sqrt{\frac{1 - r^2}{n - 2}} = \sqrt{\frac{1 - (0.870)^2}{12 - 2}} = 0.156$$

$r_{0.05(2),10} = 0.576$ (from Table 17 from *Appendix: Statistical Tables and Graphs*)

Therefore, reject H_0.

$$P < 0.001$$

Or:

$$t = \frac{r}{s_r} = \frac{0.870}{0.156} = 5.58$$

$t_{0.05(2),10} = 2.228$

Therefore, reject H_0.

$$P < 0.001 \quad [P = 0.00012]$$

Or:

$$F = \frac{1 + |r|}{1 - |r|} = \frac{1.870}{0.130} = 14.4$$

$F_{0.05(2),10,10} = 3.72$

Therefore, reject H_0.

$$P < 0.001 \quad [P = 0.00014]$$

---

where the standard error of r is calculated by Equation 3, and the degrees of freedom are $\nu = n - 2$. (Fisher, 1921, 1925b: 157).* The null hypothesis is rejected if $|t| \geq t_{\alpha(2),\nu}$.

Alternatively, this two-tailed hypothesis may be tested using

$$F = \frac{1 + |r|}{1 - |r|} \tag{5}$$

(Cacoullos, 1965), where the critical value is $F_{\alpha(2),\nu,\nu}$. (See Example 1b.) Or, critical values of $|r|$ (namely, $r_{\alpha(2),\nu}$) may be read directly from Table 17 from *Appendix: Statistical Tables and Graphs*.[†]

One-tailed hypotheses about the population correlation coefficient may also be tested by the aforementioned procedures. For the hypotheses $H_0: \rho \leq 0$ and $H_A: \rho > 0$, compute either t or F (Equations 4 or 5, respectively) and reject H_0 if r is positive and either $t \geq t_{\alpha(1),\nu}$, or $F \geq F_{\alpha(1),\nu,\nu}$, or $r \geq r_{\alpha(1),\nu}$. To test $H_0: \rho \geq 0$ vs. $H_A: \rho < 0$, reject H_0 if r is negative and either $|t| \geq t_{\alpha(1),\nu}$, or $F \geq F_{\alpha(1),\nu}$, or $|r| \geq r_{\alpha(1),\nu}$.

If we wish to test $H_0: \rho = \rho_0$ for any ρ_0 other than zero, however, Equations 4 and 5 and Table 17 from *Appendix: Statistical Tables and Graphs* are not applicable. Only for $\rho_0 = 0$ can r be considered to have come from a distribution approximated by the normal, and if the distribution of r is not normal, then the t and F statistics may not be validly employed. Fisher (1921, 1925b: 162) dealt with this problem when he proposed a transformation enabling r to be converted to a value, called z, which estimates a population parameter, ζ (lowercase Greek zeta), that *is* normally distributed. The transformation[‡] is

$$z = 0.5 \ln \left(\frac{1 + r}{1 - r} \right). \tag{9}$$

For values of r between 0 and 1, the corresponding values of Fisher's z will lie between 0 and $+\infty$; and for r's from 0 to -1, the corresponding z's will fall between 0 and $-\infty$.

---

*As an aside, t may also be computed as follows (Martín Andrés, Herranz Tejedor, and Silva Mato, 1995): Consider all N data (where $N = n_1 + n_2$) to be a sample of measurements, and associate with each datum either 0, if the datum is a value of X, or 1, if it is a value of Y; consider this set of N zeros and ones to be a second sample of data. Then calculate t for the two samples, as would be done in a two-sample t-test. This concept will be used in Section 11b.

[†]Critical values of r may also be calculated as

$$r_{\alpha,\nu} = \sqrt{\frac{t_{\alpha,\nu}^2}{t_{\alpha,\nu}^2 + \nu}}, \tag{6}$$

where α may be either one tailed or two tailed, and $\nu = n - 2$. If a regression analysis is performed, rather than a correlation analysis, the probability of rejection of $H_0: \beta = 0$ is identical to the probability of rejecting $H_0: \rho = 0$. Also, r is related to b as

$$r = \frac{s_X}{s_Y} b, \tag{7}$$

where s_X and s_Y are the standard deviations of X and Y, respectively.

[‡]z is also equal to $r + r^3/3 + r^5/5 \ldots$ and is a quantity that mathematicians recognize as the inverse hyperbolic tangent of r, namely $z = \tanh^{-1} r$. The transformation of z to r, given in Table 19 from *Appendix: Statistical Tables and Graphs*, is

$$r = \tanh z \quad \text{or} \quad r = \frac{e^{2z} - 1}{e^{2z} + 1}. \tag{8}$$

Simple Linear Correlation

For convenience, we may utilize Table 18 from *Appendix: Statistical Tables and Graphs* to avoid having to perform the computation of Equation 8 to transform r to z.[*]

For the hypothesis $H_0: \rho = \rho_0$, then we calculate a normal deviate, as

$$Z = \frac{z - \zeta_0}{\sigma_z}, \tag{10}$$

where z is the transform of r; ζ_0 is the transform of the hypothesized coefficient, ρ_0; and the standard error of z is approximated by

$$\sigma_z = \sqrt{\frac{1}{n - 3}} \tag{11}$$

(Fisher, 1925b: 162), an approximation that improves as n increases.

In Example 1a, $r = 0.870$ was calculated. If we had desired to test $H_0: \rho = 0.750$, we would have proceeded as shown in Example 2. Recall that the critical value of a normal deviate may be obtained readily from the bottom line of the t table (Table 3 from *Appendix: Statistical Tables and Graphs*), because $Z_{\alpha(2)} = t_{\alpha(2),\infty}$.

EXAMPLE 2   Testing $H_0: \rho = \rho_0$, Where $\rho_0 \neq 0$

$r = 0.870$

$n = 12$

$H_0: \rho = 0.750;$ $H_A: \rho \neq 0.750.$

$z - 0.5 \ln\left(\frac{1 + 0.870}{1 - 0.870}\right) = 1.3331$

$\zeta_0 = 0.9730$

$Z = \dfrac{z - \zeta_0}{\sqrt{\dfrac{1}{n - 3}}} = \dfrac{1.3331 \quad 0.9730}{\sqrt{\dfrac{1}{9}}} = \dfrac{0.3601}{0.3333} = 1.0803$

$Z_{0.05(2)} = t_{0.05(2),\infty} = 1.960$

Therefore, do not reject H_0.

$$0.20 < P < 0.50 \quad [P = 0.28]$$

One-tailed hypotheses may also be tested, using $Z_{\alpha(1)}$ (or $t_{\alpha(1),\infty}$) as the critical value. For $H_0: \rho \leq \rho_0$ and $H_A: \rho > \rho_0$, H_0 is rejected if $Z \geq Z_{\alpha(1)}$, and for $H_0: \rho \geq \rho_0$ versus $H_A: \rho < \rho_0$, H_0 is rejected if $Z \leq -Z_{\alpha(1)}$.

If the variables in correlation analysis have come from a bivariate normal distribution, as often may be assumed, then we may employ the aforementioned procedures, as well as those that follow. Sometimes only one of the two variables may be assumed to have been obtained randomly from a normal population. It may be possible to employ a data transformation to remedy this situation. If that cannot be done, then the hypothesis $H_0: \rho = 0$ (or its associated one-tailed hypotheses) may be tested, but none of the other testing procedures of this chapter (except for the methods

---

[*]As noted at the end of Section 7, there is a slight and correctable bias in z. Unless n is very small, however, this correction will be insignificant and may be ignored.

of Section 9) are valid. If neither variable came from a normal population and data transformations do not improve this condition, then we may turn to the procedures of Section 9.

3 CONFIDENCE INTERVALS FOR THE POPULATION CORRELATION COEFFICIENT

Confidence limits on ρ may be determined by a procedure related to Equation 5; the lower and upper confidence limits are

$$L_1 = \frac{(1 + F_\alpha)r + (1 - F_\alpha)}{(1 + F_\alpha) + (1 - F_\alpha)r} \tag{12}$$

and

$$L_2 = \frac{(1 + F_\alpha)r - (1 - F_\alpha)}{(1 + F_\alpha) - (1 - F_\alpha)r}, \tag{13}$$

respectively, where $F_\alpha = F_{\alpha(2),\nu,\nu}$ and $\nu = n - 2$ (Muddapur, 1988).* This is shown in Example 3.

Fisher's transformation may be used to approximate these confidence limits, although the confidence interval will generally be larger than that from the foregoing procedure, and the confidence coefficient may occasionally (and undesirably) be less than $1 - \alpha$ (Jeyaratnam, 1992). By this procedure, we convert r to z (using Table 18 from *Appendix: Statistical Tables and Graphs*); then the $1 - \alpha$ confidence limits may be computed for ζ:

$$z \pm Z_{\alpha(2)}\sigma_z \tag{16}$$

or, equivalently,

$$z \pm t_{\alpha(2),\infty}\sigma_z. \tag{17}$$

The lower and upper confidence limits, L_1 and L_2, are both z values and may be transformed to r values, using Table 19 from *Appendix: Statistical Tables and Graphs* or Equation 9. Example 3 demonstrates this procedure. Note that although the confidence limits for ζ are symmetrical, the confidence limits for ρ are not.

4 POWER AND SAMPLE SIZE IN CORRELATION

(a) Power in Correlation. If we test H_0: $\rho = 0$ at the α significance level, with a sample size of n, then we may estimate the probability of correctly rejecting H_0 when ρ_0 is in fact a specified value other than zero. This is done by using the Fisher z transformation for the critical value of r and for the sample r (from Table 18 from *Appendix: Statistical Tables and Graphs* or Equation 8); let us call these two transformed values z_α and z, respectively. Then, the power of the test for H_0: $\rho = 0$ is $1 - \beta(1)$, where $\beta(1)$ is the one-tailed probability of the normal deviate

$$Z_{\beta(1)} = (z - z_\alpha)\sqrt{n - 3}, \tag{18}$$

---

*Jeyaratnam (1992) asserts that the same confidence limits are obtained by

$$L_1 = \frac{r - w}{1 - rw} \quad \text{and} \tag{14}$$

$$L_2 = \frac{r + w}{1 + rw}, \tag{15}$$

where w is $r_{\alpha,\nu}$ from Equation 6 using the two-tailed $t_{\alpha(2),\nu}$.

EXAMPLE 3   Setting Confidence Limits for a Correlation Coefficient. This Example Uses the Data of Example 1a

$r = 0.870$, $n = 12$, $v = 10$, $\alpha = 0.05$

For the 95% confidence interval for ρ:

$F_\alpha = F_{0.05(2),10,10} = 3.72$; so

$$L_1 = \frac{(1 + F_\alpha)r + (1 - F_\alpha)}{(1 + F_\alpha) + (1 - F_\alpha)r} = \frac{4.11 - 2.72}{4.72 - 2.37} = 0.592$$

$$L_2 = \frac{(1 + F_\alpha)r - (1 - F_\alpha)}{(1 + F_\alpha) - (1 - F_\alpha)r} = \frac{4.11 + 2.72}{4.72 + 2.37} = 0.963.$$

For the Fisher approximation:

$r = 0.870$; therefore, $z = 1.3331$.

$$\sigma_z = \sqrt{\frac{1}{n - 3}} = 0.3333$$

$$
\begin{aligned}
\text{95\% confidence interval for } \zeta = z &\pm Z_{0.05(2)}\sigma_z \\
&= z \pm t_{0.05(2),\infty}\sigma_z \\
&= 1.3331 \pm (1.9600)(0.3333) \\
&= 1.3331 \pm 0.6533
\end{aligned}
$$

$L_1 = 0.680$; $L_2 = 1.986$

These confidence limits are in terms of z. For the 95% confidence limits for ρ, transform L_1 and L_2 from z to r: $L_1 = 0.592$, $L_2 = 0.963$.

Instead of using the Appendix table, this confidence-limit transformation from z to r may be done using Equation 9:

$$L_1 = \frac{e^{2(0.680)} - 1}{e^{2(0.680)} + 1} = \frac{2.8962}{4.8962} = 0.592$$

$$L_2 = \frac{e^{2(1.986)} - 1}{e^{2(1.986)} + 1} = \frac{52.0906}{54.0906} = 0.963.$$

(Cohen, 1988: 546), as demonstrated in Example 4. This procedure may be used for one-tailed as well as two-tailed hypotheses, so α may be either $\alpha(1)$ or $\alpha(2)$, respectively.

(b) Sample Size for Correlation Hypothesis Testing.   If the desired power is stated, then we can ask how large a sample is required to reject H_0: $\rho = 0$ if it is truly false with a specified $\rho_0 \neq 0$. This can be estimated (Cohen, 1988: 546) by calculating

$$n = \left(\frac{Z_{\beta(1)} + Z_\alpha}{\zeta_0}\right)^2 + 3, \tag{19}$$

where ζ_0 is the Fisher transformation of the ρ_0 specified, and the significance level, α, can be either one-tailed or two-tailed. This procedure is shown in Example 5a.

> **EXAMPLE 4**   Determination of Power of the Test of $H_0: \rho = 0$ in Example 1b
>
> $n = 12; \ \nu = 10$
>
> $r = 0.870$, so $z = 1.3331$
>
> $r_{0.05(2),10} = 0.576$, so $z_{0.05} = 0.6565$
>
> $Z_{\beta(1)} = (1.3331 - 0.6565)\sqrt{12 - 3}$
>
> $= 2.03$
>
> $P(Z \geq 2.03) = 0.0212 = \beta$. Therefore, the power of the test is $1 - \beta = 0.98$.

> **EXAMPLE 5a**   Determination of Required Sample Size in Testing $H_0: \rho = 0$
>
> We desire to reject $H_0: \rho = 0$ 99% of the time when $|\rho| \geq 0.5$ and the hypothesis is tested at the 0.05 level of significance. Therefore, $\beta(1) = 0.01$ and (from the last line of Table 3 from *Appendix: Statistical Tables and Graphs*) and $Z_{\beta(1)} = 2.3263$; $\alpha(2) = 0.05$ and $Z_{\alpha(2)} = 1.9600$; and, for $r = 0.5$, $z = 0.5493$.
>   Then
>
> $$n = \left(\frac{2.3263 + 1.9600}{0.5493}\right)^2 + 3 = 63.9,$$
>
> so a sample of size at least 64 should be used.

(c) Hypothesizing ρ Other Than 0.   For the two-tailed hypothesis $H_0: \rho = \rho_0$, where $\rho_0 \neq 0$, the power of the test is determined from

$$Z_{\beta(1)} = |z - z_0| - z_{\alpha(2)}\sqrt{n - 3} \tag{20}$$

instead of from Equation 18; here, z_0 is the Fisher transformation of ρ_0. One-tailed hypotheses may be addressed using $\alpha(1)$ in place of $\alpha(2)$ in Equation 20.

(d) Sample Size for Confidence Limits.   After we calculate a sample correlation coefficient, r, as an estimate of a population correlation coefficient, ρ, we can estimate how large a sample would be needed from this population to determine a confidence interval for ρ that is no greater than a specified size.

     The confidence limits in Example 5b, determined from a sample of $n = 12$, define a confidence interval having a width of $0.963 - 0.592 = 0.371$. We could ask how large a sample from this population would be needed to state a confidence interval no wider than 0.30. As shown in Example 5b, this sample size may be estimated by the iterative process of applying Equations 12 and 13. Because the desired size of the confidence interval (0.30) is smaller than the confidence-interval width obtained from the sample of size 12 in Example 3 (0.371), we know that a sample larger than 12 would be needed. Example 5b shows the confidence interval calculated for $n = 15$ (0.31), which is a little larger than desired; a confidence interval is calculated for $n = 16$, which is found to be the size desired (0.30). If the same process is used for determining the sample size needed to obtain a confidence interval no larger than 0.20, it is estimated that n would have to be at least 30.

EXAMPLE 5b   Determination of Required Sample Size in Expressing Confidence Limits for a Correlation Coefficient

If a calculated r is 0.870 (as in Example 1a), and a 95% confidence interval no wider than 0.30 is desired for estimating ρ, the following iterative process may be employed:

If $n = 15$ were used, then $\nu = 13$, and $F_{0.05(2),13,13} = 3.12$, so

$$L_1 = \frac{(1 + F_\alpha)r + (1 - F_\alpha)}{(1 + F_\alpha) + (1 - F_\alpha)r} = \frac{(4.12)(0.870) - 2.12}{4.12 - (2.12)(0.870)} = \frac{3.584 - 2.12}{4.12 - 1.844}$$

$$= \frac{1.464}{2.276} = 0.643$$

$$L_2 = \frac{(1 + F_\alpha)r - (1 - F_\alpha)}{(1 + F_\alpha) - (1 - F_\alpha)r} = \frac{(4.12)(0.870) + 2.12}{4.12 + (2.12)(0.870)} = \frac{3.584 + 2.12}{4.12 + 1.844}$$

$$= \frac{5.704}{5.964} = 0.956$$

and the width of the confidence interval is $L_2 - L_1 = 0.956 - 0.643 = 0.31$, which is a little larger than desired, so larger n is needed.

If $n = 20$ were used, then $\nu = 18$, and $F_{0.05(2),18,18} = 2.60$, so

$$L_1 = \frac{(1 + F_\alpha)r + (1 - F_\alpha)}{(1 + F_\alpha) + (1 - F_\alpha)r} = \frac{(3.60)(0.870) - 1.60}{3.60 - (1.60)(0.870)} = \frac{3.132 - 1.60}{3.60 - 1.392}$$

$$= \frac{1.532}{2.208} = 0.694$$

$$L_2 = \frac{(1 + F_\alpha)r - (1 - F_\alpha)}{(1 + F_\alpha) - (1 - F_\alpha)r} = \frac{3.132 + 1.60}{3.60 + 1.392} = \frac{4.732}{4.992} = 0.948$$

and the width of the confidence interval is $L_2 - L_1 = 0.948 - 0.694 = 0.25$, which is smaller than that desired, so a smaller n may be used.

If $n = 16$ were used, then $\nu = 14$, and $F_{0.05(2),14,14} = 2.98$, so

$$L_1 = \frac{(1 + F_\alpha)r + (1 - F_\alpha)}{(1 + F_\alpha) + (1 - F_\alpha)r} = \frac{(3.98)(0.870) - 1.98}{3.98 - (1.98)(0.870)} = \frac{3.463 - 1.98}{3.98 - 1.723}$$

$$= \frac{1.483}{2.257} = 0.657$$

$$L_2 = \frac{(1 + F_\alpha)r - (1 - F_\alpha)}{(1 + F_\alpha) - (1 - F_\alpha)r} = \frac{3.463 + 1.98}{3.98 + 1.723} = \frac{5.443}{5.703} = 0.954$$

and the width of the confidence interval is $L_2 - L_1 = 0.954 - 0.657 = 0.30$, so it is estimated that a sample size of at least 16 should be used to obtain the desired confidence interval.

To calculate the desired confidence interval using the Fisher transformation, $r = 0.870$; $z = 1.3331$ (e.g., from Table 18 from *Appendix: Statistical Tables and Graphs*); $Z_{0.05(2)} = 1.9600$.

If $n = 15$ were used, then $\sigma_z = \sqrt{\dfrac{1}{15 - 3}} = 0.2887$.

The 95% confidence interval for ζ is $1.3331 \pm (1.9600)(0.2887) = 1.3331 \pm 0.5658$.

Simple Linear Correlation

The confidence limits are

$$L_1 = 0.7672; \quad L_2 = 1.8990$$

For the 95% confidence interval for ρ, transform this L_1 and L_2 for z to L_1 and L_2 for r (e.g., using Table 19 from *Appendix: Statistical Tables and Graphs*):

$$L_1 = 0.645; \quad L_2 = 0.956$$

and the width of the confidence interval is estimated to be $0.956 - 0.645 = 0.31$, which is a little larger than that desired, so a larger n should be used.

If $n = 16$ were used, then $\sigma_z = \sqrt{\dfrac{1}{16 - 3}} = 0.2774$.

For the 95% confidence interval for $\zeta = 1.3331 \pm (1.9600)(0.2774)$
$$= 1.3331 \pm 0.5437$$

and $L_1 = 0.7894; \quad L_2 = 1.8768$

For the 95% confidence interval for ρ, transform the z'a to r':

The confidence limits are

$$L_1 = 0.658; \quad L_2 = 0.954$$

and the width of the confidence interval is estimated to be $0.954 - 0.658 = 0.30$, so it is estimated that a sample size of at least 16 should be used.

5 COMPARING TWO CORRELATION COEFFICIENTS

Hypotheses (either one-tailed or two-tailed) about two correlation coefficients may be tested by the use of

$$Z = \frac{z_1 - z_2}{\sigma_{z_1 - z_2}}, \tag{21}$$

where

$$\sigma_{z_1 - z_2} = \sqrt{\frac{1}{n_1 - 3} + \frac{1}{n_2 - 3}}. \tag{22}$$

If $n_1 = n_2$, then Equation 22 reduces to

$$\sigma_{z_1 - z_2} = \sqrt{\frac{2}{n - 3}}, \tag{23}$$

where n is the size of each sample. The use of the Fisher z transformation both normalizes the underlying distribution of each of the correlation coefficients, r_1 and r_2, and stabilizes the variances of these distributions (Winterbottom, 1979). The multisample hypothesis test recommended by Paul (1988), and presented in Section 7, may be used for the two-tailed two-sample hypothesis mentioned previously. It tends to result in a probability of Type I error that is closer to the specified α; but, as it also tends to be larger than α, I do not recommend it for the two-sample case. The preferred procedure for testing the two-tailed hypotheses, $H_0: \rho_1 = \rho_2$ versus $H_A: \rho_1 = \rho_2$, is to employ Equation 21, as shown in Example 6.* One-tailed hypotheses may be tested using one-tailed critical values, namely $Z_{\alpha(1)}$.

---

*A null hypothesis such as $H_0: \rho_1 - \rho_2 = \rho_0$, where $\rho_0 \neq 0$, might be tested by substituting $|z_1 - z_2| - \zeta_0$ for the numerator in Equation 21, but no utility for such a test is apparent.

Simple Linear Correlation

EXAMPLE 6   Testing the Hypothesis $H_0: \rho_1 = \rho_2$

For a sample of 98 bird wing and tail lengths, a correlation coefficient of 0.78 was calculated. A sample of 95 such measurements from a second bird species yielded a correlation coefficient of 0.84. Let us test for the equality of the two population correlation coefficients.

$H_0: \rho_1 = \rho_2; \quad H_A: \rho_1 \neq \rho_2$

$r_1 = 0.78 \qquad r_2 = 0.84$

$z_1 = 1.0454 \qquad z_2 = 1.2212$

$n_1 = 98 \qquad n_2 = 95$

$$Z = \frac{1.0454 - 1.2212}{\sqrt{\dfrac{1}{n_1 - 3} + \dfrac{1}{n_2 - 3}}} = \frac{-0.1758}{0.1463} = -1.202$$

$Z_{0.05(2)} = t_{0.05(2),\infty} = 1.960$

Therefore, do not reject H_0.

$$0.20 < P < 0.50 \quad [P = 0.23]$$

The common correlation coefficient may then be computed as

$$z_w = \frac{(n_1 - 3)z_1 + (n_2 - 3)z_2}{(n_1 - 3) + (n_2 - 3)} = \frac{(95)(1.0454) + (92)(1.2212)}{95 + 92} = 1.1319$$

$r_w = 0.81.$

Occasionally we want to test for equality of two correlation coefficients that are not independent. For example, if Sample 1 in Example 6 were data from a group of 98 young birds, and Sample 2 were from 95 of these birds when they were older (three of the original birds having died or escaped), the two sets of data should not be considered to be independent. Procedures for computing r_1 and r_2, taking dependence into account, are reviewed by Steiger (1980).

(a) Common Correlation Coefficient. As in Example 6, a conclusion that $\rho_1 = \rho_2$ would lead us to say that both of our samples came from the same population of data, or from two populations with identical correlation coefficients. In such a case, we may combine the information from the two samples to calculate a better estimate of a single underlying ρ. Let us call this estimate the *common*, or *weighted*, *correlation coefficient*. We obtain it by converting

$$z_w = \frac{(n_1 - 3)z_1 + (n_2 - 3)z_2}{(n_1 - 3) + (n_2 - 3)} \tag{24}$$

to its corresponding r value, r_w, as shown in Example 6. If both samples are of equal size (i.e., $n_1 = n_2$), then the previous equation reduces to

$$z_w = \frac{z_1 + z_2}{2}. \tag{25}$$

Table 19 from *Appendix: Statistical Tables and Graphs* gives the conversion of z_w to the common correlation coefficient, r_w (which estimates the common population coefficient, ρ). Paul (1988) has shown that if ρ is less than about 0.5, then a better estimate of that parameter utilizes

$$z_w = \frac{(n_1 - 1)z_1' + (n_2 - 1)z_2'}{(n_1 - 1) + (n_2 - 1)}, \tag{26}$$

where

$$z_i' = z_i - \frac{3z_i + r_i}{4(n_i - 1)} \tag{27}$$

and z_i is the z in Equation 8 (Hotelling, 1953).

We may test hypotheses about the common correlation coefficient ($H_0: \rho = 0$ versus $H_A: \rho \neq 0$, or $H_0: \rho = \rho_0$ versus $H_A: \rho \neq \rho_0$, or similar one-tailed tests) by Equation 33 or 34.

6 POWER AND SAMPLE SIZE IN COMPARING TWO CORRELATION COEFFICIENTS

The power of the preceding test for difference between two correlation coefficients is estimated as $1 - \beta$, where β is the one-tailed probability of the normal deviate calculated as

$$Z_{\beta(1)} = \frac{|z_1 - z_2|}{\sigma_{z_1 - z_2}} - Z_\alpha \tag{28}$$

(Cohen, 1988: 546–547), where α may be either one-tailed or two-tailed and where $Z_{\alpha(1)}$ or $Z_{\alpha(2)}$ is most easily read from the last line of Table 3 from *Appendix: Statistical Tables and Graphs*. Example 7 demonstrates this calculation for the data of Example 6.

EXAMPLE 7   **Determination of the Power of the Test of $H_0: \rho_1 = \rho_2$ in Example 6**

$z_1 = 1.0454 \qquad z_2 = 1.2212$

$\sigma_{z_1 - z_2} = 0.1463$

$Z_\alpha = Z_{0.05(2)} = 1.960$

$$Z_{\beta(1)} = \frac{|1.0454 - 1.2212|}{0.1463} - 1.960$$
$$= 1.202 - 1.960$$
$$= -0.76$$

From Appendix Table 2,

$$\beta = P(Z \geq -0.76) = 1 - P(Z \leq -0.76) = 1 - 0.2236 = 0.78.$$

Therefore,

$$\text{power} = 1 - \beta = 1 - 0.78 = 0.22.$$

If we state a desired power to detect a specified difference between transformed correlation coefficients, then the sample size required to reject H_0 when testing at the α level of significance is

$$n = 2\left(\frac{Z_\alpha + Z_{\beta(1)}}{z_1 - z_2}\right)^2 + 3 \tag{29}$$

(Cohen, 1988: 547). This is shown in Example 8.

The test for difference between correlation coefficients is most powerful for $n_1 = n_2$, and the proceding estimation is for a sample size of n in both samples. Sometimes the size of one sample is fixed and cannot be manipulated, and we then ask how large the second sample must be to achieve the desired power. If n_1 is fixed and n is determined by Equation 29, then (by considering n to be the harmonic mean of n_1 and n_2),

$$n_2 = \frac{n_1(n + 3) - 6n}{2n_1 - n - 3} \tag{30}$$

(Cohen, 1988: 137).*

EXAMPLE 8   Estimating the Sample Size Necessary for the Test of H_0: $\rho_1 = \rho_2$

Let us say we wish to be 90% confident of detecting a difference, $z_1 - z_2$, as small as 0.5000 when testing H_0: $\rho_1 = \rho_2$ at the 5% significance level. Then $\beta(1) = 0.10, \alpha(2) = 0.05$, and

$$n = 2\left(\frac{1.9600 + 1.2816}{0.5000}\right)^2 + 3$$
$$= 87.1.$$

So sample sizes of at least 88 should be used.

---

7   COMPARING MORE THAN TWO CORRELATION COEFFICIENTS

If k samples have been obtained and an r has been calculated for each, we often want to conclude whether or not all samples came from populations having identical ρ's. If H_0: $\rho_1 = \rho_2 = \cdots = \rho_k$ is not rejected, then all samples might be combined and one value of r calculated to estimate the single population ρ. As Example 9 shows, the testing of this hypothesis involves transforming each r to a z value. We may then calculate

$$\chi^2 = \sum_{i=1}^{k}(n_i - 3)z_i^2 - \frac{\left[\sum_{i=1}^{k}(n_i - 3)z_i\right]^2}{\sum_{i=1}^{k}(n_i - 3)}, \tag{31}$$

which may be considered to be a chi-square value with $k - 1$ degrees of freedom.[†]

(a) Common Correlation Coefficient.   If H_0 is not rejected, then all k sample correlation coefficients are concluded to estimate a common population ρ. A common r (also known as a weighted mean of r) may be obtained from transforming the

---

*If the denominator in Equation 30 is $\leq$ 0, then we must either increase n_1 or change the desired power, significance level, or detectable difference in order to solve for n_2.

[†]Equation 31 is a computational convenience for

$$\chi^2 = \sum(n_i - 3)(z_i - z_w)^2, \tag{31a}$$

where z_w is a weighted mean of z shown in Equation 32.

EXAMPLE 9   Testing a Three-Sample Hypothesis Concerning Correlation Coefficients

Given the following:

$n_1 = 24$ $n_2 = 29$ $n_3 = 32$
$r_1 = 0.52$ $r_2 = 0.56$ $r_3 = 0.87$

To test:

H_0: $\rho_1 = \rho_2 = \rho_3$.

H_A: All three population correlation coefficients are not equal.

i	r_i	z_i	z_i^2	n_i	$n_i - 3$	$(n_i - 3)z_i$	$(n_i - 3)z_i^2$
1	0.52	0.5763	0.3321	24	21	12.1023	6.9741
2	0.56	0.6328	0.4004	29	26	16.4528	10.4104
3	0.87	1.3331	1.7772	32	29	38.6599	51.5388
Sums:					76	67.2150	68.9233

$$\chi^2 = \sum (n_i - 3)z_i^2 - \frac{\left[\sum (n_i - 3)z_i\right]^2}{\sum (n_i - 3)}$$

$$= 68.9233 - \frac{(67.2150)^2}{76}$$

$$= 9.478$$

$$\nu = k - 1 = 2$$

$\chi^2_{0.05,2} = 5.991$

Therefore, reject H_0.

$$0.005 < P < 0.01 \quad [P = 0.0087]$$

If H_0 had not been rejected, it would have been appropriate to calculate the common correlation coefficient:

$$z_w = \frac{\sum (n_i - 3)z_i}{\sum (n_i - 3)} = \frac{67.2150}{76} = 0.884$$

$$r_w = 0.71.$$

weighted mean z value,

$$z_w = \frac{\sum\limits_{i=1}^{k} (n_i - 3)z_i}{\sum\limits_{i=1}^{k} (n_i - 3)}, \tag{32}$$

to its corresponding r value (let's call it r_w), as shown in Example 9. This transformation is that of Equation 8 and is given in Table 19 from *Appendix: Statistical Tables and Graphs*. If H_0 is not rejected, we may test H_0: $\rho = 0$ versus H_A: $\rho \neq 0$ by the method attributed to Neyman (1959) by Paul (1988):

$$Z = \frac{\sum_{i=1}^{k} n_i r_i}{\sqrt{N}}, \qquad (33)$$

where $N = \sum_{i=1}^{k} n_i$, and rejecting H_0 if $|Z| \geq Z_{\alpha(2)}$. For one-tailed testing, $H_0\colon \rho \leq 0$ versus $H_A\colon \rho > 0$ is rejected if $Z \geq Z_{\alpha(1)}$; and $H_0\colon \rho \geq 0$ versus $H_A\colon \rho < 0$ is rejected if $Z \leq -Z_{\alpha(1)}$.

For the hypotheses $H_0\colon \rho = \rho_0$ versus $H_A\colon \rho \neq \rho_0$, the transformation of Equation 8 is applied to convert ρ_0 to ζ_0. Then (from Paul, 1988),

$$Z = (z_w - \zeta_0)\sqrt{\sum_{i=1}^{n}(n_i - 3)} \qquad (34)$$

is computed and H_0 is rejected if $|Z| \geq Z_{\alpha(2)}$. For one-tailed testing, $H_0\colon \rho \leq \rho_0$ is rejected if $Z \geq Z_{\alpha(1)}$ or $H_0\colon \rho \geq \rho_0$ is rejected if $Z \leq -Z_{\alpha(1)}$.

If correlations are not independent, then they may be compared as described by Steiger (1980).

(b) Overcoming Bias. Fisher (1958: 205) and Hotelling (1953) have pointed out that the z transformation is slightly biased, in that each z will be a little inflated. This minor systematic error is likely to have only negligible effects on our previous considerations, but it is inclined to have adverse effects on the testing of multisample hypotheses, for in the latter situations several values of z_i, and therefore several small errors, are being summed. Such a hypothesis test and the estimation of a common correlation coefficient are most improved by correcting for bias when sample sizes are small or there are many samples in the analysis.

Several corrections for bias are available. Fisher recommended subtracting

$$\frac{r}{2(n - 1)}$$

from z, whereas Hotelling determined that better corrections to z are available, such as subtracting

$$\frac{3z + r}{4(n - 1)}.$$

However, Paul (1988) recommends a test that performs better than one employing such corrections for bias. It uses

$$\chi_P^2 = \sum_{i=1}^{k} \frac{n_i(r_i - r_w)^2}{(1 - r_i r_w)^2} \qquad (35)$$

with $k - 1$ degrees of freedom. Example 10 demonstrates this test.

If the multisample null hypothesis is not rejected, then ρ, the underlying population correlation coefficient, may be estimated by calculating z_w via Equation 32 and converting it to r_w. As an improvement, Paul (1988) determined that $n_i - 1$ should be used in place of $n_i - 3$ in the latter equation if ρ is less than about 0.5. Similarly, to compare ρ to a specified value, ρ_0, $n_i - 1$ would be used instead of $n_i - 3$ in Equation 34 if ρ is less than about 0.5.

EXAMPLE 10 The Hypothesis Testing of Example 9, Employing Correction for Bias

$$\chi_P^2 = \sum_{i=1}^{k} \frac{n_i(r_i - r_w)^2}{(1 - r_i r_w)^2}$$

$$= \frac{24(0.52 - 0.71)^2}{[1 - (0.52)(0.71)]^2} + \frac{29(0.56 - 0.71)^2}{[1 - (0.56)(0.71)]^2} + \frac{32(0.87 - 0.71)^2}{[1 - (0.87)(0.71)]^2}$$

$$= 2.1774 + 1.7981 + 5.6051$$

$$= 9.5806$$

$$\chi_{0.05,2}^2 = 5.991$$

Therefore, reject H_0.

$$0.005 < P < 0.01 \quad [P = 0.0083]$$

---

8 MULTIPLE COMPARISONS AMONG CORRELATION COEFFICIENTS

If the null hypothesis of the previous section (H_0: $\rho_1 = \rho_2 = \cdots = \rho_k$) is rejected, it is typically of interest to determine which of the k correlation coefficients are different from which others. This can be done, again using Fisher's z transformation (Levy, 1976).

We can test each pair of correlation coefficients, r_B and r_A, by a Tukey-type test, if $n_B = n_A$:

$$q = \frac{z_B - z_A}{\text{SE}}, \tag{36}$$

where

$$\text{SE} = \sqrt{\frac{1}{n - 3}} \tag{37}$$

and n is the size of each sample. If the sizes of the two samples, A and B, are not equal, then we can use

$$\text{SE} = \sqrt{\frac{1}{2}\left(\frac{1}{n_B - 3} + \frac{1}{n_A - 3}\right)}. \tag{38}$$

The appropriate critical value for this test is $q_{\alpha,\infty,k}$ (from Table 5 from *Appendix: Statistical Tables and Graphs*). This test is demonstrated in Example 11.

It is typically unnecessary in multiple comparison testing to employ the correction for bias described at the end of Section 7.

(a) Comparing a Control Correlation Coefficient to Each Other Correlation Coefficient. The foregoing methods enable us to compare each correlation coefficient with each other coefficient. If, instead, we desire only to compare each coefficient to one

Simple Linear Correlation

EXAMPLE 11   Tukey-Type Multiple Comparison Testing Among the Three Correlation Coefficients in Example 9

Samples ranked by correlation coefficient (i):	1	2	3
Ranked correlation coefficients (r_i):	0.52	0.56	0.87
Ranked transformed coefficients (z_i):	0.5763	0.6328	1.3331
Sample size (n_i):	24	29	32

Comparison B vs. A	Difference $z_B - z_A$	SE	q	$q_{0.05,\infty,3}$	Conclusion
3 vs. 1	$1.3331 - 0.5763 = 0.7568$	0.203	3.728	3.314	Reject H_0: $\rho_3 = \rho_1$
3 vs. 2	$1.3331 - 0.6328 = 0.7003$	0.191	3.667	3.314	Reject H_0: $\rho_3 = \rho_2$
2 vs. 1	$0.6328 - 0.5763 = 0.0565$	0.207	0.273	3.314	Do not reject H_0: $\rho_2 = \rho_1$

Overall conclusion: $\rho_1 = \rho_2 \neq \rho_3$

---

particular coefficient (call it the correlation coefficient of the "control" set of data), then a procedure analogous to the Dunnett test may be employed (Huitema, 1974).

Let us designate the control set of data as B, and each other group of data, in turn, as A. Then we compute

$$q = \frac{z_B - z_A}{\text{SE}}, \tag{39}$$

for each A. The appropriate standard error is

$$\text{SE} = \sqrt{\frac{2}{n-3}} \tag{40}$$

if samples A and B are of the same size, or

$$\text{SE} = \sqrt{\frac{1}{n_B - 3} + \frac{1}{n_A - 3}} \tag{41}$$

if $n_A \neq n_B$. The critical value is $q'_{\alpha(1),\infty,p}$ (from Table 6 from *Appendix: Statistical Tables and Graphs*) or $q'_{\alpha(2),\infty,p}$ (from Table 7 from *Appendix: Statistical Tables and Graphs*) for the one-tailed or two-tailed test, respectively.

(b) Multiple Contrasts Among Correlation Coefficients. Previously we introduced the concepts and procedures of multiple contrasts among means; these are multiple comparisons involving groups of means. In a similar fashion, multiple contrasts may be examined among correlation coefficients (Marascuilo, 1971: 454–455). We again employ the z transformation and calculate, for each contrast, the test statistic

$$S = \frac{\left| \sum_i c_i z_i \right|}{\text{SE}}, \tag{42}$$

Simple Linear Correlation

where

$$SE = \sqrt{\sum_i c_i^2 \sigma_{z_i}^2} \tag{43}$$

and c_i is a contrast coefficient. (For example, if we wished to test the hypothesis $H_0: (\rho_1 + \rho_2)/2 - \rho_3 = 0$, then $c_1 = \frac{1}{2}$, $c_2 = \frac{1}{2}$, and $c_3 = -1$.) The critical value for this test is

$$S_\alpha = \sqrt{\chi_{\alpha,(k-1)}^2}. \tag{19.44}*$$

9 RANK CORRELATION

If we have data obtained from a bivariate population that is far from normal, then the correlation procedures discussed thus far are generally inapplicable. Instead, we may operate with the ranks of the measurements for each variable. Two different *rank correlation* methods are commonly encountered, that proposed by Spearman (1904) and that of Kendall[†] (1938). And, these procedures are also applicable if the data are ordinal.

Example 12 demonstrates Spearman's rank correlation procedure. After each measurement of a variable is ranked, as done in previously described nonparametric testing procedures, Equation 1 can be applied to the ranks to obtain the *Spearman rank correlation coefficient*, r_s. However, a computation that is often simpler is

$$r_s = 1 - \frac{6 \sum_{i=1}^{n} d_i^2}{n^3 - n}, \tag{19.46}‡$$

---

*Because $\chi_{\alpha,\nu}^2 = \nu F_{\alpha(1),\nu,\infty}$, it is equivalent to write

$$S_\alpha = \sqrt{(k-1)F_{\alpha(1),(k-1),\infty}}, \tag{45}$$

but Equation 44 is preferable because it engenders less rounding error in the calculations.

[†]Charles Edward Spearman (1863–1945), English psychologist and statistician, an important researcher on intelligence and on the statistical field known as factor analysis (Cattell, 1978). Sir Maurice George Kendall (1907–1983), English statistician contributing to many fields (Bartholomew, 1983; David and Fuller, 2007; Ord, 1984; Stuart, 1984). Kruskal (1958) noted that the early development of the method promoted by Kendall began 41 years prior to the 1938 paper. Karl Pearson observed that Sir Francis Galton considered the correlation of ranks even before developing correlation of variables (Walker, 1929: 128).

[‡]As the sum of n ranks is $n(n + 1)/2$, Equation 1 may be rewritten for rank correlation as

$$r_s = \frac{\sum_{i=1}^{n} (\text{rank of } X_i)(\text{rank of } Y_i) - \frac{n(n+1)^2}{4}}{\sqrt{\left(\sum_{i=1}^{n} (\text{rank of } X_i)^2 - \frac{n(n+1)^2}{4}\right)\left(\sum_{i=1}^{n} (\text{rank of } Y_i)^2 - \frac{n(n+1)^2}{4}\right)}}. \tag{47}$$

Instead of using differences between ranks of pairs of X and Y, we may use the sums of the ranks for each pair, where $S_i = \text{rank of } X_i + \text{rank of } Y_i$ (Meddis, 1984: 227; Thomas, 1989):

$$r_s = \frac{6 \sum S_i^2}{n^3 - n} - \frac{7n + 5}{n - 1}. \tag{48}$$

EXAMPLE 12   Spearman Rank Correlation for the Relationship Between the Scores of Ten Students on a Mathematics Aptitude Examination and a Biology Aptitude Examination

Student (i)	Mathematics examination score (X_i)	Rank of X_i	Biology examination score (Y_i)	Rank of Y_i	d_i	d_i^2
1	57	3	83	7	−4	16
2	45	1	37	1	0	0
3	72	7	41	2	5	25
4	78	8	84	8	0	0
5	53	2	56	3	−1	1
6	63	5	85	9	−4	16
7	86	9	77	6	3	9
8	98	10	87	10	0	0
9	59	4	70	5	−1	1
10	71	6	59	4	2	4

$$n = 10 \qquad r_s = 1 - \frac{6 \sum d_i^2}{n^3 - n}$$

$$\sum d_i^2 = 72 \qquad = 1 - \frac{6(72)}{10^3 - 10}$$

$$= 1 - 0.436$$

$$= 0.564$$

To test H_0: $\rho_s = 0$; H_A: $\rho_s \neq 0$

$(r_s)_{0.05(2),10} = 0.648$ (Table 20 from *Appendix: Statistical Tables and Graphs*)

Therefore, do not reject H_0.

$$P = 0.10$$

where d_i is a difference between X and Y ranks: $d_i =$ rank of X_i − rank of Y_i.* The value of r_s, as an estimate of the population rank correlation coefficient, ρ_s, may range from −1 to +1, and it has no units; however, its value is not to be expected to be the same as the value of r that might have been calculated for the original data instead of their ranks.

Table 20 from *Appendix: Statistical Tables and Graphs* may be used to assess the significance of r_s. A comment following that table refers to approximating the exact probability of r_s. If n is greater than that provided for in that table, then r_s may be

---

*Spearman (1904) also presented a rank-correlation method, later (Spearman, 1906) called the "footrule" coefficient. Instead of using the squares of the d_i's, as r_s does, this coefficient employs the absolute values of the d_i's:

$$r_f = 1 - \frac{3 \sum |d_i|}{n^2 - 1}. \tag{48a}$$

However, r_f typically does not range from −1 to 1; its lower limit is −0.5 if n is odd and, if n is even, it is −1 when $n = 2$ and rapidly approaches −0.5 as n increases (Kendall, 1970: 32–33).

Simple Linear Correlation

used in place of r in the hypothesis testing procedures of Section 2.* If either the Spearman or the parametric correlation analysis (Section 2) is applicable, the former is $9/\pi^2 = 0.91$ times as powerful as the latter (Daniel, 1990: 362; Hotelling and Pabst, 1936; Kruskal, 1958).[†]

(a) Correction for Tied Data. If there are tied data, then they are assigned average ranks as described before and r_s is better calculated either by Equation 1 applied to the ranks (Iman and Conover, 1978) or as

$$(r_s)_c = \frac{(n^3 - n)/6 - \sum d_i^2 - \sum t_X - \sum t_Y}{\sqrt{\left[(n^3 - n)/6 - 2\sum t_X\right]\left[(n^3 - n)/6 - 2\sum t_Y\right]}} \tag{50}$$

(Kendall, 1962: 38; Kendall and Gibbons, 1990: 44; Thomas, 1989). Here,

$$\sum t_X = \frac{\sum(t_i^3 - t_i)}{12}, \tag{51}$$

where t_i is the number of tied values of X in a group of ties, and

$$\sum t_Y = \frac{\sum(t_i^3 - t_i)}{12}, \tag{52}$$

where t_i is the number of tied Y's in a group of ties; this is demonstrated in Example 13. If $\sum t_X$ and $\sum t_Y$ are zero, then Equation 50 is identical to Equation 46. Indeed, the two equations differ appreciably only if there are numerous tied data.

Computationally, it is simpler to apply Equation 1 to the ranks to obtain $(r_s)_c$ when ties are present.

(b) Other Hypotheses, Confidence Limits, Sample Size, and Power. If $n \geq 10$ and $\rho_s \leq 0.9$, then the Fisher z transformation may be used for Spearman coefficients, just as it was in Sections 2 through 6, for testing several additional kinds of hypotheses (including multiple comparisons), estimating power and sample size, and setting confidence limits around ρ_s. But in doing so it is recommended that $1.060/(n - 3)$ be used instead of $1/(n - 3)$ in the variance of z (Fieller, Hartley, and Pearson, 1957, 1961). That is,

$$(\sigma_z)_s = \sqrt{\frac{1.060}{n - 3}} \tag{53}$$

should be used for the standard error of z (instead of Equation 11).

(c) The Kendall Rank Correlation Coefficient. In addition to some rarely encountered rank-correlation procedures (e.g., see Kruskal, 1958), the Kendall

---

*In this discussion, r_s will be referred to as an unbiased estimate of a population correlation coefficient, ρ_s, although that is not strictly true (Daniel, 1990: 365; Gibbons and Chakraborti, 2003: 432; Kruskall, 1958).

[†]Zimmerman (1994b) presented a rank-correlation procedure that he asserted is slightly more powerful than the Spearman r_s method. Martín Andrés, Herranz Tejedor, and Silva Mato (1995) showed a relationship between the Spearman rank correlation and the Wilcoxon-Mann-Whitney test. The Spearman statistic is related to the coefficient of concordance, W (Section 13), for two groups of ranks:

$$W = (r_s + 1)/2. \tag{49}$$

EXAMPLE 13   The Spearman Rank Correlation Coefficient, Computed for the Data of Example 1

X	Rank of X	Y	Rank of Y	d_i	d_i^2
10.4	4	7.4	5	−1	1
10.8	8.5	7.6	7	1.5	2.25
11.1	10	7.9	11	−1	1
10.2	1.5	7.2	2.5	−1	1
10.3	3	7.4	5	−2	4
10.2	1.5	7.1	1	0.5	0.25
10.7	7	7.4	5	2	4
10.5	5	7.2	2.5	2.5	6.25
10.8	8.5	7.8	9.5	−1	1
11.2	11	7.7	8	3	9
10.6	6	7.8	9.5	−3.5	12.25
11.4	12	8.3	12	0	0

$n = 12$
$$r_s = 1 - \frac{6 \sum d_i^2}{n^3 - n}$$

$\sum d_i^2 = 42.00$
$$= 1 - \frac{6(42.00)}{1716}$$
$$= 1 - 0.147$$
$$= 0.853$$

To test H_0: $\rho_s = 0$; H_A: $\rho_s \neq 0$,

$(r_s)_{0.05(2),12} = 0.587$ (from Table 20 from *Appendix: Statistical Tables and Graphs*)

Therefore, reject H_0.
$$P < 0.001$$

To employ the correction for ties (see Equation 50):

among the X's there are two measurements of 10.2 cm and two of 10.8 cm, so

$$\sum t_X = \frac{(2^3 - 2) + (2^3 - 2)}{12} = 1;$$

among the Y's there are two measurements tied at 7.2 cm, three at 7.4 cm, and two at 7.8 cm, so

$$\sum t_Y = \frac{(2^3 - 2) + (3^3 - 3) + (2^3 - 2)}{12} = 3;$$

therefore,

$$(r_s)_c = \frac{(12^3 - 12)/6 - 42.00 - 1 - 3}{\sqrt{[(12^3 - 12)/6 - 2(1)][(12^3 - 12)/6 - 2(3)]}} = \frac{242}{284.0} = 0.852;$$

and the hypothesis test proceeds exactly as above.

rank-correlation method is often used* (see, e.g., Daniel, 1990: 365–381; Kendall and Gibbons, 1938, 1990: 3–8; Siegel and Castellan, 1988: 245–254). The sample Kendall correlation coefficient is commonly designated as τ (lowercase Greek tau, an exceptional use of a Greek letter to denote a sample statistic).[†]

The correlation statistic τ for a set of paired X and Y data is a measure of the extent to which the order of the X's differs from the order of the Y's. Its calculation will not be shown here, but this coefficient may be determined—with identical results—either from the data or from the ranks of the data.

For example, these six ranks of X's and six ranks of Y's are in exactly the same order:

X: 1   2   3   4   5   6
Y: 1   2   3   4   5   6

and τ would be calculated to be 1, just as r_s would be 1, for there is perfect agreement of the orders of the ranks of X's and Y's. However, the following Y ranks are in the reverse sequence of the X ranks:

X: 1   2   3   4   5   6
Y: 6   5   4   3   2   1

and τ would be -1, just as r_s would be, for there is an exact reversal of the relationship between the X's and Y's. And, just as with r_s, τ will be closer to zero the further the X and Y ranks are from either perfect agreement or an exact reversal of agreement; but the values of τ and r_s will not be the same except when $\tau = -1$ or 1.

The performances of the Spearman and Kendall coefficients for hypothesis testing are very similar, but the former may be a little better, especially when n is large (Chow, Miller, and Dickinson, 1974), and for a large n the Spearman measure is also easier to calculate than the Kendall. Jolliffe (1981) describes the use of the runs-up-and-down test to perform nonparametric correlation testing in situations where r_s and τ are ineffective.

10   WEIGHTED RANK CORRELATION

The rank correlation of Section 9 gives equal emphasis to each pair of data. There are instances, however, when our interest is predominantly in whether there is correlation among the largest (or smallest) ranks in the two populations. In such cases we should prefer a procedure that will give stronger weight to intersample agreement on which items have the smallest (or largest) ranks, and Quade and Salama (1992) refer to such a method as *weighted rank correlation* (a concept they introduced in Salama and Quade, 1982).

In Example 14, a study has determined the relative importance of eight ecological factors (e.g., aspects of temperature and humidity, diversity of ground cover, abundance of each of several food sources) in the success of a particular species of bird in a particular habitat. A similar study ranked the same ecological factors for a second species in that habitat, and the desire is to ask whether the same ecological factors are most important for both species. We want to ask whether there is a positive correlation between the factors most important to one species and the factors most important to the other species. Therefore, a one-tailed weighted correlation analysis is called for.

---

*The idea of this correlation measure was presented as early as 1899 in a posthumous publication of the German philosopher, physicist, and psychologist Gustav Theodor Fechner (1801–1887) (Kruskal, 1958).

[†]It is much less frequently designated as T, t, r_k, or $\hat{\tau}$.

EXAMPLE 14 **A Top-Down Correlation Analysis, Where for Each of Two Bird Species, Eight Ecological Factors Are Weighted in Terms of Their Importance to the Success of the Birds in a Given Habitat**

H_0: The same ecological factors are most important to both species.

H_A: The same ecological factors are not most important to both species.

	Rank		Savage number (S_i)		
Factor (i)	Species 1	Species 2	Species 1	Species 2	$(S_i)_1(S_i)_2$
A	1	1	2.718	2.718	7.388
B	2	2	1.718	1.718	2.952
C	3	3	1.218	1.218	1.484
D	4	7	0.885	0.268	0.237
E	5	8	0.635	0.125	0.079
F	6	6	0.435	0.435	0.189
G	7	5	0.268	0.635	0.170
H	8	4	0.125	0.885	0.111
Sum			8.002	8.002	12.610

$n = 20$

$$\sum_{i=1}^{n}(S_i)_1(S_i)_2 = 12.610$$

$$r_T = \frac{\sum_{i=1}^{n}(S_i)_1(S_i)_2 - n}{(n - S_1)}$$

$$= \frac{12.610 - 8}{8 - 2.718} = 0.873$$

$$0.005 < P < 0.01$$

A correlation analysis performed on the pairs of ranks would result in a Spearman rank correlation coefficient of $r_s = 0.548$, which is not significantly different from zero. (The one-tailed probability is $0.05 < P < 0.10$.) Iman (1987) and Iman and Conover (1987) propose weighting the ranks by replacing them with the sums of reciprocals known as Savage scores (Savage, 1956). For a given sample size, n, the ith Savage score is

$$S_i = \sum_{j=i}^{n} \frac{1}{j}. \tag{54}$$

Thus, for example, if $n = 4$, then $S_1 = 1/1 + 1/2 + 1/3 + 1/4 = 2.083, S_2 = 1/2 + 1/3 + 1/4 = 1.083, S_3 = 1/3 + 1/4 = 0.583$, and $S_4 = 1/4 = 0.250$. A check on arithmetic is that $\sum_{i=1}^{n} S_i = n$; for this example, $n = 4$ and $2.083 + 1.083 + 0.583 + 0.250 = 3.999$. Table 1 gives Savage scores for n of 3 through 20. Scores for larger n are readily computed; but, as rounding errors will be compounded in the summation, it is wise to employ extra decimal places in such calculations. If there are tied ranks, then we may use the mean of the Savage scores for the positions of the tied data. For example, if $n = 4$ and ranks 2 and 3 are tied, then use $(1.083 + 0.583)/2 = 0.833$ for both S_2 and S_3.

TABLE 1: Savage Scores, S_i, for Various Sample Sizes, n

n	i= 1	2	3	4	5	6	7	8	9	10
3	1.833	0.833	0.333							
4	2.083	1.083	0.583	0.250						
5	2.283	1.283	0.783	0.450	0.200					
6	2.450	1.450	0.950	0.617	0.367	0.167				
7	2.593	1.593	1.093	0.756	0.510	0.310	0.143			
8	2.718	1.718	1.218	0.885	0.635	0.435	0.268	0.125		
9	2.829	1.829	1.329	0.996	0.746	0.546	0.379	0.236	0.111	
10	2.929	1.929	1.429	1.096	0.846	0.646	0.479	0.336	0.211	0.100
11	3.020	2.020	1.520	1.187	0.937	0.737	0.570	0.427	0.302	0.191
12	3.103	2.103	1.603	1.270	1.020	0.820	0.653	0.510	0.385	0.274
13	3.180	2.180	1.680	1.347	1.097	0.897	0.730	0.587	0.462	0.351
14	3.252	2.251	1.752	1.418	1.168	0.968	0.802	0.659	0.534	0.423
15	3.318	2.318	1.818	1.485	1.235	1.035	0.868	0.725	0.600	0.489
16	3.381	2.381	1.881	1.547	1.297	1.097	0.931	0.788	0.663	0.552
17	3.440	2.440	1.940	1.606	1.356	1.156	0.990	0.847	0.722	0.611
18	3.495	2.495	1.995	1.662	1.412	1.212	1.045	0.902	0.777	0.666
19	3.548	2.548	2.048	1.714	1.464	1.264	1.098	0.955	0.830	0.719
20	3.598	2.598	2.098	1.764	1.514	1.314	1.148	1.005	0.880	0.769

n	i= 11	12	13	14	15	16	17	18	19	20
11	0.091									
12	0.174	0.083								
13	0.251	0.160	0.077							
14	0.323	0.232	0.148	0.071						
15	0.389	0.298	0.215	0.138	0.067					
16	0.452	0.361	0.278	0.201	0.129	0.062				
17	0.510	0.420	0.336	0.259	0.188	0.121	0.059			
18	0.566	0.475	0.392	0.315	0.244	0.177	0.114	0.056		
19	0.619	0.528	0.445	0.368	0.296	0.230	0.167	0.108	0.053	
20	0.669	0.578	0.495	0.418	0.346	0.280	0.217	0.158	0.103	0.050

The Pearson correlation coefficient of Equation 1 may then be calculated using the Savage scores, a procedure that Iman and Conover (1985, 1987) call "top-down correlation"; we shall refer to the top-down correlation coefficient as r_T. Alternatively, if there are no ties among the ranks of either of the two samples, then

$$r_T = \frac{\sum_{i=1}^{n}(S_i)_1(S_i)_2 - n}{(n - S_1)},\tag{55}$$

where $(S_i)_1$ and $(S_i)_2$ are the ith Savage scores in Samples 1 and 2, respectively; this is demonstrated in Example 14, where it is concluded that there is significant agreement between the two rankings for the most important ecological factors. (As indicated previously, if all factors were to receive equal weight in the analysis of this set of data, a nonsignificant Spearman rank correlation coefficient would have been calculated.)*

Significance testing of r_T refers to testing H_0: $\rho_T \leq 0$ against H_A: $\rho_T > 0$ and may be effected by consulting Table 21 from *Appendix: Statistical Tables and Graphs*, which gives critical values for r_T. For sample sizes greater than those appearing in this table, a one-tailed normal approximation may be employed (Iman and Conover, 1985, 1987):

$$Z = \frac{r_T}{\sqrt{n-1}}. \tag{56}$$

The top-down correlation coefficient, r_T, is 1.0 when there is perfect agreement among the ranks of the two sets of data. If the ranks are completely opposite in the two samples, then $r_T = -1.0$ only if $n = 2$; it approaches -0.645 as n increases. If we wished to perform a test that was especially sensitive to agreement at the bottom, instead of the top, of the list of ranks, then the foregoing procedure would be performed by assigning the larger Savage scores to the larger ranks.

11 CORRELATION WITH NOMINAL-SCALE DATA

(a) Both Variables Are Dichotomous. *Dichotomous* normal-scale data are data recorded in two nominal categories (e.g., observations might be recorded as male or female, dead or alive, with or without thorns). Data collected for a dichotomous variable may be presented in the form of a table with two rows and two columns (a "2 × 2 contingency table"). The data of Example 15, for instance, may be cast into a 2 × 2 table, as shown. We shall set up such tables by having f_{11} and f_{22} be the frequencies of agreement between the two variables (where f_{ij} is the frequency in row i and column j).

Many measures of association of two dichotomous variables have been suggested (e.g., Conover, 1999: Section 4.4; Everitt, 1992: Section 3.6; Gibbons and Chakraborti, 2003: Section 14.3). So-called *contingency coefficients*,[†] such as

$$\sqrt{\frac{\chi^2}{\chi^2 + n}} \tag{57}$$

and

$$\sqrt{\frac{\dfrac{\chi^2}{n}}{1 + \dfrac{\chi^2}{n}}}, \tag{57a}$$

---

*Procedures other than the use of Savage scores may be used to assign differential weights to the ranks to be analyzed; some give more emphasis to the lower ranks and some give less (Quade and Salama, 1992). Savage scores are recommended as an intermediate strategy.

[†] A term coined by Karl Pearson (Walker, 1958).

EXAMPLE 15   Correlation for Dichotomous Nominal-Scale Data. Data Are Collected to Determine the Degree of Association, or Correlation, Between the Presence of a Plant Disease and the Presence of a Certain Species of Insect

Case	Presence of plant disease	Presence of insect
1	+	+
2	+	+
3	−	−
4	−	+
5	+	+
6	−	+
7	−	−
8	+	+
9	−	+
10	−	−
11	+	+
12	−	−
13	+	+
14	−	+

The data may be tabulated in the following 2×2 contingency table:

Insect	Plant Disease		
	Present	Absent	Total
Present	6	4	10
Absent	0	4	4
Total	6	8	14

$$\phi_1 = \frac{f_{11}f_{22} - f_{12}f_{21}}{\sqrt{C_1 C_2 R_1 R_2}}$$

$$= \frac{(6)(4) - (4)(0)}{\sqrt{(6)(8)(10)(4)}}$$

$$= 0.55$$

$$Q = \frac{f_{11}f_{22} - f_{12}f_{21}}{f_{11}f_{22} + f_{12}f_{21}} = \frac{(6)(4) - (4)(0)}{(6)(4) + (4)(0)} = 1.00$$

$$r_n = \frac{(f_{11} + f_{22}) - (f_{12} - f_{21})}{(f_{11} + f_{22}) + (f_{12} + f_{21})} = \frac{(6 + 4) - (4 + 0)}{(6 + 4) + (4 + 0)} = \frac{10 - 4}{10 + 4} = 0.43$$

employ the χ^2 statistic. However, they have drawbacks, among them the lack of the desirable property of ranging between 0 and 1. [They are indeed zero when $\chi^2 = 0$ (i.e., when there is no association between the two variables), but the coefficients can never reach 1, even if there is complete agreement between the two variables.]

The Cramér, or phi, coefficient* (Cramér, 1946: 44),

$$\phi = \sqrt{\frac{\chi^2}{n}}, \tag{58}$$

does range from 0 to 1 (as does ϕ^2, which may also be used as a measure of association).[†] It is based upon χ^2 (uncorrected for continuity), as obtained from Equation 58a. Therefore, we can write

$$\chi^2 = \frac{n(f_{11}f_{22} - f_{12}f_{21})^2}{R_1 R_2 C_1 C_2}. \tag{58a}$$

$$\phi_1 = \frac{f_{11}f_{22} - f_{12}f_{21}}{\sqrt{C_1 C_2 R_1 R_2}}, \tag{59}$$

where R_i is the sum of the frequencies in row i and C_j is the sum of column j. This measure is preferable to Equation 58 because it can range from -1 to $+1$, thus expressing not only the strength of an association between variables but also the direction of the association (as does r). If $\phi = 1$, all the data in the contingency table lie in the upper left and lower right cells (i.e., $f_{12} = f_{21} = 0$). In Example 15 this would mean there was complete agreement between the presence of both the disease and the insect; either both were always present or both were always absent. If $f_{11} = f_{22} = 0$, all the data lie in the upper right and lower left cells of the contingency table, and $\phi = -1$. The measure ϕ may also be considered as a correlation coefficient, for it is equivalent to the r that would be calculated by assigning a numerical value to members of one category of each variable and another numerical value to members of the second category. For example, if we replace each "+" with 0, and each "−" with 1, in Example 15, we would obtain (by Equation 1) $r = 0.55$.[‡]

---

*Harald Cramér (1893–1985) was a distinguished Swedish mathematician (Leadbetter, 1988). This measure is commonly symbolized by the lowercase Greek phi ϕ—pronounced "fy" as in "simplify"—and is a sample statistic, not a population parameter as a Greek letter typically designates. (It should, of course, not be confused with the quantity used in estimating the power of a statistical test, which is discussed elsewhere in this book.) This measure is what Karl Pearson called "mean square contingency" (Walker, 1929: 133).

[†]ϕ may be used as a measure of association between rows and columns in contingency tables larger than 2×2, as

$$\phi = \sqrt{\frac{\chi^2}{n(k-1)}}, \tag{58b}$$

where k is the number of rows or the number of columns, whichever is smaller (Cramér, 1946: 443); ϕ^2 is also known as the *mean square contingency* (Cramér 1946: 282).

[‡]If the two rows and two columns of data are arranged so $C_2 \geq R_2$ (as is the case in Example 15), the maximum possible ϕ is

$$\phi_{max} = \sqrt{\frac{R_2 C_1}{R_1 C_2}}; \tag{59a}$$

$\phi = 0.55$ is, in fact, ϕ_{max} for marginal totals of 10, 4, 6, and 8, but if the data in Example 15 had been $f_{11} = 5, f_{12} = 5, f_{21} = 1$, and $f_{22} = 3$, ϕ would have been 0.23, resulting in $\phi/\phi_{max} = 0.42$. Some researchers have used ϕ/ϕ_{max} as an index of association, but Davenport and El-Sanhurry (1991) identified a disadvantage of doing so.

The statistic ϕ is also preferred over the previous coefficients of this section because it is amenable to hypothesis testing. The significance of ϕ (i.e., whether it indicates that an association exists in the sampled population) can be assessed by considering the significance of the contingency table. If the frequencies are sufficiently large, the significance of χ_c^2 (chi-square with the correction for continuity) may be determined. The variance of ϕ is given by Kendall and Stuart (1979: 572).

The Yule coefficient of association (Yule 1900, 1912),*

$$Q = \frac{f_{11}f_{22} - f_{12}f_{21}}{f_{11}f_{22} + f_{12}f_{21}}, \tag{60}$$

ranges from -1 (if either f_{11} or f_{22} is zero) to $+1$ (if either f_{12} or f_{21} is zero). The variance of Q is given by Kendall and Stuart (1979: 571).

A better measure is that of Ives and Gibbons (1967). It may be expressed as a correlation coefficient,

$$r_n = \frac{(f_{11} + f_{22}) - (f_{12} + f_{21})}{(f_{11} + f_{22}) + (f_{12} + f_{21})}. \tag{61}$$

The interpretation of positive and negative values of r_n (which can range from -1 to $+1$) is just as for ϕ.

The expression of significance of r_n involves statistical testing. The binomial test may be utilized, with a null hypothesis of $H_0: p = 0.5$, using cases of perfect agreement and cases of disagreement as the two categories. Alternatively, the sign test, the Fisher exact test, or the chi-square contingency test could be applied to the data.

Tetrachoic correlation is a situation where each of two nominal-scale variables has two categories because of an artificial dichotomy (Glass and Hopkins, 1996: 136–137; Howell, 2007: 284–285; Sheskin, 2004: 997–1000).[†] For example, data might be collected to ask whether there is a correlation between the height of children, recorded as "tall" or "short," and their performance on an intelligence test, recorded as "high" or "low." Underlying each of these two dichotomous variables is a spectrum of measurements, and observations are placed in the categories by an arbitrary definition of *tall*, *short*, *high*, and *low*. (If there are more than two categories of one or both variables, the term *polychoric correlation* may be used.) Therefore, the categories of X and the categories of Y may be considered to represent ordinal scales of measurement.

The tetrachoic correlation coefficient, r_t, is an estimate of what the correlation coefficient, r, would be if the continuous data (ratio-scale or interval-scale) were known for the underlying distributions; it ranges from -1 to 1. It is rarely encountered, largely because it is a very poor estimate (it has a large standard error and is adversely affected by nonnormality). The calculation of r_t, and its use in hypothesis testing, is discussed by Sheskin (2004: 998–1000).

(b) One Variable Is Dichotomous. *Point-biserial correlation* is the term used for a correlation between Y, a variable measured on a continuous scale (i.e., a ratio or

---

*Q is one of several measures of association discussed by British statistician George Udny Yule (1871–1951). He called it Q in honor of Lambert Adolphe Jacques Quetelet (1796–1874), a pioneering Belgian statistician and astronomer who was a member of more than 100 learned societies, including the American Statistical Association (of which he was the first foreign member elected after its formation in 1839); Quetelet worked on measures of association as early as 1832 (Walker, 1929: 130–131).

[†]This coefficient was developed by Karl Pearson (1901).

interval scale, not an ordinal scale), and X, a variable recorded in two nominal-scale categories.* Although this type of correlation analysis has been employed largely in the behavioral sciences (e.g., Glass and Hopkins, 1996: 364–365, 368–369; Howell, 1997: 279–283, 2007: 277–281; Sheskin, 2004: 990–993), it can also have application with biological data. If it is Y, instead of X, that has two nominal-scale categories, then logistic regression may be considered.

Example 16 utilizes point-biserial correlation to express the degree to which the blood-clotting time in humans is related to the type of drug that has been administered. In this example, data are tabulated denoting the use of one drug (drug B) by an X of 0 and the use of the other drug (drug G) by an X of 1. The dichotomy may be recorded by any two numbers, with identical results, but employing 0 and 1 provides the simplest computation.

Then a point-biserial correlation coefficient, r_{pb}, is calculated by applying Equation 1 or, equivalently, Equation 2 to the pairs of X and Y data. The sign of r_{pb} depends upon which category of X is designated as 0; in Example 16, r_{pb} is positive, but it would have been negative if drug B had been recorded as 1 and drug G as 0. The coefficient r_{pb} can range from -1 to 1, and it is zero when the means of the two groups of Y's are the same (i.e., $\overline{Y}_1 = \overline{Y}_0$; see the next paragraph).

A computation with equivalent results is

$$r_{pb} = \frac{\overline{Y}_1 - \overline{Y}_0}{s_Y} \sqrt{\frac{n_1 n_0}{N(N-1)}}, \tag{62}$$

where $\overline{Y}_0$ is the mean of all n_0 of the Y data associated with $X = 0$, $\overline{Y}_1$ is the mean of the n_1 Y data associated with $X = 1$, $N = n_0 + n_1$, and s_Y is the standard deviation of all N values of Y.[†]

By substituting r_{pb} for r and N for n, Equations 3 and 4 may be used for a point-biserial correlation coefficient, and hypothesis testing may proceed as in Section 2.

Hypothesis testing involving the population point-biserial correlation coefficient, ρ_{pb}, yields the same results as testing for the difference between two population means. If an analysis of this kind of data has been done by a t test on sample means, then determination of the point-biserial correlation coefficient may be accomplished by

$$r_{pb} = \sqrt{\frac{t^2}{t^2 + N - 2}}. \tag{63}$$

If variable X consists of more than two nominal-scale categories and Y is a continuous variable, then the expression of association between Y and X may be called a *point-polyserial correlation* (Olsson, Drasgow, and Dorans, 1982), a rarely encountered analytical situation not discussed here.

Biserial correlation involves a variable (Y) measured on a continuous scale and differs from point-biserial correlation in the nature of nominal-scale variable X (Glass and Hopkins, 1996: 134–136; Howell, 1997: 286–288, 2007: 284–285; Sheskin, 2004: 995–997). In this type of correlation, the nominal-scale categories are artificial. For

---

*This procedure was presented and named by Karl Pearson in 1901 (Glass and Hopkins, 1996: 133).

[†]Although this is a correlation, not a regression, situation, it can be noted that a regression line would run from $\overline{Y}_0$ to $\overline{Y}_1$, and it would have a slope of $\overline{Y}_1 - \overline{Y}_0$ and a Y intercept of $\overline{Y}_0$.

EXAMPLE 16   Point-Biserial Correlation, Using Data of Example 1 from *Two-Sample Hypotheses*

H_0: There is no correlation between blood-clotting time and drug. (H_0: $\rho_{pb} = 0$)
H_A: There is correlation between blood-clotting time and drug. (H_A: $\rho_{pb} \neq 0$)
For variable X, drug B is represented by $X = 0$ and drug G by $X = 1$; Y is the time (in minutes) for blood to clot.

X	Y
0	8.8
0	8.4
0	7.9
0	8.7
0	9.1
0	9.6
1	9.9
1	9.0
1	11.1
1	9.6
1	8.7
1	10.4
1	9.5

$n_0 = 6$ $n_1 = 7$ $N = 13$

$\sum X = 7$ $\sum Y = 120.70$

$\sum X^2 = 7$ $\sum Y^2 = 1129.55$

$\sum x^2 = 3.2308$ $\sum y^2 = 8.8969$

$\sum XY = 68.2$

$\sum xy = 3.2077$

$$r_{pb} = \frac{3.2077}{\sqrt{(3.2308)(8.8969)}} = 0.5983, \quad r_{pb}^2 = 0.3580$$

$$t = \frac{0.5983}{\sqrt{\dfrac{1 - 0.3580}{13 - 2}}} = \frac{0.5983}{0.2416} = 2.476$$

$t_{0.05(2),11} = 2.201$

Therefore, reject H_0.

$$0.02 < P < 0.05 \quad [P = 0.031]$$

This is the same result as obtained comparing the mean clotting times of the two drug groups.

---

example, mice in a diet experiment might be recorded as heavy or light in weight by declaring a particular weight as being the dividing point between "heavy" and "light" (and X behaves like an ordinal-scale variable).

This correlation coefficient, r_b, may be obtained by Equation 63a or 63b, as is r_{pb}, and will be larger than r_{pb} except that when r_{pb} is zero, r_b is also zero.

$$t = \frac{\bar{d}}{s_{\bar{d}}}. \tag{63a}$$

$$\overline{d} \pm t_{\alpha(2),\nu}\, s_{\overline{d}}. \tag{63b}$$

Because X represents an ordered measurement scale, $X = 0$ should be used for the smaller measurement (e.g., "light" in the previous example) and $X = 1$ for the larger measurement (e.g., "heavy"). If there are more than two ranked categories of X, the term *polyserial correlation* may be applied (Olsson, Drasgow, and Dorans, 1982).

However, the calculated biserial correlation coefficient is adversely affected if the distribution underlying variable X is not normal; indeed, with nonnormality, $|r_b|$ can be much greater than 1. The correlation r_b is an estimate of the correlation coefficient that would have been obtained if X were the measurements from the underlying normal distribution. The calculation of, and hypothesis testing with, this coefficient is discussed by Sheskin (2004: 995–996).

12 INTRACLASS CORRELATION

In some correlation situations it is not possible to designate one variable as X and one as Y. Consider the data in Example 17, where the intent is to determine whether there is a relationship between the weights of identical twins. Although the weight data clearly exist in pairs, the placement of a member in each pair in the first or in the second column is arbitrary, in contrast to paired-sample testing, where all the data in the first column have something in common and all the data in the second column have something in common. When pairs of data occur as in Example 17, we may employ *intraclass correlation*, a concept generally approached by analysis-of-variance considerations (specifically, Model II single-factor ANOVA. Aside from assuming random sampling from a bivariate normal distribution, this procedure also assumes that the population variances are equal.

If we consider each of the pairs in our example as groups in an ANOVA (i.e., $k = 7$), with each group containing two observations (i.e., $n = 2$), then we may calculate mean squares to express variability both between and within the k groups. Then the *intraclass correlation coefficient* is defined as

$$r_I = \frac{\text{groups MS} - \text{error MS}}{\text{groups MS} + \text{error MS}}, \tag{64}$$

this statistic being an estimate of the population intraclass correlation coefficient, ρ_I. To test $H_0\colon \rho_I = 0$ versus $H_0\colon \rho_I \neq 0$, we may utilize

$$F = \frac{\text{groups MS}}{\text{error MS}}, \tag{65}$$

a statistic associated with groups DF and error DF for the numerator and denominator, respectively.* If the measurements are equal within each group, then error MS $= 0$, and $r_I = 1$ (a perfect positive correlation). If there is more variability within groups than there is between groups, then r_I will be negative. The smallest it may be, however, is $-1/(n - 1)$; therefore, only if $n = 2$ (as in Example 17) can r_I be as small as -1.

We are not limited to pairs of data (i.e., situations where $n = 2$) to speak of intraclass correlation. Consider, for instance, expanding the considerations of Example 17 into a study of weight correspondence among triplets instead of twins. Indeed, n need not even be equal for all groups. We might, for example, ask whether there is a relationship among adult weights of brothers; here, some families might

---

*If desired, F may be calculated first, followed by computing

$$r_I = (F - 1)/(F + 1). \tag{66}$$

EXAMPLE 17   Intraclass Correlation

Testing for correlation between weights of members of pairs of human twins.

Group (i.e., twin)	Weight of one member of group (kg)	Weight of other member of group (kg)
1	70.4	71.3
2	68.2	67.4
3	77.3	75.2
4	61.2	66.7
5	72.3	74.2
6	74.1	72.9
7	71.1	69.5

Source of variation	SS	DF	MS
Total	220.17	13	
Groups	198.31	6	33.05
Error	21.86	7	3.12

(See for the MS computation procedures.)

$$r_1 = \frac{\text{groups MS} - \text{error MS}}{\text{groups MS} + \text{error MS}} = 0.827$$

To test $H_0: \rho_I = 0, H_A: \rho_I \neq 0$,

$$F = \frac{\text{groups MS}}{\text{error MS}} = \frac{33.05}{3.12} = 10.6$$

$$F_{0.05(1),6,7} = 3.87$$

Therefore, reject H_0.

$$0.0025 < P < 0.005 \quad [P = 0.0032]$$

---

consist of two brothers, some of three brothers, and so on. If n is not 2 for all k groups, then

$$r_I = \frac{\text{groups MS} - \text{error MS}}{\text{groups MS} + (n - 1)\text{ error MS}}, \tag{19.67}*$$

---

*Or, equivalently,

$$r_I = \frac{F - 1}{F + n - 1}. \tag{67a}$$

a calculation for which Equation 64 is the special case when $n = 2$. If all n's are not equal, then the appropriate n for use in Equation 67 may be obtained as

$$n = \frac{\sum_{i=1}^{k} n_i - \dfrac{\sum_{i=1}^{k} n_i^2}{\sum_{i=1}^{k} n_i}}{k - 1}. \tag{68}$$

Equation 65 is applicable for hypothesis testing in all the preceding cases.

If $n = 2$ in all groups, then we may set confidence limits and test hypotheses as would be done with r, by utilizing the z transformation,* with k in place of n. (See Sections 3 through 8.) However, the standard error of the resultant z_I will be

$$\sigma_{z_I} = \sqrt{\frac{1}{k - \dfrac{3}{2}}} \tag{69}$$

(Fisher 1958: 215), and the standard error for the difference between two z_i's will be

$$\sigma_{z_1 - z_2} = \sqrt{\frac{1}{k_1 - \dfrac{3}{2}} + \frac{1}{k_2 - \dfrac{3}{2}}}, \tag{70}$$

although an exact test is available (Zerbe and Goldgar, 1980).

If $n > 2$ in any groups, then the Fisher transformation is computed as

$$z_I = 0.5 \ln \frac{1 + (n - 1)r_I}{1 - r_I} \tag{71}$$

(Fisher 1958: 219), using Equation 68 if the n's are unequal. Table 18 from *Appendix: Statistical Tables and Graphs* may be used, instead of Equation 71, if r_I is first converted to

$$r' = \frac{nr_I}{2 + (n - 2)r_I} \tag{72}$$

(Rao, Mitra, and Matthai, 1966: 87), and using r' to enter the table. If Table 19 from *Appendix: Statistical Tables and Graphs* is used to convert z_I to r', then r_I is obtainable as

$$r_I = \frac{-2r'}{(n - 2)r' - n}. \tag{73}$$

---

*The correction for bias in z involves adding $1/(2k - 1)$ to z (Fisher, 1958: 216); it is typically negligible unless several z's are being summed, as in the procedure of Section 8. The result of Equation 8 may also be obtained as

$$z = 0.5 \ln F.$$

Donner and Wells (1986) concluded that these confidence limits are very good, though wider than those obtainable with another, far more complex, calculation.

Also, if $n > 2$, then

$$\sigma_{z_I} = \sqrt{\frac{n}{2(n-1)(k-2)}} \tag{74}$$

(Fisher, 1958: 219), and

$$\sigma_{z_1 - z_2} = \sqrt{\frac{\dfrac{n}{k_1 - 2} + \dfrac{n}{k_2 - 2}}{2(n-1)}} \tag{75}$$

(Zerbe and Goldgar, 1980). Nonparametric measures of intraclass correlation have been proposed (e.g., Rothery, 1979).

13 CONCORDANCE CORRELATION

If the intent of collecting pairs of data is to assess reproducibility or agreement of data sets, an effective technique is that which Lin (1989, 1992, 2000) refers to as *concordance correlation*. For example, the staff of an analytical laboratory might wish to know whether measurements of a particular substance are the same using two different instruments, or when performed by two different technicians.

In Example 18, the concentration of lead was measured in eleven specimens of brain tissue, where each specimen was analyzed by two different atomic-absorption spectrophotometers. These data are presented in Figure 2. If the scales of the two axes are the same, then perfect reproducibility of assay would be manifested by the data falling on a 45° line intersecting the origin of the graph (the line as shown in Figure 2), and concordance correlation assesses how well the data follow that 45° line.

The concordance correlation coefficient, r_c, is

$$r_c = \frac{2 \sum xy}{\sum x^2 + \sum y^2 + n(\overline{X} - \overline{Y})^2}. \tag{76}$$

This coefficient can range from -1 to $+1$, and its absolute value cannot be greater than the Pearson correlation coefficient, r; so it can be stated that $-1 \leq -|r| \leq r_c \leq |r| \leq 1$; and $r_c = 0$ only if $r = 0$.

Hypothesis testing is not recommended with r_c (Lin, 1992), but a confidence interval may be obtained for the population parameter ρ_c of which r_c is an estimate. To do so, the Fisher transformation (Equation 8) is applied to r_c to obtain a transformed value we shall call z_c; and the standard error of z_c is obtained as

$$\sigma_{z_c} = \sqrt{\frac{\dfrac{(1 - r^2) r_c^2}{(1 - r_c^2) r^2} + \dfrac{2 r_c^3 (1 - r_c) U}{r(1 - r_c^2)^2} - \dfrac{r_c^4 U^2}{2 r^2 (1 - r_c^2)^2}}{n - 2}}, \tag{77}$$

where

$$U = \frac{\sqrt{n}(\overline{X} - \overline{Y})^2}{\sqrt{\sum x^2 \sum y^2}}. \tag{78}$$

This computation is shown in Example 18.

Furthermore, we might ask whether two concordance correlations are significantly different. For example, consider that the between-instrument reproducibility analyzed in Example 18 was reported for very experienced technicians, and a set of data

Simple Linear Correlation

EXAMPLE 18 **Reproducibility of Analyses of Lead Concentrations in Brain Tissue (in Micrograms of Lead per Gram of Tissue), Using Two Different Atomic-Absorption Spectrophotometers**

Tissue sample (i)	Tissue Lead (μg/g)	
	Spectrophotometer A (X_i)	Spectrophotometer B (Y_i)
1	0.22	0.21
2	0.26	0.23
3	0.30	0.27
4	0.33	0.27
5	0.36	0.31
6	0.39	0.33
7	0.41	0.37
8	0.44	0.38
9	0.47	0.40
10	0.51	0.43
11	0.55	0.47

$n = 11$

$\sum X = 4.24$ $\sum Y = 3.67$ $\sum XY = 1.5011$

$\sum X^2 = 1.7418$ $\sum Y^2 = 1.2949$ $\sum xy = 0.08648$

$\sum x^2 = 0.10747$ $\sum y^2 = 0.07045$

$\overline{X} = 0.385$ $\overline{Y} = 0.334$

$$r_c = \frac{2\sum xy}{\sum x^2 + \sum y^2 + n(\overline{X} - \overline{Y})^2}$$

$$= \frac{2(0.08648)}{0.10747 + 0.07045 + 11(0.385 - 0.334)^2} = \frac{0.17296}{0.20653}$$

$$= 0.8375; \quad r_c^2 = 0.7014$$

$$r = \frac{\sum xy}{\sqrt{\sum x^2 \sum y^2}} = 0.9939; \quad r^2 = 0.9878$$

For $r_c = 0.8375, z_c = 1.213$ (from Table 18 from *Appendix: Statistical Tables and Graphs*, by interpolation)

$$U = \frac{\sqrt{n}(\overline{X} - \overline{Y})^2}{\sqrt{\sum x^2 \sum y^2}} = \frac{(3.31662)(0.051)^2}{\sqrt{(0.10747)(0.07045)}} = 0.09914$$

441

$$\sigma_{z_c} = \sqrt{\dfrac{\dfrac{(1 - r^2)r_c^2}{(1 - r_c^2)r^2} + \dfrac{2r_c^3(1 - r_c)U}{r(1 - r_c^2)^2} - \dfrac{r_c^4 U^2}{2r^2(1 - r_c^2)^2}}{n - 2}}$$

$$= \sqrt{\dfrac{\dfrac{(1 - 0.9878)(0.7014)}{(1 - 0.7014)(0.9878)} + \dfrac{2(0.8375)^3(1 - 0.8375)(0.09914)}{(0.9939)(1 - 0.7014)} - \dfrac{(0.8375)^4(0.09914)^2}{2(0.9878)(1 - 0.7014)}}{11 - 2}}$$

$$= \sqrt{\dfrac{0.0291 + 0.06377 - 0.002447}{9}} = 0.0245$$

95% confidence interval for ζ_c:

$$z_c \pm Z_{0.05(2)}\sigma_{z_c} = 1.213 \pm (1.960)(0.0610) = 1.213 \pm 0.120$$

$$L_1 = 1.093; \quad L_2 = 1.333.$$

For the 95% confidence limits for ρ_c, the foregoing confidence limits for ζ_c are transformed (as with Table 19 from *Appendix: Statistical Tables and Graphs*) to

$$L_1 = 0.794; \quad L_2 = 0.870.$$

was also collected for novice analysts. In order to ask whether the measure of reproducibility (namely, r_c) is different for the highly experienced and the less experienced workers, we can employ the hypothesis testing of Section 5. For this, we obtain r_c for the data from the experienced technicians (call it r_1) and another r_c (call it r_2) for the data from the novices. Then each r_c is transformed

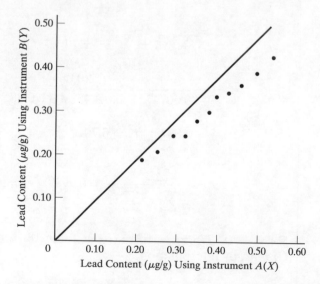

FIGURE 2: Lead concentrations in brain tissue (μg/g), determined by two different analytical instruments. The data are from Example 18 and are shown with a 45° line through the origin.

to its corresponding z_c (namely, z_1 and z_2) and the standard error to be used in Equation 21 is

$$\sigma_{z_1-z_2} = \sqrt{\sigma_{z_1}^2 + \sigma_{z_2}^2},\tag{79}$$

where each σ_z^2 is obtained as the square of the σ_{z_c} in Equation 77.

Lin (1989) has shown that this method of assessing reproducibility is superior to comparison of coefficients of variation, to the paired-t test, to regression, to Pearson correlation (Section 1), and to intraclass correlation (Section 12). And he has shown the foregoing hypothesis test to be robust with n as small as 10; however (Lin and Chinchilli, 1996), the two coefficients to be compared should have come from populations with similar ranges of data. Lin (1992) also discusses the sample-size requirement for this coefficient; and Barnhart, Haber, and Song (2002) expand concordance correlation to more than two sets of data.

14 THE EFFECT OF CODING

Except for the procedures in Sections 11 and 12, coding of the raw data will have no effect on the correlation coefficients presented in this chapter, on their z transformations, or on any statistical procedures regarding those coefficients and transformations.

EXERCISES

1. Measurements of serum cholesterol (mg/100 ml) and arterial calcium deposition (mg/100 g dry weight of tissue) were made on 12 animals. The data are as follows:

Calcium (X)	Cholesterol (Y)
59	298
52	303
42	233
59	287
24	236
24	245
40	265
32	233
63	286
57	290
36	264
24	239

   (a) Calculate the correlation coefficient.
   (b) Calculate the coefficient of determination.
   (c) Test $H_0: \rho = 0$ versus $H_A: \rho \neq 0$.
   (d) Set 95% confidence limits on the correlation coefficient.

2. Using the data from Exercise 1:
   (a) Test $H_0: \rho \leq 0$ versus $H_A: \rho > 0$.
   (b) Test $H_0: \rho = 0.50$ versus $H_A: \rho \neq 0.50$.

3. Given: $r_1 = -0.44, n_1 = 24, r_2 = -0.40, n_2 = 30$.
   (a) Test $H_0: \rho_1 = \rho_2$ versus $H_A: \rho_1 \neq \rho_2$.
   (b) If H_0 in part (a) is not rejected, compute the common correlation coefficient.

4. Given: $r_1 = 0.45, n_1 = 18, r_2 = 0.56, n_2 = 16$. Test $H_0: \rho_1 \geq \rho_2$ versus $H_A: \rho_1 < \rho_2$.

5. Given: $r_1 = 0.85, n_1 = 24, r_2 = 0.78, n_2 = 32, r_3 = 0.86, n_3 = 31$
   (a) Test $H_0: \rho_1 = \rho_2 = \rho_3$, stating the appropriate alternate hypothesis.
   (b) If H_0 in part (a) is not rejected, compute the common correlation coefficient.

6. (a) Calculate the Spearman rank correlation coefficient for the data of Exercise 1.
   (b) Test $H_0: \rho_s = 0$ versus $H_A: \rho_s \neq 0$.

7. Two different laboratories evaluated the efficacy of each of seven pharmaceuticals in treating hypertension in women, ranking them as shown below.

Drug	Lab 1 rank	Lab 2 rank
L	1	1
P	2	3
Pr	3	2
D	4	4
E	5	7
A	6	6
H	7	5

(a) Compute the top-down correlation coefficient.

(b) Test whether there is a significant agreement between the laboratories on which drugs are most effective (H_0: $\rho_T \leq 0$ versus H_A: $\rho_T > 0$).

8. To examine the proposition that the type of school college presidents lead correlates with the type of school from which they received their undergraduate education, the following data were collected and arranged in a contingency table:

College as president	Undergraduate college	
	Public	Private
Public	9	5
Private	2	7

Compute the coefficient, r_n, expressing this correlation.

9. Two samples of blood plasma from the same animal were submitted to each of four testing laboratories. The corticosterone concentrations, in grams per 100 ml, were determined as follows:

Laboratory	Sample 1	Sample 2
1	1.14	1.08
2	1.10	1.07
3	1.04	1.08
4	1.07	1.13

(a) Compute the intraclass correlation coefficient.

(b) Test whether there is a significant correlation between corticosterone determinations from the same laboratory.

10. The following avian plasma concentrations (in nanograms per milliliter) were determined by two different assay methods shortly after the blood was collected:

Blood sample	Method A	Method B
1	6.1	5.0
2	8.6	7.7
3	11.0	11.4
4	13.2	13.9
5	16.9	18.5
6	20.5	21.7
7	22.7	25.3
8	25.8	27.9
9	26.7	29.5
10	28.8	32.6
11	31.4	35.9
12	34.3	38.4

(a) Compute the concordance correlation coefficient.

(b) Compute the 95% confidence interval for the concordance correlation coefficient.

ANSWERS TO EXERCISES

1. (a) $r = 0.86$. (b) $r^2 = 0.73$. (c) H_0: $\rho = 0$; H_A: $\rho \neq 0$; $s_r = 0.16$; $t = 5.38$; as $t_{0.05(2),10} = 2.228$; reject H_0; $P < 0.001$ [$P = 0.00032$]. Or: $r = 0.86$, $r_{0.05(2),10} = 0.576$; reject H_0; $P < 0.001$. Or: $F = 13.29$, $F_{0.05(2),10,10} = 3.72$; reject H_0; $P < 0.001$. (d) $L_1 = 0.56$, $L_2 = 0.96$.

2. (a) H_0: $\rho \leq 0$; H_A: $\rho > 0$; $r = 0.86$; $t = 5.38$; $t_{0.05(1),10} = 1.812$; reject H_0; $P < 0.0005$ [$P = 0.00016$]. Or: $r_{0.05(1),10} = 0.497$; reject H_0; $P < 0.0005$. Or: $F = 13.29$; $F_{0.05(1),10,10} = 2.98$; reject H_0: $P < 0.0005$. (b) H_0: $\rho = 0.50$; H_A: $\rho \neq 0.50$; $r = 0.86$; $z = 1.2933$; $\zeta_0 = 0.5493$; $\sigma_z = 0.3333$; $Z = 2.232$; $Z_{0.05(2)} = 1.960$; reject H_0; $0.02 < P < 0.05$ [$P = 0.026$].

3. (a) H_0: $\rho_1 = \rho_2$; H_A: $\rho_1 \neq \rho_2$; $z_1 = -0.4722$. $z_2 = -0.4236$; $\sigma_{z_1-z_2} = 0.2910$; $Z = -0.167$; $Z_{0.05(2)} = 1.960$; do not reject H_0; $P > 0.50$ [$P = 0.87$]. (b) $z_w = -0.4449$; $r_w = -0.42$.

4. H_0: $\rho_1 \geq \rho_2$; H_A: $\rho_1 < \rho_2$; $z_1 = 0.4847$, $z_2 = 0.6328$; $\sigma_{z_1-z_2} = 0.3789$; $Z = -0.3909$; $Z_{0.05(1)} = 1.645$; do not reject H_0; $P > 0.25$ [$P = 0.35$].

5. (a) H_0: $\rho_1 = \rho_2 = \rho_3$; H_A: The three population correlation coefficients are not all the same; $\chi^2 = 111.6607 - (92.9071)^2/78 = 0.998$; $\chi^2_{0.05,2} = 5.991$; do not reject H_0; $0.50 < P < 0.75$ [$P = 0.61$]. $\chi^2_P = 1.095$, $0.50 < P < 0.75$ [$P = 0.58$]. (b) $z_w = 92.9071/78 = 1.1911$; $r_w = 0.83$.

6. (a) $\sum d_i^2 = 88.00$, $r_s = 0.69$; (b) H_0: $\rho_s = 0$; H_A: $\rho_s \neq 0$; as $(r_s)_{0.05(2),12} = 0.587$, reject H_0; $0.01 < P < 0.02$.

7. (a) $r_T = 0.914$; (b) reject H_0; $0.005 < P < 0.01$.

8. (a) $r_n = (16 - 7)/(16 + 7) = 0.39$. (b) H_0: There is no correlation between the type of institution a college president heads and the type

of institution he or she attended as a undergraduate; H_A: There is a correlation between the type of school headed and the type attended. By Fisher exact test (using Table 28 from *Appendix: Statistical Tables and Graphs*): $n = 23, m_1 = 9, m_2 = 11, f = 2$, critical $f_{0.05(2)} = 1$ and 7; as f is not ≤ 1 and is not ≥ 7, do not reject H_0.

9. **(a)** $r_I = (0.000946 - 0.00213)/(0.000946 + 0.001213) = -0.12$. **(b)** H_0: There is no

correlation between corticosterone determinations from the same laboratory (i.e., $\rho_I = 0$); H_A: There is no correlation between corticosterone determinations from the same laboratory (i.e., $\rho_I \neq 0$); $F = 0.000946/0.001213 = 0.78$; since $F_{0.05(1),3,4} = 6.59$, do not reject H_0; $P > 0.25$ [$P = 0.56$].

10. **(a)** $r_c = 0.9672$. **(b)** $z_c = 2.0470, r = 0.9991$, $U = 0.2502, \sigma_{z_c} = 0.2135$; for ζ_c: $L_1 = 1.6285$, $L_2 = 2.4655$; for ρ_c: $L_1 = 0.926, L_2 = 0.986$.

This page intentionally left blank

Multiple Regression and Correlation

This chapter will extend the analyses of regression and correlation relationships between two variables (*simple regression and correlation*) to regressions and correlations examining the interrelationships among three or more variables (*multiple regression and correlation*).

In *multiple regression*, one of the variables is considered to be functionally dependent upon at least one of the others. *Multiple correlation* is a situation when none of the variables is deemed to be dependent on another.*

The computations required for multiple-regression and multiple-correlation analyses would be very arduous, and in many cases prohibitive, without the computer capability that is widely available for this task. Therefore, the mathematical operations used to obtain regression and correlation coefficients, and to perform relevant hypothesis tests, will not be emphasized here. Section 1 summarizes the kinds of calculations that a computer program will typically perform and present, but that information is not necessary for understanding the statistical procedures discussed in the remainder of this chapter, including the interpretation of the results of the computer's work.

Though uncommon, there are cases where the dependent variable is recorded on a nominal scale (not on a ratio or interval scale), most often a scale with two nominal categories (e.g., male and female, infected and not infected, successful and unsuccessful). The analyses of this chapter are not applicable for such data. Instead, a procedure known as *logistic regression* may be considered.

---

*Much development in multiple correlation theory began in the late nineteenth century by several pioneers, including Karl Pearson (1857–1936) and his colleague, George Udny Yule (1871–1951) (Pearson 1967). (Pearson first called partial regression coefficients "double regression coefficients," and Yule later called them "net regression coefficients.") Pearson was the first to use the terms *multiple correlation*, in 1908, and *multiple correlation coefficient*, in 1914 (David, 1995).

Multiple regression is a major topic in theoretical and applied statistics; only an introduction is given here, and consultation with a statistical expert is often advisable.*

1 INTERMEDIATE COMPUTATIONAL STEPS

There are certain quantities that a computer program for multiple regression and/or correlation must calculate. Although we shall not concern ourselves with the mechanics of computation, intermediate steps in the calculating procedures are indicated here so the user will not be a complete stranger to them if they appear in the computer output. Among the many different programs available for multiple regression and correlation, some do not print all the following intermediate results, or they may do so only if the user specifically asks for them to appear in the output.

Consider n observations of M variables (the variables being referred to as X_1 through X_M; (see Example 1a)). If one of the M variables is considered to be dependent upon the others, then we may eventually designate that variable as Y, but the program will perform most of its computations simply considering all M variables as X's numbered 1 through M.

The sum of the observations of each of the M variables is calculated as

$$\sum_{j=1}^{n} X_{1j} \quad \sum_{j=1}^{n} X_{2j} \cdots \sum_{j=1}^{n} X_{Mj}. \tag{1}$$

For simplicity, let us refrain from indexing the $\sum$'s and assume that summations are always performed over all n sets of data. Thus, the sums of the variables could be denoted as

$$\sum X_1 \quad \sum X_2 \cdots \sum X_M. \tag{2}$$

Sums of squares and sums of cross products are calculated just as for simple regression, or correlation, for each of the M variables. The following sums, often referred to as *raw sums of squares* and *raw sums of cross products*, may be presented in computer output in the form of a matrix, or two-dimensional array:

$$
\begin{array}{ccccc}
\sum X_1^2 & \sum X_1 X_2 & \sum X_1 X_3 & \cdots & \sum X_1 X_M \\
\sum X_2 X_1 & \sum X_2^2 & \sum X_2 X_3 & \cdots & \sum X_2 X_M \\
\sum X_3 X_1 & \sum X_3 X_2 & \sum X_3^2 & \cdots & \sum X_3 X_M \\
\vdots & \vdots & \vdots & & \vdots \\
\sum X_M X_1 & \sum X_M X_2 & \sum X_M X_3 & \cdots & \sum X_M^2.
\end{array} \tag{3}
$$

As $\sum X_i X_k = \sum X_k X_i$, this matrix is said to be symmetrical about the diagonal running from upper left to lower right.* Therefore, this array, and those that follow,

---

*Greater discussion of multiple regression and correlation, often with explanation of the underlying mathematical procedures and alternate methods, can be found in many texts, such as Birkes and Dodge (1993); Chatterjee and Hadi (2006); Draper and Smith (1998); Glantz and Slinker (2001); Hair et al. (2006: Chapter 4); Howell (2007: Chapter 15); Kutner, Nachtsheim, and Neter (2004); Mickey, Dunn, and Clark (2004); Montgomery, Peck, and Vining (2006); Pedhazur (1997); Seber and Lee (2003); Tabachnik and Fidell (2001: Chapter 5); and Weisberg (2005).

*We shall refer to the values of a pair of variables as X_i and X_k.

EXAMPLE 1a The $n \times M$ Data Matrix for a Hypothetical Multiple Regression or Correlation ($n = 33$; $M = 5$)

	Variable (i)				
j	1 (°C)	2 (cm)	3 (mm)	4 (min)	5 (ml)
1	6	9.9	5.7	1.6	2.12
2	1	9.3	6.4	3.0	3.39
3	−2	9.4	5.7	3.4	3.61
4	11	9.1	6.1	3.4	1.72
5	−1	6.9	6.0	3.0	1.80
6	2	9.3	5.7	4.4	3.21
7	5	7.9	5.9	2.2	2.59
8	1	7.4	6.2	2.2	3.25
9	1	7.3	5.5	1.9	2.86
10	3	8.8	5.2	0.2	2.32
11	11	9.8	5.7	4.2	1.57
12	9	10.5	6.1	2.4	1.50
13	5	9.1	6.4	3.4	2.69
14	−3	10.1	5.5	3.0	4.06
15	1	7.2	5.5	0.2	1.98
16	8	11.7	6.0	3.9	2.29
17	−2	8.7	5.5	2.2	3.55
18	3	7.6	6.2	4.4	3.31
19	6	8.6	5.9	0.2	1.83
20	10	10.9	5.6	2.4	1.69
21	4	7.6	5.8	2.4	2.42
22	5	7.3	5.8	4.4	2.98
23	5	9.2	5.2	1.6	1.84
24	3	7.0	6.0	1.9	2.48
25	8	7.2	5.5	1.6	2.83
26	8	7.0	6.4	4.1	2.41
27	6	8.8	6.2	1.9	1.78
28	6	10.1	5.4	2.2	2.22
29	3	12.1	5.4	4.1	2.72
30	5	7.7	6.2	1.6	2.36
31	1	7.8	6.8	2.4	2.81
32	8	11.5	6.2	1.9	1.64
33	10	10.4	6.4	2.2	1.82

are sometimes presented as a half-matrix, such as

$$
\begin{array}{llll}
\sum X_1^2 & & & \\
\sum X_2 X_1 & \sum X_2^2 & & \\
\sum X_3 X_1 & \sum X_3 X_2 & \sum X_3^2 & \\
\vdots & \vdots & \vdots & \\
\sum X_M X_1 & \sum X_M X_2 & \sum X_M X_3 & \cdots & \sum X_M^2.
\end{array}
\tag{4}
$$

If a raw sum of squares, $\sum X_i^2$, is reduced by $(\sum X_i)^2/n$, we have a sum of squares that has previously been symbolized as $\sum x^2$, referring to $\sum \sum (X_{ij} - \overline{X}_i)^2$. Similarly, a raw sum of cross products, $\sum X_i X_k$, if diminished by $\sum X_i \sum X_k/n$, yields $\sum x_i x_k$, which represents $\sum (X_{ij} - \overline{X}_i)(X_{kj} - \overline{X}_k)$. These quantities are known as *corrected* sums of squares and *corrected* sums of crossproducts, respectively, and they may be presented as the following matrix:

$$
\begin{matrix}
\sum x_1^2 & \sum x_1 x_2 & \sum x_1 x_3 & \cdots & \sum x_1 x_M \\
\sum x_2 x_1 & \sum x_2^2 & \sum x_2 x_3 & \cdots & \sum x_2 x_M \\
\sum x_3 x_1 & \sum x_3 x_2 & \sum x_3^2 & \cdots & \sum x_3 x_M \\
\vdots & \vdots & \vdots & & \vdots \\
\sum x_M x_1 & \sum x_M x_2 & \sum x_M x_3 & \cdots & \sum x_M^2.
\end{matrix}
\tag{5}
$$

From Matrix 5, it is simple to calculate a matrix of simple correlation coefficients, for r_{ik} (representing the correlation between variables i and k) $= \sum x_i x_k \, ? \, \sqrt{\sum x_i^2 \sum x_k^2}$:

$$
\begin{matrix}
r_{11} & r_{12} & r_{13} & \cdots & r_{1M} \\
r_{21} & r_{22} & r_{23} & \cdots & r_{2M} \\
r_{31} & r_{32} & r_{33} & \cdots & r_{3M} \\
\vdots & \vdots & \vdots & & \vdots \\
r_{M1} & r_{M2} & r_{M3} & \cdots & r_{MM}.
\end{matrix}
\tag{6}
$$

Each element in the diagonal of this matrix (i.e., r_{ii}) is equal to 1.0, for there will always be a perfect positive correlation between a variable and itself (see Example 1b).

EXAMPLE 1b   A Matrix of Simple Correlation Coefficients, as It Might Appear as Computer Output (from the Data of Example 1a)

	1	2	3	4	5
1	1.00000	0.32872	0.16767	0.05191	−0.73081
2	0.32872	1.00000	−0.14550	0.18033	−0.21204
3	0.16767	−0.14550	1.00000	0.24134	−0.05541
4	0.05191	0.18033	0.24134	1.00000	0.31267
5	−0.73081	−0.21204	−0.05541	0.31267	1.00000

The final major manipulation necessary before the important regression or correlation statistics of the following sections can be obtained is the computation of the *inverse* of a matrix. The process of inverting a matrix will not be explained here; it is to two-dimensional algebra what taking the reciprocal is to ordinary, one-dimensional algebra.* While inverting a matrix of moderate size is too cumbersome to be performed easily by hand, it may be readily accomplished by computer. A

---

*The plural of *matrix* is *matrices*. As a shorthand notation, statisticians may refer to an entire matrix by a boldface letter, and the inverse of the matrix by that letter's reciprocal. So, Matrices

multiple-regression or correlation program may invert the corrected sum of squares and crossproducts matrix, Matrix 5, resulting in a symmetrical matrix symbolized

$$
\begin{matrix}
c_{11} & c_{12} & c_{13} & \cdots & c_{1M} \\
c_{21} & c_{22} & c_{23} & \cdots & c_{2M} \\
\vdots & \vdots & \vdots & & \vdots \\
c_{M1} & c_{M2} & c_{M3} & \cdots & c_{MM}.
\end{matrix}
\tag{7}
$$

Or the correlation matrix, Matrix 6, may be inverted, yielding a different array of values, which we may designate

$$
\begin{matrix}
d_{11} & d_{12} & d_{13} & \cdots & d_{1M} \\
d_{21} & d_{22} & d_{23} & \cdots & d_{2M} \\
d_{31} & d_{32} & d_{33} & \cdots & d_{3M} \\
\vdots & \vdots & \vdots & & \vdots \\
d_{M1} & d_{M2} & d_{M3} & \cdots & d_{MM}.
\end{matrix}
\tag{8}
$$

Computer routines might compute either Matrix 7 or 8; the choice is unimportant because the two are interconvertible:

$$
c_{ik} = \frac{d_{ik}}{\sqrt{\sum x_i^2 \sum x_k^2}},
\tag{9}
$$

or, equivalently,

$$
c_{ik} = \frac{r_{ik} d_{ik}}{\sum x_i x_k}.
\tag{10}
$$

From manipulations of these types of arrays, a computer program can derive the sample statistics and components of analysis of variance described in the following sections. If partial correlation coefficients are desired (Section 7), the matrix inversion takes place as shown. If partial regression analysis is desired (Sections 2–4), then inversion is performed only on the $M-1$ rows and $M-1$ columns corresponding to the independent variables in either Matrix 5 or 6.

2 THE MULTIPLE-REGRESSION EQUATION

A simple linear regression for a population of paired variables is the relationship

$$
Y_i = \alpha + \beta X_i.
\tag{10b}
$$

In this relationship, Y and X represent the dependent and independent variables, respectively; β is the regression coefficient in the sampled population; and α (the Y intercept) is the predicted value of Y in the population when X is zero. And the subscript i in this equation indicates the ith pair of X and Y data in the sample.

In some situations, however, Y may be considered dependent upon more than one variable. Thus,

$$
Y_j = \alpha + \beta_1 X_{1j} + \beta_2 X_{2j}
\tag{11}
$$

3, 5, and 6 might be referred to by the symbols **X**, **x**, and **r**, respectively; and Matrices 7 and 8 could be written, respectively, as $\mathbf{c} = \mathbf{x}^{-1}$ and $\mathbf{d} = \mathbf{r}^{-1}$. David (2006) gives A. C. Aitken primary credit for introducing matrix algebra into statistics in 1931.

may be proposed, implying that one variable (Y) is linearly dependent upon a second variable (X_1) and that Y is also linearly dependent upon a third variable (X_2). Here, i denotes the ith independent variable, and X_{ij} specifies the jth observation of variable i. In this particular multiple regression model, we have one dependent variable and two independent variables.* The two population parameters β_1 and β_2 are termed *partial regression coefficients*; β_1 expresses how much Y would change for a unit change in X_1, if X_2 were held constant. It is sometimes said that β_1 is a measure of the relationship of Y to X_1 after "controlling for" X_2; that is, it is a measure of the extent to which Y is related to X_1 after removing the effect of X_2. Similarly, β_2 describes the rate of change of Y as X_2 changes, with X_1 being held constant. β_1 and β_2 are called partial regression coefficients, then, because each expresses only part of the dependence relationship. The Y intercept, α, is the value of Y when *both* X_1 and X_2 are zero. Whereas Equation 10b mathematically represents a line (which may be presented on a two-dimensional graph), Equation 11 defines a plane (which may be plotted on a three-dimensional graph). A regression with m independent variables defines an m-dimensional surface, sometimes referred to as a "response surface" or "hyperplane."

The population data whose relationship is described by Equation 11 will probably not all lie exactly on a plane, so this equation may be expressed as

$$Y_j = \alpha + \beta_1 X_{1j} + \beta_2 X_{2j} + \epsilon_j; \tag{12}$$

ϵ_j, the "residual," or "error," is the amount by which Y_j differs from what is predicted by $\alpha + \beta_1 X_{1j} + \beta_2 X_{2j}$, where the sum of all ϵ's is zero, the ϵ's are assumed to be normally distributed, and each partial regression coefficient, β_i, estimates the change of Y in the population when there is a change of one unit (e.g., a change of 1 centimeter, 1 minute, or 1 milliliter) in X_i and no change in the other X's.

If we sample the population containing the three variables (Y, X_1, and X_2) in Equation 11, we can compute sample statistics to estimate the population parameters in the model. The multiple-regression function derived from a sample of data would be

$$\hat{Y}_j = a + b_1 X_{1j} + b_2 X_{2j}. \tag{13}$$

The sample statistics a, b_1, and b_2 are estimates of the population parameters α, β_1, and β_2, respectively, where each partial regression coefficient b_i is the expected change in Y in the population for a change of one unit in X_i if all of the other $m - 1$ independent variables are held constant, and a is the expected population value of Y when each X_i is zero. (Often, the sample Y intercept, a, is represented by b_0 and the population Y intercept is represented as β_0 instead of α.)

Theoretically, in multiple-regression analyses there is no limit to m, the number of independent variables (X_i) that can be proposed as influencing the dependent variable (Y), as long as $n \geq m + 2$. (There will be computational limitations, however.) The general population model, of which Equation 12 is the special case for $m = 2$, is[†]

$$Y_j = \alpha + \beta_1 X_{1j} + \beta_2 X_{2j} + \beta_3 X_{3j} + \cdots + \beta_m X_{mj} + \epsilon_j, \tag{14}$$

Dependence in a regression context refers to mathematical, not necessarily biological, dependence. Sometimes the independent variables are called "predictor" or "regressor" or "explanatory" or "exogenous" variables, and the dependent variable may be referred to as the "response" or "criterion" or "endogenous" variable.

[†]This equation reflects that multiple regression is a special case of what mathematical statisticians call the *general linear model*. Multiple correlation, simple regression and correlation, analysis of variance, and analysis of covariance are also special cases of that model.

or, more succinctly,

$$Y_j = \alpha + \sum_{i=1}^{m} \beta_j X_{ij} + \epsilon_j, \tag{15}$$

where m is the number of independent variables. This model is said to be one of *multiple linear regression* because of the linear (i.e., additive) arrangement of the parameters (α and β_i) in the model. The sample regression equation, containing the statistics used to estimate the population parameters when there are m independent variables, would be

$$\hat{Y}_j = a + b_1 X_{1j} + b_2 X_{2j} + b_3 X_{3j} + \cdots + b_m X_{mj}, \tag{16}$$

or

$$\hat{Y}_j = a + \sum_{i=1}^{m} b_i X_{ij}. \tag{17}$$

At least $m + 2$ data points are required to perform a multiple regression analysis, where m is the number of independent variables determining each data point.

The criterion for defining the "best fit" multiple regression equation is most commonly that of *least squares*,* which—as described for simple regression—represents the regression equation with the minimum residual sum of squares[†] (i.e., the minimum value of $\sum_{j=1}^{n}(Y_j - \hat{Y}_j)^2$). The idea of the least-squares fit of a plane (or hyperplane) through data is an extension of the least-squares concept regarding the fit of a line through data.

From the analysis shown in Example 1c,[‡] we arrive at a regression function having partial regression coefficients of $b_1 = -0.129$ ml/°C, $b_2 = -0.019$ ml/cm, $b_3 = 0.05$ ml/mm, $b_4 = 0.209$ ml/min, and a Y intercept of $a = 2.96$ ml.[§] Thus we can write the regression function as $\hat{Y} = 2.96 - 0.129 X_1 - 0.019 X_2 - 0.05 X_3 + 0.209 X_4$.

---

*The statistics in Equation 16, derived by the method of least squares, are known to statisticians as *best linear unbiased estimates* (BLUE) because they are unbiased estimates of the population parameters of interest, the equation is a linear combination of terms, and the statistics are "best" in the sense of having the smallest variance of any linear unbiased estimates.

[†] Another criterion that could be used—with different associated statistical procedures—is that of *least absolute deviations*, which would involve minimizing $\sum_{j=1}^{n} |Y_j - \hat{Y}_j|$ (see Birkes and Dodge, 1993; Chapter 2; Bloomfield and Steiger, 1983). This procedure may be beneficial when there are outlier data, and—as indicated in that footnote—an intermediate regression method is what is known as *M-regression*.

[‡] It should be noted that a computer program's output may display results with symbols different from those commonly found in publications such as this book. For example, n might be represented by N, t by T, and r by R; and X_1, X_2, and so on might be written as $X(1)$, $X(2)$, and so on or by $X1$, $X2$, $X3$, and so on. Numbers, especially very large or very small numbers, might be shown in "scientific notation"; for example, 0.0001234 might be displayed as 1.234×10^{-4} or $1.234 \times 10 - 4$ or 0.1234E−3. Users of computer programs should also be aware that some programs, particularly older ones, employ a small enough number of significant figures to cause sizable round-off errors to accumulate through the series of calculations noted in Section 1. Such errors may be especially severe if the variables have greatly different magnitudes or if there is considerable multicollinearity (described on Section 4).

[§] By examining the magnitude of the standard errors of the four partial regression coefficients (namely, 0.021287, 0.056278, 0.20727, and 0.067034), we observe that their second significant figures are at the third, third, second, and third decimal places, respectively, making it appropriate to state the four coefficients to those precisions.

EXAMPLE 1c   A Computer Fit of a Multiple-Regression Equation to the Data of Example 1a, Where Variable 5 Is the Dependent Variable

Regression model:

$$Y = \alpha + \beta_1 X_1 + \beta_2 X_2 + \beta_3 X_3 + \beta_4 X_4$$

For each i (where $i = 1, 2, 3, 4$),

$H_0: \beta_i = 0$

$H_A: \beta_i \neq 0$

Variable i	b_i	s_{b_i}	t	ν	b_i'
X_1	−0.12932	0.021287	−6.075	28	−0.73176
X_2	−0.018785	0.056278	−0.334	28	−0.41108
X_3	−0.046215	0.20727	−0.223	28	−0.26664
X_4	0.20876	0.067034	3.114	28	0.36451

Y intercept: $a = 2.9583$

Therefore, b_1 is an estimate of the relationship between Y and X_1 after removing the effects on Y of X_2, X_3, and X_4 (i.e., holding those three independent variables constant). So, in this example, an increase of 1° Celsius in variable 1 is predicted to be associated with a 0.129-ml decrease in volume in variable 5 (which is variable Y in this example) if there is no change in variables X_2, X_3, and X_4. Similarly, b_2 estimates the relationship between Y and X_2 after removing the effects of X_1, X_3, and X_4; and so on.

Section 4a will explain that, if independent variables are highly correlated with each other, then the interpretation of partial regression coefficients becomes questionable, as does the testing of hypotheses about the coefficients.

3   ANALYSIS OF VARIANCE OF MULTIPLE REGRESSION OR CORRELATION

A computer program for multiple regression analysis will typically include an analysis of variance (ANOVA), as shown in Example 1d, to test the null hypothesis that all partial regression coefficients (β_i) are zero against the alternate hypothesis that at least one of the β_i's is not zero. This analysis of variance is analogous to that in the case of simple regression, in that the total sum of squares and degrees of freedom are separated into two components: (1) the sum of squares and degrees of freedom due to multiple regression, and (2) the residual sum of squares and degrees of freedom; the multiple-regression mean square and the residual mean square are obtained from those quantities. The total sum of squares is an expression of the total amount of variability among the Y values (namely, $Y_j - \overline{Y}$), the regression sum of squares expresses the variability among the Y values that is attributable to the regression being fit (that is, $\hat{Y}_j - \overline{Y}$), and the residual sum of squares tells us about the amount of variability among the Y's that remains after fitting the regression ($Y_j - \hat{Y}_j$). The needed sums of squares, degrees of freedom, and means squares are summarized in Table 1. The expressions given in that table for sums of squares are the defining equations; the actual computations of these quantities may involve the use of formulas more suitable to the calculating machine.

EXAMPLE 1d   A Computer Analysis of Variance for the Multiple Regression Data of Example 1a

$H_0: \beta_1 = \beta_2 = \beta_3 = \beta_4 = 0$

$H_A: \beta_1$ and/or β_2 and/or β_3 and/or $\beta_4 \neq 0$

Source of Variation	SS	DF	MS
Total	14.747	32	
Multiple regression	9.7174	4	2.4294
Residual	5.0299	28	0.17964

$F = 13.5$, with DF of 4 and 28

$F_{0.05(1),4,28} = 2.71$, so reject H_0.

$P \ll 0.0005 \quad [P = 2.99 \times 10^{-6}]$

Coefficient of determination: $R^2 = 0.65893$

Adjusted coefficient of determination: $R_a^2 = 0.61021$

Multiple correlation coefficient: $R = 0.81175$

Standard error of estimate: $s_{Y \cdot 1,2,3,4} = 0.42384$

TABLE 1: Definitions of the Appropriate Sums of Squares, Degrees of Freedom, and Mean Squares Used in Multiple Regression or Multiple Correlation Analysis of Variance

Source of variation	Sum of squares (SS)	DF*	Mean square (MS)
Total	$\sum (Y_j - \overline{Y})^2$	$n - 1$	
Regression	$\sum (\hat{Y}_j - \overline{Y}_j)^2$	m	$\dfrac{\text{regression SS}}{\text{regression DF}}$
Residual	$\sum (Y_j - \hat{Y}_j)^2$	$n - m - 1$	$\dfrac{\text{residual SS}}{\text{residual DF}}$

*n = total number of data points (i.e., total number of Y values); m = number of independent variables in the regression model.

Note that a multiple-regression ANOVA (Table 1) becomes a simple-regression ANOVA when m, the number of independent variables, is 1.

If we assume Y to be functionally dependent on each of the X's, then we are dealing with multiple regression. If no such dependence is implied, then any of the $M = m + 1$ variables could be designated as Y for the purposes of utilizing the computer program; this is a case of *multiple correlation*. In either situation, we can test the hypothesis that there is no interrelationship among the variables, as

$$F = \frac{\text{regression MS}}{\text{residual MS}}. \tag{18}$$

The numerator and denominator degrees of freedom for this variance ratio are the regression DF and the residual DF, respectively. For multiple regression, this F tests

$$H_0: \beta_1 = \beta_2 = \cdots = \beta_M = 0,$$

which may be written as

$$H_0: \beta_i = 0 \text{ for all } i\text{'s,}$$

against

$$H_A: \beta_i \neq 0 \text{ for one or more } i\text{'s.}$$

The ratio,

$$R^2 = \frac{\text{regression SS}}{\text{total SS}} \tag{19}$$

or, equivalently,

$$R^2 = 1 - \frac{\text{residual SS}}{\text{total SS}} \tag{20}$$

is the *coefficient of determination* for a multiple regression or correlation, or the *coefficient of multiple determination*.* In a regression situation, it is an expression of the proportion of the total variability in Y that is attributable to the dependence of Y on all the X_i's, as defined by the regression model fit to the data. In the case of correlation, R^2 may be considered to be amount of variability in any one of the M variables that is accounted for by correlating it with all of the other $M - 1$ variables; the quantity $1 - R^2$ is called the *coefficient of nondetermination*, the portion of the variability in one of the variables that is *not* accounted for by its correlation with the other variables.

Healy (1984) and others caution against using R^2 as a measure of "goodness of fit" of a given regression model; and one should not attempt to employ R^2 to compare regressions with different m's and different amounts of replication. An acceptable measure of goodness of fit is what is referred to as the *adjusted coefficient of determination*[†].

$$R_a^2 = 1 - \frac{\text{residual MS}}{\text{total MS}}, \tag{22}$$

---

*R^2 may also be calculated as (Sutton, 1990)

$$R^2 = \frac{F}{F + v_2/v_1}. \tag{21}$$

Expressing R^2 may not be appropriate for regression models with no Y intercept, and various authors—and some computer programs—have used the symbol R^2 to denote somewhat different quantities (Kvålseth, 1985).

[†]Huberty and Mourad (1980) note that several adjustments to R^2 have been proposed and credit the first appearance of R_a^2, a very good one, to Ezekiel (1930: 225–226), who termed it an "index of determination" to distinguish it from the coefficient of determination. They also distinguish between a coefficient of determination when the purpose is multiple correlation (R_a^2), on the one hand, and one (which they attribute to G. E. Nicholson, in 1948, and F. M. Lord, in 1950) that they recommend where the objective is prediction via multiple regression, on the other hand:

$$R_{NL}^2 = 1 - \left(\frac{n + m + 1}{n - m - 1} \right) \left(\frac{n - 1}{n} \right) \left(1 - R^2 \right). \tag{21a}$$

The difference between R_a^2 and R_{NL}^2 increases as m increases and decreases as n increases.

which is

$$R_a^2 = 1 - \frac{n-1}{n-m-1}(1 - R^2). \tag{23}$$

R_a^2 increases only if an added X_i results in an improved fit of the regression to the data, whereas R^2 always increases with the addition of an X_i (or it is unchanged if the b_i associated with the X_i is zero). Therefore, R^2 tends to be an overestimate of the population coefficient of determination (ρ^2), with the magnitude of the overestimate greater with smaller n or larger m, and R_a^2 is a better estimate of ρ^2. Because R_a^2 is smaller than R^2, it is sometimes called the "shrunken R^2." If ρ^2 is near zero, the calculated R_a^2 may be negative (in which case it should be expressed as zero). R_a^2 is useful for comparing regression equations that have different numbers of independent variables.

The square root of the coefficient of determination is referred to as the *multiple correlation coefficient:*[*]

$$R = \sqrt{R^2}. \tag{24}$$

R is also equal to the Pearson correlation coefficient, r, for the correlation of the observed values of Y_j with the respective predicted values, $\hat{Y}_j$. For multiple correlation the F of Equation 18 allows us to draw inference about the population multiple correlation coefficient, ρ, by testing $H_0: \rho = 0$ against $H_A: \rho \neq 0$.

In a multiple-correlation analysis, Equation 18 provides the test for whether the multiple-correlation coefficient is zero in the sampled population. In the case of a multiple regression analysis, Equation 18 tests the null hypothesis of no dependence of Y on any of the independent variables, X_i; that is, $H_0: \beta_1 = \beta_2 = \cdots = \beta_m = 0$ (vs. H_A: All m population partial regression coefficients are not equal to zero). Once R^2 has been calculated, the following computation of F may be used as an alternative to Equation 18:

$$F = \left(\frac{R^2}{1 - R^2} \right) \left(\frac{\text{residual DF}}{\text{regression DF}} \right), \tag{25}$$

and F (from either Equation 18 or 25 provides a test of $H_0: \rho^2 = 0$ versus $H_A: \rho^2 \neq 0$.

The square root of the residual mean square is the *standard error of estimate* for the multiple regression:

$$s_{Y \cdot 1,2,\ldots,m} = \sqrt{\text{residual MS}}. \tag{26}$$

As the residual MS is often called the error MS, $s_{Y \cdot 1,2,\ldots,m}$ is sometimes termed the root mean square error. The subscript $(Y \cdot 1, 2, \ldots, m)$ refers to the mathematical dependence of variable Y on the independent variables 1 through m.

The addition of a variable, X_i, to a regression model increases the regression sum of squares and decreases the residual sum of squares (unless the associated b_i is zero, in which case these sums of squares are unchanged). It is important to ask, however, whether an increase in regression sum of squares is important (i.e., whether the added variable contributes useful information to our analysis). The regression degrees of

---

[*]G. U. Yule, in 1897, was the first to use R to denote the multiple correlation coefficient (Walker, 1929: 112). R. A. Fisher described the distribution of this statistic in 1928 (Lehmann, 1999). R can never be less than the coefficient of correlation between Y and any of the X_i's (Darlington, 1990: 53).

freedom also increase and the residual degrees of freedom also decrease with the addition of a variable and therefore the regression mean square might decrease and/or the residual mean square might increase, and F might be reduced. This issue is addressed in Sections 4 and 6.

(a) Assumptions of Multiple Regression Analysis. The underlying assumptions of multiple regression are analogous to those of simple regression:

1. The values of Y have come at random from the sampled population and are independent of one another.
2. For any combination of values of the X_i's in the population, there is a normal distribution of Y values. (Thus, for each of the combinations of X_i's, there is in the population a normal distribution of ϵ's.)
3. There is homogeneity of variances; that is, the variances of the population distributions of Y values for all combinations of X_i's are all equal to each other. (The residual mean square, $s^2_{Y \cdot 1,2,\ldots,m}$ is the estimate of this common variance.)
4. The independent variables, X_i's, are fixed-effects factors, the measurements of which were obtained with no error or with errors negligible compared to the magnitude of errors in measuring Y.

These assumptions do not impact the calculation of regression statistics (a, b_i, R^2), but they do underlie the performance of hypothesis testing and the expression of confidence intervals. Fortunately, regression analysis is robust to some deviation from these assumptions, especially if n is large.

Chatterjee and Hadi (2006: Chapter 4) discuss graphical examination of data for purposes including assessment of departures from assumptions, and some computer programs will provide analysis of residuals (i.e., $Y_i - \hat{Y}_i$ vs. X_i). Data transformations for variables in multiple regression may assist in meeting the regression assumptions, as in the case of simple regression (e.g., Chatterjee and Hadi, 2006: Chapter 6; Cohen et al., 2003: Section 6.4).

There are regression methods, not commonly encountered, to which the foregoing assumptions do not apply. These include nonparametric regression, least-absolute-deviations regression, and M-regression (Birkes and Dodge, 1993: Chapters 5 and 6; Cleveland, Mallows, and McRae, 1993; Draper and Smith, 1998: Chapter 25; Hollander and Wolfe, 1999: Chapter 9; Huber, 2004: Chapter 7; Kutner, Nachtsheim, and Neter, 2004: 449–558; Montgomery, Peck, and Vining, 2006: Section 7.3; Wang and Scott, 1994).

4 HYPOTHESES CONCERNING PARTIAL REGRESSION COEFFICIENTS

In employing simple regression, it is generally desired to test $H_0 : \beta = \beta_0$, a two-tailed null hypothesis where β_0 is most often zero. If, in multiple regression, Equation 18 yields a significant F (i.e., H_0: $\beta_1 = \beta_2 = \cdots = \beta_m = 0$ is rejected), then we have concluded that at least one β_i is different from zero and its associated X_i contributes to explaining Y. In that case, each of the partial regression coefficients in a multiple-regression equation may be submitted to an analogous hypothesis, H_0: $\beta_i = \beta_0$, where, again, the test is usually two-tailed and the constant is most frequently zero. For H_0: $\beta_i = 0$, Student's t may be computed as

$$t = \frac{b_i}{s_{b_i}} \tag{27}$$

We may obtain both b_i and s_{b_i} from the computer output shown in Example 1c. In the particular computer program employed for this example, the t value is also calculated for each b_i. If it had not been, then Equation 27 would have been applied. (Some computer programs present the square of this t value and call it a "partial F value.") The residual degrees of freedom are used for this test.

If the standard errors are not given by the computer program being utilized, then they may be calculated as

$$s_{b_i} = \sqrt{s^2_{Y \cdot 1,2,\ldots,m} c_{ii}}, \tag{28}$$

where $s^2_{Y \cdot 1,2,\ldots,m}$ is the square of the standard error of estimate, which is simply the residual mean square, and c_{ii} is defined in Section 1. Knowing s_{b_i}, we can obtain a $1 - \alpha$ confidence interval for a partial regression coefficient, β_i, as

$$b_i \pm t_{\alpha(2),\nu} s_{b_i}, \tag{29}$$

where ν is the residual degrees of freedom.

In general, a significant F value in testing for dependence of Y on all X_i's (by Equation 18) will be associated with significance of some of the β_i's being concluded by t-testing; but it is possible to have a significant F without any significant t's, or, in rarer cases, significant t's without a significant F (Cohen et al. 2003: 90; Cramer, 1972; Draper and Smith, 1998: 146–147; Geary and Leser, 1968). These situations can occur when there is a high degree of multicollinearity (see Section 4a), and, in general, H_0: $\beta_i = 0$ should not be tested if there is not a significant F for the multiple-regression model.

Section 6 discusses methods for concluding which of the m independent variables should be kept in the multiple-regression model and which should be deleted because they do not contribute significantly to the magnitude of Y. Those procedures may be considered if any of the partial regression coefficients are found to be nonsignificant (i.e., at least one H_0: $\beta_i = 0$ is not rejected).

Cohen and colleagues (2003: 94–95) discuss power analysis for partial regression coefficients and provide a special table for this purpose. Various rules of thumb have been presented for sample sizes desirable for testing multiple-regression hypotheses; Green (1991) critiqued several such "rules." These tend to be very general. For example, Hair et al. (2006: 194–197) provide the following recommendations for testing partial regression coefficients at the 0.05 significance level with a power of at least 0.80: a minimum n of 50, preferably 100, and a minimum n-to-i ratio of 5 : 1, preferably 15 : 1 to 20 : 1—or up to 50 : 1 if the procedure of Section 6e is used. (As with other statistical methods, specifying a smaller significance level, or a greater power, requires larger samples.) If a reasonable set of independent variables is concluded to contribute to the determination of Y, then power may not be a concern; if not, it might be wise to repeat the experiment with these recommended sample sizes. Conferring with a statistical consultant may help determine how to proceed.

(a) Multicollinearity. If independent variables, say X_1 and X_2, are highly correlated, then the partial regression coefficients associated with them (b_1 and b_2) may not be assumed to reflect the dependence of Y on X_1 or Y on X_2 that exists in the population.

Statisticians call correlation between independent variables *multicollinearity* (and sometimes, if it is between only two X_i's, it is termed *collinearity*). It is also known as *intercorrelation* or *nonorthogonality* or *illconditioning* between variables. In practice, multicollinearity is of little consequence if it is not great. But if the multicollinearity is substantial, then standard errors of the partial regression coefficients of the correlated X_i's will be large (and confidence intervals of the b_i's will be wide), significance testing will possess low power, and the interpretation of the effects of those X_i's on Y (and conclusions about the associated b_i's and t's) may be spurious or ambiguous.

Consequential multicollinearity may be suspected

- if a regression coefficient appears unreasonable (such as being unreasonable in sign or magnitude or having an insignificant t even though its X is expected to have a considerable effect on Y);
- if the F for the overall regression is significant with α much lower than the stated significance level but none of the β's are concluded to be different from zero;
- if there are significant t's without a significant F;
- if some correlation coefficients for pairs of X_i's are very high (some researchers would say >0.80 or, especially, >0.90), as observed in Matrix 6;
- if R^2 is much greater than $\displaystyle\sum_{i=1}^{m} r_{Yi}^2$, where r_{Yi}^2 represents the simple correlation of the dependent variable (Y) on an independent variable (X_i);
- if there is a great change in the b_i's associated with the other variables when a variable is added to or deleted from the regression model;
- if there is a large difference in regression coefficients upon the addition or deletion of data.

Multicollinearity is more likely with a large number of independent variables, and the adverse effect of multicollinearity may be especially pronounced if the range of any of the X_i's is narrow.* Texts such as Glantz and Slinker (2001: Chapter 5) and Hair et al. (2006: 206–207) discuss both the assessment of multicollinearity by analysis such as what is called tolerance (or its inverse, the variance inflation factor, VIF), and the reduction of multicollinearity (e.g., by deletion of one or more correlated variables from the equation).

Singularity is extreme multicollinearity, when there is a perfect correlation (i.e., $r = 1.0$ or $r = -1.0$) between two (or more) variables. In this situation, a multiple-regression analysis cannot be performed until one (or more) of the perfectly correlated variables is removed from consideration.

When multicollinearity is present, standard errors of partial regression coefficients (s_{b_i}'s) may be large, meaning that the b_i's are imprecise estimates of the relationships in the population. As a consequence, a b_i may not be declared statistically significant from zero (as by the above t test), even when Y and X_i are related in the population. With highly correlated X_i's the overall F for the regression model can be significant even when the t tests for the individual X_i's are not (Berry and Feldman, 1985: 42–43; Bertrand and Holder, 1988; Hamilton, 1987; Kendall and Stuart, 1979; 367; Routledge, 1990). An additional deleterious effect of multicollinearity is that it may lead to increased roundoff error in the computation of regression statistics.

---

*If the intercorrelation is great, we may be unable to calculate the partial regression coefficients at all, for it may not be possible to perform the matrix inversion described in Section 1.

5 STANDARDIZED PARTIAL REGRESSION COEFFICIENTS

Users of multiple-regression analysis may encounter *standardized partial regression coefficients*.* The common definition of such a coefficient employs the standard deviation of Y (namely S_Y) and of X_i (namely S_{X_i}):

$$b_i' = b_i \left(\frac{s_{X_i}}{s_Y} \right), \quad \text{or, equivalently,} \quad b_i' = b_i \sqrt{\frac{\sum x_i^2}{\sum y^2}}. \tag{30}$$

A standardized partial regression coefficient, b_i', is the partial regression coefficient that would result from using Y/s_Y in place of Y and, for that i, using $X/s_{\hat{X}}$ in place of X; or, equivalently from using $(Y - \overline{Y})/s_{\hat{Y}}$ in place of Y and, for that i, using $(X - \overline{X})/s_{\hat{X}}$ in place of X. These coefficients are sometimes reported as indicators of the relative importance of the independent variables (X_i's) in determining the value of the dependent variable Y (if the X_i's are uncorrelated).

These coefficients are unitless, so they are especially useful indicators when X_i's are on different measurement scales; a b_i' with a large absolute value is indicative of its associated X_i having a high degree of influence on Y. Many multiple-regression computer programs include standardized regression coefficients, and some also include their standard errors. A test of H_0: $\beta_i' = 0$ is typically not performed, however, for it would tell the user no more than a test performed for H_0: $\beta_i = 0$; that is, the probability associated with the former null hypothesis is equal to the probability associated with the latter. Standardized partial regression coefficients suffer from the same problems with multicollinearity as do partial regression coefficients (see Section 4a).

6 SELECTING INDEPENDENT VARIABLES

Example 1c shows the statistics for the least-squares best-fit equation for the data of Example 1a. However, although the data consisted of four independent variables, it should not be assumed that each of the four has a consequential effect on the magnitude of the dependent variable.

Challenges facing the user of multiple regression analysis include concluding which of the independent variables have a significant effect on Y in the population sampled. It is desired to employ a regression equation with as many of the independent variables as required to provide a good determination of which of these variables effect a significant change of Y in the population and to enable accurate prediction of Y. However, the resultant regression equation should comprise as few variables as are necessary for this purpose so as to minimize the time, energy, and expense expended in collecting further data or performing further calculations with the selected regression equation, to optimize statistical estimates (the variances of b_i and $\hat{Y}_j$ may increase unacceptably if nonsignificant variables are included), and, we hope, to simplify the interpretations of the resultant regression equation; a smaller number of variables will also tend to increase the precision of predicted Y's (Draper and Smith, 1998: 327).

The following statistical procedures are important if the intent of the analysis is to predict Y from a group of significantly influential X_i's. However, if the goal is to describe, and help understand, biological relationships underlying the magnitude of Y, then some analysts have argued that biological considerations, in addition to

---

*These are sometimes called beta coefficients (β_i) but should not be confused with the population parameters (β_i) estimated by b_i.

automated statistical rules, should be employed when deciding which variables to add to or delete from the regression model.

A number of procedures have been proposed to conclude which is, in some objective way, the "best" (or at least a very good) regression model. The various methods do not necessarily arrive at the same conclusions on this question, and there is not universal agreement among statisticians as to which is most advantageous. Indeed, because of drawbacks such as those noted later, some data analysts recommend against using any of them. However, inasmuch as they are commonly found in research publications and computer output, they are summarized here for the reader's benefit.

This section will discuss common methods that have been used for concluding which of the m independent variables should be included in the model, but consultation with a professional statistician may be beneficial in many cases. Deciding, by statistical processes, which of m independent variables should remain in a multiple-regression model is discussed in the references cited in the footnote at the end of the introduction to this chapter, such as in Chatterjee, Hadi, and Price (2000: Chapter 11); Draper and Smith (1998: Chapter 15); Glantz and Slinker (2001: Chapter 6); Hair et al. (2006: 209–214); Kutner, Nachtsheim, and Neter (2004: Chapter 9); and Seber and Lee (2003: Chapter 12).

Each of the methods in Sections 6b, 6c, and 6e involves more than one null hypothesis about partial regression coefficients. However, the several hypothesis tests performed on a set of data are not independent, so the probability of Type I errors may be substantially different from α, especially if the ratio of n to m is small. (See Section 4 for recommended sample sizes.) There is no consensus regarding how to correct for this, but many suggest that the same nominal significance level (α) should be used for testing each of the H_0's (though the method described later as "stepwise" might proceed otherwise, as indicated in Section 6e).

(a) Fitting All Possible Equations. One procedure would start by fitting a regression equation that contains all the independent variables. In the present example this would involve fitting an equation using all four X_i's. Then a regression fit would be calculated for each of the four different equations containing three of the four independent variables, a regression would be fit for each of the six possible equations comprising two of the X_i's, and a simple regression (that is, with one X_i) would be done using each of the four independent variables. After fitting all 15 of these regression equations, we could choose the one resulting in the lowest residual mean square, or, equivalently, the largest R_a^2 (which is preferable to using R^2) or smallest standard error of estimate.

This is often referred to as "all subsets regression." There are drawbacks to such a procedure, however. First, many regression equations must be calculated,* the number being $2^m - 1$. Thus, if $m = 5$, there would be a total of 31 regressions to be fit; if $m = 8$, then 255 regressions would be called for; if $m = 10$, the goal would be to choose among 1023 regression equations; and so on. A second difficulty with considering the very large number of all possible regressions is that of declaring an objective method for determining which among these many equations is to be considered to be the "best." Thirdly, if one regression equation is determined to be the "best" (perhaps by examining R^2, R_a^2, or $s_{Y \cdot 1,2,3,\dots,m}^2$, or by a method referred to in Section 6f), there is the challenge of concluding whether that equation is

*This calculation is the number of ways that n items can be combined, one a time, two at a time, and so on.

significantly better than the one deemed "second best." Also, this procedure may result in a regression with substantial multicollinearity.

(b) Backward Elimination of Variables. If a multiple regression equation is fitted using all m independent variables in a set of data (as done in Example 1c), then we might ask whether any of those variables have insignificant influence on Y in the sampled population and thus may be eliminated from the equation. The hypothesis H_0: $\beta_i = 0$ may be examined for each of the m partial regression coefficients. If all m of these hypothesis tests are rejected, it may be concluded that all of the X's have a significant effect on Y and none of them should be deleted from the regression model. However, if any $|t|$ values are less than the critical value, $t_{\alpha(2),\nu}$, where ν is the residual degrees of freedom ($n - m - 1$, in the model being considered at this step of the process),* then the independent variable associated with the t with the lowest absolute value is deleted from the model and a new multiple-regression equation may be fitted using the remaining $m - 1$ independent variables. The null hypothesis H_0: $\beta_i = 0$ is then tested for each partial regression coefficient in this new model, and if any of the $|t|$ values are less than the critical value, then one more variable is deleted and a new multiple-regression analysis performed.

As demonstrated in Example 1e, this procedure is repeated in what is termed a stepwise fashion, until all b_i's in the equation are concluded to estimate β_i's that are different from zero. Each time a variable is thus deleted from the regression model, the regression MS decreases slightly and the residual MS increases slightly and R^2 decreases (unless that variable's partial regression coefficient is zero, in which case there is no change).

(c) Forward Addition of Variables. Another stepwise procedure (often called forward selection) is to begin with the smallest possible regression model (i.e., one with only one independent variable; in other words, a simple regression) and gradually work up to the multiple-regression model incorporating the largest number of significantly important variables. It is first determined which is the "best" simple-regression model for the data, such as by fitting all m simple regressions and selecting the one for which b_i has the largest value of $|t|$. If none of the b_i's is significant, then it is concluded that no population relationship has been detected between Y and the X_i's and the procedure proceeds no further. If at least one b_i is significant, then a fit would be effected for each of the regressions possessing the X already selected and one of the other X's, and the equation with the largest $|t|$ associated with one of the other X's would be chosen. In a similar fashion, the "best" regression equation can be determined with one X in addition to the two already chosen, and so on. At each step, $|t|$ is compared to the critical value $t_{\alpha(2),\nu}$, where ν is the residual degrees of freedom ($n - m - 1$) at that step.†

Because the relationships among variables change as each one is added, it is not warranted to declare the importance of each variable to be indicated by the sequence in which it is added to the regression model.

(d) Backward Elimination versus Forward Addition. Mantel (1970) described how a "step-up" forward-selection process (Section 6c) can involve more computational

---

*Some computer programs express the critical value as $F_{\alpha(1),1,\nu}$, which is equal to $t^2_{\alpha(2),\nu}$ and, in the context of backward elimination, might be referred to as the "F to remove."

†If F (a "partial F," which is t^2) is used as the test statistic, some computer routines call the critical value ($F_{\alpha(1),1,\nu}$) the "F to enter."

EXAMPLE 1e   Backward Elimination of Variables in Multiple-Regression Analysis, Using the Data from Example 1a

As shown in Example 1c, the multiple regression analysis for the model $\hat{Y} = \alpha + \beta_1 X_1 + \beta_2 X_2 + \beta_3 X_3 + \beta_4 X_4$ yields the following statistics:

Variable	b_i	s_{b_i}	t	ν
X_1	−0.12932	0.021287	−6.075	28
X_2	−0.018785	0.056278	−0.334	28
X_3	−0.046215	0.20727	−0.223	28
X_4	0.20876	0.067034	3.114	28

$$a = 2.9583$$

The critical value for testing $H_0: \beta_j = 0$ against $H_A: \beta_j \neq 0$ is $t_{0.05(2),28} = 2.048$. Therefore, H_0 would be rejected for β_1 and β_4, but not for β_2 or β_3. Of the t tests for the latter two, the t for testing the significance of β_3 has the smaller absolute value. Therefore, $\beta_3 X_3$ is deleted from the model, leaving $\hat{Y} = \alpha + \beta_1 X_1 + \beta_2 X_2 + \beta_4 X_4$. The data are then subjected to a multiple-regression analysis using this model with three independent variables, and the following statistics are obtained:

Variable	b_i	s_{b_i}	t	ν
X_1	−0.13047	0.020312	−6.423	29
X_2	−0.015424	0.053325	−0.289	29
X_4	0.20450	0.063203	3.236	29

$$a = 2.6725$$

The critical value for testing the significance of these partial regression coefficients is $t_{0.05(2),29} = 2.045$. Therefore, $H_0: \beta_j = 0$ would be rejected for β_1 and for β_4, but not for β_2. Therefore, $\beta_2 X_2$ is deleted from the regression model, leaving $\hat{Y} = \alpha + \beta_1 X_1 + \beta_4 X_4$. The analysis of the data using this model, with two independent variables, yields the following statistics:

Variable	b_i	s_{b_i}	t	ν
X_1	−0.13238	0.018913	−6.999	30
X_4	0.20134	0.061291	3.285	30

$$a = 2.5520$$

The critical value for testing $H_0: \beta_j = 0$ against $H_0: \beta_j \neq 0$ is $t_{0.05(2),30} = 2.042$. Therefore, both β_1 and β_4 are concluded to be different from zero, and $\hat{Y} = 2.552 - 0.132 X_1 + 0.201 X_4$ is the final model.

effort, and is fraught with more theoretical deficiencies, than is the "step-down" backward-elimination method (Section 6b). The step-up procedure might require as many as $_{m+1}C_2$ regressions to be fit*; so if $m = 5$, there would be as many as 15 regression equations to examine, if $m = 8$, there would be as many as 36, and so on. However, the step-down method will never involve the fitting of more than m regressions. Also, forward selection will not identify situations where the addition of a significant X fails to recognize that a previously added X is no longer deemed to significant, it may fail to identify significant independent variables when multicollinearity is present (Hamilton, 1987; Mantel, 1970), and it may yield erroneous conclusions when dealing with dummy variables (described in Section 10) with more than two categories (Cohen, 1991). The backward-elimination method is generally preferred to the forward-addition process.

(e) Stepwise Regression. The procedures of Sections 6b and 6c are stepwise in their execution, but the process very commonly named stepwise is one that employs *both* the addition and the elimination of independent variables in order to conclude which of the variables should be in the multiple-regression model. The process begins as does the step-up method; but whenever an X is added, the b associated with each of the X's thus far in the model is examined to see whether it has a nonsignificant t. If any of them do, then the term with the smallest $| t |$ is eliminated at that step. No more than one X is added or removed at each step, as is the case in the step-down and step-up procedures.

Many statisticians consider this method of variable selection to be preferable to the step-up (Section 6c) or step-down (Section 6b) method, though others have serious reservations about all three of these procedures (Henderson and Denison, 1989). Some computer software for stepwise regression will allow the user to employ t (or F) with the α for adding a variable to the regression equation different from the α used to remove a variable from the model (so, for example, one might use $\alpha = 0.05$ for adding a variable and $\alpha = 0.10$ for eliminating a variable); but the α for adding should not be greater than the α for removing.

Some computer programs contain routines for performing the addition and/or elimination of variables automatically by one or more of the three stepwise procedures just described. But if a computer program does not do this, the user can determine which variable should be added or deleted at each step, and after each addition or deletion, resubmit the data for computer analysis.

(f) Other Methods. Some computer software presents other methods and criteria to select the "best" set of independent variables for a given set of data. Two such procedures employ statistics known as Mallows C_p, which is closely related to R_a^2 (Kennard, 1971),[†] and PRESS (predicted error sum of squares). These are described in references such as those cited in the footnote at the end of the introduction to this chapter.

---

*The number obtained as $_{m+1}C_2$ is called a *triangular number*. It is the sum of the consecutive integers from 1 to m and gets its name from the arrangement of objects in rows: one object in the first row, two in the second row, and so on through the mth row.

[†] C. L. Mallows introduced C_p in 1964; it was published by Gorman and Toman (1966). Mallows (1973) credited its conception to discussions with Cuthbert Daniel late in 1963, and he used the symbol C to honor the latter colleague. The symbol p is used by many authors to denote the number of independent variables (as m is used in this text).

7 PARTIAL CORRELATION

When the interest is in the relationship among all M variables, with none of them considered dependent upon the others, then the multiple-correlation coefficient, R, reflects the overall relationship of all M variables. But we may desire to examine the variables two at a time. We could calculate a simple correlation coefficient, r, for each pair of variables (i.e., what Example 1b presents to us). But the problem with considering simple correlations of all variables, two at a time, is that such correlations will fail to take into account the interactions of any of the other variables on the two in question. *Partial correlation* addresses this problem by considering the correlation between each pair of variables while holding constant the effect of each of the other variables.* Symbolically, a partial correlation coefficient for a situation considering three variables (sometimes called a first-order partial correlation coefficient) would be $r_{ik \cdot l}$, which refers to the correlation between variables i and k, considering that variable l does not change its value (i.e., we have eliminated any effect of the interaction of variable l on the relationship between variables i and k). For four variables, a partial correlation coefficient, $r_{ik \cdot lp}$ (sometimes called a second-order partial correlation coefficient), expresses the correlation between variables i and k, assuming that variables l and p were held at constant values. In general, a partial correlation coefficient might be referred to as $r_{ik \cdots}$, meaning the correlation between variables i and k, holding all other variables constant (i.e., removing, or "partialling out" the effects of the other variables).

Another way to visualize partial correlation with three variables (i.e., $M = 3$) is as follows. In a regression of variable X_i on X_l, a set of residuals $(X_i - \hat{X}_i)$ will result; and the regression of X_k on X_l will yield another set of residuals $(X_k - \hat{X}_k)$. The correlation between these two sets of residuals will be the partial correlation coefficient, $r_{ik \cdot l}$.

For three variables, partial correlation coefficients may be calculated from simple correlation coefficients as

$$r_{ik \cdot l} = \frac{r_{ik} - r_{il}r_{kl}}{\sqrt{(1 - r_{il}^2)(1 - r_{kl}^2)}}. \tag{31}$$

For more than three variables, the calculations become quite burdensome, and computer assistance is routinely employed. If a partial regression coefficient, b_i, has been obtained for the regression of Y on X_i, the partial correlation coefficient $r_{Yi \cdots}$ can be determined from the t obtained for that b_i as

$$r_{Yi \cdots} = \sqrt{\frac{t^2}{t^2 + \nu}}, \text{ where } \nu = n - M \tag{32}$$

(Algina and Seaman, 1984). So, for example (see Examples 1b and 1c),

$$r_{Y4 \cdots} = \sqrt{\frac{(3.114)^2}{(3.114)^2 + 28}} = 0.5072.$$

---

*The first (in 1892) to extend the concept of correlation to more than two variables was Francis Ysidro Edgeworth (1845–1926), a statistician and economist who was born in Ireland and spent most of his career at Oxford University (Desmond, 2000; Stigler, 1978). Karl Pearson was the first to express what we now call multiple and partial correlation coefficients; in 1897 he proposed the term *partial correlation*, in contrast to *total correlation* (i.e., what we now call simple correlation), and in preference to what G. U. Yule termed *nett* (a British variant of the word *net*) and *gross* correlation, respectively (Snedecor, 1954; Walker, 1929: 109, 111, 185).

Multiple Regression and Correlation

A computer program providing partial correlation coefficients will generally do so in the form of a matrix, such as in Example 2:

$$
\begin{matrix}
1.00 & r_{12\cdots} & r_{13\cdots} & \cdots & r_{1M\cdots} \\
r_{21\cdots} & 1.00 & r_{23\cdots} & \cdots & r_{2M\cdots} \\
r_{31\cdots} & r_{32\cdots} & 1.00 & \cdots & r_{3M\cdots} \\
\vdots & \vdots & \vdots & & \vdots \\
r_{M1\cdots} & r_{M2\cdots} & r_{M3\cdots} & \cdots & 1.00.
\end{matrix}
\tag{33}
$$

To test H_0: $\rho_{ik\cdots} = 0$, we may employ

$$
t = \frac{r_{ik\cdots}}{s_{r_{ik\cdots}}},
\tag{34}
$$

where

$$
s_{r_{ik\cdots}} = \sqrt{\frac{1 - r_{ik\cdots}^2}{n - M}}
\tag{35}
$$

and M is the total number of variables in the multiple correlation.* The statistical significance of a partial correlation coefficient (i.e., the test of H_0: $\rho_{ik\cdots} = 0$) may also be determined by employing Table 17 from *Appendix: Statistical Tables and Graphs* for $n - M$ degrees of freedom. One-tailed hypotheses may be performed as for simple correlation coefficients. If a multiple-regression and a multiple-correlation analysis were performed on the same data, the test conclusion for H_0: $\beta_i = 0$ would be identical to the test conclusion for H_0: $\rho_{ik\cdots} = 0$ (by either t testing or "partial F" testing), where variable k is the dependent variable. Hypotheses such as H_0: $\rho_{ik\cdots} = \rho_0$, or similar one-tailed hypotheses, where $\rho_0 \neq 0$, may be testing using the z transformation.

EXAMPLE 2   A Matrix of Partial Correlation Coefficients, as It Might Appear as Computer Output (from the Data of Example 1a)

	1	2	3	4	5
1	1.00000	0.19426	0.12716	0.33929	−0.75406
2	0.19426	1.00000	−0.26977	0.23500	−0.06296
3	0.12716	−0.26977	1.00000	0.26630	−0.04210
4	0.33929	0.23500	0.26630	1.00000	0.50720
5	−0.75406	−0.06296	−0.04210	0.50720	1.00000

Cohen et al. (2003: 94–95) present power estimation, with a needed table, for partial correlation. Serlin and Harwell (1993) assess several nonparametric methods for three-variable partial correlation without the assumption of normality.

---

*This test statistic may also be calculated as

$$
t = \sqrt{\frac{(n - M)(r_{ik\cdots}^2)}{1 - r_{ik\cdots}^2}}
\tag{36}
$$

(a) Semipartial Correlation.   Another correlation concept, not as commonly encountered as partial correlation, is that of *semipartial correlation* (Cohen et al., 2003: 72–73, 84–85; Howell, 2007: Section 15.7; Pedhazur, 1997: 174–180), sometimes called *part correlation*. This is the correlation between two of M variables (where $M > 2$) where the effects of all other variables are removed from only one of the two. For example, if $M = 3$, the first-order coefficient of the semipartial correlation between variables X_1 and X_2, with the influence of variable X_3 removed ("partialled out") from X_2 but not from X_1, is

$$r_{1(2.3)} = \frac{r_{12} - r_{13}r_{23}}{\sqrt{1 - r_{23}}}, \tag{36a}$$

and the second-order semipartial correlation coefficient for the relationship between X_1 and X_2, with the influence of X_3 and X_4 removed could be designated as $r_{1(2.34)}$. A generalized notation for a semipartial correlation is $r_{i(k...)}$, meaning the correlation between X_i and X_k, removing the effect on X_k of all the other variables. A simple method of calculating a semipartial correlation coefficient was given by Algina and Seaman (1984) as

$$r_{i(k...)} = t_i^2 \sqrt{\frac{\text{error MS}}{\text{total SS}}}. \tag{36b}$$

The absolute value of the coefficient of semipartial correlation between two variables is always less than the absolute value of the coefficient of partial correlation between those two variables, except that the two coefficients are equal if there is zero correlation between X_i and any variable other than X_k (Darlington, 1990: 56).

A hypothesis test for a population semipartial correlation coefficient being different from zero would exhibit the same probability as a test for the partial correlation coefficient (or the partial regression coefficient, or the standardized partial regression coefficient) being different from zero for the same two variables (Cohen et al., 2003: 89).

8   PREDICTING Y VALUES

Having fitted a multiple-regression equation to a set of data, we may desire to calculate the Y value to be expected at a particular combination of X_i values. Consider the a and b_i values determined in Example 2 for an equation of the form $\hat{Y} = a + b_1X_1 + b_4X_4$. Then the predicted value at $X_1 = 7°C$ and $X_4 = 2.0$ min, for example, would be $\hat{Y} = 2.552 - (0.132)(7) + (0.201)(2.0) = 2.03$ ml. Such predictions may be done routinely if there is a significant regression (i.e., the F from Equation 18 is significant), although, as with simple linear regression, it is unwise to predict Y for X_i's outside the ranges of the X_i's used to obtain the regression statistics.

The following is the standard error of a mean Y predicted from a multiple regression equation:

$$s_{\hat{Y}} = \sqrt{s_{Y \cdot 1,2,...,m}^2 \left[\frac{1}{n} + \sum_{i=1}^{m} \sum_{k=1}^{m} c_{ik}x_i x_k \right]}. \tag{37}$$

In this equation, $x_i = X_i - \overline{X}_i$, where X_i is the value of independent variable i at which Y is to be predicted, $\overline{X}_i$ is the mean of the observed values of variable i that were used to calculate the regression equation, and c_{ik} is from Matrix 7.

Thus, for the value of Y just predicted, we can solve Equation 37 as shown in Example 3.

EXAMPLE 3   The Standard Error of a Predicted Y

For the equation $\hat{Y} = 2.552 - 0.132X_1 + 0.201X_2$, derived from the data of Example 1a, where X_1 is the variable in column 1 of the data matrix, X_2 is the variable in column 4, and Y is the variable in column 5, we obtain the following quantities needed to solve Equation 37:

$$s^2_{Y \cdot 1,2} = 0.16844, \quad n = 33, \quad \overline{X} = 4.4546,$$

$$\overline{X}_2 = 2.5424, \quad \sum x_1^2 = 472.18, \quad \sum x_2^2 = 44.961,$$

$$d_{11} = 1.0027, \quad d_{12} = -0.052051, \quad d_{21} = -0.052051, \quad d_{22} = 1.0027.$$

By employing Equation 9, each d_{ik} is converted to a c_{ik}, resulting in

$$c_{11} = 0.0021236, \quad c_{12} = -0.00035724, \quad c_{21} = -0.00035724, \quad c_{22} = 0.022302.$$

What is the mean population value of Y at $X_1 = 7°C$ and $X_4 = 2.0$ min?

$$\hat{Y} = 2.552 - (0.132)(7) + (0.201)(2.0) = 2.030 \text{ ml}$$

What is the standard error of the mean population value of Y at $X_1 = 7°C$ and $X_4 = 2.0$ min? [Equation 37 is used.]

$$\begin{aligned}
s^2_{\hat{Y}} &= 0.16844 \left[\frac{1}{33} + (0.0021236)(7 - 4.4546)^2 \right. \\
&\quad + (-0.00035724)(7 - 4.4546)(2.0 - 2.5424) \\
&\quad + (-0.00035724)(2.0 - 2.5424)(7 - 4.4546) \\
&\quad \left. + (0.022302)(2.0 - 2.5424)^2 \right] \\
&= 0.16844 \left(\frac{1}{33} + 0.0213066 \right) \\
&= 0.008693 \text{ ml}^2 \\
s_{\hat{Y}} &= \sqrt{0.008693 \text{ ml}^2} = 0.093 \text{ ml}
\end{aligned}$$

As $t_{0.05(2),30} = 2.042$, the 95% prediction interval for the predicted Y is $2.030 \pm (2.042)(0.093)$ ml $= 2.030 \pm 0.190$ ml.

What is the predicted value of one additional Y value taken from the population at $X_1 = 7°C$ and $X_4 = 2.0$ min?

$$\hat{Y} = 2.552 - (0.132)(7) + (0.201)(2.0) = 2.030 \text{ ml}$$

What is the standard error of the predicted value of one additional Y value taken from the population at $X_1 = 7°C$ and $X_4 = 2.0$ min? [Equation 39 is used.]

$$\begin{aligned}
s_{\hat{Y}} &= \sqrt{0.16844 \left[1 + \frac{1}{33} + 0.0213066 \right]} \\
&= 0.421 \text{ ml}
\end{aligned}$$

As $t_{0.05(2),30} = 2.042$, the 95% prediction interval for the preceding predicted $\hat{Y}$ is $2.03 \pm (2.042)(0.421)$ ml $= 2.03 \pm 0.86$ ml.

What is the predicted value of the mean of 10 additional values of Y taken from the population at $X_1 = 7°C$ and $X_4 = 2.0$ min?

$$\hat{Y} = 2.552 - (0.132)(7) + (0.201)(2.0) = 2.030 \text{ ml}$$

What is the standard error of the predicted value of the mean of 10 additional values of Y taken from the population at $X_1 = 7°C$ and $X_4 = 2.0$ min? [Equation 40 is used.]

$$s_{\hat{Y}} = \sqrt{0.16844 \left[\frac{1}{10} + \frac{1}{33} + 0.0213066 \right]}$$

$$= 0.16 \text{ ml}$$

As $t_{0.05(2),30} = 2.042$, the 95% prediction interval for the predicted Y is $2.03 \pm (2.042)(0.16)$ ml $= 2.03 \pm 0.33$ ml.

A special case of Equation 37 is where each $X_i = 0$. The Y in question is then the Y intercept, a, and

$$s_a = \sqrt{s^2_{Y \cdot 1,2,\ldots,m} \left[\frac{1}{n} + \sum_{i=1}^{m} \sum_{k=1}^{m} c_{ik} \overline{X}_i \overline{X}_k \right]}. \tag{38}$$

To predict the value of Y that would be expected if one additional set of X_i were obtained, we may use Equation 16, and the standard error of this prediction is

$$(s_{\hat{Y}})_1 = \sqrt{s^2_{Y \cdot 1,2,\ldots,m} \left[1 + \frac{1}{n} + \sum_{i=1}^{m} \sum_{k=1}^{m} c_{ik} x_i x_k \right]}, \tag{39}$$

as Example 3 shows. This situation is a special case of predicting the mean Y to be expected from obtaining p additional sets of X_i, where the X_1's in all sets are equal, the X_2's in all sets are equal, and so on. Such a calculation is performed in Example 3, using

$$(s_{\hat{Y}})_p = \sqrt{s^2_{Y \cdot 1,2,\ldots,m} \left[\frac{1}{p} + \frac{1}{n} + \sum_{i=1}^{m} \sum_{k=1}^{m} c_{ik} x_i x_k \right]}. \tag{40}$$

Adding an independent variable, X_i, to a regression model increases each of the standard errors, $s_{\hat{Y}}$, in this section. Therefore, it is desirable to be assured that all variables included are important in predicting $\hat{Y}$ (see Section 6).

9 TESTING DIFFERENCE BETWEEN TWO PARTIAL REGRESSION COEFFICIENTS

If two partial regression coefficients, b_i and b_k, have the same units of measurement, it may occasionally be of interest to test $H_0: \beta_i - \beta_k = \beta_0$. This can be done by using

$$t = \frac{|b_i - b_k| - \beta_0}{s_{b_i - b_k}}. \tag{41}$$

Multiple Regression and Correlation

When $\beta_0 = 0$ is hypothesized, this may be written as

$$t = \frac{b_i - b_k}{s_{b_i} - b_k} \tag{42}$$

and the null hypothesis can be written as H_0: $\beta_i = \beta_k$. The standard error of the difference between two partial regression coefficients is*

$$s_{b_i - b_k} = \sqrt{s^2_{Y \cdot 1,2,\dots,m}[c_{ii} + c_{kk} - 2c_{ik}]} \tag{43}$$

and the degrees of freedom for this test are $n - m - 1$.

Testing other hypotheses about partial regression coefficients is discussed by Chatterjee, Hadi, and Price (2006: Section 3.9).

10 "DUMMY" VARIABLES

It is sometimes useful to introduce into a multiple regression model one or more additional variables in order to account for the effects of one or more nominal-scale variables on the dependent variable, Y. For example, we might be considering fitting the model $\hat{Y}_j = a + b_1 X_{1j} + b_2 X_{2j}$, where Y is diastolic blood pressure in a species of bear, X_1 is age, and X_2 is body weight. In addition, we might be interested in determining the effect (if any) of the animal's sex on blood pressure. Our regression model could then be expanded to $\hat{Y}_j = a + b_1 X_{1j} + b_2 X_{2j} + b_3 X_{3j}$, where X_3 is a "dummy variable," or "indicator variable," with one of two possible values: for example, set $X_3 = 0$ if the data are for a male and $X_3 = 1$ if the data are for a female. By using this dummy variable, we can test whether sex is a significant determinant of blood pressure (by the considerations of Section 4 for testing H_0: $\beta_3 = 0$). If it is, then the use of the model with all three independent variables will yield significantly more accurate Y values than the preceding model with only two independent variables, if the regression equation is used for predicting blood pressure.

If there are three levels of the nominal-scale variable, then two dummies would be needed in the regression model. For example, if we were considering the blood pressure of both sexes and of three subspecies of this bear species, then we might fit the model $\hat{Y}_j = a + b_1 X_{1j} + b_2 X_{2j} + b_3 X_{3j} + b_4 X_{4j} + b_5 X_{5j}$, where X_1, X_2, and X_3 are as before and X_4 and X_5 specify the subspecies. For example, subspecies 1 could be denoted by $X_4 = 0$ and $X_5 = 0$, subspecies 2 by $X_4 = 0$ and $X_5 = 1$, and subspecies 3 by $X_4 = 1$ and $X_5 = 0$. When L levels (i.e., nominal scale categories) of a variable are to be represented by dummy variables, $L - 1$ dummy variables are required. So, in the preceding examples, when $L = 2$ sexes, 1 dummy variable is needed; when $L = 3$ subspecies, 2 dummy variables must be used. Each dummy variable is set to either 0 or 1 for each Y (e.g., 0 or 1 for sex; and 0&0, 0&1, or 1&0 for subspecies), and, for a given Y, the sum of the 0's and 1's may not exceed 1 (so, for example, a dummy two-variable combination of 0&0, 0&1, or 1&0 is acceptable, but 1&1 is not). Further considerations of dummy variables are found in Chatterjee and Hadi (2006: Chapter 5), Draper and Smith (1998: Chapter 14), Hardy (1993), and Pedhazur (1997: 343–360).

When $L > 2$, it is inadvisable to employ stepwise regression by the forward-selection process of Section 6c (Cohen, 1991). If the dependent variable, Y, is the

*This could also be written as

$$s_{b_i - b_k} = \sqrt{s^2_{b_i} + s^2_{b_k} + 2s^2_{Y \cdot 1,2,\dots,m} c_{ik}}. \tag{43a}$$

dummy variable, appropriate procedures are more complicated and may involve the use of what is known as *logistic regression*.

11 INTERACTION OF INDEPENDENT VARIABLES

It may be proposed that two or more independent variables interact in affecting the dependent variable, Y, a concept encountered when discussing factorial analysis of variance. For example, we may propose this regression model:

$$Y_j = \alpha + \beta_1 X_{1j} + \beta_2 X_{2j} + \beta_3 X_{1j} X_{2j} + \epsilon_j. \tag{44}$$

The regression analysis would proceed by treating $X_1 X_2$ as a third independent variable (i.e., as if it were X_3); and rejecting H_0: $\beta_3 = 0$ would indicate a significant interaction between X_1 and X_2, meaning that the magnitude of the effect of X_1 on Y is dependent upon X_2 and the magnitude of the effect of X_2 on Y is dependent upon X_1. By using linear-regression equations that include interaction terms, a great variety of analysis-of-variance experimental designs can be analyzed (even those with unequal replication per cell), and this is a technique employed by some computer programs. Many ramifications of interactions in multiple regression are covered by Aiken and West (1991) and in many of the texts cited in the footnote at the end of the introduction to this chapter. Interaction, the joint effect on Y of two or more X's, should not be confused with correlation among X's ("multicollinearity," discussed in Section 4a).

12 COMPARING MULTIPLE REGRESSION EQUATIONS

Often we want to determine whether the multiple regressions from two or more sets of data, all containing the same variables, are estimating the same population regression function. We may test the null hypothesis that all the sample regression equations estimate the same population regression model. For a total of k regressions, the pooled residual sum of squares, SS_p, is the sum of all k residual sums of squares; and the pooled residual degrees of freedom, DF_p, is the sum of all k residual degrees of freedom. We then can combine the data from all k regressions and calculate a regression for this totality of data. The resulting total residual sum of squares and total degrees of freedom will be referred to as SS_t and DF_t, respectively.

The test of the null hypothesis (that there is a single set of population parameters underlying all k sample regressions) is

$$F = \frac{\dfrac{SS_t - SS_p}{(m+1)(k-1)}}{\dfrac{SS_p}{DF_p}}, \tag{45}$$

a statistic with $(m+1)(k-1)$ and DF_p degrees of freedom, Example 4 demonstrates this procedure.

We may also employ the concept of parallelism in multiple regression as we did in simple regression. A simple linear regression may be represented as a line on a two-dimensional graph, and two such lines are said to be parallel if the vertical distance between them is constant for all values of the independent variable, meaning that the regression coefficients (i.e., slopes) of the two lines are the same. A multiple regression with two independent variables may be visualized as a plane in three-dimensional space. Two planes are parallel if the vertical distance between them is the same for all combinations of the independent variables, in which case each of the partial regression coefficients for one regression is equal to the corresponding coefficient of the second regression, with only the Y intercepts possibly differing.

EXAMPLE 4   Comparing Multiple Regressions

Let us consider three multiple regressions, each fitted to a different sample of data, and each containing the same dependent variable and the same four independent variables. (Therefore, $m = 4$ and $k = 3$.) The residual sums of squares from each of the regressions are 437.8824, 449.2417, and 411.3548, respectively.

If the residual degrees of freedom for each of the regressions are 41, 32, and 38, respectively (that is, the three sample sizes were 46, 37, and 43, respectively), then the pooled residual sum of squares, SS_p, is 1298.4789, and the pooled residual degrees of freedom, DF_p, is 111.

Then, we combine the 126 data from all three samples and fit to these data a multiple regression having the same variables as the three individual regressions fitted previously. From this multiple regression let us say we have a total residual sum of squares, SS_t, of 1577.3106. The total residual degrees of freedom, DF_t, is 121.

Then we test H_0: All three sample regression functions estimate the same population regression, against H_A: All three sample regression functions do not estimate the same population regression:

$$F = \frac{\dfrac{SS_t - SS_p}{(m + 1)(k - 1)}}{\dfrac{SS_p}{DF_p}}$$

$$= \frac{\dfrac{1577.3106 - 1298.4789}{(5)(2)}}{\dfrac{1298.4789}{111}}$$

$$= 2.38.$$

The degrees of freedom associated with F are 10 and 111.

Since $F_{0.05(1),10,111} \cong 1.93$, reject H_0.

$$0.01 < P < 0.025 \quad [P = 0.013]$$

In general, two or more multiple regressions are said to be parallel if they all have the same β_1, β_2, β_3, and so on. The residual sums of squares for all k regressions are summed to give the pooled residual sum of squares, SS_p; the pooled residual degrees of freedom are

$$DF_p = \sum_{i=1}^{k} n_i - k(m + 1). \tag{46}$$

Additionally, we calculate a residual sum of squares for the "combined" regression in the following manner. Each element in a corrected sum-of-squares and sum-of-cross-products matrix (Matrix 5) is formed by summing all those elements from the k regressions. For example, element $\sum x_1^2$ for the combined regression is formed as

$(\sum x_1^2)_1 + (\sum x_1^2)_2 + (\sum x_1^2)_3 + \cdots + (\sum x_1^2)_k$, and element $\sum x_1 x_2$ is formed as $(\sum x_1 x_2)_1 + (\sum x_1 x_2)_2 + \cdots + (\sum x_1 x_2)_k$. The residual sum of squares obtained from the multiple regression analysis using the resulting matrix is the "common" residual sum of squares, SS_c; the degrees of freedom associated with it are

$$DF_c = \sum_{i=1}^{k} n_i - k - m. \tag{47}$$

Then the null hypothesis of all k regressions being parallel is tested by

$$F = \frac{\dfrac{SS_c - SS_p}{k - 1}}{\dfrac{SS_p}{DF_p}}, \tag{48}$$

with $k - 1$ and DF_p degrees of freedom.

If the null hypothesis is not rejected, we conclude that the independent variables affect the dependent variable in the same manner in all k regressions; we also conclude that all k regressions are parallel. Now we may ask whether the elevations of the k regressions are all the same. Here we proceed by an extension of the method you previously studied. The data for all k regressions are pooled together and one overall regression is fitted. The residual sum of squares of this regression is the total residual sum of squares, SS_t, which is associated with degrees of freedom of

$$DF_t = \sum_{i=1}^{k} n_t - m - 1. \tag{49}$$

(The latter degrees of freedom do not enter the calculation of F.)

Then the hypothesis of no difference among the k elevations is tested by

$$F = \frac{\dfrac{SS_t - SS_c}{k - 1}}{\dfrac{SS_c}{DF_c}}, \tag{50}$$

with $k - 1$ and DF_c degrees of freedom.

13 MULTIPLE REGRESSION THROUGH THE ORIGIN

As an expansion of the simple linear regression model presented, we might propose a multiple regression model where $\alpha = 0$; that is, when all $X_i = 0$, then $Y = 0$:

$$\hat{Y}_j = \beta_1 X_{1j} + \beta_2 X_{2j} + \cdots + \beta_m X_{mj}. \tag{51}$$

This will be encountered only rarely in biological work, but it is worth noting that some multiple-regression computer programs are capable of handling this model.* Striking differences in the computer output will be that total $DF = n$, regression $DF = m$ (the number of parameters in the model), and residual $DF = n - m$. Also, an *inverse pseudocorrelation matrix* may appear in the computer output in place of an inverse correlation or inverse sum-of-squares and sum-of-cross-products matrix. This

---

*Hawkins (1980) explains how a regression can be fitted through the origin using the output from a computer program for fitting a regression not assumed to pass through the origin.

regression model is legitimate only if each variable (i.e., Y and each X_i) is measured on a ratio scale.

14 NONLINEAR REGRESSION

Regression models such as

$$Y_i = \alpha + \beta X_i, \tag{13a}$$

$$Y_j = \alpha + \beta_1 X_{1j} + \beta_2 X_{2j} + \cdots + \beta_m X_{mj}, \tag{14}$$

or

$$Y_i = \alpha + \beta_1 X_i + \beta_2 X_i^2 + \cdots + \beta_m X_i^m \tag{14a}$$

are more completely symbolized as

$$Y_i = \alpha + \beta X_i + \epsilon_i, \tag{52}$$

$$Y_j = \alpha + \beta_1 X_{1j} + \beta_2 X_{2j} + \cdots + \beta_m X_{mj} + \epsilon_j, \tag{53}$$

or

$$Y_i = \alpha + \beta_1 X_i + \beta_2 X_i^2 + \cdots + \beta_m X_i^m + \epsilon_i, \tag{54}$$

respectively, where ϵ is the *residual* (or "error"), the difference between the value of Y predicted from the equation and the true value of Y in the population. All three of the preceding regression models are termed *linear* models because their parameters (i.e., α, β, and ϵ) appear in an additive fashion. However, cases do arise where the investigator wishes to fit to the data a model that is nonlinear with regard to its parameters. Such models might be those such as "exponential growth,"

$$Y_i = \alpha \beta^{X_i} + \epsilon_i \tag{55}$$

or

$$Y = \alpha e^{\gamma X_i} + \epsilon_i; \tag{56}$$

"exponential decay,"

$$Y_i = \alpha \beta^{-X_i} + \epsilon_i \tag{57}$$

or

$$Y_i = \alpha e^{-\gamma X_i} + \epsilon_i; \tag{58}$$

"asymptotic regression,"

$$Y_i = \alpha - \beta \delta^{X_i} + \epsilon_i \tag{59}$$

or

$$Y_i = \alpha - \beta(e^{-\gamma X_i}) + \epsilon_i; \tag{60}$$

or "logistic growth,"

$$Y_i = \frac{\alpha}{1 + \beta \delta^{X_i}} + \epsilon_i; \tag{61}$$

where the various Greek letters are parameters in the model. (See Snedecor and Cochran, 1989: 399, for graphs of such functions.) Other nonlinear models would be those in which the residuals were not additive, but, for example, might be multiplicative:

$$Y_i = \beta X_i \epsilon_i. \tag{62}$$

Sometimes a nonlinear model may be transformed into a linear one. For example, we may transform

$$Y_i = \alpha X_i^{\beta} \epsilon_i \tag{63}$$

by taking the logarithm of each side of the equation, acquiring a model that is linear in its parameters:

$$\log Y_i = \log \alpha + \beta \log X_i + \log \epsilon_i. \tag{64}$$

Transformations must be employed with careful consideration, however, so that the assumption of homogeneity of variance is not violated.

Biologists at times wish to fit nonlinear equations, some much more complex than the examples given, and computer programs are available for many of them. Such programs fall into two general groups. First are programs written to fit a particular model or a family of models, and the use of the program is little if any more complicated than the use of a multiple-linear-regression program. Second are general programs that can handle any of a wide variety of models. To use the latter type of program, however, requires the user to submit a good deal of information, perhaps the partial derivatives of the regression function with respect to each parameter in the model (thus, consulting with a statistician would be in order).

Nonlinear regression programs typically involve some sort of an iterative procedure, *iteration* being the utilization of a set of parameter estimates to arrive at a set of somewhat better parameter estimates, using the new estimates to derive better estimates, and so on. Thus, many of these programs require the user to submit initial estimates of (i.e., to guess the values of) the parameters in the model being fitted.

The program output for a nonlinear regression analysis is basically similar to much of the output from multiple-linear-regression analyses. Most importantly, the program should provide estimates of the parameters in the model (i.e., the statistics in the regression equation), the standard error of each of these statistics, and an analysis-of-variance summary including at least the regression and residual SS and DF. If regression and residual MS are not presented in the output, they may be calculated by dividing the appropriate SS by its associated DF. An F test of significance of the entire regression (or correlation) and the coefficient of determination may be obtained by means of Equations 18 and 19, respectively. Testing whether a parameter in the model is equal to a hypothesized value may be effected by a t test similar to those previously used for simple and partial regression coefficients (e.g., Section 4). Kvålseth (1985) and others warn that the computation of R^2 may be inappropriate in nonlinear regression.

Further discussions of nonlinear regression are found in Bates and Watts (1988), Berry and Feldman (1985: 51–64), Seber and Wild (1989), Snedecor and Cochran (1989: Chapter 19), and some of the books cited in the footnote at the end of the introduction to this chapter.

15 DESCRIPTIVE VERSUS PREDICTIVE MODELS

Often, it is hoped that a regression model implies a biological dependence (i.e., a cause and effect) in nature, and that this dependence is supported by the mathematical relationship described by the regression equation. However, regression equations are at times useful primarily as a means of predicting the value of a variable, if the values of a number of associated variables are known. For example, we may desire to predict the weight (call it variable Y) of a mammal, given the length of the femur (variable X). Perhaps a polynomial regression such as

$$\hat{Y}_i = a + b_1 X_i + b_2 X_i^2 + b_3 X_i^3 + b_4 X_i^4 \tag{65}$$

might be found to fit the data rather well. Or perhaps we wish to predict a man's blood pressure (call it variable Y) as accurately as we can by using measurements of his weight (variable W), his age (variable A), and his height (variable H). By deriving additional regression terms composed of combinations and powers of the three measured independent variables, we might conclude the statistical significance of each term in an equation such as

$$\hat{Y}_i = a + b_1 W_i + b_2 A_i + b_3 H_i + b_4 W_i^2 + b_5 H_i^2 + b_6 W_i^3 \tag{66}$$
$$+ b_7 W_i A_i + b_8 H_i A_i + b_9 W_i^3 A_i.$$

Equations such as 65 and 66 might have statistically significant partial regression coefficients. They might also have associated with them small standard errors of estimate, meaning that the standard error of predicted Y_i's (and, therefore, the prediction intervals) would be small. Thus, these would be good regression equations for purposes of prediction; but this does not imply that the fourth power of femur length has any natural significance in determining mammal weights, or that terms such as $H_i A_i$ or $W_i^3 A_i$ have any *biological* significance relative to human blood pressure.

To realize a regression function that describes underlying biological phenomena, the investigator must possess a good deal of knowledge about the interrelationships in nature among the variables in the model. Is it indeed reasonable to assume underlying relationships to be linear, or is there a logical basis for seeking to define a particular nonlinear relationship? (For example, forcing a linear model to fit a set of data in no way "proves" that the underlying biological relationships are, in fact, linear.) Are the variables included in the model meaningful choices? (For example, we might find a significant regression of variable A on variable B, whereas a third variable, C, is actually causing the changes in both A and B.) Statistical analysis is only a tool; it cannot be depended upon when applied to incomplete or fallacious biological information.

16 CONCORDANCE: RANK CORRELATION AMONG SEVERAL VARIABLES

The concept of nonparametric analysis of the correlation between two variables can be expanded to consider association among more than two. Such multivariate association is measurable nonparametrically by a statistic known as *Kendall's coefficient of concordance** (Kendall and Gibbons, 1990: Chapter 6; Kendall and Babbington Smith, 1939.)[†] To demonstrate, let us examine whether there is concordance (i.e., association) among the magnitudes of wing, tail, and bill lengths in birds of a particular species. Example 5 shows such data, for which we determine the ranks for each of the three variables.

---

*Maurice George Kendall (1907–1983), English statistician.

[†]Wallis (1939) introduced this statistic independently, calling it the "correlation ratio," and designating it by η_r^2 (where η is the lowercase Greek eta).

Multiple Regression and Correlation

EXAMPLE 5   Kendall's Coefficient of Concordance

H_0: In the sampled population, there is no association among the three variables (wing, tail, and bill lengths).

H_0: In the sampled population, there is a relationship among wing, tail, and bill lengths.

Birds (i)	Wing Length (cm) Data	Ranks	Tail Length (cm) Data	Ranks	Bill Length (mm) Data	Ranks	Sums of ranks (R_i)
1	10.4	4	7.4	5	17	5.5	14.5
2	10.8	8.5	7.6	7	17	5.5	21
3	11.1	10	7.9	11	20	9.5	30.5
4	10.2	1.5	7.2	2.5	14.5	2	6
5	10.3	3	7.4	5	15.5	3	11
6	10.2	1.5	7.1	1	13	1	3.5
7	10.7	7	7.4	5	19.5	8	20
8	10.5	5	7.2	2.5	16	4	11.5
9	10.8	8.5	7.8	9.5	21	11	29
10	11.2	11	7.7	8	20	9.5	28.5
11	10.6	6	7.8	9.5	18	7	22.5
12	11.4	12	8.3	12	22	12	36

$M = 3$
$n = 12$

Without correction for ties:

$$W = \frac{\sum R_i^2 - \frac{\left(\sum R_i\right)^2}{n}}{\frac{M^2(n^3 - n)}{12}}$$

$$= \frac{(14.5^2 + 21^2 + 30.5^2 + \cdots + 36^2) - \frac{(14.5 + 21 + 30.5 + \cdots + 36)^2}{12}}{\frac{3^2(12^3 - 12)}{12}}$$

$$= \frac{5738.5 - \frac{(234)^2}{12}}{\frac{15444}{12}}$$

$$= \frac{1175.5}{1287} = 0.913$$

$\chi_r^2 = M(n - 1)W$

$= (3)(12 - 1)(0.913)$

$= 30.129$

478

From Table 14 from *Appendix: Statistical Tables and Graphs*, $(X_r^2)_{0.05,3,12} = 6.167$.

Reject H_0: $P \ll 0.001$.

Incorporating the correction for ties:

In group 1 (wing length): there are 2 data tied at 10.2 cm

(i.e., $t_1 = 2$); there are 2 data tied at 10.8 cm (i.e., $t_2 = 2$).

In group 2 (tail length): there are 2 data tied at 7.2 cm

(i.e., $t_3 = 2$); there are 3 data tied at 7.4 cm (i.e., $t_4 = 3$); there are 2 data tied at 7.8 cm (i.e., $t_5 = 2$).

In group 3 (bill length): there are 2 data tied at 17 mm (i.e., $t_6 = 2$); there are 2 data tied at 20 mm (i.e., $t_7 = 2$).

Considering all seven groups of ties,

$$\sum t = \sum_{i=1}^{7} (t_i^3 - t_i)$$
$$= (2^3 - 2) + (2^3 - 2) + (2^3 - 2) + (3^3 - 3)$$
$$+ (2^3 - 2) + (2^3 - 2) + (2^3 - 2) = 60$$

and

$$W_c = \frac{1175.5}{\dfrac{15444 - 3(60)}{12}} = \frac{1175.5}{1272} = 0.924.$$

Then, to test the significance of W_c:

$$(\chi_r^2)_c = M(n - 1)W_c$$
$$= (3)(12 - 1)(0.924) = 30.492.$$

For these data, the same conclusion is reached with W_c as with W, namely: Reject H_0; and $P \ll 0.001$.

Several computational formulas for the coefficient of concordance are found in various texts. Two that are easy to use are

$$W = \frac{\sum (R_i - \overline{R})^2}{\dfrac{M^2(n^3 - n)}{12}} \tag{67}$$

and, equivalently,

$$W = \frac{\sum R_i^2 - \dfrac{\left(\sum R_i\right)^2}{n}}{\dfrac{M^2(n^3 - n)}{12}}, \tag{68}$$

where M is the number of variables being correlated, and n is the number of data per variable. The numerators of Equations 67 and 68 are simply the sum of squares of the n rank sums, R_i.*

The value of W may range from 0 (when there is no association and, consequently, the R_i's are equal and the sum of squares of R_i is zero) to 1 (when there is complete agreement among the rankings of all n groups and there is the maximum possible sum of squares for M variables). In Example 5 there is a very high level of concordance ($W = 0.913$), indicating that a bird with a large measurement for one of the variables is likely to have a large measurement for each of the other two variables.

We can ask whether a calculated sample W is significant; that is, whether it represents an association different from zero in the population of data that was sampled (Kendall and Gibbons, 1990: 224–227). The latter authors give tables of probabilities of W, but a simple way to assess the significance of W without such tables is to use the relationship between this coefficient and the Friedman X_r^2. Using the notation from the present section (Kendall and Babbington Smith, 1939),

$$\chi_r^2 = M(n - 1)W. \tag{69}$$

Thus, we can convert a calculated W to its equivalent χ_r^2 and then employ our table of critical values of χ_r^2 (Table 14 from *Appendix: Statistical Tables and Graphs*). This is demonstrated in Example 5. If either n or M is larger than that found in this table, then χ_r^2 may be assumed to be approximated by χ^2 with $n - 1$ degrees of freedom, and Table 1 from *Appendix: Statistical Tables and Graphs* is used.

(a) The Coefficient of Concordance with Tied Ranks. If there are tied ranks within any of the M groups, then mean ranks are assigned as in previous discussions. Then W is computed with a correction for ties,

$$W_c = \frac{\sum R_i^2 - \dfrac{\left(\sum R_i\right)^2}{n}}{\dfrac{M^2(n^3 - n) - M\sum t}{12}}, \tag{70}$$

where

$$\sum t = \sum_{i=1}^{m}(t_i^3 - t_i), \tag{71}$$

t_i is the number of ties in the ith group of ties, and m is the number of groups of tied ranks.† This computation of W_c is demonstrated in Example 5. W_c will not differ appreciably from W unless the numbers of tied data are great.

(b) The Coefficient of Concordance for Assessing Agreement. A common use of Kendall's coefficient of concordance is to express the intensity of agreement among several rankings. In Example 6, each of the three ten-year-old girls has been asked to rank the palatability of six flavors of ice cream. We wish to ask whether ten-year-old girls, in the population from which this sample came, agree upon the rankings.

---

*Kendall and Gibbons (1990: 123) present W with this correlation for continuity, noting that it does not appreciably alter the resultant W: Subtract 1 from the numerator and add 2 to the denominator of Equation 67 or 68.

†As in Equation 70, when ties are present, the denominator in Equations 68 and 69 would incorporate the subtraction of $M\sum t$ prior to dividing by 12.

EXAMPLE 6   Kendall's Coefficient of Concordance Used to Assess Agreement

Each of three girls ranked her taste preference for each of six flavors of ice cream (chocolate-chip, chocolate, spumoni, vanilla, butter-pecan, Neapolitan.)

H_0:   There is no agreement in flavor preference.

H_A: There is agreement in flavor preference.

	Flavors (i)					
Girl	CC	C	S	V	BP	N
1	5	1	3	2	4	6
2	6	2	3	1	5	4
3	6	3	2	1	4	5
Rank sum (R_i)	17	6	8	4	13	15	$\sum R_i = 63$

$M = 3$

$n = 6$

$$W = \dfrac{\sum R_i^2 - \dfrac{\left(\sum R_i\right)^2}{n}}{\dfrac{M^2(n^3 - n)}{12}}$$

$$= \dfrac{17^2 + 6^2 + 8^2 + 4^2 + 13^2 + 15^2 - \dfrac{63^2}{6}}{\dfrac{3^2(6^3 - 6)}{12}} = \dfrac{137.50}{157.50} = 0.873$$

$$\chi_r^2 = M(n - 1)W = (3)(6 - 1)(0.873) = 13.095$$

Using Table 14 from *Appendix: Statistical Tables and Graphs*, $(\chi_r^2)_{0.05,3,6} = 7.000$. Therefore, reject H_0. The conclusion is that there is agreement in flavor preferences.

$$P < 0.001$$

(c) The Relationship Between W and r_s.   Not only Kendall's W related to Friedman's X_r^2 (Equation 69), but it is related to the mean value of all possible Spearman rank correlation coefficients that would be obtained from all possible pairs of variables. These correlation coefficients may be listed in a matrix array:

$$
\begin{matrix}
(r_s)_{11} & (r_s)_{12} & (r_s)_{13} & \cdots & (r_s)_{1M} \\
(r_s)_{21} & (r_s)_{22} & (r_s)_{23} & \cdots & (r_s)_{2M} \\
(r_s)_{31} & (r_s)_{32} & (r_s)_{33} & \cdots & (r_s)_{3M} \\
\vdots & \vdots & \vdots & & \vdots \\
(r_s)_{M1} & (r_s)_{M2} & (r_s)_{M3} & \cdots & (r_s)_{MM},
\end{matrix}
\tag{72}
$$

a form similar to that of Matrix 6. As in Matrix 6, each element of the diagonal, $(r_s)_{ii}$, is equal to 1.0, and each element below the diagonal is duplicated above the diagonal, as $(r_s)_{ik} = (r_s)_{ki}$. There are $M!/[2(M-2)!]$ different r_s's possible for M variables.*

In Example 5, we are speaking of three r_s's: $(r_s)_{12}$, the r_s for wing length and tail length; $(r_s)_{13}$, the r_s for wing and bill lengths; and $(r_s)_{23}$, the r_s for tail and bill lengths. The Spearman rank correlation coefficient matrix, using correction for ties, would be

$$
\begin{matrix}
1.000 & & \\
0.852 & 1.000 & \\
0.917 & 0.890 & 1.000.
\end{matrix}
$$

For Example 6, the r_s matrix would be

$$
\begin{matrix}
1.000 & & \\
0.771 & 1.000 & \\
0.771 & 0.886 & 1.000.
\end{matrix}
$$

Denoting the mean of r_s as $\overline{r_s}$, the relationship with W (if there are no tied ranks) is

$$W = \frac{(M-1)\overline{r_s} + 1}{M};$$ (73)

therefore,

$$\overline{r_s} = \frac{MW - 1}{M - 1}.$$ (74)

If there are ties, then the preceding two equations relate W_c and $(\overline{r_s})_c$ in the same fashion as W and $\overline{r_s}$ are related. While the possible range of W is 0 to 1, $\overline{r_s}$ may range from $-1/(M-1)$ to 1. For Example 5, $(\overline{r_s})_c = (0.852 + 0.917 + 0.890)/3 = 0.886$, and Equation 73 yields $W = 0.924$. And for Example 6, $\overline{r_s} = 0.809$, and Equation 73 gives $W = 0.873$.

If $M = 2$ (i.e., there are only two variables, or rankings, being correlated), then either r_s or W might be computed; and

$$W = \frac{\overline{r_s} + 1}{2},$$ (75)

and

$$r_s = 2W_c - 1.$$ (76)

When $M = 2$, the use of r_s is preferable, for there are more thorough tables of critical values available.

If significant concordance is concluded for each of two groups of data, we may wish to ask if the agreement within each group is the same for both groups. For example, the data in Example 6 are for ice cream flavor preference as assessed by girls, and we might have a similar set of data for the preference exhibited by boys of the same age for these same flavors; and if there were significant concordance among girls as well as significant agreement among boys, we might wish to ask whether the consensus among girls is the same as that among boys. A test for this purpose was presented by Schucany and Frawley (1973), with elaboration by Li and Schucany (1975). However, the hypothesis test is not always conclusive with regard

*That is, M things taken two at a time.

to concordance between two groups and it has received criticism by Hollander and Sethuraman (1978), who proposed a different procedure. Serlin and Marascuilo (1983) reexamined both approaches as well as multiple comparison testing.

(d) Top-Down Concordance. The weighted-correlation procedure called "top-down correlation" is a two-sample test allowing us to give emphasis to those items ranked high (or low). An analogous situation can occur when there are more than two groups of ranks. For example, for the data of Example 6 we might have desired to know whether the girls in the sampled population agree on the most favored ice-cream flavors, with our having relatively little interest in whether they agree on the least appealing flavors. As with the correlation situation, we may employ the Savage scores, S_i, of Equation 76a, and a concordance test statistic is

$$S_i = \sum_{j=i}^{n} \frac{1}{j}. \tag{76a}$$

$$C_T = \frac{1}{M^2(n - S_1)} \left(\sum_{i=1}^{n} R_i^2 - M^2 n \right), \tag{77}$$

the significance of which may be assessed by

$$\chi_T^2 = M(n - 1)C_T, \tag{78}$$

by comparing it to the chi-square distribution with $n - 1$ degrees of freedom (Iman and Conover, 1987). Here, n and M are as in the preceding concordance computations: Each of M groups has n ranks. R_i is the sum of the Savage scores, across the M groups, at rank position i; and S_1 is Savage score 1. This is demonstrated in Example 7. In this example, it is concluded that there is agreement among the girls regarding the most tasty ice cream flavors. We could instead have asked whether there was agreement as to the least tasty flavors. This would have been done by assigning Savage scores in reverse order (i.e., $S_1 = 2.450$ assigned to rank 6, S_2 to rank 5, and so on). If this were done we would have found that $C_T = 0.8222$ and $\chi_T^2 = 12.333$, which would have resulted in a rejection of the null hypothesis of no agreement regarding the least liked flavors ($0.025 < P < 0.05$; $P = 0.030$).

EXAMPLE 7   Top-down Concordance, Using the Data of Example 6 to Ask Whether There Was Significant Agreement Among Children Regarding the Most Desirable Ice Cream Flavors. The Table of Data Shows the Savage Scores in Place of the Ranks of Example 6.

H_0: There is no agreement regarding the most preferred flavors.
H_A: There is agreement regarding the most preferred flavors.

	Flavors (i)					
Girl	CC	C	S	V	BP	N
1	0.367	2.450	0.950	1.450	0.617	0.167
2	0.167	1.450	0.950	2.450	0.367	0.617
3	0.167	0.950	1.450	2.450	0.617	0.367
R_i	0.701	4.850	3.350	6.350	1.601	1.151

$$C_T = \frac{1}{M^2(n - S_1)} \left(\sum_{i=1}^{n} R_i^2 - M^2 n \right)$$

$$= \frac{1}{3^2(6 - 2.450)} \left[0.701^2 + 4.850^2 + 3.350^2 + 6.350^2 \right.$$

$$\left. + 1.601^2 + 1.151^2 - (3^2)(6) \right]$$

$$= 0.03130[79.4469 - 54] = 0.03130(25.4469) = 0.7965$$

$$\chi_T^2 = 3(6 - 1)C_T$$

$$= (15)(0.7965) = 11.948$$

$$\nu = n - 1 = 5$$

$$\chi_{0.05,5}^2 = 11.070$$

Reject H_0.

$$0.025 < P < 0.05 \quad [P = 0.036]$$

EXERCISES

1. Given the following data:

Y (g)	X_1 (m)	X_2 (cm)	X_3 (m²)	X_4 (cm)
51.4	0.2	17.8	24.6	18.9
72.0	1.9	29.4	20.7	8.0
53.2	0.2	17.0	18.5	22.6
83.2	10.7	30.2	10.6	7.1
57.4	6.8	15.3	8.9	27.3
66.5	10.6	17.6	11.1	20.8
98.3	9.6	35.6	10.6	5.6
74.8	6.3	28.2	8.8	13.1
92.2	10.8	34.7	11.9	5.9
97.9	9.6	35.8	10.8	5.5
88.1	10.5	29.6	11.7	7.8
94.8	20.5	26.3	6.7	10.0
62.8	0.4	22.3	26.5	14.3
81.6	2.3	37.9	20.0	0.5

(a) Fit the multiple regression model $Y = \alpha + \beta_1 X_1 + \beta_2 X_2 + \beta_3 X_3 + \beta_4 X_4$ to the data, computing the sample partial regression coefficients and Y intercept.

(b) By analysis of variance, test the hypothesis that there is no significant multiple regression relationship.

(c) If H_0 is rejected in part (b), compute the standard error of each partial regression coefficient and test each H_0: $\beta_i = 0$.

(d) Calculate the standard error of estimate and the coefficient of determination.

(e) What is the predicted mean population value of Y at $X_1 = 5.2$ m, $X_2 = 21.3$ cm, $X_3 = 19.7$ m², and $X_4 = 12.2$ cm?

(f) What are the 95% confidence limits for the $\hat{Y}$ of part (e)?

(g) Test the hypothesis that the mean population value of Y at the X_i's stated in part (e) is greater than 50.0 g.

2. Subject the data of Exercise 1 to a stepwise regression analysis.

3. Analyze the five variables in Exercise 1 as a multiple correlation.

(a) Compute the multiple-correlation coefficient.

(b) Test the null hypothesis that the population multiple-correlation coefficient is zero.

(c) Compute the partial correlation coefficient for each pair of variables.

(d) Determine which of the calculated partial correlation coefficients estimate population partial correlation coefficients that are different from zero.

4. The following values were obtained for three multiple regressions of the form $\hat{Y} = a + b_1 X_1 + b_2 X_2 + b_3 X_3$. Test the null hypothesis that each of the three sample regressions estimates the same population regression function.

Regression	Residual sum of squares	Residual degrees of freedom
1	44.1253	24
2	56.7851	27
3	54.4288	21
All data combined	171.1372	

5. Each of five research papers was read by each of four reviewers. Each reviewer then ranked the quality of the five papers, as follows:

ANSWERS TO EXERCISES

1. (a) $\hat{Y} = -30.14 + 2.07X_1 + 2.58X_2 + 0.64X_3 + 1.11X_4$.   **(b)** H_0: No population regression; H_A: There is a population regression; $F = 90.2$, $F_{0.05(1),4,9} = 3.63$, reject H_0, $P \ll 0.0005$ $[P = 0.00000031]$.   **(c)** H_0: $\beta_1 = 0$, H_A: $\beta_1 \neq 0$; $t_{0.05(2),9} = 2.262$; "*" below denotes significance:

i	b_i	s_{b_i}	$t = \dfrac{b_i}{s_{b_i}}$	Conclusion
1	2.07	0.46	4.50*	Reject H_0.
2	2.58	0.74	3.49*	Reject H_0.
3	0.64	0.46	1.39	Do not reject H_0
4	1.11	0.76	1.46	Do not reject H_0.

(d) $s_{Y \cdot 1,2,3,4} = 3.11$ g; $R^2 = 0.9757$.   **(e)** $\overline{Y} = 61.73$ g.   **(f)** $s_{\hat{Y}} = 2.9549$ g, $L_1 = 55.0$ g, $L_2 = 68.4$ g.   **(g)** H_0: $\mu_Y \leq 50.0$ g, H_A: $\mu_Y > 50.0$ g, $t = 3.970$; $t_{0.05(1),9} = 1.833$, reject H_0; $0.001 < P < 0.0025$ $[P = 0.0016]$.

2. (1) With X_1, X_2, X_3, and X_4 in the model, see Exercise 1c.   (2) Delete X_3. With X_1, X_2, and X_4 in the model, $t_{0.05(2),10} = 2.228$ and:

i	b_i	t
1	1.48	9.15*
2	1.73	4.02*
4	0.21	0.50
	$a = 16.83$	

(3) Delete X_4. With X_1 and X_2 in the model, $t_{0.05(2),11} = 2.201$ and:

	Papers				
	1	2	3	4	5
Reviewer 1	5	4	3	1	2
Reviewer 2	4	5	3	2	1
Reviewer 3	5	4	1	2	3
Reviewer 4	5	3	2	4	1

(a) Calculate the Kendall coefficient of concordance.

(b) Test whether the rankings by the four reviewers are in agreement.

i	b_i	t
1	1.48	9.47*
2	1.53	13.19*
	$a = 24.96$	

(4) Therefore, the final equation is $\hat{Y} = 24.96 + 1.48X_1 + 1.53X_2$.

3. (a) $R = 0.9878$.   **(b)** $F = 90.2$, $F_{0.05(1),4,9} = 3.63$, reject H_0: There is no population correlation among the five variables; $P \ll 0.0005$ $[P = 0.00000031]$.   **(c)** Partial correlation coefficients:

	1	2	3	4	5
1	1.0000				
2	−0.9092*	1.0000			
3	−0.8203*	−0.8089*	1.0000		
4	−0.7578*	−0.9094*	−0.8724*	1.0000	
5	0.8342*	0.7583*	0.4183	0.4342	1.0000

(d) From Table 17 from *Appendix: Statistical Tables and Graphs*, $r_{0.05(2),9} = 0.602$, and the significant partial correlation coefficients are indicated with asterisks in part (c).

4. H_0: Each of the three sample regressions estimates the same population regression; H_A: Each of the three sample regressions does not estimate the same population regression; $F = 0.915$; as $F_{0.05(1),8,72} = 2.07$, do not reject H_0; $P > 0.25$ $[P = 0.51]$.

5. (a) $W = 0.675$.   **(b)** H_0: There is no agreement among the four faculty reviewers; H_A: There is agreement among the four faculty reviewers; $\chi_r^2 = 10.800$; $(\chi_r^2)_{0.05,4,5} = 7.800$; reject H_0; $0.005 < P < 0.01$.

This page intentionally left blank

Polynomial Regression

From Chapter 21 of *Biostatistical Analysis*, Fifth Edition, Jerrold H. Zar. Copyright © 2010 by
Pearson Education, Inc. Publishing as Pearson Prentice Hall. All rights reserved.

Polynomial Regression

A specific type of multiple regression is that concerning a *polynomial* expression:

$$Y_i = \alpha + \beta_1 X_i + \beta_2 X_i^2 + \beta_3 X_i^3 + \cdots + \beta_m X_i^m + \epsilon_i, \tag{1}$$

a model with parameters estimated in the expression

$$\hat{Y}_i = a + b_1 X_i + b_2 X_i^2 + b_3 X_i^3 + \cdots + b_m X_i^m, \tag{2}$$

for which a more concise symbolism is

$$\hat{Y}_i = a + \sum_{j=1}^{m} b_j X_i^j. \tag{3}$$

If $m = 1$, then the polynomial regression reduces to a simple linear regression.

As shown in Example 1, a polynomial equation such as Equation 2 deals with only two variables: the dependent variable, Y, and the independent variable, X. Additional terms in the polynomial equation consist of powers of X as if they are additional independent variables. That is, Equation 2 may be expressed as $\hat{Y}_j = a + b_1 X_{1j} + b_2 X_{2j} + b_3 X_{3j} + \cdots + b_m X_{mj}$, where, corresponding to the terms in Equation 2, X_{1j} is X_j, X_{2j} is X_j^2, X_{3j} is X_j^3, and so on, and X_{mj} is X_j^m.

The highest power in a polynomial equation, m, is known as the *degree* or *order* of the equation. There may be an underlying biological relationship warranting description by a polynomial model, but this is unlikely to involve an equation with an exponent larger than 2 or 3. The more common objective of polynomial regression, especially when $m > 2$, is to obtain an equation with which to predict the population value of Y at a specified X.

Polynomial regression is discussed in greater detail in Cohen et al. (2003: Section 6.2), von Eye and Schuster (1998: Chapter 7), and some of the books noted in the introduction to Chapter 20 (e.g., Draper and Smith, 1998: Chapter 12; Glantz and Slinker, 2001: 91–96; Kutner, Nachtsheim, and Neter, 2004: Section 8.1).

1   POLYNOMIAL CURVE FITTING

A polynomial equation may be analyzed by submitting values of Y, X, X^2, X^3, and so on to multiple regression computer programs.* There are also computer programs

---

*Serious rounding errors can readily arise when dealing with powers of X_i, and these problems can often be reduced by coding. A commonly recommended coding is to subtract $\overline{X}$ (i.e., to use $X_i - \overline{X}$ in place of X_i); this is known as *centering* the data (e.g., Cohen et al., 2003: Section 6.2.3; Ryan, 1997: Sections 3.2.4 and 4.2.1). Coding should be attempted with rounding-error in polynomial regression.

EXAMPLE 1 Stepwise Polynomial Regression

The following shows the results of a polynomial-regression analysis, by forward addition of terms, of data collected from a river, where X is the distance from the mouth of the river (in kilometers) and Y is the concentration of iron in the water (in micrograms per liter).

X (km)	Y (μg/L)
1.22	40.9
1.34	41.8
1.51	42.4
1.66	43.0
1.72	43.4
1.93	43.9
2.14	44.3
2.39	44.7
2.51	45.0
2.78	45.1
2.97	45.4
3.17	46.2
3.32	47.0
3.50	48.6
3.53	49.0
3.85	49.7
3.95	50.0
4.11	50.8
4.18	51.1

$n = 19$

First, a linear regression is fit to the data ($m = 1$), resulting in

$a = 37.389$, $b = 3.1269$, and $s_b = 0.15099$.

To test $H_0\colon \beta = 0$ against $H_A\colon \beta \neq 0$, $t = \frac{b}{s_b} = 20.709$, with $\nu = 17$.

As $t_{0.05(2),17} = 2.110$, H_0 is rejected.

Then, a quadratic (second-power) regression is fit to the data ($m = 2$), resulting in

$$a = 40.302, \quad b_1 = 0.66658, \quad s_{b_1} = 0.91352$$
$$b_2 = 0.45397, \quad s_{b_2} = 0.16688.$$

To test $H_0\colon \beta_2 = 0$ against $H_A\colon \beta_2 \neq 0$, $t = 2.720$, with $\nu = 16$.
As $t_{0.05(2),16} = 2.120$, H_0 is rejected.

Then, a cubic (third-power) regression is fit to the data ($m = 3$), resulting in

$$a = 32.767, \quad b_1 = 10.411, \quad s_{b_1} = 3.9030$$
$$b_2 = -3.3868, \quad s_{b_2} = 1.5136$$
$$b_3 = 0.47011, \quad s_{b_3} = 0.18442.$$

To test H_0: $\beta_3 = 0$ against H_A: $\beta_3 \neq 0, t = 2.549$, with $\nu = 15$.
As $t_{0.05(2),15} = 2.131, H_0$ is rejected.

Then, a quartic (fourth-power) regression is fit to the data ($m = 4$), resulting in

$$a = 6.9265, \quad \begin{array}{ll} b_1 = 55.835, & s_{b_1} = 12.495 \\ b_2 = -31.487, & s_{b_2} = 7.6054 \\ b_3 = 7.7625, & s_{b_3} = 1.9573 \\ b_4 = -0.67507, & s_{b_4} = 0.18076. \end{array}$$

To test H_0: $\beta_4 = 0$ against H_A: $\beta_4 \neq 0, t = 3.735$, with $\nu = 14$.
As $t_{0.05(2),14} = 2.145, H_0$ is rejected.

Then, a quintic (fifth-power) regression is fit to the data ($m = 5$), resulting in

$$a = 36.239, \quad \begin{array}{ll} b_1 = -9.1615, & s_{b_1} = 49.564 \\ b_2 = 23.387, & s_{b_2} = 41.238 \\ b_3 = -14.346, & s_{b_3} = 16.456 \\ b_4 = 3.5936, & s_{b_4} = 3.1609 \\ b_5 = -0.31740, & s_{b_5} = 0.23467. \end{array}$$

To test H_0: $\beta_5 = 0$ against H_A: $\beta_5 \neq 0, t = 1.353$, with $\nu = 13$.
As $t_{0.05(2),13} = 2.160$, do not reject H_0.

Therefore, it appears that a quartic polynomial is an appropriate regression function for the data. But to be more confident, we add one more term beyond the quintic to the model (i.e., a sextic, or sixth-power, polynomial regression is fit to the data; $m = 6$), resulting in

$$a = 157.88, \quad \begin{array}{ll} b_1 = -330.98, & s_{b_1} = 192.28 \\ b_2 = 364.04, & s_{b_2} = 201.29 \\ b_3 = -199.36, & s_{b_3} = 108.40 \\ b_4 = 58.113, & s_{b_4} = 31.759 \\ b_5 = -8.6070, & s_{b_5} = 4.8130 \\ b_6 = 0.50964, & s_{b_6} = 0.29560. \end{array}$$

To test H_0: $\beta_6 = 0$ against H_A: $\beta_6 \neq 0, t = 1.724$, with $\nu = 12$.
As $t_{0.05(2),12} = 2.179$, do not reject H_0.

In concluding that the quartic regression is a desirable fit to the data, we have $\hat{Y} = 6.9265 + 55.835X - 31.487X^2 + 7.7625X^3 - 0.67507X^4$. See Figure 1 for graphical presentation of the preceding polynomial equations.

that will perform polynomial regression with the input of only Y and X data (with the program calculating the powers of X instead of the user having to submit them as computer input).

The power, m, for fitting a polynomial to the data may be no greater than $n - 1$*; but m's larger than 4 or 5 are very seldom warranted.

The appropriate maximum m may be determined in one of two ways. One is the backward-elimination multiple-regression procedure. This would involve beginning with the highest-order term (the term with the largest m) in which we have any

---

*If $m = n - 1$, the curve will fit perfectly to the data (i.e., $R^2 = 1$). For example, it can be observed that for two data ($n = 2$), a linear regression line ($m = 1$) will pass perfectly through the two data points; for $n = 3$, the quadratic curve from a second-order polynomial regression ($m = 2$) will fit perfectly through the three data points; and so on.

interest. But, except for occasional second- or third-order equations, this m is difficult to specify meaningfully before the analysis.

The other procedure, which is more commonly used, is that of forward-selection multiple regression. A simple linear regression ($\hat{Y}_i = a + bX_i$) is fit to the data as in Figure 1a. Then a second-degree polynomial (known as a *quadratic* equation, $\hat{Y}_i = a + b_1X_i + b_2X_i^2$) is fit, as shown in Figure 1b. The next step would be to fit a third-degree polynomial (called a *cubic* equation, $\hat{Y}_i = a + b_1X_i + b_2X_i^2 + b_3X_i^3$), and the stepwise process of adding terms could continue beyond that. But at each step we ask whether adding the last term significantly improved the polynomial-regression equation. This "improvement" may be assessed by the t test for H_0: $\beta_j = 0$, where b_j, the sample estimate of β_j, is the partial-regression coefficient in the last term added.*

At each step of adding a term, rejection of H_0: $\beta = 0$ for the last term added indicates that the term significantly improves the model; and it is recommended practice that, at each step, each previous (i.e., lower-order) term is retained even if its b is no longer significant. If the H_0 is *not* rejected, then the final model might be expressed without the last term, as the equation assumed to appropriately describe the mathematical relationship between Y and X. But, as done in Example 1, some would advise carrying the analysis one or two terms beyond the point where the preceding H_0 is not rejected, to reduce the possibility that significant terms are being neglected inadvertently. For example, it is possible to not reject H_0: $\beta_3 = 0$, but by testing further to reject H_0: $\beta_4 = 0$.

After arriving at a final equation in a polynomial regression analysis, it may be desired to predict values of Y at a given value of X. The precision of a predicted $\hat{Y}$ (expressed by a standard error or confidence interval) may also be computed. Indeed, prediction is often the primary goal of a polynomial regression and biological interpretation is generally difficult, especially for $m > 2$.

It is very dangerous to extrapolate by predicting Y's beyond the range of the observed X's, and this is even more unwise than in the case of simple regression or other multiple regression. It should also be noted that use of polynomial regression can be problematic, especially for m larger than 2, because X_i is correlated with powers of X_i (i.e., with X_i^2, X_i^3, and so on), so the analysis may be very adversely affected by multicollinearity.

The concept of polynomial regression may be extended to the study of relationships of Y to more than one independent variable. For example, equations such as these may be analyzed by considering them to be multiple regressions:

$$\hat{Y} = a + b_1X_1 + b_2X_1^2 + b_3X_2 + b_4X_1X_2$$
$$\hat{Y} = a + b_1X_1 + b_2X_1^2 + b_3X_2 + b_4X_2^2 + b_5X_1X_2.$$

---

*This hypothesis may also be tested by

$$F = \frac{(\text{Regression SS for model of degree } m) - (\text{Regression SS for model of degree } m - 1)}{\text{Residual MS for the model of degree } m},$$

(4)

with a numerator DF of 1 and a denominator DF that is the residual DF for the m-degree model, and this gives results the same as from the t test.

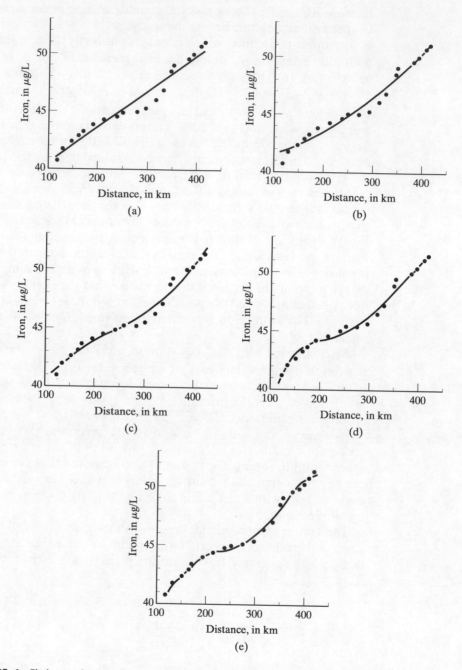

FIGURE 1: Fitting polynomial regression models. Each of the following regressions is fit to the 19 data points of Example 1. (a) Linear: $\hat{Y} = 37.389 + 3.1269X$. (b) Quadratic: $\hat{Y} = 40.302 + 0.66658X + 0.45397X^2$. (c) Cubic: $\hat{Y} = 32.767 + 10.411X - 3.3868X^2 + 0.47011X^3$. (d) Quartic: $\hat{Y} = 6.9265 + 55.835X - 31.487X^2 + 7.7625X^3 - 0.67507X^4$. (e) Quintic $\hat{Y} = 36.239 - 9.1615X + 23.387X^2 - 14.346X^3 + 3.5936X^4 - 0.31740X^5$. The stepwise analysis of Example 1 concludes that the quartic equation provides the appropriate fit; that is, the quintic expression does not provide a significant improvement in fit over the quartic.

Polynomial Regression

In these examples, the term X_1X_2 represents interaction between the two independent variables. Because there is more than one independent variable, there is no clear sequence of adding one term at a time in a forward-selection procedure, and some other method would have to be employed to strive for the best set of terms to compose the multiple-regression model.

2 QUADRATIC REGRESSION

The most common polynomial regression is the second-order, or *quadratic*, regression:

$$Y_i = \alpha + \beta_1 X_i + \beta_2 X_i^2 + \epsilon_i \tag{5}$$

with three population parameters, α, β_1, and β_2, to be estimated by three regression statistics, a, b_1, and b_2, respectively, in the quadratic equation

$$\hat{Y}_i = a + b_1 X_i + b_2 X_i^2. \tag{6}$$

The geometric shape of the curve represented by Equation 6 is a *parabola*. An example of a quadratic regression line is shown in Figure 2. If b_2 is negative as shown in Figure 2, the parabola will be concave downward. If b_2 is positive (as shown in Figure 1b), the curve will be concave upward. Therefore, one-tailed hypotheses may be desired: Rejection of H_0: $\beta_2 \geq 0$ would conclude a parabolic relationship in the population that is concave downward ($\beta_2 < 0$), and rejecting H_0: $\beta_2 \leq 0$ would indicate the curve is concave upward in the population ($\beta_2 > 0$).

(a) Maximum and Minimum Values of Y_i. A common interest in polynomial regression analysis, especially where $m = 2$ (quadratic), is the determination of a maximum or minimum value of Y_i (Bliss 1970: Section 14.4; Studier, Dapson, and Bigelow, 1975). A maximum value of Y_i is defined as one that is greater than those Y_i's that are close to it; and a minimum Y_i is one that is less than the nearby Y_i's. If, in a quadratic regression (Equation 6), the coefficient b_2 is negative, then there will be a maximum, as shown in Figure 2. If b_2 is positive, there will be a minimum (as is implied in Figure 1b). It may be desired to determine what the maximum or minimum value of Y_i is and what the corresponding value of X_i is.

The maximum or minimum of a quadratic equation is at the following value of the independent variable:

$$\hat{X}_0 = \frac{-b_1}{2b_2}. \tag{7}$$

Placing $\hat{X}_0$ in the quadratic equation (Equation 6), we find that

$$\hat{Y}_0 = a - \frac{b_1^2}{4b_2}. \tag{8}$$

Thus, in Figure 2, the maximum is at

$$\hat{X}_0 = \frac{-17.769}{2(-7.74286)} = 1.15 \text{ hr},$$

at which

$$\hat{Y}_0 = 1.39 - \frac{(17.769)^2}{4(-7.74286)} = 11.58 \text{ mg}/100 \text{ ml}.$$

Polynomial Regression

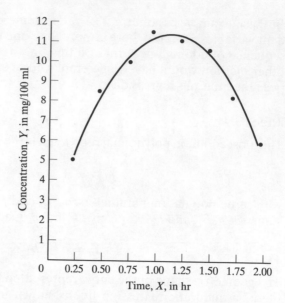

FIGURE 2: Quadratic fit to eight data points resulting in the equation $\hat{Y}_i = 1.39 + 17.769X_i - 7.74286X_i^2$.

A confidence interval for a maximum or minimum $\hat{Y}_0$ may be computed.

EXERCISES

1. The following measurements are the concentrations of leaf stomata (in numbers of stomata per square millimeter) and the heights of leaves above the ground (in centimeters). Subject the data to a polynomial regression analysis by stepwise addition of terms.

Y	X
(number/mm^2)	(cm)
4.5	21.4
4.4	21.7
4.6	22.3
4.7	22.9
4.5	23.2
4.4	23.8
4.5	24.8
4.2	25.4
4.4	25.9
4.2	27.2
3.8	27.4
3.4	28.0
3.1	28.9
3.2	29.2
3.0	29.8

2. Consider the following data, where X is temperature (in degrees Celsius) and Y is the concentration of a mineral in insect hemolymph (in millimoles per liter).

X	Y
($^\circ$C)	(mmole/L)
3.0	2.8
5.0	4.9
8.0	6.7
14.0	7.6
21.0	7.2
25.0	6.1
28.0	4.7

(a) Fit a quadratic equation to these data.

(b) Test for significance of the quadratic term.

(c) Estimate the mean population value of $\hat{Y}_i$ at $X_i = 10.0°C$ and compute the 95% confidence interval for the estimate.

(d) Determine the values of X and Y at which the quadratic function is maximum.

ANSWERS TO EXERCISES

1. In each step, H_0: $\beta_i = 0$ versus H_A: $\beta_i \neq 0$ is tested, where i is the highest term in the polynomial expression. An asterisk indicates H_0 is rejected. (1) Linear regression: $\hat{Y} = 8.8074 - 0.18646X$; $t = 7.136^*$; $t_{0.05(2),13} = 2.160$. (2) Quadratic regression: $\hat{Y} = -14.495 + 1.6595X - 0.036133X^2$; $t = 5.298^*$; $t_{0.05(2),12} = 2.179$. (3) Cubic regression: $\hat{Y} = -33.810 + 3.9550X - 0.12649X^2 + 0.0011781X^3$; $t = 0.374$; $t_{0.05(2),11} = 2.201$. (4) Quartic regression: $\hat{Y} = 525.30 - 84.708X + 5.1223X^2 - 0.13630X^3 + 0.0013443X^4$; $t = 0.911$; $t_{0.05(2),10} = 2.28$.

Therefore, the quadratic expression is concluded to be the "best."

2. **(a)** $\hat{Y} = 1.00 + 0.851X - 0.0259X^2$. **(b)** H_0: $\beta_2 = 0$; H_A: $\beta_2 \neq 0$; $F = 69.4$; $F_{0.05(1),1,4} = 7.71$; reject H_0; $0.001 < P < 0.0025$ [$P = 0.0011$]. **(c)** $\hat{Y} = 6.92$ eggs/cm^2; $s_{\hat{Y}} = 0.26$ eggs/cm^2; 95% confidence interval $- 6.92 \pm 0.72$ eggs/cm^2. **(d)** $\hat{X}_0 = 16.43°C$; $\hat{Y}_0 = 7.99$ eggs/ cm^2. **(e)** For $\hat{X}_0$: 95% confidence interval $= 16.47 \pm 0.65°C$; for $\hat{Y}_0$: 95% confidence interval $= 7.99 \pm 0.86$ eggs/cm^2.

This page intentionally left blank

Testing for Goodness of Fit

From Chapter 22 of *Biostatistical Analysis*, Fifth Edition, Jerrold H. Zar. Copyright © 2010 by
Pearson Education, Inc. Publishing as Pearson Prentice Hall. All rights reserved.

Testing for Goodness of Fit

This chapter concentrates on some statistical methods designed for use with nominal-scale data. As nominal data are counts of items or events in each of several categories, procedures for their analysis are sometimes referred to as *enumeration statistical methods*. This chapter deals with methods that address how well a sample of observations from a population of data conforms to the population's distribution of observations expressed by a null hypothesis. These procedures, which compare frequencies in a sample to frequencies hypothesized in the sampled population, are called *goodness-of-fit tests*. In testing such hypotheses, the widely used chi-square statistic* (χ^2) will be discussed, as will the more recently developed log-likelihood ratio introduced in Section 7. Goodness of fit for ordered categories (as contrasted with nominal-scale categories) is addressed by the Kolmogorov-Smirnov test of 8 or by the Watson test.

---

*The symbol for chi-square is χ^2, where the Greek lowercase letter chi (χ) is pronounced as the "ky" in "sky". Some authors use the notation X^2 instead of χ^2, which avoids employing a Greek letter for something other than a population parameter; but this invites confusion with the designation of X^2 as the square of an observation, X; so the symbol χ^2 will be used in this text. Karl Pearson (1900) pioneered the use of this statistic for goodness-of-fit analysis, and David (1995) credits him with the first use of the terms *chi-squared* and *goodness of fit* at that time. Pearson and R. A Fisher subsequently expanded the theory and application of chi-square (Lancaster, 1969: Chapter 1). *Chi-squared* is the term commonly preferred to *chi-square* by British writers.

Karl Pearson (1857–1936) was a remarkable British mathematician. Walker (1958) notes that Pearson has been referred to as "the founder of the science of statistics"; she called Pearson's development of statistical thinking and practice "an achievement of fantastic proportions" and said of his influence on others: "Few men in all the history of science have stimulated so many other people to cultivate and enlarge the fields they had planted." Karl Pearson, Walter Frank, and Francis Galton founded the British journal *Biometrika*, which was first issued in October 1901 and which still influences statistics in many areas. Pearson edited this journal for 35 years, succeeded for 30 years by his son, Egon Sharpe Pearson, himself a powerful contributor to statistical theory and application (see Bartlett, 1981).

1 CHI-SQUARE GOODNESS OF FIT FOR TWO CATEGORIES

It is often desired to obtain a sample of nominal scale data and to infer whether the population from which it came conforms to a specified distribution. For example, a plant geneticist might raise 100 progeny from a cross that is hypothesized to result in a 3:1 phenotypic ratio of yellow-flowered to green-flowered plants. Perhaps this sample of 100 is composed of 84 yellow-flowered plants and 16 green-flowered plants, although the hypothesis indicates an expectation of 75 yellow- and 25 green-flowered plants. The sampled population is the flower colors of all possible offspring from parent plants of the kind used in the experiment. The question of interest, then, is whether the observed frequencies (84 and 16) deviate significantly from the frequencies (75 and 25) expected from sampling this population.

The following chi-square statistic may be used as a measure of how much an observed sample distribution of nominal-scale data differs from a hypothesized distribution:

$$\chi^2 = \sum_{i=1}^{k} \frac{(f_i - \hat{f}_i)^2}{\hat{f}_i}. \tag{1}*$$

Here, f_i is the frequency (that is, the number of counts) observed in category i, $\hat{f}_i$ is the frequency expected in category i if the null hypothesis is true,[†] and the summation is performed over all k categories of data. For the aforementioned flower-color data, which are in two categories, Example 1 shows the two observed frequencies (f_1 and f_2), the two expected frequencies ($\hat{f}_1$ and $\hat{f}_2$), and the null and alternate hypotheses (H_0 and H_A). The expected frequency, $\hat{f}_i$, for each category may be calculated by multiplying the total number of observations, n, by the proportion of the total that the null hypothesis specifies for each category. Therefore, for the two flower colors in this example, $\hat{f}_1 = (100)(\frac{3}{4}) = 75$ and $\hat{f}_2 = (100)(\frac{1}{4}) = 25$.

Examining Equation 1 shows that larger disagreement between observed and expected frequencies (i.e., larger $f_i - \hat{f}_i$ values) will result in a larger χ^2 value. Thus, this type of calculation is referred to as a measure of *goodness of fit* (although it might better have been named a measure of "poorness of fit"). A calculated χ^2 value can be as small as zero, in the case of a perfect fit (i.e., each f_i value equals its corresponding $\hat{f}_i$), or very large if the fit is very bad; it can never be a negative value.

It is fundamentally important to appreciate that the chi-square statistic is calculated using the actual frequencies observed. It is not valid to convert the data to percentages

---

*Equation 1 can be rewritten as

$$\chi^2 = \sum_{i=1}^{k} \frac{f_i^2}{\hat{f}_i} - n, \tag{2}$$

where n is the sum of all the f_i's, namely the total number of observations in the sample. Although this formula renders the calculation of χ^2 a little easier, it has the disadvantage of not enabling us to examine each contribution to χ^2 [i.e., each $(f_i - \hat{f}_i)^2/\hat{f}_i$], and, as shown in Section 4, such an examination is an aid in determining how we might subdivide an overall chi-square analysis into component chi-square analyses for additional data collection. Thus, Equation 2 is seldom encountered.

[†]The symbol $\hat{f}$ is pronounced "f hat."

EXAMPLE 1   Calculation of Chi-Square Goodness of Fit, of Data Consisting of the Colors of 100 Flowers, to a Hypothesized Color Ratio of 3 : 1

H_0: The sample data came from a population having a 3 : 1 ratio of yellow to green flowers.

H_A: The sample data came from a population not having a 3 : 1 ratio of yellow to green flowers.

The data recorded are the 100 observed frequencies, f_i, in each of the two flower-color categories, with the frequencies expected under the null hypothesis, $\hat{f}_i$, in parentheses.

	Category (flower color)		
	Yellow	*Green*	*n*
f_i	84	16	100
$(\hat{f}_i)$	(75)	(25)	

Degrees of freedom $= \nu = k - 1 = 2 - 1 = 1$

$$\chi^2 = \sum \frac{(f_i - \hat{f}_i)^2}{\hat{f}_i} = \frac{(84 - 75)^2}{75} + \frac{(16 - 25)^2}{25}$$

$$= \frac{9^2}{75} + \frac{9^2}{25}$$

$$= 1.080 + 3.240$$

$$= 4.320$$

$\chi^2_{0.05,1} = 3.841$

Therefore, reject H_0.

$$0.025 < P < 0.05 \quad [P = 0.038]$$

An improved procedure is presented in Section 2 (Example 2).

---

and attempt to submit the percentages to Equation 1. An additional consideration in calculating chi-square is described in Section 2.

Critical values of χ^2 are given in Table 1 from *Appendix: Statistical Tables and Graphs*. For chi-square goodness-of-fit testing, the degrees of freedom, ν, are $k - 1$, so in the present example $\nu = 2 - 1 = 1$, and the first line of Table 1 from *Appendix: Statistical Tables and Graphs* is consulted to decide whether the null hypothesis, H_0, should be rejected. As in most hypothesis testing, a calculated χ^2 greater than or equal to the critical value causes rejection of H_0.

In Example 1, χ^2 is calculated to be 4.320 and the critical value is $\chi^2_{0.05,1} = 3.841$. This means that the probability of obtaining a sample of data diverging at least this far from the hypothesized distribution, *if* the null hypothesis is true, is less than 0.05. Therefore, if testing is being performed at the 5% significance level, H_0 is rejected and declared not to be a true statement about the distribution of flower colors in the sampled population. Indeed, examination of the first line of Table 1 from

Appendix: Statistical Tables and Graphs indicates that this probability lies between 0.025 and 0.05 (which we can express as $0.025 < P < 0.05$).*

The numbers of items in the two categories may be expressed as proportions (or percentages): In Example 1, yellow-flowered plants compose 0.84 (i.e., 84%) of the sample, and 0.16 (i.e., 16%) are green flowered.

2   CHI-SQUARE CORRECTION FOR CONTINUITY

Chi-square values obtained from actual data, using Equation 1, belong to a discrete, or discontinuous, distribution, in that they can take on only certain values. For instance, in Example 1 we calculated a chi-square value of 4.320 for $f_1 = 84, f_2 = 16, \hat{f}_1 = 75$, and $\hat{f}_2 = 25$. If we had observed $f_1 = 83$ and $f_2 = 17$, the calculated chi-square value would have been $(83 - 75)^2/75 + (17 - 25)^2/25 = 0.8533 + 2.5600 = 3.413$; for $f_1 = 82$ and $f_2 = 18, \chi^2 = 2.613$; and so on. These chi-square values obviously form a discrete distribution, for results between 4.320 and 3.413 or between 3.413 and 2.613 are not possible with the given $\hat{f}_i$ values. However, the theoretical χ^2 distribution, from which Table 1 from *Appendix: Statistical Tables and Graphs* is derived, is a continuous distribution; that is, for example, all values of χ^2 between 2.613 and 4.320 are possible. Thus, our need to determine the probability of a calculated χ^2 can be met only approximately by consulting Table 1 from *Appendix: Statistical Tables and Graphs*, and our conclusions are not taking place exactly at the level of α which we set. This situation would be unfortunate were it not for the fact that the approximation is a very good one, except when $\nu = 1$ (and in the instances described in Section 5). In the case of $\nu = 1$, it is usually recommended to use the *Yates correction for continuity* (Yates, 1934),[†] where the absolute value of each deviation of f_i from $\hat{f}_i$ is reduced by 0.5. That is,

$$\chi_c^2 = \sum_{i=1}^{2} \frac{(|f_i - \hat{f}_i| - 0.5)^2}{\hat{f}_i}, \tag{3}$$

where χ_c^2 denotes the chi-square value calculated with the correction for continuity.

This correction is demonstrated in Example 2, which presents the determination of χ_c^2 for the data of Example 1. For this example, the use of χ_c^2 yields the same conclusion as is arrived at without the correction for continuity, but this will not always be the case. Without the continuity correction, the calculated χ^2 may be inflated enough to cause us to reject H_0, whereas the corrected χ_c^2 value might not. In other words, not correcting for continuity may cause us to commit the Type I error with a probability greater than the stated α. The Yates correction should routinely be used when $\nu = 1$; it is not applicable for $\nu > 1$. For very large n, the effect of discontinuity is small, even for $\nu = 1$, and in such cases the Yates correction will

---

*Some calculators and computer programs have the capability of determining the exact probability of a given χ^2. For the present example, we would thereby find that $P(\chi^2 \geq 4.320) = 0.038$.

[†] Although English statistician Frank Yates (1902–1994) deserves the credit for suggesting this correction for chi-square testing, it had previously been employed in other statistical contexts (Pearson, 1947). R. A. Fisher associated it with Yates's name in 1936 (David, 1995). This was one of many important contributions Yates made over a distinguished 59-year publishing career, and he was also one of the earliest users of electronic computers to summarize and analyze data (Dyke, 1995). The correction should not be applied in the very rare situations where the numerator of χ^2 is *increased*, instead of decreased, by its use (that is, when $|f_i - \hat{f}_i| < 0.25$).

change the calculated chi-square very little. Its use remains appropriate with $\nu = 1$, however, regardless of n.

EXAMPLE 2   Chi-Square Goodness of Fit, Using the Yates Correction for Continuity

For the hypothesis and data of Example 1:

	Category (flower color)		
	Yellow	Green	n
f_i	84	16	100
$(\hat{f_i})$	(75)	(25)	

$\nu = k - 1 = 2 - 1 = 1$

$$\chi_c^2 = \sum_{i=1}^{2} \frac{(|f_i - \hat{f_i}| - 0.5)^2}{\hat{f_i}} = \frac{(|84 - 75| - 0.5)^2}{75} + \frac{(|16 - 25| - 0.5)^2}{25}$$

$$= 0.9633 + 2.8900 = 3.853$$

$\chi_{0.05,1}^2 = 3.841.$
Therefore, reject H_0.

$$0.025 < P < 0.05 \quad [P = 0.0497]$$

For $k = 2$, if H_0 involves a 1 : 1 ratio,

$$\chi^2 = \frac{(f_1 - f_2)^2}{n} \tag{4}$$

may be used in place of Equation 1, and

$$\chi_c^2 = \frac{(|f_1 - f_2| - 1)^2}{n} \tag{5}$$

may be used instead of Equation 3. In these two shortcut equations, $\hat{f_1}$ and $\hat{f_2}$ need not be calculated, thus avoiding the concomitant rounding errors.

If, when $\nu = 1$, the chi-square calculation is performed by a computer, the user should be aware whether the continuity correction is employed.

3 CHI-SQUARE GOODNESS OF FIT FOR MORE THAN TWO CATEGORIES

Example 1 demonstrated chi-square goodness-of-fit testing when there are two categories of data (i.e., $k = 2$). This kind of analysis may be extended readily to sets of data with larger numbers of categories, as Example 3 exemplifies. Here, 250 plants were examined ($n = 250$), and their seeds were classified into four categories ($k = 4$). The calculated χ^2, using Equation 1, is 8.972. (This text will routinely express a calculated chi-square to at least three decimal places, for that is the accuracy of the table of critical values, Table 1 from *Appendix: Statistical Tables and Graphs*. Therefore, to help avoid rounding errors, intermediate calculations, including those of $\hat{f_i}$, will be performed to four or more decimal places.)

EXAMPLE 3   Chi-square Goodness of Fit for $k = 4$

H_0: The sample comes from a population having a $9 : 3 : 3 : 1$ ratio of yellow-smooth to yellow-wrinkled to green-smooth to green-wrinkled seeds.

H_A: The sample comes from a population not having a $9 : 3 : 3 : 1$ ratio of the above four seed phenotypes.

The sample data are recorded as observed frequencies, f_i, with the frequencies expected under the null hypothesis, $\hat{f}_i$, in parentheses.

	Yellow smooth	Yellow wrinkled	Green smooth	Green wrinkled	n
f_i	152	39	53	6	250
$(\hat{f}_i)$	(140.6250)	(46.8750)	(46.8750)	(15.6250)	

$\nu = k - 1 = 3$

$$\chi^2 = \frac{11.3750^2}{140.6250} + \frac{7.8750^2}{46.8750} + \frac{6.1250^2}{46.8750} + \frac{9.6250^2}{15.6250}$$

$$= 0.9201 + 1.3230 + 0.8003 + 5.9290$$

$$= 8.972$$

$\chi^2_{0.05,3} = 7.815$

Therefore, reject H_0.

$$0.025 < P < 0.05 \quad [P = 0.030]$$

It has already been pointed out that larger χ^2 values will result from larger differences between f_i and $\hat{f}_i$, but large calculated χ^2 values may also simply be the result of a large number of classes of data, because the calculation involves the summing over all classes. Thus, in considering the significance of a calculated χ^2, the value of k must be taken into account. What is done is to consider the degrees of freedom* (ν). For the chi-square goodness-of-fit testing discussed in this chapter, $\nu = k - 1$. Thus, in Example 3 $\nu = 4 - 1 = 3$, while the calculated χ^2 is 8.972. Entering Table 1 from *Appendix: Statistical Tables and Graphs* in the row for $\nu = 3$, it is seen that $P(\chi^2 \geq 7.815) = 0.05$ and $P(\chi^2 \geq 9.348) = 0.025$. Therefore, $0.025 < P(\chi^2 \geq 8.972) < 0.05$; and, if testing at the 5% significance level, we would reject the null hypothesis that the sample came from a population having a $9 : 3 : 3 : 1$ ratio of yellow-smooth to yellow-wrinkled to green-smooth to green-wrinkled seeds. The tabled critical values may be denoted as $\chi^2_{\alpha,\nu}$; thus, for example, we can write $\chi^2_{0.05,3} = 7.815$ and $\chi^2_{0.025,3} = 0.348$.

When we say the degrees of freedom are $k - 1$, we are stating that, given the frequencies in any $k - 1$ of the categories, we can determine the frequency in the remaining category. This is so because n is known, and the sum of the frequencies in all k categories equals n. In other words, there is "freedom" to assign frequencies to only $k - 1$ categories. It may also be noted that the degrees of freedom are k minus

---

*This term was introduced by R. A. Fisher, in 1922, while discussing contingency tables (David, 1995).

the number of sample constants used to calculate the expected frequencies. In the present examples, only one constant, n, was so used, so $v = k - 1$.

4 SUBDIVIDING CHI-SQUARE GOODNESS OF FIT

In Example 3, the chi-square analysis detected a difference between the observed and expected frequencies too great to be attributed to chance, and the null hypothesis was rejected. This conclusion may be satisfactory in some instances, but in many cases the investigator will wish to perform further analysis.

For the example under consideration, the null hypothesis is that the sample came from a population having a $9 : 3 : 3 : 1$ phenotypic ratio. If the chi-square analysis had not led to a rejection of the hypothesis, we would proceed no further. But since H_0 was rejected, we may wish to ask whether the significant disagreement between observed and expected frequencies was concentrated in certain of the categories, or whether the difference was due to the effects of the data in all of the classes. The four individual contributions to the chi-square value are 0.9201, 1.3230, 0.8003, and 5.9290; and the contribution resulting from the last class (the green-wrinkled seeds) contributes a relatively large amount to the size of the calculated χ^2. Thus we see that the nonconformity of the sample frequencies to those expected from a population with a $9 : 3 : 3 : 1$ ratio is due largely to the magnitude of the discrepancy between f_4 and $\hat{f}_4$.

This line of thought can be examined as shown in Example 4. First, we test H_0: f_1, f_2, and f_3 came from a population having a 9:3:3 ratios with H_A: The frequencies in the first three categories came from a population having a phenotypic ratio other than $9 : 3 : 3$. This null hypothesis is not rejected, indicating that the frequencies in the first three categories conform acceptably well to those predicted by H_0. Then we can test the frequency of green-wrinkled seeds against the combined frequencies for the other three phenotypes, under the null hypothesis of a $1 : 15$ ratio. The calculated χ^2 value causes us to reject this hypothesis, however, and we draw the conclusion that the nonconformity of the data in Example 3 to the hypothesized frequencies is due primarily to the observed frequency of green-wrinkled seeds. In the latter hypothesis test, χ_c^2 is employed instead of χ^2 because $v = 1$.

EXAMPLE 4   Chi-Square Goodness of Fit, Subdividing the Chi-Square Analysis of Example 3

H_0: The sample came from a population with a $9 : 3 : 3$ ratio of the first three phenotypes in Example 2.

H_A: The sample came from a population not having a $9 : 3 : 3$ ratio of the first three phenotypes in Example 2.

	Seed Characteristics			
	Yellow smooth	Yellow wrinkled	Green smooth	n
f_i	152	39	53	244
$(\hat{f}_i)$	(146.4000)	(48.8000)	(48.8000)	

$v = k - 1 = 2$

$$\chi^2 = \frac{5.6000^2}{146.4000} + \frac{-9.8000^2}{48.8000} + \frac{4.2000^2}{48.8000}$$

$$= 0.2142 + 1.9680 + 0.3615$$

$$= 2.544$$

$\chi^2_{0.05,2} = 5.991$

Therefore, do not reject H_0.

$$0.25 < P < 0.50 \quad [P = 0.28]$$

H_0: The sample came from a population with a $1:15$ ratio of green-wrinkled to other seed phenotypes.

H_A: The sample came from a population not having the $1:15$ ratio stated in H_0.

	Seed Characteristics		
	Green wrinkled	*Others*	*n*
f_i	6	244	250
$(\hat{f}_i)$	*(15.6250)*	*(234.3750)*	

$\nu = k - 1 = 1$

$$\chi^2_c = \sum_{i=1}^{2} \frac{(|f_i - \hat{f}_i| - 0.5)^2}{\hat{f}_i} = \frac{(9.6250 - 0.5)^2}{15.6250} + \frac{(9.6250 - 0.5)^2}{234.3750}$$

$$= 5.3290 + 0.3553 - 5.684$$

$\chi^2_{0.05,1} = 3.841$

Therefore, reject H_0.

$$0.01 < P < 0.025 \quad [P = 0.017]$$

Note: It is not proper to test statistical hypotheses that were stated *after* examining the data to be tested. Therefore, the analyses described in this section should be considered only a guide to developing hypotheses that subdivide a goodness-of-fit analysis. And the newly proposed hypotheses should then be stated in advance of their being tested with a new set of data.

5 CHI-SQUARE GOODNESS OF FIT WITH SMALL FREQUENCIES

In order for us to assign a probability to the results of a chi-square goodness-of-fit test, and thereby assess the statistical significance of the test, the calculated χ^2 must be a close approximation to the theoretical distribution that is summarized in Table 1 from *Appendix: Statistical Tables and Graphs*. This approximation is quite acceptable as long as the expected frequencies are not too small. If $\hat{f}_i$ values are very small, however, the calculated χ^2 is biased in that it is larger than the theoretical χ^2 it is supposed to estimate, and there is a tendency to reject the null hypothesis with a probability greater than α. This is undesirable, and statisticians have attempted to define in a convenient manner what would constitute $\hat{f}_i$'s that are "too small."

For decades a commonly applied general rule was that no expected frequency should be less than 5.0,* even though it has long been known that it is tolerable to have a few $\hat{f}_i$'s considerably smaller than that (e.g., Cochran, 1952, 1954). By a review of previous recommendations and an extensive empirical analysis, Roscoe and Byars (1971) reached conclusions that provide less restrictive guidelines for chi-square goodness-of-fit testing. They and others have found that the test is remarkably robust when testing for a uniform distribution — that is, for H_0: In the population, the frequencies in all k categories are equal — in which case $\hat{f}_i = n/k$. In this situation, it appears that it is acceptable to have expected frequencies as small as 1.0 for testing at α as small as 0.05, or as small as 2.0 for α as small as 0.01. The chi-square test works nearly as well when there is moderate departure from a uniform distribution in H_0, and the average expected frequencies may be as small as those indicated for a uniform distribution. And even with great departure from uniform, it appears that the average expected frequency (i.e., n/k) may be as small as 2.0 for testing at α as low as 0.05 and as small as 4.0 for α as small as 0.01. Koehler and Larntz (1980) suggested that the chi-square test is applicable for situations where $k \geq 3, n \geq 10$, and $n^2/k \geq 10$. Users of goodness-of-fit testing can be comfortable if their data fit both the Roscoe and Byars and the Koehler and Larntz guidelines. These recommendations are for situations where there are more than two categories. If $k = 2$, then it is wise to have $\hat{f}_i$'s of at least 5.0, or to use the binomial test as indicated in the next paragraph.

The chi-square calculation can be employed if the data for the classes with offensively low $\hat{f}_i$ values are simply eliminated from H_0 and the subsequent analysis. Or, certain of the classes of data might be meaningfully combined so as to result in all $\hat{f}_i$ values being large enough to proceed with the analysis. Such modified procedures are not to be recommended as routine practice, however. Rather, the experimenter should strive to obtain a sufficiently large n for the analysis to be performed. When $k = 2$ and each f_i is small, the use of the binomial test is preferable to chi-square analysis. [Similarly, use of the multinomial, rather than the binomial, distribution is appropriate when $k > 2$ and the f_i's are small; however, this is a tedious procedure and will not be demonstrated here (Radlow and Alf, 1975).]

6 HETEROGENEITY CHI-SQUARE TESTING FOR GOODNESS OF FIT

It is sometimes the case that a number of sets of data are being tested against the same null hypothesis, and we wish to decide whether we may combine all of the sets in order to perform one overall chi-square analysis. As an example, let us examine some of the classic data of Gregor Mendel[†](1865). In one series of 10 experiments,

---

*Some statisticians have suggested lower limits as small as 1.0 and others recommend limits as large as 20.0 (as summarized by Cressie and Read, 1989; Tate and Hyer, 1973), with lower $\hat{f}_i$'s acceptable in some cases where the $\hat{f}_i$'s are all equal.

†Born Johann Mendel (1822–1884), he was an Augustinian monk (taking the name *Gregor* when entering the monastery), an Austrian schoolteacher, and a pioneer in biological experimentation and its quantitative analysis — although his data have been called into question by statisticians (Edwards, 1986; Fisher, 1936). His research was unappreciated until sixteen years after his death.

Mendel obtained pea plants with either yellow or green seeds, with the frequency of yellow-seed plants and green-seed plants shown in Example 5.* The data from each of the 10 samples are tested against the null hypothesis that there is a 3-to-1 ratio of plants with yellow seeds to plants with green seeds in the population from which the sample came. H_0 is not rejected in any of the 10 experiments, so it is reasonable to test a null hypothesis examining heterogeneity, that all 10 samples could have come from the same population (or from more than one population having the same ratios). This new hypothesis may be tested by the procedure called *heterogeneity chi-square* analysis (sometimes referred to as "interaction" chi-square analysis or even "homogeneity analysis"). In addition to performing the 10 separate chi-squares tests, we total all 10 f_i values and total all 10 $\hat{f}_i$ values and perform a chi-square test on these totals. But in totaling these values, commonly called *pooling* them, we must assume that all ten samples came from the same population (or from populations having identical seed-color ratios). If this assumption is true, we say that the samples are *homogeneous*. If this assumption is false, the samples are said to be *heterogeneous*, and the chi-square analysis on the pooled data would not be justified. So we are faced with the desirability of testing for heterogeneity, using the null hypothesis that the samples could have come from the same population (i.e., H_0: The samples are homogeneous).

Testing for heterogeneity among replicated goodness-of-fit tests is based on the fact that the sum of chi-square values is itself a chi-square value. If the samples are indeed homogeneous, then the total of the individual chi-square values should be close to the chi-square for the total frequencies. In Example 5, the total chi-square is 7.1899, with a total of 10 degrees of freedom; and the chi-square of the totals is 0.1367, with 1 degree of freedom. The absolute value of the difference between these two chi-squares is itself a chi-square (called the *heterogeneity chi-square*), 7.053, with $\nu = 10 - 1 = 9$.

Consulting Table 1 from *Appendix: Statistical Tables and Graphs*, we see that for the heterogeneity chi-square, $\chi^2_{0.05,9} = 16.9$, so H_0 is not rejected. Thus we conclude that the 10 samples could have come from the same population and that their frequencies might justifiably be pooled. The Yates correction for continuity may not be applied in a heterogeneity chi-square analysis (Cochran, 1942; Lancaster, 1949). But if we conclude that the sample data may be pooled, we should then analyze these

---

EXAMPLE 5   Heterogeneity Chi-Square Analysis

The data are the number of pea plants with yellow seeds and the number with green seeds, in each of 10 plant-breeding experiments.

The null hypothesis for each experiment is that the population sampled has a 3 : 1 ratio of plants with yellow seeds to plants with green seeds.

The null hypothesis for the heterogeneity chi-square test is that all 10 samples of data came from the same population (or from populations with the same ratios).

For each experiment, the observed frequencies, f_i, are given, with the frequencies, $\hat{f}_i$, predicted by the null hypothesis within parentheses.

---

*These ten data sets come from what Mendel (1865) collectively called "Experiment 2," in which, he reports, "258 plants yielded 8023 seeds, 6022 yellow and 2001 green; their ratio, therefore, is 3.01 : 1." (He was expressing that 6022/2001 = 3.01.)

Experiment	Plants with yellow seeds	Plants with green seeds	Total plants (n)	Uncorrected chi-square*	ν
1	25 (27.0000)	11 (9.0000)	36	0.5926	1
2	32 (29.2500)	7 (9.7500)	39	1.0342	1
3	14 (14.2500)	5 (4.7500)	19	0.0175	1
4	70 (72.7500)	27 (24.2500)	97	0.4158	1
5	24 (27.7500)	13 (9.2500)	37	2.0270	1
6	20 (19.5000)	6 (6.5000)	26	0.0513	1
7	32 (33.7500)	13 (11.2500)	45	0.3630	1
8	44 (39.7500)	9 (13.2500)	53	1.8176	1
9	50 (48.0000)	14 (16.0000)	64	0.3333	1
10	44 (46.5000)	18 (15.5000)	62	0.5376	1

Total of chi-squares				7.1899	10
Chi-square of totals (i.e., pooled chi-square)	355 (358.5000)	123 (119.5000)	478	0.1367	1
Heterogeneity chi-square				7.0532	9

$\chi^2_{0.05,9} = 16.919$.

Do not reject the homogeneity null hypothesis. $0.50 < P < 0.75$ $[P = 0.63]$

---

* In heterogeneity analysis, chi-square is computed *without* correction for continuity.

Pooled chi-square with continuity correction: $\chi^2_c = 0.1004$ and $\chi^2_{0.05,1} = 3.841$. Do not reject H_0 of 3 : 1 ratio. $0.50 < P < 0.75$ $[P = 0.75]$

pooled data using the correction for continuity. Thus, for Example 5, $\chi^2_c = 0.128$, rather than $\chi^2 = 0.137$, should be used, once it has been determined that the samples are homogeneous and the data may be pooled. Heterogeneity testing may also be done using the log-likelihood statistic, G (Section 7), instead of χ^2.

Example 6 demonstrates how we can be misled by pooling heterogeneous samples without testing for acceptable homogeneity. If the six samples shown were pooled and a chi-square computed ($\chi^2 = 0.2336$), we would not reject the null hypothesis. But such a procedure would have ignored the strong indication obtainable

EXAMPLE 6   Hypothetical Data for Heterogeneity Chi-Square Analysis, Demonstrating Misleading Results from the Pooling of Heterogeneous Samples

H_0: The sample population has a 1 : 1 ratio of right- to left-handed men.

H_A: The sampled population does not have a 1 : 1 ratio of right- to left-handed men.

Sample frequencies observed, f_i, are listed, with the frequencies predicted by H_0 $(\hat{f_i})$ in parentheses.

Sample	Right-handed	Left-handed	n	Uncorrected chi-square	ν
1	3 (7.0000)	11 (7.0000)	14	4.5714	1
2	4 (8.0000)	12 (8.0000)	16	4.0000	1
3	5 (10.0000)	15 (10.0000)	20	5.0000	1
4	14 (9.0000)	4 (9.0000)	18	5.5556	1
5	13 (8.5000)	4 (8.5000)	17	4.7647	1
6	17 (11.0000)	5 (11.0000)	22	6.5455	1
Total of chi-squares				30.4372	6
Chi-square of totals (i.e., pooled chi-square)	56 (53.5000)	51 (53.5000)	107	0.2336	1
Heterogeneity chi-square				30.2036	5

$\chi^2_{0.05,5} = 11.070$.

Reject H_0 for homogeneity. $P < 0.001$ $[P = 0.000013]$

Therefore, we are not justified in performing a goodness-of-fit analysis on the pooled data.

from the heterogeneity analysis ($P < 0.001$) that the samples came from more than one population. The appearance of the data in this example suggests that Samples 1, 2, and 3 came from one population, and Samples 4, 5, and 6 came from another, possibilities that can be reexamined with new data.

It is also important to realize that the pooling of homogeneous data can, in same cases, result in a more powerful analysis. Example 7 presents hypothetical data for four replicate chi-square analyses. None of the individual chi-square tests detects a significant deviation from the null hypothesis; but on pooling them, the chi-square test performed on the larger number of data does reject H_0. The nonsignificant

EXAMPLE 7   Hypothetical Data for Heterogeneity Chi-Square Analysis, Demonstrating How Nonsignificant Sample Frequencies Can Result in Significant Pooled Frequencies

For each sample, and for the pooled sample:

H_0: The sampled population has equal frequencies of right- and left-handed men.

H_A: The sampled population does not have equal frequencies of right- and left-handed men.

For heterogeneity testing:

H_0: All the samples came from the same population.

H_A: The samples came from at least two different populations.

For each sample, the observed frequencies, f_i, are given, together with the expected frequencies, $\hat{f}_i$, in parentheses.

Sample	Right-handed	Left-handed	n	Uncorrected chi-square	ν
1	15 (11.0000)	7 (11.0000)	22	2.9091	1
2	16 (12.0000)	8 (12.0000)	24	2.6667	1
3	12 (8.5000)	5 (8.5000)	17	2.8824	1
4	13 (9.0000)	5 (9.0000)	18	3.5556	1
Total of chi-squares				12.0138	4
Chi-square of totals (pooled chi-square)	56 (40.5000)	25 (40.5000)	81	11.8642	1
Heterogeneity chi-square				0.1496	3

$\chi^2_{0.05,3} = 7.815$.

The homogeneity H_0 is not rejected. $0.975 < P < 0.99$ $[P = 0.985]$

Therefore, we are justified in pooling the four sets of data. On doing so, $\chi^2_c = 11.111$, DF $= 1$, $P = 0.00086$, H_0 is rejected.

heterogeneity chi-square shows that we are justified in pooling the replicates in order to analyze a single set of data with a large n.

7   THE LOG-LIKELIHOOD RATIO FOR GOODNESS OF FIT

The *log-likelihood ratio* is applicable to goodness-of-fit analysis in circumstances having data for which chi-square may be employed. The log-likelihood ratio,*

---

*Proposed by Wilks (1935), based upon concepts of Neyman and Pearson (1928a, 1928b). This procedure, often referred to simply as the *likelihood ratio* (abbreviated LR), considers the ratio between two likelihoods (i.e., probabilities). Referring to Example 3, one likelihood is the likelihood of the population containing the same proportions that the sample has of the data

$\sum f_i \ln(f_i/\hat{f}_i)$, may also be written as $\sum f_i \ln f_i - \sum f_i \ln \hat{f}$. Twice this quantity, a value called G, approximates the χ^2 distribution.* Thus†

$$G = 2\sum f_i \ln \frac{f_i}{\hat{f}_i} \quad \text{or} \quad G = 4.60517 \sum f_i \log \frac{f_i}{\hat{f}_i}, \qquad (6)$$

or, equivalently,

$$G = 2\left[\sum f_i \ln f_i - \sum f_i \ln \hat{f}_i\right] \quad \text{or} \quad G = 4.60517\left[\sum f_i \log f_i - \sum f_i \log \hat{f}_i\right] \qquad (7)$$

is applicable as a test for goodness of fit, utilizing Table 1 from *Appendix: Statistical Tables and Graphs* with the same degrees of freedom as would be used for chi-square testing. Example 8 demonstrates the G test for the data of Example 3. In this case, the same conclusion is reached using G and χ^2, but this will not always be so.

EXAMPLE 8 Calculation of the G Statistic for the Log-Likelihood Ratio Goodness-of-Fit Test. The Data and the Hypotheses Are Those of Example 3

	Yellow smooth	Yellow wrinkled	Green smooth	Green wrinkled	n
f_i	152	39	53	6	250
$(\hat{f}_i)$	(140.6250)	(46.8750)	(46.8750)	(15.6250)	

$\nu = k - 1 = 3$

$$G = 4.60517\left[\sum f_i \log f_i - \sum f_i \log \hat{f}_i\right]$$

$= 4.60517[(152)(2.18184) + (39)(1.59106) + (53)(1.72428)$

$\quad + (6)(0.77815) - (152)(2.14806) \quad (39)(1.67094)$

$\quad - (53)(1.67094) - (6)(1.19382)]$

$= 4.60517[331.63968 + 62.05134 + 91.38684 + 4.66890$

$\quad - 326.50512 - 65.16666 - 88.55982 - 7.16292]$

$= 4.60517[2.35224]$

$= 10.832$‡

$\chi^2_{0.05,3} = 7.815$

Therefore, reject H_0.

$$0.01 < P < 0.025 \quad [P = 0.013]$$

---

‡Using natural logarithms (see Equations 6 or 7) yields the same value of G.

---

in the four categories. And the other is the likelihood of the population containing the proportions, in the four categories, that are stated in the null hypothesis. The ratio of the first likelihood to the second will be larger for greater departures of population proportions from the proportions observed in the sample.

*G also appears in the literature written as G^2 and occasionally as *likelihood ratio* χ^2; it is sometimes referred to as a measure of *deviance*.

†"ln" refers to natural logarithm (in base e) and "log" to common logarithm (in base 10). Many modern calculators can employ either.

Williams (1976) recommended G be used in preference to χ^2 whenever any $|f_i - \hat{f}_i| \geq \hat{f}_i$. The two methods often yield the same conclusions, especially when n is large; when they do not, some statisticians prefer G; others recommend χ^2, for while G may result in a more powerful test in some cases, χ^2 tends to provide a test that operates much closer to the stated level of α (e.g., Chapman, 1976; Cressie and Read, 1989; Hutchinson, 1979; Larntz, 1978; Lawal, 1984; Moore, 1986; Rudas, 1986), with the probability of a Type I error often far above α when employing G.

When $\nu = 1$, the Yates correction for continuity is applied in a fashion analogous to that in chi-square analysis in Section 2. The procedure is to make each f_i closer to $\hat{f}_i$ by 0.5 and to apply Equation 7 (or Equation 6) using these modified f_i's. This is demonstrated in Example 9.

EXAMPLE 9 The G Test for Goodness of Fit for Two Categories, for the Hypotheses and Data of Example 1

(a) Without the Yates correction for continuity:

	Category (flower color)		
	Yellow	Green	n
f_i	84	16	100
$(\hat{f}_i)$	(75)	(25)	

$\nu = k - 1 = 2 - 1 = 1$

$$G = 4.60517[(84)(1.92428) + (16)(1.20412) - (84)(1.87506)$$
$$- (16)(1.39794)]$$
$$= 4.60517[1.03336] = 4.759$$

$\chi^2_{0.05,1} = 3.841$
Therefore, reject H_0.

$$0.025 < P < 0.05 \quad [P = 0.029]$$

(b) With the Yates correction for continuity:

	Category (flower color)		
	Yellow	Green	n
f_i	84	16	100
$(\hat{f}_i)$	(75)	(25)	
Modified f_i	83.5	16.5	

$\nu = k - 1 = 2 - 1 = 1$

$$G_c = 4.60517[(83.5)(1.92169) + (16.5)(1.21748) - (83.5)(1.87506)$$
$$- (16.5)(1.39794)]$$
$$= 4.60517[0.916015] = 4.218$$

$$\chi^2_{0.05,1} = 3.841$$

Therefore, reject H_0.

$$0.025 < P < 0.05 \quad [P = 0.040]$$

8 KOLMOGOROV-SMIRNOV GOODNESS OF FIT

This chapter has thus far dealt with goodness-of-fit tests applicable to nominal-scale data. This section will present goodness-of-fit testing for data measured on a ratio, interval, or ordinal scale.

Example 10 presents data that are measurements of the height above the ground at which each of 15 moths was found on the trunk of a 25-meter-tall tree. For each height, X_i, the observed frequency is f_i, which is the number of moths found at that height. The Kolmogorov-Smirnov goodness-of-fit test (Kolmogorov, 1933; Smirnov, 1939a, 1939b), also called the Kolmogorov-Smirnov one-sample test, examines how well an observed cumulative frequency distribution conforms to an

---

EXAMPLE 10 Two-Tailed Kolmogorov-Smirnov Goodness of Fit for Continuous Ratio-Scale Data, Vertical Distribution of Moths on a Tree Trunk

H_0: Moths are distributed uniformly from ground level to height of 25 m.

H_A: Moths are not distributed uniformly from ground level to height of 25 m.

Each X_i is a height (in meters) at which a moth was observed on the tree trunk.

i	X_i	f_i	F_i	rel F_i	rel $\hat{F}_i$	D_i	D'_i
1	1.4	1	1	0.0667	0.0560	0.0107	0.0560
2	2.6	1	2	0.1333	0.1040	0.0293	0.0373
3	3.3	1	3	0.2000	0.1320	0.0680	0.0013
4	4.2	1	4	0.2667	0.1680	0.0987	0.0320
5	4.7	1	5	0.3333	0.1880	0.1453	0.0787
6	5.6	2	7	0.4667	0.2240	0.2427	0.1093
7	6.4	1	8	0.5333	0.2560	0.2773	0.2107
8	7.7	1	9	0.6000	0.3080	0.2920	0.2253
9	9.3	1	10	0.6667	0.3720	0.2947	0.2280
10	10.6	1	11	0.7333	0.4240	0.3093	0.2427
11	11.5	1	12	0.8000	0.4600	0.3400	0.2733
12	12.4	1	13	0.8667	0.4960	0.3707	0.3040
13	18.6	1	14	0.9333	0.7440	0.1893	0.1227
14	22.3	1	15	1.0000	0.8920	0.1080	0.0413

$n = 15$

$\max D_i = D_{12} = |0.8667 - 0.4960| = |0.3707| = 0.3707$

$\max D'_i = D'_{12} = |0.8000 - 0.4960| = |0.3040| = 0.3040$

$D = 0.3707$

$D_{0.05(2),15} = 0.33760$

Therefore, reject H_0.

$$0.02 < P < 0.05$$

---

expected frequency distribution.* The test considers how likely it is to obtain the observed data distribution at random from a population having the distribution specified in the null hypotheses.

For the test applicable to continuous data (i.e., ratio-scale or interval-scale data), the observed frequencies are arranged in ascending order and each cumulative observed frequency, F_i, is obtained as the sum of the observed frequencies from f_1 up to and including f_i. (For example, F_{10} is the sum of f_1 through f_{10}.) And from these cumulative frequencies the *cumulative relative observed frequencies* are determined as

$$\text{rel } F_i = \frac{F_i}{n}, \tag{8}$$

where n, which is $\sum f_i$, is the number of data in the sample. Thus, rel F_i is simply the proportion of the data that are measurements $\leq X_i$. For the data being discussed, n is 15, so, for example, rel $F_{10} = 11/15 = 0.7333$.

Then, for each X_i, the *cumulative relative expected frequency*, $\hat{F}_i$, is calculated as follows (where *expected* refers to the distribution specified in the null hypothesis). In Example 10, H_0 proposes a uniform distribution of moths over the heights 0 to 25 meters, so rel $\hat{F}_i = X_i/25$ m (for example, $\hat{F}_{10} = 10.6$ m/25 m $= 0.4240$). If, in a similar study, the null hypothesis were a uniform distribution over heights 1 to 25 m from the ground, then rel $\hat{F}_i$ would be $(X_i - 1)/24$ m.

The test statistic, D, for the Kolmogorov-Smirnov goodness-of-fit is obtained by first calculating both

$$D_i = |\text{rel } F_i - \text{rel } \hat{F}_i| \tag{9}$$

and

$$D'_i = |\text{rel } F_{i-1} - \text{rel } \hat{F}_i| \tag{10}$$

for each i. For the data under consideration, for example, $D_{10} = |0.7333 - 0.4240| = 0.3093$ and $D'_{10} = |0.6667 - 0.4240| = 0.2427$. In using Equation 10 it is important to know that $F_0 = 0$, so $D'_1 = \text{rel } \hat{F}_1$ (and in Example 10, $D'_1 = |0 - 0.0560| = 0.0560$). Then the test statistic is

$$D = \max[(\max D_i), (\max D'_i)], \tag{11}$$

which means "D is the largest value of D_i or the largest value of D'_i, whichever is larger." Critical values for this test statistic are referred to as $D_{\alpha,n}$ in Table 9 from *Appendix: Statistical Tables and Graphs*. If $D \geq D_{\alpha,n}$, then H_0 is rejected at the α level of significance.

Figure 1 demonstrates why it is necessary to examine both D_i and D'_i in comparing an observed to a hypothesized cumulative frequency distribution for continuous data. (See also D'Agostino and Noether, 1973; Fisz, 1963: Section 12.5A;

---

*The name of the test honors the two Russian mathematicians who developed its underlying concepts and procedures: Andrei Nikolaevich Kolmogorov (1903–1987) and Nikolai Vasil'evich Smirnov (1900–1906). Körner (1996: 190) reported that "Kolmogorov worked in such a large number of mathematical fields that eleven experts were required to describe his work for his London Mathematical Society obituary." Kolmogorov originated the test for the one-sample situation discussed here, and Smirnov described a two-sample test to assess how well two observed cumulative frequency distributions represent population distributions that coincide. (See, e.g., Daniel, 1990: Section 8.3; Hollander and Wolfe, 1999: Section 5.4; Siegel and Castellan, 1988: 144–151, 166; and Sprent and Smeeton, 2001: 185–187, for discussion of the Kolmogorov-Smirnov two-sample test.)

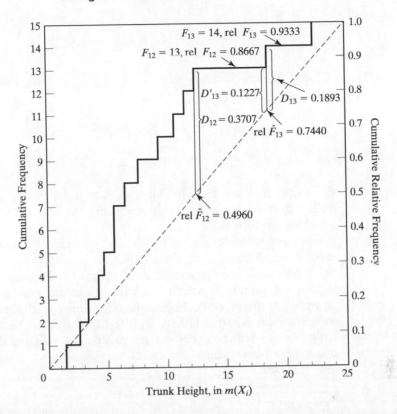

FIGURE 1: Graphical representation of Example 10, Kolmogorov-Smirnov goodness-of-fit testing for continuous data. The solid line plots the observed frequencies, and the dashed line shows the expected frequencies.

Gibbons and Chakraborti, 2003: Section 4.3.) What is sought is the maximum deviation between the observed distribution, F (which looks like a staircase when graphed), and the hypothesized distribution, $\hat{F}$. (The distribution of $\hat{F}$ appears as a straight line if H_0 proposes a uniform distribution; but other distributions, such as a normal distribution, may be hypothesized.) For each $\hat{F}_i$, we must consider the vertical distance $D_i = |F_i - \hat{F}_i|$, which occurs at the left end of a step, as well as the vertical distance $D'_i = |F_{i-1} - \hat{F}_i|$, which is at the right end of a step.

A lesser-known, but quite good, alternative for the Kolmogorov-Smirnov test for goodness of fit of continuous data is the Watson goodness-of-fit test. It is especially suited for data on a circular scale, but it is applicable as well to data on a linear scale such as in the present section. Other alternatives have been proposed, including those that have special emphasis on the differences in the tails of the distributions (Calitz, 1987).

(a) Correction for Increased Power. For small sample sizes (say, $n \leq 25$), the power of Kolmogorov-Smirnov testing can be increased impressively by employing the correction expounded by Harter, Khamis, and Lamb (1984) and Khamis (1990, 1993). For each i, Equation 8 is modified to

$$\text{rel } \mathscr{F}_i = \frac{F_i}{n + 1} \tag{12}$$

and

$$\text{rel } \mathscr{F}_i' = \frac{F_i - 1}{n - 1}.$$

(13)

Then, differences analogous to D_i and D_i' of Equations 9 and 10, respectively, are obtained as

$$D_{0,i} = \left| \text{rel } \mathscr{F}_i' - \text{rel } \hat{F}_i \right|$$

(14)

$$D_{1,i} = \left| \text{rel } \mathscr{F}_i' - \text{rel } \hat{F}_i \right|.$$

(15)

For these two statistics, the subscripts 0 and 1 are denoted as δ (lowercase Greek delta), so the developers of this procedure call it the δ-corrected Kolmogorov-Smirnov goodness-of-fit test.

The test statistic is either max $D_{0,i}$ or max $D_{1,i}$, whichever leads to the higher level of significance (i.e., the smaller probability). Table 10 from *Appendix: Statistical Tables and Graphs* gives critical values for $D_{\delta,n}$ for various levels of α. This test is demonstrated in Example 11. Although in this example the conclusion is the same as with the uncorrected Kolmogorov-Smirnov test (Example 10), this is not always so. However, Khamis (1990) reported that if $n > 20$, the results of this corrected Kolmogorov-Smirnov method are practically indistinguishable from those from the uncorrected procedure, and in such cases either the uncorrected or corrected test may be used.

EXAMPLE 11 δ-corrected Kolmogorov-Smirnov Goodness of Fit

The hypotheses and data are those of Example 10.

i	X_i	F_i	rel $\hat{F}_i$	rel $\mathscr{F}_i$	$D_{0,i}$	rel $\mathscr{F}_i'$	$D_{1,i}$
1	1.4	1	0.0560	0.0625	0.0065	0.0000	0.0560
2	2.6	2	0.1040	0.1250	0.0210	0.0714	0.0326
3	3.3	3	0.1320	0.1875	0.0555	0.1429	0.0109
4	4.2	4	0.1680	0.2500	0.0820	0.2143	0.0463
5	4.7	5	0.1880	0.3125	0.1245	0.2857	0.0977
6	5.6	7	0.2240	0.4375	0.2135	0.4286	0.2046
7	6.4	8	0.2560	0.5000	0.2440	0.5000	0.2440
8	7.7	9	0.3080	0.5625	0.2545	0.5714	0.2634
9	9.3	10	0.3720	0.6250	0.2530	0.6429	0.2709
10	10.6	11	0.4240	0.6875	0.2635	0.7143	0.2903
11	11.5	12	0.4600	0.7500	0.2900	0.7857	0.3257
12	12.4	13	0.4960	0.8125	0.3165	0.8571	0.3611
13	18.6	14	0.7440	0.8750	0.1310	0.9286	0.1846
14	22.3	15	0.8920	0.9375	0.0455	1.0000	0.1080

$n = 15$

max $D_{0,i} = D_{0,12} = 0.3165$, which has a probability of $0.05 < P < 0.10$

max $D_{1,i} = D_{1,12} = 0.3611$, which has a probability of $0.02 < P < 0.05$

Therefore, reject H_0; $0.02 < P < 0.05$.

Testing for Goodness of Fit

The δ-corrected procedure can result in a test with a Type I error slightly greater than α; Khamis (2000) presented an adjustment to remedy this. Feltz (1998) discussed similar corrections that have more power in some circumstances.

(b) Sample Size Required. When it is planned to apply a Kolmogorov-Smirnov test to continuous data, it may be asked how large a sample is needed to be able to detect a significant difference of a given magnitude between an observed and a hypothesized cumulative frequency distribution. All that need be done is to seek the desired minimum detectable difference in the body of the table of critical values of D (Table 9 from *Appendix: Statistical Tables and Graphs*), for the selected significance level, α. For example, to be able to detect a difference as small as 0.30 between an observed and a hypothesized cumulative relative frequency distribution, at a significance level of 0.05, a sample of of at least 20 would be needed, for $D_{0.05,19} = 0.30143$, which is larger than 0.30, and $D_{0.05,20} = 0.29408$, which is smaller than 0.30. If the desired difference is not in the table, then the nearest smaller one is used. Thus, for a study such as that in Example 10, it is estimated that at least 20 moths would have to be observed to be able to detect, at the 5% significance level, a difference between the cumulative frequency distributions — a difference of either D or D' — as small as 0.30.

If the desired detectable difference is beyond the $D_{\alpha,n}$ values in Table 9 from *Appendix: Statistical Tables and Graphs* (i.e., the difference is $<D_{\alpha,160}$), then we know that the required sample size is greater than 160. This sample size may be estimated by employing the values of d_α at the end of Table 9 from *Appendix: Statistical Tables and Graphs*.* If we wish to detect a difference as small as Δ, then the sample size should be at least[†]

$$n = \frac{d_\alpha^2}{\Delta^2}. \tag{17}$$

For example, if the collector of data in Example 11 had desired to be able to detect a difference, D_i or D_i', as small as 0.10, a sample size of at least 185 moth observations should have been obtained, for

$$n = \frac{(1.35810)^2}{(0.10)^2}$$
$$= 184.4.$$

(c) Discrete or Grouped Data. Ordinal data, such as in Example 12, are not measurements on a continuous scale and should not be analyzed by the Kolmogorov-Smirnov procedures discussed previously. But the following method is applicable.

Example 12 shows the results of an experiment in which cats were given a choice of five foods, identical in all respects except moisture content. A total of 40 observations were recorded. The experiment was performed in a fashion that ensured that all 40 were independent; this was done by using 40 cats, each given a choice among the five foods (not, for example, by using eight cats with each cat being given five opportunities to choose among the food types), and the cats were subjected to the experiment one at a time, so no individual's actions would influence another's.

---

*These values at the end of Appendix Table 9 are

$$d_\alpha = \sqrt{\frac{-\ln \alpha}{2}}. \tag{16}$$

[†] Δ is the capital Greek letter delta.

517

EXAMPLE 12   **Kolmogorov-Smirnov Goodness-of-Fit Test for Discrete Ordered Data**

H_0: Cats have no preference along a food-moisture gradient.

H_A: Cats do have preference along a food-moisture gradient.

Food Moisture (i)	f_i	$\hat{f}_i$	F_i	$\hat{F}_i$	d_i
1 (driest)	5	8	5	8	3
2	6	8	11	16	5
3	7	8	18	24	6
4	10	8	28	32	4
5 (moistest)	12	8	40	40	0

$n = 40; \; k = 5$

$d_{max} = d_3 = 6$

$(d_{max})_{0.05, 5, 40} = 8$

Therefore, do not reject H_0.

$$[0.10 < P < 0.20]$$

The food moisture is expressed on an ordinal scale, for although we can say that food 1 is drier than food 2 and food 2 is drier than food 3, we cannot say that the difference in moisture between foods 1 and 2 is quantitatively equal to the difference between foods 2 and 3. That is, we can speak only of relative magnitudes, and not quantitative measurements, of the foods' moisture contents.

The null hypothesis of equal preference for the five food types could be tested by chi-square goodness of fit (Section 3), and this would be appropriate if the five foods were nominal-scale categories (for example, different brands or different recipes). But the present data are in categories that have a rational order, and the null hypothesis is that there is no preference along the gradient of food moisture (that is, no preference among the five moisture categories arranged in ascending order).

The data are observed frequencies, f_i, namely the numbers of animals choosing each of the five food types. The expected frequencies, $\hat{f}_i$, are the numbers expected if the null hypothesis is true. In the present example, hypothesizing no preferred food type, a uniform distribution (i.e., a frequency of eight in each of the five categories) would be expected.

For the Kolmogorov-Smirnov goodness-of-fit test, cumulative observed frequencies (F_i) and cumulative expected frequencies ($\hat{F}_i$) are calculated for categories 1 through k. (In Example 12, $k = 5$.) The cumulative frequency for category i is the sum of all frequencies from categories 1 through i (in Example 12, the frequencies for food as moist as, or moister than, i).

For each category, i, the absolute value of difference between the two cumulative frequency distributions is determined:

$$d_i = |F_i - \hat{F}_i| \; . \tag{18}$$

The largest d_i is the test statistic; let us call it d_{max}.

Critical values of $d_{\max}$ are found in Table 8 from *Appendix: Statistical Tables and Graphs* (which requires that in the experiment n, the total number of data, is a multiple of k, the number of categories).* Also, the tabled critical values are for situations where all of the expected frequencies, $\hat{f}_i$, are equal, but the table also works well for unequal $\hat{f}_i$ if the inequality is not great (Pettitt and Stephens, 1977).

The $d_{\max}$ procedure is also appropriate when data are recorded on a continuous scale but are grouped into broad categories on that scale so $f_i > 1$ for several f_i's. For such data, or for ordinal data, the test appropriate for ungrouped continuous data (using D) is conservative (e.g., Noether, 1963; Pettitt and Stephens, 1977), meaning that the testing is occurring at an α smaller — perhaps much smaller — than that stated, and the probability of a Type II error is inflated; that is, the power of the test is reduced. Therefore, use of $d_{\max}$ is preferred to using D for grouped or ordinal data.[†]

Example 13 shows how the data of Example 10 would look had the investigator recorded them in 5-meter ranges of trunk heights. Note that power is lost (and H_0 is not rejected) by grouping the data, and grouping should be avoided or minimized whenever possible.

When applicable (that is, when the categories are ordered), the Kolmogorov-Smirnov test is more powerful than the chi-square test when n is small or when $\hat{f}_i$

EXAMPLE 13   Kolmogorov-Smirnov Goodness-of-Fit Test for Continuous, But Grouped, Data

The hypotheses are as in Example 10, with that example's data recorded in 5-meter segments of tree height (where, for example, 5–10 m denotes a height of at least 5 m but less than 10 m).

Trunk height i	(X_i)	f_i	$\hat{f}_i$	F_i	$\hat{F}_i$	d_i
1	0–5 m	5	3	5	3	2
2	5–10 m	5	3	10	6	4
3	10–15 m	5	3	13	9	4
4	15–20 m	1	3	14	12	2
5	20–25 m	1	3	15	15	0

$n = 15$; $k = 5$

$d_{\max} = 4$

$(d_{\max})_{0.05,5,15} = 5$

Therefore, do not reject H_0.

$$[0.10 < P < 0.20]$$

---

*If n is not evenly divisible by k, then, conservatively, the critical value for the nearest larger n in the table may be used. (However, that critical value might not exist in the table.)

[†]The first footnote of this section refers to the Kolmogorov-Smirnov two-sample test, which also yields conservative results if applied to discrete data (Noether, 1963).

values are small, and often in other cases.* Another advantage of the Kolmogorov-Smirnov test over chi-square is that it is not adversely affected by small expected frequencies (see Section 5).

EXERCISES

1. Consult Table 1 from *Appendix: Statistical Tables and Graphs*
   (a) What is the probability of computing a χ^2 at least as large as 3.452 if DF $= 2$ and the null hypothesis is true?
   (b) What is $P(\chi^2 \geq 8.668)$ if $v = 5$?
   (c) What is $\chi^2_{0.05,4}$?
   (d) What is $\chi^2_{0.01,8}$?

2. Each of 126 individuals of a certain mammal species was placed in an enclosure containing equal amounts of each of six different foods. The frequency with which the animals chose each of the foods was:

Food item (i)	f_i
N	13
A	26
W	31
G	14
M	28
C	14

   (a) Test the hypothesis that there is no preference among the food items.
   (b) If the null hypothesis is rejected, ascertain which of the foods are preferred by this species.

3. A sample of hibernating bats consisted of 44 males and 54 females. Test the hypothesis that the hibernating population consists of equal numbers of males and females.

4. In attempting to determine whether there is a 1 : 1 sex ratio among hibernating bats, samples were taken from four different locations in a cave:

Location	Males	Females
V	44	54
D	31	40
E	12	18
M	15	16

By performing a heterogeneity chi-square analysis, determine whether the four samples may justifiably be pooled. If they may, pool them and retest the null hypothesis of equal sex frequencies.

5. Test the hypothesis and data of Exercise 2 using the log-likelihood G.

6. A straight line is drawn on the ground perpendicular to the shore of a body of water. Then the locations of ground arthropods of a certain species are measured along a 1-meter-wide band on either side of the line. Use the Kolmogorov-Smirnov procedure on the following data to test the null hypothesis of uniform distribution of this species from the water's edge to a distance of 10 meters inland.

Distance from water (m)	Number observed	Distance from from (m)	Number observed
0.3	1	3.4	1
0.6	1	4.1	1
1.0	1	4.6	1
1.1	1	4.7	1
1.2	1	4.8	1
1.4	1	4.9	1
1.6	1	4.9	1
1.9	1	5.3	1
2.1	1	5.8	1
2.2	1	6.4	1
2.4	1	6.8	1
2.6	1	7.5	1
2.8	1	7.7	1
3.0	1	8.8	1
3.1	1	9.4	1

7. For a two-tailed Kolmogorov-Smirnov goodness-of-fit test with continuous data at the 5% level of significance, how large a sample is necessary to detect a difference as small as 0.25 between cumulative relative frequency distributions?

*A chi-square goodness-of-fit test performed on the data of Example 12, for H_0: There is no preference among the five food categories, would disregard the order of the categories and would yield $\chi^2 = 4.250$; and the log-likelihood goodness of fit would result in $G = 4.173$. Each of those statistics would be associated with a probability between 0.25 and 0.50.

8. A bird feeder is placed at each of six different heights. It is recorded which feeder was selected by each of 18 cardinals. Using the Kolmogorov-Smirnov procedure for discrete data, test the null hypothesis that each feeder height is equally desirable to cardinals.

Feeder height	Number observed
1 (lowest)	2
2	3
3	3
4	4
5	4
6 (highest)	2

ANSWERS TO EXERCISES

1. **(a)** For $v = 2$, $P(\chi^2 \geq 3.452)$ is between 0.10 and 0.25 $[P = 0.18]$; **(b)** For $v = 5$, $0.10 < (\chi^2 \geq 8.668) < 0.25 [P = 0.12]$; **(c)** $\chi^2_{0.05,4} = 9.488$; **(d)** $\chi^2_{0.01,8} = 20.090$.

2. **(a)** $\chi^2 = 16.000$, $v = 5$, $0.005 < P < 0.01$ $[P = 0.0068]$. As $P < 0.05$, reject H_0 of equal food item preference. **(b)** By grouping food items N, G, and C; $n = 41$, and for H_0: Equal food preference, $\chi^2 = 0.049$, $v = 2$, $0.975 < P < 0.99$ $[P = 0.98]$; as $P > 0.05$, H_0 is not rejected. By grouping food items A, W, and M; $n = 85$, and for H_0: Equal food preference, $\chi^2 = 0.447$, $v = 2$, $0.75 < P < 0.90 [P = 0.80]$; as $P > 0.05$, H_0 is not rejected. By considering food items N, G, and C as one group and items A, W, and M as a second group, and H_0: Equal preference for the two groups, $\chi^2_c = 14.675$, $v = 1$, $P < 0.001$ $[P = 0.00013]$; H_0 is rejected.

3. $\chi^2_c = 0.827$, $v = 1$, $0.25 < P < 0.50 [P = 0.36]$. As $P > 0.05$, do not reject H_0: The population consists in equal numbers of males and females.

4.

Location	Males	Females	χ^2	v
1	44	54	1.020	1
2	31	40	1.141	1
3	12	18	1.200	1
4	15	16	0.032	1
Total of chi-squares			3.393	4
Pooled chi-square 102	128		2.939	1
Heterogeneity chi-square			0.454	3

$$0.90 < P < 0.95$$

Because P(heterogeneity χ^2) > 0.05, the four samples may be pooled with the following results: $\chi^2_c = 2.717$, $v = 1$, $0.05 < P < 0.10 [P = 0.099]$; $P > 0.05$, so do not reject H_0: Equal numbers of males and females in the population.

5. $G = 16.188$, $v = 5$, $0.005 < P < 0.01 [P = 0.0063]$; $P < 0.05$, so reject H_0 of no difference in food preference.

6. H_0: There is a uniform distribution of the animals from the water's edge to a distance of 10 meters upland; max $D_i = 0.24333$, max $D'_i = 0.2033$, $D = 0.2433$; $D_{0.05(2),31} = 0.23788$; reject H_0, $0.02 < P < 0.05$.

7. $D_{0.05,27} = 0.25438$ and $D_{0.05,28} = 0.24993$, so a sample size of at least 28 is called for.

8. $d_{max} = 1$; $(d_{max})_{0.05,6,18} = 6$; do not reject H_0: The feeders are equally desirable to the birds; $P > 0.50$.

This page intentionally left blank

Contingency Tables

Contingency Tables

Enumeration data may be collected simultaneously for two nominal-scale variables. These data may be displayed in what is known as a *contingency table*, where the r rows of the table represent the r categories of one variable and the c columns indicate the c categories of the other variable; thus, there are rc "cells" in the table. (This presentation of data is also known as a *cross tabulation* or *cross classification*.)

Example 1a is of a contingency table of two rows and four columns, and may be referred to as a 2 × 4 ("two by four") table having $(2)(4) = 8$ cells. A sample of 300 people has been obtained from a specified population (let's say members of an actors' professional association), and the variables tabulated are each person's sex and each person's hair color. In this 2 × 4 table, the number of people in the sample with each of the eight combinations of sex and hair color is recorded in one of the eight cells of the table. These eight data could also be recorded in a 4 × 2 contingency table, with the four hair colors appearing as rows and the two sexes as columns, and that would not change the statistical hypothesis tests or the conclusions that result from them. As with previous statistical tests, the total number of data in the sample is designated as n.

EXAMPLE 1   A 2 × 4 Contingency Table for Testing the Independence of Hair Color and Sex in Humans

(a) H_0:   Human hair color is independent of sex in the population sampled.

H_A: Human hair color is not independent of sex in the population sampled.

$\alpha = 0.05$

Sex	Hair color				Total
	Black	*Brown*	*Blond*	*Red*	
Male	32	43	16	9	$100 (= R_1)$
Female	55	65	64	16	$200 (= R_2)$
Total	87 $(= C_1)$	108 $(= C_2)$	80 $(= C_3)$	25 $(= C_4)$	$300 (= n)$

Contingency Tables

(b) The observed frequency, f_{ij}, in each cell is shown, with the frequency expected if H_0 is true (i.e., $\hat{f}_{ij}$) in parentheses.

| | **Hair color** | | | | |
Sex	Black	Brown	Blond	Red	**Total**
Male	32 (29.0000)	43 (36.0000)	16 (26.6667)	9 (8.3333)	100 ($= R_1$)
Female	55 (58.0000)	65 (72.0000)	64 (53.3333)	16 (16.6667)	200 ($= R_2$)
Total	87 ($= C_1$)	108 ($= C_2$)	80 ($= C_3$)	25 ($= C_4$)	300 ($= n$)

$$\chi^2 = \sum\sum \frac{(f_{ij} - \hat{f}_{ij})^2}{\hat{f}_{ij}}$$

$$= \frac{(32 - 29.0000)^2}{29.0000} + \frac{(43 - 36.0000)^2}{36.0000} + \frac{(16 - 26.6667)^2}{26.6667}$$

$$+ \frac{(9 - 8.3333)^2}{8.3333} + \frac{(55 - 58.0000)^2}{58.0000} + \frac{(65 - 72.0000)^2}{72.0000}$$

$$+ \frac{(64 - 53.3333)^2}{53.3333} + \frac{(16 - 16.6667)^2}{16.6667}$$

$$= 0.3103 + 1.3611 + 4.2667 + 0.0533 + 0.1552 + 0.6806 + 2.1333$$

$$+ 0.0267 = 8.987$$

$$\nu = (r - 1)(c - 1) = (2 - 1)(4 - 1) = 3$$

$$\chi^2_{0.05,3} = 7.815$$

Therefore, reject H_0.

$$0.025 < P < 0.05 \quad [P = 0.029]$$

The hypotheses to be tested in this example may be stated in any of these three ways:

H_0: In the sampled population, a person's hair color is independent of that person's sex (that is, a person's hair color is not associated with the person's sex), and

H_A: In the sampled population, a person's hair color is not independent of that person's sex (that is, a person's hair color is associated with the person's sex), *or*

H_0: In the sampled population, the ratio of males to females is the same for people having each of the four hair colors, and

H_A: In the sampled population, the ratio of males to females is not the same for people having each of the four hair colors; *or*

525

H_0: In the sampled population, the proportions of people with the four hair colors is the same for both sexes, and

H_A: In the sampled population, the proportions of people with the four hair colors is not the same for both sexes.

In order to test the stated hypotheses, the sample of data in this example could have been collected in a variety of ways:

- It could have been stipulated, in advance of collecting the data, that a specified number of males would be taken at random from all the males in the population and a specified number of females would be taken at random from all the females in the population. Then the hair color of the people in the sample would be recorded for each sex. That is what was done for Example 1a, where it was decided, before the data were collected, that the sample would consist of 100 males and 200 females.

- It could have been stipulated, in advance of collecting the data, that a specified number of people with each hair color would be taken at random from all persons in the population with that hair color. Then the sex of the people in the sample would be recorded for each hair color.

- It could have been stipulated, in advance of collecting, that a sample of n people would be taken at random from the population, without specifying how many of each sex would be in the sample or how many of each hair color would be in the sample. Then the sex and hair color of each person would be recorded.

For most contingency-table situations, the same statistical testing procedure applies to any one of these three methods of obtaining the sample of n people, and the same result is obtained. However, when dealing with the smallest possible contingency table, namely one with only two rows and two columns (Section 3), an additional sampling strategy may be encountered that calls for a different statistical procedure.

Section 8 will introduce procedures for analyzing contingency tables of more than two dimensions, where frequencies are tabulated simultaneously for more than two variables.

1 CHI-SQUARE ANALYSIS OF CONTINGENCY TABLES

The most common procedure for analyzing contingency table data uses the chi-square statistic.* Recall that for the computation of chi-square one utilizes observed and expected frequencies (and never proportions or percentages). For the goodness-of-fit analysis, f_i denoted the frequency observed in category i of the variable under study. In a contingency table, we have two variables under consideration, and we denote an observed frequency as f_{ij}. Using the double subscript, f_{ij} refers to the frequency observed in row i and column j of the contingency table. In Example 1, the value in row 1 column 1 is denoted as f_{11}, that in row 2 column 3 as f_{23}, and so on. Thus, $f_{11} = 32, f_{12} = 43, f_{13} = 16, \ldots, f_{23} = 64$, and $f_{24} = 16$.

The total frequency in row i of the table is denoted as R_i and is obtained as $R_i = \sum_{j=1}^{c} f_{ij}$. Thus, $R_1 = f_{11} + f_{12} + f_{13} + f_{14} = 100$, which is the total number of males in the sample, and $R_2 = f_{21} + f_{22} + f_{23} + f_{24} = 200$, which is the total number of females in the sample. The column totals, C_j, are obtained by analogous

---

*The early development of chi-square analysis of contingency tables is credited to Karl Pearson (1904) and R. A. Fisher (1922). In 1904, Pearson was the first to use the term "contingency table" (David, 1995).

summations: $C_j = \sum_{i=1}^{r} f_{ij}$. For example, the total number of blonds in the sample data is $C_3 = \sum_{i=1}^{2} f_{i3} = f_{13} + f_{23} = 80$, the total number of redheads is $C_4 = \sum_{i=1}^{2} f_{i4} = 25$, and so on. The total number of observations in all cells of the table is called the grand total and is $\sum_{i=1}^{r} \sum_{j=1}^{c} f_{ij} = f_{11} + f_{12} + f_{13} + \cdots + f_{23} + f_{24} = 300$, which is n, the size of our sample. The computation of the grand total may be written in several other notations: $\sum_i \sum_j f_{ij}$ or $\sum_{i,j} f_{ij}$, or simply $\sum \sum f_{ij}$. When no indices are given on the summation signs, we assume that the summation of all values in the sample is desired.

The most common calculation of chi-square analysis of contingency tables is

$$\chi^2 = \sum \sum \frac{(f_{ij} - \hat{f}_{ij})^2}{\hat{f}_{ij}}. \tag{1}$$

In this formula, similar to the equation for chi-square goodness of fit, $\hat{f}_{ij}$ refers to the frequency expected in a row i column j if the null hypothesis is true.* If, in Example 1a, hair color is in fact independent of sex, then $\frac{100}{300} = \frac{1}{3}$ of all black-haired people would be expected to be males and $\frac{200}{300} = \frac{2}{3}$ would be expected to be females. That is, $\hat{f}_{11} = \frac{100}{300}(87) = 29$ (the expected number of black-haired males), $\hat{f}_{21} = \frac{200}{300}(87) = 58$ (the expected number of black-haired females), $\hat{f}_{12} = \frac{100}{300}(108) = 36$ (the expected number of brown-haired males), and so on.

This may also be explained by the probability rule: The probability of two independent events occurring at once is the product of the probabilities of the two events. Thus, if having black hair is independent of being male, then the probability of a person being both black-haired and male is the probability of a person being black-haired multiplied by the probability of a person being male, namely $\left(\frac{87}{300}\right) \times \left(\frac{100}{300}\right)$, which is 0.0966667. This means that the expected number of black-haired males in a sample of 300 is $(0.0966667)(300) = 29.0000$. In general, the frequency expected in a cell of a contingency table is

$$\hat{f}_{ij} = \left(\frac{R_i}{n}\right)\left(\frac{C_j}{n}\right)n, \tag{3}$$

which reduces to the commonly encountered formula,

$$\hat{f}_{ij} = \frac{(R_i)(C_j)}{n}, \tag{4}$$

---

*The following are mathematically equivalent to Equation 1 for contingency tables:

$$\chi^2 = \sum \sum \frac{f_{ij}^2}{\hat{f}_{ij}} - n \tag{2}$$

and

$$\chi^2 = n\left(\sum \sum \frac{f_{ij}^2}{R_i C_j} - 1\right). \tag{2a}$$

These formulas are computationally simpler than Equation 1, the latter not even requiring the calculation of expected frequencies; however, they do not allow for the examination of the contributions to the computed chi-square, the utility of which will be seen in Section 6.

and it is in this way that the $\hat{f}_{ij}$ values in Example 1b were obtained. Note that we can check for arithmetic errors in our calculations by observing that $R_i = \sum_{j=1}^{c} f_{ij} = \sum_{j=1}^{c} \hat{f}_{ij}$ and $C_j = \sum_{i=1}^{r} f_{ij} = \sum_{i=1}^{r} \hat{f}_{ij}$. That is, the row totals of the expected frequencies equal the row totals of the observed frequencies, and the column totals of the expected frequencies equal the column totals of the observed frequencies.

Once χ^2 has been calculated, its significance can be ascertained from Table 1 from *Appendix: Statistical Tables and Graphs*, but to do so we must determine the degrees of freedom of the contingency table.

The degrees of freedom for a chi-square calculated from contingency-table data are*

$$\nu = (r - 1)(c - 1). \tag{5}$$

In Example 1, which is a 2×4 table, $\nu = (2 - 1)(4 - 1) = 3$. The calculated statistic is 9.987 and the critical value is $\chi^2_{0.05,3} = 7.815$, so the null hypothesis is rejected.

It is good to calculate expected frequencies and other intermediate results to at least four decimal places and to round to three decimal places after arriving at the value of χ^2. Barnett and Lewis (1994: 431–440) and Simonoff (2003: 228–234) discuss outliers in contingency-table data.

(a) Comparing Proportions. Hypotheses for data in a contingency table with only two rows (or only two columns) often refer to ratios or proportions. In Example 1, the null hypothesis could have been stated as, "In the sampled population, the sex ratio is the same for each hair color" or as "In the sampled population, the proportion of males is the same for each hair color." The comparison of two proportions is discussed in Sections 3b and 10.

2 VISUALIZING CONTINGENCY-TABLE DATA

Among the ways to present contingency-table data in graphical form is a method known as a *mosaic* display.[†]

Nominal-scale data were presented in a bar graph. The categories of the nominal-scale variable appear on one axis of the graph (typically the horizontal axis), and the number of observations is on the other axis. The lengths of the bars in the graph

---

*In the early days of contingency-table analysis, K. Pearson and R. A. Fisher disagreed vehemently over the appropriate degrees of freedom to employ; Fisher's (1922) view has prevailed (Agresti, 2002: 622; Savage, 1976), as has his use of the term *degrees of freedom*.

[†]The current use of mosaic displays is attributed to Hartigan and Kleiner (1981). In an historical review of rectangular presentations of data, Friendly (2002) credits the English astronomer Edmond (a.k.a. Edmund) Halley (1656–1742), famous for his 1682 observation of the comet that bears his name, with the first use of rectangular areas in the data representation for two independent variables (which, however, were not variables for a contingency table). Further developments in the visual use of rectangular areas took place in France and Germany in the early 1780s; a forerunner of mosaic graphs was introduced in 1844 by French civil engineer Charles Joseph Minard (1791–1870), and what resembled the modern mosaic presentation was first used in 1877 by German statistician Georg von Mayr (1841–1925). In 1977, French cartographer Jacques Bertin (1918–) used graphs very similar to the mosaics of Hartigan and Kleiner.

Contingency Tables

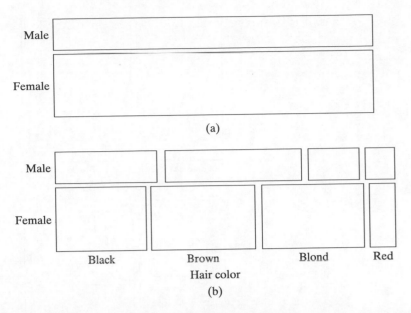

(a)

Black   Brown   Blond   Red
Hair color

(b)

FIGURE 1: A mosaic display for the contingency-table data of Example 1. (a) The first step displays two horizontal bars of equal width with the height of one of them representing the number of males and the height of the other representing the number of females in the sample. (b) The second step divides each of the two horizontal bars into four tiles, with the width of each tile depicting the frequency in the sample of a hair color among the individuals of one of the sexes.

are representations of the frequencies of occurrence of observations in the data categories; and, when bars are of equal width, the areas of the bars also depict those frequencies.

Figure 1 demonstrates visualizing the data in Example 1 and shows the two-step process of preparing a mosaic display. The first step is to prepare Figure 1a, which is a graph reflecting the numbers of males and females in the sample of data described in Example 1. Of the 300 data, 100 are males and 200 are females, so the bar for females is two times as high as the bar for males. (The bars are graphed horizontally to reflect the rows in the Example 1 contingency table, but they could have been drawn vertically instead.) The bars are drawn with equal widths, so their areas also express visually the proportion of the 300 data in each sex category, with the lower (female) bar having two times the area of the upper (male) bar.

The second step, shown in Figure 1b, is to divide each sex's horizontal bar into four segments representing the relative frequencies of the four hair colors within that sex. For example, black hair among males was exhibited by $32/100 = 0.32$ of the males in the sample, so black is depicted by a bar segment that is 32% of the width of the male bar; and $16/200 = 0.08$ of the sample's females had red hair, so the red-hair segment for females is 8% of the width of the female bar. These bar segments are often referred to as *tiles*, and there will be a tile for each of the $r \times c$ cells in the contingency table. Mosaic displays are usually, but not necessarily, drawn with small gaps between adjacent tiles.

If the boundaries of the tiles for the two bars were perfectly aligned vertically, that would indicate that the $r \times c$ frequencies were in perfect agreement with the

529

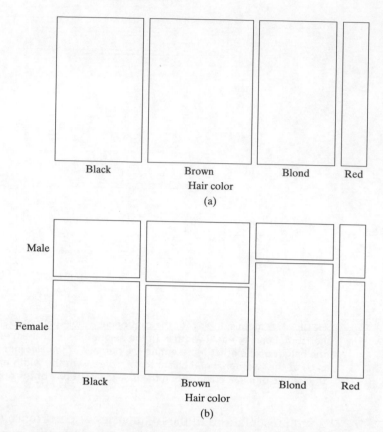

FIGURE 2: A mosaic display for the contingency-table data of Example 1. (a) The first step displays four vertical bars of equal height, one for each of the hair colors in the sample, with the width of the bars expressing the relative frequencies of the hair colors. (b) The second step divides each of the four vertical bars into two tiles, with the length of each tile depicting the frequency of a members of a sex among the individuals of one of the hair colors.

null hypothesis. The more out of alignment the tiles are, the less likely it is that the sampled population conforms to that specified in H_0.

In Figure 1, the data of Example 1 were displayed graphically by showing the frequency of each hair color within each sex. Alternatively, the data could be presented as the frequency of each sex for each hair color. This is shown in Figure 2. In Figure 2a, the widths of the four vertical bars represent the relative frequencies of the four hair colors in the sample, and Figure 2b divides each of those four bars into two segments (tiles) with sizes reflecting the proportions of males and females with each hair color.

Either graphical depiction (Figure 1b or 2b) is legitimate, with the choice depending upon the visual emphasis the researcher wants to give to each of the variables.

Mosaic displays may also be presented for contingency tables having more than two rows *and* more than two columns (such as Exercise 4 at the end of this chapter). Friendly (1994, 1995, 1999, 2002) described how the interpretation of mosaic graphs can be enhanced by shading or coloring the tiles to emphasize the degree to which observed frequencies differ from expected frequencies in the cells of the contingency table; and mosaic presentations are also used for contingency tables with more than two dimensions (which are introduced in Section 8).

3 2 × 2 CONTINGENCY TABLES

The smallest possible contingency table is that consisting of two rows and two columns. It is referred to as a 2 × 2 ("two by two") table or a fourfold table, and it is often encountered in biological research. By Equation 5, the degrees of freedom for 2 × 2 tables is $(2 - 1)(2 - 1) = 1$.

The information in a 2 × 2 contingency table may be displayed as

$$
\begin{array}{cc|c}
f_{11} & f_{12} & R_1 \\
f_{21} & f_{22} & R_2 \\
\hline
C_1 & C_2 & n
\end{array}
$$

where f_{ij} denotes the frequency observed in row i and column j, R_i is the sum of the two frequencies in row i, C_j is the sum of the two frequencies in column j, and n is the total number of data in the sample. (The sample size, n, is the sum of all four of the f_{ij}'s, is the sum of the two row totals, and is the sum of the two column totals.) The row totals, R_1 and R_2, are said to occupy one margin of the table, and the column totals, C_1 and C_2, are said to occupy an adjacent margin of the table.

There are different experimental designs that result in data that can be arranged in contingency tables, depending upon the nature of the populations from which the samples come. As described by Barnard (1947) and others, these can be categorized on the basis of whether the marginal totals are set by the experimenter before the data are collected.

(a) No Margin Fixed. There are situations where only the size of the sample (n) is declared in advance of data collection, and neither the row totals nor the column totals are prescribed.* In Example 2a, the experimenter decided that the total number of data in the sample would be $n = 70$, but there was no specification prior to the data collection of what the total number of boys, of girls, of right-handed children, or of left-handed children would be. A sample of 70 was taken at random from a population of children (perhaps of a particular age of interest), and then the numbers of right-handed boys, right-handed girls, left-handed boys, and left-handed girls were recorded as shown in this example. The statistical analysis shown in Example 2b will be discussed in Section 3d.

EXAMPLE 2   A 2 × 2 Contingency Table with No Fixed Margins

(a) H_0:   In the sampled population, handedness is independent of sex.
       H_A: In the sampled population, handedness is not independent of sex.

   $\alpha = 0.05$

	Boys	Girls	Total
Left-handed	6	12	18
Right-handed	28	24	52
Total	34	36	70

---

*This kind of experimental design is sometimes referred to as a double dichotomy or as representing a multinomial sampling distribution, and the resulting test as a test of association or test of independence.

(b) Using Equation 6 (Equation 1 could also be used, with the same result),

$$\chi^2 = \frac{n(f_{11}f_{22} - f_{12}f_{21})^2}{R_1 R_2 C_1 C_2}$$

$$= \frac{70\,[(6)(24) - (12)(28)]^2}{(18)(52)(34)(36)}$$

$$= 2.2524.$$

$\nu = 1;\ \chi^2_{0.05,1} = 3.841$

Therefore, do not reject H_0.

$$0.10 < P < 0.25 \quad [P = 0.22]$$

(b) One Margin Fixed. Some experimental designs not only specify the sample size, n, but also indicate — prior to collecting data — how many data in the sample will be in each row (or how many will be in each column).* Thus, in Example 2a, it could have been declared, before counting how many children were in each of the four categories, how many boys would be taken at random from all the boys in the population and how many girls would be taken at random from the girls in the population. Or the column totals might have been fixed, stating how many right-handed children and how many left-handed children would be selected from their respective populations.

Another example of a contingency table with one pair of marginal totals fixed is shown in Example 3a. In this study, it was decided to collect, at random, 24 mice of species 1 and 25 of species 2, and the researcher recorded the number of mice of each species that were infected with a parasite of interest.

EXAMPLE 3   A 2 x 2 Contingency Table with One Fixed Margin

(a) H_0: The proportion of the population infected with an intestinal parasite is the same in two species of mouse.

H_A: The proportion of the population infected with an intestinal parasite is not the same in two species of mouse.

$\alpha = 0.05$

	Species 1	Species 2	Total
With parasite	18	10	28
Without parasite	6	15	21
Total	24	25	49

---

*This experimental design is often called a comparative trial, the resulting test a test of homogeneity, and the underlying distributions binomial distributions.

(b) Using Equation 6 (Equation 1 could also be used, with the same result),

$$\chi^2 = \frac{n(f_{11}f_{22} - f_{12}f_{21})^2}{R_1 R_2 C_1 C_2}$$

$$= \frac{49\left[(18)(15) - (10)(6)\right]^2}{(28)(21)(24)(25)}$$

$$= 6.1250.$$

$$0.01 < P < 0.025 \quad [P = 0.013]$$

If one margin is fixed, hypotheses might be expressed in terms of proportions. For Example 3a, the null hypothesis could be stated as H_0: In the sampled population, the proportion of infected mice is the same in species 1 and species 2. The statistical analysis shown in Example 3b will be discussed in Section 3d.

(c) Both Margins Fixed. In some cases (which are very uncommon), both margins in the contingency table are fixed.* That is, R_1, R_2, C_1, C_2, and n are all set before the collection of data.

Data for such a 2×2 table are shown in Example 4a, where an ecologist wanted to compare the ability of two species of snails to tolerate the current of a stream and adhere to the stream's substrate. The researcher labeled 30 snails that were clinging to the bottom of the stream, 19 of them selected at random from a population of snails of one species and 11 selected at random from a population of snails of a second species. These 30 individuals were then observed as the current washed over them, and it was decided before the experiment began that data collection would end when more than half of the 30 (that is, 16) yielded to the current and were swept downstream.

EXAMPLE 4   A 2 × 2 Contingency Table with Two Fixed Margins

(a) H_0: The ability of snails to resist the current is no different between the two species.

H_A: The ability of snails to resist the current is different between the two species.

$\alpha = 0.05$

The four marginal totals are set before performing the experiment, and the four cell frequencies are collected from the experiment.

	Resisted	Yielded	
Species 1	12	7	19
Species 2	2	9	11
	14	16	30

---

*The sampling in this experimental design comes from what is known as a hypergeometric distribution, and the experimental design is sometimes called an independence trial.

(b) Using Equation 7 (Equation 1 could also be used, with the same result, if $f_{ij} - \hat{f}_{ij}$ is replaced by $|f_{ij} - \hat{f}_{ij}| - 0.5$), the chi-square with the Yates correction for continuity is

$$\chi_c^2 = \frac{n\left(|f_{11}f_{22} - f_{12}f_{21}| - \dfrac{n}{2}\right)^2}{R_1 R_2 C_1 C_2}$$

$$= \frac{30\left[|(12)(9) - (7)(2)| - \dfrac{30}{2}\right]^2}{(19)(11)(14)(16)}$$

$$= 3.999.$$

$\nu = 1$

$\chi_{0.05,1}^2 = 3.841.$

Therefore, reject H_0.

$$0.025 < P < 0.05 \quad [P = 0.046]$$

(c) Using Equation 7b, the chi-square with the Cochran-Haber correction for continuity is calculated as follows:

$m_1 = R_2 = 11, \; m_2 = C_1 = 14$

$\hat{f} = m_1 m_2 / n = (11)(14)/30 = 5.13$

$f = f_{21} = 2; \; d = |f - \hat{f}| = |2 - 5.13| = 3.13$

$2\hat{f} = 2(5.13) = 10.26;$

As $f < 2\hat{f}, D = 3.0$

$$\chi_H^2 = \frac{n^3 D^2}{R_1 R_2 C_1 C_2}$$

$$= \frac{(30)^3 (3.0)^2}{(19)(11)(14)(16)}$$

$$= 5.191.$$

As $\chi_{0.05,1}^2 = 3.841$, reject H_0.

$$0.01 < P < 0.025 \quad [P = 0.023]$$

Thus, prior to collecting the data, the number of snails of each species was decided upon (as 19 and 11), *and* the total numbers of snails dislodged by the current (16 and 14) were specified. The statistical analysis demonstrated in Example 4b will be discussed in Section 3d.

(d) Analysis of 2 × 2 Contingency Tables. Contingency-table hypotheses may be examined by chi-square, as shown in Section 1, calculating χ^2 with Equation 1 with the expected frequencies $(\hat{f}_{ij})$ obtained via Equation 4. However, for a 2 × 2 table, the following is a simpler computation,* for it does not require that the expected frequencies be determined, and it avoids rounding error associated with calculating $\hat{f}_{ij}$ and $f_{ij} - \hat{f}_{ij}$:

$$\chi^2 = \frac{n(f_{11}f_{22} - f_{12}f_{21})^2}{R_1 R_2 C_1 C_2}. \tag{6}$$

As with goodness of fit, chi-square values that are calculated come from a discrete distribution, but they are to be compared (such as by Table 1 from *Appendix: Statistical Tables and Graphs*) to chi-square values from a continuous distribution. Thus, statisticians may recommend that a correction for continuity be applied when $v = 1$ (which is the case when dealing with a 2×2 contingency table). More than 20 continuity corrections have been proposed; the most commonly considered is the Yates (1934) correction[†] (as was used for goodness of fit), which is the modification of Equation 1 by substituting $|f_{ij} - \hat{f}_{ij}| - 0.5$ for $f_{ij} - \hat{f}_{ij}$ or, equivalently, using the following instead of Equation 6:

$$\chi_c^2 = \frac{n\left(|f_{11}f_{22} - f_{12}f_{21}| - \frac{n}{2}\right)^2}{R_1 R_2 C_1 C_2}. \tag{7}$$

This is the calculation employed in Example 4b, and its use approximates the two-tailed Fisher exact test.

Haber (1980) showed that there are other correction methods that often perform better than that of Yates, which tends to be conservative (in that it has a probability less than α of a Type I error and has lower power than a nonconservative test). He proposed using a procedure based on a principle expounded by Cochran (1942, 1952). In the Cochran-Haber method (demonstrated in Example 4c), the smallest of the four expected frequencies is determined; using Equation 4, this frequency is

$$\hat{f} = \frac{m_1 m_2}{n}, \tag{7a}$$

where m_1 is the smallest of the four marginal totals and m_2 is the smaller of the two totals in the other margin. In Example 4, the smallest marginal total, m_1, is 11, which is a row total; and m_2 is, therefore, the smaller of the two column totals, namely 14. Then the absolute difference between this expected frequency ($\hat{f}$) and its corresponding observed frequency (f) is $d = |f - \hat{f}|$; and

- If $f \leq 2\hat{f}$, then define D = the largest multiple of 0.5 that is $<d$; and
- If $f > 2\hat{f}$, then define $D = d - 0.5$.

The chi-square with the Cochran-Haber correction is

$$\chi_H^2 = \frac{n^3 D^2}{R_1 R_2 C_1 C_2}. \tag{7b}$$

---

*Richardson (1994) attributed Equation 6 to Fisher (1922). Upton (1982) reported a "slight" improvement if $n - 1$ is employed in place of n.

[†]Pearson (1947) points out that Yates's use of this correction for chi-square analysis was employed as early as 1921 for other statistical purposes. The continuity correction for 2×2 tables should *not* be used in the very rare instances that its inclusion increases, instead of decreases, the numerator (that is, when $|f_{11} - f_{22}| < n/2$).

If $f > 2\hat{f}$, then the Cochran-Haber-corrected chi-square (χ_H^2) is the same as the chi-square with the Yates correction (χ_c^2). Also, if either $C_1 = C_2$ or $R_a = R_2$, then $\chi_H^2 = \chi_c^2$.

A great deal has been written about 2×2 contingency-table testing.* For example, it has been reported that the power of chi-square testing increases with larger n or with more similarity between the two totals in a margin, and that the difference between results using chi-square and a continuity-corrected chi-square is less for large n.

In addition, many authors have reported that, for 2×2 tables having no fixed margin or only one fixed margin, χ_c^2 provides a test that is very, very conservative (that is, the probability of a Type I error is far less than that indicated by referring to the theoretical chi-square distribution—such as in Table 1 from *Appendix: Statistical Tables and Graphs*), with relatively low power; and they recommend that it should not be used for such sets of data. The use of χ^2 instead of χ_c^2 will occasionally result in a test that is somewhat liberal (i.e., the probability of a Type I error is a little greater than that indicated by the chi-square distribution, though it will typically be closer to the latter distribution than χ_c^2 will be); this liberalism is more pronounced when the two row totals are very different or the two column totals are very different.

For many decades there has been debate and disagreement over the appropriate statistical procedure for each of the aforementioned three sampling models for data in a 2×2 contingency table, with arguments presented on both theoretical and empirical grounds. There is still no consensus, and some believe there never will be,[†] but there is significant agreement on the following:

- If the 2×2 table has no margin fixed or only one margin fixed, then use χ^2. This is demonstrated in Examples 2b and 3b.
- If the 2×2 table has both margins fixed, then use χ_c^2 or χ_H^2, as demonstrated in Example 4, or use the Fisher exact test. As noted after Equation 7b, there are situations in which χ_c^2 and χ_H^2 are equal; otherwise, χ_H^2 is routinely a better approximation of the Fisher exact test and is preferred to χ_c^2.

Computer software may present χ^2 or a continuity-corrected χ_c^2, or both, and the user must decide which one of these two test statistics to use (such as by the guidelines just given).

---

*This paragraph and the next are a summary of the findings in many publications, such as those cited in the footnote that follows this one.

[†]Those promoting the analysis of any of the three models by using chi-square with the Yates correction for continuity (χ_c^2), or the Fisher exact test, include Camilli (1990), Cox (1984), Fisher (1935), Kendall and Stuart (1979), Martín Andrés (1991), Mehta and Hilton (1993), Upton (1992), and Yates (1984). Among those concluding that χ_c^2 should not be employed for all three models are Barnard (1947, 1979); Berkson (1978); Camilli and Hopkins (1978); Conover (1974); D'Agostino, Chase, and Belanger (1988); Garside and Mack (1976); Grizzle (1967); Haber (1980, 1982, 1987, 1990); Haviland (1990); Kempthorne (1979); Kroll (1989); Liddell (1976); Parshall and Kromrey (1996); Pearson (1947); Plackett (1964); Richardson (1990, 1994); Starmer, Grizzle, and Sen (1974); Storer and Kim (1990); and Upton (1982). Other procedures for testing 2×2 tables have been proposed (e.g., see Martín Andrés and Silva Mato (1994); Martín Andrés and Tapia Garcia (2004); and Overall, Rhoades, and Starbuck (1987)).

(e) One-Tailed Testing. The preceding hypotheses are two-tailed, which is the typical situation. However, one-tailed hypotheses (where a one-tailed hypothesis is specified before data are collected) are possible for data in 2×2 tables. In Example 2, the hypotheses could have been stated as follows:

H_0: In the sampled population, the proportion of left-handed children is the same or greater for boys compared to girls.

H_A: In the sampled population, the proportion of left-handed children is less for boys than for girls.

If the direction of the difference in the sample is that indicated in the null hypothesis (i.e., if $f_{11}/C_1 \geq f_{12}/C_2$), then H_0 cannot be rejected and the one-tailed analysis proceeds no further. However, if the direction of the difference in the sample is not in the direction of the null hypothesis (as in Example 2, where $6/34 < 12/36$), then it can be asked whether that difference is likely to indicate a difference in that direction in the population. In this example, one-tailed hypotheses could also have been stated as follows:

H_0: In the sampled population, the proportion of boys is the same or less for left-handed compared to right-handed children.

H_A: In the sampled population, the proportion of boys is greater for left-handed than for right-handed children.

This would ask whether the sample proportion f_{11}/R_1 (namely, $6/18$) resulted from a population proportion less than or equal to the population proportion estimated by f_{21}/R_2 (i.e., $28/52$).

Consistent with the preceding recommendations for two-tailed hypothesis testing, the following can be advised for one-tailed testing: For 2×2 tables in which no margin or only one margin is fixed, test by using one-half of the chi-square probability (for example, employing the critical value $\chi^2_{0.10,1}$ for testing at $\alpha = 0.05$), by dividing the resultant P by 2, or by using one-tailed values for Z. For tables with two fixed margins, the Fisher exact test is the preferred method of analysis, though if $R_1 = R_2$ or $C_1 = C_2$, we may calculate χ^2_c or, preferably, χ^2_H, and proceed as indicated previously for situations with one fixed margin. If neither $R_1 = R_2$ nor $C_1 = C_2$, using χ^2_H or χ^2_c yields a very poor approximation to the one-tailed Fisher exact test and is not recommended.

4 CONTINGENCY TABLES WITH SMALL FREQUENCIES

We previously discussed bias in chi-square goodness-of-fit testing when expected frequencies are "too small." As with goodness-of-fit testing, for a long time many statisticians (e.g., Fisher, 1925b) advised that chi-square analysis of contingency tables be employed only if each of the expected frequencies was at least 5.0 — even after there was evidence that such analyses worked well with smaller frequencies (e.g., Cochran, 1952, 1954). The review and empirical analysis of Roscoe and Byars (1971) offer more useful guidelines. Although smaller sample sizes are likely to work well, a secure practice is to have the mean expected frequency be at least 6.0 when testing with α as small as 0.05, and at least 10.0 for $\alpha = 0.01$. Requiring an average expected frequency of at least 6 is typically less restrictive than stipulating that each $\hat{f}_{ij}$ be at least 5. Since the mean expected frequency is n/rc, the minimum sample size for testing at the 0.05

significance level should be at least $n = 6rc = 6(2)(4) = 48$ for a 2×4 contingency table (such as in Example 1) and at least $6(2)(2) = 24$ for a 2×2 table (as in Exercises 2, 3, 4, and 5).

If any of the expected frequencies are smaller than recommended, then one or more rows or columns containing an offensively low $\hat{f}_{ij}$ might be discarded, or rows or columns might be combined to result in $\hat{f}_{ij}$'s of sufficient magnitude. However, such practices are not routinely advised, for they disregard information that can be important to the hypothesis testing. When possible, it is better to repeat the experiment with a sufficiently large n to ensure large enough expected frequencies. Some propose employing the log-likelihood ratio of Section 7 as a test less affected than chi-square by low frequencies, but this is not universally suggested. If both margins are fixed in a 2×2 contingency table, then the Fisher exact test is highly recommended when frequencies are small.

5 HETEROGENEITY TESTING OF 2 × 2 TABLES

Testing for heterogeneity of replicate samples in goodness-of-fit analysis was discussed previously. An analogous procedure may be used with contingency-table data, as demonstrated in Example 5. Here, data set 1 is the data from Example 2, and each of the three other sets of data is a sample obtained by the same data-collection procedure for the purpose of testing the same hypothesis. Heterogeneity testing asks whether all four of the data sets are likely to have come from the same population of data. In this example, a calculation of χ^2 was done, as in Section 3a, for each of the four contingency tables; and H_0 was not rejected for any of the data sets. This failure to reject H_0 might reflect low power of the test due to small sample sizes, so it would be helpful to use the heterogeneity test to conclude whether it would be reasonable to combine the four sets of data and perform a more powerful test of H_0 with the pooled number of data.

EXAMPLE 5   A Heterogeneity Chi-Square Analysis of Four 2 × 2 Contingency Tables, Where Data Set 1 Is That of Example 2

(a) H_0: In the sampled population, handedness is independent of sex.
H_A: In the sampled population, handedness is not independent of sex.
$\alpha = 0.05$

Data Set 1

From the data of Example 2, $\chi^2 = 2.2523$, DF = 1, $0.10 < P < 0.25$.

Data Set 2

	Boys	Girls	Total
Left-handed	4	7	11
Right-handed	25	13	38
Total	29	20	49

$\chi^2 = 3.0578$, DF = 1, $0.05 < P < 0.10$

Data Set 3

	Boys	Girls	Total
Left-handed	7	10	17
Right-handed	27	18	45
Total	34	28	62

$\chi^2 = 1.7653, \ DF = 1, \ 0.10 < P < 0.25$

Data Set 4

	Boys	Girls	Total
Left-handed	4	7	11
Right-handed	22	14	36
Total	26	21	47

$\chi^2 = 2.0877, \ DF = 1, \ 0.10 < P < 0.25$

(b) H_0: The four samples are homogeneous.

H_A: The four samples are heterogeneous.

Data Sets 1–4 Pooled

	Boys	Girls	Total
Left-handed	21	36	57
Right-handed	102	69	171
Total	123	105	228

$\chi^2 = 8.9505$
$DF = 1$

χ^2 for Data Set 1:	2.2523	DF = 1
χ^2 for Data Set 2:	3.0578	DF = 1
χ^2 for Data Set 3:	1.7653	DF = 1
χ^2 for Data Set 4:	2.0877	DF = 1
Total chi-square:	9.1631	DF = 4
Chi-square of pooled data:	8.9505	DF = 1
Heterogeneity chi-square	0.2126	DF = 3

For heterogeneity testing (using $\chi^2 = 0.2126$):

$\chi^2_{0.05,3} = 7.815$.

Therefore, do not reject H_0.

$$0.975 < P < 0.99 \quad [P = 0.98]$$

(c) H_0: In the sampled population, handedness is independent of sex.

H_A: In the sampled population, handedness is not independent of sex.

$\alpha = 0.05$

	Data Sets 1–4 Pooled		
	Boys	*Girls*	*Total*
Left-handed	21	36	57
Right-handed	102	69	171
Total	123	105	228

$$\chi^2_{0.05,1} = 3.841$$

$$\chi^2 = 8.9505; \text{ therefore, reject } H_0.$$

$$0.001 < P < 0.005 \ [P = 0.0028]$$

In the test for heterogeneity, chi-square is calculated for each of the samples; these four separate χ^2 values are shown in Example 5a, along with the χ^2 for the contingency table formed by the four sets of data combined. The χ^2 values for the four separate contingency tables are them summed (to obtain what may be called a total chi-square, which is 9.1631), and the degrees of freedom for the four tables are also summed (to obtain a total DF, which is 4), as shown in Example 5b. The test for heterogeneity employs a chi-square value that is the absolute difference between the total chi-square and the chi-square from the table of combined data, with degrees of freedom that are the difference between the total degrees of freedom and the degrees of freedom from the table of combined data. In the present example, the heterogeneity χ^2 is 0.2126, with 3 degrees of freedom. That chi-square is associated with a probability much greater than 0.05, so H_0 is not rejected and it is concluded that the data of the four samples may be combined.

Example 5c considers the contingency table formed by combining the data of all four of the original tables and tests the same hypothesis of independence that was tested for each of the original tables. When the heterogeneity test fails to reject H_0, pooling of the data is generally desirable because it allows contingency-table analysis with a larger n.

Heterogeneity testing with 2×2 tables is performed without the chi-square correction for continuity, except when both margins are fixed, in which case χ^2_c is used for the combined data while χ^2 is used for all other steps in the analysis (Cochran, 1942; Lancaster, 1949). The heterogeneity test may also be performed for contingency tables with more than two rows or columns. To test for heterogeneity, the log-likelihood ratio, G (Section 7), may be used instead of χ^2.

6 SUBDIVIDING CONTINGENCY TABLES

In Example 1, the analysis of a 2×4 contingency table, it was concluded that there was a significant difference in human hair-color frequencies between males and females. Expressing the percent males and percent females in each column, as in Example 6a, and examining Figures 1 and 2 shows that the proportion of males in the blond column is prominently less than in the other columns. (Examining the data in this fashion can be helpful, although frequencies, not proportions, are used for the hypothesis test.)

EXAMPLE 6a   The Data of Example 1, Where for Each Hair Color the Percent Males and Percent Females Are Indicated

Sex	Hair color				Total
	Black	Brown	Blond	Red	
Male	32 (37%)	43 (40%)	16 (20%)	9 (36%)	100
Female	55 (63%)	65 (60%)	64 (80%)	16 (64%)	200
Total	87	108	80	25	300

In Example 1, the null hypothesis that the four hair colors are independent of sex was rejected.

Thus, it might be suspected that the significant χ^2 calculated in Example 1 was due largely to the frequencies in column 3 of the table. To pursue that supposition, the data in column 3 may be momentarily ignored and the remaining 2×3 table considered; this is done in Example 6b. The nonsignificant χ^2 for this table supports the null hypothesis that these three hair colors are independent of sex in the population from which the sample came. Then, in Example 6c, a 2×2 table is formed by considering blond versus all other hair colors combined. For this table, the null hypothesis of independence is rejected.

EXAMPLE 6b   The 2×3 Contingency Table Formed from Columns 1, 2, and 4 of the Original 2×4 Table. $\hat{f}_{ij}$ Values for the Cells of the 2×3 Table Are Shown in Parentheses

H_0: The occurrence of black, brown, and red hair is independent of sex.
H_A: The occurrence of black, brown, and red hair is not independent of sex.

$\alpha = 0.05$

Sex	Hair color			Total
	Black	Brown	Red	
Male	32 (33.2182)	43 (41.2364)	9 (9.5455)	84
Female	55 (53.7818)	65 (66.7636)	16 (15.4545)	136
Total	87	108	25	220

$\chi^2 = 0.245$ with DF $= 2$

$\chi^2_{0.05,2} = 5.991$

Therefore, do not reject H_0.

$$0.75 < P < 0.90 \quad [P = 0.88]$$

EXAMPLE 6c **The 2 x 2 Contingency Table Formed by Combining Columns 1, 2, and 4 of the Original Table**

H_0: Occurrence of blond and nonblond hair color is independent of sex.

H_A: Occurrence of blond and nonblond hair color is not independent of sex.

$\alpha = 0.05$

	Hair color		
Sex	*Blond*	*Nonblond*	**Total**
Male	16	84	100
Female	64	136	200
Total	80	220	300

$\chi^2 = 8.727$

DF $= 1$

$\chi^2_{0.05,1} = 3.841$

Therefore, reject H_0.

$$0.001 < P < 0.005 \quad [P = 0.0036]$$

By the described series of subdivisions and column combinations of the original contingency table, we see evidence suggesting that, among the four hair colors in the population, blond occurs between the sexes with relative frequencies different from those of the other colors. However, it is not strictly proper to test statistical hypotheses developed after examining the data to be tested. Therefore, the analysis of a subdivided contingency table should be considered only as a guide to developing hypotheses. Hypotheses suggested by this analysis then can be tested by obtaining a new set of data from the population of interest and stating those hypotheses in advance of the testing.

7 THE LOG-LIKELIHOOD RATIO FOR CONTINGENCY TABLES

The G statistic (sometimes called G^2) was presented as an alternative to chi-square for goodness-of-fit testing. The G test may also be applied to contingency tables

Contingency Tables

(Neyman and Pearson, 1928a, 1928b; Wilks, 1935), where

$$G = 2 \left[\sum_i \sum_j f_{ij} \ln \left(\frac{f_{ij}}{\hat{f}_{ij}} \right) \right], \tag{8}$$

which, without the necessity of calculating expected frequencies, may readily be computed as

$$G = 2 \left[\sum_i \sum_j f_{ij} \ln f_{ij} - \sum_i R_i \ln R_i - \sum_j C_j \ln C_j + n \ln n \right]. \tag{9}$$

If common logarithms (denoted by "log") are used instead of natural logarithms (indicated as "ln"), then use 4.60517 instead of 2 prior to the left bracket. Because G is approximately distributed as χ^2, Table 1 from *Appendix: Statistical Tables and Graphs* may be used with $(r - 1)(c - 1)$ degrees of freedom. In Example 7, the contingency table of Example 1 is analyzed using the G statistic, with very similar results.

EXAMPLE 7   The G Test for the Contingency Table Data of Example 1

H_0:   Hair color is independent of sex.

H_A: Hair color is not independent of sex.

$\alpha = 0.05$

	Hair color				
Sex	*Black*	*Brown*	*Blond*	*Red*	**Total**
Male	32	43	16	9	100
Female	55	65	64	16	200
Total	87	108	80	25	300

$G = 4.60517 \left[\sum \sum f_{ij} \log f_{ij} - \sum R_i \log R_i - \sum C_j \log C_j + n \log n \right]$

$\quad = 4.60517[(32)(1.50515) + (43)(1.63347) + (16)(1.20412) + (9)(0.95424)$

$\quad\quad + (55)(1.74036) + (65)(1.81291) + (64)(1.80618) + (16)(1.20412)$

$\quad\quad - (100)(2.00000) - (200)(2.30103) - (87)(1.93952)$

$\quad\quad - (108)(2.03342) - (80)(1.90309) - (25)(1.39794) + (300)(2.47712)]$

$\quad = 4.60517(2.06518)$

$\quad = 9.510$ with DF $= 3$

$\chi^2_{0.05,3} = 7.815$

Therefore, reject H_0.

$$0.01 < P < 0.025 \quad [P = 0.023]$$

In the case of a 2×2 table, the Yates correction for continuity (see Sections 3c and 3d) is applied by making each f_{ij} 0.5 closer to $\hat{f}_{ij}$. This may be accomplished (without calculating expected frequencies) as follows: If $f_{11}f_{22} - f_{12}f_{21}$ is negative, add 0.5 to f_{11} and f_{22} and subtract 0.5 from f_{12} and f_{21}; if $f_{11}f_{22} - f_{12}f_{21}$ is positive, subtract 0.5 from f_{11} and f_{22} and add 0.5 to f_{12} and f_{21}; then Equation 8 or 9 is applied using these modified values of f_{11}, f_{12}, f_{21}, and f_{22}.

Williams (1976) recommended that G be used in preference to χ^2 whenever $| f_{ij} - \hat{f}_{ij} | \geq \hat{f}_{ij}$ for any cell. Both χ^2 and G commonly result in the same conclusion for the hypothesis test, especially when n is large. When they do not, some statisticians favor employing G, and its use is found in some research reports and computer software. However, many others (e.g., Agresti, 2002: 24, 396; Agresti and Yang, 1987; Berry and Mielke, 1988; Hosmane, 1986; Hutchinson, 1979; Koehler, 1986; Larntz, 1978; Margolin and Light, 1974; Stelzl, 2000; Upton, 1982) have concluded that the χ^2 procedure is preferable to G; and generally it more closely refers to the probability of a Type I error.

8 MULTIDIMENSIONAL CONTINGENCY TABLES

Thus far, this chapter has considered two-dimensional contingency tables (tables with rows and columns as the two dimensions), where each of the two dimensions represents a nominal-scale variable. However, categorical data may also be collected and tabulated with respect to three or more nominal-scale variables, resulting in what are called multidimensional contingency tables — that is, tables with three or more dimensions (e.g., see Christensen, 1990; Everitt, 1992: Chapter 4; Fienberg, 1970, 1980; Goodman, 1970; Simonoff, 2003: Chapter 8). An example would be data from a study similar to that in Example 1, but where eye color is a third variable — in addition to the variables hair color and sex.

As the number of dimensions increases, so does the complexity of the analysis, and various interactions of variables are potentially of interest. Multidimensional contingency tables may be analyzed by extensions of the χ^2 and G testing discussed earlier in this chapter, as will be indicated in this section. Computer-program libraries often include provision for the analysis of such tables, including by utilizing what are known as *log-linear models*,* a large body of statistical procedures (e.g., see Everitt, 1992: Chapter 5; Fienberg, 1970, 1980; Howell, 2007: Chapter 17; Kennedy, 1992; Knoke and Burke, 1980; Tabachnik and Fidell, 2001: Chapter 7).

Figure 3 shows a three-dimensional contingency table. The three "rows" are species, the four "columns" are geographic locations, and the two "tiers" (or "layers") are presence and absence of a disease. If a sample is obtained containing individuals of these species, from these locations, and with and without the disease in question, then observed frequencies can be recorded in the 24 cells of this $3 \times 4 \times 2$ contingency table. We shall refer to the observed frequency in row i, column j, and tier l as f_{ijl}. We shall refer to the number of rows, columns, and tiers as r, c, and t, respectively. The sum of the frequencies in row i will be designated R_i, the sum in column j as C_j, and the sum in tier l as T_l. Friendly (1994, 1999), Hartigan and Kleiner (1981, 1984), and

---

*Log-linear models are mathematical representations that also underlie analysis of variance and multiple regression. The term *log-linear model* was introduced in 1969 by Y. M. M. Bishop and S. E. Fienberg (David, 1995).

Contingency Tables

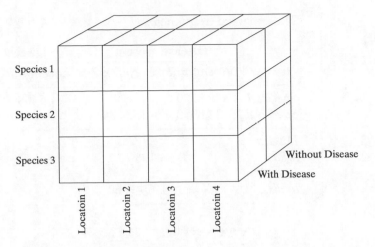

FIGURE 3: A three-dimensional contingency table, where the three rows are species, the four columns are locations, and the two tiers are occurrence of a disease. An observed frequency, f_{ijl}, will be recorded in each combination of row, column, and tier.

Simonoff (2003: 329) discuss mosaic displays for contingency tables with more than two dimensions, and such graphical presentations can make multidimensional contingency table data easier to visualize and interpret than if they are presented only in tabular format.

Example 8 presents a $2 \times 2 \times 2$ contingency table where data (f_{ijl}) are collected as described previously, but only for two species and two locations. Note that throughout the following discussions the sum of the expected frequencies for a given row, column, or tier equals the sum of the observed frequencies for that row, column, or tier.

EXAMPLE 8   Test for Mutual Independence in a 2 x 2 x 2 Contingency Table

H_0: Disease occurrence, species, and location are all mutually independent in the population sampled.

H_A: Disease occurrence, species, and location are not all mutually independent in the population sampled.

The observed frequencies (f_{ijl}):

	Disease present		**Disease absent**		**Species totals**
	Location 1	*Location 2*	*Location 1*	*Location 2*	($r = 2$)
Species 1	44	12	38	10	$R_1 = 104$
Species 2	28	22	20	18	$R_2 = 88$
Disease totals ($t = 2$):	$T_1 = 106$		$T_2 = 86$		*Grand total:*
Location totals: ($c = 2$):	$C_1 = 130, C_2 = 62$				$n = 192$

The expected frequencies $(\hat{f}_{ijl})$:

	Disease present		Disease absent		
	Location 1	Location 2	Location 1	Location 2	Species totals
Species 1	38.8759	18.5408	31.5408	15.0425	$R_1 = 104$
Species 2	32.8950	15.6884	26.6884	12.7283	$R_2 = 88$
Disease totals:	$T_1 = 106$		$T_2 = 86$		Grand total:
Location totals:	$C_1 = 130, C_2 = 62$				$n = 192$

$$\chi^2 = \sum \sum \sum \frac{(f_{ijl} - \hat{f}_{ijl})^2}{\hat{f}_{ijl}}$$

$$\chi^2 = \frac{(44 - 38.8759)^2}{38.8759} + \frac{(12 - 18.5408)^2}{18.5408} + \frac{(38 - 31.5408)^2}{31.5408}$$

$$+ \frac{(10 - 15.0425)^2}{15.0425} + \frac{(28 - 32.8950)^2}{32.8950} + \frac{(22 - 15.6884)^2}{15.6884}$$

$$+ \frac{(20 - 26.6884)^2}{26.6884} + \frac{(18 - 12.7283)^2}{12.7283}$$

$$= 0.6754 + 2.3075 + 1.3228 + 1.6903 + 0.7284 + 2.5392$$

$$+ 1.6762 + 2.1834$$

$$= 13.123$$

$$\nu = rct - r - c - t + 2 = (2)(2)(2) - 2 - 2 - 2 + 2 = 4$$

$$\chi^2_{0.05,4} = 9.488$$

Reject H_0.

$$0.01 < P < 0.025 \quad [P = 0.011]$$

(a) Mutual Independence. We can test more than one null hypothesis using multidimensional contingency-table data. An overall kind of hypothesis is that which states mutual independence among all the variables. Another way of expressing this H_0 is that there are no interactions (either three-way or two-way) among any of the variables. For this hypothesis, the expected frequency in row i, column j, and tier l is

$$\hat{f}_{ijl} = \frac{R_i C_j T_l}{n^2}, \tag{10}$$

where n is the total of all the frequencies in the entire contingency table.

In Example 8 this null hypothesis would imply that presence or absence of the disease occurred independently of species and location. For three dimensions, this null hypothesis is tested by computing

$$\chi^2 = \sum_{i=1}^{r}\sum_{j=1}^{c}\sum_{l=1}^{t} \frac{(f_{ijl} - \hat{f}_{ijl})^2}{\hat{f}_{ijl}}, \tag{11}$$

which is a simple extension of the chi-square calculation for a two-dimensional table (by Equation 1). The degrees of freedom for this test are the sums of the degrees of freedom for all interactions:

$$\nu = (r-1)(c-1)(t-1) + (r-1)(c-1) + (r-1)(t-1) + (c-1)(t-1), \tag{12}$$

which is equivalent to

$$\nu = rct - r - c - t + 2. \tag{13}$$

(b) Partial Independence.   If the preceding null hypothesis is not rejected, then we conclude that all three variables are mutually independent and the analysis proceeds no further. If, however, H_0 is rejected, then we may test further to conclude between which variables dependencies and independencies exist. For example, we may test whether one of the three variables is independent of the other two, a situation known as *partial independence.*[*]

For the hypothesis of rows being independent of columns and tiers, we need total frequencies for rows and total frequencies for combinations of columns and tiers. Designating the total frequency in column j and tier l as $(CT)_{jl}$, expected frequencies are calculated as

$$\hat{f}_{ijl} = \frac{R_i(CT)_{jl}}{n}, \tag{14}$$

and Equation 11 is used with degrees of freedom

$$\nu = (r-1)(c-1)(t-1) + (r-1)(c-1) + (r-1)(t-1), \tag{15}$$

which is equivalent to

$$\nu = rct - ct - r + 1. \tag{16}$$

For the null hypothesis of columns being independent of rows and tiers, we compute expected frequencies using column totals, C_j, and the totals for row and tier combinations, $(RT)_{il}$:

$$\hat{f}_{ijl} = \frac{C_j(RT)_{il}}{n}, \tag{17}$$

and

$$\nu = rct - rt - c + 1. \tag{18}$$

And, for the null hypothesis of tiers being independent of rows and columns, we use tier totals, T_l, and the totals for row and column combinations, $(RC)_{ij}$:

$$\hat{f}_{ijl} = \frac{T_l(RC)_{ij}}{n}; \tag{19}$$

$$\nu = rct - rc - t + 1. \tag{20}$$

---

[*] A different hypothesis is that of *conditional independence*, where two of the variables are said to be independent in each level of the third (but each may have dependence on the third). This is discussed in the references cited at the beginning of this section.

In Example 9, all three pairs of hypotheses for partial independence are tested. In one of the three (the last), H_0 is not rejected; thus we conclude that presence of disease is independent of species and location. However, the hypothesis test of Example 8 concluded that all three variables are not independent of each other. Therefore, we suspect that species and location are not independent. The independence of these two variables may be tested using a two-dimensional contingency table, as described earlier, in Section 3, and demonstrated in Example 10. In the present case, the species-location interaction is tested by way of a 2×2 contingency table, and we conclude that these two factors are not independent (i.e., species occurrence depends on geographic location).

In general, hypotheses to be tested should be stated before the data are collected. But the hypotheses proposed in Example 10 were suggested *after* the data were examined. Therefore, instead of accepting the present conclusion of the analysis in Example 10, such a conclusion should be reached by testing this pair of hypotheses upon obtaining a new set of data from the population of interest and stating the hypotheses in advance of the testing.

EXAMPLE 9   Test for Partial Independence in a 2 x 2 x 2 Contingency Table. As the H_0 of Overall Independence Was Rejected in Example 8, We May Test the Following Three Pairs of Hypotheses

H_0: Species is independent of location and disease.

H_A: Species is not independent of location and disease.

The expected frequencies ($\hat{f}_{ijl}$):

| | **Disease present** | | **Disease absent** | | |
	Location 1	*Location 2*	*Location 1*	*Location 2*	**Species totals**
Species 1	39.0000	18.4167	31.4167	15.1667	$R_1 = 104$
Species 2	33.0000	15.5833	26.5833	12.8333	$R_2 = 88$
Location and disease totals:	$(CT)_{11}$ $= 72$	$(CT)_{12}$ $= 34$	$(CT)_{21}$ $= 58$	$(CT)_{22}$ $= 28$	*Grand total:* $n = 192$

$$\chi^2 = \frac{(44 - 39.0000)^2}{39.0000} + \frac{(12 - 18.4167)^2}{18.4167} + \frac{(38 - 31.4167)^2}{31.4167}$$

$$+ \cdots + \frac{(18 - 12.8333)^2}{12.8333}$$

$$= 0.6410 + 2.2357 + 1.3795 + 1.7601 + 0.7576 + 2.6422$$

$$+ 1.6303 + 2.0801$$

$$= 13.126$$

$$\nu = rct - ct - r + 1 = (2)(2)(2) - (2)(2) - 2 + 1 = 3$$
$$\chi^2_{0.05,3} = 7.815$$

Reject H_0. Species is not independent of location and presence of disease.

$$0.005 < P < 0.001 \quad [P = 0.0044]$$

H_0: Location is independent of species and disease.

H_A: Location is not independent of species and disease.

The expected frequencies ($\hat{f}_{ijl}$):

	Disease present		Disease absent		
	Species 1	Species 2	Species 1	Species 2	Location totals
Location 1	37.91677	33.8542	32.5000	25.7292	$C_1 = 130$
Location 2	18.0833	16.1458	15.5000	12.2708	$C_2 = 62$
Species and disease totals:	$(RT)_{11}$ $= 56$	$(RT)_{12}$ $= 50$	$(RT)_{21}$ $= 48$	$(RT)_{22}$ $= 38$	Grand total: $n = 192$

$$\chi^2 = \frac{(44 - 37.9167)^2}{37.9167} + \frac{(28 - 33.8542)^2}{33.8542} + \cdots + \frac{(18 - 12.2708)^2}{12.2708}$$

$$= 0.9760 + 1.0123 + 0.9308 + 1.2757 + 2.0464 + 2.1226$$

$$+ 1.9516 + 2.6749$$

$$= 12.990$$

$$\nu = rct - rt - c + 1 = (2)(2)(2) - (2)(2) - 2 + 1 = 3$$

$$\chi^2_{0.05,3} = 7.815$$

Reject H_0. Location is not independent of species and presence of disease.

$$0.001 < P < 0.005 \quad [P = 0.0047]$$

H_0: Presence of disease is independent of species and location.

H_A: Presence of disease is not independent of species and location.

The expected frequencies ($\hat{f}_{ijl}$):

	Species 1		Species 2		
	Location 1	Location 2	Location 1	Location 2	Disease totals
Disease present	45.2708	12.1458	26.5000	22.0833	$T_1 = 106$
Disease absent	36.7292	9.8542	21.5000	17.9167	$T_2 = 86$
Species and location totals:	$(RC)_{11}$ $= 82$	$(RC)_{12}$ $= 22$	$(RC)_{21}$ $= 48$	$(RC)_{22}$ $= 40$	Grand total: $n = 192$

$$\chi^2 = \frac{(44 - 45.2708)^2}{45.2708} + \frac{(12 - 12.1458)^2}{12.1458} + \cdots + \frac{(18 - 17.9167)^2}{17.9167}$$

$$= 0.0357 + 0.0018 + 0.0849 + 0.0003 + 0.0440 + 0.0022$$

$$+ 0.1047 + 0.0004$$

$$= 0.274$$

$$\nu = rct - rc - t + 1 = (2)(2)(2) - (2)(2) - 2 + 1 = 3$$
$$\chi^2_{0.05,3} = 7.815$$
Do not reject H_0.

$$0.95 < P < 0.975 \quad [P = 0.96]$$

EXAMPLE 10   Test for Independence of Two Variables, Following Tests for Partial Dependence

The hypothesis test of Example 8 concluded that all three variables are not mutually independent, while the last test in Example 9 concluded that presence of disease is independent of species and location. Therefore, it is desirable (and permissible) to test the following two-dimensional contingency table:

H_0: Species occurrence is independent of location.

H_A: Species occurrence is not independent of location.

	Location 1	Location 2	Total
Species 1	82	22	104
Species 2	48	40	88
Total	130	62	192

$$\chi^2 = 12.874$$
$$\nu = (r - 1)(c - 1) = 1$$
$$\chi^2_{0.05,1} = 3.841$$
Reject H_0.

$$P < 0.001 \quad [P = 0.00033]$$

(c) The Log-Likelihood Ratio. The log-likelihood ratio of Section 7 can be expanded to contingency tables with more than two dimensions. While some authors have chosen this procedure over chi-square testing and it is found in some statistical computer packages, others (e.g., Haber, 1984; Hosmane, 1987; Koehler, 1986; Larntz, 1978; Rudas, 1986; and Stelzl, 2000) have concluded that χ^2 is preferable. With χ^2 in contrast to G, the probability of a Type I error is generally closer to α.

EXERCISES

1. Consider the following data for the abundance of a certain species of bird.
  (a) Using chi-square, test the null hypothesis that the ratio of numbers of males to females was the same in all four seasons.
  (b) Apply the G test to that hypothesis.

Sex	Spring	Summer	Fall	Winter
Males	163	135	71	43
Females	86	77	40	38

2. The following data are frequencies of skunks found with and without rabies in two different geographic areas.
  (a) Using chi-square, test the null hypothesis that the incidence of rabies in skunks is the same in both areas.
  (b) Apply the G test to that hypothesis.

Area	With rabies	Without rabies
E	14	29
W	12	38

3. Data were collected as in Exercise 2, but with the additional tabulation of the sex of each skunk recorded, as follows. Test for mutual independence; and, if H_0 is rejected, test for partial independence.

	With rabies		Without rabies	
Area	Male	Female	Male	Female
E	42	33	55	63
W	84	51	34	48

4. A sample of 150 was obtained of men with each of three types of cancer, and the following data are the frequencies of blood types for the men.
  (a) Using chi-square, test the null hypothesis that, in the sampled population, the frequency distribution of the three kinds of cancer is the same for men with each of the four blood types (which is the same as testing the H_0 that the frequency distribution of the four blood types is the same in men with each of the three kinds of cancer).
  (b) Apply the G test to the same hypothesis.

	Blood type				
Cancer type	O	A	B	AB	Total
Colon	61	65	18	6	$R_1 = 150$
Lung	69	57	15	9	$R_2 = 150$
Prostate	73	60	12	5	$R_3 = 150$

ANSWERS TO EXERCISES

1. (a) $\hat{f}_{11} = 157.1026, \hat{f}_{12} = 133.7580, \hat{f}_{13} = 70.0337, \hat{f}_{14} = 51.1057, \hat{f}_{21} = 91.8974, \hat{f}_{22} = 78.2420, \hat{f}_{23} = 40.9663, \hat{f}_{24} = 29.8943, R_1 = 412, R_2 = 241, C_1 = 249, C_2 = 212, C_3 = 111, C_4 = 81, n = 653;$ $\chi^2 = 0.2214 + 0.0115 + 0.0133 + 1.2856 + 0.3785 + 0.0197 + 0.0228 + 2.1978 = 4.151;$ $\nu = (2 - 1)(4 - 1) = 3; \chi^2_{0.05,3} = 7.815, 0.10 < P(\chi^2 \geq 4.156) < 0.25 [P = 0.246]; P > 0.05,$ do not reject H_0. **(b)** $G = 4.032, \nu = 3, \chi^2_{0.05,3} = 7.815, 0.25 < P(\chi^2 \geq 4.032) < 0.50 [P = 0.26]; P > 0.05,$ do not reject H_0.

2. (a) $f_{11} = 14, f_{12} = 29, f_{21} = 12, f_{22} = 38, R_1 = 43, R_2 = 50, C_1 = 26, C_2 = 67, n = 93; \chi^2 = 0.8407, \nu = 1; \chi^2_{0.05,1} = 3.841, 0.25 < P(\chi^2 \geq 0.8407) < 0.50;$ as $P > 0.05,$ do not reject $H_0 [P = 0.36].$ **(b)** $G = 0.8395, \nu = 1; \chi^2_{0.05,1} = 3.841, 0.25 < P(\chi^2 \geq 0.8395) < 0.50;$ as $P > 0.05,$ do not reject $H_0 [P = 0.36].$

3. H_0: Sex, area, and occurrence of rabies are mutually independent; $\chi^2 = 33.959; \nu = 4; \chi^2_{0.05,4} = 9.488;$ reject $H_0; P < 0.001 [P = 0.00000076].$ H_0: Area is independent of sex and rabies; $\chi^2 = 23.515; \nu = 3; \chi^2_{0.05,3} = 7.815;$ reject $H_0; P < 0.001 [P = 0.000032].$ H_0: Sex is independent of area and rabies; $\chi^2 = 11.130; \nu = 3;$ reject $H_0; 0.01 < P < 0.25 [P = 0.011].$ H_0: Rabies is independent of area and sex; $\chi^2 = 32.170; \nu = 3;$ reject $H_0; P < 0.001 [P = 0.00000048].$

4. (a) $R_1 = R_2 = R_3 = 150, C_1 = 203, C_2 = 182, C_3 = 45, C_4 = 20, n = 450; \hat{f}_{11} = \hat{f}_{21} = \hat{f}_{31} = 67.6667, \hat{f}_{12} = \hat{f}_{22} = \hat{f}_{32} = 60.6667, \hat{f}_{13} = \hat{f}_{23} = \hat{f}_{33} = 15.0000, \hat{f}_{14} = \hat{f}_{24} = \hat{f}_{34} = 6.6667; \chi^2 = 4.141; \nu = (2)(3) = 6; \chi^2_{0.05,6} = 12.592; 0.50 < P < 0.75 [P = 0.66];$ do not reject H_0. **(b)** $G = 4.141,$ same probability and conclusion as part (a).

This page intentionally left blank

Dichotomous Variables

From Chapter 24 of *Biostatistical Analysis*, Fifth Edition, Jerrold H. Zar. Copyright © 2010 by
Pearson Education, Inc. Publishing as Pearson Prentice Hall. All rights reserved.

Dichotomous Variables

This chapter will concentrate on nominal-scale data that come from a population with only two categories. As examples, members of a mammal litter might be classified as male or female, victims of a disease as dead or alive, trees in an area as "deciduous" or "evergreen," or progeny as color-blind or not color-blind. A nominal-scale variable having two categories is said to be *dichotomous*. Such variables have already been discussed in the context of goodness of fit and contingency tables.

The proportion of the population belonging to one of the two categories is denoted as p (here departing from the convention of using Greek letters for population parameters). Therefore, the proportion of the population belonging to the second class is $1 - p$, and the notation $q = 1 - p$ is commonly employed. For example, if 0.5 (i.e., 50%) of a population were male, then we would know that 0.5 (i.e., $1 - 0.5$) of the population were female, and we could write $p = 0.5$ and $q = 0.5$; if 0.4 (i.e., 40%) of a population were male, then 0.6 (i.e., 60%) of the population were female, and we could write $p = 0.4$ and $q = 0.6$.

If we took a random sample of ten from a population where $p = q = 0.5$, then we might expect that the sample would consist of five males and five females. However, we should not be too surprised to find such a sample consisting of six males and four females, or four males and six females, although neither of these combinations would be expected with as great a frequency as samples possessing the population sex ratio of 5 : 5. It would, in fact, be possible to obtain a sample of ten with nine males and one female, or even one consisting of all males, but the probabilities of such samples being encountered by random chance are relatively low.

If we were to obtain a large number of samples from the population under consideration, the frequency of samples consisting of no males, one male, two males, and so on would be described by the *binomial distribution* (sometimes referred to as the "Bernoulli distribution"*). Let us now examine binomial probabilities.

1 BINOMIAL PROBABILITIES

Consider a population consisting of two categories, where p is the proportion of individuals in one of the categories and $q = 1 - p$ is the proportion in the other. Then the probability of selecting at random from this population a member of the first category is p, and the probability of selecting a member of the second category is q.[†]

For example, let us say we have a population of female and male animals, in proportions of $p = 0.4$ and $q = 0.6$, respectively, and we take a random sample of two individuals from the population. The probability of the first being a female is p (i.e., 0.4) and the probability of the second being a female is also p. As the probability of two independent (i.e., mutually exclusive) events both occurring is the product of the probabilities of the two separate events, the probability of having two females in a sample of two is $(p)(p) = p^2 = 0.16$; the probability of the sample of two consisting of two males is $(q)(q) = q^2 = 0.36$.

What is the probability of the sample of two consisting of one male and one female? This could occur by the first individual being a female and the second a male (with a probability of pq) or by the first being a male and the second a female (which would occur with a probability of qp). The probability of either of two mutually exclusive outcomes is the sum of the probabilities of each outcome, so the probability of one female and one male in the sample is $pq + qp = 2pq = 2(0.4)(0.6) = 0.48$. Note that $0.16 + 0.36 + 0.48 = 1.00$.

Now consider another sample from this population, one where $n = 3$. The probability of all three individuals being female is $ppp = p^3 = (0.4)^3 = 0.064$. The probability of two females and one male is ppq (for a sequence of ♀♀♂) + pqp (for ♀♂♀) + qpp (for ♂♀♀), or $3p^2q = 3(0.4)^2(0.6) = 0.288$. The probability of one female and two males is pqq (for ♀♂♂) + qpq (for ♂♀♂) + qqp (for ♂♂♀), or $3pq^2 = 3(0.4)(0.6)^2 = 0.432$. And, finally, the probability of all three being males is $qqq = q^3 = (0.6)^3 = 0.216$. Note that $p^3 + 3p^2q + 3pq^2 + q^3 = 0.064 + 0.288 + 0.432 + 0.216 = 1.000$ (meaning that there is a 100% probability — that is, it is certain — that the three animals will be in one of these three combinations of sexes).

If we performed the same exercise with $n = 4$, we would find that the probability of four females is $p^4 = (0.4)^4 = 0.0256$, the probability of three females (and one male) is $4p^3q = 4(0.4)^3(0.6) = 0.1536$, the probability of two females is $6p^2q^2 = 0.3456$,

---

*The binomial formula in the following section was first described, in 1676, by English scientist-mathematician Sir Isaac Newton (1642–1727), more than 10 years after he discovered it (Gullberg, 1997: 776). Its first proof, for positive integer exponents, was given by the Swiss mathematician Jacques (also known as Jacob, Jakob, or James) Bernoulli (1654–1705), in a 1713 posthumous publication; thus, each observed event from a binomial distribution is sometimes called a Bernoulli trial. Jacques Bernoulli's nephew, Nicholas Bernoulli (1687–1759), is given credit for editing that publication and writing a preface for it, but Hald (1984) explains that in 1713 Nicholas also presented an improvement to his uncle's binomial theorem. David (1995) attributes the first use of the term *binomial distribution* to G. U. Yule, in 1911.

[†]This assumes "sampling with replacement." That is, each individual in the sample is taken at random from the population and then is returned to the population before the next member of the sample is selected at random. Sampling without replacement is discussed in Section 2. If the population is very large compared to the size of the sample, then sampling with and without replacement are indistinguishable in practice.

the probability of one female is $4pq^3 = 0.3456$, and the probability of no females (i.e., all four are male) is $q^4 = 0.1296$. (The sum of these five terms is 1.0000, a good arithmetic check.)

If a random sample of size n is taken from a binomial population, then the probability of X individuals being in one category (and, therefore, $n - X$ individuals in the second category) is

$$P(X) = \binom{n}{X} p^X q^{n-X}. \tag{1}$$

In this equation, $p^X q^{n-X}$ refers to the probability of sample consisting of X items, each having a probability of p, and $n - X$ items, each with probability q. The *binomial coefficient*,

$$\binom{n}{X} = \frac{n!}{X!(n - X)!}, \tag{2}$$

is the number of ways X items of one kind can be arranged with $n - X$ items of a second kind, or, in other words, it is $_nC_X$, the number of possible *combinations* of n items divided into one group of X items and a second group of $n - X$ items. Therefore, Equation 1 can be written as

$$P(X) = \frac{n!}{X!(n - X)!} p^X q^{n-X}. \tag{3}$$

Thus, $\binom{n}{X} p^X q^{n-X}$ is the Xth term in the expansion of $(p + q)^n$, and Table 1 shows this expansion for powers up through 6. Note that for any power, n, the sum of the two exponents in any term is n. Furthermore, the first term will always be p^n, the second will always contain $p^{n-1}q$, the third will always contain $p^{n-2}q^2$, and so on, with the last term always being q^n. The sum of all the terms in a binomial expansion will always be 1.0, for $p + q = 1$, and $(p + q)^n = 1^n = 1$.

As for the coefficients of these terms in the binomial expansion, the Xth term of the nth power expansion can be calculated by Equation 3. Furthermore, the examination of these coefficients as shown in Table 2 has been deemed interesting for centuries. This arrangement is known as *Pascal's triangle.** We can see from this triangular

TABLE 1: Expansion of the Binomial, $(p + q)^n$

n	$(p + q)^n$
1	$p + q$
2	$p^2 + 2pq + q^2$
3	$p^3 + 3p^2q + 3pq^2 + q^3$
4	$p^4 + 4p^3q + 6p^2q^2 + 4pq^3 + q^4$
5	$p^5 + 5p^4q + 10p^3q^2 + 10p^2q^3 + 5pq^4 + q^5$
6	$p^6 + 6p^5q + 15p^4q^2 + 20p^3q^3 + 15p^2q^4 + 6pq^5 - q^6$

---

*Blaise Pascal (1623–1662), French mathematician and physicist and one of the founders of probability theory (in 1654, immediately before abandoning mathematics to become a religious recluse). He had his triangular binomial coefficient derivation published in 1665, although knowledge of the triangular properties appears in Chinese writings as early as 1303 (Cajori, 1954; David, 1962; Gullberg 1997: 141; Struik, 1967: 79). Pascal also invented (at age 19) a mechanical adding and subtracting machine which, though patented in 1649, proved too expensive to be practical to construct (Asimov, 1982: 130–131). His significant contributions to the study of fluid pressures have been honored by naming the international unit of pressure the pascal, which is a pressure of one

TABLE 2: Binomial Coefficient, $_nC_X$

n	$X = 0$	1	2	3	4	5	6	7	8	9	10	Sum of coefficients
1	1	1										$2 = 2^1$
2	1	2	1									$4 = 2^2$
3	1	3	3	1								$8 = 2^3$
4	1	4	6	4	1							$16 = 2^4$
5	1	5	10	10	5	1						$32 = 2^5$
6	1	6	15	20	15	6	1					$64 = 2^6$
7	1	7	21	35	35	21	7	1				$128 = 2^7$
8	1	8	28	56	70	56	28	8	1			$256 = 2^8$
9	1	9	36	84	126	126	84	36	9	1		$512 = 2^9$
10	1	10	45	120	210	252	210	120	45	10	1	$1024 = 2^{10}$

array that any binomial coefficient is the sum of two coefficients on the line above it, namely,

$$\binom{n}{X} = \binom{n-1}{X-1} + \binom{n-1}{X}. \tag{4}$$

This can be more readily observed if we display the triangular array as follows:

```
            1
         1     1
       1    2     1
     1    3     3    1
   1   4    6     4    1
 1   5   10    10    5    1
```

Also note that the sum of all coefficients for the nth power binomial expansion is 2^n. Table 26a from *Appendix: Statistical Tables and Graphs* presents binomial coefficients for much larger n's and X's, and they will be found useful later in this chapter.

Thus, we can calculate probabilities of category frequencies occurring in random samples from a binomial population. If, for example, a sample of five (i.e., $n = 5$) is taken from a population composed of 50% males and 50% females (i.e., $p = 0.5$ and $q = 0.5$) then Example 1 shows how Equation 3 is used to determine the probability of the sample containing 0 males, 1 male, 2 males, 3 males, 4 males, and 5 males. These probabilities are found to be 0.03125, 0.15625, 0.31250, 0.31250, 0.15625, and 0.03125, respectively. This enables us to state that if we took 100 random samples of five animals each from the population, about three of the samples [i.e., $(0.03125)(100) = 3.125$ of them] would be expected to contain all females, about 16 [i.e., $(0.15625)(100) = 15.625$] to contain one male and four females, 31 [i.e., $(0.31250)(100)$] to consist of two males and three females, and so on. If we took 1400 random samples of five, then $(0.03125)(1400) = 43.75$ [i.e., about 44] of them would be expected to contain all females, and so on. Figure 1a shows graphically

newton per square meter (where a newton — named for Sir Isaac Newton — is the unit of force representing a one-kilogram mass accelerating at the rate of one meter per second per second). Pascal is also the name of a computer programming language developed in 1970 by Niklaus Wirth. The relationship of Pascal's triangle to $_nC_X$ was first published in 1685 by the English mathematician John Wallis (1616–1703) (David, 1962: 123–124).

EXAMPLE 1   Computing Binomial Probabilities, $P(X)$, Where $n = 5$, $p = 0.5$, and $q = 0.5$ (Following Equation 3)

X	$P(X)$
0	$\dfrac{5!}{0!5!}(0.5^0)(0.5^5) = (1)(1.0)(0.03125) = 0.03125$
1	$\dfrac{5!}{1!4!}(0.5^1)(0.5^4) = (5)(0.5)(0.0625) = 0.15625$
2	$\dfrac{5!}{2!3!}(0.5^2)(0.5^3) = (10)(0.25)(0.125) = 0.31250$
3	$\dfrac{5!}{3!2!}(0.5^3)(0.5^2) = (10)(0.125)(0.25) = 0.31250$
4	$\dfrac{5!}{4!1!}(0.5^4)(0.5^1) = (5)(0.0625)(0.5) = 0.15625$
5	$\dfrac{5!}{5!0!}(0.5^5)(0.5^0) = (1)(0.03125)(1.0) = 0.03125$

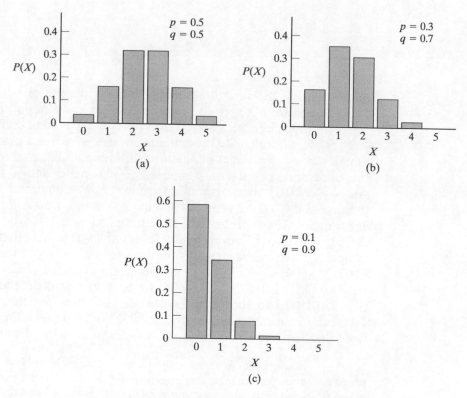

FIGURE 1: The binomial distribution, for $n = 5$. (a) $p = q = 0.5$. (b) $p = 0.3$, $q = 0.7$. (c) $p = 0.1$, $q = 0.9$. These graphs were drawn utilizing the proportions given by Equation 1.

the binomial distribution for $p = q = 0.5$, for $n = 5$. Note, from Figure 1a and Example 1, that when $p = q = 0.5$ the distribution is symmetrical [i.e., $P(0) = P(n)$, $P(1) = P(n - 1)$, etc.], and Equation 3 becomes

$$P(X) = \frac{n!}{X!(n - X)!} 0.5^n. \tag{5}$$

Table 26b from *Appendix: Statistical Tables and Graphs* gives binomial probabilities for $n = 2$ to $n = 20$, for $p = 0.5$.

Example 2 presents the calculation of binomial probabilities for the case where $n = 5$, $p = 0.3$, and $q = 1 - 0.3 = 0.7$. Thus, if we were sampling a population consisting of 30% males and 70% females, 0.16807 (i.e., 16.807%) of the samples would be expected to contain no males, 0.36015 to contain one male and four females, and so on. Figure 1b presents this binomial distribution graphically, whereas Figure 1c shows the distribution where $p = 0.1$ and $q = 0.9$.

EXAMPLE 2   Computing Binomial Probabilities, $P(X)$, Where $n = 5$, $p = 0.4$, and $q = 0.7$ (Following Equation 3)

X	$P(X)$
0	$\dfrac{5!}{0!5!}(0.3^0)(0.7^5) = (1)(1.0)(0.16807) = 0.16807$
1	$\dfrac{5!}{1!4!}(0.3^1)(0.7^4) = (5)(0.3)(0.2401) = 0.36015$
2	$\dfrac{5!}{2!3!}(0.3^2)(0.7^3) = (10)(0.09)(0.343) = 0.30870$
3	$\dfrac{5!}{3!2!}(0.3^3)(0.7^2) = (10)(0.027)(0.49) = 0.13230$
4	$\dfrac{5!}{4!1!}(0.3^4)(0.7^1) = (5)(0.0081)(0.7) = 0.02835$
5	$\dfrac{5!}{5!0!}(0.3^5)(0.7^0) = (1)(0.00243)(1.0) = 0.00243$

For calculating binomial probabilities for large n, it is often convenient to employ logarithms. For this reason, Table 40 from *Appendix: Statistical Tables and Graphs*, a table of logarithms of factorials, is provided. Alternatively, it is useful to note that the denominator of Equation 3 cancels out much of the numerator, so that it is possible to simplify the computation of $P(X)$, especially in the tails of the distribution (i.e., for low X and for high X), as shown in Example 3. If p is very small, then the use of the Poisson distribution should be considered.[*]

The mean of a binomial distribution of counts X is

$$\mu_X = np, \tag{6}$$

---

[*]Raff (1956) and Molenaar (1969a, 1969b) discuss several approximations to the binomial distribution, including the normal and Poisson distributions.

EXAMPLE 3   Computing Binomial Probabilities, $P(X)$, with $n = 400$, $p = 0.02$, and $q = 0.98$

(Many calculators can operate with large powers of numbers; otherwise, logarithms may be used.)

X	$P(X)$
0	$\dfrac{n!}{0!(n-0)!}p^0 q^{n-0} = q^n = 0.98^{400} = 0.00031$
1	$\dfrac{n!}{1!(n-1)!}p^1 q^{n-1} = npq^{n-1} = (400)(0.02)(0.98^{399}) = 0.00253$
2	$\dfrac{n!}{2!(n-2)!}p^2 q^{n-2} = \dfrac{n(n-1)}{2!}p^2 q^{n-2}$
	$= \dfrac{(400)(399)}{2}(0.02^2)(0.98^{398}) = 0.01028$
3	$\dfrac{n!}{3!(n-3)!}p^3 q^{n-3} = \dfrac{n(n-1)(n-2)}{3!}p^3 q^{n-3}$
	$= \dfrac{(400)(399)(398)}{(3)(2)}(0.02^3)(0.98^{397}) = 0.02784$

and so on.

---

the variance* of X is

$$\sigma_X^2 = npq, \tag{8}$$

and the standard deviation of X is

$$\sigma_X = \sqrt{npq}. \tag{9}$$

Thus, if we have a binomially distributed population where p (e.g., the proportion of males) $= 0.5$ and q (e.g., the proportion of females) $= 0.5$ and we take 10 samples from that population, the mean of the 10 X's (i.e., the mean number of males per sample) would be expected to be $np = (10)(0.05) = 5$ and the standard deviation of the 10 X's would be expected to be $\sqrt{npq} = \sqrt{(10)(0.5)(0.5)} = 1.58$. Our concern typically is with the distribution of the expected probabilities rather than the expected X's, as will be explained in Section 3.

2 THE HYPERGEOMETRIC DISTRIBUTION

Binomial probabilities (Section 1) may result from what is known as "sampling with replacement." This means that after an item is randomly removed from the

---

*A measure of symmetry for a binomial distribution is

$$\gamma_1 = \frac{q-p}{\sqrt{npq}}, \tag{7}$$

so it can be seen that $\gamma_1 = 0$ only when $p = q = 0.05$, $\gamma_1 > 0$ implies a distribution skewed to the right (as in Figures 1b and 1c) and $\gamma_1 < 0$ indicates a distribution skewed to the left.

population to be part of the sample it is returned to the population before randomly selecting another item for inclusion in the sample. (This assumes that after the item is returned to the population it has the same chance of being selected again as does any other member of the population; in many biological situations — such as catching a mammal in a trap — this is not so.) Sampling with replacement ensures that the probability of selecting an item belonging to a specific one of the binomial categories remains constant. If sampling from an actual population is performed without replacement, then selecting an item from the first category reduces p and increases q (and, if the selected item were from the second category, then q would decrease and p would increase). Binomial probabilities may also arise from sampling "hypothetical" populations, such as proportions of heads and tails from all possible coin tosses or of males and females in all possible fraternal twins.

Probabilities associated with sampling without replacement follow the *hypergeometric distribution* instead of the binomial distribution. The probability of obtaining a sample of n items from a hypergeometric distribution, where the sample consists of X items in one category and $n - X$ items in a second category, is

$$P(X) = \frac{\binom{N_1}{X}\binom{N_2}{n-X}}{\binom{N_T}{n}} \tag{10}$$

$$= \frac{N_1! N_2! n! (N_T - n)!}{X!(N_1 - X)!(n - X)!(N_2 - n + X)! N_T!}. \tag{11}$$

Here, N_T is the total number of items in the population, N_1 in category 1 and N_2 in category 2. For example, we could ask what the probability is of forming a sample consisting of three women and two men by taking five people at random from a group of eight women and six men. As $N_1 = 8, N_2 = 6, N_T = 14, n = 5$, and $X = 3$, the probability is

$$P(X) = \frac{\binom{8}{3}\binom{6}{2}}{\binom{14}{5}}$$

$$= \frac{8!\, 6!\, 5!\, X\, 9!}{3!\, 5!\, 2!\, 4!\, 14!}$$

$$= \frac{(8 \cdot 7 \cdot 6 \cdot 5 \cdot 4 \cdot 3 \cdot 2)(6 \cdot 5 \cdot 4 \cdot 3 \cdot 2)(5 \cdot 4 \cdot 3 \cdot 2)(9 \cdot 8 \cdot 7 \cdot 6 \cdot 5 \cdot 4 \cdot 3 \cdot 2)}{(3 \cdot 2)(5 \cdot 4 \cdot 3 \cdot 2)(2)(4 \cdot 3 \cdot 2)(14 \cdot 13 \cdot 12 \cdot 11 \cdot 10 \cdot 9 \cdot 8 \cdot 7 \cdot 6 \cdot 5 \cdot 4 \cdot 3 \cdot 2)}$$

$$= 0.4196.$$

If the population is very large compared to the size of the sample, then the result of sampling with replacement is indistinguishable from that of sampling without replacement, and the hypergeometric distribution approaches — and is approximated by — the binomial distribution. Table 3 compares the binomial distribution with $p = 0.01$ and $n = 100$ to three hypergeometric distributions with the same p and n but with different population sizes. It can be seen that for larger N_T the hypergeometric is closer to the binomial distribution.

TABLE 3: The Hypergeometric Distribution Where N_1, the Number of Items of One Category, Is 1% of the Population Size, N_T; and the Binomial Distribution with $p = 0.01$; the Sample Size, n, Is 100 in Each Case

X	$P(X)$ for hypergeometric: $N_T = 1000, N_1 = 10$	$P(X)$ for hypergeometric: $N_T = 2000, N_1 = 20$	$P(X)$ for hypergeometric: $N_T = 5000, N_1 = 50$	$P(X)$ for binomial: $p = 0.01$
0	0.34693	0.35669	0.36235	0.36603
1	0.38937	0.37926	0.37347	0.36973
2	0.19447	0.18953	0.18670	0.18486
3	0.05691	0.05918	0.06032	0.06100
4	0.01081	0.01295	0.01416	0.01494
5	0.00139	0.00211	0.00258	0.00290
6	0.00012	0.00027	0.00038	0.00046
> 6	0.00000	0.00001	0.00004	0.00008
Total	1.00000	1.00000	1.00000	1.00000

3 SAMPLING A BINOMIAL POPULATION

Let us consider a population of N individuals: Y individuals in one category and $N - Y$ in the second category. Then the proportion of individuals in the first category is

$$p = \frac{Y}{N} \tag{12}$$

and the proportion in the second is

$$q = 1 - p \quad \text{or} \quad q = \frac{N - Y}{N}. \tag{13}$$

If a sample of n observations is taken from this population, with replacement, and X observations are in one category and $n - X$ are in the other, then the population parameter p is estimated by the sample statistic

$$\hat{p} = \frac{X}{n}, \tag{14}$$

which is the proportion of the sample that is in the first category.* The estimate of q is

$$\hat{q} = 1 - p \quad \text{or} \quad \hat{q} = \frac{n - X}{n}, \tag{15}$$

which is the proportion of the sample occurring in the second category. In Example 4 we have $X = 4$ and $n = 20$, so $\hat{p} = 4/20 = 0.20$ and $\hat{q} = 1 - p = 0.80$.

If our sample of 20 were returned to the population (or if the population were extremely large), and we took another sample of 20, and repeated this multiple sampling procedure many times, we could obtain many calculations of $\hat{p}$, each estimating the population parameter p. If, in the population, $p = 0$, then obviously any sample from that population would have $\hat{p} = 0$; and if $p = 1.0$, then each and

---

*Placing the symbol "^" above a letter is statistical convention for denoting an estimate of the quantity which that letter denotes. Thus, $\hat{p}$ refers to an estimate of p, and the statistic $\hat{q}$ is a sample estimate of the population parameter q. Routinely, $\hat{p}$ is called "p hat" and $\hat{q}$ is called "q hat."

EXAMPLE 4   Sampling a Binomial Population

From a population of male and female spiders, a sample of 20 is taken, which contains 4 males and 16 females.

$$n = 20$$
$$X = 4$$

By Equation 14,

$$\hat{p} = \frac{X}{n} = \frac{4}{20} = 0.20.$$

Therefore, we estimate that 20% of the population are males and, by Equation 15,

$$\hat{q} = 1 - \hat{p} = 1 - 0.20 = 0.80$$

or

$$\hat{q} = \frac{n - X}{n} = \frac{20 - 4}{20} = \frac{16}{20} = 0.80,$$

so we estimate that 80% of the population are females.

The variance of the estimate $\hat{p}$ (or of $\hat{q}$) is, by Equation 17,

$$s_{\hat{p}}^2 = \frac{\hat{p}\hat{q}}{n - 1} = \frac{(0.20)(0.80)}{20 - 1} = 0.008421.$$

If we consider that the sample consists of four 1's and sixteen 0's, then $\sum X = 4$, $\sum X^2 = 4$, and the variance of the twenty 1's and 0's is, $s^2 = (4 - 4^2/20)/(20 - 1) = 0.168421$, and the variance of the mean, is $s_{\overline{X}}^2 = 0.168421/20 = 0.008421$.

The standard error (or standard deviation) of $\hat{p}$ (or of $\hat{q}$) is, by Equation 21, $s_{\hat{p}} = \sqrt{0.008421} = 0.092.$

---

every $\hat{p}$ would be 1.0. However, if p is neither 0 nor 1.0, then all the many samples from the population would not have the same values of $\hat{p}$. The variance of all possible $\hat{p}$'s is

$$\sigma_{\hat{p}}^2 = \frac{pq}{n}, \tag{16}$$

which can be estimated from our sample as

$$s_{\hat{p}}^2 = \frac{\hat{p}\hat{q}}{n - 1}. \tag{17}$$

This variance is essentially a variance of means. The variance of $\hat{q}$ is the same as the variance of $\hat{p}$; that is

$$\sigma_{\hat{q}}^2 = \sigma_{\hat{p}}^2 \tag{18}$$

and

$$s_{\hat{q}}^2 = s_{\hat{p}}^2. \tag{19}$$

The standard error of $\hat{p}$ (or of $\hat{q}$), also called the standard deviation, is

$$\sigma_{\hat{p}} = \sqrt{\frac{pq}{n}}, \tag{20}$$

which is estimated from a sample as*

$$s_{\hat{p}} = \sqrt{\frac{\hat{p}\hat{q}}{n-1}}. \tag{21}$$

The possible values of $\sigma_{\hat{p}}^2, \sigma_{\hat{q}}^2, \sigma_{\hat{p}},$ and $\sigma_{\hat{q}}$ range from a minimum of zero when either $\hat{p}$ or $\hat{q}$ is zero, to a maximum when $\hat{p} = \hat{q} = 0.5$; and $s_{\hat{p}}^2, s_{\hat{q}}^2, s_{\hat{p}},$ and $s_{\hat{q}}$ can range from a minimum of zero when either $\hat{p}$ or $\hat{q}$ is zero, to a maximum when $\hat{p} = \hat{q} = 0.5$.

(a) Sampling Finite Populations[†]. If n is a substantial portion of the entire population of size N, and sampling is without replacement, then a finite population correction is called for in estimating $\sigma_{\hat{p}}^2$ or $\sigma_{\hat{p}}$:

$$s_{\hat{p}}^2 = \frac{\hat{p}\hat{q}}{n-1}\left(1 - \frac{n}{N}\right) \tag{23}$$

and

$$s_{\hat{p}} = \sqrt{\frac{\hat{p}\hat{q}}{n-1}\left(1 - \frac{n}{N}\right)}, \tag{24}$$

when n/N is called the *sampling fraction*, and $1 - n/N$ is the *finite population correction*, the latter also being written as $(N - n)/N$. As N becomes very large compared to n, Equation 23 approaches Equation 17 and Equation 24 approaches 21.

We can estimate Y, the total number of occurrences in the population in the first category, as

$$\hat{Y} = \hat{p}N; \tag{25}$$

and the variance and standard error of this estimate are

$$s_{\hat{Y}}^2 = \frac{N(N-n)\hat{p}\hat{q}}{n-1} \tag{26}$$

and

$$s_{\hat{Y}} = \sqrt{\frac{N(N-n)\hat{p}\hat{q}}{n-1}}, \tag{27}$$

respectively.

---

*We often see

$$s_{\hat{p}} = \sqrt{\frac{\hat{p}\hat{q}}{n}} \tag{22}$$

used to estimate $\sigma_{\hat{p}}$. Although it is an underestimate, when n is large the difference between Equations 21 and 22 is slight.

[†]These procedures are from Cochran (1977: 52). When sampling from finite populations, the data follow the hypergeometric (Section 2), rather than the binomial, distribution.

4 GOODNESS OF FIT FOR THE BINOMIAL DISTRIBUTION

(a) When p Is Hypothesized to Be Known. In some biological situations the population proportions, p and q, might be postulated, as from theory. For example, theory might tell us that 50% of mammalian sperm contain an X chromosome, whereas 50% contain a Y chromosome, and we can expect a 1 : 1 sex ratio among the offspring. We may wish to test the hypothesis that our sample came from a binomially distributed population with equal sex frequencies. We may do this as follows, by goodness-of-fit testing.

Let us suppose that we have tabulated the sexes of the offspring from 54 litters of five animals each (Example 5). Setting $p = q = 0.5$, the proportion of each possible litter composition can be computed by the procedures of Example 1, using Equation 3, or they can be read directly from Table 26b from *Appendix: Statistical Tables and Graphs*. From these proportions, we can tabulate expected frequencies, and then can subject observed and expected frequencies of each type of litter to a chi-square goodness-of-fit analysis, with $k - 1$ degrees of freedom (k being the number of classes of X). In Example 5, we do not reject the null hypothesis, and therefore we conclude that the sampled population is binomial with $p = 0.5$.

EXAMPLE 5   Goodness of Fit of a Binomial Distribution, When p Is Postulated

The data consist of observed frequencies of females in 54 litters of five offspring per litter. $X = 0$ denotes a litter having no females, $X = 1$ a litter having one female, and so on; f is the observed number of litters, and $\hat{f}$ is the number of litters expected if the null hypothesis is true. Computation of the values of $\hat{f}$ requires the values of $P(X)$, as obtained in Example 1.

H_0: The sexes of the offspring reflect a binomial distribution with $p = q = 0.5$.

H_A: The sexes of the offspring do not reflect a binomial distribution with $p = q = 0.5$.

X_i	f_i	$\hat{f}_i$
0	3	$(0.03125)(54) = 1.6875$
1	10	$(0.15625)(54) = 8.4375$
2	14	$(0.31250)(54) = 16.8750$
3	17	$(0.31250)(54) = 16.8750$
4	9	$(0.15625)(54) = 8.4375$
5	1	$(0.03125)(54) = 1.6875$

$$\chi^2 = \frac{(3 - 1.6875)^2}{1.6875} + \frac{(10 - 8.4375)^2}{8.4375} + \frac{(14 - 16.8750)^2}{16.8750}$$

$$+ \frac{(17 - 16.8750)^2}{16.8750} + \frac{(9 - 8.4375)^2}{8.4375} + \frac{(1 - 1.6875)^2}{1.6875}$$

$$= 1.0208 + 0.2894 + 0.4898 + 0.0009 + 0.0375 + 0.2801 = 2.1185$$

$$\nu = k - 1 = 6 - 1 = 5$$

$$\chi^2_{0.05,5} = 11.070$$

Therefore, do not reject H_0.

$$0.75 < P < 0.90 \quad [P = 0.83]$$

To avoid bias in this chi-square computation, no expected frequency should be less than 1.0 (Cochran, 1954). If such small frequencies occur, then frequencies in the appropriate extreme classes of X may be pooled to arrive at sufficiently large $\hat{f}_i$ values. Such pooling was not necessary in Example 5, as no $\hat{f}_i$ was less than 1.0. But it will be shown in Example 6.

EXAMPLE 6   Goodness of Fit of a Binomial Distribution, When p Is Estimated from the Sample Data

The data consist of observed frequencies of left-handed persons in 75 samples of eight persons each. $X = 0$ denotes a sample with no left-handed persons, $X = 1$ a sample with one left-handed person, and so on; f is the observed number of samples, and $\hat{f}$ is the number of samples expected if the null hypothesis is true. Each $\hat{f}$ is computed by multiplying 75 by $P(X)$, where $P(X)$ is obtained from Equation 3 by substituting $\hat{p}$ and $\hat{q}$ for p and q, respectively.

H_0: The frequencies of left- and right-handed persons in the population follow a binomial distribution.

H_A: The frequencies of left- and right-handed persons in the population do not follow a binomial distribution.

$$\overline{X} = \frac{\sum f_i X_i}{\sum f_i} = \frac{96}{75} = 1.2800$$

$$\hat{p} = \frac{\overline{X}}{n} = \frac{1.2800}{8} = 0.16 = \text{probability of a person being left-handed}$$

$$\hat{q} = 1 - \hat{p} = 0.84 = \text{probability of a person being right-handed}$$

X_i	f_i	$f_i X_i$	$\hat{f}_i$
0	21	0	$\frac{8!}{0!8!}(0.16^0)(0.84^8)(75) = (0.24788)(75) = 18.59$
1	26	26	$(0.37772)(75) = 28.33$
2	19	38	$(0.25181)(75) = 18.89$
3	6	18	$(0.09593)(75) = 7.19$
4	2	8	$(0.02284)(75) = 1.71$
5	0	0	$(0.00348)(75) = 0.26$
6	1 } 3	6	$(0.00033)(75) = 0.02$ } 1.99
7	0	0	$(0.00002)(75) = 0.00$
8	0	0	$(0.00000)(75) = 0.00$
	75	96	

Note: The extremely small $\hat{f}$ values of 0.00, 0.00, 0.02, and 0.26 are each less than 1.00. So they are combined with the adjacent $\hat{f}$ of 1.71. This results in an $\hat{f}$ of 1.99 for a corresponding f of 3.

$$\sum f_i = 75$$

$$\sum f_i X_i = 96$$

$$\chi^2 = \frac{(21 - 18.59)^2}{18.59} + \frac{(26 - 28.33)^2}{28.33} + \frac{(19 - 18.89)^2}{18.89} + \frac{(6 - 7.19)^2}{7.19}$$

$$+ \frac{(3 - 1.99)^2}{1.99}$$

$$= 1.214$$

$$\nu = k - 2 = 5 - 2 = 3$$

$$\chi^2_{0.05,3} = 7.815$$

Therefore, do not reject H_0.

$$0.50 < P < 0.75 \quad [P = 0.7496]$$

The G statistic may be calculated in lieu of chi-square, with the summation being executed over all classes except those where not only $f_i = 0$ but also all more extreme f_i's are zero. The Kolmogorov-Smirnov statistic could also be used to determine the goodness of fit. Heterogeneity testing may be performed for several sets of data hypothesized to have come from a binomial distribution.

If the preceding null hypothesis had been rejected, we might have looked in several directions for a biological explanation. The rejection of H_0 might have indicated that the population p was, in fact, not 0.5. Or, it might have indicated that the underlying distribution was not binomial. The latter possibility may occur when membership of an individual in one of the two possible categories is dependent upon another individual in the sample. In Example 5, for instance, identical twins (or other multiple identical births) might have been a common occurrence in the species in question. In that case, if one member of a litter was found to be female, then there would be a greater-than-expected chance of a second member of the litter being female.

(b) When p Is Not Assumed to Be Known. Commonly, we do not postulate the value of p in the population but estimate it from a sample of data. As shown in Example 7, we may do this by calculating

$$\hat{p} = \frac{\sum_{i=1}^{k} f_i X_i \Big/ \sum_{i=1}^{k} f_i}{n}. \tag{28}$$

It then follows that $\hat{q} = 1 - \hat{p}$.

The values of $\hat{p}$ and $\hat{q}$ may be substituted in Equation 3 in place of p and q, respectively. Thus, expected frequencies may be calculated for each X, and a chi-square goodness-of-fit analysis may be performed as it was in Example 5. In such a procedure, however, ν is $k - 2$ rather than $k - 1$, because two constants

(n and $\hat{p}$) must be obtained from the sample, and ν is, in general, determined as k minus the number of such constants. The G statistic may be employed when p is not known, but the Kolmogorov-Smirnov test is very conservative in such cases and should be avoided.

The null hypothesis for such a test would be that the sampled population was distributed binomially, with the members of the population occuring independently of one another.

5 THE BINOMIAL TEST AND ONE-SAMPLE TEST OF A PROPORTION

With the ability to determine binomial probabilities, a simple procedure may be employed for goodness-of-fit testing of nominal data distributed between two categories. This method is especially welcome as an alternative to chi-square goodness of fit where the expected frequencies are small. If p is very small, then the Poisson distribution may be used; and it is simpler to employ when n is very large. Because the binomial distribution is discrete, this procedure is conservative in that the probability of a Type I error is $\leq \alpha$.

(a) One-Tailed Testing.   Animals might be introduced one at a time into a passageway at the end of which each has a choice of turning either to the right or to the left. A substance, perhaps food, is placed out of sight to the left or right; the direction is randomly determined (as by the toss of a coin). We might state a null hypothesis, that there is no tendency for animals to turn in the direction of the food, against the alternative, that the animals prefer to turn toward the food. If we consider p to be the probability of turning toward the food, then the hypothesis (one-tailed) would be stated as $H_0: p \leq 0.5$ and $H_A: p > 0.5$, and such an experiment might be utilized, for example, to determine the ability of the animals to smell the food. We may test H_0 as shown in Example 7. In this procedure, we determine the probability of obtaining, at random, a distribution of data deviating as much as, or more than, the observed data. In Example 7, the most likely distribution of data in a sample of twelve from a population where p, in fact, was 0.5, would be six left and six right. The samples deviating from a $6:6$ ratio even more than our observed sample (having a $10:2$ ratio) would be those possessing eleven left, one right, and twelve left, zero right.

EXAMPLE 7   A One-Tailed Binomial Test

Twelve animals were introduced, one at a time, into a passageway at the end of which they could turn to the left (where food was placed out of sight) or to the right. We wish to determine if these animals came from a population in which animals would choose the left more often than the right (perhaps because they were able to smell the food).

Thus, $n = 12$, the number of animals; X is the number of animals turning left; and p is the probability of animals in the sampled population that would turn left.

$H_0: p \leq 0.5$ and $H_A: p > 5$

In this example, $P(X)$ is obtained either from Table 26b from *Appendix: Statistical Tables and Graphs* or by Equation 3.

(a) The test using binomial probabilities

X	$P(X)$
0	0.00024
1	0.00293
2	0.01611
3	0.05371
4	0.12085
5	0.19336
6	0.22559
7	0.19336
8	0.12085
9	0.05371
10	0.01611
11	0.00293
12	0.00024

On performing the experiment, ten of the twelve animals turned to the left and two turned to the right. If H_0 is true, $P(X \geq 10) = 0.01611 + 0.00293 + 0.00024 = 0.01928$. As this probability is less than 0.05, reject H_0.

(b) The test using a confidence limit

This test could also be performed by using the upper confidence limit of p as a critical value. For example, by Equation 35, with a one-tailed F,

$$X = pn = (0.5)(12) = 6,$$

$$\nu_1' = 2(6 + 1) = 14,$$

$$\nu_2' = 2(12 - 6) = 12,$$

$$F_{0.05(1),14,12} = 2.64, \text{ and}$$

$$L_2 = \frac{(6 + 1)(2.64)}{12 - 6 + (6 + 1)(2.64)} = 0.755.$$

Because the observed $\hat{p}$ (namely $X/n = 10/12 = 0.833$) exceeds the critical value (0.755), we reject H_0.

(c) A simpler alternative

Table 27 from *Appendix: Statistical Tables and Graphs* can be consulted for $n = 12$ and $\alpha(1) = 0.05$ and to find an upper critical value of $n - C_{0.05(1),n} = 12 - 2 = 10$. As an X of 10 falls within the range of $\hat{p} \geq n - C_{0.05(1),12}, n, H_0$ is rejected; and (by examining the column headings) $0.01 < P(X \geq 10) < 0.25$.

The general one-tailed hypotheses are $H_0: p \leq p_0$ and $H_A: p > p_0$, or $H_0: p \geq p_0$ and $H_A: p < p_0$, where p_0 need not be 0.5. The determination of the probability of $\hat{p}$

as extreme as, or more extreme than, that observed is shown in Example 8, where the expected frequencies, $P(X)$, are obtained either from Table 26b from *Appendix: Statistical Tables and Graphs* or by Equation 3. If the resultant probability is less than or equal to α, then H_0 is rejected. A simple procedure for computing this P when $p_0 = 0.5$ is shown in Section 6.

Alternatively, a critical value for the one-tailed binomial test may be found using the confidence-limit determinations of Section 8. This is demonstrated in Example 7b, using a one-tailed F for the confidence interval presented in Section 8a. If $H_A: p > 0.5$ (as in Example 7), then Equation 35 is used to obtain the upper critical value as the critical value for the test. If the alternative hypothesis had been $H_A: p > 0.5$, then Equation 29 would have been appropriate, calculating the lower confidence limit to be considered the critical value. Or the needed upper or lower confidence limit could be obtained using a one-tailed Z for the procedure of Section 8c. Employing the confidence limits of Section 8b is not recommended.

A simpler alternative, demonstrated as Example 8c, is to consult Table 27 from *Appendix: Statistical Tables and Graphs* to obtain the lower confidence limit, $C_{\alpha(1),n}$, for one-tailed probabilities, $\alpha(1)$. If $H_A: p < 0.5$, then H_0 is rejected if $X \leq C_{\alpha(1),n}$, where $X = np$. If $H_A: p > 0.5$ (as in Example 7), then H_0 is rejected if $X > n - C_{\alpha(1),n}$.

(b) Two-Tailed Testing. The preceding experiment might have been performed without expressing an interest specifically in whether the animals were attracted toward the introduced substance. Thus, there would be no reason for considering a preference for only one of the two possible directions, and we would be dealing with two-tailed hypotheses, $H_0: p = 0.5$ and $H_A: p \neq 0.5$. The testing procedure would be identical to that in Example 7, except that we desire to know $P(X \leq 2 \text{ or } X \geq 10)$. This is the probability of a set of data deviating *in either direction* from the expected as much as or more than those data observed. This is shown in Example 8. The general two-tailed hypotheses are $H_0: p = p_0$ and $H_A: p \neq p_0$. If $p_0 = 0.5$, a simplified computation of P is shown in Equation 5.

Instead of enumerating the several values of $P(X)$ required, we could determine critical values for the two-tailed binomial test as the two-tailed confidence limits described in Section 8. If the observed $\hat{p}$ lies outside the interval formed by L_1 and L_2, then H_0 is rejected. This is demonstrated in Example 8b, using the confidence limits of Section 8a. If the hypothesized p is 0.5, then ν_1 and ν_2 are the same for L_1 and L_2; therefore, the required critical value of F is the same for both confidence limits. This is shown in Example 8a, where $\nu_1 = \nu_1'$, $\nu_2 = \nu_2'$, and $F_{0.05(2),14,12}$ is used for both L_1 and L_2. The confidence-limit calculation of Section 8c can also be used, but employing Section 8b is not recommended.

A simpler two-tailed binomial test is possible using Table 27 from *Appendix: Statistical Tables and Graphs*, as demonstrated in Example 8c. If the observed count, $X = pn$, is either $\leq C_{\alpha(2),\nu}$ or $\geq n - C_{\alpha(2),n}$, then H_0 is rejected.

(c) Normal and Chi-Square Approximations. Some researchers have used the normal approximation to the binomial distribution to perform the two-tailed test (for $H_0: p = p_0$ versus $H_A: p \neq p_0$) or the one-tailed test (either $H_0: p \leq p_0$ versus $H_A: p > p_0$, or $H_0: p \geq p_0$ versus $H_A: p < p_0$). The test statistic is

$$Z = \frac{X - np_0}{\sqrt{np_0 q_0}}, \tag{29}$$

Dichotomous Variables

EXAMPLE 8   A Two-Tailed Binomial Test

The experiment is as described in Example 7, except that we have no a priori interest in the animals' turning either toward or away from the introduced substance.

$$H_0: \ p = 0.5$$
$$H_A: \ p \neq 0.5$$

$n = 12$

(a) The probabilities of X, for $X = 0$ through $X = 12$, are given in Example 7.

$$P(X \geq 10 \text{ or } X \leq 2) = 0.01611 + 0.00293 + 0.00024 + 0.01611$$
$$+ 0.00293 + 0.00024$$
$$= 0.03856$$

As this probability is less than 0.05, reject H_0.

(b) Alternatively, this test could be performed by using the confidence limits as critical values. By Equations 34 and 35, we have

$$X = pn = (0.5)(12) = 6,$$

and for L_1 we have

$$\nu_1 = 2(12 - 6 + 1) = 14$$

$$\nu_2 = 2(6) = 12$$

$$F_{0.05(2),14,12} = 3.21$$

$$L_1 = \frac{6}{6 + (12 - 6 + 1)(3.21)} = 0.211,$$

and for L_2 we have

$$\nu_1' = 2(6 + 1) = 14$$

$$\nu_2' = 2(12 - 6) = 12$$

$$F_{0.05(2),14,12} = 3.21$$

$$L_2 = \frac{(6 + 1)(3.21)}{12 - 6 + (6 + 1)(3.21)} = 0.789.$$

As the observed $\hat{p}$ (namely $X/n = 10/12 = 0.833$) lies outside the range of 0.211 to 0.789, we reject H_0.

(c) A simpler procedure uses Table 27 from *Appendix: Statistical Tables and Graphs* to obtain critical values of $C_{0.05(2),6} = 2$ and $n - C_{0.05(2),6} = 12 - 2 = 10$. As $X = 10$, H_0 is rejected; and $0.02 < P(X \leq 2 \text{ or } X > 10) < 0.05$.

---

where $X = n\hat{p}$, the number of observations in one of the two categories, and np_0 is the number of observations expected in that category if H_0 is true. Equivalently, this may be expressed as

$$Z = \frac{p - p_0}{\sqrt{p_0 q_0 / n}}. \tag{30}$$

The two-tailed null hypothesis, $H_0 : p = p_0$ (tested in Example 9), is rejected if $|Z| > Z_{\alpha(2)}$; the one-tailed $H_0 : p \leq p_0$ is rejected if $|Z| > Z_{\alpha(1)}$, and $H_0 : p \geq p_0$ is rejected if $|Z| < Z_{\alpha(1)}$.

EXAMPLE 9   The Normal Approximation to the Binomial Test

For the normal approximation to the binomial test of Example 8, using the data of Example 7, $H_0: p = 0.5$; $H_A: p \neq 0.5$; $p_0 = 0.5$; $n = 12$; $X = 10$; $\hat{p} = 10/12 = 0.8333$.

Using Equation 29,

$$Z = \frac{X - np_0}{\sqrt{np_0 q_0}} = \frac{10 - 6}{\sqrt{(12)(0.5)(0.5)}} = \frac{4}{1.7321} = 2.3098$$

$$Z_{0.05(2)} = 1.9600$$

Therefore, reject H_0.

$$0.02 < P < 0.05 \quad [0.021]$$

$$Z_c = \frac{|X - np_0| - 0.5}{\sqrt{np_0 q_0}} = \frac{|10 - 6| - 0.5}{\sqrt{(12)(0.5)(0.5)}} = \frac{3.5}{1.7321} = 2.0207$$

$$0.02 < P < 0.05 \quad [0.043]$$

Using Equation 30,

$$Z = \frac{p - p_0}{\sqrt{p_0 q_0 / n}} = \frac{0.8333 - 0.5000}{\sqrt{(0.5)(0.5/12)}} = \frac{0.3333}{0.0208} = 2.3093$$

$$0.02 < P < 0.05 \quad [0.021]$$

$$Z_c = \frac{|p - p_0| - 0.5/n}{\sqrt{p_0 q_0 / n}} = \frac{|0.8333 - 0.5000| - 0.5/12}{\sqrt{(0.5)(0.5/12)}}$$

$$= \frac{0.3333 - 0.0417}{0.0208} = 2.0208$$

$$0.02 < P < 0.05 \quad [0.043]$$

The test may also be performed with a correction for continuity, by bringing np_0 closer by 0.5 to X in Equation 29 or by bringing p_0 nearer by $0.5/n$ to p in Equation 30. This is shown in Example 9, where the probability using Z_c is seen to be closer than $P(Z)$ to the probability determined by the binomial test.

The two-tailed (but not the one-tailed) test can be effected as a chi-square goodness of fit. For two-tailed hypotheses, Z testing is equivalent to chi-square testing, for $Z^2 = \chi^2$ (and, if $Z_c^2 = \chi_c^2$). However, Ramsey and Ramsey (1988) concluded that the approximation is not as powerful as the binomial test.

The normal approximation to the binomial test and the chi-square goodness-of-fit test does not work well when n is small. $\hat{f}_1$ and $\hat{f}_2$ should be at least 5 for the chi-square procedure; the equivalent statement for the normal approximation to the binomial test is that $p_0 n$ and $q_0 n$ should be at least 5.

6 THE SIGN TEST

For two samples where interval- or ratio scale data occur in pairs, hypotheses of no difference between the means of the two samples may be tested by the t test. If paired data are measured on an ordinal scale, the nonparametric *sign test** can be useful; it may be employed whenever the Wilcoxon paired-sample test is appropriate, although it is not as powerful as the latter. The sign test may also be used with interval- or ratio-scale data, but in those circumstances it is not as powerful as the paired t test and it does not express hypotheses in terms of population means.

The actual differences between members of a pair are not utilized in the sign test; only the direction (or sign) of each difference is tabulated. In Example 10, all that need be recorded is whether each hindleg length is greater than, equal to, or less than its corresponding foreleg length; we do this by recording $+$, 0, or $-$, respectively. We then ask what the probability is of the observed distribution, or a more extreme distribution, of $+$ and $-$ signs if the null hypothesis is true. (A difference of zero is deleted from the analysis, so n is here defined as the number of differences having a sign.[†]) The analysis proceeds as a binomial test with H_0: $p = 0.5$, and the null hypothesis tested is, essentially, that in the population the median difference is zero (i.e., the population frequencies of positive differences and negative differences are the same), but it differs from the median test in that the data in the two samples are paired.

In performing a binomial test with $p_0 = q_0 = 0.5$, which is always the case with the sign test, the exact probability, P, may be obtained by the following simple considerations. As introduced in Equation 5, for a given n the probability of a specified X is 0.5^n times the binomial coefficient. And binomial coefficients are defined in Equation 2 and presented in Table 2 and Table 26a from *Appendix: Statistical Tables and Graphs*. In performing the binomial or sign test, we sum binomial terms in one or both tails of the distribution, and if $p_0 = q_0 = 0.5$, then this is the same as multiplying 0.5^n by the sum of the binomial coefficients in the one or two tails. Examining Example 10, what is needed is the sum of the probabilities in the two tails defined by $X \leq 2$ and $X \geq 8$. Thus, we may sum the coefficients, $_{10}C_X$, for $X \leq 2$ and $X \geq 8$, and multiply that sum by 0.5^{10} or, equivalently, divide the sum by 2^{10}. For this example, the binomial coefficients are 1, 10, 45, 45, 10, and 1, so the probability of H_0 being a true statement about the sampled population is

$$\frac{1 + 10 + 45 + 45 + 10 + 1}{2^{10}} = \frac{112}{1024} = 0.1094.$$

---

*The sign test was first employed by Scottish physician and mathematician John Arbuthnott (1667–1735), and his 1710 publication is perhaps the earliest report of something resembling a statistical hypothesis (Noether, 1984).

[†]Methods have been described specifically for situations where there are many differences of zero (Coakley and Heise, 1996; Fong et al., 2003; Wittkowski, 1998).

EXAMPLE 10	The Sign Test for the Paired-Sample Data.		

Deer	Hindleg length (cm)	Foreleg length (cm)	Difference
1	142	138	+
2	140	136	+
3	144	147	−
4	144	139	+
5	142	143	−
6	146	141	+
7	149	143	+
8	150	145	+
9	142	136	+
10	148	146	+

H_0: There is no difference between hindleg and foreleg length in deer. ($p = 0$)

H_A: There is a difference between hindleg and foreleg length in deer. ($p \neq 0$)

$n = 10$, and there are 8 positive differences and 2 negative differences.

Using Table 26b from *Appendix: Statistical Tables and Graphs* for $n = 10$ and $p = 0.50$,

$$P(X \leq 2 \text{ or } X \geq 8)$$
$$= 0.04395 + 0.00977 + 0.00098 + 0.04395 + 0.00977 + 0.00098$$
$$= 0.1094.$$

As the probability is greater than 0.05, do not reject H_0.

Using binomial coefficients,

$$\frac{1 + 10 + 45 + 45 + 10 + 1}{2^{10}} = \frac{112}{1024} = 0.1094.$$

Using Table 27 from *Appendix: Statistical Tables and Graphs* for $n = 10$, the critical values are $C_{0.05(2),10} = 1$ and $n - C_{0.05(2),10} = 10 - 1 = 9$. As neither $X = 2$ nor $X = 8$ is as small as 1 or as large as 9, H_0 is not rejected; and by consulting Table 27 from *Appendix: Statistical Tables and Graphs*, we state $0.10 < P < 0.20$.

This calculation can be more accurate than summing the six individual binomial probabilities, as shown in Example 10, for it avoids rounding error.

An extension of the sign test to nonparametric testing for blocked data from more than two groups is found in the form of the Friedman test.

It is noted that the Wilcoxon paired-sample test can be applied to hypotheses expressing differences of specified magnitude. The sign test can be used in a similar fashion. For instance, it can be asked whether the hindlegs in the population sampled in Example 10 are 3 cm longer than the lengths of the forelegs. This can be done by applying the sign test after subtracting 3 cm from each hindleg length in the sample (or adding 3 cm to each foreleg length).

7 POWER, DETECTABLE DIFFERENCE, AND SAMPLE SIZE FOR THE BINOMIAL AND SIGN TESTS

The power of, and required sample size for, the binomial test may be determined by examining the cumulative binomial distribution. As the sign test is essentially a binomial test with p hypothesized to be 0.5 (see Section 6), its power and sample size may be assessed in the same manner as for the binomial test.

(a) Power of the Test. If a binomial test is performed at significance level α and with sample size n, we can estimate the power of the test (i.e., the probability of correctly rejecting H_0) as follows. First we determine the critical value(s) of X for the test. For a one-tailed test of $H_0: p \leq p_0$ versus $H_A: p > p_0$, the critical value is the smallest value of X for which the probability of that X or a larger X is $\leq \alpha$. (In Example 7a this is found to be $X = 10$.) For a one-tailed test of $H_0: p \leq p_0$ versus $H_A: p > p_0$, the critical value is the largest X for which the probability of that X or of a smaller X is $\leq \alpha$. Then we examine the binomial distribution for the observed proportion, $\hat{p}$, from our sample. The power of the test is $\geq$ the probability of an X at least as extreme as the critical value referred to previously.* This is demonstrated in Example 11.

EXAMPLE 11   Determination of the Power of the One-Tailed Binomial Test of Example 7a

In Example 7, $H_0: p \leq 0.5$ and $\hat{p} = X/n = 10/12 = 0.833$. And $X = 10$ is the critical value, because $P(X \geq 10) < 0.05$, but $P(X \geq 11) > 0.05$.

Using Equation 3, $P(X)$ for X's of 10 through 12 is calculated for the binomial distribution having $p = 0.833$ and $n = 12$:

X	$P(X)$
10	0.296
11	0.269
12	0.112

Thus, the power when performing this test on a future set of such data is estimated to be $\geq 0.296 + 0.269 + 0.112 = 0.68$.

For a two-tailed test of $H_0: p = p_0$ versus $H_A: p \neq p_0$, there are two critical values of X, one that cuts off $\alpha/2$ of the binomial distribution in each tail. Knowing these two X's, we examine the binomial distribution for $\hat{p}$, and the power of the test is the probability in the latter distribution that X is at least as extreme as the critical values. This is demonstrated in Example 12.

Cohen (1988: Section 5.4) presents tables to estimate sample size requirements in the sign test.

(b) Normal Approximation for Power. Having performed a binomial test or a sign test, it may be estimated what the power would be if the same test were

---

*If the critical X delineates a probability of exactly α in the tail of the distribution, then the power is equal to that computed; if the critical value defines a tail of less than α, then the power is greater than that calculated.

EXAMPLE 12   Determination of the Power of the Two-Tailed Sign Test of Example 10

In Example 10, H_0: $p = 0.50$, the critical values are 1 and 9, and $\hat{p} = X/n = 8/10 = 0.800$.

Using Equation 3, $P(X)$ is calculated for all X's equal to or more extreme than the critical values, for the binomial distribution having $p = 0.800$ and $n = 10$:

X	$P(X)$
0	0.000
1	0.000
9	0.269
10	0.112

Therefore, the power of performing this test on a future set of data is $\geq$ 0.000 + 0.000 + 0.269 + 0.112 = 0.38.

---

performed on a future set of data taken from the same population. As noted earlier, this is not an estimate of the power of the test already performed; it is an estimate of the power of this test performed on a new set of data obtained from the same population. It has been noted in the preceding discussions of the binomial distribution that normal approximations to that distribution are generally poor and inadvisable. However, rough estimates of power are often sufficient in planning data collection. If n is not small and the best estimate of the population proportion, p, is not near 0 or 1, then an approximation of the power of a binomial or sign test can be calculated as

$$\text{power} = P\left[Z \leq \frac{p_0 - p}{\sqrt{\dfrac{pq}{n}}} - Z_{\alpha(2)}\sqrt{\frac{p_0 q_0}{pq}}\right] + P\left[Z \geq \frac{p_0 - p}{\sqrt{\dfrac{pq}{n}}} + Z_{\alpha(2)}\sqrt{\frac{p_0 q_0}{pq}}\right]$$

$$(31)$$

(Marascuilo and McSweeney, 1977: 62). Here p_0 is the population proportion in the hypothesis to be tested, $q_0 = 1 - p_0$, p is the true population proportion (or our best estimate of it), $q = 1 - p$, $Z_{\alpha(2)} = t_{\alpha(2),\infty}$, and the probabilities of Z are found in Table 2 from *Appendix: Statistical Tables and Graphs*. This is demonstrated in Example 13.

For the one-tailed test, H_0: $p \leq p_0$ versus H_A: $p > p_0$, the estimated power is

$$\text{power} = P\left[Z \geq \frac{p_0 - p}{\sqrt{\dfrac{pq}{n}}} + Z_{\alpha(1)}\sqrt{\frac{p_0 q_0}{pq}}\right];$$

$$(31a)$$

EXAMPLE 13   Estimation of Power in a Two-Tailed Binomial Test, Using the Normal Approximation

To test $H_0: p = 0.5$ versus $H_A: p \neq 0.5$, using $\alpha = 0.05$ (so $Z_{0.05(2)} = 1.9600$) and a sample size of 50, when p in the population is actually 0.5:
Employing Equation 31,

$$\text{power} = P\left[Z \leq \frac{p_0 - p}{\sqrt{\frac{pq}{n}}} - Z_{\alpha(2)}\sqrt{\frac{p_0 q_0}{pq}}\right]$$

$$+ P\left[Z \geq \frac{p_0 - p}{\sqrt{\frac{pq}{n}}} + Z_{\alpha(2)}\sqrt{\frac{p_0 q_0}{pq}}\right]$$

$$= P\left[Z \leq \frac{0.5 - 0.4}{\sqrt{\frac{(0.4)(0.6)}{50}}} - Z_{\alpha(2)}\sqrt{\frac{(0.5)(0.5)}{(0.4)(0.6)}}\right]$$

$$+ P\left[Z \geq \frac{0.5 - 0.4}{\sqrt{\frac{(0.4)(0.6)}{50}}} + Z_{\alpha(2)}\sqrt{\frac{(0.5)(0.5)}{(0.4)(0.6)}}\right]$$

$$= P[Z \leq 1.4434 - (1.9600)(1.0206)]$$
$$+ P[Z \geq 1.4434 + (1.9600)(1.0206)]$$
$$= P[Z \leq 1.4434 - 2.0004] + P[Z \leq 1.4434 + 2.0004]$$
$$= P[Z \leq -0.56] + P[Z \geq 3.44] = P[Z \geq 0.56] + P[Z \geq 3.44]$$
$$= 0.29 + 0.00 = 0.29.$$

and for the one-tailed hypotheses, $H_0: p \geq p_0$ versus $H_A: p < p_0$,

$$\text{power} = P\left[Z \leq \frac{p_0 - p}{\sqrt{\frac{pq}{n}}} - Z_{\alpha(1)}\sqrt{\frac{p_0 q_0}{pq}}\right]. \tag{31b}$$

(c) Sample Size Required and Minimum Detectable Difference.   Prior to designing an experiment, an estimate of the needed sample size may be obtained by specifying

α and the minimum difference between p_0 and p that is desired to be detected with a given power.

If p is not very near 0 or 1, this may be done with a normal approximation, with the understanding that this will result only in a rough estimate. Depending upon the hypothesis to be tested, Equation 31, 31a, or 31b may be used to estimate power for any sample size (n) that appears to be a reasonable guess. If the calculated power is less than desired, then the calculation is repeated with a larger n; if it is greater than that desired, the calculation is repeated with a smaller n. This repetitive process (called iteration) is performed until the specified power is obtained from the equation, at which point n is the estimate of the required sample size.

An estimate of the required n may be obtained without iteration, when one-tailed testing is to be performed. Equations 31a and 31b can be rearranged as follows (Simonoff, 2003, 59–60):

For $H_0: p \le p_0$ versus $H_A: p > p_0$,

$$n = \left(\frac{Z_{\alpha(1)}\sqrt{p_0 q_0} - Z_{\beta(1)}\sqrt{pq}}{p_0 - p} \right)^2 ; \tag{32}$$

and for $H_0: p \ge p_0$ versus $H_A: p < p_0$,

$$n = \left(\frac{Z_{\alpha(1)}\sqrt{p_0 q_0} + Z_{\beta(1)}\sqrt{pq}}{p_0 - p} \right)^2 . \tag{32a}$$

Levin and Chen (1999) have shown that these two equations provide values of n that tend to be underestimates, and they present estimates that are often better.

If p is not very near 0 or 1, we can specify α, power, and n, and employ iteration to obtain a rough estimate of the smallest difference between p_0 and p that can be detected in a future experiment.

A reasonable guess of this minimum detectable difference, $p_0 - p$, may be inserted into Equation 31, 31a, or 31b (depending upon the hypothesis to be tested), and the calculated power is then examined. If the power is less than that specified, then the calculation is performed again, using a larger value of $p_0 - p$; if the calculated power is greater than the specified power, the computation is performed again, using a smaller $p_0 - p$. This iterative process, involving increasing or decreasing $p_0 - p$ in the equation, is repeated until the desired power is achieved, at which point the $p_0 - p$ used to calculate that power is the minimum detectable difference sought.

Or, to estimate the minimum detectable difference for one-tailed testing, either Equation 32 or 32a (depending upon the hypotheses) may be rearranged to give

$$p_0 - p = \frac{Z_{\alpha(1)}\sqrt{p_0 q_0} - Z_{\beta(1)}\sqrt{pq}}{\sqrt{n}}, \tag{33}$$

or

$$p_0 - p = \frac{Z_{\alpha(1)}\sqrt{p_0 q_0} + Z_{\beta(1)}\sqrt{pq}}{\sqrt{n}}, \tag{33a}$$

respectively.

8 CONFIDENCE LIMITS FOR A POPULATION PROPORTION

Confidence intervals for the binomial parameter, p, can be calculated by a very large number of methods.* Among them are the following:

(a) Clopper-Pearson Interval. A confidence interval for p may be computed (Agresti and Coull, 1998; Bliss, 1967: 199–201; Brownlee, 1965: 148–149; Fleiss, Levin, and Paik, 2003: 25) using a relationship between the F distribution and the binomial distribution (Clopper and Pearson, 1934). As demonstrated in Example 14a, the lower confidence limit for p is

$$L_1 = \frac{X}{X + (n - X + 1)F_{\alpha(2),\nu_1,\nu_2}},$$ (34)

where $\nu_1 = 2(n - X + 1)$ and $\nu_2 = 2X$. And the upper confidence limit for p is

$$L_2 = \frac{(X + 1)F_{\alpha(2),\nu'_1,\nu'_2}}{n - X + (X + 1)F_{\alpha(2),\nu'_1,\nu'_2}},$$ (35)

with $\nu'_1 = 2(X + 1)$, which is the same as $\nu_2 + 2$, and $\nu'_2 = 2(n - X)$, which is equal to $\nu_1 - 2$.

The interval specified by L_1 and L_2 is one of many referred to as an "exact" confidence interval, because it is based upon an exact distribution (the binomial distribution) and not upon an approximation of a distribution. But it is not exact in the sense of specifying an interval that includes p with a probability of exactly $1 - \alpha$. Indeed, the aforementioned interval includes p with a probability of *at least* $1 - \alpha$, and the probability might be much greater than $1 - \alpha$. (So a confidence interval calculated in this fashion using $\alpha = 0.05$, such as in Example 14a, will contain p with a probability of 95% *or greater*.) Because this interval tends to be larger than necessary for $1 - \alpha$ confidence, it is said to be a *conservative* confidence interval (although the conservatism is less when n is large).

(b) Wald Interval. This commonly encountered approximation for the confidence interval, based upon the normal distribution,[†] is shown in Exercise 14b:

$$\hat{p} \pm Z_{\alpha(2)}\sqrt{\frac{\hat{p}\hat{q}}{n}}.$$ (36)

But this approximation can yield unsatisfactory results, especially when p is near 0 or 1 or when n is small. (Although the approximation improves somewhat as n or $\hat{p}\hat{q}$ increases, it still performs less well than the method discussed in Section 8c.)

One problem with this confidence interval is that it overestimates the precision of estimating p (and thus is said to be "liberal." That is, the interval includes p less

---

*Many of these are discussed by Agresti and Coull (1998); Blyth (1986); Böhning (1994); Brown, Cai, and DasGupta (2002); Fujino (1980); Newcombe (1998a); and Vollset (1993).

†Brownlee (1965: 136) credits Abraham de Moivre as the first to demonstrate, in 1733, the approximation of the binomial distribution by the normal distribution. Agresti (2002: 15) refers to this as one of the first confidence intervals proposed for any parameter, citing Laplace (1812: 283).

EXAMPLE 14 **Determination of 95% Confidence Interval for the Binomial Population Parameter, p**

One hundred fifty birds were randomly collected from a population, and there were 65 females in the sample. What proportion of the population is female?

$$n = 150, \quad X = 65$$

$$\hat{p} = \frac{X}{n} = \frac{65}{150} = 0.4333, \quad \hat{q} = 1 - \hat{p} = 1 - 0.4333 = 0.5667$$

(a) The Clopper-Pearson Confidence Interval:

For the lower 95% confidence limit,

$$\nu_1 = 2(n - X + 1) = 2(150 - 65 + 1) = 172$$

$$\nu_2 = 2X = 2(65) = 130$$

$$F_{0.05(2),172,130} \approx F_{0.05(2),140,120} = 1.42$$

$$L_1 = \frac{X}{X + (n - X + 1)F_{0.05(2),172,130}}$$

$$\approx \frac{65}{65 + (150 - 65 + 1)(1.42)} = 0.347.$$

For the upper 95% confidence limit,

$$\nu_1' = 2(X + 1) = 2(65 + 1) = 132$$

$$\text{or } \nu_1' = \nu_2 + 2 = 130 + 2 = 132$$

$$\nu_2' = 2(n - X) = 2(150 - 65) = 170$$

$$\text{or } \nu_2' = \nu_1 - 2 = 172 - 2 = 170$$

$$F_{0.05(2),132,170} \approx F_{0.05(2),120,160} = 1.39$$

$$L_2 = \frac{(X + 1)F_{0.05(2),132,170}}{n - X + (X + 1)F_{0.05(2),132,170}}$$

$$\approx \frac{(65 + 1)(1.39)}{150 - 65 + (65 + 1)(1.39)} = 0.519.$$

Therefore, we can state the 95% confidence interval as

$$P(0.347 \leq p \leq 0.519) = 0.95,$$

which is to say that there is 95% confidence that the interval between 0.347 and 0.519 includes the population parameter p.

Note: In this example, the required critical values of F have degrees of freedom (172 and 130 for L_1, and 132 and 170 for L_2) that are not in Table 4 from *Appendix: Statistical Tables and Graphs.* So the next lower available degrees of freedom were used, which is generally an acceptable procedure. Exact critical values from an appropriate computer program are $F_{0.05(2),172,130} = 1.387$ and $F_{0.05(2),132,170} = 1.376$, yielding $L_1 = 0.353$ and $L_2 = 0.514$, which are results very similar to those just given.

(b) The Wald Confidence Interval:

$$Z_{0.05(2)} = 1.9600$$

$$\hat{p} \pm Z_{\alpha(2)}\sqrt{\frac{\hat{p}\hat{q}}{n}}$$

$$= 0.4333 \pm 1.9600\sqrt{\frac{(0.4333)(0.5667)}{150}} = 0.4333 \pm 0.0793$$

$$L_1 = 0.354, \quad L_2 = 0.513$$

(c) The Adjusted Wald Confidence Interval:

$$Z_{0.05(2)}^2 = 1.9600^2 = 3.8416; \quad Z_{0.05(2)}^2/2 = 1.9600^2/2 = 1.9208$$

$$\text{adjusted } X \text{ is } \tilde{X} = X + Z_{0.05(2)}^2/2 = 65 + 1.9208 = 66.9208$$

$$\text{adjusted } n \text{ is } \tilde{n} = n + Z_{0.05(2)}^2 = 150 + 3.8416 = 153.8416$$

$$\text{adjusted } \hat{p} \text{ is } \tilde{p} = \frac{\tilde{X}}{\tilde{n}} = \frac{66.9208}{153.8416} = 0.4350, \quad \hat{q} = 0.5650$$

$$95\% \text{ confidence interval for } p \text{ is } \tilde{p} \pm Z_{0.05(2)}\sqrt{\frac{\tilde{p}\tilde{q}}{\tilde{n}}}$$

$$= 0.4350 \pm 1.9600\sqrt{\frac{(0.4350)(0.5650)}{153.8416}} = 0.4350 \pm 0.0783$$

$$L_1 = 0.357, \quad L_2 = 0.513$$

than $1 - \alpha$ of the time (e.g., less than 95% of the time when $\alpha = 0.05$). Another objection is that the calculated confidence interval is always symmetrical around $\hat{p}$ (that is, it has a lower limit as far from $\hat{p}$ as the upper limit is), although a binomial distribution is skewed (unless $p = 0.50$; see Figure 1). This forced symmetry can result in a calculated L_1 that is less than 0 or an L_2 greater than 1, which would be unreasonable. Also, when $\hat{p}$ is 0 or 1, no confidence interval is calculable; the upper and lower confidence limits are both calculated to be $\hat{p}$.

Though it is commonly encountered, many authors* have noted serious disadvantages of Equation 36 (even with application of a continuity correction), have strongly discouraged its use, and have discussed some approximations that perform much better.

(c) Adjusted Wald Interval. A very simple, and very good, modification of the Wald interval (called an "adjusted Wald interval" by Agresti and Caffo, 2000, and Agresti

*See, for example; Agresti and Caffo (2000); Agresti and Coull (1998); Blyth (1986); Brown, Cai, and DasGupta (2001, 2002); Fujino (1980); Newcombe (1998a); Schader and Schmid (1990); and Vollset (1993). Fleiss, Levin, and Paik (2003: 28–29) present a normal approximation more complicated than, but apparently more accurate than, Equation 36.

and Coull, 1998) substitutes* $\tilde{X} = X + Z^2_{\alpha(2)}/2$ for X and $\tilde{n} = n + Z^2_{\alpha(2)}$ for n in Equation 36,[†] as shown in Example 14c; then, $\tilde{p} = \tilde{X}/\tilde{n}$, and $\tilde{q} = 1 - \tilde{p}$.

The probability of this confidence interval containing the population parameter is closer to $1 - \alpha$, although that probability may be a little below or a little above $1 - \alpha$ in individual cases. And neither L_1 nor L_2 is as likely as with the nonadjusted Wald interval to appear less than 0 or greater than 1.

(d) Which Interval to Use.   Because n is fairly large in Example 14, and $\hat{p}$ and $\hat{q}$ are close to 0.5, the confidence limits do not vary appreciably among the aforementioned procedures. However, the following general guidelines emerge from the many studies performed on confidence intervals for the binomial parameter, p, although there is not unanimity of opinion:

- If it is desired that there is probability of at least $1 - \alpha$ that the interval from L_1 to L_2 includes p, even if the interval might be very conservative (i.e., the probability might be much greater than $1 - \alpha$), then the Clopper-Pearson interval (Section 8a) should be used.

- If it is desired that the probability is close to $1 - \alpha$ that the interval from L_1 to L_2 includes p, even though it might be either a little above or a little below $1 - \alpha$, then the adjusted Wald interval (Section 8c) is preferable. Another approximation is that of Wilson (1927) and is sometimes called the "score interval" (with different, but equivalent, formulas given by Agresti and Coull, 2000 and Newcombe, 1998a); and it yields results similar to those of the adjusted Wald interval.

- The nonadjusted Wald interval (Section 8b) should not be used.

- None of these confidence-interval calculations works acceptably when $X = 0$ (that is, when $\hat{p} = 0$) or when $X = 1$ (that is, when $\hat{p} = 1$). The following exact confidence limits (Blyth, 1986; Fleiss, Levin, and Paik, 2003: 23; Sprent and Smeeton, 2001: 81; Vollset, 1993) should be used in those circumstances[‡]:

$$\text{If } X = 0: \qquad L_1 = 0 \text{ and } L_2 = 1 - \sqrt[n]{\alpha(1)} \tag{37}$$

$$\text{If } X = 1: \qquad L_1 = \sqrt[n]{\alpha(1)} \text{ and } L_2 = 1. \tag{38}$$

One-tailed confidence intervals can be determined via the the use of a one-tailed critical value for F or Z.

---

*The symbol above the X and n is called a tilde (pronounced "til-duh"); $\tilde{X}$ is read as "X tilde" and $\tilde{n}$ as "n tilde."

[†]For a 95% confidence interval, $Z_{\alpha(2)} = 1.9600$, $Z^2_{0.05(2)} = 3.8416$, and $Z^2_{0.05(2)}/2 = 1.9208$; so it is often recommended to simply use $X + 2$ in place of X and $n + 4$ in place of n, which yields very good results. For the data of Example 14, this would give the same confidence limits as in Example 14c: $L_1 = 0.357$ and $L_2 = 0.513$.

[‡]The notation $\sqrt[n]{X}$ represents the nth root of X, which may also be written as $X^{1/n}$, so $\sqrt[n]{\alpha(1)}$ may be seen written as $[\alpha(1)]^{1/n}$. It can also be noted that negative exponents represent reciprocals: $X^{-a} = 1/X^a$. This modern notation for fractional and negative powers was introduced by Sir Isaac Newton (1642–1727) in 1676, although the notion of fractional powers was conceived much earlier, such as by French writer Nicole Oresme (ca. 1323–1382) (Cajori, 1928–1929: Vol. I: 91, 354, 355).

There are computer programs and published tables that provide confidence limits for p, but users of them should be confident in the computational method employed to obtain the results.

(e) Confidence Limits with Finite Populations. It is usually considered that the size of a sampled population (N) is *very much* larger than the size of the sample (n), as if the population size was infinite. If, however, n is large compared to N (i.e., the sample is a large portion of the population), it is said that the population is finite. As n approaches N, the estimation of p becomes more accurate, and the calculation of confidence limits by the adjusted Wald method (or by the Wald method) improves greatly by converting the lower confidence limit, L_1, to $(L_1)_c$ and the upper confidence limit, L_2, to $(L_2)_c$, as follows (Burstein, 1975):

$$(L_1)_c = \frac{\tilde{X} - 0.5}{\tilde{n}} - \left(\frac{\tilde{X} - 0.5}{\tilde{n}} - L_1\right)\sqrt{\frac{N - \tilde{n}}{N - 1}} \tag{39}$$

$$(L_2)_c = \frac{\tilde{X} + \tilde{X}/\tilde{n}}{\tilde{n}} + \left(L_2 - \frac{\tilde{X} + \tilde{X}/\tilde{n}}{\tilde{n}}\right)\sqrt{\frac{N - \tilde{n}}{N - 1}}. \tag{40}$$

The more $(N - n)/(N - 1)$ differs from 1.0, the more the confidence limits from Equations 39 and 40 will be preferable to those that do not consider the sample population to be finite.

(f) Sample Size Requirements. A researcher may wish to estimate, for a given $\hat{p}$, how large a sample is necessary to produce a confidence interval of a specified width.

Section 8b presents a procedure (although often a crude one) for calculating a confidence interval for p when p is not close to 0 or 1. That normal approximation can be employed, with $\hat{p}$ and $\hat{q}$ obtained from an existing set of data, to provide a rough estimate of the number of data (n) that a future sample from the same population must contain to obtain a confidence interval where both L_1 and L_2 are at a designated distance, δ, from p:

$$n = \frac{Z_{0.05(2)}^2 \hat{p}\hat{q}}{\delta^2} \tag{41}$$

(Cochran, 1977: 75–76; Hollander and Wolfe, 1999: 30), where $Z_{0.05(2)}$ is a two-tailed normal deviate. If we do not have an estimate of p, then a conservative estimate of the required n can be obtained by inserting 0.5 for $\hat{p}$ and for $\hat{q}$ in Equation 41.

If the sample size, n, is not a small portion of the population size, N, then the required sample size is smaller than the n determined by Equation 41 and can be estimated as

$$m = \frac{n}{1 + (n - 1)/N} \tag{42}$$

(Cochran, 1977: 75–76).

For the confidence intervals of Sections 8a and 8c, a much better estimate of the needed sample size may be obtained by iteration. For a given $\hat{p}$, a value of n may be proposed, from Equation 41 or otherwise. (Equation 41 will yield an underestimate of n.) Then the confidence limits are calculated. If the confidence interval is wider than desired, perform the calculation again with a larger n; and if it is narrower than desired, try again with a smaller n. This process can be repeated until an estimate of the required n is obtained for the interval width desired.

9 CONFIDENCE INTERVAL FOR A POPULATION MEDIAN

The confidence limits for a population median* may be obtained by considering a binomial distribution with $p = 0.5$. The procedure thus is related to the binomial and sign tests in earlier sections of this chapter and may conveniently use Table 27 from *Appendix: Statistical Tables and Graphs*. That table gives $C_{\alpha,n}$, and from this we can state the confidence interval for a median to be

$$P(X_i \leq \text{population median} \leq X_j) \geq 1 - \alpha, \tag{43}$$

where

$$i = C_{\alpha(2),n} + 1 \tag{44}$$

and

$$j = n - C_{\alpha(2),n} \tag{45}$$

(e.g., MacKinnon, 1964), if the data are arranged in order of magnitude (so that X_i is the smallest measurement and X_j is the largest). The confidence limits, therefore, are $L_1 = X_i$ and $L_2 = X_j$. Because of the discreteness of the binomial distribution, the confidence will typically be a little greater than the $1 - \alpha$ specified. This procedure is demonstrated in Example 14a.

EXAMPLE 14a   A Confidence Interval for a Median

Let us determine a 95% confidence interval for the median of the population from which each of the two sets of data in Example 3 from *Measures of Central Tendency* came, where the population median was estimated to be 40 mo for species A and 52 mo for species B.

For species $A, n = 9$, so (from the before-mentioned Table 27) $C_{0.05(2),9} = 1$ and $n - C_{0.05(2),9} = 9 - 1 = 8$. The confidence limits are, therefore, X_i and X_j, where $i = 1 + 1 = 2$ and $j = 8$; and we can state

$$P(X_2 \leq \text{population median} \leq X_8) \geq 0.95$$

or

$$P(36 \text{ mo} \leq \text{population median} \leq 43 \text{ mo}) \geq 0.95.$$

For species $B, n = 10$, and Table 27 *Appendix: Statistical Tables and Graphs* informs us that $C_{0.05(2),10} = 1$; therefore, $n - C_{0.05(2),10} = 10 - 1 = 9$. The confidence limits are X_i and X_j, where $i = 1 + 1 = 2$ and $j = 9$; thus,

$$P(X_2 \leq \text{population median} \leq X_9) \geq 0.95.$$

or

$$P(36 \text{ mo} \leq \text{population median} \leq 69 \text{ mo}) \geq 0.95.$$

Hutson (1999) discussed calculation of confidence intervals for quantiles other than the median.

---

*Such confidence intervals were first discussed by William R. Thompson in 1936 (Noether, 1984).

(a) A Large-Sample Approximation. For samples larger than appearing in Table 27 *Appendix: Statistical Tables and Graphs*, an excellent approximation of the lower confidence limit (based on Hollander and Wolfe, 1999: Section 3.6), is derived from the normal distribution as

$$L_1 = X_i, \tag{46}$$

where

$$i = \frac{n - Z_{\alpha(2)}\sqrt{n}}{2} \tag{47}$$

rounded to the nearest integer, and $Z_{\alpha(2)}$ is the two-tailed normal deviate read from Table 2 *Appendix: Statistical Tables and Graphs*. (Recall that $Z_{\alpha(2)} = t_{\alpha(2),\infty}$, and so may be read from the last line of Table 3 *Appendix: Statistical Tables and Graphs*.) The upper confidence limit is

$$L_2 = X_{n-i+1}. \tag{48}$$

By this method we approximate a confidence interval for the population median with confidence $\geq 1 - \alpha$.

10 TESTING FOR DIFFERENCE BETWEEN TWO PROPORTIONS

Two proportions may be compared by casting the underlying data in a 2×2 contingency table and considering that one margin of the table is fixed. For example, the column totals (the total data for each species) are fixed and the proportion of mice afflicted with the parasite are $\hat{p}_1 = 18/24 = 0.75$ for species 1 and $\hat{p}_2 = 10/25 = 0.40$ for species 2. The null hypothesis ($H_0: p_1 = p_2$) may be tested using the normal distribution (as shown in Example 15), by computing

$$Z = \frac{\hat{p}_1 - \hat{p}_2}{\sqrt{\dfrac{\overline{p}\,\overline{q}}{n_1} + \dfrac{\overline{p}\,\overline{q}}{n_2}}}. \tag{49}$$

Here, $\overline{p}$ is the proportion of parasitized mice obtained by pooling all n data:

$$\frac{X_1 + X_2}{n_1 + n_2} \tag{50}$$

or, equivalently, as

$$\frac{n_1\hat{p}_1 + n_2\hat{p}_2}{n_1 + n_2}, \tag{51}$$

and $\overline{q} = 1 - \overline{p}$.

A null hypothesis may propose a difference other than zero between two proportions. With p_0 the specified difference, $H_0: |p_1 - p_2| = p_0$ may be tested by replacing the numerator* of Equation 49 with $|p_1 - p_2| - p_0$.

One-tailed testing is also possible (with $p_0 = 0$ or $p_0 \neq 0$) and is effected in a fashion analogous to that used for testing difference between two means. This is demonstrated in Example 15b, where the alternate hypothesis is that the parasite infects a higher proportion of fish in species 1 than in species 2, and also in Example 15c.

---

*If p_0 is not zero in H_0 and H_A, in some cases a somewhat more powerful test has been reported if the denominator of Equation 49 is replaced by $\sqrt{\hat{p}_1\hat{q}_1/n_1 + \hat{p}_2\hat{q}_2/n_2}$ (Agresti, 2002: 77; Eberhardt and Fligner, 1977).

EXAMPLE 15   Testing for Difference Between Two Proportions

Using the data of Example 3 from *Contingency Tables*,

$$\hat{p}_1 = X_1/n_1 = 18/24 = 0.75$$

$$\hat{p}_2 = X_2/n_2 = 10/25 = 0.40$$

$$\bar{p} = (18 + 10)/(24 + 25) = 28/49 = 0.5714$$

$$\bar{q} = 1 - 0.5714 = 0.4286.$$

(a) The two-tailed test for $H_0\colon p_1 = p_2$ versus $H_A\colon p_1 \neq p_2$

$$Z = \frac{\hat{p}_1 - \hat{p}_2}{\sqrt{\dfrac{\bar{p}\,\bar{q}}{n_1} + \dfrac{\bar{p}\,\bar{q}}{n_2}}}$$

$$= \frac{0.75 - 0.40}{\sqrt{\dfrac{(0.5714)(0.4286)}{24} + \dfrac{(0.5714)(0.4286)}{25}}}$$

$$= \frac{0.35}{\sqrt{0.1414}} = 2.4752$$

For $\alpha = 0.05$, $Z_{0.05(2)} = 1.9600$ and H_0 is rejected.

$$0.01 < P < 0.02 \quad [P = 0.013]$$

(b) The one-tailed test for $H_0\colon p_1 \leq p_2$ versus $H_A\colon p_1 > p_2$

For $\alpha = 0.05$, $Z_{0.05(1)} = 1.6449$. Because $Z > 1.6449$ and the difference (0.35) is in the direction of the alternate hypothesis, H_0 is rejected.

$$0.005 < P < 0.01 \quad [P = 0.007]$$

(c) The one-tailed test for $H_0\colon p_1 \geq p_2$ versus $H_A\colon p_1 < p_2$

For $\alpha = 0.05$, $Z_{0.05(1)} = 1.6449$. Because $Z > 1.6449$ but the difference (0.35) is not in the direction of the alternate hypothesis, H_0 is not rejected.

Important note: Three pairs of hypotheses are tested in this example. This is done only for demonstration of the test, for in practice it would be proper to test only one of these pairs of hypotheses for a given set of data. The decision of which one of the three pairs of hypotheses to use should be made on the basis of the biological question being asked and is to be made *prior to* the collection of the data.

11 CONFIDENCE LIMITS FOR THE DIFFERENCE BETWEEN PROPORTIONS

When a proportion ($\hat{p}_1$) is obtained by sampling one population, and a proportion ($\hat{p}_2$) is from another population, confidence limits for the difference between the two population proportions ($p_1 - p_2$) can be calculated by many methods.* The most common are these:

(a) Wald Interval. A confidence interval may be expressed in a fashion analogous to the Wald interval for a single proportion (which is discussed in Section 8b):

$$\hat{p}_1 - \hat{p}_2 \pm Z_{\alpha(2)} \sqrt{\frac{\hat{p}_1 \hat{q}_1}{n_1} + \frac{\hat{p}_2 \hat{q}_2}{n_2}}, \tag{52}$$

where $\hat{p}_1 = X_1/n_1, \hat{p}_2 = X_2/n_2, \hat{q}_1 = 1 - \hat{p}_1$, and $\hat{q}_2 = 1 - \hat{p}_2$.

Though commonly used, this calculation of confidence limits yields poor results, even when sample sizes are large. These confidence limits include $\hat{p}_1 - \hat{p}_2$ less than $1 - \alpha$ of the time (e.g., less than 95% of the time when expressing a 95% confidence interval), and thus they are said to be "liberal." The confidence limits include $\hat{p}_1 - \hat{p}_2$ *much* less than $1 - \alpha$ of the time when $\hat{p}_1$ and $\hat{p}_2$ are near 0 or 1. Also, Equation 52 produces confidence limits that are always symmetrical around $\hat{p}_1 - \hat{p}_2$, whereas (unless $\hat{p}_1 - \hat{p}_2 = 0.50$) the distance between the lower confidence limit (L_1) and $\hat{p}_1 - \hat{p}_2$ should be different than the distance between $\hat{p}_1 - \hat{p}_2$ and the upper limit (L_2). This unrealistic symmetry can produce a calculated L_1 that is less than 0 or greater than 1, which would be an unreasonable result. In addition, when both $\hat{p}_1$ and $\hat{p}_2$ are either 0 or 1, a confidence interval cannot be calculated by this equation. Therefore, this calculation (as is the case of Equation 36 in Section 8b) generally is not recommended.

(b) Adjusted Wald Interval. Agresti and Caffo (2000) have shown that it is far preferable to employ an "adjusted" Wald interval (analogous to that in the one-sample situation of Section 8c), where Equation 52 is employed by substituting[†] $\tilde{X}_i = X_i + Z_{\alpha(2)}^2/4$ for X_i and $\tilde{n}_i = n_i Z_{\alpha(2)}^2/2$ for n_i. As shown in Example 15a, the adjusted confidence interval is obtained by using $\tilde{p}_i = \tilde{X}_i/\tilde{n}_i$ in place of $\hat{p}_i$ in Equation 52. This adjusted Wald confidence interval avoids the undesirable severe liberalism obtainable with the unadjusted interval, although it can be slightly conservative (i.e., have a probability of a little greater than $1 - \alpha$ of containing $p_1 - p_2$) when $\tilde{p}_1$ and $\tilde{p}_2$ are both near 0 or 1.

Newcombe (1998b) discussed a confidence interval that is a modification of the one-sample interval based upon Wilson (1927) and mentioned in Section 8d. It is said to produce results similar to those of the adjusted Wald interval.

(c) Sample Size Requirements. If the statistics $\hat{p}_1$ and $\hat{p}_2$ are obtained from sampling two populations, it may be desired to estimate how many data must be collected from those populations to calculate a confidence interval of specified width for $p_1 - p_2$. The following may be derived from the calculation of the Wald interval

---

*A large number of them are discussed by Agresti and Caffo (2000), Agresti and Coull (1998), Blyth (1986), Hauck and Anderson (1986), Newcombe (1998b), and Upton (1982).

[†]For a 95% confidence interval, $Z_{\alpha(2)}^2/4 = (1.9600)^2/4 = 0.9604$ and $Z_{\alpha(2)}^2/2 = (1.9600)^2/2 = 1.9208$, so using $X_i + 1$ in place of X_i and $n_i + 2$ in place of n_i yields very good results.

EXAMPLE 15a   Confidence Interval for the Difference Between Two Population Proportions

For the 95% adjusted Wald confidence interval using the data,

$$Z_{0.05(2)} = 1.9600,$$

$$\text{so } (1.9600)^2/4 = 0.9604 \text{ and } (1.9600)^2/2 = 1.9208$$

$X_1 = 18, \text{ so } \tilde{X}_1 = 18 + 0.9604 = 18.9604$

$X_2 = 10, \text{ so } \tilde{X}_2 = 10 + 0.9604 = 10.9604$

$n_1 = 24, \text{ so } \tilde{n}_1 = 24 + 1.9208 = 25.9208$

$n_2 = 25, \text{ so } \tilde{n}_2 = 25 + 1.9208 = 26.9208$

$\tilde{p}_1 = 18.9604/25.9208 = 0.7315 \text{ and } \tilde{q}_1 = 1 - 0.7315 = 0.2685$

$\tilde{p}_2 = 10.9604/26.9208 = 0.4071 \text{ and } \tilde{q}_1 = 1 - 0.4071 = 0.5929$

$$95\% \text{ CI for } p_1 - p_2 = \tilde{p}_1 - \tilde{p}_2 \pm Z_{0.05(2)}\sqrt{\frac{\tilde{p}_1\tilde{q}_1}{\tilde{n}_1} + \frac{\tilde{p}_2\tilde{q}_2}{\tilde{n}_2}}$$

$$= 0.7315 - 0.4071 \pm 1.9600\sqrt{\frac{(0.7315)(0.2685)}{25.9208} + \frac{(0.4071)(0.5929)}{26.9208}}$$

$$= 0.3244 \pm 1.9600\sqrt{0.0076 + 0.0090} = 0.3244 \pm 0.2525$$

$L_1 = 0.07; \quad L_2 = 0.58.$

(Section 11a) for equal sample sizes ($n = n_1 = n_2$):

$$n = Z_{\alpha(2)}^2 \left[\frac{\hat{p}_1\hat{q}_1 + \hat{p}_2\hat{q}_2}{(\hat{p}_1 - \hat{p}_2)^2} \right]. \tag{53}$$

This is an underestimate of the number of data needed, and a better estimate may be obtained by iteration, using the adjusted Wald interval (Section 11b). For the given $\hat{p}_1$ and $\hat{p}_2$, a speculated value of n is inserted into the computation of the adjusted Wald interval in place of n_1 and n_2. If the calculated confidence interval is wider than desired, the adjusted Wald calculation is performed again with a larger n; if it is narrower than desired, the calculation is executed with a smaller n. This process is repeated until n is obtained for the interval width desired.

The iteration using the adjusted Wald equation may also be performed with one of the future sample sizes specified, in which case the process estimates the size needed for the other sample.

12   POWER, DETECTABLE DIFFERENCE, AND SAMPLE SIZE IN TESTING DIFFERENCE BETWEEN TWO PROPORTIONS

(a) Power of the Test.   If the test of $H_0: p_1 = p_2$ versus $H_A: p_1 \neq p_2$ is to be performed at the α significance level, with n_1 data in sample 1 and n_2 data in sample 2, and if the two samples come from populations actually having proportions of p_1

and p_2, respectively, then an estimate of power is

$$\text{power} = P\left[Z \le \frac{-Z_{\alpha(2)}\sqrt{\overline{p}\,\overline{q}/n_1 + \overline{p}\,\overline{q}/n_2} - (p_1 - p_2)}{\sqrt{p_1 q_1/n_1 + p_2 q_2/n_2}}\right]$$

$$+ P\left[Z \ge \frac{Z_{\alpha(2)}\sqrt{\overline{p}\,\overline{q}/n_1 + \overline{p}\,\overline{q}/n_2} - (p_1 - p_2)}{\sqrt{p_1 q_1/n_1 + p_2 q_2/n_2}}\right] \tag{54}$$

(Marascuilo and McSweeney, 1977: 111), where

$$\overline{p} = \frac{n_1 p_1 + n_2 p_2}{n_1 + n_2}, \tag{55}$$

$$q_1 = 1 - p_1, \tag{56}$$

$$q_2 = 1 - p_2, \tag{57}$$

and

$$\overline{q} = 1 - \overline{p}. \tag{58}$$

The calculation is demonstrated in Example 16.

For the one-tailed test of $H_0: p_1 \ge p_2$ versus $H_A: p_1 < p_2$, the estimated power is

$$\text{power} = P\left[Z \le \frac{-Z_{\alpha(1)}\sqrt{\overline{p}\,\overline{q}/n_1 + \overline{p}\,\overline{q}/n_2} - (p_1 - p_2)}{\sqrt{p_1 q_1/n_1 + p_2 q_2/n_2}}\right]; \tag{59}$$

and for the one-tailed hypotheses, $H_0: p_1 \le p_2$ versus $H_A: p_1 > p_2$,

$$\text{power} = P\left[Z \ge \frac{Z_{\alpha(1)}\sqrt{\overline{p}\,\overline{q}/n_1 + \overline{p}\,\overline{q}/n_2} - (p_1 - p_2)}{\sqrt{p_1 q_1/n_1 + p_2 q_2/n_2}}\right]. \tag{60}$$

These power computations are based on approximations to the Fisher exact test (Section 16) and tend to produce a conservative result. That is, the power is likely to be greater than that calculated.

(b) Sample Size Required and Minimum Detectable Difference. Estimating the sample size needed in a future comparison of two proportions, with a specified power, has been discussed by several authors,* using a normal approximation. Such an estimate may also be obtained by iteration analogous to that of Section 7c.

---

*See, for example, Casagrande, Pike, and Smith (1978); Cochran and Cox (1957: 27); and Fleiss, Levin, and Paik (2003: 72). Also, Hornick and Overall (1980) reported that the following computation of Cochran and Cox (1957: 27) yields good results and appears not to have a tendency to be conservative:

$$n = \left(Z_\alpha + Z_{\beta(1)}\right)^2 \Big/ \left[2\left(\arcsin\sqrt{p_1} - \arcsin\sqrt{p_2}\right)\right], \tag{61}$$

where the arcsines are expressed in radians.

EXAMPLE 16   Estimation of Power in a Two-Tailed Test Comparing Two Proportions

We propose to test $H_0: p_1 = p_2$ versus $H_A: p_1 \neq p_2$, with $\alpha = 0.05$, $n_1 = 50$, and $n_2 = 45$, where in the sampled populations $p_1 = 0.75$ and $p_2 = 0.50$. The power of the test can be estimated as follows.

We first compute (by Equation 55):

$$\bar{p} = \frac{(50)(0.75) + (45)(0.50)}{50 + 45} = 0.6316 \quad \text{and} \quad \bar{q} = 1 - \bar{p} = 0.3684.$$

Then

$$\frac{\bar{p}\,\bar{q}}{n_1} = \frac{(0.6316)(0.3684)}{50} = 0.0047; \quad \frac{\bar{p}\,\bar{q}}{n_2} = 0.0052;$$

$$\frac{p_1 q_1}{n_1} = \frac{(0.75)(0.25)}{50} = 0.0038; \quad \frac{p_2 q_2}{n_2} = \frac{(0.50)(0.50)}{45} = 0.0056;$$

$$Z_{0.05(2)} = 1.9600;$$

and, using Equation 64,

$$\text{power} = P\left[Z \leq \frac{-1.9600\sqrt{0.0047 + 0.0052} - (0.75 - 0.50)}{\sqrt{0.0038 + 0.0056}} \right]$$

$$+ P\left[Z \geq \frac{1.9600\sqrt{0.0047 + 0.0052} - (0.75 - 0.50)}{\sqrt{0.0038 + 0.0056}} \right]$$

$$= P(Z \leq -4.59) + P(Z \geq -0.57)$$

$$= P(Z \geq -4.59) + [1 - P(Z \geq 0.57)]$$

$$= 0.0000 + [1.0000 - 0.2843]$$

$$= 0.72.$$

The estimation procedure uses Equation 54, 59, or 60, depending upon the null hypothesis to be tested. The power is thus determined for the difference $(p_1 - p_2)$ desired to be detected between two population proportions. This is done using equal sample sizes $(n_1 = n_2)$ that are a reasonable guess of the sample size that is required from each of the two populations. If the power thus calculated is less than desired, then the calculation is repeated using a larger sample size. If the calculated power is greater than the desired power, the computation is performed again but using a smaller sample size. Such iterative calculations are repeated until the specified power has been obtained, and the last n used in the calculation is the estimate of the sample size required in each of the two samples.

The samples from the two populations should be of the same size $(n_1 = n_2)$ for the desired power to be calculated with the fewest total number of data $(n_1 + n_2)$. Fleiss, Tytun, and Ury (1980), Levin and Chen (1999), and Ury and Fleiss (1980) discuss the estimation of n_1 and n_2 when acquiring equal sample sizes is not practical.

In a similar manner, n, α, and power may be specified, and the minimum detectable difference ($p_1 - p_2$) may be estimated. This is done by iteration, using Equation 54, 59, or 60, depending upon the null hypothesis to be tested. A reasonable guess of the minimum detectable difference can be entered into the equation and, if the calculated power is less than that desired, the computation is repeated with a larger ($p_1 - p_2$); if the calculated power is greater than that desired, the computation is repeated by inserting a smaller $p_1 - p_2$ into the equation; and when the desired power is obtained from the equation, the $p_1 - p_2$ last used in the calculation is an expression of the estimated minimum detectable difference.

Ury (1982) described a procedure for estimating one of the two population proportions if n, α, the desired power, and the other population proportion are specified.

13 COMPARING MORE THAN TWO PROPORTIONS

Comparison of proportions may be done by contingency-table analysis. For example, the null hypothesis could be stated as, "The proportions of males and females are the same among individuals of each of the four hair colors."

Alternatively, an approximation related to the normal approximation is applicable (if n is large and neither p nor q is very near 1). Using this approximation, one tests $H_0: p_1 = p_2 = \cdots = p_k$ against the alternative hypothesis that all k proportions are not the same, as

$$\chi^2 = \sum_{i=1}^{k} \frac{(X_i - n_i\overline{p})^2}{n_i\overline{p}\,\overline{q}} \tag{62}$$

(Pazer and Swanson, 1972: 187–190). Here,

$$\overline{p} = \frac{\displaystyle\sum_{i=1}^{k} X_i}{\displaystyle\sum_{i=1}^{k} n_i} \tag{63}$$

is a pooled proportion, $\overline{q} = 1 - \overline{q}$, and χ^2 has $k - 1$ degrees of freedom. Example 17 demonstrates this procedure, which is equivalent to χ^2 testing of a contingency table with two rows (or two columns).

We can instead test whether k p's are equal not only to each other but to a specified constant, p_0 (i.e., $H_0: p_1 = p_2 = \cdots = p_k = p_0$). This is done by computing

$$\chi^2 = \sum_{i=1}^{k} \frac{(X_i - n_i p_0)^2}{n_i p_0 (1 - p_0)}, \tag{64}$$

which is then compared to the critical value of χ^2 for k (rather than $k - 1$) degrees of freedom (Kulkarni and Shah, 1995, who also discuss one-tailed testing of H_0, where H_A is that $p_i \neq p_0$ for at least one i).

EXAMPLE 17 **Comparing Four Proportions, Using the Data of Example 1 from *Contingency Tables***

$$n_1 = 87, \; X_1 = 32, \; \hat{p}_1 = \frac{32}{87} = 0.368, \; \hat{q}_1 = 0.632$$

$$n_2 = 108, \; X_2 = 43, \; \hat{p}_2 = \frac{43}{108} = 0.398, \; \hat{q}_2 = 0.602$$

$$n_3 = 80, \; X_3 = 16, \; \hat{p}_3 = \frac{16}{80} = 0.200, \; \hat{q}_3 = 0.800$$

$$n_4 = 25, \; X_4 = 9, \; \hat{p}_4 = \frac{9}{25} = 0.360, \; \hat{q}_4 = 0.640$$

$$\bar{p} = \frac{\sum X_i}{\sum n_i} = \frac{32 + 43 + 16 + 9}{87 + 108 + 80 + 25} = \frac{100}{300} = \frac{1}{3}$$

$$\bar{q} = 1 - \bar{p} = \frac{2}{3}$$

$$\chi^2 = \sum \frac{(X_i - n_i \bar{p})^2}{n_i \bar{p} \, \bar{q}}$$

$$= \frac{\left[132 - (87) \left(\frac{1}{3} \right) \right]^2}{(87) \left(\frac{1}{3} \right) \left(\frac{2}{3} \right)} + \frac{\left[43 - (108) \left(\frac{1}{3} \right) \right]^2}{(108) \left(\frac{1}{3} \right) \left(\frac{2}{3} \right)} + \frac{\left[16 - (80) \left(\frac{1}{3} \right) \right]^2}{(80) \left(\frac{1}{3} \right) \left(\frac{2}{3} \right)}$$

$$+ \frac{\left[9 - (25) \left(\frac{1}{3} \right) \right]^2}{(25) \left(\frac{1}{3} \right) \left(\frac{2}{3} \right)}$$

$$= 0.4655 + 2.0417 + 6.4000 + 0.0800$$

$$= 8.987$$

$$\nu = k - 1 = 4 - 1 = 3$$

$$\chi^2_{0.05,3} = 7.815$$

Therefore, reject H_0.

$$0.025 < P < 0.05 \quad [P = 0.029]$$

Note how the two procedures yield the same results for contingency tables with two rows or two columns.

Finally, it should be noted that comparing several p's yields the same results as if one compared the associated q's.

14 MULTIPLE COMPARISONS FOR PROPORTIONS

(a) Comparisons of All Pairs of Proportions. If the null hypothesis $H_0: p_1 = p_2 = \cdots = p_k$ (see Section 13) is rejected, then we may desire to determine specifically which population proportions are different from which others. The following procedure (similar to that of Levy, 1975a) allows for testing analogous to the Tukey test. An angular transformation of each sample proportion is to be used. If $\hat{p}$, but not X and n, is known, then Equation 64a may be used. If, however, X and n are known, then either Equation 64b or 64c is preferable. (The latter two equations give similar results, except for small or large $\hat{p}$, where Equation 64c is probably better.)

$$p' = \arcsin \sqrt{\hat{p}}, \tag{64a}$$

$$p' = \arcsin \sqrt{\frac{X + \frac{3}{8}}{n + \frac{3}{4}}}. \tag{64b}$$

$$p' = \frac{1}{2}\left[\arcsin \sqrt{\frac{X}{n + 1}} + \arcsin \sqrt{\frac{X + 1}{n + 1}} \right], \tag{64c}$$

As shown in Example 18, the multiple comparison procedure is similar to that the Tukey test. The standard error for each comparison is, in degrees,[*]

$$SE = \sqrt{\frac{820.70}{n + 0.5}} \tag{65}$$

---

[*]The constant 820.70 square degrees results from $(180°/2\pi)^2$, which follows from the variances reported by Anscombe (1948) and Freeman and Tukey (1950).

EXAMPLE 18 **Tukey-Type Multiple Comparison Testing Among the Four Proportions of Example 17**

Samples ranked by proportion (i):	3	4	1	2
Ranked sample proportions ($p_i = X_i/n_i$):	16/80	9/25	32/87	43/108
	= 0.200	= 0.360	= 0.368	= 0.398
Ranked transformed proportions (p'_i, in degrees):	26.85	37.18	37.42	39.18

Comparison B vs. A	Difference $p'_B - p'_A$	SE	q	$q_{0.05,\infty,4}$	Conclusion
2 vs. 3	$39.18 - 26.85 = 12.33$	2.98	4.14	3.633	Reject H_0: $p_2 = p_3$
2 vs. 4	$39.18 - 37.18 = 2.00$	4.46	0.45	3.633	Do not reject H_0: $p_2 = p_4$
2 vs. 1	Do not test				
1 vs. 3	$37.42 - 26.85 = 10.57$	3.13	3.38	3.633	Do not reject H_0: $p_1 = p_3$
1 vs. 4	Do not test				
4 vs. 3	Do not test				

Overall conclusion: $p_4 = p_1 = p_2$ and $p_3 = p_4 = p_1$, which is a kind of ambiguous result. By chi-square analysis it was concluded that $p_3 \neq p_4 = p_1 = p_2$; it is likely that the present method lacks power for this set of data.

For sample 3, for example, $X/(n + 1) = 16/81 = 0.198$ and $(X + 1)/(n + 1) = 17/81 = 0.210$, so $p'_3 = \frac{1}{2}[\arcsin\sqrt{0.198} + \arcsin\sqrt{0.210}] = \frac{1}{2}[26.4215 + 27.2747] = 26.848$. If we use Table 24 from *Appendix: Statistical Tables and Graphs* to obtain the two needed arcsines, we have $p'_3 = \frac{1}{2}[26.42 + 27.27] = 26.845$.

---

if the two samples being compared are the same size, or

$$\text{SE} = \sqrt{\frac{410.35}{n_A + 0.5} + \frac{410.35}{n_B + 0.5}} \tag{66}$$

if they are not. The critical value is $q_{\alpha,\infty,k}$.

Use of the normal approximation to the binomial is possible in multiple-comparison testing (e.g., Marascuilo, 1971: 380–382); but the preceding procedure is preferable, even though it — and the methods to follow in this section — may lack desirable power.

(b) Comparison of a Control Proportion to Each Other Proportion. A procedure analogous to the Dunnett test may be used as a multiple comparison test where instead of comparing all pairs of proportions we desire to compare one proportion (designated as the "control") to each of the others. Calling the control group B, and each other group, in turn, A, we compute the Dunnett test statistic:

$$q = \frac{p'_B - p'_A}{\text{SE}}. \tag{67}$$

Here, the proportions have been transformed as earlier in this section, and the appropriate standard error is

$$SE = \sqrt{\frac{1641.40}{n + 0.5}} \qquad (68)$$

if Samples A and B are the same size, or

$$SE = \sqrt{\frac{820.70}{n_A + 0.5} + \frac{820.70}{n_B + 0.5}} \qquad (69)$$

if $n_A \neq n_B$. The critical value is $q'_{\alpha(1),\infty,p}$ or $q'_{\alpha(2),\infty,p}$ for one-tailed or two-tailed testing, respectively.

(c) Multiple Contrasts Among Proportions. The Scheffé procedure for multiple contrasts among means may be adapted to proportions by using angular transformations as done earlier in this section. For each contrast, we calculate

$$S = \frac{\left| \sum_i c_i p'_i \right|}{SE}, \qquad (70)$$

where

$$SE = \sqrt{820.70 \sum_i \frac{c_i^2}{n_i + 0.5}}, \qquad (71)$$

and c_i is a contrast coefficient. For example, if we wished to test the hypothesis H_0: $(p_1 + p_2 + p_4)/3 - p_3 = 0$, then $c_1 = \frac{1}{3}, c_2 = \frac{1}{3}, c_3 = -1$, and $c_4 = \frac{1}{3}$.

15 TRENDS AMONG PROPORTIONS

In a $2 \times c$ contingency table (2 rows and c columns), the columns may have a natural quantitative sequence. For example, they may represent different ages, different lengths of time after a treatment, different sizes, different degrees of infection, or different intensities of a treatment. In Example 19, the columns represent three age classes of women, and the data are the frequencies with which the 104 women in the sample exhibit a particular skeletal condition. The chi-square contingency-table analysis tests the null hypothesis that the occurrence of this condition is independent of age class (i.e., that the population proportion, p, of women with the condition is the same for all three age classes). It is seen, in Example 19a, that this hypothesis is rejected, so we conclude that in the sampled population there is a relationship between the two variables (age class and skeletal condition).

EXAMPLE 19   Testing for Linear Trend in a 2 × 3 Contingency Table. The Data Are the Frequencies of Occurrence of a Skeletal Condition in Women, Tabulated by Age Class

	Age class			
	Young	*Medium*	*Older*	*Total*
Condition present	6	16	18	40
Condition absent	22	28	14	64
Total	28	44	32	104
$\hat{p}_j$	0.2143	0.3636	0.5625	
X_j	−1	0	1	

(a) Comparison of proportions

H_0: In the sampled population, the proportion of women with this condition is the same for all three age classes.

H_A: In the sampled population, the proportion of women with this condition is not the same for all three age classes.

$$\hat{f}_{11} = (40)(28)/104 = 10.7692, \ \hat{f}_{12} = (40)(44)/104 = 16.9231, \ldots,$$
$$\hat{f}_{23} = (64)(28)/104 = 19.6923$$

$$\chi^2 = \sum_{i=1}^{r} \sum_{j=1}^{c} \frac{(f_{ij} - \hat{f}_{ij})^2}{\hat{f}_{ij}}$$

$$= \frac{(6 - 10.7692)^2}{10.7692} + \frac{(16 - 16.9231)^2}{16.9231} + \cdots + \frac{(14 - 19.6923)^2}{19.6923}$$

$$= 2.1121 + 0.0504 + 2.6327 + 1.3200 + 0.0315 + 1.6454$$

$$= 7.7921$$

(b) Test for trend

H_0: In the sampled population, there is a linear trend among these three age categories for the proportion of women with this condition.

H_A: In the sampled population, there is not a linear trend among these three age categories for the proportion of women with this condition.

$$\chi_t^2 = \frac{n}{R_1 R_2} \cdot \frac{\left(n \sum_{j=1}^{c} f_{1,j} X_j - R_1 \sum_{j=1}^{c} C_j X_j \right)^2}{n \sum_{j=1}^{c} C_j X_j^2 - \left(\sum_{j=1}^{c} C_j X_j \right)^2}$$

$$= \frac{104}{(40)(64)} \cdot \frac{\left\{ \begin{array}{l} 104[(6)(-1) + (16)(0) + (18)(1)] \\ -40[(28)(-1) + (44)(0) + (32)(1)] \end{array} \right\}^2}{\begin{array}{l} 104[(28)(-1^2) + (44)(0^2) + (32)(1^2)] \\ -[(28)(-1) + (44)(0) + (32)(1)]^2 \end{array}}$$

$$= 0.04062 \cdot \frac{(1248 - 160)^2}{6240 - 16}$$

$$= (0.04062)(190.1902) = 7.726$$

	Chi-square	ν	P
Total	$\chi^2 = 7.792$	2	$0.01 < P < 0.025$ $[P = 0.020]$
Linear trend	$\chi_t^2 = 7.726$	1	
Departure from linear trend	$\chi_d^2 = 0.066$	1	$0.75 < P < 0.90$ $[P = 0.80]$

In addition, we may ask whether the difference among the three age classes follows a linear trend; that is, whether there is either a greater occurrence of the condition of interest in women of greater age or a lesser occurrence with greater age. The question of linear trend in a $2 \times n$ contingency table may be addressed by the method promoted by Armitage (1955, 1971: 363–365) and Armitage, Berry, and Matthews (2002: 504–509). To do so, the magnitudes of ages expressed by the age classes may be designated by consecutive equally spaced ordinal scores: X. For example, the "young," "medium," and "older" categories in the present example could be indicated by X's of 1, 2, and 3; or by 0, 1, and 2; or by -1, 0, and 1; or by 0.5, 1, and 1.5; or by 3, 5, and 7; and so on. The computation for trend is made easier if the scores are consecutive integers centered on zero, so scores of -1, 0, and 1 are used in Example 19. (If the number of columns, c, were 4, then X's such as -2, -1, 1, and 2, or 1, 0, 2, and 3 could be used.)

The procedure divides the contingency-table chi-square into component parts, somewhat as sum-of-squares partitioning is done in analysis of variance. The chi-square may be referred to as the total chi-square, a portion of which can be identified as being due to a linear trend:

$$\text{chi-square for linear trend} = \chi_t^2 = \frac{n}{R_1 R_2} \cdot \frac{\left(n\sum_{j=1}^{c} f_{1,j} X_j - R_1 \sum_{j=1}^{c} C_j X_j \right)^2}{n\sum_{j=1}^{c} C_j X_j^2 - \left(\sum_{j=1}^{c} C_j X_j \right)^2}, \qquad (72)$$

and the remainder is identified as not being due to a linear trend:

$$\text{chi-square for departure from linear trend} = \chi_d^2 = \chi^2 - \chi_t^2. \qquad (73)$$

Associated with these three chi-square values are degrees of freedom of $c - 1$ for χ^2, 1 for χ_t^2 and $c - 2$ for χ_d^2.* This testing for trend among proportions is more powerful than the chi-square test for difference among the c proportions, so a trend might be identified even if the latter chi-square test concludes no significant difference among the proportions.

If, in Example 19, the data in the second column (16 and 28) were for older women and the data in the third column (18 and 14) were for medium-aged women, the total chi-square would have been the same, but χ_t^2 would have been 0.899 $[P = 0.34]$, and it would have been concluded that there was no linear trend with age.

In Example 19, the presence of a physical condition was analyzed in reference to an ordinal scale of measurement ("young," "medium," and "older"). In other situations, an interval or ratio scale may be encountered. For example, the three columns might have represented age classes of known quantitative intervals. If the three categories were equal intervals of "20.0 − −39.9 years," "40.0–59.9 years," and "60.0 − −79.9 years," then the X's could be set as equally spaced values (for example, as −1, 0, and 1) the same as in Example 19. However, if the intervals were unequal in size, such as "20.0 − −29.9 years," "30.0 − −49.9 years," and "50.0 − −79.9 years," then the X's should reflect the midpoints of the intervals, such as by 25, 40, and 65; or (with the subtraction of 25 years) by 0, 15, and 40; or (with subtraction of 40 years) by −15, 0, and 25.

16 THE FISHER EXACT TEST

In the discussion of 2×2 contingency tables, we described contingency tables that have two fixed margins, and we recommended analyzing such tables using a contingency-corrected chi-square (χ_c^2 or χ_H^2), or a procedure known as the *Fisher exact test*.[†] The test using chi-square corrected for continuity is an approximation of the Fisher exact test, with χ_H^2 the same as χ_c^2 or routinely a better approximation than χ_c^2.

The Fisher exact test is based upon hypergeometric probabilities (see Section 2). The needed calculations can be tedious, but some statistical computer programs (e.g., Zar, 1987, and some statistical packages) can perform the test. Although this text recommends this test only for 2×2 tables having both margins fixed, some researchers use it for the other kinds of 2×2 tables.

---

*Armitage (1955) explained that this procedure may be thought of as a regression of the sample proportions, $\hat{p}_j$, on the ordinal scores, X_j, where the $\hat{p}_j$'s are weighted by the column totals, C_j; or as a regression of n pairs of Y and X, where Y is 1 for each of the observations in row 1 and is 0 for each of the observations in row 2.

[†]Named for Sir Ronald Aylmer Fisher (1890–1962), a monumental statistician recognized as a principal founder of modern statistics, with extremely strong influence in statistical theory and methods, including many areas of biostatistics (see, for example, Rao, 1992). At about the same time he published this procedure (Fisher, 1934: 99–101; 1935), it was also presented by Yates (1934) and Irwin (1935), so it is sometimes referred to as the Fisher-Yates test or Fisher-Irwin test. Yates (1984) observed that Fisher was probably aware of the exact-test procedure as early as 1926. Although often referred to as a statistician, R. A. Fisher also had a strong reputation as a biologist (e.g., Neyman, 1967), publishing — from 1912 to 1962 — 140 papers on genetics as well as 129 on statistics and 16 on other topics (Barnard, 1990).

The probability of a given 2×2 table is

$$P = \frac{\binom{R_1}{f_{11}} \binom{R_2}{f_{21}}}{\binom{n}{C_1}}, \qquad (74)$$

which is identical to

$$P = \frac{\binom{C_1}{f_{11}} \binom{C_2}{f_{12}}}{\binom{n}{R_1}}. \qquad (75)$$

From Equation 11, both Equations 74 and 75 reduce to

$$P = \frac{R_1! \, R_2! \, C_1! \, C_2!}{f_{11}! \, f_{21}! \, f_{12}! \, f_{22}! \, n!}, \qquad (76)$$

and it will be seen that there is advantage in expressing this as

$$P = \frac{\dfrac{R_1! \, R_2! \, C_1! \, C_2!}{n!}}{f_{11}! \, f_{12}! \, f_{21}! \, f_{22}!}. \qquad (77)$$

(a) One-Tailed Testing. If species 1 is naturally found in more rapidly moving waters, it would be reasonable to propose that it is better adapted to resist current, and the test could involve one-tailed hypotheses: H_0: The proportion of snails of species 1 resisting the water current is no greater than (i.e., less than or equal to) the proportion of species 2 withstanding the current, and H_A: The proportion of snails of species 1 resisting the current is greater than the proportion of species 2 resisting the current. The Fisher exact test proceeds as in Example 20.

EXAMPLE 20   A One-Tailed Fisher Exact Test, Using the Data of Example 4 from *Contingency Tables*

H_0: The proportion of snails of species 1 able to resist the experimental water current is no greater than the proportion of species 2 snails able to resist the current.

H_A: The proportion of snails of species 1 able to resist the experimental water current is greater than the proportion of species 2 snails able to resist the current.

	Resisted	Yielded	
Species 1	12	7	19
Species 2	2	9	11
	14	16	30

Expressing the proportion of each species resisting the current in the sample,

	Resisted	*Yielded*	*Total*
Species 1	0.63	0.37	1.00
Species 2	0.18	0.82	1.00

The sample data are in the direction of H_A, in that the species 1 sample has a higher proportion of resistant snails than does the species 2 sample. But are the data significantly in that direction? (If the data were not in the direction of H_A, the conclusion would be that H_0 cannot be rejected, and the analysis would proceed no further.) The probability of the observed table of data is

$$P = \frac{\dfrac{R_1!\,R_2!\,C_1!\,C_2!}{n!}}{f_{11}!\,f_{12}!\,f_{21}!\,f_{22}!}$$

$$= \frac{\dfrac{19!\,11!\,14!\,16!}{30!}}{12!\,7!\,2!\,9!}$$

$$= \text{antilog}\,[(\log 19! + \log 11! + \log 14! + \log 16! - \log 30!)$$
$$- (\log 12! + \log 7! + \log 2! + \log 9!)]$$

$$= \text{antilog}\,[(17.08509 + 7.60116 + 10.94041 + 13.32062$$
$$- 32.42366) - (8.68034 + 3.70243 + 0.30103$$
$$+ 5.55976)]$$

$$= \text{antilog}\,[16.52362 - 18.24356]$$

$$= \text{antilog}\,[-1.71994]$$

$$= \text{antilog}\,[0.28006 - 2.00000]$$

$$= 0.01906.$$

There are two tables with data more extreme than the observed data; they are as follows:

Table A:

13	6	19
1	10	11
14	16	30

$$P = \frac{\dfrac{19!\,11!\,14!\,16!}{30!}}{13!\,6!\,1!\,10!}$$

$$= \text{antilog}\,[16.52362 - (\log 13! + \log 6! + \log 1! + \log 10!)]$$

$$= \text{antilog}\,[-2.68776]$$

$$= 0.00205$$

Table B:

14	5	19
0	11	11
14	16	30

$$P = \frac{\dfrac{19!\,11!\,14!\,16!}{30!}}{14!\,5!\,0!\,11!}$$

$$= \text{antilog}\,[16.52362 - (\log 14! + \log 5! + \log 0! + \log 11!)]$$

$$= \text{antilog}\,[-4.09713]$$

$$= 0.00008$$

To summarize the probability of the original table and of the two more extreme tables (where f_0 in each table is the smallest of the four frequencies in that table),

	f_0	P
Original table	2	0.01906
More extreme table A	1	0.00205
More extreme table B	0	0.00008
Entire tail		0.02119

Therefore, if the null hypothesis is true, the probability of the array of data in the observed table or in more extreme tables is 0.02119. As this probability is less than 0.05, H_0 is rejected.

Note that if the hypotheses had been H_0: Snail species 2 has no greater ability to resist current than species 1 and H_A: Snail species 2 has greater ability to resist current than species 1, then we would have observed that the sample data are *not* in the direction of H_A and would not reject H_0, without even computing probabilities.

Instead of computing this exact probability of H_0, we may consult Table 28 from *Appendix: Statistical Tables and Graphs*, for $n = 30$, $m_1 = 11$, $m_2 = 14$; and the one-tailed critical values of f, for $\alpha = 0.05$, are 2 and 8. As the observed f in the cell corresponding to $m_1 = 11$ and $m_2 = 14$ is 2, H_0 may be rejected.

The probability of the observed contingency table occurring by chance, given the row and column totals, may be computed using Equation 76 or 77. Then the probability is calculated for each possible table having observed data more extreme that those of the original table. If the smallest observed frequency in the original table is designated as f_0 (which is 2 in Example 20), the more extreme tables are those that have smaller values of f_0 (which would be 1 and 0 in this example). (If the smallest observed frequency occurs in two cells of the table, then f_0 is designated to be the one with the smaller frequency diagonally opposite of it.)

The null hypothesis is tested by examining the sum of the probabilities of the observed table and of all the more extreme tables. This procedure yields the exact probability (hence the name of the test) of obtaining this set of tables by chance if the null hypothesis is true; and if this probability is less than or equal to the significance level, α, then H_0 is rejected.

Note that the quantity $R_1! R_2! C_1! C_2!/n!$ appears in each of the probability calculations using Equation 76 and therefore need be computed only once. It is only the value of $f_{11}! f_{12}! f_{21}! f_{22}!$ that needs to be computed anew for each table. To undertake these computations, the use of logarithms is advised for all but the smallest tables; and Table 40 from *Appendix: Statistical Tables and Graphs* provides logarithms of factorials. It is also obvious that, unless the four cell frequencies are small, this test calculation is tedious without a computer.

An alternative to computing the exact probability in the Fisher exact test of 2×2 tables is to consult Table 28 from *Appendix: Statistical Tables and Graphs* to obtain critical values with which to test null hypotheses for n up to 30. We examine the four marginal frequencies, $R_1, R_2, C_1,$ and C_2; and we designate the smallest of the four as m_1. If m_1 is a row total, then we call the smaller of the two column totals m_2; if m_1 is a column total, then the smaller row total is m_2. In Example 20, $m_1 = R_2$ and $m_2 = C_1$; and the one-tailed critical values in Table 28 from *Appendix: Statistical Tables and Graphs*, for $\alpha = 0.05$, are 2 and 8. The observed frequency in the cell corresponding to marginal totals m_1 and m_2 is called f; and if f is equal to or more extreme than 2 or 8 (i.e., if $f \leq 2$ or $f \geq 8$), then H_0 is rejected. However, employing tables of critical values results in expressing a range of probabilities associated with H_0; and a noteworthy characteristic of the exact test — namely the exact probability — is absent.

Bennett and Nakamura (1963) published tables for performing an exact test of 2×3 tables where the three column (or row) totals are equal and n is as large as 60. Computer programs have been developed to perform exact testing of $r \times c$ tables where r and/or c is greater than 2.

Feldman and Kluger (1963) demonstrated a simpler computational procedure for obtaining the probabilities of tables more extreme than those of the observed table. It will not be presented here because the calculations shown on this section and in Section 16c are straightforward and because performance of the Fisher exact test is so often performed via computer programs.

(b) Two-Tailed Testing. For data in a 2×2 contingency table, the Fisher exact test may also be used to test two-tailed hypotheses, particularly when both margins of the table are fixed. What is needed is the sum of the probabilities of the observed table and of all tables more extreme in the same direction as the observed data. This is the probability obtained for the one-tailed test shown in Example 20. If either $R_1 = R_2$ or $C_1 = C_2$, then the two-tailed probability is two times the one-tailed probability. Otherwise, it is not, and the probability for the second tail is computed as follows.*

Again designating f_0 to be the smallest of the four observed frequencies and m_1 to be the smallest of the four marginal frequencies in the original table, a 2×2 table is formed by replacing f_0 with $m_1 - 0$, and this is the most extreme table in the second tail. This is shown as Table C in Example 21. The probability of that table is calculated with Equation 76 or 77; if it is greater than the probability of the original table, then the two-tailed probability equals the one-tailed probability and the computation is complete. If the probability of the newly formed table is not greater than that of the original table, then it contributes to the probability of the second tail and the calculations continue. The probability of the next less extreme table is

---

*Some (e.g., Dupont, 1986) recommend that the two-tailed probability should be determined as two times the one-tailed probability. Others (e.g., Lloyd, 1998) argue against that calculation, and that practice is not employed here; and it can be noted that the second tail may be much smaller than the first and such a doubling procedure could result in a computed two-tailed probability that is greater than 1.

EXAMPLE 21   A Two-Tailed Fisher Exact Test, Using the Data and Hypotheses of Example 4 from *Contingency Tables*

The probability of the observed table was found, in Example 20, to be 0.01906, and the one-tailed probability was calculated to be 0.02119. In determining the one-tailed probability, the smallest cell frequency (f_0) in the most extreme table (Table B) was 0, and the smallest marginal frequency (m_1) was 11. So $m_1 - f_0 = 11 - 0 = 11$ is inserted in place of f_0 to form the most extreme table in the opposite tail:

Table C:

3	16	19
11	0	11
14	16	30

the probability of which is

$$P = \frac{\dfrac{19!\,11!\,14!\,16!}{30!}}{3!\,16!\,11!\,0!}$$

$$= 0.00000663, \text{ which is rounded to } 0.00001.$$

The less extreme tables that are in the second tail, and have probabilities less than the probability of the observed table, are these two:

Table D:

4	15	19	
10	1	11	$P = 0.00029$
14	16	30	cumulative $P = 0.00001 + 0.00029 = 0.00030$

Table E:

5	14	19	
9	2	11	$P = 0.00440$
14	16	30	cumulative $P = 0.00030 + 0.00440 = 0.00470$

The next less extreme table is this:

Table F:

6	13	19	
8	3	11	$P = 0.03079$
14	16	30	cumulative $P = 0.00470 + 0.03079 = 0.03549$

The Table F cumulative probability (0.03549) is larger than the probability of the original table (0.02119), so the Table F probability is not considered a relevant part of the second tail. The second tail consists of the following:

	f_0	P
Table C	3	0.00001
Table D	4	0.00029
Table E	5	0.00440
Entire second tail		0.00470

and the two-tailed P is, therefore, $0.02119 + 0.00470 = 0.02589$. As this is less than 0.05, we may reject H_0. Note that χ^2_H in Example has a probability close to that of this Fisher exact test.

If using Table 28 from *Statistical Tables and Graphs*, $n = 30, m_1 = 11, m_2 = 14$, and the f corresponding to m_1 and m_2 in the observed table is 2. As the two-tailed critical values of f, for $\alpha = 0.05$, are 2 and 9, H_0 is rejected.

determined; that table (Table D in Example 21) has cell frequency f_0 increased by 1, keeping the marginal frequencies the same. The two probabilities calculated for the second tail are summed and, if the sum is no greater than the probability of the original table, that cell frequency is again increased by 1 and a new probability computed. This process is continued as long as the sum of the probabilities in that tail is no greater than the probability of the original table.

(c) Probabilities Using Binomial Coefficients. Ghent (1972), Leslie (1955), Leyton (1968), and Sakoda and Cohen (1957) have shown how the use of binomial coefficients can eliminate much of the laboriousness of Fisher-exact-test computations, and Ghent (1972) and Carr (1980) have expanded these considerations to tables with more than two rows and/or columns. Using Table 26a from *Statistical Tables and Graphs*, this computational procedure requires much less effort than the use of logarithms of factorials, and it is at least as accurate. It may be employed for moderately large sample sizes, limited by the number of digits on one's calculator.

Referring back to Equation 75, the probability of a given 2×2 table is seen to be the product of two binomial coefficients divided by a third. The numerator of Equation 75 consists of one binomial coefficient representing the number of ways C_1 items can be combined f_{11} at a time (or f_{21} at a time, which is equivalent) and a second coefficient expressing the number of ways C_2 items can be combined f_{12} at a time (or, equivalently, f_{22} at a time). And the denominator denotes the number of ways n items can be combined R_1 at a time (or R_2 at a time). Table 26a from *Appendix: Statistical Tables and Graphs* provides a large array of binomial coefficients, and the proper selection of those required leads to simple computation of the probability of a 2×2 table.

The procedure is demonstrated in Example 22, for the data in Example 20. Consider the first row of the contingency table and determine the largest f_{11} and the smallest f_{12} that are possible without exceeding the row totals and column totals. These are $f_{11} = 14$ and $f_{12} = 5$, which sum to the row total of 19. (Other frequencies, such as 15 and 4, also add to 19, but the frequencies in the first column are limited to 14.) In a table where $f_{11} < f_{12}$, switch the two columns of the 2×2 table before performing these calculations.

EXAMPLE 22   The Fisher Exact Tests of Examples 20 and 21, Employing the Binomial-Coefficient Procedure

The observed 2 × 2 contingency table is

12	7	19
2	9	11
14	16	30

The top-row frequencies, f_{11} and f_{12}, of all contingency tables possible with the observed row and column totals, and their associated binomial coefficients and coefficient products, are as follows. The observed contingency table is indicated by "*".

f_{11}	f_{12}	$C_1 = 14$	$C_2 = 16$		Coefficient product	
14	5	1.	×	4,368.	=	4,368
13	6	14.	×	8,008.	=	112,112
12*	7*	91.	×	11,440.	=	1,041,040*	1,157,520
11	8	364.	×	12,870.	=	4,684,680
10	9	1,001.	×	11,440.	=	11,451,440
9	10	2,002.	×	8,008.	=	16,032,016
8	11	3,003.	×	4,368.	=	13,117,104
7	12	3,432.	×	1,820.	=	6,246,240
6	13	3,003.	×	560.	=	1,681,680
5	14	2,002.	×	120.	=	240,240
4	15	1,001.	×	16.	=	16,016	256,620
3	16	364.	×	1.	=	364

54,627,300

One-tailed probability: $\dfrac{1,157,520}{54,627,300} = 0.02119$

Probability associated with the opposite tail: $\dfrac{256,620}{54,627,300} = 0.00470$

Two-tailed probability: 0.02589

We shall need to refer to the binomial coefficients for what Table 26a from *Statistical Tables and Graphs* refers to as $n = 14$ and $n = 16$, for these are the two column totals (C_1 and C_2) in the contingency table in Example 20. We record, from Table 26a from *Appendix: Statistical Tables and Graphs*, the binomial coefficient for $n = C_1 = 14$ and $X = f_{11} = 14$ (which is 1), the coefficient for $n = C_2 = 16$ and $X = f_{12} = 5$ (which is 4,368), and the product of the two coefficients (which is 4,368).

Then we record the binomial coefficients of the next less extreme table; that is, the one with $f_{11} = 13$ and $f_{12} = 6$ (that is, coefficients of 14 and 8,008) and their product (i.e., 112,112). This process is repeated for each possible table until f_{11} can be no smaller (and f_{12} can be no larger): that is, $f_{11} = 3$ and $f_{12} = 16$. The sum of all the coefficient products (54,627,300 in this example) is the number of ways n things may be combined R_1 at a time (where n is the total of the frequencies in the 2×2 table); this is the binomial coefficient for n (the total frequency) and $X = R_1$ (and in the present example this coefficient is $_{30}C_{11} = 54,627,300$). Determining this coefficient is a good arithmetic check against the sum of the products of the several coefficients of individual contingency tables.

Dividing the coefficient product for a contingency table by the sum of the products yields the probability of that table. Thus, the table of observed data in Example 20 has $f_{11} = 14$ and $f_{12} = 5$, and we may compute $1,041,040/54,627,300 = 0.01906$, exactly the probability obtained in Example 20 using logarithms of factorials. The probability of the one-tailed test employs the sum of those coefficient products equal to or smaller than the product for the observed table and in the same tail as the observed table. In the present example, this tail would include products 4,368, 112,112, and 1,041,040, the sum of which is 1,157,520, and $1,157,520/54,627,300 = 0.02119$, which is the probability calculated in Example 20. To obtain the probability for the two-tailed test, we add to the one-tailed probability the probabilities of all tables in the opposite tail that have coefficient products equal to or less than that of the observed table. In our example these products are 240,240, 16,016, and 364, their sum is 256,620, and $256,620/54,627,300 = 0.00470$; the probabilities of the two tails are 0.02119 and 0.00470, which sum to the two-tailed probability of 0.02589 (which is what was calculated in Example 21).

17 PAIRED-SAMPLE TESTING OF NOMINAL-SCALE DATA

(a) **Data in a 2 × 2 Table.** Nominal-scale data may come from paired samples. A 2×2 table containing data that are *dichotomous* (i.e., the nominal-scale variable has two possible values) may be analyzed by the *McNemar test* (McNemar, 1947).

For example, assume that we wish to test whether two skin lotions are equally effective in relieving a poison-ivy rash. Both of the lotions might be tested on each of 50 patients with poison-ivy rashes on both arms, by applying one lotion to one arm and the other lotion to the other arm (using, for each person, a random selection of which arm gets which lotion). The results of the experiment can be summarized in a table such as in Example 23, where the results for each of the 50 patients consist of a pair of data (i.e., the outcomes of using the two lotions on each patient). As with other 2×2 tables, the datum in row i and column j will be designated as f_{ij}. Thus, $f_{11} = 12$, $f_{12} = 5$, $f_{21} = 11$, and $f_{22} = 22$; and the total of the four frequencies is $n = 50$. The two-tailed null hypothesis is that, in the sampled population of people who might be treated with these two medications, the proportion of them that would obtain relief from lotion A (call it p_1) is the same as the proportion receiving relief from lotion B (call it p_2); that is, $H_0: p_1 = p_2$ (vs. $H_A: p_1 \neq p_2$). In the sample, 12 patients (f_{11}) experienced relief from both lotions and 22 (f_{22}) had relief from neither lotion. The proportion of people in the sample who experienced relief from lotion A is $\hat{p}_1 = (f_{11} + f_{21})/n = (12 + 11)/50 = 0.46$, and the proportion benefiting from lotion B is $\hat{p}_2 = (f_{11} + f_{12})/n = (12 + 5)/50 = 0.34$. The sample estimate of $p_1 - p_2$ is

$$\hat{p}_1 - \hat{p}_2 = \frac{f_{11} + f_{21}}{n} - \frac{f_{11} + f_{12}}{n} = \frac{f_{11}}{n} + \frac{f_{21}}{n} - \frac{f_{11}}{n} - \frac{f_{12}}{n} = \frac{f_{21}}{n} - \frac{f_{12}}{n}.$$

$$(78)$$

That is, of the four data in the 2×2 table, only f_{12} and f_{21} are needed to test the hypotheses. The test is essentially a goodness-of-fit procedure where we ask whether the ratio of f_{12} to f_{21} departs significantly from $1 : 1$. Thus, the hypotheses could also be stated as H_0: $\psi = 1$ and H_A: $\psi \neq 1$, where* ψ is the population ratio estimated by f_{12}/f_{21}.

The goodness-of-fit test for this hypothesis it is readily performed via

$$\chi^2 = \frac{(f_{12} - f_{21})^2}{f_{12} + f_{21}}, \tag{79}$$

which is equivalent to using the normal deviate for a two-tailed test:

$$Z = \frac{|f_{12} - f_{21}|}{\sqrt{f_{12} + f_{21}}}. \tag{80}$$

Because χ^2 and Z are continuous distributions and the data to be analyzed are counts (i.e., integers), some authors have employed corrections for continuity. A common one is the Yates correction for continuity. This is accomplished, with

$$\chi_c^2 = \frac{|(f_{12} - f_{21}| - 1)^2}{f_{12} + f_{21}}, \tag{81}$$

which is the same as employing

$$Z_c = \frac{|f_{12} - f_{21}| - 1}{\sqrt{f_{12} + f_{21}}}. \tag{82}$$

The calculation of χ_c^2 for this test is demonstrated in Example 23.

A McNemar test using χ^2 operates with the probability of a Type I error much closer to α than testing with χ_c^2, although that probability will occasionally be a little greater than α. That is, the test can be liberal, rejecting H_0 more often than it should at the α level of significance. Use of χ_c^2 will routinely result in a test that is conservative, rejecting H_0 less often than it should and having less power than employing χ^2. Often, as in Example 23, the same conclusion is reached using the test with and without a correction for continuity. Bennett and Underwood (1970) and others have advised that the continuity correction should not generally be used.

Because this test employs only two of the four tabulated data (f_{12} and f_{21}), the results are the same regardless of the magnitude of the other two counts (f_{11} and f_{22}), which are considered as tied data and ignored in the analysis. If f_{12} and f_{21} are small, this test does not work well. If $f_{12} + f_{21} \leq 10$, the binomial test (Section 5) is recommended, with $n = f_{12} + f_{21}$ and $X = f_{12}$ (or f_{21}).

Another type of data amenable to McNemar testing results from the situation where experimental responses are recorded before and after some event, in which case the procedure may be called the "McNemar test for change." For example, we might record whether students saying the plan to pursue a career in microbiology,

---

*The symbol ψ is the lowercase Greek letter psi.

EXAMPLE 23   McNemar's Test for Paired-Sample Nominal Scale Data

H_0: The proportion of persons experiencing relief is the same with both lotions (i.e., $H_0: p_1 = p_2$).

H_A: The proportion of persons experiencing relief is not the same with both lotions (i.e., $H_0: p_1 \neq p_2$).

$\alpha = 0.05$
$n = 50$

| | Lotion A | |
Lotion B	Relief	No relief
Relief	12	5
No relief	11	22

$$\chi^2 = \frac{(f_{12} - f_{21})^2}{(f_{12} - f_{21})} = \frac{(5-11)^2}{5+11} = 2.250$$

$\nu = 1$, $\chi^2_{0.05,1} = 3.841$. Therefore, do not reject H_0.

$$0.10 < P < 0.25 \quad [P = 0.13]$$

Alternatively, and with the same result,

$$Z = \frac{|f_{12} - f_{21}|}{\sqrt{f_{12} + f_{21}}} = \frac{|5-1|}{\sqrt{5+11}} = 1.500,$$

$Z_{0.05(2)} = 1.900$. Therefore, do not reject H_0.

$$0.10 < P < 0.20 \quad [P = 0.13]$$

With a correction for continuity,

$$\chi^2_c = \frac{(|f_{12} - f_{21}|-1)^2}{(f_{12} - f_{21})} = \frac{(|5-11|-1)^2}{5+11} = 1.562.$$

Do not reject H_0.

$$0.10 < P < 0.25 \quad [P = 0.21]$$

Alternatively, and with the same result,

$$Z_c = \frac{|f_{12} - f_{21}|-1}{\sqrt{f_{12} + f_{21}}} = \frac{|5-1|-1}{\sqrt{5+11}} = 1.250.$$

Do not reject H_0.

$$0.20 < P < 0.50 \quad [P = 0.21]$$

---

before and after an internship experience in microbiology laboratory. The column headings could be "yes" and "no" before the internship, and the row designations then would be "yes" and "no" after the internship.

The McNemar test should not be confused with 2×2 contingency-table analysis. Contingency-table data are analyzed using a null hypothesis of independence between rows and columns, whereas in the case of data subjected to the McNemar test, there is intentional association between the row and column data.

(b) The One-Tailed McNemar Test. Using the normal deviate (Z) as the test statistic, one-tailed hypotheses can be tested. So, for example, the hypotheses for a poison ivy-treatment experiment could be H_0: The proportion of people experiencing relief with lotion A is not greater than (i.e., is less than or equal to) the proportion having relief with lotion B, versus H_A: The proportion of people experiencing relief with lotion A is greater than the proportion obtaining relief with lotion B. And H_0 would be rejected if Z (or Z_c, if using the continuity correction) were greater than or equal to $Z_{\alpha(1)}$ and $f_{21} > f_{12}$.

(c) Power and Sample Size for the McNemar Test. The ability of the McNemar test to reject a null hypothesis, when the hypothesis is false, may be estimated by computing

$$Z_{\beta(1)} = \frac{\sqrt{n}\sqrt{p}\,(\psi - 1) - Z_\alpha\sqrt{\psi + 1}}{\sqrt{(\psi + 1) - p(\psi - 1)^2}} \tag{83}$$

(Connett, Smith, and McHugh, 1987). Here, n is the number of pairs to be used (i.e., $n = f_{11} + f_{12} + f_{21} + f_{22}$); p is an estimate, as from a pilot study, of the proportion f_{12}/n or f_{21}/n, whichever is smaller; ψ is the magnitude of difference desired to be detected by the hypothesis test, expressed as the ratio in the population of either f_{12} to f_{21}, or f_{21} to f_{12}, whichever is larger; and Z_α is $Z_{\alpha(2)}$ or $Z_{\alpha(1)}$, depending upon whether the test is two-tailed or one-tailed, respectively. Then, using Appendix Tabl 2 or the last line in Table 3 from *Appendix: Statistical Tables and Graphs* (i.e., for t with $\nu = \infty$), determine $\beta(1)$; and the estimated power of the test is $1 - \beta(1)$. This estimation procedure is demonstrated in Example 24.

Similarly, we can estimate the sample size necessary to perform a McNemar test with a specified power:

$$n = \frac{\left[Z_\alpha\sqrt{\psi + 1} + Z_{\beta(1)}\sqrt{(\psi + 1) - p(\psi - 1)^2}\right]^2}{p(\psi - 1)^2} \tag{84}$$

(Connett, Smith, and McHugh, 1987). This is demonstrated in Example 25

(d) Data in Larger Tables. The McNemar test may be extended to square tables larger than 2×2 (Bowker, 1948; Maxwell, 1970). What we test is whether the upper right corner of the table is symmetrical with the lower left corner. This is done by ignoring the data along the diagonal containing f_{ii} (i.e., row 1, column 1; row 2, column 2; etc.). We compute

$$\chi^2 = \sum_{i=1}^{r} \sum_{j>i} \frac{(f_{ij} - f_{ji})^2}{f_{ij} + f_{ji}}, \tag{85}$$

where, as before, f_{ij} is the observed frequency in row i and column j, and the degrees of freedom are

$$\nu = \frac{r(r - 1)}{2}, \tag{86}$$

EXAMPLE 24 Determination of Power of the McNemar Test

Considering the data of Example 23 to be from a pilot study, what would be the probability of rejecting H_0 if 200 pairs of data were used in a future study, if the test were performed at the 0.05 level of significance and if the population ratio of f_{21} to f_{12} were at least 2?

From the pilot study, using $n = 51$ pairs of data, $f_{12/n} = 6/51 = 0.1176$ and $f_{21/n} = 10/51 = 0.1961$; so $p = 0.1176$. We specify $\alpha(2) = 0.05$, so $Z_{0.05(2)} = 1.9600$ (from the last line of Table 3 from *Appendix: Statistical Tables and Graphs*). And we also specify a new sample size of $n = 200$ and $\psi = 2$. Therefore,

$$Z_{\beta(1)} = \frac{\sqrt{n}\sqrt{p}(\psi - 1) - Z_{\alpha(2)}\sqrt{\psi + 1}}{\sqrt{(\psi + 1) - p(\psi - 1)^2}}$$

$$= \frac{\sqrt{200}\sqrt{0.1176}(2 - 1) - 1.9600\sqrt{2 + 1}}{\sqrt{(2 + 1) - 0.1176(2 - 1)^2}}$$

$$= \frac{(14.1421)(0.3429)(1) - 1.9600(1.7321)}{\sqrt{3 - (0.1176)(1)}}$$

$$= \frac{4.8493 - 3.3949}{\sqrt{2.8824}} = \frac{1.4544}{1.6978} = 0.86.$$

From Table 2 from *Appendix: Statistical Tables and Graphs*, if $Z_{\beta(1)} = 0.86$, then $\beta(1) = 0.19$; therefore, power [i.e., $1 - \beta(1)$] is $1 - 0.19 = 0.81$.

From Table 3 from *Appendix: Statistical Tables and Graphs*, if $Z_{\beta(1)}$ [i.e., $t_{\beta(1),\infty}$] is 0.86, then $\beta(1)$ lies between 0.25 and 0.10, and the power [i.e., $1 - \beta(1)$] lies between 0.75 and 0.90. [$\beta(1) = 0.19$ and power $= 0.81$.]

---

and where r is the number of rows (or, equivalently, the number of columns) in the table of data. This is demonstrated in Example 26.

Note that Equation 85 involves the testing of a series of 1 : 1 ratios by what is essentially an expansion of Equation 79. Each of these 1 : 1 ratios derives from a unique pairing of the r categories taken two at a time. Recall that the number of ways that r items can be combined two at a time is $_rC_2 = r!/[2(r - 2)!]$. So, in Example 26, where there are three categories, there are $_3C_2 = 3!/[2(3 - 2)!] = 3$ pairings, resulting in three terms in the χ^2 summation. If there were four categories of religion, then the summation would involve $_4C_2 = 4!/[2(4 - 2)!] = 6$ pairings, and 6 χ^2 terms; and so on. For data of this type in a 2 × 2 table, Equation 85 becomes Equation 79, and Equation 86 yields $v = 1$.

(e) Testing for Effect of Treatment Order. If two treatments are applied sequentially to a group of subjects, we might ask whether the response to each treatment depended on the order in which the treatments were administered. For example, suppose we have two medications for the treatment of poison-ivy rash, but, instead of the situation in Example 23, they are to be administered orally rather than by external

---

EXAMPLE 25   Determination of Sample Size for the McNemar Test

Considering the data of Example 23 to be from a pilot study, how many pairs of data would be needed to have a 90% probability of rejecting the two-tailed H_0 if a future test were performed at the 0.05 level of significance and the ratio of f_{21} to f_{12} in the population were at least 2?

As in Example 24, $p = 0.1176$ and $Z_{\alpha(2)} = 1.9600$. In addition, we specify that $\psi = 2$ and that the power of the test is to be 0.90 [so $\beta(1) = 0.10$]. Therefore, the required sample size is

$$n = \frac{\left[Z_{\alpha(2)}\sqrt{\psi + 1} + Z_{\beta(1)}\sqrt{(\psi + 1) - p(\psi - 1)^2} \right]^2}{p(\psi - 1)^2}$$

$$= \frac{\left[1.9600\sqrt{2 + 1} + 1.2816\sqrt{(2 + 1) - (0.1176)(2 - 1)^2} \right]^2}{(0.1176)(2 - 1)^2}$$

$$= \frac{[1.9600(1.7321) + 1.2816(1.6978)]^2}{0.1176} = \frac{(5.5708)^2}{0.1176} = 263.9.$$

Therefore, at least 264 pairs of data should be used.

---

application to the skin. Thus, in this example, both arms receive medication at the same time, but the oral medications must be given at different times.

Gart (1969a) provides the following procedure to test for the difference in response between two sequentially applied treatments and to test whether the order of application had an effect on the response. The following 2×2 contingency table is used to test for a treatment effect:

	Order of Application of Treatments A and B		
	A, then B	B, then A	Total
Response with first treatment	f_{11}	f_{12}	R_1
Response with second treatment	f_{21}	f_{22}	R_2
Total	C_1	C_2	n

By redefining the rows, the following 2×2 table may be used to test the null hypothesis of no difference in response due to order of treatment application:

	Order of Application of Treatments A and B		
	A, then B	B, then A	Total
Response with treatment A	f_{11}	f_{12}	R_1
Response with treatment B	f_{21}	f_{22}	R_2
Total	C_1	C_2	n

EXAMPLE 26   McNemar's Test for a 3 × 3 Table of Nominal-Scale Data

H_0: Of men who adopt a religion different from that of their fathers, a change from one religion to another is as likely as a change from the latter religion to the former.

H_A: Of men who adopt a religion different from that of their fathers, a change from one religion to another is not as likely as a change from the latter religion to the former.

Man's	Man's Father's Religion		
Religion	*Protestant*	*Catholic*	*Jewish*
Protestant	173	20	7
Catholic	15	51	2
Jewish	5	3	24

$r = 3$

$$\chi^2 = \sum_{i=1}^{r}\sum_{j>i} \frac{(f_{ij} - f_{ji})^2}{f_{ij} + f_{ji}}$$

$$= \frac{(f_{12} - f_{21})^2}{f_{12} - f_{21}} + \frac{(f_{13} - f_{31})^2}{f_{13} - f_{31}} + \frac{(f_{23} - f_{32})^2}{f_{23} - f_{32}}$$

$$= \frac{(20 - 15)^2}{20 + 15} + \frac{(7 - 5)^2}{7 + 5} + \frac{(2 - 3)^2}{2 + 3}$$

$$= 0.7143 + 0.3333 + 0.2000$$

$$= 1.248$$

$\nu = \frac{r(r-1)}{2} = \frac{3(2)}{2} = 3$

$\chi^2_{0.05,3} = 7.815$

Do not reject H_0.

$$0.50 < P < 0.75 \ [P = 0.74]$$

These two contingency tables have one fixed margin (the column totals are fixed) and they may be tested by chi-square, which is shown in Example 27.

EXAMPLE 27   Gart's Test for Effect of Treatment and Treatment Order

H_0: The two oral medications have the same effect on relieving poison-ivy rash.

H_A: The two oral medications do not have the same effect on relieving poison-ivy rash.

	Order of Application of Medications A and B		
	A, then B	B, then A	Total
Response with 1st medication	14	6	20
Response with 2nd medication	4	12	16
Total	18	18	36

$$\chi^2 = \frac{n(f_{11}f_{22} - f_{12}f_{21})^2}{R_1 R_2 C_1 C_2}$$

$$= \frac{36\left[(14)(12) - (6)(4)\right]^2}{(20)(16)(18)(18)} = 7.200.$$

$\chi^2_{0.05,1} = 3.841$; reject H_0.

That is, it is concluded that there is a difference in response to the two medications, regardless of the order in which they are administered.

$$0.005 < P < 0.01 \quad [P = 0.0073]$$

H_0: The order of administration of the two oral medications does not affect their abilities to relieve poison-ivy rash.

H_A: The order of administration of the two oral medications does affect their abilities to relieve poison-ivy rash.

	Order of Application of Medications A and B		
	A, then B	B, then A	Total
Response with medication A	14	12	26
Response with medication B	4	6	10
Total	18	18	36

$$\chi^2 = \frac{36\left[(14)(6) - (12)(4)\right]^2}{(26)(10)(18)(18)} = 0.554$$

$\chi^2_{0.05,1} = 3.841$; do not reject H_0.

That is, it is concluded that the effects of the two medications are not affected by the order in which they are administered.

$$0.25 < P < 0.50 \quad [P = 0.46]$$

18 LOGISTIC REGRESSION

Previous discussions of regression considered data measured on a continuous (i.e., ratio or interval) scale, where there are measurements of a dependent variable (Y) associated with measurements of one or more independent variables (X's). However, there are situations (commonly involving clinical, epidemiological, or sociological data) where the dependent variable is measured on a nominal scale; that is, where the data are in two or more categories. For example, a sample of men could be examined for the presence of arterial plaque, and this information is recorded together with the age of each man. The mean age of men with plaque and the mean of those without plaque could be compared via a two-sample t test. But the data can be analyzed in a different fashion, deriving a quantitative expression of the relationship between the presence of plaque and the age of the subject and allowing for the prediction of the probability of plaque at a specified age.

Regression data with Y recorded on a dichotomous scale do not meet the assumptions of the previously introduced regression methods, assumptions such as that the Y's and residuals (ϵ's) have come from a normal distribution at each value of X and have the same variance at all values of X. So another statistical procedure must be sought. The most frequently employed analysis of such data is *logistic regression*.*

A brief introduction to logistic regression is given here.[†] Because of the intense calculations required, users of this regression technique will depend upon computer programs, which are found in several statistical computer packages, and will typically benefit from consultation with a statistician familiar with the procedures.

(a) Simple Logistic Regression. The simplest — and most common — logistic regression situation is where the categorical data (Y) are binomial, also known as *dichotomous* (i.e., the data consist of each observation recorded as belonging in one of two categories). Each value of Y is routinely recorded as "1" or "0" and might, for example, refer to a characteristic as being "present" (1) or "absent" (0), or to subjects being "with" (1) or "without" (0) a disease.[‡]

Logistic regression considers the probability (p) of encountering a Y of 1 at a given X in the population that was sampled. So, for example, p could be the probability of

---

*A similar procedure, but one that is less often preferred, is known as *discriminant analysis*. What statisticians refer to as the *general linear model* underlies several statistical techniques, including analysis of variance, analysis of covariance, multivariate analysis of variance, linear regression, logistic regression, and discriminant analysis.

[†]Logistic regression is a wide-ranging topic and is covered in portions of comprehensive books on regression (e.g., Chatterjee and Hadi, 2006: Chapter 12; Glantz and Slinker, 2001: Chapter 12; Hair et al., 2006: Chapter 5; Kutner, Nachtsheim, and Neter, 2004: Chapter 14; Meyers, Gamst, and Guarino, 2006: Chapter 6A and 6B; Montgomery, Peck, and Vining, 2001: Section 14.2; Pedhazur, 1997: Chapter 17; Vittinghoff et al., 2005: Chapter 6); in books on the analysis of categorical data (e.g., Agresti, 2002: Chapters 5 and 6; Agresti, 2007: Chapters 4 and 5; Fleiss, Levin, and Paik, 2003: Chapter 11); and in works that concentrate specifically on logistic regression (e.g., Garson, 2006; Hosmer and Lemeshow, 2000; Kleinbaum and Klein, 2002; Menard, 2002; Pampel, 2000).

[‡]Designating observations of Y as "0" or "1" is thus an example of using a "dummy variable," first described in Section 20.10. Any two integers could be used, but 0 and 1 are almost always employed, and this results in the mean of the Y's being the probability of $Y = 1$.

Dichotomous Variables

encountering a member of the population that has a specified characteristic present, or the proportion that has a specified disease.

The logistic regression relationship in a population is

$$p = \frac{e^{\alpha + \beta X}}{1 + e^{\alpha + \beta X}}, \tag{87}$$

This equation may also be written, equivalently, using the abbreviation "exp" for an exponent on e:

$$p = \frac{\exp(\alpha + \beta X)}{1 + \exp(\alpha + \beta X)}, \tag{87a}$$

or, equivalently, as

$$p = \frac{1}{1 + e^{-(\alpha + \beta X)}} \text{ or } p = \frac{1}{1 + \exp[-(\alpha + \beta X)]}. \tag{87b}$$

The parameter α is often seen written as β_0.

The sample regression equations corresponding to these expressions of population regression are, respectively,

$$\hat{p} = \frac{e^{a + bX}}{1 + e^{a + bX}}, \tag{88}$$

$$\hat{p} = \frac{\exp(a + bX)}{1 + \exp(a + bX)}, \tag{88a}$$

$$\hat{p} = \frac{1}{1 + e^{-(a + bX)}} \text{ or } \hat{p} = \frac{1}{1 + \exp[-(a + bX)]}; \tag{88b}$$

and a is often written as b_0.

Logistic regression employs the concept of the *odds* of an event namely the probability of the event occurring expressed relative to the probability of the event not occurring. Using the designations $p = P(Y = 1)$ and $q = 1 - p = P(Y = 0)$, the odds can be expressed by these four equivalent statements:

$$\frac{P(Y = 1)}{1 - P(Y = 1)} \text{ or } \frac{P(Y = 1)}{P(Y = 0)} \text{ or } \frac{p}{1 - p} \text{ or } \frac{p}{q}, \tag{89}$$

and the fourth will be employed in the following discussion.

A probability, p, must lie within a limited range ($0 \leq p \leq 1$).* However, odds, p/q, have no upper limit. For example, if $p = 0.1$, odds $= 0.1/0.9 = 0.11$; if $p = 0.5$, odds $= 0.5/0.5 = 1$; if $p = 0.9$, odds $= 0.9/0.1 = 9$; if $p = 0.97$, odds $= 0.97/0.03 = 32.3$; and so on. Expanding the odds terminology, if, for example, a population contains 60% females and 40% males, it is said that the odds are $0.60/0.40 = $ "6 to 4" or "1.5 to 1" *in favor of* randomly selecting a female from the population, or 1.5 to 1 *against* selecting a male.

In order to obtain a linear model using the regression terms α and βX, statisticians utilize the natural logarithm of the odds, a quantity known as a *logit*; this is sometimes

---

*If a linear regression were performed on p versus X, predicted values of $\hat{p}$ could be less than 0 or greater than 1, an untenable outcome. This is another reason why logistic regression should be used when dealing with a categorical dependent variable.

referred to as a "logit transformation" of the dependent variable:

$$\text{logit} = \ln(\text{odds}) = \ln\left(\frac{p}{q}\right). \tag{90}$$

For $0 < \text{odds} < 1.0$, the logit is a negative number (and it becomes farther from 0 the closer the odds are to 0); for odds $= 0.5$, the logit is 0; and for odds > 0, the logit is a positive number (and it becomes farther from 0 the closer the odds are to 1).

Using the logit transformation, the population linear regression equation and sample regression equations are, respectively,

$$\text{logit for } p = \alpha + \beta X \tag{91}$$

and

$$\text{logit for } \hat{p} = a + bX. \tag{92}$$

This linear relationship is shown in Figure 2b. Determining a and b for Equation 92 is performed by an iterative process termed *maximum likelihood estimation*, instead of by the least-squares procedure used in the linear regressions. The term maximum likelihood refers to arriving at the a and b that are most likely to estimate the population parameters underlying the observed sample of data.

As with the regression procedures previously discussed, the results of a logistic regression analysis will include a and b as estimates of the population parameters α and β, respectively. Computer output will routinely present these, with the standard error of b (namely, s_b), confidence limits for β, and a test of H_0: $\beta = 0$. In logistic regression there are no measures corresponding to the coefficient of determination (r^2) in linear regression analysis, but some authors have suggested statistics to express similar concepts.

For a logistic relationship, a plot of p versus X will display an S-shaped curve rising from the lower left to the upper right, such as that in Figure 2a. In such a graph, p is near zero for very small values of X, it increases gradually as X increases, then it increases much more rapidly with further increase in X, and then it increases at a slow rate for larger X's, gradually approaching 1.0. Figure 2a is a graph for a logistic equation with $\alpha = -2.0$ and $\beta = 1.0$. The graph would shift to the left by 1 unit of X for each increase of α by 1, and the rise in p would be steeper with a larger β. If β were negative instead of positive, the curve would be a reverse S shape, rising from the lower right to the upper left of the graph, instead of from the lower left to the upper right.*

For a 1-unit increase in X, the odds increase by a factor of e^β. If, for example, $\beta = 1.0$, then the odds for $X = 4$ would be $e^{1.0}$ (namely, 2.72) times the odds for $X = 3$. And a one-unit increase in X will result in a unit increase in the logit of p. So, if $\beta = 1.0$, then the logit for $X = 4$ would be 1.0 larger than the logit for $X = 3$. If β is negative, then the odds and logit decrease instead of increase.

Once the logistic-regression relationship is determined for a sample of data, it may be used to predict the probability of X for a given X. Equation 92 yields a logit of p for the specified X; then

$$\text{odds} = e^{\text{logit}} \tag{93}$$

---

*Another representation of binary Y data in an S-shaped relationship to X is that of *probits*, based upon a cumulative normal distribution. The use of logits is simpler and is preferred by many.

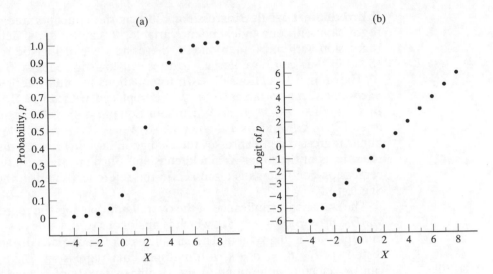

FIGURE 2: Logistic regression, where $\alpha = -2.0$ and $\beta = 1.0$. (a) The relationship of $p = 1/[1 + e^{-(\alpha + \beta X)}]$ to X. (b) The relationship of the logit of p to X.

and

$$\hat{p} = \frac{\text{odds}}{1 + \text{odds}}. \tag{94}$$

If n is large, the hypothesis $H_0: \beta = 0$ may be tested using

$$Z = \frac{b}{s_b}, \tag{95}$$

which is analogous to the linear-regression test of that hypothesis using t (Section 17.3b). This is known as the Wald test*; sometimes, with exactly the same result, the test statistic used is $\chi^2 = Z^2$, with 1 degree of freedom. In further analogy with linear regression, confidence limits for β are obtainable as

$$b + Z_{\alpha(2)} s_b. \tag{96}$$

Assessing $H_0: \beta = 0$ and expressing a confidence interval for β may also be done using a computer-intensive process known as likelihood-ratio, or log-likelihood, testing. It has been noted (e.g., by Hauck and Donner, 1977; Menard, 2002: 43; Pampel, 2000: 30) that when b is large, its standard error (s_b) is inflated, thereby increasing the probability of a Type II error in a Wald test (and thus decreasing the power of the test). Therefore, it is sometimes recommended that the likelihood-ratio test be used routinely in preference to the commonly encountered Wald test. For the same reason, likelihood-ratio confidence limits can be recommended over those obtained by Equation 96. The interpretation of logistic-regression coefficients is discussed more fully in the references cited in the second footnote in Section 18.

In recommending an adequate number of data for logistic-regression analysis, some authors have suggested that n be large enough that there are at least 10 observations of $Y = 1$ and at least 10 of $Y = 0$.

---

*Named for Hungarian-born, Vienna-educated, American mathematician and econometrician Abraham Wald (1902–1950).

(b) Multiple Logistic Regression. Just as the concepts and procedures of linear regression with one independent variable, X, can be expanded into those of linear regression with more than one X, the basic ideas of logistic regression with one X (Section 18a) can be enlarged to those of logistic regression with more than one X. Expanding Equations 87–87b to equations for multiple logistic regression with m independent variables, $\alpha + \beta X$ is replaced with $\alpha + \Sigma_{i=1}^{m} \beta_i X_i$, which is $\alpha + \beta_1 X_1 + \beta_2 X_2 + \cdots + \beta_m X_m$; and, in Equations 88–88b, $a + bX$ is replaced with $a + \Sigma_{i=1}^{m} b_i X_i$, which is $a + b_1 X_1 + b_2 X_2 + \cdots + b_m X_m$. Analogous to multiple linear regression, β_i expresses the change in ln(logit) for a 1-unit change in X_i, with the effects of the other X_i's held constant. Further interpretation of logistic partial-regression coefficients is discussed in the references cited in the second footnote of Section 18.

The statistical significance of the overall multiple logistic regression model is tested via $H_0: \beta_1 = \beta_2 = \cdots = \beta_m$, which is analogous to the analysis-of-variance testing in multiple linear regression. The significance of each partial-regression coefficient is tested via $H_0: \beta_i = 0$. As with multiple liner regression, $H_0: \beta_1 = \beta_2 = \cdots = \beta_m$ can be rejected while none of the significance tests of individual partial-regression coefficients result in rejection of $H_0: \beta_i = 0$ (especially when the conservative Wald test is used).

The standardized partial-regression coefficients in multiple linear regression cannot be computed for logistic regression, but some authors have proposed similar coefficients. Also, there is no perfect logistic-regression analog to coefficients of determination (R^2 and R_a^2). Several measures have been suggested by various authors to express the concept of such a coefficient; sometimes they are referred to as "pseudo-R^2" values (and some do not have 1.0 as their maximum). Just as with multiple linear-regression analysis, multiple logistic regression is adversely affected by multicolinearity. Logistic analysis does not work well with small numbers of data, and some authors recommend that the sample be large enough that there are at least $10m$ 0's and at least $10m$ 1's for the dependent variable. And, as is the case of multiple linear regression, multiple logistic regression is adversely affected by outliers.

(c) Other Models of Logistic Regression. Though not commonly encountered, the dependent variable can be one that has more than two categories. This is known as a *polytomous* (or "polychotomous" or "multinomial") variable. Also, the dependent variable can be one that is measured on an ordinal scale but recorded in nominal-scale categories. For example, subjects could be classified as "underweight," "normal weight," and "overweight."

Logistic regression may also be performed when an independent variable (X) is recorded on a dichotomous nominal scale. For example, the dependent variable could be recorded as hair loss ($Y = 1$) or no hair loss ($Y = 0$) in men, with the independent variable (X) being exposed ($X = 1$) or not exposed ($X = 0$) to a certain drug or radiation treatment. Such data can be subjected to a 2×2 contingency-table analysis, where the null hypothesis is that the proportion (p) of men with loss of hair is the same in the treated and nontreated groups. Using logistic regression, however, the null hypothesis is that hair loss is statistically dependent upon receiving the treatment and the relationship between p and whether treatment was applied is quantified.

In multiple logistic regression, one or more of the X's may be recorded on a dichotomous scale. So, for example, Y could be recorded as hair loss or no hair loss, X_1 could be age (measured on a continuous scale), and X_2 could be sex (a dichotomous variable: male or female). Indeed, using dummy variables, X's can be recorded on a nominal scale with more than two categories. If all m of the independent variables are nominal and Y is dichotomous, the data could be arranged in a $2 \times m$ contingency table, but different hypotheses would be tested thereby.

It is also possible to perform stepwise multiple logistic regression in order to determine which of the X's should be in the final regression model, logistic regression with polynomial terms and logistic regression with interaction of independent variables.

EXERCISES

1. If, in a binomial population, $p = 0.3$ and $n = 6$, what proportion of the population does $X = 2$ represent?

2. If, in a binomial population, $p = 0.22$ and $n = 5$, what is the probability of $X = 4$?

3. Determine whether the following data, where $n = 4$, are likely to have come from a binomial population with $p = 0.25$:

X	f
0	30
1	51
2	33
3	10
4	2

4. Determine whether the following data, where $n = 4$, are likely to have come from a binomial population:

X	f
0	20
1	41
2	33
3	11
4	4

5. A randomly selected male mouse of a certain species was placed in a cage with a randomly selected male mouse of a second species, and it was recorded which animal exhibited dominance over the other. The experimental procedure was performed, with different pairs of animals, a total of twenty times, with individuals from species 1 being dominant six times and those from species 2 being dominant fourteen times. Test the null hypothesis that there is no difference in the ability of members of either species to dominate.

6. A hospital treated 412 skin cancer patients over a period of time. Of these, 197 were female. Using the normal approximation to the binomial test, test the hypothesis that equal numbers of males and females seek treatment for skin cancer.

7. Test the null hypothesis, using the binomial test normal approximation.

8. Ten students were given a mathematics aptitude test in a quiet room. The same students were given a similar test in a room with background music. Their performances were as follows. Using the sign test, test the hypothesis that the music has no effect on test performance.

Student	Score without music	Score with music
1	114	112
2	121	122
3	136	141
4	102	107
5	99	96
6	114	109
7	127	121
8	150	146
9	129	127
10	130	128

9. Estimate the power of the hypothesis test of Exercise 5 if $\alpha = 0.05$.

10. Using the normal approximation, estimate the power of the hypothesis test of Exercise 6 if $\alpha = 0.05$.

11. In a random sample of 30 boys, 18 have curly hair. Determine the 95% confidence limits for the proportion of curly-haired individuals in the population of boys that was sampled.

(a) Determine the Clopper-Pearson interval.

(b) Determine the Wald interval.

(c) Determine the adjusted Wald interval.

12. In a random sample of 1215 animals, 62 exhibited a certain genetic defect. Determine the 95% confidence interval for the proportion of the population displaying this defect.

(a) Determine the Clopper-Pearson interval.

(b) Determine the Wald interval.

(c) Determine the adjusted Wald interval.

13. From this sample of 14 measurements, determine the 90% confidence limits for the population median running speed of monkeys: 28.3, 29.1, 29.5, 20.1, 30.2, 31.4, 32.2, 32.8, 33.1, 33.2, 33.6, 34.5, 34.7, and 34.8 km/hr.

14. Using the data of Exercise 3 from *Contingency Tables*, test $H_0: p_1 = p_2$ versus $H_A: p_1 \neq p_2$.

15. Using the data of Example 3 from *Contingency Tables*, determine the 95% adjusted Wald confidence limits for $p_1 - p_2$.

16. Using the data of Exercise 1 from *Contingency Tables*, test the null hypothesis that there is the same proportion of males in all four seasons.

17. If the null hypothesis in Exercise 16 is rejected, perform a Tukey-type multiple-comparison test to conclude which population proportions are different from which.

18. A new type of heart valve has been developed and is implanted in 63 dogs that have been raised on various levels of exercise. The numbers of valve transplants that succeed are tabulated as follows.

(a) Is the proportion of successful implants the same for dogs on all exercise regimens?

(b) Is there a trend with amount of exercise in the proportion of successful implants?

Implant	**Amount of exercise**				
	None	Slight	Moderate	Vigorous	Total
Successful	8	9	17	14	48
Unsuccessful	7	3	3	2	15
Total	15	12	20	16	63

19. In investigating the cold tolerance of adults of a species of tropical butterfly, 68 of the butterflies (32 females and 36 males) were subjected to a cold temperature until half of the 68 had died. Twenty of the females survived, as did 14 of the males, with the data tabulated as follows:

	Females	Males	
Alive	20	14	34
Dead	12	22	34
	32	36	68

Prior to performing the experiment and collecting the data, it was stated that H_0: females are as likely as males to survive the experimental temperature, and H_A: Females and males are not equally likely to survive.

(a) Use the Fisher exact test for the one-tailed hypotheses.

(b) Use chi-square with the Yates correction for continuity (Sections 3c and 3d from *Contingency Tables*) for the two-tailed hypotheses.

(c) Use chi-square with the Cochran-Haber correction for continuity (Sections 3c and 3d from *Contingency Tables*) for the two-tailed hypotheses.

(d) Use the Fisher exact test for the two-tailed hypotheses.

20. Thirteen snakes of species S and 17 of species E were placed in an enclosure containing 14 mice of species M and 16 of species U. Each of the 30 snakes ate one of the 30 mice, and the following results were recorded:

	Snakes S	Snakes E	
Mice M	3	11	14
Mice U	10	6	16
	13	17	30

Prior to performing the experiment and collecting the data, it was decided whether the interest was in a one-tailed or two-tailed test. The one-tailed test hypotheses would be H_0: Under the conditions of this experiment, snakes of species E are not more likely than species S to eat mice of species M (i.e., they are less likely or equally likely to do so), and H_A: Snakes of species E are more likely to eat mice of species M. The two-tailed hypotheses would be H_0: Under the conditions of this experiment, snakes of species S and species E are equally likely to eat mice of species M, and H_A: Snakes of species S and species E are not equally likely to eat mice of species M.

(a) Use the Fisher exact test for the one-tailed hypotheses.

(b) Use chi-square with the Yates correction for continuity for the two-tailed hypotheses.

(c) Use chi-square with the Cochran-Haber correction for continuity.

(d) Use the Fisher exact test for the two-tailed hypotheses.

21. One hundred twenty-two pairs of brothers, one member of each pair overweight and the other of normal weight, were examined for presence of varicose veins. Use the McNemar test for the data below to test the hypothesis that there is no relationship between being overweight and developing varicose veins (i.e., that the same proportion of overweight men as normal weight men possess

varicose veins). In the following data tabulation, "v.v." stands for "varicose veins."

Normal Weight	Overweight With v.v.	Overweight Without v.v.	
With v.v.	19	5	
Without v.v.	12	86	$n = 122$

ANSWERS TO EXERCISES

1. $P(X = 2) = 0.32413$.

2. $P(X = 4) = 0.00914$.

3. H_0: The sampled population is binomial with $p = 0.25$; H_A: The sampled population is not binomial with $p = 0.25$; $\sum f_i = 126$; $F_1 = (0.31641)(126) = 39.868$, $F_2 = 53.157$, $F_3 = 26.578$, $F_4 = 5.907$, $F_5 = 0.493$; combine F_4 and F_5 and combine f_4 and f_5; $\chi^2 = 11.524$, $\nu = k - 1 = 3$, $\chi^2_{0.05, 3} = 7.815$; reject H_0; $0.005 < P < 0.01$ [$P = 0.0092$].

4. H_0: The sampled population is binomial; H_A: The sampled population is not binomial; $\hat{p} = \frac{156}{109}/4 = 0.3578$; $\chi^2 = 3.186$, $\nu = k - 2 = 3$, $\chi^2_{0.05, 3} = 7.815$, do not reject H_0; $0.25 < P < 0.50$ [$P = 0.36$].

5. H_0: $p = 0.5$; H_A: $p \neq 0.5$; $n = 20$; $P(X \leq 6$ or $X \geq 14) = 0.11532$; since this probability is greater than 0.05, do not reject H_0.

6. H_0: $p = 0.5$; H_A: $p \neq 0.5$; $\hat{p} = \frac{197}{412} = 0.4782$; $Z = -0.888$, $Z_c = 0.838$; $Z_{0.05(2)} = t_{0.05(2), \infty} = 1.960$; therefore, do not reject H_0; $P \approx 0.37$ [$P = 0.40$].

7. H_0: $p = 0.5$; H_A: $p \neq 0.5$; $X = 44$; $Z = -1.0102$; $Z_{0.05(2)} = t_{0.05(2), \infty} = 1.960$; do not reject H_0; $0.20 < P < 0.50$ [$P = 0.30$].

8. H_0: $p = 0.5$; H_A: $p \neq 0.5$; number of positive differences = 7; for $n = 10$ and $p = 0.5$; $P(X \leq 3$ or $X \geq 7) = 0.34378$; since this probability is greater than 0.05, do not reject H_0.

9. $n = 20$, $p = 0.50$; critical values are 5 and 15; $\hat{p} = 6/20 = 0.30$, power = $0.00080 + 0.00684 + 0.02785 + 0.07160 + 0.13042 + 0.17886 + 0.00004 + 0.00001 = 0.42$.

10. $p_0 = 0.50$, $p = 0.4782$, $n = 412$; power = $P(Z < -1.08) + P(Z > 2.84) = 0.1401 + 0.0023 = 0.14$.

11. $X = 18$, $n = 30$, $\hat{p} = 0.600$ **(a)** $F_{0.05(2), 26, 36} \approx F_{0.05(2), 26, 35} = 2.04$, $L_1 = 0.404$; $F_{0.05(2), 38, 24} \approx F_{0.05(2), 30, 24} = 2.21$, $L_2 = 0.778$. Using exact probabilities of F: $F_{0.05(2), 26, 36} = 2.025$, $L_1 = 0.406$; $F_{0.05(2), 28, 34} = 2.156$, $L_2 = 0.773$.
(b) $Z_{0.05(2)} = 1.9600$, confidence interval is

$0.600 \pm (1.9600)(0.0894)$, $L_1 = 0.425$, $L_2 = 0.775$. **(c)** $\tilde{X} = 19.92$, $\tilde{n} = 33.84$, $\tilde{p} = 0.589$, confidence interval is $0.589 \pm (1.9600)(0.0846)$, $L_1 = 0.423$, $L_2 = 0.755$.

12. $X = 62$, $n = 1215$, $\hat{p} = 0.051$ **(a)** $F_{0.05(2), 2308, 124} \approx F_{0.05(2), \infty, 120} = 1.31$, $L_1 = 0.039$; $F_{0.05(2), 126, 2306} \approx F_{0.05(2), 120, \infty} = 1.27$, $L_2 = 0.065$. Using exact probabilities of F: $F_{0.05(2), 2308, 124} = 1.312$, $L_1 = 0.039$; $F_{0.05(2), 126, 2306} = 1.271$, $L_2 = 0.061$. **(b)** $Z_{0.05(2)} = 1.9600$, confidence interval is $0.051 \pm (1.9600)(0.0631)$, $L_1 = 0.039$, $L_2 = 0.063$. **(c)** $\tilde{X} = 63.92$, $\tilde{n} = 1218.84$, $\tilde{p} = 0.052$. confidence interval is $0.052 \pm (1.9600)(0.00638)$, $L_1 = 0.039$, $L_2 = 0.065$.

13. sample median = 32.5 km/hr, $i = 4$, $j = 1$, $P(29.5$ km/hr $\leq$ population median $\leq$ 33.6 km/hr$) = 0.90$.

14. $\hat{p}_1 = 0.7500$, $\hat{p}_2 = 0.4000$, $\bar{p} = 0.5714$, $\bar{q} = 0.4286$, SE $= 0.1414$, $Z = 2.475$, $0.01 < P < 0.02$, reject H_0 [$P = 0.013$].

15. $\tilde{X}_1 = 18.96$, $\tilde{n}_1 = 25.92$, $\tilde{p}_1 = 0.7315$, $\tilde{X}_2 = 10.96$, $\tilde{n}_2 = 26.92$, $\tilde{p}_2 = 0.4071$, SE $= 0.1286$, 95% confidence interval $= 32.444 \pm 0.2521$, $L_1 = 0.372$, $L_2 = 0.576$.

16. H_0: $p_1 = p_2 = p_3 = p_4$, H_A: All four population proportions are not equal; $X_1 = 163$, $X_2 = 135$, $X_3 = 71$, $X_4 = 43$, $n_1 = 249$, $n_2 = 212$, $n_3 = 111$, $n_4 = 81$; $\hat{p}_1 = 0.6546$, $\hat{p}_2 = 0.6368$, $\hat{p}_3 = 0.6396$, $\hat{p}_4 = 0.5309$; $\bar{p} = 412/653 = 0.6309$, $\chi^2 = 0.6015 + 0.0316 + 0.0364 + 3.4804 = 4.150$, $\chi^2_{0.05, 3} = 7.815$; do not reject H_0; $0.10 < P < 0.25$ [$P = 0.246$].

17. H_0: $p_1 = p_2 = p_3 = p_4$ is not rejected. So multiple-comparison testing is not done.

18. (a) P of original table $= 0.02965$; P of next more extreme table (i.e., where $f_{11} = 21) = 0.01037$; P of next more extreme table (i.e., where $f_{11} = 22) = 0.00284$; and so on with total P for that tail $= 0.0435$; H_0 is rejected. **(b)** $\chi^2_c = 2.892$, $0.05 < P < 0.10$ [$P = 0.089$], H_0 is not rejected. **(c)** $\chi^2_H = 2.892$, $0.05 < P < 0.10$ [$P = 0.089$], H_0 is not rejected. **(d)** Since $R_1 = R_2$, the two-tailed P is 2 times the one-tailed P; two-tailed $P = 2(0.0435) = 0.087$, H_0 is not rejected.

19. (a) P of original table $= 0.02034$; P of next most extreme table (i.e., where $f_{11} = 2$) $= 0.00332$; P of next most extreme table (i.e., where $f_{11} = 1$) $= 0.00021$; and so on, with a total P for that tail $= 0.02787$, H_0 is rejected. **(b)** $\chi_c^2 = 3.593$, $0.05 < P < 0.10$ $[P = 0.057]$, H_0 is not rejected. **(c)** $\chi_H^2 = 4.909$, $0.025 < P < 0.05$ $[P = 0.027]$, and H_0 is not rejected. **(d)** For the most extreme table in the tail opposite from that in part **(a)**, $f_{11} = 13$ and $f_{12} = 1$, $P = 0.00000$; for next more extreme table, $f_{11} = 12$, $P = 0.00001$; for the next more extreme table, $f_{11} = 11$, $P = 0.00027$;

and so on through each of the tables in this tail with a probability less than 0.02034; sum of the tables' probabilities in the second tail $= 0.00505$; sum of the probabilities of the two tails $= 0.02787 + 0.00505 = 0.03292$; H_0 is rejected.

20. H_0: There is no difference in frequency of occurrence of varicose veins between overweight and normal weight men; H_A: There is a difference in frequency of occurrence of varicose veins between overweight men and normal weight men; $f_{11} = 19$, $f_{12} = 5$, $f_{21} = 12$, $f_{22} = 86$, $n = 122$; $\chi_c^2 = 2.118$; $\chi_{0.05,1}^2 = 3.841$; do not reject H_0; $0.10 < P < 0.25$ $[P = 0.15]$.

Testing for Randomness

A *random distribution* of objects in space is one in which each one of equal portions of the space has the same probability of containing an object, and the occurrence of an object in no way influences the occurrence of any of the other objects. A biological example in one-dimensional space could be the linear distribution of blackbirds along the top of a fence, an example in two-dimensional space could be the distribution of cherry trees in a forest, and an example in three-dimensional space could be the distribution of unicellular algae in water.* A random distribution of events in time is one in which each period of time of given length (e.g., an hour or a day) has an equal chance of containing an event, and the occurrence of any one of the events is independent of the occurrence of any of the other events. An example of events in periods of time could be the numbers of heart-attack patients entering a hospital each day.

1   POISSON PROBABILITIES

The *Poisson distribution*[†] is important in describing *random* occurrences when the probability of an occurrence is small. The terms of the Poisson distribution are

$$P(X) = \frac{e^{-\mu}\mu^X}{X!}$$

(1a)

or, equivalently,

$$P(X) = \frac{\mu^X}{e^{\mu}X!},$$

(1b)

---

*An extensive coverage of the description and analysis of spatial pattern is given by Upton and Fingleton (1985).

†Also known as *Poisson's law* and named for Siméon Denis Poisson (1781–1840), a French mathematician, astronomer, and physicist (Féron, 1978). He is often credited with the first report of this distribution in a 1837 publication. However, Dale (1989) reported that it appeared earlier in an 1830 memoir of an 1829 presentation by Poisson, and Abraham de Moivre (1667–1754) apparently described it in 1718 (David, 1962: 168; Stigler, 1982). It was also described independently by others, including "Student" (W. S. Gosset, 1976–1937) during 1906–1909 (Boland, 2000; Haight, 1967: 117). Poisson's name might have first been attached to this distribution, in contrast to his being merely cited, by H.E. Soper in 1914 (David, 1995).

From Chapter 25 of *Biostatistical Analysis*, Fifth Edition, Jerrold H. Zar. Copyright © 2010 by Pearson Education, Inc. Publishing as Pearson Prentice Hall. All rights reserved.

where $P(X)$ is the probability of X occurrences in a unit of space (or time) and μ is the population mean number of occurrences in the unit of space (or time). Thus,

$$P(0) = e^{-\mu}, \tag{2}$$

$$P(1) = e^{-\mu}\mu, \tag{3}$$

$$P(2) = \frac{e^{-\mu}\mu^2}{2}, \tag{4}$$

$$P(3) = \frac{e^{-\mu}\mu^3}{(3)(2)}, \tag{5}$$

$$P(4) = \frac{e^{-\mu}\mu^4}{(4)(3)(2)}, \tag{6}$$

and so on, where $P(0)$ is the probability of no occurrences in the unit space, $P(1)$ is the probability of exactly one occurrence in the unit space, and so on. Figure 1 presents some Poisson probabilities graphically.

In calculating a series of Poisson probabilities, as represented by the preceding five equations, a simple computational expedient is available:

$$P(X) = \frac{P(X-1)\mu}{X}. \tag{7}$$

Example 1 demonstrates these calculations for predicting how many plants will have no beetles, how many will have one beetle, how many will have two beetles, and so on, if 80 beetles are distributed randomly among 50 plants.

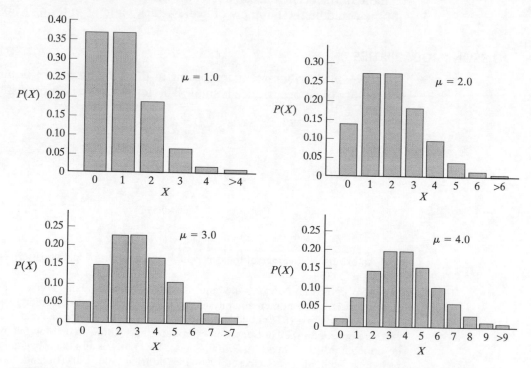

FIGURE 1: The Poisson distribution for various values of μ. These graphs were prepared by using Equation 1.

EXAMPLE 1   Frequencies from a Poisson Distribution

There are 50 plants in a greenhouse. If 80 leaf-eating beetles are introduced and they randomly land on the plants, what are the expected numbers of beetles per plant?

$$n = 50$$

$$\mu = \frac{80 \text{ beetles}}{50 \text{ plants}} = 1.6 \text{ beetles/plant}$$

Using Equation 2, $P(0) = e^{-\mu} = e^{-1.6} = 0.20190$; thus, 20.190% of the plants — that is, $(0.20190)(50) = 10.10$ (about 10) — are expected to have no beetles. The probabilities of plants with $X_i = 1, 2, 3, \ldots$ beetles are as follows, using Equation 7 (although Equation 1a or 1b could be used instead):

Number of beetles X	Poisson probability $P(X)$	Estimated number of plants	
		$\hat{f} = [P(X)][n]$	$\hat{f}$ rounded
0	0.20190	$(0.20190)(50) = 10.10$	10
1	$(0.20190)(1.6)/1 = 0.32304$	$(0.32304)(50) = 16.15$	16
2	$(0.32304)(1.6)/2 = 0.25843$	$(0.25843)(50) = 12.92$	13
3	$(0.25843)(1.6)/3 = 0.13783$	$(0.13783)(50) = 6.89$	7
4	$(0.13783)(1.6)/4 = 0.05513$	$(0.05513)(50) = 2.76$	3
5	$(0.05513)(1.6)/5 = 0.01764$	$(0.01764)(50) = 0.88$	1
≤ 5	0.99397	49.70	50
≥ 6	$1.00000 - 0.99397 = 0.00603$	$(0.00603)(50) = 0.30$	0
		50.00	50

The Poisson distribution is appropriate when there is a small probability of a single event, as reflected in a small μ, and this distribution is very similar to the binomial distribution where n is large and p is small. For example, Table 1 compares the Poisson distribution where $\mu = 1$ with the binomial distribution where $n = 100$ and $p = 0.01$ (and, therefore, $\mu = np = 1$). Thus, the Poisson distribution has importance in describing binomially distributed events having low probability. Another interesting property of the Poisson distribution is that $\sigma^2 = \mu$; that is, the variance and the mean are equal.

2 CONFIDENCE LIMITS FOR THE POISSON PARAMETER

Confidence limits for the Poisson distribution parameter, μ, may be obtained as follows. The lower $1 - \alpha$ confidence limit is

$$L_1 = \frac{\chi^2_{(1-\alpha/2),\, \nu}}{2}, \tag{8}$$

where $\nu = 2X$; and the upper $1 - \alpha$ confidence limit is

$$L_2 = \frac{\chi^2_{\alpha/2,\, \nu}}{2}, \tag{9}$$

TABLE 1: The Poisson Distribution Where $\mu = 1$ Compared with the Binomial Distribution Where $n = 100$ and $p = 0.01$ (i.e., with $\mu = 1$) and the Binomial Distribution Where $n = 10$ and $p = 0.1$ (i.e., with $\mu = 1$)

X	$P(X)$ for Poisson: $\mu = 1$	$P(X)$ for binomial: $n = 100, p = 0.01$	$P(X)$ for binomial: $n = 10, p = 0.1$
0	0.36788	0.36603	0.34868
1	0.36788	0.36973	0.38742
2	0.18394	0.18486	0.19371
3	0.06131	0.06100	0.05740
4	0.01533	0.01494	0.01116
5	0.00307	0.00290	0.00149
6	0.00050	0.00046	0.00014
7	0.00007	0.00006	0.00001
>7	0.00001	0.00002	0.00000
Total	1.00000	1.00000	1.00001

where $\nu = 2(X + 1)$ (Pearson and Hartley, 1966: 81). This is demonstrated in Example 2. L_1 and L_2 are the confidence limits for the population mean and for the population variance. Confidence limits for the population standard deviation, σ, are simply the square roots of L_1 and L_2. The confidence limits, L_1 and L_2 (or their square roots), are not symmetrical around the parameter to which they refer. This procedure is a fairly good approximation. If confidence limits are desired to be accurate to more decimal places than given by the available critical values of χ^2, we may engage in the more tedious process of examining the tails of the Poisson distribution (e.g., see Example 3) to determine the value of X that cuts off $\alpha/2$ of each tail. Baker (2002) and Schwertman and Martinez (1994) discuss several approximations to L_1 and L_2, but the best of them require more computational effort than do the exact limits given previously, and they are all poor estimates of those limits.

EXAMPLE 2   Confidence Limits for the Poisson Parameter

An oak leaf contains four galls. Assuming that there is a random occurrence of galls on oak leaves in the population, estimate with 95% confidence the mean number of galls per leaf in the population.

The population mean, μ, is estimated as $X = 4$ galls/leaf.

The 95% confidence limits for μ are

$$L_1 = \frac{\chi^2_{(1-\alpha/2),\,\nu}}{2}, \quad \text{where} \quad \nu = 2X = 2(4) = 8$$

$$L_1 = \frac{\chi^2_{0.975,\,8}}{2} = \frac{2.180}{2} = 1.1 \text{ galls/leaf}$$

$$L_2 = \frac{\chi^2_{\alpha/2,\,\nu}}{2}, \quad \text{where} \quad \nu = 2(X + 1) = 2(4 + 1) = 10$$

$$L_2 = \frac{\chi^2_{0.025,\,10}}{2} = \frac{20.483}{2} = 10.2 \text{ galls/leaf}$$

Therefore, we can state

$$P(1.1 \text{ galls/leaf} \leq \mu \leq 10.2 \text{ galls/leaf}) \geq 0.95$$

and

$$P(1.1 \text{ galls/leaf} \leq \sigma^2 \leq 10.2 \text{ galls/leaf}) \geq 0.95;$$

and, using the square roots of L_1 and L_2,

$$P(1.0 \text{ galls/leaf} \leq \sigma \leq 3.2 \text{ galls/leaf}) \geq 0.95.$$

3 GOODNESS OF FIT FOR THE POISSON DISTRIBUTION

The goodness of fit of a set of data to the Poisson distribution is a test of the null hypothesis that the data are distributed randomly within the space that was sampled. This may be tested by chi-square, as was done with the binomial distribution. When tabulating the observed frequencies (f_i) and the expected frequencies ($\hat{f}_i$), the frequencies in the tails of the distribution should be pooled so no $\hat{f}_i$ is less than 1.0 (Cochran, 1954). The degrees of freedom are $k - 2$ (where k is the number of categories of X remaining after such pooling). Example 3 fits a set of data to a Poisson distribution, using the sample mean, $\overline{X}$, as an estimate of μ in Equations 2 and 7. The G statistic may be used for goodness-of-fit analysis instead of chi-square. It will give equivalent results when n/k is large; if n/k is very small, G is preferable to χ^2 (Rao and Chakravarti, 1956).

If μ were known for the particular population sampled, or if it were desired to assume a certain value of μ, then the parameter would not have to be estimated by $\overline{X}$, and the degrees of freedom for χ^2 for G goodness-of-fit testing would be $k - 1$. For example, if the 50 plants in Example 1 were considered the only plants of interest and 80 beetles were distributed among them, μ would be $80/50 = 1.6$. Then the observed number of beetles per plant could be counted and those numbers compared to the expected frequencies ($\hat{f}$) determined in Example 1 for a random distribution. It is only when this parameter is specified that the Kolmogorov-Smirnov goodness-of-fit procedure may be applied (Massey, 1951).

EXAMPLE 3   Fitting the Poisson Distribution

Thirty plots of ground were examined within an abandoned golf course, each plot being the same size; a total of 74 weeds were counted in the 30 plots. The frequency f is the number of plots found to contain X weeds; $P(X)$ is the probability of X weeds in a plot if the distribution of weeds is random within the golf course.

H_0: The weeds are distributed randomly.

H_A: The weeds are not distributed randomly.

$n = 30$

$\overline{X} = \dfrac{74 \text{ weeds}}{30 \text{ plots}} = 2.47 \text{ weeds/plot}$

$P(0) = e^{-\overline{X}} = e^{-2.47} = 0.08458$

X	f	fX	$P(X)$	$\hat{f} = [P(X)][n]$
0	2	0	0.08458	2.537
1	1	1	$(0.08458)(2.47)/1 = 0.20891$	6.267
2	13	26	$(0.20891)(2.47)/2 = 0.25800$	7.740
3	10	30	$(0.25800)(2.47)/3 = 0.21242$	6.373
4	3	12	$(0.21242)(2.47)/4 = 0.13116$	3.935
5	1	5	$(0.13116)(2.47)/5 = 0.06479$	1.944
6	0	0	$(0.06479)(2.47)/6 = 0.02667$	0.800
	30	74	0.98653	29.596

The last $\hat{f}$ calculated (for $X = 6$) is less than 1.0, so the calculation of $\hat{f}$'s proceeds no further. The sum of the seven calculated $\hat{f}$'s is 25.596, so $P(X > 6) = 30 - 25.596 = 0.404$ and the $\hat{f}$'s of 0.800 and 0.404 are summed to 1.204 to obtain an $\hat{f}$ that is no smaller than 1.0.*

Then the chi-square goodness of fit would proceed:

X:	0	1	2	3	4	5	≥ 6	n
f:	2	1	13	10	3	1	0	30
$\hat{f}$:	2.537	6.267	7.740	6.373	3.935	1.944	1.204	

$$\chi^2 = \frac{(2 - 2.537)^2}{2.537} + \frac{(1 - 6.267)^2}{6.267} + \frac{(13 - 7.740)^2}{7.740}$$

$$+ \frac{(10 - 6.373)^2}{6.373} + \frac{(3 - 3.935)^2}{3.935} + \frac{(1 - 1.944)^2}{1.944} + \frac{(0 - 1.204)^2}{1.204}$$

$$= 0.114 + 4.427 + 3.575 + 2.064 + 0.222 + 0.458 + 1.204$$

$$= 12.064$$

$$\nu = k - 2 = 7 - 2 = 5$$

$$\chi^2_{0.05,5} = 11.070.$$

Therefore, reject H_0.

$$0.025 < P < 0.05$$

---

*$P(X) > 6$ could also have been obtained by adding all of the $P(X)$'s in the preceding table, which would result in a sum of 0.98653; and $(0.98653)(30) = 29.596$.

The null hypothesis in Poisson goodness-of-fit testing is that the distribution of objects in space (or events in time) is random.

- A *random* distribution of objects in a space is one in which each object has the same probability of occurring in each portion of the space; that is, the occurrence of each object is independent of the occurrence of any other object. There are these two kinds of deviation from randomness that will cause rejection of the null hypothesis:

- A *uniform* distribution of objects in a space is one in which there is equal distance between adjacent objects, as if they are repelling each other.
- A *contagious* distribution* (also referred to as a "clumped," "clustered," "patchy," or "aggregated" distribution) in a space is one in which objects are more likely than in a random distribution to occur in the vicinity of other objects, as if they are attracting each other.

Figures 2 and 3 show examples of these three kinds of distributions.

If a population has a random (Poisson) distribution, the variance is the same as the mean: that is, $\sigma^2 = \mu$ and $\sigma^2/\mu = 1.0$.[†] If the distribution is more uniform than random (said to be "underdispersed"), $\sigma^2 < \mu$ and $\sigma^2/\mu < 1.0$; and if the distribution is distributed contagiously ("overdispersed"), $\sigma^2 > \mu$ and $\sigma^2/\mu > 1.0$.

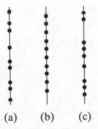

(a)   (b)   (c)

FIGURE 2: Distributions in one-dimensional space (i.e., along a line): (a) random (Poisson), in which $\sigma^2 = \mu$; (b) uniform, in which $\sigma^2 < \mu$; (c) contagious, in which $\sigma^2 > \mu$.

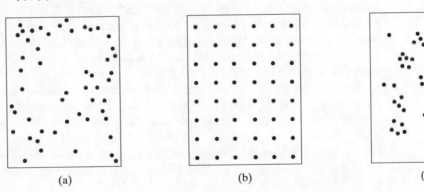

(a)   (b)   (c)

FIGURE 3: Distributions in two-dimensional space: (a) random (Poisson), in which $\sigma^2 = \mu$; (b) uniform, in which $\sigma^2 < \mu$; (c) contagious, in which $\sigma^2 > \mu$.

---

*A mathematical distribution that is sometimes used to describe contagious distributions of biological data is the *negative binomial distribution*, which is described, for example, by Ludwig and Reynolds (1988: 24–26, 32–35) and Pielou (1977: 278–281), and by Ross and Preece (1985), who credit a 1930 French paper by G. Polya with the first use of the term *contagious* in this context. David (1995) reported that "negative binomial distribution" is a term first used by M. Greenwood and G. U. Yule, in 1920.

†Although a population with a random distribution will always have its mean equal to its variance, Hurlbert (1990), Pielou (1967: 155), and others have emphasized that not every population with $\mu = \sigma^2$ has a random distribution.

The investigator generally has some control over the size of the space, or the length of the time interval, from which counts are recorded. So a plot size twice as large as that in Example 3 might have been used, in which case each f would most likely have been twice the size as in this example, with $\overline{X}$ of 4.94, instead of 2.47. In analyses using the Poisson distribution, it is desirable to use a sample distribution with a fairly small mean — let us say certainly below 10, preferably below 5, and ideally in the neighborhood of 1. If the mean is too large, then the Poisson too closely resembles the binomial, as well as the normal, distribution. If it is too small, however, then the number of categories, k, with appreciable frequencies will be too small for sensitive analysis.

Graphical testing of goodness of fit is sometimes encountered. The reader may consult Gart (1969b) for such considerations.

4   THE POISSON DISTRIBUTION FOR THE BINOMIAL TEST

The binomial test was introduced as a goodness-of-fit test for counts in two categories. If n is large, the binomial test is unwieldy. If p is small, it may be convenient to use the Poisson distribution, for it becomes very similar to the binomial distribution at such p's.

(a) One-Tailed Testing.   Let us consider the following example. It is assumed (as from a very large body of previous information) that a certain type of genetic mutation naturally occurs in an insect population with a frequency of 0.0020 (i.e., on average in 20 out of 10,000 insects). On exposing a large number of these insects to a particular chemical, we wish to ask whether that chemical increases the rate of this mutation. Thus, we state $H_0: p \leq 0.0020$ and $H_A: p > 0.0020$. (The general one-tailed hypotheses of this sort would be $H_0: p \leq p_0$ and $H_A: p > p_0$, where p_0 is the proportion of interest in the statistical hypotheses. If we had reason to ask whether some treatment reduced the natural rate of mutations, then the one-tailed test would have used $H_0: p \geq p_0$ and $H_A: p < p_0$.)

As an example, if performing this exposure experiment for the hypotheses $H_0: p_0 \leq 0.0020$ and $H_A: p_0 > 0.0020$ yielded 28 of the mutations of interest in 8000 insects observed, then the sample mutation rate is $\hat{p} = X/n = 28/8000 = 0.0035$. The question is whether the rate of 0.0035 is significantly greater than 0.0020. If we conclude that there is a low probability (i.e., $\leq \alpha$) of a sample rate being at least as large as 0.0035 when the sample is taken at random from a population having a rate of 0.0020, then H_0 is to be rejected.

The hypotheses could also be stated in terms of numbers, instead of proportions, as $H_0: \mu \leq \mu_0$ and $H_A: \mu > \mu_0$, where $\mu_0 = p_0 n$ (which is $0.0020 \times 8000 = 16$ in this example).

By substituting $p_0 n$ for μ in Equation 1, we determine the probability of observing $X = 28$ mutations if our sample came from a population with $p_0 = 0.0020$. To test the hypothesis at hand, we determine the probability of observing $X \geq 28$ mutations in a sample. (If the alternate hypothesis being considered were $H_A: p < p_0$, then we would compute the probability of mutations less than or equal to the number observed.) If the one-tailed probability is less than or equal to α, then H_0 is rejected at the α level of significance. This process is shown in Examples 4 and 5a.

EXAMPLE 4 Poisson Probabilities for Performing the Binomial Test with a Very Small Proportion

$p_0 = 0.0020$

$n = 8000$

We substitute $p_0 n = (0.0020)(8000) = 16$ for μ in Equation 1 to compute the following*:

For lower tail of distribution			For upper tail of distribution		
X	$P(X)$	Cumulative $P(X)$	X	$P(X)$	Cumulative $P(X)$
0	0.00000	0.00000	23	0.02156	0.05825
1	0.00000	0.00000	24	0.01437	0.03669
2	0.00001	0.00001	25	0.00920	0.02232
3	0.00008	0.00009	26	0.00566	0.01312
4	0.00031	0.00040	27	0.00335	0.00746
5	0.00098	0.00138	28	0.00192	0.00411
6	0.00262	0.00400	29	0.00106	0.00219
7	0.00599	0.00999	30	0.00056	0.00113
8	0.01199	0.02198	31	0.00029	0.00057
9	0.02131	0.04329	32	0.00015	0.00028
10	0.03410	0.07739	33	0.00007	0.00013
			34	0.00003	0.00006
			35	0.00002	0.00003
			36	0.00001	0.00001
			37	0.00000	0.00000

The cumulative probability is the probability in the indicated tail. For example, $P(X \leq 10) = 0.07739$ and $P(X \geq 25) = 0.02232$. This series of computations terminates when we reach a $P(X)$ that is zero to the number of decimal places used.

---

*For example, using Equation 1a, $P(X = 28) = \dfrac{e^{-16} 16^{28}}{28!} = 0.00192$; and $P(X = 29) = \dfrac{e^{-16} 16^{29}}{29!} = 0.00106$.

(b) Two-Tailed Testing. If there is no reason, a priori, to hypothesize that a change in mutation rate would be in one specified direction (e.g., an increase) from the natural rate, then a two-tailed test is appropriate. The probability of the observed number of mutations is computed as shown in Example 4. Then we calculate and sum all the probabilities (in both tails) that are equal to or smaller than that of the observed. This is demonstrated in Example 5b.

(c) Power of the Test. Recall that the power of a statistical test is the probability of that test rejecting a null hypothesis that is in fact a false statement about the

EXAMPLE 5a   A One-Tailed Binomial Test for a Proportion from a Poisson Population, Using the Information of Example 4

H_0: $p \leq 0.0020$
H_A: $p > 0.0020$
  $\alpha = 0.05$
  $n = 8000$
  $X = 28$
  $p_0 n = (0.0020)(8000) = 16$

Therefore, we could state

H_0: $\mu \leq 16$

H_A: $\mu > 16$.

From Example 4, we see that $P(X = 28) = 0.00192$ and $P(X \geq 28) = 0.00411$. As $0.00411 < 0.05$, reject H_0.

EXAMPLE 5b   A Two-Tailed Binomial Test for a Proportion from a Poisson Population, Using the Information of Example 4

H_0: $p = 0.0020$

H_A: $p \neq 0.0020$

  $\alpha = 0.05$

  $n = 8000$

  $X = 28$

  $p_0 n = (0.0020)(8000) = 16$

Therefore, we can state

H_0: $\mu = 16$

H_A: $\mu \neq 16$.

From Example 4, we see that $P(X = 28) = 0.00192$.

The sum of the probabilities in one tail that are ≤ 0.00192 is 0.00411; the sum of the probabilities in the other tail that are ≤ 0.00192 is 0.00138. Therefore, the probability of obtaining these data from a population where H_0 is true is $0.00411 + 0.00138 = 0.00549$.

As $0.00549 < 0.05$, reject H_0.

---

population. We can determine the power of the preceding test when it is performed with a sample size of n at a significance level of α. For a one-tailed test, we first determine the critical value of X (i.e., the smallest X that delineates a proportion of the Poisson distribution $\leq \alpha$). Examining the distribution of Example 4, for example, for $\alpha = 0.05$, we see that the appropriate X is 24 [for $P(X \geq 24) = 0.037$), while $P(X \geq 23) = 0.058$]. We then examine the Poisson distribution having the sample X replace μ in Equation 1. The power of the test is $\geq$ the probability of an X at least as extreme as the critical value of X.*

For a two-tailed hypothesis, we identify one critical value of X as the smallest X that cuts off $\leq \alpha/2$ of the distribution in the upper tail and one as the largest X that cuts off $\leq \alpha/2$ of the lower tail. In Example 6, these two critical values for $\alpha = 0.05$ (i.e., $\alpha/2 = 0.025$) are $X = 25$ and $X = 8$ [as $P(X \geq 25) = 0.022$ and $P(X \leq 8) = 0.022$]. Then we examine the Poisson distribution having the sample X replace μ in Equation 1. As shown in Example 6, the power of the two-tailed test is at least as large as the probability of X in the latter Poisson distribution being more extreme than either of the critical values. That is, power $\geq P(X \geq$ upper critical value$) + P(X \leq$ lower critical value).

5 COMPARING TWO POISSON COUNTS

If we have two counts, X_1 and X_2, each from a population with a Poisson distribution, we can ask whether they are likely to have come from the same population (or from populations with the same mean). The test of H_0: $\mu_1 = \mu_2$ (against H_A: $\mu_1 \neq \mu_2$) is related to the binomial test with $p = 0.50$ (Przyborowski and Wilenski, 1940; Pearson and Hartley, 1966: 78–79), so that Table 27 from *Appendix: Statistical Tables and Graphs* can be utilized, using $n = X_1 + X_2$. For the two-tailed test, H_0 is rejected if either X_1 or X_2 is $\leq$ the critical value, $C_{\alpha(2),n}$. This is demonstrated in Example 7.

For a one-tailed test of H_0: $\mu_1 \leq \mu_2$ against H_A: $\mu_1 > \mu_2$, we reject H_0 if $X_1 > X_2$ and $X_2 \leq C_{\alpha(1),n}$, where $n = X_1 + X_2$. For H_0: $\mu_1 \geq \mu_2$ and H_A: $\mu_1 < \mu_2$, H_0 is rejected if $X_1 < X_2$ and $X_1 \leq C_{\alpha(1),n}$, where $n = X_1 + X_2$.

This procedure results in conservative testing, and if n is at least 5, then a normal approximation should be used (Detre and White, 1970; Przyborowski and Wilenski, 1940; Sichel, 1973). For the two-tailed test,

$$Z = \frac{|X_1 - X_2|}{\sqrt{X_1 + X_2}} \tag{10}$$

is considered a normal deviate, so the critical value is $Z_{\alpha(2)}$ (which can be read as $t_{\alpha(2),\infty}$ at the end of Table 3 from *Appendix: Statistical Tables and Graphs*). This is demonstrated in Example 7.

For a one-tailed test,

$$Z = \frac{X_1 - X_2}{\sqrt{X_1 - X_2}}. \tag{11}$$

For H_0: $\mu_1 \leq \mu_2$ versus H_A: $\mu_1 > \mu_2$, H_0 is rejected if $X_1 > X_2$ and $Z \geq Z_{\alpha(1)}$. For H_0: $\mu_1 \geq \mu_2$ versus H_A: $\mu_1 < \mu_2$, H_0 is rejected if $X_1 < X_2$ and $Z \leq -Z_{\alpha(1)}$.

The normal approximation is sometimes seen presented with a correction for continuity, but Pirie and Hamdan (1972) concluded that this produces results that are excessively conservative and the test has very low power.

An alternative normal approximation, based on a square-root transformation (Anscombe, 1948) is

$$Z = \left| \sqrt{2X_1 + \frac{3}{4}} - \sqrt{2X_2 + \frac{3}{4}} \right| \tag{12}$$

---

*If the critical value delineates exactly α of the tail of the Poisson distribution, then the test's power is exactly what was calculated; if the critical value cuts off $< \alpha$ of the tail, then the power is $>$ that calculated.

EXAMPLE 6 **Estimation of the Power of the Small-Probability Binomial Tests of Examples 5a and 5b, Using $\alpha = 0.05$**

Substituting $X = 28$ for μ in Equation 1, we compute the following*:

For lower tail of distribution			For upper tail of distribution		
X	$P(X)$	Cumulative $P(X)$	X	$P(X)$	Cumulative $P(X)$
0	0.000	0.000	24	0.060	0.798
1	0.000	0.000	25	0.067	0.738
2	0.000	0.000	26	0.072	0.671
3	0.000	0.000	27	0.075	0.599
3	0.000	0.000	28	0.075	0.524
5	0.000	0.000	29	0.073	0.449
6	0.000	0.000	30	0.068	0.376
7	0.000	0.000	31	0.061	0.308
8	0.000	0.000	32	0.054	0.247
			33	0.045	0.193
			34	0.037	0.148
			35	0.030	0.111
			36	0.023	0.081
			37	0.018	0.058
			38	0.013	0.040
			39	0.009	0.027
			40	0.007	0.018
			41	0.004	0.011
			42	0.003	0.007
			43	0.002	0.004
			44	0.001	0.002
			45	0.001	0.001
			46	0.000	0.000

The critical value for the one-tailed test of Example 5a is $X = 24$. The power of this test is $>P(X \geq 24)$ in the preceding distribution. That is, the power is >0.798.

The critical values for the two-tailed test of Example 5b are 25 and 8. The power of this test is $>P(X \geq 25) + P(X \leq 8) = 0.738 + 0.000$. That is, the power is >0.738.

---

*For example, using Equation 1a, $P(X = 24) = \dfrac{e^{-28}28^{24}}{24!} = 0.06010$; and $P(X = 25) = \dfrac{e^{-28}28^{25}}{25!} = 0.06731$.

(Best, 1975). It may be used routinely in place of Equation 10, and it has superior power when testing at $\alpha < 0.05$. Equation 12 is for a two-tailed test; for one-tailed testing, use

$$Z = \sqrt{2X_1 + \frac{3}{4}} - \sqrt{2X_2 + \frac{3}{4}}, \tag{13}$$

with the same procedure to reject H_0 as with Equation 11.

EXAMPLE 7   A Two-Sample Test with Poisson Data

One fish is found to be infected with 13 parasites and a second fish with 220. Assuming parasites are distributed randomly among fish, test whether these two fish are likely to have come from the same population. (If the two were of different species, or sexes, then we could ask whether the two species, or sexes, are equally infected.) The test is two-tailed for hypotheses H_0: $\mu_1 = \mu_2$ and H_A: $\mu_1 \neq \mu_2$.

Using Table 27 from *Appendix: Statistical Tables and Graphs* for $n = X_1 + X_2 = 13 + 22 = 35$, we find a critical value of $C_{0.05(2),35} = 11$. Because neither X_1 nor X_2 is ≤ 11, H_0 is not rejected. Using the smaller of the two X's, we conclude that the probability is between 0.20 and 0.50 that a fish with 13 parasites and one with 22 parasites come from the same Poisson population (or from two Poisson populations having the same mean).

Using the normal approximation of Equation 10,

$$Z = \frac{|X_1 - X_2|}{\sqrt{X_1 + X_2}} = \frac{|13 - 22|}{\sqrt{13 + 22}} = \frac{9}{5.916} = 1.521$$

$$Z_{0.05(2)} = t_{0.05(2),\infty} = 1.960.$$

Therefore, do not reject H_0.

$$0.10 < P < 0.20 \quad [P = 0.13]$$

---

6   SERIAL RANDOMNESS OF NOMINAL-SCALE CATEGORIES

Representatives of two different nominal-scale categories may appear serially in space or time, and their randomness of occurrence may be assessed as in the following example. Members of two species of antelopes are observed drinking along a river, and their linear order is as shown in Example 8. We may ask whether the sequence of occurrence of members of the two species is random (as opposed to the animals either forming groups with individuals of the same species or shunning members of the same species). A sequence of like elements, bounded on either side by either unlike elements or no elements, is termed a *run*. Thus, any of the following arrangements of five members of antelope species A and seven members of species B would be considered to consist of five runs:

BAABBBAAABBB,   or   *BBAAAABBBBBAB,*   or   *BABAAAABBBBB,*   or   *BAAABAABBBBB,*   and so on.

To test the null hypothesis of randomness, we may use the *runs test*.* If n_1 is the total number of elements of the first category (in the present example, the number of antelope of species A), n_2 the number of antelope of species B, and u the number of runs in the entire sequence, then the critical values, $u_{\alpha(2),n_1,n_2}$, can be read from Table 29 from *Appendix: Statistical Tables and Graphs* for cases where both $n_1 \leq 30$ and $n_2 \leq 30$. The critical values in this table are given in pairs; if the u in the sample is $\leq$ the first member of the pair *or* $\geq$ the second, then H_0 is rejected.

---

*From its inception the runs test has also been considered to be a nonparametric test of whether two samples come from the same population (e.g., Wald and Wolfowitz, 1940), but as a two-sample test it has very poor power and the Mann-Whitney test is preferable.

> **EXAMPLE 8   The Two-Tailed Runs Test with Elements of Two Kinds**
>
> Members of two species of antelopes (denoted as species A and B) are drinking along a river in the following order: *AABBAABBBBAAABBBBAABBB*.
>
> H_0: The distribution of members of the two species along the river is random.
> H_A: The distribution of members of the two species along the river is not random.
>
> For species A, $n_1 = 9$; for species B, $n_2 = 13$; and $u = 9$.
>
> $u_{0.05(2),9,13} = 6$ and 17 (from Appendix Table 29)
>
> As u is neither ≤ 6 nor ≥ 17, do not reject H_0.
>
> $$0.10 \leq P \leq 0.20$$

The power of the runs test increases with sample size.* Although Table 29 from *Appendix: Statistical Tables and Graphs* cannot be employed if either n_1 or n_2 is larger than 30, for such samples the distribution of u approaches normality with a mean of

$$\mu_u = \frac{2n_1 n_2}{N} + 1 \tag{14}$$

and a standard deviation of

$$\sigma_u = \sqrt{\frac{2n_1 n_2(2n_1 n_2 - N)}{N^2(N - 1)}}, \tag{15}$$

where $N = n_1 + n_2$ (Brownlee, 1965: 226–230; Wald and Wolfowitz, 1940). And the statistic

$$Z_c = \frac{|u - \mu_u| - 0.5}{\sigma_u} \tag{16}$$

may be considered a normal deviate, with $Z_{\alpha(2)}$ being the critical value for the test. (The 0.5 in the numerator of Z_c is a correction for continuity.)

Using Equation 16, the runs test may be extended to data with more than two categories (Wallis and Roberts, 1956: 571), for, in general,

$$\mu_u = \frac{N(N + 1) - \sum n_i^2}{N} \tag{17}$$

and

$$\sigma_u = \sqrt{\frac{\sum n_i^2 \left[\sum n_i^2 + N(N + 1)\right] - 2N \sum n_i^3 - N^3}{N^2(N - 1)}}, \tag{18}$$

where n_i is the number of items in category i, N is the total number of items (i.e., $N = \sum n_i$), and the summations are over all categories. (For two categories,

---

*Mogull (1994) has shown that the runs test should not used in the unusual case of a sample consisting entirely of runs of two (for example, a sample consisting of *BBAABBAABBAABB*). In such situations the runs test is incapable of concluding departures from randomness; it has very low power, and the power *decreases* with increased sample size.

Equations 17 and 18 are equivalent to Equations 14 and 15, respectively.) O'Brien (1976) and O'Brien and Dyck (1985) present a runs test, for two or more categories, that utilizes more information from the data, and is more powerful, than the above procedure.

(a) One-Tailed Testing. There are two ways in which a distribution of nominal-scale categories can be nonrandom: (a) The distribution may have fewer runs than would occur at random, in which case the distribution is more clustered, or contagious, than random; (b) the distribution may have more runs than would occur at random, indicating a tendency toward a uniform distribution.

To test for the one-tailed situation of contagion, we state H_0: The elements in the population are not distributed contagiously, versus H_A: The elements in the population are distributed contagiously; and H_0 would be rejected at the $\alpha(1)$ significance level if $u \leq$ the lower of the pair of critical values in Table 29 from *Appendix: Statistical Tables and Graphs*. Thus, had the animals in Example 8 been arranged $AAAAABBBBBBAAAABBBBBBB$, then $u = 4$ and the one-tailed 5% critical value would be the lower value of $u_{0.05(1),9,13}$, which is 7; as $4 < 7$, H_0 is rejected and the distribution is concluded to be clustered. In using the normal approximation, H_0 is rejected if $Z_c \geq Z_{\alpha(1)}$ and $u \leq \mu_u$.

To test for uniformity, we use H_0: The elements in the population are not uniformly distributed versus H_A: The elements in the population are uniformly distributed. If $u \geq$ the upper critical value in Table 29 from *Appendix: Statistical Tables and Graphs* for $\alpha(1)$, then H_0 is rejected. If the animals in Example 8 had been arranged as $ABABABBABABBABABBABBAB$, then $u = 18$, which is greater than the upper critical value of $u_{0.05(1),9,13}$ (which is 16); therefore, H_0 would have been rejected. If the normal approximation were used, H_0 would be rejected if $Z_c \geq Z_{\alpha(1)}$ and $u \geq \mu_u$.

(b) Centrifugal and Centripetal Patterns. Occasionally, nonrandomness in the sequential arrangement of two nominal-scale categories is characterized by one of the categories being predominant toward the ends of the series and the other toward the center. In the following sequence, for example,

$$AAAABAABBBBBBABBBBBAABBAAAA$$

the A's are more common toward the termini of the sequence, and the B's are more common toward the center of the sequence. Such a situation might be the pattern of two species of plants along a line transect from the edge of a marsh, through the center of the marsh, to the opposite edge. Or we might observe the occurrence of diseased and healthy birds in a row of cages, each cage containing one bird. Ghent (1993) refers to this as a *centrifugal* pattern of A's and a *centripetal* pattern of B's and presents a statistical test to detect such distributions of observations.

7 SERIAL RANDOMNESS OF MEASUREMENTS: PARAMETRIC TESTING

Biologists may encounter continuous data that have been collected serially in space or time. For example, rates of conduction might be measured at successive lengths along a nerve. A null hypothesis of no difference in conduction rate as one examines successive portions essentially is stating that all the measurements obtained are a random sample from a population of such measurements.

Example 9 presents data consisting of dissolved oxygen measurements of a water solution determined on the same instrument every five minutes. The desire

---

EXAMPLE 9   The Mean Square Successive Difference Test

An instrument for measuring dissolved oxygen is used to record a measurement every five minutes from a container of lake water. It is desired to know whether the differences in measurements are random or whether they are systematic. (If the latter, it could be due to the dissolved oxygen content in the water changing, or the instrument's response changing, or to both.) The data (in ppm) are as follows, recorded in the sequence in which they were obtained: 9.4, 9.3, 9.3, 9.2, 9.3, 9.2, 9.1, 9.3, 9.2, 9.1, 9.1.

H_0:   Consecutive measurements obtained on the lake water with this instrument have random variability.

H_A: Consecutive measurements obtained on the lake water with this instrument have nonrandom variability and are serially correlated.

$n = 11$

$s^2 = 0.01018 \ (\text{ppm})^2$

$$s_*^2 = \frac{(9.3 - 9.4)^2 + (9.3 - 9.3)^2 + (9.2 - 9.3)^2 + \ldots + (9.1 - 9.1)^2}{2(11 - 1)}$$

$$= 0.00550$$

$$C = 1 - \frac{0.00550}{0.01018} = 1 - 0.540 = 0.460$$

$C_{0.05,11} = 0.452$

Therefore, reject H_0.

$$0.025 < P < 0.05$$

---

is to conclude whether fluctuations in measurements are random or whether they indicate a nonrandom instability in the measuring device (or in the solution). The null hypothesis that the sequential variability among measurements is random may be subjected to the *mean square successive difference* test, a test that assumes normality in the underlying distribution. In this procedure, we calculate the sample variance, s^2, which is an estimate of the population variance, σ^2:

$$s^2 = \frac{\sum\limits_{i=1}^{n}(X_i - \overline{X})^2}{n - 1}, \tag{17}$$

or

$$s^2 = \frac{\sum\limits_{i=1}^{n} X_i^2 - \dfrac{\left(\sum\limits_{i=1}^{n} X_i\right)^2}{n}}{n - 1}. \tag{18}$$

If the null hypothesis is true, then another estimate of σ^2 is

$$s_*^2 = \frac{\sum\limits_{i=1}^{n-1}(X_{i+1} - X_i)^2}{2(n - 1)} \tag{19}$$

(von Neumann et al., 1941). Therefore, the ratio s_*^2/s^2 should equal 1 when H_0 is true. Using Young's (1941) notation, the test statistic is

$$C = 1 - \frac{s_*^2}{s^2},$$
(20)

and if this value equals or exceeds the critical value $C_{a,n}$, in Table 30 from *Appendix: Statistical Tables and Graphs*, we reject the null hypothesis of serial randomness.* The mean square successive difference test considers the one-tailed alternate hypothesis that measurements are serially correlated.

For n larger than those in Table 30 from *Appendix: Statistical Tables and Graphs*, the hypothesis may be tested by a normal approximation:

$$Z = \frac{C}{\sqrt{\dfrac{n-2}{n^2-1}}}$$
(22)

(von Neumann et al., 1941), with the value of the calculated Z being compared with the critical value of $Z_{\alpha(1)} = t_{\alpha(1),\infty}$. This approximation is very good for $\alpha = 0.05$, for n as small as 10; for $\alpha = 0.10, 0.25$, or 0.025, for n as small as 25; and for $\alpha = 0.01$ and 0.005, for n of at least 100.

8 SERIAL RANDOMNESS OF MEASUREMENTS: NONPARAMETRIC TESTING

If we do not wish to assume that a sample of serially obtained measurements came from a normal population, then the procedure of Section 7 should not be employed. Instead, there are nonparametric methods that address hypotheses about serial patterns.

(a) Runs Up and Down: Two-Tailed Testing. We may wish to test the null hypothesis that successive directions of change in serial data tend to occur randomly, with the alternate hypothesis stating the directions of change occur *either* in clusters (that is, where an increase from one datum to another is likely to be followed by another increase, and a decrease in the magnitude of the variable is likely to be followed by another decrease) *or* with a tendency toward regular alternation of increases and decreases (i.e., an increase is likely to be followed by a decrease, and vice versa). In the series of n data, we note whether datum $i + 1$ is larger than datum i (and denote this as a positive change, indicated as "+") or is smaller than datum i (which is referred to as a negative change, indicated as "−"). By a nonparametric procedure presented by Wallis and Moore (1941), the series of +'s and −'s is examined and we determine the number of runs of +'s and −'s, calling this number u as we did in Section 6. Table 31 from *Appendix: Statistical Tables and Graphs* presents pairs of critical values for u, where for the two-tailed test for deviation from randomness one would reject H_0 if u were *either* $\leq$ the first member of the pair *or* $\geq$ the second member of the pair.

---

*C can be computed as

$$C = 1 - \frac{\displaystyle\sum_{i=1}^{n-1}(X_i - X_{i+1})^2}{2(\text{SS})},$$
(21)

For sample sizes larger than those in Table 31 from *Appendix: Statistical Tables and Graphs*, a normal approximation may be employed (Edgington, 1961; Wallis and Moore, 1941) using Equation 16, where

$$\mu_u = \frac{2n - 1}{3} \tag{23}$$

and

$$\sigma_u = \sqrt{\frac{16n - 29}{90}}. \tag{24}$$

The runs-up-and-down test may be used for ratio, interval, or ordinal data and is demonstrated in Example 10. It is most powerful when no adjacent data are the same; if there are identical adjacent data, as in Example 10, then indicate the progression from each adjacent datum to the next as "0" and determine the mean of all the u's that would result from all the different conversions of the 0's to either +'s or −'s. Levene (1952) discussed the power of this test.

EXAMPLE 10   Testing of Runs Up and Down

Data are measurements of temperature in a rodent burrow at noon on successive days.

H_0: The successive positive and negative changes in temperature measurements are random.

H_A: The successive positive and negative changes in the series of temperature measurements are not random.

Day	Temperature (°C)	Difference
1	20.2	
2	20.4	+
3	20.1	−
4	20.3	+
5	20.5	+
6	20.7	+
7	20.5	−
8	20.4	−
9	20.8	+
10	20.8	0
11	21.0	+
12	21.7	+

$n = 12$

If the difference of 0 is counted as +, then $u = 5$; if the 0 is counted as −, then $u = 7$; mean $u = 6$.

For $\alpha = 0.05$, the critical values are $u_{0.05(2),12} = 4$ and 11.

H_0 is not rejected; $0.25 < P \leq 0.50$.

(b) Runs Up and Down: One-Tailed Testing. In a fashion similar to that in Section 6a, a one-tailed test would address one of two situations. One is where H_0: In the sampled population, the successive positive and negative changes in the series of data are not clustered (i.e., are not contagious), and H_A: In the sampled population, the successive positive and negative changes in the series of data are clustered (i.e., are contagious). For this test, H_0 would be rejected if $u \leq$ the first member of the pair of one-tailed critical values in Table 31 from *Appendix: Statistical Tables and Graphs*; if the test is performed using the normal approximation, H_0 would be rejected if $|Z_c \geq Z_{\alpha(1)}|$ and $u \leq \mu_u$. The other one-tailed circumstance is where H_0: In the sampled population, the successive positive and negative changes in the series of data do not alternate regularly (i.e., are not uniform), versus H_A: In the sampled population, the series of data do alternate regularly (i.e., are uniform). H_0 would be rejected if $u \geq$ the second member of the pair of one-tailed critical values; if the test uses the normal approximation, H_0 would be rejected if $Z_c \geq Z_{\alpha(1)}$ and $u \geq \mu_u$.

(c) Runs Above and Below the Median. Another method of assessing randomness of ratio-, interval-, or ordinal-scale measurements examines the pattern of their distribution with respect to the set of data. We first determine the median of the sample. Then we record each datum as being either above ($+$) or below ($-$) the median. If a sample datum is equal to the median, it is discarded from the analysis. We then record u, the number of runs, in the resulting sequence of $+$'s and $-$'s. The test then proceeds as the runs test of Section 6. This is demonstrated, for two-tailed hypotheses, in Example 11; common one-tailed hypotheses are those inquiring into contagious (i.e., clumped) distributions of data above or below the median.

EXAMPLE 11   Runs Above and Below the Median

The data and hypotheses are those of Example 10.

The median of the 12 data is determined to be 20.3°C.

The sequence of data, indicating whether they are above ($+$) or below ($-$) the median, is $-\ -\ -\ -\ 0 + 0 - + + + ++$.

For the runs test, $n_1 = 5, n_2 = 5, u = 4$.

The critical values are $u_{0.05(2),5,5} = 2$ and 10; therefore, do not reject H_0; $0.20 < P \leq 0.50$.

Although the test for runs up and down and the test for runs above and below the median may both be considered nonparametric alternatives to the parametric means square successive difference test of Section 7, the latter runs test often resembles the parametric test more than the former runs test does. The test for runs up and down works well to detect long-term trends in the data, unless there are short-term random fluctuations superimposed upon those trends. The other two tests tend to perform better in detecting long-term patterns in the presence of short-term randomness (A. W. Ghent, personal communication).

EXERCISES

1. If, in a Poisson distribution, $\mu = 1.5$, what is $P(0)$? What is $P(5)$?

2. A solution contains bacterial viruses in a concentration of 5×10^8 bacterial-virus particles per milliliter. In the same solution are 2×10^8 bacteria per milliliter. If there is a random distribution of virus among the bacteria,
   (a) What proportion of the bacteria will have no virus particles?
   (b) What proportion of the bacteria will have virus particles?
   (c) What proportion of the bacteria will have at least two virus particles?
   (d) What proportion of the bacteria will have three virus particles?

3. Fifty-seven men were seated in an outdoor area with only their arms exposed. After a period of time, the number of mosquito bites (X) on each man's arms was recorded, as follows, where f is the number of men with X bites. Test the null hypothesis that mosquitoes bite these men at random.

X	f
0	8
1	17
2	18
3	11
4	3
≥ 5	0

4. We wish to compile a list of certain types of human metabolic diseases that occur in more than 0.01% of the population. A random sample of 25,000 infants reveals five infants with one of these diseases. Should that disease be placed on our list?

5. A biologist counts 112 diatoms in a milliliter of lake water, and 134 diatoms are counted in a milliliter of a second collection of lake water. Test the hypothesis that the two water collections came from the same lake (or from lakes with the same mean diatom concentrations).

6. An economic entomologist rates the annual incidence of damage by a certain beetle as mild (M) or heavy (H). For a 27-year period he records the following: $H\,M\,M\,M\,H\,H\,M\,M\,M\,H\,M\,H\,H\,H\,H\,M\,M\,H$ $H\,H\,H\,M\,M\,H\,H\,M\,M\,M\,M$. Test the null hypothesis that the incidence of heavy damage occurs randomly over the years.

7. The following data are the magnitudes of fish kills along a certain river (measured in kilograms of fish killed) over a period of years. Test the null hypothesis that the magnitudes of the fish kills were randomly distributed over time.

Year	Kill (kg)
1955	147.4
1956	159.8
1957	155.2
1958	161.3
1959	173.2
1960	191.5
1961	198.2
1962	166.0
1963	171.7
1964	184.9
1965	177.6
1966	162.8
1967	177.9
1968	189.6
1969	206.9
1970	221.5

8. Analyze the data of Exercise 7 nonparametrically to test for serial randomness.

ANSWERS TO EXERCISES

1. If $\mu = 1.5$, $P(X = 0) = 0.2231$ and $P(X = 5) = 0.0141$.

2. $\mu = \frac{5}{2} = 2.5$ viruses per bacterium.
   (a) $P(X = 0) = 0.0821$. (b) $P(X > 0) = 1.0000 - P(X = 0) = 1.0000 - 0.0821 = 0.9197$.
   (c) $P(X \geq 2) = 1.0000 - P(X = 0) - P(X = 1) = 1.0000 - 0.0821 - 0.2052 = 0.7127$.
   (d) $P(X = 3) = 0.2138$.

3. H_0: Biting mosquitoes select the men randomly; H_A: Biting mosquitoes do not select the mean randomly. $\overline{X} = \sum f_i X_i = 98/57 = 1.7193$;

   $\chi^2 = 3.060$, $\nu = 6 - 2 = 4$, $\chi^2_{0.05,4} = 7.815$; do not reject H: $0.50 < P < 0.75$ $[P = 0.55]$

4. $H_0: p \leq 0.00010$; $H_A: p > 0.00010$; $p_0 = 0.00010$; $n = 25,000$; $p_0 n = 2.5$; $X = 5$; $P(X \geq 5) = 0.1087$; do not reject H_0; do not include this disease on the list.

5. $H_0: \mu_1 = \mu_2$; $H_A: \mu_1 \neq \mu_2$; $X_1 = 112$, $X_2 = 134$; $Z = 1.40$; $Z_{0.05(2)} = 1.9600$; do not reject H_0; $0.10 < P < 0.20$ $[P = 0.16]$.

6. H_0: The incidence of heavy damage is random over the years; H_A: The incidence of heavy damage is not random over the years; $n_1 = 14$,

$n_2 = 13, u = 12, u_{0.05, 14, 13} = 9$ and 20. As 12 is neither ≤ 9 nor ≥ 20; do not reject H_0; $P = 0.50$.

7. H_0: The magnitude of fish kills is randomly distributed over time; H_A: The magnitude of fish kills is not randomly distributed over time; $n = 16$,

$s^2 = 400.25, s_*^2 = 3126.77/30 = 104.22$; $C = 0.740, C_{0.05, 16} = 0.386$; reject H_0; $P < 0.0005$.

8. H_0: The data are sequentially random; H_A: The data are not sequentially random; $n = 16, u = 7$; critical values $= 6$ and 14; do not reject H_0; $0.05 < P \leq 0.10$.

This page intentionally left blank

Circular Distributions: Descriptive Statistics

1 DATA ON A CIRCULAR SCALE

An interval scale of measurement was defined as a scale with equal intervals but with no true zero point. A special type of interval scale is a circular scale, where not only is there no true zero, but any designation of high or low values is arbitrary. A common example of a circular scale of measurement is compass direction (Figure 1a), where a circle is said to be divided into 360 equal intervals, called degrees,* and for which the zero point is arbitrary. There is no physical justification for a direction of north to be designated 0 (or 360) degrees, and a direction of 270° cannot be said to be a "larger" direction than 90°.[†]

Another common circular scale is time of day (Fig. 1b), where a day is divided into 24 equal intervals, called hours, but where the designation of midnight as the zero or starting point is arbitrary. One hour of a day corresponds to 15° (i.e., 360°/24) of a circle, and 1° of a circle corresponds to four minutes of a day. Other time divisions, such as weeks and years (see Figure 1c), also represent circular scales of measurement.

---

*A degree is divided into 60 minutes (i.e., $1° = 60'$) and a minute into 60 seconds ($1' = 60''$). A number system based upon 60 is termed *sexagesimal*, and we owe the division of the circle into 360 degrees — and the 60-minute hour and 60-second minute — to the ancient Babylonians (about 3000 years ago). The use of the modern symbols (° and ′ and ″) appears to date from the 1570s (Cajori, 1928–1929, Vol. II: 146).

[†]Occasionally one will encounter angular measurements expressed in radians instead of in degrees. A radian is the angle that is subtended by an arc of a circle equal in length to the radius of the circle. As a circle's circumference is 2π times the radius, a radian is $360°/2\pi = 180°/\pi = 57.29577951°$ (or 57 deg, 17 min, 44.8062 sec). The term *radian* was first used, in 1873, by James Thomson, brother of Baron William Thomson (Lord Kelvin), the famous Scottish mathematician and physicist (Cajori, 1928–1929, Vol. II: 147). A direction measured clockwise, from 0° at north, is called an *azimuth*. Rarely, a direction is recorded as an angular measurement called a *grad*; a right angle (90°) is divided into 100 grads, so a grad is 0.9 of a degree.

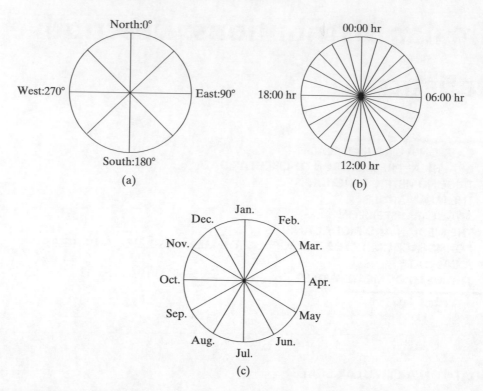

FIGURE 1: Common circular scales of measurement. (a) Compass directions. (b) Times of day. (c) Days of the year (with the first day of each month shown).

In general, X time units may be converted to an angular direction (a, in degrees), where X has been measured on a circular scale having k time units in the full cycle:

$$a = \frac{(360°)(X)}{k}.\tag{1}$$

For example, to convert a time of day (X, in hours) to an angular direction, $k = 24$ hr; to convert a day of the week to an angular direction, number the seven days from some arbitrary point (e.g., Sunday = day 1) and use Equation 1 with $k = 7$; to convert the Xth day of the year to an angular direction, $k = 365$ (or, $k = 366$ in a leap year); to convert a month of the year, $k = 12$; and so on.* Such conversions are demonstrated in Example 1.

Data from circular distributions generally may *not* be analyzed using the statistical methods presented earlier in this text. This is so for theoretical reasons as well as for empirically obvious reasons stemming from the arbitrariness of the zero point on the circular scale. For example, consider three compass directions — 10°, 30°, and 350°,

---

*Equation 1 gives angular directions corresponding to the ends of time periods (e.g., the end of the Xth day of the year). If some other point in a time period is preferred, the equation can be adjusted accordingly. For example, noon can be considered on the Xth day of the year by using $X-0.5$ in place of X. If the same point is used in each time period (e.g., always using either noon or midnight), then the statistical procedures of this and the following chapter will be unaffected by the choice of point. (However, graphical procedures, as in Section 2, will of course be affected, in the form of a rotation of the graph if Equation 1 is adjusted. If, for example, we considered noon on the Xth day of the year, the entire graph would be rotated about half a degree counterclockwise.)

EXAMPLE 1   Conversions of Times Measured on a Circular Scale to Corresponding Angular Directions

By Equation 1,

$$a = \frac{(360°)(X)}{k}.$$

1. Given a time of day of 06:00 hr (which is one-fourth of the 24-hour clock and should correspond, therefore, to one-fourth of a circle),

$$X = 6 \text{ hr}, k = 24 \text{ hr, and}$$

$$a = (360°)(6 \text{ hr})/24 \text{ hr} = 90°.$$

2. Given a time of day of 06:15 hr,

$$X = 6.25 \text{ hr}, k = 24 \text{ hr, and}$$

$$a = (360°)(6.25 \text{ hr})/24 \text{ hr} = 93.75°.$$

3. Given the 14th day of February, being the 45th day of the year,

$$X = 45 \text{ days}, k = 365 \text{ days, and}$$

$$a = (360°)(45 \text{ days})/365 \text{ days} = 44.38°.$$

---

for which we wish to calculate an arithmetic mean. The arithmetic mean calculation of $(10° + 30° + 350°)/3 = 390°/3 = 130°$ is clearly absurd, for all data are northerly directions and the computed mean is southeasterly.

This chapter introduces some basic considerations useful in calculating descriptive statistics for circular data.[*] Statistical methods have also been developed for data that occur on a sphere (which are of particular interest to earth scientists).[‡]

2   GRAPHICAL PRESENTATION OF CIRCULAR DATA

Circular data are often presented as a scatter diagram, where the scatter is shown on the circumference of a circle. Figure 2 shows such a graph for the data of Example 2. If frequencies of data are too large to be plotted conveniently on a scatter diagram, then a bar graph, or histogram, may be drawn. This is demonstrated in Figure 3, for the data presented in Example 3. Recall that in a histogram, the length, as well as the area, of each bar is an indication of the frequency observed at each plotted value of the variable. Occasionally, as shown in Figure 4, a histogram is seen presented with sectors, instead of bars, composing the graph; this is sometimes called a *rose diagram*. Here, the radii forming the outer boundaries

---

[*]More extensive reviews of methods for circular data include Batschelet[†](1965, 1972, 1981), Fisher (1993), Jammalamadaka and SenGupta (2001), Mardia (1972a, 1981), and Mardia and Jupp (2000).

[†]Edward Batschelet (1914–1979), Swiss biomathematician, was one of the most influential writers in developing, explaining, and promulgating circular statistical methods, particularly among biologists.

[‡]Notable discussions of the statistical analysis of sperical data are as follows: Batschelet (1981: Chapter 11); Fisher, Lewis, and Embleton (1987); Mardia (1972a: Chapters 8 and 9); Mardia and Jupp (2000: Chapters 9, 10, etc.); Upton and Fingleton (1989: Chapter 10); and Watson (1983).

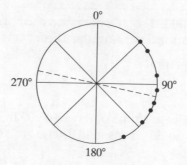

FIGURE 2: A circular scatter diagram for the data of Example 2. (The dashed line defines the median as explained in Section 6.)

EXAMPLE 2   A Sample of Circular Data. These Data Are Plotted in Figure 2

Eight trees are found leaning in the following compass directions: 45°, 55°, 81°, 96°, 110°, 117°, 132°, 154°.

---

of the sectors are proportional to the frequencies being represented, but the areas of the sectors are not. Since it is likely that the areas will be judged by eye to represent the frequencies, the reader of the graph is being misled, and this type of graphical presentation is not recommended. However, a true-area rose diagram can be obtained by plotting the square roots of frequencies as radii.*

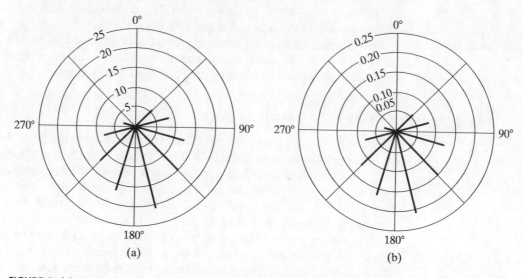

(a)   (b)

FIGURE 3: (a) Circular histogram for the data of Example 3 where the concentric circles represent frequency increments of 5. (b) A relative frequency histogram for the data of Example 3 with the concentric circles representing relative frequency increments of 0.05.

---

*The earliest user of rose diagrams was the founder of modern nursing and pioneer social and health statistician, Florence Nightingale (1820–1910), in 1858. She employed true-area colored diagrams, which she termed "coxcombs," to indicate deaths from various causes over months of the year (Fisher, 1993: 5–6). (Nightingale gained much fame for her work with the British Army during the Crimean War.)

EXAMPLE 3   A Sample of Circular Data, Presented as a Frequency Table, Where a_i Is an Angle and f_i Is the Observed Frequency of a_i. These Data Are Plotted in Figure 3

a_i (deg)	f_i	Relative f_i
0–30	0	0.00
30–60	6	0.06
60–90	9	0.09
90–120	13	0.12
120–150	15	0.14
150–180	22	0.21
180–210	17	0.16
210–240	12	0.11
240 270	8	0.08
270–300	3	0.03
300–330	0	0.00
330–360	0	0.00
$n = 105$		Total $= 1.00$

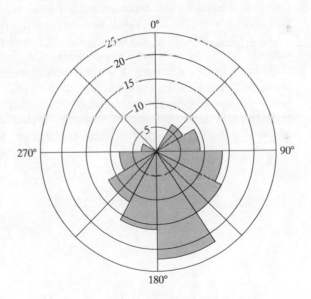

FIGURE 4: A rose diagram of the data of Example 3, utilizing sectors instead of bars. This procedure is not recommended unless square roots of the frequencies are employed (see Section 2).

Another manner of expressing circular frequency distributions graphically is shown in Figure 5. Here, the length of each bar of the histogram represents a frequency, as in Figure 3(a), but the bars extend from the circumference of a circle instead of from the center. In addition, an arrow extending from the circle's center toward the circumference indicates both the direction and the length of the mean vector, and this expresses visually both the mean angle and a measure of data concentration (as explained in Sections 4 and 5).

FIGURE 5: Circular histogram for the data of Example 3, including an arrow depicting the mean angle (*a*) and a measure of dispersion (*r*).

A histogram of circular data can also be plotted as a linear histogram, with degrees on the horizontal axis and frequencies (or relative frequencies) on the vertical axis. But the impression on the eye may vary with the arbitrary location of the origin of the horizontal axis, and (unless the range of data is small — say, no more than 180°) the presentation of Figure 3 or Figure 5 is preferable.

3 TRIGONOMETRIC FUNCTIONS

A great many of the procedures that follow in this chapter and the next require the determination of basic trigonometric functions. Consider that a circle (perhaps representing a compass face) is drawn on rectangular coordinates (as on common graph paper) with the center as the origin (i.e., zero) of both a vertical X axis and a horizontal Y axis; this is what is done in Figure 6.

There are two methods that can be used to locate any point on a plane (such as a sheet of paper). One is to specify X and Y. However, with circular data it is conventional to use a vertical, instead of a horizontal, X axis. This second method specifies both the angle, *a*, with respect to some starting direction (say, clockwise from the top of the X axis, namely "north") and the straight-line distance, *r*, from some reference

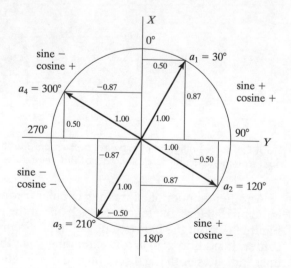

FIGURE 6: A unit circle, showing four points and their polar (*a* and *r*) and rectangular (*X* and *Y*) coordinates.

point (the center of the circle). This pair of numbers, a and r, is known as the "polar coordinates" of a point.* Thus, for example, in Figure 6, point 1 is uniquely identified by polar coordinates $a = 30°$ and $r = 1.00$, point 2 by $a = 120°$ and $r = 1.00$, and so on. If the radius of the circle is specified to be 1 unit, as in Figure 6, the circle is called a *unit circle*.

If a is negative, it is expressing a direction *counterclockwise* from zero. It may be added to 360° to yield the equivalent positive angle; thus, for example, $-60° = 360° - 60° = 300°$. An angle greater than 360° is equivalent to the number of degrees by which it exceeds 360° or a multiple of 360°. So, for example, $450° = 450° - 360° = 90°$ and $780° = 780° - 360° - 360° = 60°$.

The first-mentioned method of locating points on a graph referred to the X and Y axes. By this method, point 1 in Figure 6 is located by the "rectangular coordinates" $X = 0.87$ and $Y = 0.50$, point 2 by $X = -0.50$ and $Y = 0.87$, point 3 by $X = -0.87$ and $Y = -0.50$, and point 4 by $X = 0.50$ and $Y = -0.87$. The *cosine* (abbreviated "cos") of an angle is defined as the ratio of the X and the r associated with the circular measurement:

$$\cos a = \frac{X}{r}, \tag{2}$$

while the *sine* (abbreviated "sin") of the angle is the ratio of the associated Y and r:

$$\sin a = \frac{Y}{r}. \tag{3}$$

Thus, for example, the sine of a_1 in Figure 6 is $\sin 30° = 0.50/1.00 = 0.50$, and its cosine is $\cos 30° = 0.87/1.00 = 0.87$. Also, $\sin 120° = 0.87/1.00 = 0.87$, $\cos 120° = -0.50/1.00 = -0.50$, and so on. Sines and cosines (two of the most used "trigonometric[†] functions") are readily available in published tables, and many electronic calculators give them (and sometimes convert between polar and rectangular coordinates as well). The sines of 0° and 180° are zero, angles between 0° and 180° have sines that are positive, and the sines are negative for $180° < a < 360°$. The cosine is zero for 90° and 270°, with positive cosines obtained for $0° < a < 90°$ and for $270° < a < 360°$, and negative cosines for angles between 90° and 270°.

A third trigonometric function is the *tangent*:[‡]

$$\tan a = \frac{Y}{X} = \frac{\sin a}{\cos a}. \tag{5}$$

On the circle, two different angles have the same sine, two have the same cosine, and two have the same tangent:

$$\sin a = \sin (180° - a)$$
$$\cos a = \cos (360° - a)$$
$$\tan a = \tan (180° + a).$$

---

*This use of the symbol r has no relation to the r that denotes a sample correlation coefficient.

[†]*Trigonometry* refers, literally, to the measurement of triangles (such as the triangles that emanate from the center of the circle in Figure 6).

[‡]The angle having a tangent of Y/X is known as the arctangent (abbreviated *arctan*) of Y/X. As noted at the end of Section 3, a given tangent can be obtained from either of two different angles. If $X \geq 0$, then $a = \arctan(Y/X)$; if $X < 0$, then $a = \arctan(Y/X) + 180°$. The *cotangent* is the reciprocal of the tangent, namely

$$\cot a = \frac{X}{Y} = \frac{\cos a}{\sin a}. \tag{4}$$

We shall see later that rectangular coordinates, X and Y, may also be used in conjunction with mean angles just as they are with individual angular measurements.[*]

4   THE MEAN ANGLE

If a sample consists of n angles, denoted as a_1 through a_n, then the mean of these angles, $\bar{a}$, is to be an estimate of the mean angle, μ_a in the sampled population. To compute the sample mean angle, $\bar{a}$, we first consider the rectangular coordinates of the mean angle:

$$X = \frac{\sum_{i=1}^{n} \cos a_i}{n} \qquad (6)$$

and

$$Y = \frac{\sum_{i=1}^{n} \sin a_i}{n}. \qquad (7)$$

Then, the quantity

$$r = \sqrt{X^2 + Y^2} \qquad (8)$$

is computed;[†] this is the length of the mean vector, which will be further discussed in Section 5. The value of $\bar{a}$ is determined as the angle having the following cosine and sine:

$$\cos \bar{a} = \frac{X}{r} \qquad (9)$$

and

$$\sin \bar{a} = \frac{Y}{r}. \qquad (10)$$

Example 4 demonstrates these calculations. It is also true that

$$\tan \bar{a} = \frac{Y}{X} = \frac{\sin \bar{a}}{\cos \bar{a}}. \qquad (11)$$

If $r = 0$, the mean angle is undefined and we conclude that there is no mean direction.

If the circular data are times instead of angles, then the mean time corresponding to the mean angle may be determined from a manipulation of Equation 1:

$$\overline{X} = \frac{ka}{360°}. \qquad (12)$$

---

[*]Over time, many different symbols and abbreviations have been used for trigonometric functions. The abbreviations *sin.* and *tan.* were established in the latter half of the sixteenth century, and the periods were dropped early in the next century (Cajori, 1928–1929, Vol. II: 150, 158). The cosine was first known as the "sine of the complement"—because the cosine of a equals the sine of $90° - a$ for angles from $0°$ to $90°$—and the English writer E. Gunter changed "complementary sine" to "cosine" and "complementary tangent" to "cotangent" in 1620 (ibid.: 157).

[†]This use of the symbol r has no relation to the r that denotes a sample correlation coefficient.

EXAMPLE 4 **Calculating the Mean Angle for the Data of Example 2**

a_i (deg)	$\sin a_i$	$\cos a_i$
45	0.70711	0.70711
55	0.81915	0.57358
81	0.98769	0.15643
96	0.99452	−0.10453
110	0.93969	−0.34202
117	0.89101	−0.45399
132	0.74315	−0.66913
154	0.43837	−0.89879

$$\sum \sin a_i = 6.52069 \qquad \sum \cos a_i = -1.03134$$

$$Y = \frac{\sum \sin a_i}{n} \qquad\qquad X = \frac{\sum \cos a_i}{n}$$

$$= 0.81509 \qquad\qquad = -0.12892$$

$n = 8$

$$r = \sqrt{X^2 + Y^2} = \sqrt{(-0.12892)^2 + (0.81509)^2} = \sqrt{0.68099} = 0.82522$$

$$\cos \bar{a} = \frac{X}{r} = \frac{-0.12892}{0.82522} = -0.15623$$

$$\sin \bar{a} - \frac{Y}{r} = \frac{0.81509}{0.82522} = 0.98772$$

The angle with this sine and cosine is $\bar{a} = 99°$.

So, to determine a mean time of day, $\overline{X}$, from a mean angle, $\bar{a}$, $\overline{X} = (24 \text{ hr})(\bar{a})/360°$. For example, a mean angle of 270° on a 24-hour clock corresponds to $\overline{X} = (24 \text{ hr})(270°)/360° = 18{:}00$ hr (also denoted as 6:00 P.M.).

If the sine of a is S, then it is said that the *arcsine* of S is a; for example, the sine of 30° is 0.50, so the arcsine of 0.50 is 30°. If the cosine of a is C, then the *arccosine* of C is a; for example, the cosine of 30° is 0.866, so the arccosine of 0.866 is 30°. And, if the tangent of a is T, then the *arctangent* of T is a; for example, the tangent of 30° is 0.577, so the arctangent of 0.577 is 30°*

$$P(X) = \frac{P(X-1)\mu}{X}. \tag{12a}$$

(a) Grouped Data. Circular data are often recorded in a frequency table (as in Example 3). For such data, the following computations are convenient alternatives

---

*The arcsine is also referred to as the "inverse sine" and can be abbreviated "arcsin" or $\sin^{-1}$; the arccosine can be designated as "arccos" or $\cos^{-1}$, and the arctangent as "arctan" or $\tan^{-1}$.

to Equations 6 and 12a, respectively:

$$X = \frac{\sum f_i \cos a_i}{n} \qquad (13)$$

$$Y = \frac{\sum f_i \sin a_i}{n} \qquad (14)$$

In these equations, a_i is the midpoint of the measurement interval recorded (e.g., $a_2 = 45°$ in Example 3, which is the midpoint of the second recorded interval, $30 - 60°$), f_i is the frequency of occurrence of data within that interval (e.g., $f_2 = 6$ in that example), and $n = \sum f_i$. Example 5 demonstrates the determination of $\bar{a}$ for the grouped data (where f_i is not 0) of Example 3.

EXAMPLE 5 **Calculating the Mean Angle for the Data of Example 3**

a_i	f_i	$\sin a_i$	$f_i \sin a_i$	$\cos a_i$	$f_i \cos a_i$
45°	6	0.70711	4.24266	0.70711	4.24266
75°	9	0.96593	8.69337	0.25882	2.32938
105°	13	0.96593	12.55709	-0.25882	-3.36466
135°	15	0.70711	10.60665	-0.70711	-10.60665
165°	22	0.25882	5.69404	-0.96593	-21.25046
195°	17	-0.25882	-4.39994	-0.96593	-16.42081
225°	12	-0.70711	-8.48532	-0.70711	-8.48532
255°	8	-0.96593	-7.72744	-0.25882	-2.07056
285°	3	-0.96593	-2.89779	0.25882	0.77646

$$n = 105 \qquad \sum f_i \sin a_i = 18.28332 \qquad \sum f_i \cos a_i = -54.84996$$

$$Y = \frac{\sum f_i \sin a_i}{n} \qquad\qquad X = \frac{\sum f_i \cos a_i}{n}$$

$$= 0.17413 \qquad\qquad = -0.52238$$

$$r = \sqrt{X^2 + Y^2} = \sqrt{(-0.52238)^2 + (0.17413)^2} = 0.55064$$

$$\cos \bar{a} = \frac{X}{r} = \frac{-0.52238}{0.55064} = -0.94868$$

$$\sin \bar{a} = \frac{Y}{r} = \frac{0.17413}{0.55064} = 0.31623$$

The angle with this cosine and sine is $\bar{a} = 162°$.

There is a bias in computing r from grouped data, in that the result is too small. A correction for this is available (Batschelet, 1965: 16–17, 1981: 37–40; Mardia, 1972a: 78–79; Mardia and Jupp, 2000: 23; Upton and Fingleton, 1985: 219), which may be applied when the distribution is unimodal and does not deviate greatly from symmetry. For data grouped into equal intervals of d degrees each,

$$r_c = cr, \qquad (15)$$

where r_c is the corrected r, and c is a correction factor,

$$c = \frac{\dfrac{d\pi}{360°}}{\sin\left(\dfrac{d}{2}\right)}.\tag{16}$$

The correction is insignificant for intervals smaller than 30°. This correction is for the quantity r; the mean angle, $\bar{a}$, requires no correction for grouping.

5 ANGULAR DISPERSION

When dealing with circular data, it is desirable to have a measure, analogous to those for a linear scale, to describe the dispersion of the data.

We can define the *range* in a circular distribution of data as the smallest arc (i.e., the smallest portion of the circle's circumference) that contains all the data in the distribution. For example, in Figure 7a, the range is zero; in Figure 7b, the shortest arc is from the data point at 38° to the datum at 60°, making the range 22°; in Figure 7c, the data are found from 10° to 93°, with a range of 83°; in Figure 7d, the data run from 322° to 135°, with a range of 173°; in Figure 7e, the shortest arc containing all the data is that running clockwise from 285° to 171°, namely an arc of 246°; and in Figure 7f, the range is 300°. For the data of Example 4, the range is 109° (as the data run from 45° to 154°).

Another measure of dispersion is seen by examining Figure 7; the value of r (by Equation 8, and indicated by the length of the broken line) varies inversely with the amount of dispersion in the data. Therefore, r is a measure of concentration. It has no units and it may vary from 0 (when there is so much dispersion that a mean angle cannot be described) to 1.0 (when all the data are concentrated at the same direction). (An r of 0 does not, however, necessarily indicate a uniform distribution. For example, the data of Figure 8 would also yield $r = 0$). A line specified by both its direction and length is called a *vector*, so r is sometimes called the length of the mean vector.

The mean on a linear scale was noted to be the center of gravity of a group of data. Similarly, the tip of the mean vector (i.e., the quantity r), in the direction of the mean angle ($\bar{a}$) lies at the center of gravity. (Consider that each circle in Figure 7 is a disc of material of negligible weight, and each datum is a dot of unit weight. The disc, held parallel to the ground, would balance at the tip of the arrow in the figure. In Figure 7f, $r = 0$ and the center of gravity is the center of the circle.)

Because r is a measure of concentration, $1-r$ is a measure of dispersion. Lack of dispersion would be indicated by $1-r = 0$, and maximum dispersion by $1-r = 1.0$. As a measure of dispersion reminiscent of those for linear data, Mardia (1972a: 45), Mardia and Jupp (2000: 18), and Upton and Fingleton (1985: 218) defined *circular variance*:

$$S^2 = 1 - r.\tag{17}$$

Batschelet (1965, 1981: 34) defined *angular variance:*

$$s^2 = 2(1 - r)\tag{18}$$

as being a closer analog to linear variance. While S^2 may range from 0 to 1, and s^2 from 0 to 2, an S^2 of 1 or an s^2 of 2 does not necessarily indicate a uniform distribution of data around the circle because, as noted previously, $r = 0$ does not necessarily indicate

Circular Distributions: Descriptive Statistics

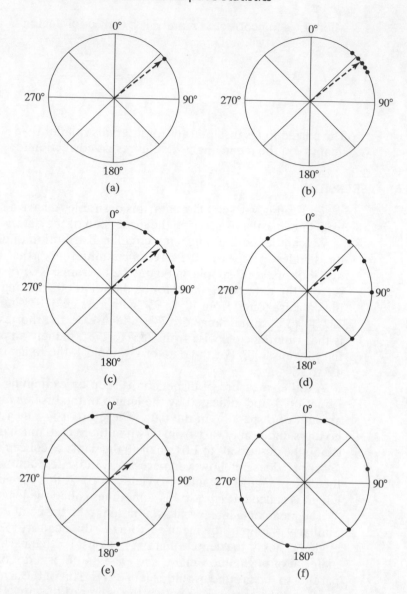

FIGURE 7: Circular distributions with various amounts of dispersion. The direction of the broken-line arrow indicates the mean angle, which is 50° in each case, and the length of the arrow expresses r (Equation 8), a measure of concentration. The magnitude of r varies inversely with the amount of dispersion, and that the values of s and s_0 vary directly with the amount of dispersion. (a) $r = 1.00, s = 0°, s_0 = 0°$. (b) $r = 0.99, s = 8.10°, s_0 = 8.12°$. (c) $r = 0.90, s = 25.62°, s_0 = 26.30°$. (d) $r = 0.60, s = 51.25°, s_0 = 57.91°$. (e) $r = 0.30, s = 67.79°, s_0 = 88.91$. (f) $r = 0.00, s = 81.03°, s_0 = \infty$. (By the method, the magnitude of r is statistically significant in Figs. a, b, and c, but not in d, e, and f.)

a uniform distribution. The variance measure

$$s_0^2 = -2\ln r \tag{19}$$

is a statistic that ranges from 0 to ∞ (Mardia, 1972a: 24). These three dispersion measures are in radians squared. To express them in degrees squared, multiply each by $(180°/\pi)^2$.

Measures analogous to the linear standard deviation include the "mean angular deviation," or simply the *angular deviation*, which is

$$s = \frac{180°}{\pi}\sqrt{2(1 - r)}, \tag{20}$$

in degrees.* This ranges from a minimum of zero (e.g., Fig. 7a) to a maximum of 81.03° (e.g., Fig. 7f).[†] Mardia (1972a: 24, 74) defines *circular standard deviation* as

$$s_0 = \frac{180°}{\pi}\sqrt{-2\ln r} \tag{21}$$

degrees; or, employing common, instead of natural, logarithms:

$$s_0 = \frac{180°}{\pi}\sqrt{-4.60517\log r} \tag{22}$$

degrees. This is analogous to the standard deviation, s, on a linear scale in that it ranges from zero to infinity (see Fig. 7). For large r, the values of s and s_0 differ by no more than 2 degrees for r as small as 0.80, by no more than 1 degree for r as small as 0.87, and by no more than 0.1 degree for r as small as 0.97. It is intuitively reasonable that a measure of angular dispersion should have a finite upper limit, so s is the dispersion measure preferred in this text. Tables 32 and 33 from *Appendix: Statistical Tables and Graphs* convert r to s and s_0, respectively. If the data are grouped, then s and s_0 are biased in being too high, so r_c (by Equation 15) can be used in place of r. For the data of Example 4 (where $r = 0.82522$), $s = 34°$ and $s_0 = 36°$; for the data of Example 5 (where $r = 0.55064$), $s = 54°$ and $s_0 = 63°$.

Dispersion measures analogous to the linear mean deviation utilize absolute deviations of angles from the mean or median (e.g., Fisher, 1993: 36).

Measures of symmetry and kurtosis on a circular scale, analogous to those that may be calculated on a linear scale, are discussed by Batschelet (1965: 14–15, 1981: 34–44); Mardia (1972a: 36–38, 74–76), and Mardia and Jupp (2000: 22, 31, 145–146).

6 THE MEDIAN AND MODAL ANGLES

In a fashion analogous to considerations for linear scales of measurement, we can determine the sample median and mode of a set of data on a circular scale.

To find the *median angle*, we first determine which diameter of the circle divides the data into two equal-sized groups. The median angle is the angle indicated by that diameter's radius that is nearer to the majority of the data points. If n is even, the median is nearly always midway between two of the data. In Example 2, a diameter extending from 103° to 289° divides the data into two groups of four each (as indicated by the dashed line in Fig. 2). The data are concentrated around 103°, rather than 289°, so the sample median is 103°. If n is odd, the median will almost always be one of the data points or 180° opposite from one. If the data in Example 2 had been seven in number — with the 45° lacking — then the diameter line would have run through 110° and 290°, and the median would have been 110°.

Though uncommon, it is possible for a set of angular data to have more than one angle fit this definition of median. In such a case, Otieno and Anderson-Cook (2003)

---

*Simply delete the constant, $180°/\pi$, in this and in the following equations if the measurement is desired in radians rather than degrees.

[†]This is a range of 0 to 1.41 radians.

recommend calculating the estimate of the population median as the mean of the two or more medians fitting the definition. Mardia (1972a: 29–30) shows how the median is estimated when it lies within a group of tied data. If a sample has the data equally spaced around the circle (as in Fig. 7f), then the median, as well as the mean, is undefined.

The *modal angle* is defined as is the mode for linear scale data. Just as with linear data, there may be more than one mode or there may be no modes.

7   CONFIDENCE LIMITS FOR THE POPULATION MEAN AND MEDIAN ANGLES

The confidence limits of the mean of angles may be expressed as

$$\bar{a} \pm d. \tag{23}$$

That is, the lower confidence limit is $L_1 = \bar{a} - d$ and the upper confidence limit is $L_2 = \bar{a} + d$. For n as small as 8, the following method may be used (Upton, 1986). For $r \leq 0.9$,

$$d = \arccos\left[\frac{\sqrt{\dfrac{2n(2R^2 - n\chi^2_{\alpha,1})}{4n - \chi^2_{\alpha,1}}}}{R}\right] \tag{24}$$

and for $r \geq 0.9$,

$$d = \arccos\left[\frac{\sqrt{n^2 - (n^2 - R^2)e^{\chi^2_{\alpha,1}/n}}}{R}\right], \tag{25}$$

where

$$R = nr. \tag{26}$$

This is demonstrated in Example 6.* As this procedure is only approximate, d — and confidence limits — should not be expressed to fractions of a degree. This procedure is based on the von Mises distribution, a circular analog of the normal distribution.[†] Batschelet (1972: 86; Zar, 1984: 665–666) presents nomograms that yield similar results.

---

*As shown in this example, a given cosine is associated with two different angles: a and $360° - a$; the smaller of the two is to be used.

[†]Richard von Mises (1883–1953), a physicist and mathematician, was born in the Austro-Hungarian Empire and moved to Germany, Turkey, and the United States because of two world wars (Geiringer, 1978). He introduced this distribution (von Mises, 1918), and it was called "circular normal" by Gumbel, Greenwood, and Durand (1953), and later by others, because of its similarity to the linear-scale normal distribution. It is described mathematically by Batschelet (1981: 279–282), Fisher (1993: 48–56), Jammalamadaka and SenGupta (2001: 35–42), Mardia (1972a: 122–127), Mardia and Jupp (2000: 36, 68–71, 85–88, 167–173), and Upton and Fingleton (1985: 277–229).

EXAMPLE 6   The 95% Confidence Interval for the Data of Example 4

$n = 8$

$\bar{a} = 99°$

$r = 0.82522$

$R = nr = (8)(0.82522) = 6.60108$

$\chi^2_{0.05,1} = 3.841$

Using Equation 24:

$$d = \arccos \left[\frac{\sqrt{\dfrac{2n\left(2R^2 - n\chi^2_{\alpha,1}\right)}{4n - \chi^2_{\alpha,1}}}}{R} \right]$$

$$= \arccos \left[\frac{\sqrt{\dfrac{2(8)[2(6.60108)^2 - 8(3.841)]}{4(8) - 3.841}}}{6.60108} \right]$$

$$= \arccos 0.85779$$

$$= 31° \text{ or } 360° - 31° = 229°.$$

The 95% confidence interval is $99° \pm 31°$; $L_1 = 68°$ and $L_2 = 130°$.

The median is determined as in Section 6. Then the data are numbered 1 through n, with 1 representing the datum farthest from the median in a counterclockwise direction and n denoting the datum farthest in a clockwise direction.

8   AXIAL DATA

Although not common, circular data may be encountered that are bimodal and have their two modes diametrically opposite on the circle. An example of such data is shown in Figure 8, where there is a group of seven angular data opposite a group of eight data (with the data shown as small black circles). Such measurements are known as *axial data*, and it is desirable to calculate the angle that best describes the circle diameter that runs through the two groups of data. Determining the mean angle ($\bar{a}$) of the diameter in one direction means that the mean angle of the diameter in the other direction is $\bar{a} + 180°$.

These 15 measurements are given in Example 7 and resulted from the following experiment: A river flows in a generally southeasterly–northwesterly direction. Fifteen fish, of a species that prefers shallow water at river edges, were released in the middle of the river. Then it was recorded which direction from the point of release each of the fish traveled.

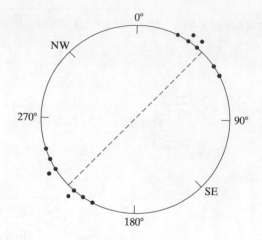

FIGURE 8: A bimodal circular distribution, showing the data of Example 7.

EXAMPLE 7   The Data of Fig. 8, and Their Axial Mean

a_i (degrees)	modulo $2a_i$ (degrees)	$\sin 2a_i$	$\cos 2a_i$
35	70	0.93964	0.34202
40	80	0.98481	0.17365
40	80	0.98481	0.17365
45	90	1.00000	0.00000
45	90	1.00000	0.00000
55	110	0.93969	−0.34202
60	120	0.86602	−0.50000
215	70	0.93964	0.34202
220	80	0.98481	0.17365
225	90	1.00000	0.00000
225	90	1.00000	0.00000
235	110	0.93969	−0.34202
235	110	0.93969	−0.34202
240	120	0.86602	−0.50000
245	130	0.76604	−0.64279

$$n = 15 \qquad \Sigma \sin 2a_i \qquad \Sigma \cos 2a_i$$
$$= 14.15086 \qquad = -1.46386$$
$$Y = -0.94339 \qquad X = -0.09759$$

$r = 0.94842$

$\sin \bar{a} = 0.99470$

$\cos \bar{a} = -0.10290$

The angle ($2\bar{a}$) with this sine and cosine is 95.9°; so $\bar{a} = 95.9°/2 = 48°$.
Also: $\tan 2\bar{a} = Y/X = -9.66687$; and, since $C < 0$, $\bar{a} = \arctan -9.66687 + 180° = -84.1° + 180° = 95.9°$; so $\bar{a} = 48°$.

The statistical procedure is to calculate the mean angle (Section 4) after doubling each of the data (that is, to find the mean of $2a_i$). It will be noted that doubling of an angle greater than 180° will result in an angle greater than 360°; in that case, 360° is subtracted from the doubled angle. (This results in angles that are said to be "modulo 360°.") The doubling of angles for axial data is also appropriate for calculating other statistics, such as those in Sections 5–7.

The mean angle of 48° determined in Example 7 indicates that a line running from 48° to 228° (that is, from 48° to 48°+180°) is the axis of the bimodal data (shown as a dashed line in Fig. 8).

9 THE MEAN OF MEAN ANGLES

If a mean is determined for each of several groups of angles, then we have a set of mean angles. Consider the data in Example 8. Here, a mean angle, $\bar{a}$, has been calculated for each of k samples of circular data, using the procedure of Section 4. If, now, we desire to determine the grand mean of these several means, it is not appropriate to consider each of the sample means as an angle and employ the method of Section 4. To do so would be to assume that each mean had a vector length, r, of 1.0 (i.e., that an angular deviation, s, of zero was the case in each of the k samples), a most unlikely situation. Instead, we shall employ the procedure promulgated by Batschelet* (1978, 1981: 201–202), whereby the grand mean has rectangular coordinates

$$\overline{X} = \frac{\sum_{j=1}^{k} X_j}{k} \tag{27}$$

and

$$\overline{Y} = \frac{\sum_{j=1}^{k} Y_j}{k}, \tag{28}$$

where X_j and Y_j are the quantities X and Y, respectively, applying Equations 6 and 7 to sample j; k is the total number of samples. If we do not have X and Y for each sample, but we have $\bar{a}$ and r (polar coordinates) for each sample, then

$$\overline{X} = \frac{\sum_{j=1}^{k} r_j \cos \bar{a}_j}{k} \tag{29}$$

and

$$\overline{Y} = \frac{\sum_{j=1}^{k} r_j \sin \bar{a}_j}{k}. \tag{30}$$

---

*Batschelet (1981: 198) refers to statistical analysis of a set of angles as a first-order analysis and the analysis of a set of mean angles as a second-order analysis.

Having obtained $\overline{X}$ and $\overline{Y}$, we may substitute them for X and Y, respectively, in Equations 8, 9, and 10 (and 11, if desired) in order to determine $\overline{a}$, which is the grand mean. For this calculation, all n_j's (sample sizes) should be equal, although unequal sample sizes do not appear to seriously affect the results (Batschelet, 1981: 202).

Figure 9 shows the individual means and the grand mean for Example 8. (By the hypothesis testing, we would conclude that there is in this example no significant mean direction for Samples 5 and 7. However, the data from these two samples should not be deleted from the present analysis.)

Batschelet (1981: 144, 262–265) discussed confidence limits for the mean of mean angles.

EXAMPLE 8   The Mean of a Set of Mean Angles

Under particular light conditions, each of seven butterflies is allowed to fly from the center of an experimental chamber ten times. From the procedures of Section 4, the values of $\overline{a}$ and r for each of the seven samples of data are as follows.

$$k = 7; \quad n = 10$$

Sample (j)	$\overline{a}_j$	r_j	$X_j = r_j \cos \overline{a}_j$	$Y_j = r_j \sin \overline{a}_j$
1	160°	0.8954	−0.84140	0.30624
2	169	0.7747	−0.76047	0.14782
3	117	0.4696	−0.21319	0.41842
4	140	0.8794	−0.67366	0.56527
5	186	0.3922	−0.39005	−0.04100
6	134	0.6952	−0.48293	0.50009
7	171	0.3338	−0.32969	0.05222
			−3.69139	1.94906

$$\overline{X} = \frac{\sum r_j \cos \overline{a}_j}{k} = \frac{-3.69139}{7} = -0.52734$$

$$\overline{Y} = \frac{\sum r_j \sin \overline{a}_j}{k} = \frac{1.94906}{7} = 0.27844$$

$$r = \sqrt{\overline{X}^2 + \overline{Y}^2} = \sqrt{0.35562} = 0.59634$$

$$\cos \overline{a} = \frac{\overline{X}}{r} = \frac{-0.52734}{0.59634} = -0.88429$$

$$\sin \overline{a} = \frac{\overline{Y}}{r} = \frac{0.27844}{0.59634} = 0.46691$$

Therefore, $\overline{a} = 152°$.

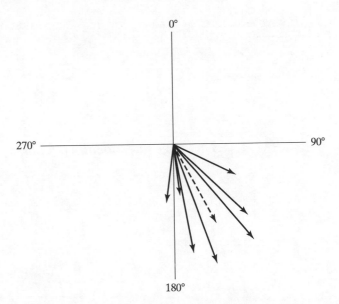

FIGURE 9: The data of Example 8. Each of the seven vectors in this sample is itself a mean vector. The mean of these seven means is indicated by the broken line.

EXERCISES

1. Twelve nests of a particular bird species were recorded on branches extending in the following directions from the trunks of trees:

Direction		Frequency
N:	0°	2
NE:	45°	4
E:	90°	3
SE:	135°	1
S:	180°	1
SW:	225°	1
W:	270°	0
NW:	315°	0

(a) Compute the sample mean direction.
(b) Compute the angular deviation for the data.

(c) Determine 95% confidence limits for the population mean.
(d) Determine the sample median direction.

2. A total of 15 human births occurred as follows:

1:15 A.M.	4:40 A.M.	5:30 A.M.	6:50 A.M.
2:00 A.M.	11:00 A.M.	4:20 A.M.	5:10 A.M.
4:30 A.M.	5:15 A.M.	10:30 A.M.	8:55 A.M.
6:10 A.M.	2:45 A.M.	3:10 A.M.	

(a) Compute the mean time of birth.
(b) Compute the angular deviation for the data.
(c) Determine 95% confidence limits for the population mean time.
(d) Determine the sample median time.

ANSWERS TO EXERCISES

1. $n = 12$, $Y = 0.48570$, $X = 0.20118$, $r = 0.52572$ ($c = 1.02617$, $r_c = 0.53948$). **(a)** $\bar{a} = 68°$.
(b) $s = 56°$ (using correction for grouping, $s = 55°$), $s' = 65°$ (using correction for grouping, $s' = 64°$). **(c)** $68° \pm 47°$ (using correction for grouping, $68° \pm 46°$). **(d)** median = 67.5°.
2. $n = 15$, $Y = 0.76319$, $X = 0.12614$, $r = 0.77354$. **(a)** $a = 5:22$ A.M. **(b)** $s = 2:34$ hr. **(c)** 5:22 hr $\pm$ 1:38 hr. **(d)** median = 5:10 A.M.

This page intentionally left blank

Circular Distributions: Hypothesis Testing

Circular Distributions: Hypothesis Testing

Armed with the information contained in the basic statistics of circular distributions (primarily $\bar{a}$ and r), we can now examine a number of methods for testing hypotheses about populations measured on a circular scale.

1   TESTING SIGNIFICANCE OF THE MEAN ANGLE

(a) The Rayleigh Test for Uniformity.   We can place more confidence in $\bar{a}$ as an estimate of the population mean angle, μ_a, if s is small, than if it is large. This is identical to stating that $\bar{a}$ is a better estimate of μ_a if r is large than if r is small. What is desired is a method of asking whether there is, in fact, a mean direction for the population of data that were sampled, for even if there is no mean direction (i.e., the circular distribution is uniform) in the population, a random sample might still display a calculable mean. The test we require is that concerning H_0: The sampled population is uniformly distributed around a circle versus H_A: The population is not a uniform circular distribution. This may be tested by the *Rayleigh test**. As circular uniformity implies there is no mean direction, the Rayleigh test may also be said to test H_0: $\rho = 0$ versus H_A: $\rho \neq 0$, where ρ is the population mean vector length.

---

*Named for Lord Rayleigh [John William Strutt, Third Baron Rayleigh (1842–1919)], a physicist and applied mathematician who gained his greatest fame for discovering and isolating the chemical element argon (winning him the Nobel Prize in physics in 1904), although some of his other contributions to physics were at least as important (Lindsay, 1976). He was a pioneering worker with directional data beginning in 1880 (Fisher, 1993: 10; Moore, 1980; Rayleigh, 1919).

The Rayleigh test asks how large a sample r must be to indicate confidently a nonuniform population distribution. A quantity referred to as "Rayleigh's R" is obtainable as

$$R = nr, \tag{1}$$

and the so-called "Rayleigh's z" may be utilized for testing the null hypothesis of no population mean direction:

$$z = \frac{R^2}{n} \quad \text{or} \quad z = nr^2. \tag{2}$$

Table 34 from *Appendix: Statistical Tables and Graphs* presents critical values of $z_{\alpha,n}$. Also an excellent approximation of the probability of Rayleigh's R is*

$$P = \exp\left[\sqrt{1 + 4n + 4(n^2 - R^2)} - (1 + 2n)\right] \tag{4}$$

(derived from Greenwood and Durand, 1955). This calculation is accurate to three decimal places for n as small as 10 and to two decimal places for n as small as 5.[†] The Rayleigh test assumes sampling from a von Mises distribution, a circular analog of the linear normal distribution.

If H_0 is rejected by Rayleigh's test, we may conclude that there is a mean population direction (see Example 1), and if H_0 is not rejected, we may conclude the population distribution to be uniform around the circle; but only if we may assume that the population distribution does not have more than one mode.

EXAMPLE 1   Rayleigh's Test for Circular Uniformity, Applied to the Data of Example 2 from *Circular Distributions: Descriptive Statistics*

These data are plotted in Figure 2 from *Circular Distributions: Descriptive Statistics*.

H_0: $\rho = 0$ (i.e., the population is uniformly distributed around the circle).

H_A: $\rho \neq 0$ (i.e., the population is not distributed uniformly around the circle).

Following Example 4 from *Appendix: Statistical Tables and Graphs*:

$n = 8$

$r = 0.82522$

$R = nr = (8)(0.82522) = 6.60176$

$z = \dfrac{R^2}{n} = \dfrac{(6.60176)^2}{8} = 5.448.$

Using Table 34 from *Appendix: Statistical Tables and Graphs*, $z_{0.05,8} = 2.899$. Reject H_0. $0.001 < P < 0.002$

---

*Recall the following notation:

$$\exp[C] = e^C. \tag{3}$$

[†]A simpler, but less accurate, approximation for P is to consider $2z$ as a chi-square with 2 degrees of freedom (Mardia, 1972a: 113; Mardia and Jupp, 2000: 92). This is accurate to two decimal places for n as small as about 15.

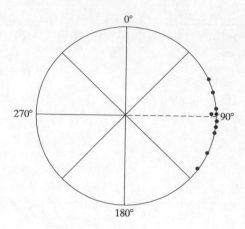

FIGURE 1: The data for the *V* test of Example 2. The broken line indicates the expected mean angle (94°).

Axially bimodal data can be transformed into unimodal data, thereafter to be subjected to Rayleigh testing and other procedures requiring unimodality. What is known as "Rao's spacing test" (Batschelet, 1981: 66–69; Rao, 1976) is particularly appropriate when circular data are neither unimodal nor axially bimodal, and Russell and Levitin (1994) have produced excellent tables for its use.

(b) Modified Rayleigh Test for Uniformity versus a Specified Mean Angle. The Rayleigh test looks for any departure from uniformity. A modification of that test (Durand and Greenwood, 1958; Greenwood and Durand, 1955) is available for use when the investigator has reason to propose, *in advance*, that if the sampled distribution is not uniform it will have a specified mean direction. In Example 2 (and presented graphically in Figure 1), ten birds were released at a site directly west of their home. Therefore, the statistical analysis may include the suspicion that such birds will tend to fly directly east (i.e., at an angle of 90°). The testing procedure considers H_0: The population directions are uniformly distributed versus H_A: The directions in the population are not uniformly distributed and $\mu_a = 90°$. By using additional information, namely the proposed mean angle, this test is more powerful than Rayleigh's test (Batschelet, 1972; 1981: 60).

The preceding hypotheses are tested by a modified Rayleigh test that we shall refer to as the *V test*, in which the test statistic is computed as

$$V = R\cos(\bar{a} - \mu_0),\qquad(5)$$

where μ_0 is the mean angle proposed. The significance of V may be ascertained from

$$u = V\sqrt{\frac{2}{n}}.\qquad(6)$$

Table 35 from *Appendix: Statistical Tables and Graphs* gives critical values of $u_{\alpha,n}$, a statistic which, for large sample sizes, approaches a one-tailed normal deviate, $Z_{\alpha(1)}$, especially in the neighborhood of probabilities of 0.05. If the data are grouped, then R may be determined from r_c rather than from r.

(c) One-Sample Test for the Mean Angle. The Rayleigh test and the *V* test are nonparametric methods for testing for uniform distribution of a population

EXAMPLE 2   The V Test for Circular Uniformity Under the Alternative of Nonuniformity and a Specified Mean Direction

H_0: The population is uniformly distributed around the circle (i.e., H_0: $\rho = 0$).

H_A: The population is not uniformly distributed around the circle (i.e., H_A: $\rho \neq 0$), but has a mean of 90°.

a_i (deg)	$\sin a_i$	$\cos a_i$
66	0.91355	0.40674
75	0.96593	0.25882
86	0.99756	0.06976
88	0.99939	0.03490
88	0.99939	0.03490
93	0.99863	0.05234
97	0.99255	0.12187
101	0.98163	0.19081
118	0.88295	0.46947
130	0.76604	0.64279

$n = 10 \quad \sum \sin a_i = 9.49762 \quad \sum \cos a_i = -0.67216$

$$Y = \frac{9.49762}{10} = 0.94976$$

$$X = -\frac{0.67216}{10} = -0.06722$$

$$r = \sqrt{(-0.06722)^2 + (0.94976)^2} = 0.95214$$

$$\sin \bar{a} = \frac{Y}{r} = 0.99751$$

$$\cos \bar{a} = \frac{X}{r} = -0.07060$$

$\bar{a} = 94°$.

$R = (10)(0.95214) = 9.5214$

$V = R \cos(94° - 90°)$

$= 9.5214 \cos(4°)$

$= (9.5214)(0.99756)$

$= 9.498$

$u = V\sqrt{\frac{2}{n}}$

$= (9.498)\sqrt{\frac{2}{10}}$

$= 4.248$

Using Table 35 from *Appendix: Statistical Tables and Graphs*, $u_{0.05,10} = 1.648$. Reject H_0. $P < 0.0005$.

of data around the circle. (See Batschelet, 1981: Chapter 4, for other tests of the null hypothesis of randomness.) If it is desired to test whether the population mean angle is equal to a specified value, say μ_0, then we have a one-sample test situation analogous to that of the one-sample t test for data on a linear scale. The hypotheses are

$$H_0: \mu_a = \mu_0$$

and

$$H_A: \mu_a \neq \mu_0,$$

and H_0 is tested simply by observing whether μ_0 lies within the $1 - \alpha$ confidence interval for μ_a. If μ_a lies outside the confidence interval, then H_0 is rejected. Example 3 demonstrates the hypothesis-testing procedure.*

EXAMPLE 3   The One-Sample Test for the Mean Angle, Using the Data of Example 2

H_0: The population has a mean of 90° (i.e., $\mu_a = 90°$).

H_A: The population mean is not 90° (i.e., $\mu_a \neq 90°$).

The computation of the following is given in Example 2:

$r = 0.95$

$\bar{a} = 94°$

For $\alpha = 0.05$ and $n = 10$:

$R = nr = (10)(0.95) = 9.5$

$\chi^2_{0.05,1} = 3.841$

$$d = \arccos\left[\frac{\sqrt{n^2 - (n^2 - R^2)e^{\chi^2_{\alpha,1}/n}}}{R}\right]$$

$$= \arccos\left[\frac{\sqrt{10^2 - (10^2 - 9.5^2)e^{3.841/10}}}{9.5}\right]$$

$$= \arccos[0.9744]$$

$$= 13°, \text{ or } 360° - 13° = 347°.$$

Thus, the 95% confidence interval for μ_a is 94° $\pm$ 13°.

As this confidence interval does contain the hypothesized mean ($\mu_0 = 90°$), do not reject H_0.

---

*For demonstration purposes (Examples 2 and 3) we have applied the V test and the one-sample test for the mean angle to the same set of data. In practice this would not be done. Deciding which test to employ would depend, respectively, on whether the intention is to test for circular uniformity or to test whether the population mean angle is a specified value.

2 TESTING SIGNIFICANCE OF THE MEDIAN ANGLE

(a) The Hodges-Ajne Test for Uniformity.   A simple alternative to the Rayleigh test (Section 1) is the so-called *Hodges-Ajne test*,* which does not assume sampling from a specific distribution. This is called an "omnibus test" because it works well for unimodal, bimodal, and multimodal distributions. If the underlying distribution is that assumed by the Rayleigh test, then the latter procedure is the more powerful.

Given a sample of circular data, we determine the smallest number of data that occur within a range of 180°. As shown in Example 4, this is readily done by drawing a line through the center of the circle (i.e., drawing a diameter) and rotating that line around the center until there is the greatest possible difference between the numbers of data on each side of the line. If, for example, the diameter line were vertical (i.e., through 0° and 180°), there would be 10 data on one side of it and 14 on the other; if the line were horizontal (i.e., through 90° and 270°), then there would be 3.5 points on one side and 20.5 points on the other; and if the diameter were rotated slightly counterclockwise from horizontal (shown as a dashed line in the figure in Example 4), then there would be 3 data on one side and 21 on the other, and no line will split the data with fewer data on one side and more on the other. The test statistic, which we shall call m, is the smallest number of data that can be partitioned on one side of a diameter; in Example 4, $m = 3$.

The probability of an m at least this small, under the null hypothesis of circular uniformity, is

$$P = \frac{(n - 2m)\binom{n}{m}}{2^{n-1}} = \frac{(n - 2m)\dfrac{n!}{m!(n - m)!}}{2^{n-1}} \tag{7}$$

(Hodges, 1955), using the binomial coefficient notation. Instead of computing this probability, we may refer to Table 36 from *Appendix: Statistical Tables and Graphs*, which gives critical values for m as a function of n and α. (It can be seen from this table that in order to test at the 5% significance level, we must have a sample of at least nine data.) For $n > 50$, P may be determined by the following approximation:

$$P \approx \frac{\sqrt{2\pi}}{A} \exp\left[\frac{-\pi^2}{8A^2}\right], \tag{8}$$

where

$$A = \frac{\pi\sqrt{n}}{2(n - 2m)} \tag{9}$$

(Ajne, 1968); the accuracy of this approximation is indicated at the end of Table 36 from *Appendix: Statistical Tables and Graphs*.

(b) Modified Hodges-Ajne Test for Uniformity versus a Specified Angle.   Just as (in Section 1b) the V test is a modification of the Rayleigh test to test for circular uniformity against an alternative that proposes a specified angle, a test presented by Batschelet (1981: 64–66) is a modification of the Hodges-Ajne test to test

---

*This procedure was presented by Ajne (1968). Shortly thereafter, Bhattacharyya and Johnson (1969) showed that his test is identical to a test given by Hodges (1955) for a different purpose.

EXAMPLE 4   The Hodges-Ajne Test for Circular Uniformity

H_0: The population is uniformly distributed around the circle.

H_A: The population is not uniformly distributed around the circle.

This sample of 24 data is collected: 10°, 15°, 25°, 30°, 30°, 30°, 35°, 45°, 50°, 60°, 75°, 80°, 100°, 110°, 255°, 270°, 280°, 280°, 300°, 320°, 330°, 350°, 350°, 355°.

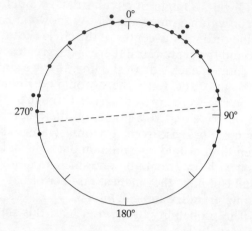

$n = 24; m = 3$

For $\alpha = 0.05$, the critical value (from Table 36 from *Appendix: Statistical Tables and Graphs*) is $m_{0.05,24} = 4$; reject H_0. $0.002 < P \le 0.005$.

$$\text{Exact probability} = P = \frac{(n - 2m)\binom{n}{m}}{2^{n-1}} = \frac{(n - 2m)\dfrac{n!}{m!(n - m)!}}{2^{n-1}}$$

$$= \frac{(24 - 6)\dfrac{24!}{3!\,21!}}{2^{23}} = 0.0043$$

For comparison, the Rayleigh test for these data would yield $\bar{a} = 12°$, $r = 0.563, R = 13.513, z = 7.609, P < 0.001$.

nonparametrically for uniformity against an alternative that specifies an angle. For the Batschelet test, we count the number of data that lie within ±90° of the specified angle; let us call this number m' and the test statistic is

$$C = n - m'. \tag{10}$$

We may proceed by performing a two-tailed binomial test, with $p = 0.5$ and with C counts in one category and m' counts in the other. As shown in the figure in Example 5, this may be visualized as drawing a diameter line perpendicular to the radius extending in the specified angle and counting the data on either side of that line.

(c) A Binomial Test.   A nonparametric test to conclude whether the population median angle is equal to a specified value may be performed as follows. Count the number of observed angles on either side of a diameter through the hypothesized angle and subject these data to the binomial test, with $p = 0.5$.

EXAMPLE 5   The Batschelet Test for Circular Uniformity

H_0: The population is uniformly distributed around the circle.

H_A: The population is not uniformly distributed around the circle, but is concentrated around 45°.

The data are those of Example 4.

$n = 24; p = 0.5; m' = 19; C = 5$

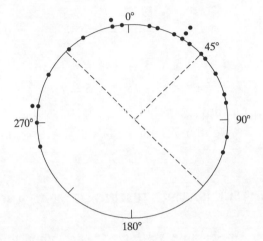

Using Table 27 from *Appendix: Statistical Tables and Graphs*, $C_{0.05(2),24} = 6$, reject H_0, $0.005 < P \leq 0.01$; by the procedure shown, the exact probability would be $P = 0.00661$.

3 TESTING SYMMETRY AROUND THE MEDIAN ANGLE

The symmetry of a distribution around the median may be tested nonparametrically using the Wilcoxon paired-sample test (also known as the Wilcoxon signed-rank test). For each angle (X_i) we calculate the deviation of X_i from the median (i.e., $d_i = X_i -$ median), and we then analyze the d_i's. This is shown in Example 6 for a two-tailed test, where H_0: The underlying distribution is not skewed from the median. A one-tailed test could be used to ask whether the distribution was skewed in a specific direction from the median. (T_- would be the test statistic for H_0: The distribution is not skewed clockwise from the median, and T_+ would be the test statistic for H_0: The distribution is not skewed counterclockwise from the median.)

EXAMPLE 6   Testing for Symmetry Around the Median Angle, for the Data of Example 6

H_0: The underlying distribution is symmetrical around the median.

H_A: The underlying distribution is not symmetrical around the median.

For the 8 data below, the median is 161.5°.

Using the Wilcoxon signed-rank test:

X_i	$d_i =$ X_i- median	Rank of $\|d_i\|$	Signed rank of $\|d_i\|$
97°	−64.5°	8	−8
104°	−57.5°	7	−7
121°	−40.5°	5	−5
159°	−2.5°	1.5	−1.5
164°	2.5°	1.5	1.5
172°	10.5°	3	3
195°	33.5°	4	4
213°	51.5°	6	6

$T_+ = 1.5 + 3 + 4 + 6 = 14.5$

$T_- = 8 + 7 + 5 + 1.5 = 21.5$

$T_{0.05(2),8} = 3$

Neither T_+ nor T_- is $< T_{0.05(2),8}$, so do not reject H_0. $P > 0.50$

4 TWO-SAMPLE AND MULTISAMPLE TESTING OF MEAN ANGLES

(a) Two-Sample Testing. It is common to consider the null hypothesis $H_0: \mu_1 = \mu_2$, where μ_1 and μ_2 are the mean angles for each of two circular distributions (see Example 7). Watson and Williams (1956, with an improvement by Stephens, 1972) proposed a test that utilizes the statistic

$$F = K\frac{(N - 2)(R_1 + R_2 - R)}{N - R_1 - R_2},$$

(11)

where $N = n_1 + n_2$. In this equation, R is Rayleigh's R calculated by Equation 1 with the data from the two samples being combined; R_1 and R_2 are the values of Rayleigh's R for the two samples considered separately. K is a factor, obtained from Table 37 from *Appendix: Statistical Tables and Graphs*, that corrects for bias in the F calculation; in that table we use the weighted mean of the two vector lengths for the column headed r:

$$r_w = \frac{n_1 r_1 + n_2 r_2}{N} = \frac{R_1 + R_2}{N}.$$

(12)

The critical value for this test is $F_{\alpha(1),1,N-2}$. Alternatively,

$$t = \sqrt{K\frac{(N - 2)(R_1 + R_2 - R)}{N - R_1 - R_2}}$$

(13)

may be compared with $t_{\alpha(2),N-2}$. This test may be used for r_w as small as 0.75, if $N/2 \geq 25$. (Batschelet, 1981: 97, 321; Mardia 1972a: 155; Mardia and Jupp, 2000: 130). The underlying assumptions of the test are discussed at the end of this section.

EXAMPLE 7   The Watson-Williams Test for Two Samples

H_0: $\mu_1 = \mu_2$

H_A: $\mu_1 \neq \mu_2$

Sample 1			Sample 2		
a_i (deg)	$\sin a_i$	$\cos a_i$	a_i (deg)	$\sin a_i$	$\cos a_i$
94	0.99756	−0.06976	77	0.97437	0.22495
65	0.90631	0.42262	70	0.93969	0.34202
45	0.70711	0.70711	61	0.87462	0.48481
52	0.78801	0.61566	45	0.70711	0.70711
38	0.61566	0.78801	50	0.76604	0.64279
47	0.73135	0.68200	35	0.57358	0.81915
73	0.95630	0.29237	48	0.74314	0.66913
82	0.99027	0.13917	65	0.90631	0.42262
90	1.00000	0.00000	36	0.58779	0.80902
40	0.64279	0.76604			
87	0.99863	0.05234			

$n_1 = 11$	$\sum \sin a_i$ $= 9.33399$	$\sum \cos a_i$ $= 4.39556$	$n_2 = 9$	$\sum \sin a_i$ $= 7.07265$	$\sum \cos a_i$ $= 5.12160$

$Y = 0.84854, \quad X = 0.39960$ $\qquad\qquad$ $Y = 0.78585, \quad X = 0.56907$

$r_1 = 0.93792$ $\qquad\qquad\qquad\qquad$ $r_2 = 0.97026$

$\sin \bar{a}_1 = 0.90470$ $\qquad\qquad\qquad$ $\sin \bar{a}_2 = 0.80994$

$\cos \bar{a}_1 = 0.42605$ $\qquad\qquad\qquad$ $\cos \bar{a}_2 = 0.58651$

$\bar{a}_1 = 65°$ $\qquad\qquad\qquad\qquad$ $\bar{a}_2 = 54°$

$R_1 = 10.31712$ $\qquad\qquad\qquad\qquad$ $R_2 = 8.73234$

By combining the twenty data from both samples:

$$\sum \sin a_i = 9.33399 + 7.07265 = 16.40664$$

$$\sum \cos a_i = 4.39556 + 5.12160 = 9.51716$$

$$N = 11 + 9 = 20$$

$$Y = \frac{16.40664}{20} = 0.82033$$

$$X = \frac{9.51716}{20} = 0.47586$$

$$r = 0.94836$$

$$R = 18.96720$$

$$r_w = \frac{10.31712 + 8.73234}{20} = 0.952; K = 1.0251$$

$$F = K \frac{(N-2)(R_1 + R_2 - R)}{N - R_1 - R_2}$$

$$= (1.0351) \frac{(20-2)(10.31712 + 8.73234 - 18.96720)}{20 - 10.31712 - 8.73234}$$

$$= (1.0351) \frac{1.48068}{0.95054}$$

$$= 1.61$$

$F_{0.05(1),1,18} = 4.41$.

Therefore, do not reject H_0.

$$0.10 < P < 0.25 \quad [P = 0.22]$$

Thus, we conclude that the two sample means estimate the same population mean, and the best estimate of this population mean is obtained as

$$\sin \bar{a} = \frac{Y}{r} = 0.86500$$

$$\cos \bar{a} = \frac{X}{r} = 0.50177$$

$$\bar{a} = 60°.$$

The data may be grouped as long as the grouping interval is $\leq 10°$. See Batschelet (1972; 1981: Chapter 6) for a review of other two-sample testing procedures. Mardia (1972a: 156–158) and Mardia and Jupp (2000: 130) give a procedure for an approximate confidence interval for $\mu_1 - \mu_2$.

(b) Multisample Testing. The Watson-Williams test can be generalized to a multisample test for testing $H_0: \mu_1 = \mu_2 = \cdots = \mu_k$, a hypothesis reminiscent of analysis of variance considerations for linear data. In multisample tests with circular data (such as Example 8),

$$F = K \frac{(N-k)\left(\sum_{j=1}^{k} R_j - R\right)}{(k-1)\left(N - \sum_{j=1}^{k} R_j\right)}. \tag{14}$$

Here, k is the number of samples, R is the Rayleigh's R for all k samples combined, and $N = \sum_{j=1}^{n} n_j$. The correction factor, K, is obtained from Table 37 from *Appendix: Statistical Tables and Graphs*, using

$$r_w = \frac{\sum_{j=1}^{k} n_j r_j}{N} = \frac{\sum_{j=1}^{k} R_j}{N}. \tag{15}$$

The critical value for this test is $F_{\alpha(1),k-1,N-k}$. Equation 15 (and, thus this test) may be used for r_w as small as 0.45 if $N/k \geq 6$ (Batschelet, 1981: 321; Mardia, 1972a: 163; Mardia and Jupp, 2000: 135). If the data are grouped, the grouping interval should be

EXAMPLE 8   The Watson-Williams Test for Three Samples

H_0: All three samples are from populations with the same mean angle.

H_A: All three samples are not from populations with the same mean angle.

Sample 1			Sample 2		
a_1 (deg)	$\sin a_i$	$\cos a_i$	a_i (deg)	$\sin a_i$	$\cos a_i$
135	0.70711	−0.70711	150	0.50000	−0.86603
145	0.57358	−0.81915	130	0.76604	−0.64279
125	0.81915	−0.57358	175	0.08716	−0.99619
140	0.64279	−0.76604	190	−0.17365	−0.98481
165	0.25882	−0.96593	180	0.00000	−1.00000
170	0.17365	−0.98481	220	−0.64279	−0.76604

$n_1 = 6$ $\quad \sum \sin a_i$ $\quad \sum \cos a_i$ $\qquad n_2 = 6$ $\quad \sum \sin a_i$ $\quad \sum \cos a_i$

$\qquad\qquad = 3.17510 \qquad = -4.81662 \qquad\qquad\qquad = 0.53676 \qquad = -5.25586$

$\qquad\qquad \bar{a}_1 = 147° \qquad\qquad\qquad\qquad\qquad\qquad \bar{a}_2 = 174°$

$\qquad\qquad r_1 = 0.96150 \qquad\qquad\qquad\qquad\qquad\qquad r_2 = 0.88053$

$\qquad\qquad R_1 = 5.76894 \qquad\qquad\qquad\qquad\qquad\qquad R_2 = 5.28324$

Sample 3		
a_i (deg)	$\sin a_i$	$\cos a_i$
140	0.64279	−0.76604
165	0.25882	−0.96593
185	−0.08715	−0.99619
180	0.00000	−1.00000
125	0.81915	−0.57358
175	0.08716	−0.99619
140	0.64279	−0.76604

$n_3 = 7$ $\quad \sum \sin a_i$ $\quad \sum \cos a_i$

$\qquad\qquad = 2.36356 \qquad = -6.06397$

$\qquad\qquad \bar{a}_3 = 159°$

$\qquad\qquad r_3 = 0.92976$

$\qquad\qquad R_3 = 6.50832$

$k = 3$

$N = 6 + 6 + 7 = 19$

For all 19 data:

$$\sum \sin a_i = 3.17510 + 0.53676 + 2.36356 = 6.07542$$

$$\sum \cos a_i = -4.81662 - 5.25586 - 6.06397 = -16.13645$$

$Y = 0.31976$

$X = -0.84929$

$$r = 0.90749$$

$$R = 17.24231$$

$$r_w = \frac{5.76894 + 5.28324 + 6.50832}{19} = 0.924$$

$$F = K \frac{(N - k)(\sum R_j - R)}{(k - 1)(N - \sum R_j)}$$

$$= (1.0546) \frac{(19 - 3)(5.76894 + 5.28324 + 6.50832 - 17.24231)}{(3 - 1)(19 - 5.76894 - 5.28324 - 6.50832)}$$

$$= (1.0546) \frac{5.09104}{2.87900}$$

$$= 1.86$$

$$\nu_1 = k - 1 = 2$$

$$\nu_2 = N - k = 16$$

$$F_{0.05(1),2,16} = 3.63.$$

Therefore, do not reject H_0.

$$0.10 < P < 0.25 \quad [P = 0.19]$$

Thus, we conclude that the three sample means estimate the same population mean, and the best estimate of that population mean is obtained as

$$\sin \bar{a} = \frac{Y}{r} = 0.35236$$

$$\cos \bar{a} = \frac{X}{r} = -0.93587$$

$$\bar{a} = 159°.$$

no larger than 10°. Upton (1976) presents an alternative to the Watson-Williams test that relies on χ^2, instead of F, but the Watson-Williams procedure is a little simpler to use.

The Watson-Williams tests (for two or more samples) are parametric and assume that each of the samples came from a population conforming to what is known as the von Mises distribution, a circular analog to the normal distribution of linear data. In addition, the tests assume that the population dispersions are all the same. Fortunately, the tests are robust to departures from these assumptions. But if the underlying assumptions are known to be violated severely (as when the distributions are not unimodal), we should be wary of their use. In the two-sample case, the nonparametric test of Section 5 is preferable to the Watson-Williams test when the assumptions of the latter are seriously unmet.

Stephens (1982) developed a test with characteristics of a hierarchical analysis of variance of circular data, and Harrison and Kanji (1988) and Harrison, Kanji, and Gadsden (1986) present two-factor ANOVA (including the randomized block design).

Batschelet (1981: 122–126), Fisher (1993: 131–133), Jammalamadake and SenGupta (2001: 128–130), Mardia (1972a: 158–162, 165–166), and Stephens (1972) discuss testing of equality of population concentrations.

5 NONPARAMETRIC TWO-SAMPLE AND MULTISAMPLE TESTING OF ANGLES

If data are grouped — Batschelet (1981: 110) recommends a grouping interval larger than 10° — then contingency-table analysis may be used as a two-sample test. The runs test of Section 18 may be used as a two-sample test, but it is not as powerful for that purpose as are the procedures below, and it is best reserved for testing the hypothesis in Section 18.

(a) Watson's Test. Among the nonparametric procedures applicable to two samples of circular data (e.g., see Batschelet, 1972, 1981: Chapter 6; Fisher, 1993: Section 5.3; Mardia, 1972a: Section 2.4; Mardia and Jupp, 2000: Section 8.3) are the median test of Section 6 and the Watson test.

The Watson test, a powerful procedure developed by Watson* (1962), is recommended in place of the Watson-Williams two-sample test of Section 4 when at least one of the sampled populations is not unimodal or when there are other considerable departures from the assumptions of the latter test. It may be used on grouped data if the grouping interval is no greater than 5° (Batschelet, 1981: 115).

The data in each sample are arranged in ascending order, as demonstrated in Example 9. For the two sample sizes, n_1 and n_2, let us denote the ith observation in Sample 1 as a_{1i} and the jth datum in Sample 2 as a_{2j}. Then, for the data in Example 9, $a_{11} = 35°$, $a_{21} = 75°$, $a_{12} = 45°$, $a_{22} = 80°$, and so on. The total number of data is $N = n_1 + n_2$. The cumulative relative frequencies for the observations in Sample 1 are i/n_1, and those for Sample 2 are j/n_2. As shown in the present example, we then define values of d_k (where k runs from 1 through N) as the differences between the two cumulative relative frequency distributions. The test statistic, called the Watson U^2, is computed as

$$U^2 = \frac{n_1 n_2}{N^2} \sum_{k=1}^{N} (d_k - \bar{d})^2, \tag{16a}$$

where $\bar{d} = \Sigma d_k / N$; or, equivalently, as

$$U^2 = \frac{n_1 n_2}{N^2} \left[\sum_{k=1}^{N} d_k^2 - \frac{\left(\sum_{k=1}^{N} d_k \right)^2}{N} \right]. \tag{16b}$$

Critical values of U^2_{α,n_1,n_2} are given in Table 38a from *Appendix: Statistical Tables and Graphs* bearing in mind that $U^2_{\alpha,n_1,n_2} = U^2_{\alpha,n_2,n_1}$.

Watson's U^2 is especially useful for circular data because the starting point for determining the cumulative frequencies is immaterial. It may also be used in any situation with linear data that are amenable to Mann-Whitney testing, but it is generally not recommended as a substitute for the Mann-Whitney test; the latter is easier to perform and has access to more extensive tables of critical values, and the former may declare significance because group dispersions are different.

---

*Geoffrey Stuart Watson (1921–1998), outstanding Australian-born statistician (Mardia, 1998).

EXAMPLE 9   Watson's U^2 Test for Nonparametric Two-Sample Testing

H_0: The two samples came from the same population, or from two populations having the same direction.

H_A: The two samples did not come from the same population, or from two populations having the same directions.

i	Sample 1 a_{1i} (deg)	$\dfrac{i}{n_1}$	j	Sample 2 a_{2j} (deg)	$\dfrac{j}{n_2}$	$d_k = \dfrac{i}{n_1} - \dfrac{j}{n_2}$	d_k^2
1	35	0.1000			0.0000	0.1000	0.0100
2	45	0.2000			0.0000	0.2000	0.0400
3	50	0.3000			0.0000	0.3000	0.0900
4	55	0.4000			0.0000	0.4000	0.1600
5	60	0.5000			0.0000	0.5000	0.2500
6	70	0.6000			0.0000	0.6000	0.3600
		0.6000	1	75	0.0909	0.5091	0.2592
		0.6000	2	80	0.1818	0.4182	0.1749
7	85	0.7000			0.1818	0.5182	0.2685
		0.7000	3	90	0.2727	0.4273	0.1826
8	95	0.8000			0.2727	0.5273	0.2780
		0.8000	4	100	0.3636	0.4364	0.1904
9	105	0.9000			0.3636	0.5364	0.2877
		0.9000	5	110	0.4546	0.4454	0.1984
10	120	1.0000			0.4546	0.5454	0.2975
		1.0000	6	130	0.5455	0.4545	0.2066
		1.0000	7	135	0.6364	0.3636	0.1322
		1.0000	8	140	0.7273	0.2727	0.0744
		1.0000	9	150	0.8182	0.1818	0.0331
		1.0000	10	155	0.9091	0.0909	0.0083
		1.0000	11	165	1.0000	0.0000	0.0000
	$n_1 = 10$			$n_2 = 11$		$\sum d_k = 7.8272$	$\sum d_k^2 = 3.5018$

$N = n_1 + n_2 = 21$

$$U^2 = \frac{n_1 n_2}{N^2}\left[\sum d_k^2 - \frac{\left(\sum d_k\right)^2}{N}\right]$$

$$= \frac{(10)(11)}{21^2}\left[3.5018 - \frac{(7.8272)^2}{21}\right]$$

$$= 0.1458$$

$U^2_{0.05,10,11} = 0.1856$

Do not reject H_0.

$0.10 < P < 0.20$

(b) Watson's Test with Ties. If there are some tied data (i.e., there are two or more observations having the same numerical value), then the Watson two-sample test is modified as demonstrated in Example 10. We define t_{1i} as the number of data in

---

EXAMPLE 10   Watson's U^2 Test for Data Containing Ties

H_0: The two samples came from the same population, or from two populations having the same directions.

H_A: The two samples did not come from the same population, or from two populations having the same directions.

i	a_{1i}	t_{1i}	m_{1i}	$\dfrac{m_{1i}}{n_1}$	j	a_{2j}	t_{2j}	m_{2j}	$\dfrac{m_{2j}}{n_2}$	$d_k = \dfrac{m_{1i}}{n_1} - \dfrac{m_{2j}}{n_2}$	d_k^2	t_k
				0.0000	1	30°	1	1	0.1000	−0.1000	0.0100	1
				0.0000	2	35	1	2	0.2000	−0.2000	0.0400	1
1	40°	1	1	0.0833					0.2000	−0.1167	0.0136	1
2	45	1	2	0.1667					0.2000	−0.0333	0.0011	1
3	50	1	3	0.2500	3	50	1	3	0.3000	−0.0500	0.0025	2
4	55	1	4	0.3333					0.3000	0.0333	0.0011	1
				0.3333	4	60	1	4	0.4000	−0.0667	0.0044	1
				0.3333	5	65	2	6	0.6000	−0.2677	0.0711	2
5	70	1	5	0.4167					0.6000	−0.1833	0.0336	1
				0.4167	6	75	1	7	0.7000	−0.2833	0.0803	1
6	80	2	7	0.5833	7	80	1	8	0.8000	−0.2167	0.0470	3
				0.5833	8	90	1	9	0.9000	−0.3167	0.1003	1
7	95	1	8	0.6667					0.9000	−0.2333	0.0544	1
				0.6667	9	100	1	10	1.0000	−0.3333	0.1111	1
8	105	1	9	0.7500					1.0000	−0.2500	0.0625	1
9	110	2	11	0.9167					1.0000	−0.0833	0.0069	2
10	120	1	12	1.0000					1.0000	0.0000	0.0000	1

$$n_1 = 12 \qquad\qquad n_2 = 10 \qquad\qquad \sum t_k d_k = -3.5334 \qquad \sum t_k d_k^2 = -0.8144$$

$N = 12 + 10 = 22$

$$U^2 = \frac{n_1 n_2}{N^2}\left[\sum t_k d_k^2 - \frac{\left(\sum t_k d_k\right)^2}{N}\right]$$

$$= \frac{(12)(10)}{22^2}\left[0.8144 - \frac{(-3.5334)^2}{22}\right]$$

$$= 0.0612$$

$U^2_{0.05,10,12} = 0.2246$

Do not reject H_0.

$P > 0.50$

Sample 1 with a value of a_{1i} and t_{2j} as the number of data in Sample 2 that have a value of a_{2j}. Additionally, m_{1i} and m_{2j} are the cumulative number of data in Samples 1 and 2, respectively; so the cumulative relative frequencies are m_{1i}/n_1 and m_{2j}/n_2, respectively. As in Section 5a, d_k represents a difference between two cumulative distributions; and each t_k is the total number of data, in both samples, at each a_{ij}. The test statistic is

$$U^2 = \frac{n_1 n_2}{N^2} \left[\sum_{k=1}^{N} t_k d_k^2 - \frac{\left(\sum_{k=1}^{N} t_k d_k \right)^2}{N} \right]. \tag{17}$$

(c) Wheeler and Watson Test.   Another nonparametric test for the null hypothesis of no difference between two circular distributions is one presented by Wheeler and Watson (1964; developed independently by Mardia, 1967). This procedure ranks all N data and for each a calculates what is termed a *uniform score* or *circular rank*:

$$d = \frac{(360°)(\text{rank of } a)}{N}. \tag{18}$$

This spaces all of the data equally around the circle. Then

$$C_i = \sum_{j=1}^{n_i} \cos d_j \tag{19}$$

and

$$S_i = \sum_{j=1}^{n_i} \sin d_j, \tag{20}$$

where i refers to *either* sample 1 or 2; it does not matter which one of the two samples is used for this calculation. The test statistic is

$$W = \frac{2(N-1)(C_i^2 + S_i^2)}{n_1 n_2}. \tag{21}$$

Critical values of W have been published for some sample sizes (Batschelet, 1981: 344; Mardia, 1967; Mardia and Jupp, 2000: 375–376; Mardia and Spurr, 1973). It has also been shown that W approaches a χ^2 distribution with 2 degrees of freedom for large N. This approximation works best for significance levels no less than 0.025 and has been deemed acceptable by Batschelet (1981: 103) if N is larger than 17, by Fisher (1993: 123) if n_1 and n_2 are each at least 10, and by Mardia and Spurr (1973) if N is greater than 20. This approximation should not be used if there are tied data or if the two sample dispersions are very different (Batschelet 1981: 103). An approximation related to the F distribution has been proposed (Mardia, 1967; Mardia and Jupp, 2000: 148–149) as preferable for some sample sizes.

   This test is demonstrated in Example 11. This example shows C_i and S_i for each of the two samples, but C and S are only needed from one of the samples in order to perform the test.

EXAMPLE 11   The Wheeler and Watson Two-Sample Test for the Data of Example 9

H_0: The two samples came from the same population, or from two populations having the same directions.

H_A: The two samples did not come from the same population, or from two populations having the same directions.

$n_1 = 10, n_2 = 11$, and $N = 21$

$$\frac{360°}{N} = \frac{360°}{21} = 17.1429°$$

Sample 1			Sample 2		
Direction (degrees)	Rank of direction	Circular rank (degrees)	Direction (degrees)	Rank of direction	Circular rank (degrees)
35	1	17.14			
45	2	34.29			
50	3	51.43			
55	4	68.57			
60	5	85.71			
70	6	102.86			
			75	7	120.00
			80	8	137.14
85	9	154.29			
			90	10	171.43
95	11	188.57			
			100	12	205.71
105	13	222.86			
			110	14	240.00
120	15	257.14			
			130	16	274.29
			135	17	291.43
			140	18	308.57
			150	19	325.71
			160	20	342.86
			165	21	360.00

$C_1 = -0.2226$ $C_2 = 0.2226$

$S_1 = 3.1726$ $S_2 = -3.1726$

$$W = \frac{2(N-1)(C_1^2 + S_1^2)}{n_1 n_2}$$

$$= \frac{2(21-1)[(-0.2226)^2 + (3.1726)^2]}{(10)(11)} = 3.678$$

$\nu = 2$

$\chi^2_{0.05,2} = 5.991$

Do not reject H_0.

$0.10 < P < 0.25$

(b) Multisample Testing. The Wheeler and Watson test may also be applied to more than two samples. The procedure is as before, where all N data from all k samples are ranked and (by Equation 18) the circular rank, d, is calculated for each datum. Equations 19 and 20 are applied to each sample, and

$$W = 2 \sum_{i=1}^{k} \left[\frac{C_i^2 + S_i^2}{n_i} \right].$$ (22)

Some critical values for $k = 3$ have been published (Batschelet, 1981: 345; Mardia, 1970b; Mardia and Spurr, 1973). For large sample sizes, W may be considered to approximate χ^2 with $2(k - 1)$ degrees of freedom. This approximation is considered adequate by Fisher (1993: 123) if each n_i is at least 10 and by Mardia and Spurr (1973) if N is greater than 20.

Maag (1966) extended the Watson U^2 test to $k > 2$, but critical values are not available. Comparison of more than two medians may be effected by the procedure in Section 6.

If the data are in groups with a grouping interval larger than 10, then an $r \times c$ contingency table analysis may be performed, for r samples in c groups.

6 TWO-SAMPLE AND MULTISAMPLE TESTING OF MEDIAN ANGLES

The following comparison of two or more medians is presented by Fisher (1933: 114), who states it as applicable if each sample size is at least 10 and all data are within 90° of the grand median (i.e., the median of all N data from all k samples). If we designate m_i to be the number of data in sample i that lie between the grand median and the grand median $- 90°$, and $M = \sum_{i=1}^{k} m_i$, then

$$\frac{N^2}{M(N - M)} \sum_{i=1}^{k} \frac{m_i^2}{n_i} - \frac{NM}{N - M}$$ (23)

is a test statistic that may be compared to χ^2 with $k - 1$ degrees of freedom.*

If "H_0: All k population medians are equal" is not rejected, then the grand median is the best estimate of the median of each of the k populations.

7 TWO-SAMPLE AND MULTISAMPLE TESTING OF ANGULAR DISTANCES

Angular distance is simply the shortest distance, in angles, between two points on a circle. For example, the angular distance between 95° and 120° is 25°, between 340° and 30° is 50°, and between 190° and 5° is 175°. In general, we shall refer to the angular distance between angles a_1 and a_2 as $d_{a_1-a_2}$. (Thus, $d_{95°-120°} = 25°$, and so on.)

---

*The same results are obtained if m_i is defined as the number of data in sample i that lie between the grand median and the grand median $+ 90°$.

Angular distances are useful in drawing inferences about departures of data from a specified direction. We may observe travel directions of animals trained to travel in a particular compass direction (perhaps "homeward"), or of animals confronted with the odor of food coming from a specified direction. If dealing with times of day we might speak of the time of a physiological or behavioral activity in relation to the time of a particular stimulus.

(a) Two-Sample Testing. If a specified angle (e.g., direction or time of day) is μ_0, we may ask whether the mean of a sample of data, $\bar{a}$, significantly departs from μ_0 by testing the one-sample hypothesis, $H_0: \mu_a = \mu_0$, as explained in Section 1c. However, we may have two samples, Sample 1 and Sample 2, each of which has associated with it a specified angle of interest, μ_1 and μ_2, respectively (where μ_1 and μ_2 need not be the same). We may ask whether the angular distances for Sample 1 $(d_{a_{1i}-\mu_1})$ are significantly different from those for Sample 2 $(d_{a_{2i}-\mu_2})$. As shown in Example 12, we can rank the angular distances of both samples combined and then perform a Mann-Whitney test. This was suggested by Wallraff (1979).

EXAMPLE 12   Two-Sample Testing of Angular Distances

Birds of both sexes are transported away from their homes and released, with their directions of travel tabulated. The homeward direction for each sex is 135°.

H_0: Males and females orient equally well toward their homes.

H_A: Males and females do not orient equally well toward their homes.

Males			Females		
Direction traveled	Angular distance	Rank	Direction traveled	Angular distance	Rank
145°	10°	6	160°	25°	12.5
155	20	11	135	0	1
130	5	2.5	145	10	6
145	10	6	150	15	9.5
145	10	6	125	10	6
160	25	12.5	120	15	9.5
140	5	2.5			
		46.5			44.5

For the two-tailed Mann-Whitney test:

$n_1 = 7, R_1 = 46.5$

$n_2 = 6, R_2 = 44.5$

$U = n_1 n_2 + \dfrac{n_1(n_1 + 1)}{2} - R_1 = (7)(6) + \dfrac{7(8)}{2} - 46.5 = 23.5$

$U' = n_1 n_2 - U = (7)(6) - 23.5 = 18.5$

$U_{0.05(2),7,6} = U_{0.05(2),6,7} = 36.$

Do not reject H_0.

$$P > 0.20$$

The procedure could be performed as a one-tailed instead of a two-tailed test, if there were reason to be interested in whether the angular distances in one group were greater than those in the other.

(b) Multisample Testing. If more than two samples are involved, then the angular deviations of all of them are pooled and ranked, whereupon the Kruskal-Wallis test may be applied, followed if necessary by nonparametric multiple-comparison testing.

8 TWO-SAMPLE AND MULTISAMPLE TESTING OF ANGULAR DISPERSION

The Wallraff (1979) procedure of analyzing angular distances (Section 7) may be applied to testing for angular dispersion. The angular distances of concern for Sample 1 are $d_{a_{1i}-\bar{a}_1}$ and those for Sample 2 are $d_{a_{2i}-\bar{a}_2}$. Thus, just as measures of dispersion for linear data may refer to deviations of the data from their mean, here we consider the deviations of circular data from their mean.

The angular distances of the two samples may be pooled and ranked for application of the Mann-Whitney test, which may be employed for either two-tailed (Example 13) or one-tailed testing.

EXAMPLE 13   Two-Sample Testing for Angular Dispersion

The times of day that males and females are born are tabulated. The mean time of day for each sex is determined (to the nearest 5 min). (For males, $\bar{a}_1 = 7{:}55$ A.M.; for females, $\bar{a}_2 = 8{:}15$ A.M.)

H_0: The times of day of male births are as variable as the times of day of female births.

H_A: The times of day of male births do not have the same variability as the times of day of female births.

	Male			Female	
Time of day	*Angular distance*	*Rank*	*Time of day*	*Angular distance*	*Rank*
05:10 hr	2:45 hr	11	08:15 hr	0:00 hr	1
06:30	1:25	4	10:20	2:05	8.5
09:40	1:45	6	09:45	1:30	5
10:20	2:25	10	06:10	2:05	8.5
04:20	3:35	13	04:05	4:10	14
11:15	3:20	12	07:50	0:25	2
			09:00	0:45	3
			10:10	1:55	7
		$R_1 = 56$			$R_2 = 49$

For the two-tailed Mann-Whitney test:

$$n_1 = 6, R_1 = 56$$

$$n_2 = 8, R_2 = 49$$

$$U = n_1 n_2 + \frac{n_1(n_1 + 1)}{2} - R_1 = (6)(8) + \frac{6(7)}{2} - 56 = 13$$

$$U' = n_1 n_2 - U = (6)(8) - 13 = 35$$

$$U_{0.05(2),6,8} = 40.$$

Do not reject H_0.

$$P = 0.20$$

If we wish to compare the dispersions of more than two samples, then the aforementioned Mann-Whitney procedure may be expanded by using the Kruskal-Wallis test, followed if necessary by nonparametric multiple comparisons.

9 PARAMETRIC ANALYSIS OF THE MEAN OF MEAN ANGLES

A set of n angles, a_i, has a mean angle, $\bar{a}$, and an associated mean vector length, r. This set of data may be referred to as a *first-order sample*. A set of k such means may be referred to as a *second-order sample*. We can also test the statistical significance of a mean of means.

Assuming that the second-order sample comes from a bivariate normal distribution (i.e., a population in which the X_j's follow a normal distribution, and the Y_j's are also normally distributed), a testing procedure due to Hotelling* (1931) may be applied.

The sums of squares and crossproducts of the k means are

$$\sum x^2 = \sum X_j^2 - \frac{\left(\sum X_j\right)^2}{k}, \tag{24}$$

$$\sum y^2 = \sum Y_j^2 - \frac{\left(\sum Y_j\right)^2}{k}, \tag{25}$$

and

$$\sum xy = \sum X_j Y_j - \frac{\sum X_j \sum Y_j}{k}, \tag{26}$$

where $\sum$ in each instance refers to a summation over all k means (i.e., $\sum = \sum_{j=1}^{k}$).

Then, we can test the null hypothesis that there is no mean direction (i.e., $H_0 : \rho = 0$) in the population from which the second-order sample came by using as a test statistic

$$F = \frac{k(k-2)}{2} \left[\frac{\bar{X}^2 \sum y^2 - 2\bar{X}\bar{Y} \sum xy + \bar{Y}^2 \sum x^2}{\sum x^2 \sum y^2 - \left(\sum xy\right)^2} \right], \tag{27}$$

with the critical value being the one-tailed F with degrees of freedom of 2 and $k - 2$ (Batschelet, 1978; 1981: 144–150). This test is demonstrated in Example 14.

This test assumes the data are not grouped. The assumption of bivariate normality is a serious one. Although the test appears robust against departures due to kurtosis,

---

*Harold Hotelling (1895–1973), American mathematical economist and statistician. He owed his life, and thus the achievements of an impressive career, to a zoological mishap. While attending the University of Washington he was called to military service in World War I and appointed to care for mules. One of his charges (named Dynamite) broke Hotelling's leg, thus preventing the young soldier from accompanying his division to France where the unit was annihilated in battle (Darnell, 1988).

EXAMPLE 14   The Second-Order Analysis for Testing the Significance of the Mean of the Sample Means in Example 8 from *Circular Distributions: Descriptive Statistics*

H_0: There is no mean population direction (i.e., $\rho = 0$).

H_A: There is a mean population direction (i.e., $\rho \neq 0$).

$k = 7, \quad \overline{X} = -0.52734, \quad \overline{Y} = 0.27844$

$$\sum X_j = -3.69139, \quad \sum X_j^2 = 2.27959, \quad \sum x^2 = 0.33297$$

$$\sum Y_j = 1.94906, \quad \sum Y_j^2 = 0.86474, \quad \sum y^2 = 0.32205$$

$$\sum X_j Y_j = -1.08282, \quad \sum xy = -0.05500$$

$$F = \frac{7(7-2)}{2}\left[\frac{(-0.52734)^2(0.32205) - 2(-0.52734)(0.27844)(-0.05500) + (0.27844)^2(0.33297)}{(0.33297)(0.32205) - (-0.05500)^2}\right]$$

$$= 16.66$$

$F_{0.05(1),2,5} = 5.79$

Reject H_0.

$$0.005 < P < 0.01$$

And, from *Circular Distributions: Descriptive Statistics*, we see that the population mean angle is estimated to be 152°.

the test may be badly affected by departures due to extreme skewness, rejecting a true H_0 far more often than indicated by the significance level, α (Everitt, 1979; Mardia, 1970a).

10   NONPARAMETRIC ANALYSIS OF THE MEAN OF MEAN ANGLES

The Hotelling testing procedure of Section 9 requires that the k $\overline{X}$'s come from a normal distribution, as do the k $\overline{Y}$'s. Although we may assume the test to be robust to some departure from this bivariate normality, there may be considerable nonnormality in a sample, in which case a nonparametric method is preferable.

Moore (1980) has provided a nonparametric modification of the Rayleigh test, which can be used to test a sample of mean angles; it is demonstrated in Example 15. The k vector lengths are ranked, so that r_1 is the smallest and r_k is the largest. We shall call the ranks i (where i ranges from 1 through k) and compute

$$X = \frac{\sum_{i=1}^{k} i \cos \overline{a}_i}{k} \tag{28}$$

$$Y = \frac{\sum_{i=1}^{k} i \sin \overline{a}_i}{k} \tag{29}$$

$$R' = \sqrt{\frac{X^2 + Y^2}{k}}. \tag{30}$$

EXAMPLE 15   Nonparametric Second-Order Analysis for Significant Direction in the Sample of Means of Example from *Circular Distributions: Descriptive Statistics*

H_0: The population from which the sample of means came is uniformly distributed around the circle (i.e., $\rho = 0$).

H_A: The population of means is not uniformly distributed around the circle (i.e., $\rho \neq 0$).

Sample rank (i)	r_i	a_i	$\cos \bar{a}_i$	$\sin \bar{a}_i$	$i \cos \bar{a}_i$	$i \sin \bar{a}_i$
1	0.3338	171°	−0.98769	0.15643	−0.98769	0.15643
2	0.3922	186	−0.99452	−0.10453	−1.98904	−0.20906
3	0.4696	117	−0.45399	0.04570	−1.36197	2.67302
4	0.6962	134	−0.69466	0.71934	−2.77863	2.87736
5	0.7747	169	−0.98163	0.19081	−4.90814	0.95404
6	0.8794	140	−0.76604	0.64279	−4.59627	3.85673
7	0.8954	160	−0.93969	0.34202	−6.57785	2.39414
					−23.19959	12.70266

$k = 7$

$$X = \frac{\sum i \cos \bar{a}_i}{k} = \frac{-23.19959}{7} = -3.31423$$

$$Y = \frac{\sum i \sin \bar{a}_i}{k} = \frac{12.70266}{7} = 1.81467$$

$$R' = \sqrt{\frac{X^2 + Y^2}{k}} = \sqrt{\frac{(-3.31423)^2 + (1.81467)^2}{7}} = \sqrt{2.03959} = 1.428$$

$R'_{0.05,7} = 1.066$

Therefore, reject H_0.

$P < 0.001$

The test statistic, R', is then compared to the appropriate critical value, $R'_{\alpha,n}$, in Table 39 from *Statistical Tables and Graphs*.

(a) Testing with Weighted Angles. The Moore modification of the Rayleigh test can also be used when we have a sample of angles, each of which is weighted. We may then perform the ranking of the angles by the weights, instead of by the vector lengths, r. For example, the data could be ranked by the amount of leaning. Or, if we are recording the direction each of several birds flies from a release point, the weights could be the distances flown. (If the birds disappear at the horizon, then the weights of their flight angles are all the same.)

11   PARAMETRIC TWO-SAMPLE ANALYSIS OF THE MEAN OF MEAN ANGLES

Batschelet (1978, 1981: 150–154) explained how the Hotelling (1931) procedure of Section 9 can be extended to consider the hypothesis of equality of the means of two populations of means (assuming each population to be bivariate normal). We proceed as in Section 9, obtaining an $\overline{X}$ and $\overline{Y}$ for each of the two samples ($\overline{X}_1$ and $\overline{Y}_1$ for Sample 1, and $\overline{X}_2$ and $\overline{Y}_2$ for Sample 2). Then, we apply Equations 24, 25, and 26 to each of the two samples, obtaining $\left(\sum x^2\right)_1$, $\left(\sum xy\right)_1$, and $\left(\sum y^2\right)_1$ for Sample 1, and $\left(\sum x^2\right)_2$, $\left(\sum xy\right)_2$, and $\left(\sum y^2\right)_2$ for Sample 2.

Then we calculate

$$\left(\sum x^2\right)_c = \left(\sum x^2\right)_1 + \left(\sum x^2\right)_2; \tag{31}$$

$$\left(\sum y^2\right)_c = \left(\sum y^2\right)_1 + \left(\sum y^2\right)_2; \tag{32}$$

$$\left(\sum xy\right)_c = \left(\sum xy\right)_1 + \left(\sum xy\right)_2; \tag{33}$$

and the null hypothesis of the two population mean angles being equal is tested by

$$F = \frac{N-3}{2\left(\dfrac{1}{k_1} + \dfrac{1}{k_2}\right)}\left[\frac{(\overline{X}_1 - \overline{X}_2)^2\left(\sum y^2\right)_c - 2(\overline{X}_1 - \overline{X}_2)(\overline{Y}_1 - \overline{Y}_2)\left(\sum xy\right)_c + (\overline{Y}_1 - \overline{Y}_2)^2\left(\sum x^2\right)_c}{\left(\sum x^2\right)_c\left(\sum y^2\right)_c - \left(\sum xy\right)_c^2}\right],$$

$$\tag{34}$$

where $N = k_1 + k_2$, and F is one-tailed with 2 and $N - 3$ degrees of freedom. This test is shown in Example 16, using the data of Figure 2.

EXAMPLE 16 Parametric Two-Sample Second-Order Analysis for Testing the Difference Between Mean Angles

We have two samples, each consisting of mean directions and vector lengths, as shown in Figure 2. Sample 1 is the data from Example 14, where

$$k_1 = 7; \quad \overline{X}_1 = -0.52734; \quad \overline{Y}_1 = 0.27844; \quad \bar{a}_1 = 152°;$$

$$\left(\sum x^2\right)_1 = 0.33297; \quad \left(\sum y^2\right)_1 = 0.32205; \quad \left(\sum xy\right)_1 = -0.05500.$$

Sample 2 consists of the following 10 data:

j	$\bar{a}_j$	r_j
1	115°	0.9394
2	127	0.6403
3	143	0.3780
4	103	0.6671
5	130	0.8210
6	147	0.5534
7	107	0.8334
8	137	0.8139
9	127	0.2500
10	121	0.8746

Applying the calculations of Example 14, we find

$$k_2 = 10; \quad \sum r_j \cos a_j = -3.66655; \quad \sum r_j \sin a_j = 5.47197;$$

$$\overline{X}_2 = -0.36660; \quad \overline{Y}_2 = 0.54720; \quad \bar{a}_2 = 124°.$$

$$\left(\sum x^2\right)_2 = 0.20897; \quad \left(\sum y^2\right)_2 = 0.49793; \quad \left(\sum xy\right)_2 = -0.05940.$$

Then, we can test

H_0: $\mu_1 = \mu_2$ (The means of the populations from which these two samples came are equal.)

H_A: $\mu_1 \neq \mu_2$ (The two population means are not equal.)

$N = 7 + 10$

$$\left(\sum x^2\right)_c = 0.33297 + 0.20897 = 0.54194$$

$$\left(\sum y^2\right)_c = 0.32205 + 0.49793 = 0.81998$$

$$\left(\sum xy\right)_c = -0.05500 + (-0.05940) = -0.11440$$

$$F = \frac{(17 - 3)}{2\left(\frac{1}{7} + \frac{1}{10}\right)}$$

$$\times \left[\frac{\begin{array}{l} [-0.52734 - (-0.36660)]^2(0.81998) \\ \quad - 2[-0.52734 - (-0.36660)](0.27844 - 0.54720)(-0.11440) \\ \quad + (0.27844 - 0.54720)^2(0.54194) \end{array}}{(0.54194)(0.81998) - (-0.11440)^2} \right]$$

$= 4.69$

$F_{0.05(1),2,14} = 3.74.$

Reject H_0.

$0.025 < P < 0.05 \quad [P = 0.028]$

The two-sample Hotelling test is robust to departures from the normality assumption (far more so than is the one-sample test of Section 9), the effect of nonnormality being slight conservatism (i.e., rejecting a false H_0 a little less frequently than indicated by the significance level, α) (Everitt, 1979). The two samples should be of the same size, but departure from this assumption does not appear to have serious consequences (Batschelet, 1981: 202).

12 NONPARAMETRIC TWO-SAMPLE ANALYSIS OF THE MEAN OF MEAN ANGLES

The parametric test of Section 11 is based on sampled populations being bivariate normal and the two populations having variances and covariances in common, unlikely assumptions to be strictly satisfied in practice. While the test is rather robust to departures from these assumptions, employing a nonparametric test may be employed to assess whether two second-order populations have the same directional orientation.

Batschelet (1978; 1981: 154–156) presented the following nonparametric procedure (suggested by Mardia, 1967) as an alternative to the Hotelling test of Section 11. First compute the grand mean vector, pooling all data from both samples. Then, the X coordinate of the grand mean is subtracted from the X coordinate of each of the data in both samples, and the Y of the grand mean is subtracted from the Y of each of the data. (This maneuver determines the direction of each datum from

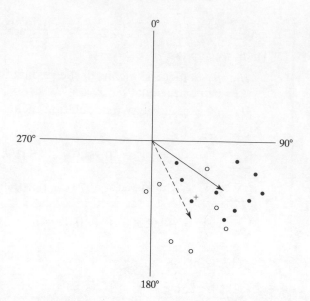

FIGURE 2: The data of Example 16. The open circles indicate the ends of the seven mean vectors of Sample 1, with the mean of these seven indicated by the broken-line vector. The solid circles indicate the 10 data of Sample 2, with their mean shown as a solid-line vector. (The "+" indicates the grand mean vector of all seventeen data, which is used in Example 17.)

the grand mean.) As shown in Example 17, the resulting vectors are then tested by a nonparametric two-sample test (as in Section 5). This procedure requires that the data not be grouped.

EXAMPLE 17   Nonparametric Two-Sample Second-Order Analysis, Using the Data of Example 16

H_0: The two samples came from the same population, or from two populations with the same directions.

H_A: The two samples did not come from the same population, nor from two populations with the same directions.

Total number of vectors $= 7 + 10 = 17$

To determine the grand mean vector (which is shown in Figure 2):

$$\sum r_j \cos a_j = (-3.69139) + (-3.66655) = -7.35794$$

$$\sum r_j \sin a_j = 1.94906 + 5.47197 = 7.42103$$

$$\overline{X} = \frac{-7.35794}{17} = -0.43282$$

$$\overline{Y} = \frac{7.42103}{17} = 0.43653$$

$\overline{X}$ and $\overline{Y}$ are all that we need to define the grand mean; however, if we wish we can also determine the length and direction of the grand mean vector:

$$r = \sqrt{\overline{X}^2 + \overline{Y}^2} = \sqrt{(-0.43282)^2 + (0.43653)^2} = 0.61473$$

$$\cos \overline{a} = \frac{-0.43282}{0.61473} = -0.70408$$

$$\sin \overline{a} = \frac{0.43653}{0.61473} = 0.71012$$
$$\overline{a} = 135°.$$

Returning to the hypothesis test, we subtract the foregoing $\overline{X}$ from the X, and the $\overline{Y}$ from the Y, for each of the 17 data, arriving at 17 new vectors, as follows:

Sample 1

Datum	X	$X - \overline{X}$	Y	$Y - \overline{Y}$	New a
1	−0.84140	−0.40858	0.30624	−0.13029	184°
2	−0.76047	−0.32765	0.14782	−0.28871	210
3	−0.21319	0.21963	0.41842	−0.01811	20
4	−0.67366	−0.24084	0.56527	0.12874	137
5	−0.39005	0.04277	−0.04100	−0.47753	276
6	−0.48293	−0.05011	0.50009	0.06356	107
7	−0.32969	0.10313	0.05222	−0.38431	290

Sample 2

Datum	X	$X - \overline{X}$	Y	$Y - \overline{Y}$	New a
1	−0.39701	0.03581	0.85139	0.41485	86°
2	−0.38534	0.04748	0.51137	0.07484	75
3	−0.30188	0.13084	0.22749	−0.20904	320
4	−0.15006	0.28276	0.65000	0.21347	48
5	−0.52773	−0.09491	0.62892	0.19239	108
6	−0.46412	−0.03130	0.30140	−0.13513	230
7	−0.24366	0.18916	0.79698	0.36045	68
8	−0.59525	−0.16243	0.55508	0.11855	127
9	−0.15045	0.28237	0.19966	−0.23687	334
10	−0.45045	−0.01763	0.74968	0.31315	92

Now, using Watson's two-sample test (Section 5) on these new angles:

i	a_{1i}	i/n_1	j	a_{2j}	j/n_2	d_k	d_k^2
					Sample 2		
	Sample 1		Sample 2				
1	20	0.1429			0.0000	0.1429	0.0204
		0.1429	1	48	0.1000	0.0429	0.0018
		0.1429	2	68	0.2000	−0.0571	0.0033
		0.1429	3	75	0.3000	−0.1571	0.0247
		0.1429	4	86	0.4000	−0.2571	0.0661
		0.1429	5	92	0.5000	−0.3571	0.1275
2	107	0.2857			0.5000	−0.2143	0.0459
		0.2857	6	108	0.6000	−0.3143	0.0988
		0.2857	7	127	0.7000	−0.4143	0.1716
3	137	0.4286			0.7000	−0.2714	0.0737
4	184	0.5714			0.7000	−0.1286	0.0165
5	210	0.7143			0.7000	0.0143	0.0002
		0.7143	8	230	0.8000	−0.0857	0.0073
6	276	0.8571			0.8000	0.0571	0.0033

7	290	1.0000			0.8000	0.2000	0.0400
		1.0000	9	320	0.9000	0.1000	0.0100
		1.0000	10	334	1.0000	0.0000	0.0000

$n_1 = 7$		$n_2 = 10$				Σd_k	Σd_k^2
						$= -1.6998$	$= 0.7111$

$$N = 7 + 10 = 17$$

$$U^2 = \frac{n_1 n_2}{N}\left[\Sigma d_k^2 - \frac{(\Sigma d_k)^2}{N}\right]$$

$$= \frac{(7)(10)}{17^2}\left[0.7111 - \frac{(-1.6998)^2}{17}\right]$$

$$= 0.1311$$

$U_{0.05,7,10}^2 = 0.1866$

Do not reject H_0.

$$0.10 < P < 0.20$$

13 PARAMETRIC PAIRED-SAMPLE TESTING WITH ANGLES

For linear data, we two samples having paired data can be reduced to a one-sample test employing the differences between members of pairs.

Circular data in two samples might also be paired, in which case the one-sample Hotelling test of Section 9 may be used after forming a single sample of data from the differences between the paired angles. If a_{ij} is the jth angle in the ith sample, then a_{1j} and a_{2j} are a pair of data. A single set of rectangular coordinates, X's and Y's, is formed by computing

$$X_j = \cos a_{2j} - \cos a_{1j} \tag{35}$$

and

$$Y_j = \sin a_{2j} - \sin a_{1j}. \tag{36}$$

Then the procedure of Section 9 may be applied, as shown in Example 18 (where k is the number of pairs).

EXAMPLE 18   The Hotelling Test for Paired Samples of Angles

Ten birds are marked for individual identification, and we record on which side of a tree each bird sits to rest in the morning and in the afternoon. We wish to test the following.

H_0: The side of a tree on which birds sit is the same in the morning and in the afternoon.

H_A: The side of a tree on which birds sit is not the same in the morning and in the afternoon.

Bird (j)	Morning Direction (a_{1j})	$\sin a_{1j}$	$\cos a_{1j}$	Afternoon Direction (a_{2j})	$\sin a_{2j}$	$\cos a_{2j}$	Difference Y_j	X_j	$X_j Y_j$
1	105°	0.9659	−0.2588	205°	−0.4226	−0.9063	−1.3885	−0.6475	0.8991
2	120	0.8660	−0.5000	210	−0.5000	−0.8660	−1.3660	−0.3660	0.5000
3	135	0.7071	−0.7071	235	−0.8192	−0.5736	−1.5263	0.1335	−0.2038
4	95	0.9962	−0.0872	245	−0.9063	−0.4226	−1.9025	−0.3354	0.6381
5	155	0.4226	−0.9063	260	−0.9848	−0.1736	−1.4074	0.7327	−1.0312
6	170	0.1736	−0.9848	255	−0.9659	−0.2588	−1.1395	0.7260	−0.8273
7	160	0.3420	−0.9397	240	−0.8660	−0.5000	−1.2080	0.4397	−0.5312
8	155	0.4226	−0.9063	245	−0.9063	−0.4226	−1.3289	0.4837	−0.6428
9	120	0.8660	−0.5000	210	−0.5000	−0.8660	−1.3660	−0.3660	0.5000
10	115	0.9063	−0.4226	200	−0.3420	−0.9397	−1.2483	−0.5171	0.6455

$k = 10$

$\overline{X} = 0.0284$

$\overline{Y} = -1.3881$

$\Sigma Y_j = -13.8814 \qquad \Sigma X_j = 0.2836$

$\Sigma Y_j^2 = 19.6717 \qquad \Sigma X_j^2 = 2.5761 \qquad \Sigma X_j Y_j = -0.0536$

$$\sum x^2 = 2.5761 - \frac{(0.2836)^2}{10} = 2.5681$$

$$\sum y^2 = 19.6717 - \frac{(-13.8814)^2}{10} = 0.4024$$

$$\sum xy = -0.0536 - \frac{(0.2836 - 13.8814)}{10} = 0.3402$$

$$F = \frac{k(k-2)}{2} \left[\frac{\overline{X}^2 \Sigma y^2 - 2\overline{X}\,\overline{Y}\Sigma xy + \overline{Y}^2 \Sigma x^2}{\Sigma x^2 \Sigma y^2 - \left(\Sigma xy\right)^2} \right]$$

$$F = \frac{10(10-2)}{2} \left[\frac{\begin{array}{c}(0.0284)^2(0.4024) - 2(0.0284)(-1.3881) \\ \times\, (0.3402) + (-1.3881)^2(2.5681)\end{array}}{(2.5681)(0.4024) - (0.3402)^2} \right]$$

$$= 217$$

$F_{0.05(1),2,8} = 4.46.$

Reject H_0.

$$P \ll 0.0005 \quad [P = 0.00000011]$$

If each member of a pair of data is a mean angle ($\bar{a}$) from a sample, with an associated vector length (r), then we are dealing with a second-order analysis. The aforementioned Hotelling test may be applied if the following computations are used in place of Equations 35 and 36, respectively:

$$X_j = r_{2j} \cos \bar{a}_{2j} - r_{1j} \cos \bar{a}_{1j} \tag{37}$$

$$Y_j = r_{2j} \sin \bar{a}_{2j} - r_{ij} \sin \bar{a}_{1j}. \tag{38}$$

14 NONPARAMETRIC PAIRED-SAMPLE TESTING WITH ANGLES

Circular data in a paired-sample experimental design may be tested nonparametrically by forming a single sample of the paired differences, which can then be subjected to the Moore test of Section 10. We calculate rectangular coordinates (X_j and Y_j) for each paired difference, as done in Equations 35 and 36. Then, for each of the j paired differences, we compute

$$r_j = \sqrt{X_j^2 + Y_j^2}, \tag{39}$$

$$\cos a_j = \frac{X_j}{r_j}, \tag{40}$$

$$\sin a_j = \frac{Y_j}{r_j}. \tag{41}$$

Then the values of r_j are ranked, with ranks (i) running from 1 through n, and we complete the analysis using Equations 28, 29, and 30 and Table 39 from *Appendix: Statistical Tables and Graphs*, substituting n for k. The procedure is demonstrated in Example 19.

If each member of a pair of circular-scale data is a mean angle, $\bar{a}_j$, with an associated vector length, r_j, then we modify the preceding analysis. Calculate X_j and Y_j by Equations 37 and 38, respectively, instead of by Equations 35 and 36, respectively. Then apply Equations 39 through 41 and Equations 28 through 30 to complete the analysis (where k is the number of paired means).

15 PARAMETRIC ANGULAR CORRELATION AND REGRESSION

Correlation involving angular data may be of two kinds: Either both variables are measured on a circular scale (a situation sometimes termed "angular-angular," or "spherical," correlation), or one variable is on a circular scale with the other measured on a linear scale (sometimes called an "angular-linear," or "cylindrical," correlation). The study of biological rhythms deals essentially with the rhythmic dependence (i.e., regression) of a linear scale variable (e.g., a measure of biological activity, such as body temperature) on a circular scale variable (namely, time).

(a) Angular-Angular Correlation. Correlation measures developed for correlation between two angular variables were for years characterized by serious deficiencies, such as not distinguishing between positive and negative relationships (e.g., see the review by Jupp and Mardia, 1980). However, Fisher and Lee (1983) presented a correlation coefficient analogous to the familiar parametric correlation

EXAMPLE 19   The Moore Test for Paired Data on a Circular Scale of Measurement

Ten birds are marked for individual identification, and we record on which side of a tree each bird sits to rest in the morning and in the afternoon. (The data are the same as in Example 18.) We wish to test the following:

H_0: The side of a tree on which birds sit is the same in the morning and in the afternoon.

H_A: The side of a tree on which birds sit is not the same in the morning and in the afternoon.

Bird (j)	Morning Direction (a_{1j})	$\sin a_{1j}$	$\cos a_{1j}$	Afternoon Direction (a_{2j})	$\sin a_{2j}$	$\cos a_{2j}$	Y_j	X_j	r_j	$\sin a_j$	$\cos a_j$	Rank of r_j (i)
1	105°	0.9659	−0.2588	205°	−0.4226	−0.9063	−1.3885	0.6475	1.5321	0.9063	0.4226	7.5
2	120	0.8660	−0.5000	210	−0.5000	−0.8660	−1.3660	−0.3660	1.4142	−0.9659	−0.2588	4.5
3	135	0.7071	−0.7071	235	−0.8192	−0.5736	−1.5263	0.1335	1.5321	−0.9962	0.0871	7.5
4	95	0.9962	−0.0872	245	−0.9063	−0.4226	−1.9025	−0.3354	1.9318	−0.9848	−0.1736	10
5	155	0.4226	−0.9063	260	−0.9848	−0.1736	−1.4074	0.7327	1.5867	−0.8870	0.4618	9
6	170	0.1736	−0.9848	255	−0.9659	−0.2588	−1.1395	0.7260	1.3511	−0.8434	0.5373	2
7	160	0.3420	−0.9397	240	−0.8660	−0.5000	−1.2080	0.4397	1.2855	−0.9397	0.3420	1
8	155	0.4226	−0.9063	245	−0.9063	−0.4226	−1.3289	0.4837	1.4152	−0.9390	0.3418	6
9	120	0.8660	−0.5000	210	−0.5000	−0.8660	−1.3660	−0.3660	1.4142	−0.9659	−0.2588	4.5
10	115	0.9063	−0.4226	200	−0.3420	−0.9397	−1.2483	−0.5171	1.3512	−0.9238	−0.3827	3

$n = 10$

$$X = \frac{\sum_{i=1}^{n} i \cos a_i}{n} = \frac{1(0.3420) + 2(0.5373) + \cdots + 10(-0.1736)}{10}$$

$$= -0.0106$$

$$Y = \frac{\sum_{i=1}^{n} i \sin a_i}{n} = \frac{1(-0.9397) + 2(-0.8434) + \cdots + 10(-0.9848)}{10}$$

$$= -5.1825$$

$$R' = \sqrt{\frac{X^2 + Y^2}{n}} = \sqrt{\frac{(-0.0106)^2 + (-5.1825)^2}{10}} = \sqrt{2.685842} = 1.639$$

$R'_{0.05,10} = 1.048.$

Reject H_0.

$$P < 0.001$$

coefficient*; it is[†]

$$r_{aa} = \frac{\displaystyle\sum_{i=1}^{n-1}\sum_{j=i+1}^{n} \sin(a_i - a_j)\sin(b_i - b_j)}{\sqrt{\displaystyle\sum_{i=1}^{n-1}\sum_{j=i+1}^{n} \sin^2(a_i - a_j) \sum_{i=1}^{n-1}\sum_{j=i+1}^{n} \sin^2(b_i - b_j)}}, \tag{43}$$

where the ith pair of data is denoted as a_i, b_i[‡].

Upton and Fingleton (1989: 303) gave a relatively simple method to test the significance of r_{aa} — that is, to test whether the sample of data came from a population having a correlation coefficient, ρ_{aa}, different from zero. The procedure involves computing r_{aa} an additional n times for the sample, each time eliminating a different one of the n pairs of a and b data.[§]

Then a mean and variance of these n additional r_{aa}'s is calculated (let's call the mean $\bar{r}_{aa}$, and the variance $s^2_{r_{aa}}$); and confidence limits for ρ_{aa} are obtained as

$$L_1 = n r_{aa} - (n-1)\bar{r}_{aa} - Z_{\alpha(2)}\sqrt{\frac{s^2_{r_{aa}}}{n}} \tag{45}$$

and

$$L_2 = n r_{aa} - (n-1)\bar{r}_{aa} + Z_{\alpha(2)}\sqrt{\frac{s^2_{r_{aa}}}{n}}. \tag{46}$$

If the confidence interval (i.e., the interval between L_1 and L_2) does *not* include zero, then H_0: $\rho_{aa} = 0$ is rejected in favor of H_A: $\rho_{aa} \neq 0$. The computation of r_{aa}, and testing its significance, is shown in Example 20.

---

*Results may be obtained by

$$r = \frac{\displaystyle\sum_{i=1}^{n-1}\sum_{j=i+1}^{n} (X_i - X_j)(Y_i - Y_j)}{\sqrt{\displaystyle\sum_{i=1}^{n-1}\sum_{j=i+1}^{n} (X_i - X_j)^2 \sum_{i=1}^{n-1}\sum_{j=i+1}^{n} (Y_i - Y_j)^2}}. \tag{42}$$

[†]The notation "$\sin^2(a_i - a_j)$" means "$[\sin(a_i - a_j)]^2$."
[‡]Fisher (1993: 151) gives an alternate computation of r_{aa} as

$$\frac{4\left[\left(\displaystyle\sum_{i=1}^{n} \cos a_i \cos b_i\right)\left(\displaystyle\sum_{i=1}^{n} \sin a_i \sin b_i\right) - \left(\displaystyle\sum_{i=1}^{n} \cos a_i \sin b_i\right)\left(\displaystyle\sum_{i=1}^{n} \sin a_i \cos b_i\right)\right]}{\sqrt{\left[n^2 - \left(\displaystyle\sum_{i=1}^{n} \cos(2a_i)\right)^2 - \left(\displaystyle\sum_{i=1}^{n} \sin(2a_i)\right)^2\right]\left[n^2 - \left(\displaystyle\sum_{i=1}^{n} \cos(2b_i)\right)^2 - \left(\displaystyle\sum_{i=1}^{n} \sin(2b_i)\right)^2\right]}}. \tag{44}$$

[§]This involves what statisticians call the *jackknife* technique (introduced by Quenouille, 1956), named in 1964 by R. G. Miller (David, 1995).

EXAMPLE 20   Angular-Angular Correlation

We wish to assess the relationship between the orientation of insects and the direction of a light source.

H_0: $\rho_{aa} = 0$; H_A: $\rho_{aa} \neq 0$

i	Insect a_i	Light b_i
1	145°	120°
2	190°	180°
3	310°	330°
4	210°	225°
5	80°	55°

$n = 5$; the computations proceed as follows:

i	j	$a_i - a_j$	$b_i - b_j$	$\sin(a_i - a_j)$	$\sin(b_i - b_j)$	$\sin(a_i - a_j) \times \sin(b_i - b_j)$	$\sin^2(a_i - a_j)$	$\sin^2(b_i - b_j)$
1	2	−45°	−60°	−0.70711	−0.86603	0.61237	0.50000	0.75001
1	3	−165°	−210°	−0.25882	0.50000	−0.12941	0.06699	0.25000
1	4	−65°	−105°	−0.90631	−0.96593	0.87543	0.82140	0.93302
1	5	65°	65°	0.90631	0.90631	0.82140	0.82140	0.82140
2	3	−120°	−150°	−0.86603	−0.50000	0.43302	0.75001	0.25000
2	4	−20°	−45°	−0.34202	−0.70711	0.24185	0.11698	0.50000
2	5	110°	125°	0.93969	0.81915	0.76975	0.88302	0.67101
3	4	100°	105°	0.98481	0.96593	0.95126	0.96985	0.93302
3	5	230°	275°	−0.76604	−0.99619	0.76312	0.58682	0.99239
4	5	130°	170°	0.76604	0.17365	0.13302	0.58682	0.03015
					Sum:	5.47181	6.10329	6.13100

$$r_{aa} = \frac{\sum_{i=1}^{n-1} \sum_{j=i+1}^{n} \sin(a_i - a_j) \sin(b_i - b_j)}{\sqrt{\sum_{i=1}^{n-1} \sum_{j=i+1}^{n} \sin^2(a_i - a_j) \sum_{i=1}^{n-1} \sum_{j=i+1}^{n} \sin^2(b_i - b_j)}}$$

$$= \frac{5.47181}{\sqrt{(6.10329)(6.13100)}} = \frac{5.47181}{\sqrt{37.41927}} = \frac{5.47181}{6.11713} = 0.8945$$

Five r_{aa}'s computed for the above data, each with a different pair of data deleted:

i deleted:	1	2	3	4	5
r_{aa}	0.90793	0.87419	0.92905	0.89084	0.87393

$\bar{r}_{aa} = 0.89519$; $s^2_{r_{aa}} = 0.0005552$

$nr_{aa} - (n-1)\bar{r}_{aa} = (5)(0.8945) - (5-1)(0.89519) = 0.8917$

$$Z_{0.05(2)}\sqrt{\frac{s_{r_{aa}}^2}{n}} = 1.9600\sqrt{\frac{0.0005552}{5}} = 1.9600(0.0105) = 0.0206$$

$$L_1 = 0.8917 - 0.0206 = 0.8711$$

$$L_2 = 0.8917 + 0.0206 = 0.9123.$$

As this confidence interval does not encompass zero, reject H_0.

(b) Angular-Linear Correlation. Among the procedures proposed for correlating an angular variable (a) with a linear variable (X) is one by Mardia (1976). Using Equation 46a,

$$r = \frac{\sum xy}{\sqrt{\sum x^2 \sum y^2}} \tag{46a}$$

determine coefficients for the correlation between X and the sine of a (call it r_{XS}), the correlation between X and the cosine of a (call it r_{XC}), and the correlation between the cosine and the sine of a (call it r_{CS}). Then, the angular-linear correlation coefficient is

$$r_{al} = \sqrt{\frac{r_{XC}^2 + r_{XS}^2 - 2r_{XC}r_{XS}r_{CS}}{1 - r_{CS}^2}}. \tag{47}$$

For angular-linear correlation, the correlation coefficient lies between 0 and 1 (i.e., there is no negative correlation). If n is large, then the significance of the correlation coefficient may be assessed by comparing nr_{al}^2 to χ_2^2 (Batschelet, 1981: 193). This procedure is shown in Example 21. It is not known how large n must be for the chi-square approximation to give good results, and Fisher (1993: 145) recommends a different (laborious) method for assessing significance of the test statistic.

EXAMPLE 21   Angular-Linear Correlation

For a sampled population of animals, we wish to examine the relationship between distance traveled and direction traveled.

$$H_0: \rho_{al} = 0; \quad H_A: \rho_{al} \neq 0$$

$$\alpha = 0.05$$

i	X distance (km)	a_i direction (deg)	$\sin a_i$	$\cos a_i$
1	48	190	−0.17364	−0.98481
2	55	160	0.34202	−0.93969
3	26	210	−0.50000	−0.86603
4	23	225	−0.70711	−0.70711
5	22	220	−0.64279	−0.76604
6	62	140	0.64279	−0.76604
7	64	120	0.86603	−0.50000

$n = 7$

$$\sum X_i = 300 \text{ kilometers}; \quad \sum X_i^2 = 14{,}958 \text{ km}^2$$

$$\sum \sin a_i = -0.17270 \text{ degrees}; \quad \sum \sin^2 a_i = 2.47350 \text{ deg}^2$$

$$\sum \cos a_i = -5.52972 \text{ degrees}; \quad \sum \cos^2 a_i = 4.52652 \text{ deg}^2$$

"Sum of squares" of $X = \sum x^2 = 14{,}958 - 300^2/7 = 2100.86 \text{ km}^2$

"Sum of squares" of $\sin a_i = 2.47350 - (-0.17270)^2/7 = 2.46924 \text{ deg}^2$

"Sum of squares" of $\cos a_i = 4.52652 - (-5.52972)^2/7 = 0.15826 \text{ deg}^2$

"Sum of cross products" of X and $\cos a_i$ = $(48)(-0.98481) + (55)$ $(-0.93969) + \cdots + (64)(-0.50000) - (300)(-5.52972)/7 = -234.08150$ $- (-236.98800) = 2.90650 \text{ deg-km}$

"Sum of cross products" of X and $\sin a_i = (48)(-0.17364) + (55)(0.34202)$ $+ \cdots + (64)(0.86603) - (300)(-0.17270)/7 = 62.35037 - (-7.40143)$ $= 69.75180 \text{ deg-km}$

"Sum of cross products" of $\cos a_i$ and $\sin a_i = (-0.98481)(-0.17364)$ $+ (-0.93969)(0.34202) + \cdots + (-0.50000)(0.86603) - (-5.52972)$ $(-0.17270)/7 = 0.34961 - 0.13643 = -0.21318 \text{ deg-km}$

$r_{XC} = 0.15940; \quad r_{XC}^2 = 0.02541$

$r_{XS} = 0.96845; \quad r_{XS}^2 = 0.93789$

$r_{CS} = 0.34104; \quad r_{CS}^2 = 0.11630$

$$r_{al}^2 = \frac{r_{XC}^2 + r_{XS}^2 - 2r_{XC}r_{XS}r_{CS}}{1 - r_{CS}^2}$$

$$= \frac{0.02541 + 0.93789 - 2(0.15940)(0.96845)(0.34104)}{1 - 0.11630}$$

$$= \frac{0.85801}{0.88370} = 0.97093$$

$r_{al} = \sqrt{0.97093} = 0.9854$

$nr_{al}^2 = (7)(0.97093) = 6.797.$

$\chi_{0.05,2}^2 = 5.991$; reject H_0; $\quad 0.025 < P < 0.05$

(c) Regression. Linear-circular regression, in which the dependent variable (Y) is linear and the independent variable (a) circular, may be analyzed as

$$Y_i = b_0 + b_1 \cos a_i + b_2 \sin a_i \tag{48}$$

(Fisher, 1993: 139–140), where b_0 is the Y-intercept and b_1 and b_2 are partial regression coefficients.

In circular-linear regression, where a is the dependent variable and Y the independent variable, the situation is rather more complicated and is discussed by Fisher (1993: 155–168). Regression where both the dependent and independent variables are on a circular scale (angular-angular regression), or where there is a circular dependent variable and both circular and linear independent variables, has received little attention in the statistical literature (Fisher, 1993: 168; Lund, 1999).

(d) Rhytomometry. The description of biological rhythms may be thought of as a regression (often called *periodic regression*) of the linear variable on time (a circular variable). Excellent discussions of such regression are provided by Batschelet (1972; 1974; 1981: Chapter 8), Bliss (1970: Chapter 17), Bloomfield (1976), and Nelson et al. (1979).

The *period*, or length, of the cycle* is often stated in advance. Parameters to be estimated in the regression are the *amplitude* of the rhythm (which is the range from the minimum to the maximum value of the linear variable)[†] and the *phase angle*, or *acrophase*, of the cycle (which is the point on the circular time scale at which the linear variable is maximum). If the period is also a parameter to be estimated, then the situation is more complex and one may resort to the broad area of *time series analysis* (e.g., Fisher, 1993: 172–189). Some biological rhythms can be fitted, by least-squares regression, by a sine (or cosine) curve; and if the rhythm does not conform well to such a symmetrical functional relationship, then a "harmonic analysis" (also called a "Fourier analysis") may be employed.

16 NONPARAMETRIC ANGULAR CORRELATION

(a) Angular-Angular Correlation. A nonparametric correlation procedure proposed by Mardia (1975) employs the ranks of circular measurements as follows. If pair i of circular data is denoted by measurements a_i and b_i, then these two statistics are computed:

$$r' = \frac{\left\{\sum_{i=1}^{n} \cos[C(\text{rank of } a_i - \text{rank of } b_i)]\right\}^2 + \left\{\sum_{i=1}^{n} \sin[C(\text{rank of } a_i - \text{rank of } b_i)]\right\}^2}{n^2} \tag{49}$$

$$r'' = \frac{\left\{\sum_{i=1}^{n} \cos[C(\text{rank of } a_i + \text{rank of } b_i)]\right\}^2 + \left\{\sum_{i=1}^{n} \sin[C(\text{rank of } a_i + \text{rank of } b_i)]\right\}^2}{n^2}, \tag{50}$$

where

$$C = \frac{360°}{n}; \tag{51}$$

and Fisher and Lee (1982) showed that

$$(r_{aa})_s = r' - r'' \tag{52}$$

---

*A rhythm with one cycle every twenty-four hours is said to be "circadian" (from the Latin *circa*, meaning "about" and *diem*, meaning "day"); a rhythm with a seven-day period is said to be "circaseptan"; a rhythm with a fourteen-day period is "circadiseptan"; one with a period of one year is "circannual" (Halberg and Lee, 1974).

[†]The amplitude is often defined as half this range.

is, for circular data, analogous to the Spearman rank correlation coefficient. For n of 8 or more, we may calculate

$$(n - 1)(r_{aa})_s$$

and compare it to the critical value of

$$A + B/n,$$

using A and B from Table 1 (which yields excellent approximations to the values given by Fisher and Lee, 1982). This procedure is demonstrated in Example 22.

TABLE 1: Constants A and B for Critical Values for Nonparametric Angular-Angular Correlation

$\alpha(2)$:	0.20	0.10	0.05	0.02	0.01
$\alpha(1)$:	0.10	0.05	0.025	0.01	0.005
A:	1.61	2.30	2.99	3.91	4.60
B:	1.52	2.00	2.16	1.60	1.60

EXAMPLE 22   Nonparametric Angular-Angular Correlation

For a population of birds sampled, we wish to correlate the direction toward which they attempt to fly in the morning with that in the evening.

H_0: $(\rho_{aa})_s = 0$; H_A: $(\rho_{aa})_s \neq 0$

$\alpha = 0.05$

Bird i	Direction Evening a_i	Morning b_i	Rank of a_i	Rank of b_i	Rank difference	Rank sum
1	30°	60°	4.5	5	−0.5	9.5
2	10°	50°	2	4	−2	6
3	350°	10°	8	2	6	10
4	0°	350°	1	8	−7	9
5	340°	330°	7	7	0	14
6	330°	0°	6	1	5	7
7	20°	40°	3	3	0	6
8	30°	70°	4.5	6	−1.5	10.5

$n = 8$; $C = 360°/n = 45°$

$$r' = \frac{\left\{\sum_{i=1}^{n}\cos[C(\text{rank of } a_i - \text{rank of } b_i)]\right\}^2 + \left\{\sum_{i=1}^{n}\sin[C(\text{rank of } a_i - \text{rank of } b_i)]\right\}^2}{n^2}$$

$$= \Big(\{\cos[45°(-0.5)] + \cos[45°(-2)] + \cdots + \cos[45°(-1.5)]\}^2$$

$$+\{\sin[45°(-0.5)] + \sin[45°(-1)] + \cdots + \sin[45°(-1.5)]\}^2\Big)/8^2$$

$$= 0.3654$$

$$r'' = \frac{\left\{\sum_{i=1}^{n}\cos[C(\text{rank of }a_i+\text{rank of }b_i)]\right\}^2 + \left\{\sum_{i=1}^{n}\sin[C(\text{rank of }a_i+\text{rank of }b_i)]\right\}^2}{n^2}$$

$$= \Big(\{\cos[45°(9.5)] + \cos[45°(6)] + \cdot\cdot + \cos[45°(10.5)]\}^2$$

$$+\{\sin[45°(9.5)] + \sin[45°(6)] + \cdots + \sin[45°(10.5)]\}^2\Big) \big/ 8^2$$

$$= 0.0316$$

$$(r_{aa})_s = r' - r'' = 0.3654 - 0.0316 = 0.3338$$

$$(n - 1)(r_{aa})_s = (8 - 1)(0.3338) = 2.34$$

For $\alpha(2) = 0.05$ and $n = 8$, the critical value is estimated to be

$$A + B/n = 2.99 + 2.16/8 = 3.26.$$

As $2.34 < 3.26$, do not reject H_0.

We may also compute critical values for other significance levels:

for $\alpha(2) = 0.20 : 1.61 + 1.52/8 = 1.80$;

for $\alpha(2) = 0.10 : 2.30 + 2.00/8 = 2.55$;

therefore, $0.10 < P < 0.20$.

Fisher and Lee (1982) also described a nonparametric angular-angular correlation that is analogous to the Kendall rank correlation (see also Upton and Fingleton, 1989).

(b) Angular-Linear Correlation. Mardia (1976) presented a ranking procedure for correlation between a circular and a linear variable, which is analogous to the Spearman rank correlation. (See also Fisher, 1993: 140–141; Mardia and Jupp, 2000: 246–248.)

17 GOODNESS-OF-FIT TESTING FOR CIRCULAR DISTRIBUTIONS

Either χ^2 or G may be used to test the goodness of fit of a theoretical to an observed circular frequency distribution. The procedure is to determine each expected frequency, $\hat{f}_i$, corresponding to each observed frequency, f_i, in each category, i. For the data of Example 3, for instance, we might hypothesize a uniform distribution of data among the 12 divisions of the data. The test of this hypothesis is presented in Example 23. Batschelet (1981: 72) recommends grouping the data so that no expected frequency is less than 4 in using chi-square. All of the k categories do not have to be the same size. If they are (as in Example 23, where each category is 30° wide), Fisher (1993: 67) recommends that n/k be at least 2.

EXAMPLE 23   Chi-Square Goodness of Fit for the Circular Data of Example 3

H_0: The data in the population are distributed uniformly around the circle.

H_A: The data in the population are not distributed uniformly around the circle.

a_i (deg)	f_i	$\hat{f}_i$
0–30	0	8.7500
30–60	6	8.7500
60–90	9	8.7500
90–120	13	8.7500
120–150	15	8.7500
150–180	22	8.7500
180–210	17	8.7500
210–240	12	8.7500
240–270	8	8.7500
270–300	3	8.7500
300–330	0	8.7500
330–360	0	8.7500

$k = 12 \qquad n = 105$

$\hat{f}_i = 105/12 = 8.7500$ for all i

$$\chi^2 = \frac{(0 \quad 8.7500)^2}{8.7500} + \frac{(6 \quad 8.7500)^2}{8.7500}$$
$$+ \frac{(9 - 8.7500)^2}{8.7500} + \cdots + \frac{(0 - 8.7500)^2}{8.7500}$$
$$= 8.7500 + 0.8643 + 0.0071 + \cdots + 8.7500$$
$$= 66.543$$

$\nu = k - 1 = 11$

$\chi^2_{0.05,11} = 19.675$

Reject H_0. $P \ll 0.001$ $[P = 0.00000000055]$

Recall that goodness-of-fit testing by the chi-square or G statistic does not take into account the sequence of categories that occurs in the data distribution. The Kolmogorov-Smirnov test was introduced in preference to chi-square when the categories of data are, in fact, ordered. Unfortunately, the Kolmogorov-Smirnov test yields different results for different starting points on a circular scale; however, a modification of this test by Kuiper (1960) provides a goodness-of-fit test, the results of which are unrelated to the starting point on a circle.

If data are not grouped, the Kuiper test is preferred to the chi-square procedure. It is discussed by Batschelet (1965: 26–27; 1981: 76–79), Fisher (1993: 66–67), Mardia (1972a: 173–180), Mardia and Jupp (2000: 99–103). Among others, another goodness-of-fit test applicable to circular distributions of ungrouped data (Watson, 1961, 1962, 1995), is often referred to as the *Watson one-sample U^2 test*, which is demonstrated in Example 24. To test the null hypothesis of uniformity, first transform each angular measurement, a_i, by dividing it by 360°:

$$u_i = \frac{a_i}{360°}.$$ (53)

Then the following quantities are obtained for the set of n values of u_i : $\sum u_i, \sum u_i^2, \bar{u}$, and $\sum i u_i$. The test statistic, called "Watson's U^2," is

$$U^2 = \sum u_i^2 - \frac{\left(\sum u_i\right)^2}{n} - \frac{2}{n} \sum i u_i + (n+1)\,\bar{u} + \frac{n}{12}$$ (54)

(Mardia, 1972a: 182; Mardia and Jupp, 2000: 103–105). Critical values for this test are $U^2_{\alpha,n}$ in Table 38b from *Appendix: Statistical Tables and Graphs*. Kuiper's test and Watson's test appear to be very similar in power (Stephens, 1969b; Mardia and Jupp, 2000: 115). Lockhart and Stephens (1985) discussed the use of Watson's U^2 for goodness of fit to the von Mises distribution and provided tables for that application.

EXAMPLE 24   Watson's Goodness-of-Fit Testing Using the Data of Example

H_0: The sample data come from a population uniformly distributed around the circle.

H_A: The sample data do not come from a population uniformly distributed around the circle.

i	a_i	u_i	u_i^2	iu_i
1	45°	0.1250	0.0156	0.1250
2	55°	0.1528	0.0233	0.3056
3	81°	0.2250	0.0506	0.6750
4	96°	0.2667	0.0711	1.0668
5	110°	0.3056	0.0934	1.5280
6	117°	0.3250	0.1056	1.9500
7	132°	0.3667	0.1345	2.5669
8	154°	0.4278	0.1830	3.4224

$n = 8$ $\quad \sum u_i = 2.1946 \quad \sum u_i^2 = 0.6771 \quad \sum i u_i = 11.6397$

$$\bar{u} = \frac{\sum u_i}{n} = \frac{2.1946}{8} = 0.2743$$

$$U^2 = \sum u_i^2 - \frac{\left(\sum u_i\right)^2}{n} - \frac{2}{n}\sum iu_i + (n + 1)\bar{u} + \frac{n}{12}$$

$$= 0.6771 - \frac{(2.1946)^2}{8} - \frac{2}{8}(11.6397) + (8 + 1)(0.2743) + \frac{8}{12}$$

$$= 0.6771 - 0.6020 - 2.9099 + 2.4687 + 0.6667$$

$$= 0.3006$$

$$U^2_{0.05,7} = 0.179$$

Therefore, reject H_0. $0.002 < P < 0.005$.

18 SERIAL RANDOMNESS OF NOMINAL-SCALE CATEGORIES ON A CIRCLE

When dealing with the occurrence of members of two nominal-scale categories along a linear space or time, the runs test is appropriate. A runs test is also available for spatial or temporal measurements that are on a circular scale. This test may also be employed as a two-sample test, but the tests of Sections 4 and 5 are more powerful for that purpose; the circular runs test is best reserved for testing the hypothesis of random distribution of members of two categories around a circle.

We define a run on a circle as a sequence of like elements, bounded on each side by unlike elements. We let n_1 be the total number of elements in the first category, n_2 the number of elements in the second category, and u the number of runs in the entire sequence of elements. For the runs test on a linear scale, the number of runs may be even or odd; however, on a circle the number of runs is always even: half of the runs (i.e., $u/2$) consist of elements belonging to one of the categories, and there are also $u/2$ runs of elements of the other category.

The null hypothesis may be tested by analysis of the following 2×2 contingency table (Stevens, 1939), where $u' = u/2$:

u'	$n_1 - u'$	n_1
$n_2 - u'$	$u' - 1$	$n_2 - 1$
n_2	$n_1 - 1$	$n_1 + n_2 - 1$

This should be done by the Fisher exact test, as demonstrated in Example 25. For that test, m_1, m_2, n, and f are as defined at the end of Table 28 from *Appendix: Statistical Tables and Graphs*. For two-tailed testing, as in Example 25, the second pair of critical values in that table are used.

Circular Distributions: Hypothesis Testing

EXAMPLE 25   The Two-Tailed Runs Test on a Circle

Members of two antelope species (referred to as species A and B) are observed drinking on the shore of a pond in the following sequence:

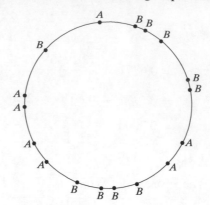

H_0: The distribution of members of the two species around the pond is random.

H_A: The distribution of members of the two species around the pond is not random.

$n_1 = 7, n_2 = 10, u = 6, u' = 3$

3	4	7
7	2	9
10	6	16

To use Table 28 from *Appendix: Statistical Tables and Graphs*, $m_1 = 6$, $m_2 = 7$, $f = 4$, $n = 17$. For a two-tailed test, the critical values of f for $\alpha = 0.05$ are 0 and 5. Therefore, we do not reject H_0; $P \geq 0.20$.

If one or both sample sizes exceed those in Table 28 from *Appendix: Statistical Tables and Graphs*, then this 2×2 contingency table may be subjected to analysis by chi-square, but a correction for continuity should be used. Ghent and Zar (1992) discuss normal approximations for circular runs testing.

Although this discussion and Example 25 depict a distribution around a circle, the testing procedure is appropriate if the physical arrangement of observations is in the shape of an ellipse, a rectangle, or any other closed figure — however irregular — providing that the figure is everywhere wider than the spacing of the elements along its periphery; and it may also be used for data that are conceptually circular, such as clock times or compass directions.

(a) One-Tailed Testing.   For one-tailed testing we use the first pair of critical values in Table 28 from *Appendix: Statistical Tables and Graphs*. We can test specifically whether the population is nonrandom due to clustering (also known as contagion) in the following manner. We state H_0: In the population the members of each of the two groups are not clustered (i.e., not distributed contagiously) around the circle and

H_A: In the population the members of each of the two groups are clustered (i.e., distributed contagiously) around the circle; and if $f \leq$ the first member of the pair of one-tailed critical values, then H_0 is rejected. In using a normal approximation, this H_0 is rejected if $Z \geq Z_{\alpha(1)}$ and $u' \leq \mu_{u'}$.

If our interest is specifically whether the population distribution is nonrandom owing to a tendency toward being uniform, then we state H_0: In the population the members of each of the two groups do not tend to be uniformly distributed around the circle versus H_A: In the population the members of each of the two groups tend to be uniformly distributed around the circle; and if $f \geq$ the second member of the one-tailed pair of critical values then H_0 is rejected. If a normal approximation is employed, this H_0 is rejected if $Z \geq Z_{\alpha(1)}$ and $u' \geq \mu_{u'}$.

EXERCISES

1. Consider the data below. Test the null hypothesis that the population is distributed uniformly around the circle (i.e., $\rho = 0$).

   Twelve nests of a particular bird species were recorded on branches extending in the following directions from the trunks of trees:

Direction		Frequency
N:	0°	2
NE:	45°	4
E:	90°	3
SE:	135°	1
S:	180°	1
SW:	225°	1
W:	270°	0
NW:	315°	0

   (a) Compute the sample mean direction.
   (b) Compute the angular deviation for the data.
   (c) Determine 95% confidence limits for the population mean.
   (d) Determine the sample median direction.

   Consider the data below. Test the null hypothesis that time of birth is distributed uniformly around the clock (i.e., $\rho = 0$).

2. A total of 15 human births occurred as follows:

1:15 A.M.	4:40 A.M.	5:30 A.M.	6:50 A.M.
2:00 A.M.	11:00 A.M.	4:20 A.M.	5:10 A.M.
4:30 A.M.	5:15 A.M.	10:30 A.M.	8:55 A.M.
6:10 A.M.	2:45 A.M.	3:10 A.M.	

   (a) Compute the mean time of birth.
   (b) Compute the angular deviation for the data.
   (c) Determine 95% confidence limits for the population mean time.
   (d) Determine the sample median time.

3. Trees are planted in a circle to surround a cabin and protect it from prevailing west (i.e., 270°) winds. The trees suffering the greatest wind damage are the eleven at the following directions.

   (a) Using the V test, test the null hypothesis that tree damage is independent of wind direction, versus the alternate hypothesis that tree damage is concentrated around 270°.

   (b) Test H_0: $\mu_a = 270°$ vs. H_A: $\mu_a \neq 270°$.

285°	295°	335°
240	275	260
280	310	300
255	260	

4. Test nonparametrically for uniformity, using the data of Exercise 1.

5. Test nonparametrically for the data and experimental situation of Exercise 3.

6. The direction of the spring flight of a certain bird species was recorded as follows in eight individuals released in full sunlight and seven individuals released under overcast skies:

Sunny	Overcast
350°	340°
340	305
315	255
10	270
20	305
355	320
345	335
360	

   Using the Watson-Williams test, test the null hypothesis that the mean flight direction in this species is the same under both cloudy and sunny skies.

7. Using the data of Exercise 6, test nonparametrically the hypothesis that birds of the species under consideration fly in the same direction under sunny as well as under cloudy skies.

8. Times of arrival at a feeding station of members of three species of hummingbirds were recorded as follows:

Species 1	Species 2	Species 3
05:40 hr	05:30 hr	05:35 hr
07:15	07:20	08:10
09:00	09:00	08:15
11:20	09:40	10:15
15:10	11:20	14:20
17:25	15:00	15:35
	17:05	16:05
	17:20	
	17:40	

Test the null hypothesis that members of all three species have the same mean time of visiting the feeding station.

9. For the data in Exercise 6, the birds were released at a site from which their home lies due north (i.e., in a compass direction of 0°). Test whether birds orient homeward better under sunny skies than under cloudy skies.

10. For the data in Exercise 6, test whether the variability in flight direction is the same under both sky conditions.

11. The following data, for each of nine experimental animals, are the time of day when body temperature is greatest and the time of day when heart rate is greatest.

(a) Determine and test the correlation of these two times of day.

(b) Perform nonparametric correlation analysis on these data.

Animal	Time of day	
i	Body temperature a_i	Heart rate b_i
1	09:50	10:40
2	10:20	09:30
3	11:40	11:10
4	08:40	08:30
5	09:10	08:40
6	10:50	09:10
7	13:20	12:50
8	13:10	13:30
9	12:40	13:00

12. For a sample of nine human births, the following are the times of day of the births and the ages of the mothers. Test whether there is correlation between these two variables.

Birth i	Age (yr) X_i	Time of day (hr:min) a_i
1	23	06:25
2	22	07:20
3	19	07:05
4	25	08:15
5	28	15:40
6	24	09:25
7	31	18:20
8	17	07:30
9	27	16:10

13. Eight men (M) and eight women (W) were asked to sit around a circular conference table; they did so in the following configuration (see figure). Test whether there is evidence that members of the same sex tend to sit next to each other.

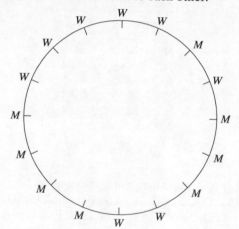

ANSWERS TO EXERCISES

1. H_0: $\rho = 0$; H_A: $\rho \neq 0$; $r = 0.526$; $R = 6.309$; $z = 3.317$, $z_{0.05,12} = 2.932$; reject H_0; $0.02 < P < 0.05$.

2. H_0: $\rho = 0$, H_A: $\rho \neq 0$; $r = 0.774$; $R = 11.603$; $z = 8.975$, $z_{0.05,15} = 2.945$; reject H_0; $P < 0.001$.

3. H_0: $\rho = 0$; H_A: $\rho \neq 0$; $n = 11$, $Y = -0.88268$, $X = 0.17138$, $r = 0.89917$, $\bar{a} = 281°$, $R = 9.891$, $\mu_0 = 270°$. **(a)** $V = 9.709$, $u = 4.140$, $u_{0.05,11} = 1.648$; reject H_0; $P < 0.0005$. **(b)** H_0: $\mu_a = 270°$, H_A: $\mu_a \neq 270°$, 95% confidence interval for $\mu_a = 281° \pm 19°$, so do not reject H_0.

4. $n = 12$, $m = 2$, $m_{0.05,12} = 0$, do not reject H_0; $0.20 < P \leq 0.50$.

5. $n = 11$, $m' = 0$, $C = 11$, $C_{0.05(2),11} = 1$, reject H_0; $P < 0.001$.

6. H_0: Mean flight direction is the same under the two sky conditions; H_A: Mean flight direction is not the same under the two sky conditions; $n_1 = 8$, $n_2 = 7$, $R_1 = 7.5916$, $R_2 = 6.1130$, $\bar{a}_1 = 352°$, $\bar{a}_2 = 305°$, $N = 15$, $r_w = 0.914$, $R = 12.5774$; $F = 12.01$, $F_{0.05(1),1,13} = 4.67$; reject H_0; $0.0025 < P < 0.005$ $[P = 0.004]$.

7. H_0: The flight direction is the same under the two sky conditions; H_A: The flight direction is not the same under the two sky conditions; $n_1 = 8$, $n_2 = 7$, $N = 15$; $\sum d_k = -2.96429$. $\sum d_k^2 = 1.40243$, $U^2 = 0.2032$, $U_{0.05,8,7}^2 = 0.1817$; do not reject H_0; $0.02 < P < 0.05$.

8. H_0: Members of all three hummingbird species have the same mean time of feeding at the feeding station; H_A: Members of all three species do not have the same mean time of feeding at the feeding station; $n_1 = 6$, $n_2 = 9$, $n_3 = 7$, $N = 22$; $R_1 = 2.965$, $R_2 = 3.938$, $R_3 = 3.868$; $\bar{a}_1 = 10{:}30$ hr, $\bar{a}_2 = 11{:}45$ hr, $\bar{a}_3 = 11{:}10$ hr; $r_w = 0.490$, $F =$

0.206, $F_{0.05(1),2,19} = 3.54$; do not reject H_0; $P > 0.25$ $[P = 0.82]$. Therefore, all three $\bar{a}_i$'s estimate the same μ_a, the best estimate of which is 11:25 hr.

9. H_0: Birds do not orient better when skies are sunny than when cloudy; H_A: Birds do orient better when skies are sunny than when cloudy. Angular distances for group 1 (sunny): 10, 20, 45, 10, 20, 5, 15, and 0°; for group 2 (cloudy): 20, 55, 105, 90, 55, 40, and 25°. For the one-tailed Mann-Whitney test: $n_1 = 8$, $n_2 = 7$, $R_1 = 40$, $U = 52$, $U_{0.05(1),8,7} = 43$; reject H_0; $P = 0.0025$.

10. H_0: Variability in flight direction is the same under both sky conditions; H_A: Variability in flight direction is not the same under both sky conditions; $\bar{a}_1 = 352°$, $\bar{a}_2 = 305°$; angular distances for group 1 (sunny): 2, 12, 37, 18, 28, 3, 7, and 8°, and for group 2 (cloudy): 35, 0, 50, 35, 0, 15, and 30°; for the two-tailed Mann-Whitney test: $R_1 = 58$, $U = 34$, $U' = 22$, $U_{0.05(2),8,7} = 46$; reject H_0; $P < 0.001$.

11. **(a)** H_0: $\rho_{aa} = 0$; H_A: $\rho_{aa} \neq 0$; $r_{aa} = 0.9244$; $\bar{r}_{aa} = 0.9236$; $s_{r_{aa}}^2 = 0.0004312$; $L_1 = 0.9169$, $L_2 = 0.9440$; reject H_0. **(b)** H_0: $(\rho_{aa})_s = 0$, H_A: $(\rho_{aa}) \neq 0$; $r' = 0.453$, $r'' = 0.009$, $(r_{aa})_s = 0.365$; $(n - 1)(r_{aa})_s = 2.92$, for $\alpha(2) = 0.05$ the critical value is 3.23; do not reject H_0; for $\alpha(2) = 0.10$, the critical value is 2.52, so $0.05 < P < 0.10$.

12. H_0: $\rho_{al} = 0$; H_A: $\rho_{al} \neq 0$; $r_{al} = 0.833$, $nr_{al}^2 = 6.24$, $\chi_{0.05,2}^2 = 5.991$, reject H_0.

13. H_0: The distribution is not contagious; H_A: The distribution is contagious; $n_1 = 8$, $n_2 = 8$, $u = 6$, $u' = 3$; using Table 28 from *Appendix: Statistical Tables and Graphs*: $m_1 = 7$, $m_2 = 7$, $f = 2$, $n = 15$, critical values are 1 and 6, so do not reject H_0; $P \geq 0.50$.

This page intentionally left blank

Literature Cited

Literature Cited

ACTON, F. S. 1966. *Analysis of Straight Line Data.* Dover, New York. 267 pp.

ADRIAN, Y. E. O. 2006. *The Pleasures of Pi,e and Other Interesting Numbers.* World Scientific Publishing, Singapore. 239 pp.

AGRESTI, A. 1992. A survey of exact inference for contingency tables. *Statist. Sci.* 7: 131–153. *Also* Comment, by E. J. Bedrick and J. R. Hill, *ibid.* 7: 153–157; D. E. Duffy, *ibid.* 7: 157–160; L. D. Epstein and S. E. Fienberg, *ibid.* 7: 160–163; S. Kreiner, Exact inference in multidimensional tables. *ibid.* 7: 163–165; D. Y. Lin and L. J. Wei, *ibid.* 7: 166–167; C. R. Mehta, An interdisciplinary approach to exact inference in contingency tables. *ibid.* 167–170; S. Suissa, *ibid.* 7: 170–172; Rejoinder, by A. Agresti, *ibid.* 7: 173–177.

AGRESTI, A. 2002. *Categorical Data Analysis.* Wiley-Interscience, New York. 710 pp.

AGRESTI, A. 2007. *An Introduction to Categorical Data Analysis.* 2nd ed. Wiley-Interscience, New York. 372 pp.

AGRESTI, A. and B. CAFFO. 2000. Simple and effective confidence intervals for proportions and differences of proportions result from adding two successes and two failures. *Amer. Statist.* 54: 280–288.

AGRESTI, A. and B. A. COULL. 1998. Approximate is better than "exact" for interval estimation of binomial proportions. *Amer. Statist.* 52: 119–126.

AGRESTI, A. and M.-C. YANG. 1987. An empirical investigation of some effects of sparseness in contingency tables. *Computa. Statist. Data Anal.* 5: 9–21.

AIKEN, L. S. and S. G. WEST. 1991. *Multiple Regression: Testing and Interpreting Interactions.* Sage Publications, Newbury Park, CA. 212 pp.

AJNE, B. 1968. A simple test for uniformity of a circular distribution. *Biometrika* 55: 343–354.

AKRITAS, M. G. 1990. The rank transform method in some two-factor designs. *J. Amer. Statist. Assoc.* 85: 73–78.

ALGINA, J. and S. SEAMAN. 1984. Calculation of semipartial correlations. *Educ. Psychol. Meas.* 44: 547–549.

ANDERSON, N. H. 1961. Scales and statistics: parametric and non-parametric. *Psychol. Bull.* 58: 305–316.

ANDRÉ, D. 1883. Sur le nombre de permutations de *n* éléments qui présentent *S* séquences. *Compt. Rend. (Paris)* 97: 1356–1358. [Cited in Bradley, 1968: 281.]

ANDREWS, F. C. 1954. Asymptotic behavior of some rank tests for analysis of variance. *Ann. Math. Statist.* 25: 724–735.

ANSCOMBE, F. J. 1948. The transformation of Poisson, binomial, and negative binomial data. *Biometrika* 35: 246–254.

ARMITAGE, P. 1955. Tests for linear trends in proportions and frequencies. *Biometrics* 11: 375–386.

ARMITAGE, P. 1971. *Statistical Methods in Medical Research.* John Wiley, New York. 504 pp.

ARMITAGE, P. 1985. Biometry and medical statistics. *Biometrics* 41: 823–833.

ARMITAGE, P., G. BERRY, and J. N. S. MATTHEWS. 2002. *Statistical Methods in Medical Research.* Blackwell Science, Malden, MA. 817 pp.

ASIMOV, I. 1982. *Asimov's Biographical Encyclopedia of Science and Technology.* 2nd rev. ed. Doubleday, Garden City, NY. 941 pp.

Literature Cited

BAHADUR, R. R. 1967. Rates of convergence of estimates and test statistics. *Ann. Math. Statist.* 38: 303–324.

BAKER, L. 2002. A comparison of nine confidence intervals for a Poisson parameter when the expected number of events is ≤ 2 5. *Amer. Statist.* 56: 85–89.

BALL, W. W. R. 1935. *A Short Account of the History of Mathematics.* Macmillan, London. 522 pp.

BARNARD, G. A. 1947. Significance tests for 2×2 tables. *Biometrika* 34: 123–138.

BARNARD, G. A. 1979. In contradiction to J. Berkson's dispraise: Conditional tests can be more efficient. *J. Statist. Plan. Inf.* 2: 181–187.

BARNARD, G. A. 1984. Discussion of Dr. Yate's paper. *J. Roy. Statist. Soc.* Ser. A, 147: 449–450.

BARNARD, G. 1990. Fisher: A retrospective. *CHANCE* 3: 22–28.

BARNETT, V. and T. LEWIS. 1994. *Outliers in Statistical Data.* 3rd ed. John Wiley, Chichester, NY. 584 pp.

BARNHART, H. X., M. HABER, and J. SONG. 2002. Overall concordance correlation coefficient for evaluating agreement among multiple observers. *Biometrics* 58: 1020–1027.

BARR, D. R. 1969. Using confidence intervals to test hypothses. *J. Qual. Technol.* 1: 256–258.

BARTHOLOMEW, D. J. 1983. Sir Maurice Kendall FBA. *Statistician* 32: 445–446.

BARTLETT, M. S. 1936. The square root transformation in analysis of variance. *J. Royal Statist. Soc. Suppl.* 3: 68–78.

BARTLETT, M. S. 1937a. Some examples of statistical methods of research in agriculture and applied biology. *J. Royal Statist. Soc. Suppl.* 4: 137–170.

BARTLETT, M. S. 1937b. Properties of sufficiency and statistical tests. *Proc. Roy. Statist. Soc.* Ser. A, 160: 268–282.

BARTLETT, M. S. 1939. A note on tests of significance in multivariate analysis. *Proc. Cambridge Philos. Soc.* 35: 180–185.

BARTLETT, M. S. 1947. The use of transformations. *Biometrics* 3: 39–52.

BARTLETT, M. S. 1965. R. A. Fisher and the last 50 years of statistical methodology. *J. Amer. Statist. Assoc.* 6: 395–409.

BARTLETT, M. S. 1981. Egon Sharpe Pearson, 1895–1980. *Biometrika* 68: 1–12.

BASHARIN, G. P. 1959. On a statistical estimate for the entropy of a sequence of independent variables. *Theory Prob. Appl.* 4: 333–336.

BATES, D. M. and D. G. WATTS. 1988. *Nonlinear Linear Regression Analysis and Its Applications.* John Wiley, New York. 365 pp.

BATSCHELET, E. 1965. *Statistical Methods for the Analysis of Problems in Animal Orientation and Certain Biological Rhythms.* American Institute of Biological Sciences, Washington, DC. 57 pp.

BATSCHELET, E. 1972. Recent statistical methods for orientation data, pp. 61–91. (With discussion.) *In* S. R. Galler, K. Schmidt-Koenig, G. J. Jacobs, and R. E. Belleville (eds.), *Animal Orientation and Navigation.* National Aeronautics and Space Administration, Washington, DC.

BATSCHELET, E. 1974. Statistical rhythm evaluations, pp. 25–35. *In* M. Ferin, F. Halberg, R.M. Richart, and R. L. Vande Wiele (eds.), *Biorhythms and Human Reproduction.* John Wiley, New York.

BATSCHELET, E. 1976. *Mathematics for Life Scientists.* 2nd ed. Springer-Verlag, New York. 643 pp.

BATSCHELET, E. 1978. Second-order statistical analysis of directions, pp. 1–24. *In* K. Schmidt-Koenig and W. T. Keeton (eds.), *Animal Migration, Navigation, and Homing.* Springer-Verlag, Berlin.

BATSHCELET, E. 1981. *Circular Statistics in Biology.* Academic Press, New York. 371 pp.

BAUSELL, R. B. and Y.-F. LI. 2002. *Power Analysis for Experimental Research.* Cambridge University Press, Cambridge, UK. 363 pp.

Literature Cited

BEALL, G. 1940. The transformation of data from entomological field experiments. *Can. Entomol.* 72: 168.

BEALL, G. 1942. The transformation of data from entomological field experiments so that the analysis of variance becomes applicable. *Biometrika* 32: 243–262.

BECKMANN, P. 1977. *A History of π*. 4th ed. Golem Press, Boulder, Colorado. 202 pp.

BECKMAN, R. J. and R. D. COOK. 1983. Outlier ....s. *Technometrics* 25: 119–149, and discussion and response.

BEHRENS, W. V. 1929. Ein Beitrag zur Fehlerberechnung bei weinige Beobachtungen. *Landwirtsch Jahrbücher* 68: 807–837.

BENARD, A. and P. VAN ELTEREN. 1953. A generalization of the method of m rankings. *Indagatones Mathematicae* 15: 358–369.

BENNETT, B. M. and E. NAKAMURA. 1963. Tables for testing significance in a 2×3 contingency table. *Technometrics* 5: 501–511.

BENNETT, B. M. and R. E. UNDERWOOD. 1970. On McNemar's test for the 2×2 table and its power function. *Biometrics* 26: 339–343.

BERKSON, J. 1978. In dispraise of the exact test. *J. Statist. Plan. Infer.* 2:27–42.

BERNHARDSON, C. S. 1975. Type I error rates when multiple comparison procedures follow a significant F test of ANOVA. *Biometrics* 31: 229–232.

BERRY, K. J. and P. W. MIELKE JR. 1988. Monte Carlo comparisons of the asumptotic chi-square and likelihood-ratio test for sparse $r \times c$ tables. *Psychol. Bull.* 103: 256–264.

BERRY, W. D. and S. FELDMAN. 1985. *Multiple Regression in Practice*. Sage Publications, Beverly Hills, CA. 95 pp.

BERTRAND, P. V. and R. L. HOLDER. 1988. A quirk in multiple regression: the whole regression can be greater than the sum of its parts. *Statistician* 37: 371–374.

BEST, D. J. 1975. The difference between two Poisson expectations. *Austral. J. Statist.* 17: 29–33.

BEST, D. J. and D. E. ROBERTS. 1975. The percentage points of the χ^2 distribution. *J. Royal Statist. Soc. Ser. C. Appl. Statist.* 24: 385–388.

BEST, D. J. and J. C. W. RAYNER. 1987. Welch's approximate solution for the Behrens-Fisher problem. *Technometrics* 29: 205–210.

BHATTACHARJEE, G. P. 1970. The incomplete gamma interval. *J. Royal Statist. Soc. Ser. C. Appl. Statist.* 19: 285–287.

BHATTACHARYYA, G. K. and R. A. JOHNSON. 1969. On Hodges's bivariate sign test and a test for uniformity of a circular distribution. *Biometrika* 56: 446–449.

BIRKES, D. and Y. DODGE. 1993. *Alternative Methods of Regression*. John Wiley, New York. 228 pp.

BIRNBAUM, Z. W. and F. H. TINGEY. 1951. One-sided confidence contours for probability distribution functions. *Ann. Math. Statist.* 22: 592–596.

BISSELL, A. F. 1992. Lines through the origin—is NO INT the answer? *J. Appl. Statist.* 19: 192–210.

BLAIR, R. C. and J. J. HIGGINS. 1980a. A comparison of the power of Wilcoxon's rank-sum test statistic to that of Student's t statistic under various nonnormal distributions. *J. Educ. Statist.* 5: 309–335.

BLAIR, R. C. and J. J. HIGGINS. 1980b. The power of t and Wilcoxon statistics. A comparison. *Eval. Rev.* 4: 645–656.

BLAIR, R. C. and J. J. HIGGINS. 1985. Comparison of the power of the paired samples t test to that of Wilcoxon's signed-ranks test under various population shapes. *Psychol. Bull.* 97: 119–128.

BLAIR, R. C., J. J. HIGGINS, and W. D. S. SMITLEY. 1980. On the relative power of the U and t tests. *Brit. J. Math. Statist. Psychol.* 38: 114–120.

BLAIR, R. C., S. S. SAWILOWSKY, and J. J. HIGGINS. 1987. Limitations of the rank transform statistic in tests for interactions. *Communic. Statist. — Simula.* 16: 1133–1145.

Literature Cited

BLATNER, D. 1997. *The Joy of* π. Walker, New York. 130 pp.

BLISS, C. I. 1967. *Statistics in Biology,* Vol. 1. McGraw-Hill, New York. 558 pp.

BLISS, C. I. 1970. *Statistics in Biology,* Vol. 2. McGraw-Hill, New York. 639 pp.

BLOOMFIELD, P. 1976. *Fourier Analysis of Time Series: An Introduction.* John Wiley, New York. 258 pp.

BLOOMFIELD, P. and W. L. STEIGER. 1983. *Least Absolute Deviations: Theory, Applications, and Algorithms.* Birkhäuser, Boston, MA. 349 pp.

BLYTH, C. R. 1986. Approximate binomial confidence limits. *J. Amer. Statist. Assoc.* 81: 843–855.

BOEHNKE, K. 1984. *F*- and *H*-test assumptions revisited. *Educ. Psychol. Meas.* 44: 609–615.

BÖHNING, D. 1994. Better approximate confidence intervals for a binomial parameter. *Can. J. Statis.* 22: 207–218.

BOLAND, P. J. 1984. A biographical glimpse of William Sealy Gosset. *Amer. Statist.* 38: 179–183.

BOLAND, P. J. 2000. William Sealy Gosset — alias 'Student' 1876–1937, pp. 105–122. *In* Houston, K. (ed.), *Creators of Mathematics: The Irish Connection,* University College Dublin Press, Dublin.

BONEAU, C. A. 1960. The effects of violations of assumptions underlying the *t* test. *Psychol. Bull.* 57: 49–64.

BONEAU, C. A. 1962. A comparison of the power of the *U* and *t* tests. *Psychol. Rev.* 69: 246–256.

BOOMSMA, A. and I. W. MOLENAAR. 1994. Four electronic tables for probability distributions. *Amer. Statist.* 48: 153–162.

BOWKER, A. H. 1948. A test for symmetry in contingency tables. *J. Amer. Statist. Assoc.* 43: 572–574.

BOWLEY, A. L. 1920. *Elements of Statistics.* 4th ed. Charles Scribner's Sons, New York. 459 pp. + tables.

BOWMAN, K. O., K. HUTCHESON, E. P. ODUM, and L. R. SHENTON. 1971. Comments on the distribution of indices of diversity, pp. 315–366. *In* G. P. Patil, E. C. Pielou, and W. E. Waters (eds.), *Statistical Ecology,* Vol. 3. Many Species Populations, Ecosystems, and Systems Analysis. Pennsylvania State University Press, University Park.

BOWMAN, K. O. and L. R. SHENTON. 1975. Omnibus test contours for departures from normality based on $\sqrt{b_1}$ and b_2. *Biometrika* 62: 243–250.

BOWMAN, K. O. and L. R. SHENTON. 1986. Moment ($\sqrt{b_1}$, b_2) techniques, pp. 279–329. *In* R. B. D'Agostino and M. A. Stephens (eds.), *Goodness-of-Fit Techniques.* Marcel Dekker, New York.

BOX, G. E. P. 1949. A general distribution theory for a class of likelihood criteria. *Biometrika* 36: 317–346.

BOX, G. E. P. 1950. Problems in the analysis of growth and linear curves. *Biometrics* 6: 362–389.

BOX, G. E. P. 1953. Non-normality and tests on variances. *Biometrika* 40: 318–335.

BOX, G. E. P. and S. L. ANDERSON. 1955. Permutation theory in the derivation of robust criteria and the study of departures from assumption. *J. Royal Statist. Soc.* B17: 1–34.

BOX, G. E. P. and D. R. COX. 1964. An analysis of transformations. *J. Royal Statist. Soc.* B26: 211–243.

BOX, J. F. 1978. *R. A. Fisher: The Life of a Scientist.* John Wiley, New York. 512 pp.

BOZIVICH, H., T. A. BANCROFT, and H. O. HARTLEY. 1956. Power of analysis of variance test procedures for certain incompletely specified models. *Ann. Math. Statist.* 27: 1017–1043.

BRADLEY, R. A. and M. HOLLANDER. 1978. Wilcoxon, Frank, pp. 1245–1250. *In* Kruskal and Tanur (1978).

BRAY, J. S. and S. E. MAXWELL. 1985. *Multivariate Analysis of Variance.* Sage Publications, Beverly Hills, CA. 80 pp.

BRILLOUIN, L. 1962. *Science and Information Theory.* Academic Press, New York. 351 pp.

BRITS, S. J. M. and H. H. LEMMER. 1990. An adjusted Friedman test for the nested design. *Communic. Statist. — Theor. Meth.* 19: 1837–1855.

Literature Cited

BROWER, J. E., J. H. ZAR, and C. N. VON ENDE. 1998. *Field and Laboratory Methods for General Ecology*. 4th ed. McGraw-Hill, Boston. 273 pp.

BROWN, L. D., T. T. CAI, and A. DASGUPTA. 2001. Interval estimation for a binomial proportion. *Statist. Sci.* 16: 101–117. Comment by A. Agresti and B. A. Coull, *ibid.* 16: 117–120; G. Casella, *ibid.* 16: 120–122; C. Corcoran and C. Mehta, *ibid.* 16: 122–124; M. Ghosh, *ibid.* 16: 124–125; T. J. Santner, *ibid.* 16: 126–128. Rejoinder by L. D. Brown, T. T. Cai, and A. DasGupta, *ibid.* 16: 128–133.

BROWN, L. D., T. T. CAI, and A. DASGUPTA. 2002. Confidence intervals for a binomial proportion and asymptotic expansions. *Ann. Statist.* 30: 160–201.

BROWN, M. B. and A. B. FORSYTHE. 1974a. The small size sample behavior of some statistics which test the equality of several means. *Technometrics* 16: 129–132.

BROWN, M. B. and A. B. FORSYTHE. 1974b. The ANOVA and multiple comparisons for data with heterogeneous variances. *Biometrics* 30: 719–724.

BROWN, M. B. and A. B. FORSYTHE. 1974c. Robust tests for the equality of variance. *J. Amer. Statist. Assoc.* 69: 364–367.

BROWNE, R. H. 1979. On visual assessment of the significance of a mean difference. *Biometrics* 35: 657–665.

BROWNLEE, K. A. 1965. *Statistical Theory and Methodology in Science and Engineering*. 2nd ed. John Wiley, New York. 590 pp.

BRUNNER, E. and N. NEUMANN. 1986. Rank tests in 2×2 designs. *Statist. Neerland.* 40: 251–272. [Cited in McKean and Vidmar, 1994.]

BUCKLE, N., C. KRAFT, and C. VAN EEDEN. 1969. An approximation to the Wilcoxon-Mann-Whitney distribution. *J. Amer. Statist. Assoc.* 64: 591–599.

BUDESCU, D. V. and M. I. APPELBAUM. 1981. Variance stablizing transformations and the power of the F test. *J. Educ. Statist.* 6: 55–74.

BÜNING, H. 1997. Robust analysis of variance. *J. Appl. Statist.* 24: 319–332.

BURR, E. J. 1964. Small-sample distributions of the two-sample Cramér-von Mises' W^2 and Watson's U^2. *Ann. Math. Statist.* 35: 1091–1098.

BURSTEIN, H. 1971. *Attribute Sampling. Tables and Explanations*. McGraw-Hill, New York. 464 pp.

BURSTEIN, H. 1975. Finite population correction for binomial confidence limits. *J. Amer. Statist. Assoc.* 70: 67–69.

BURSTEIN, H. 1981. Binomial 2×2 test for independent samples with independent proportions. *Communic. Statist. — Theor. Meth.* A10: 11–29.

CACOULLOS, T. 1965. A relation between the t and F distributions. *J. Amer. Statist. Assoc.* 60: 528–531.

CAJORI, F. 1928–1929. *A History of Mathematical Notations*. Vol. I: *Notations in Elementary Mathematics*, 451 pp. Vol. II: *Notations Mainly in Higher Mathematics*, 367 pp. Open Court Publishing, LaSalle, Illinois. As one volume: Dover, New York, 1993.

CAJORI, F. 1954. Binomial formula, pp. 588–589. In *The Encyclopaedia Britannica*, Vol. 3. Encyclopaedia Britannica Inc., New York.

CALITZ, F. 1987. An alternative to the Kolmogorov-Smirnov test for goodness of fit. *Communic. Statist. — Theor. Meth.* 16: 3519–3534.

CAMILLI, G. 1990. The test of homogeneity for 2×2 contingency tables: A review of and some personal opinions on the controversy. *Psychol. Bull.* 108: 135–145.

CAMILLI, G. and K. D. HOPKINS. 1978. Applicability of chi-square to 2×2 contingency tables with small expected frequencies. *Psychol. Bull.* 85: 163–167.

CAMILLI, G. and K. D. HOPKINS. 1979. Testing for association in 2×2 contingency tables with very small sample sizes. *Psychol. Bull.* 86: 1011–1014.

CARMER, S. G. and M. R. SWANSON. 1973. An evaluation of ten pairwise multiple comparison procedures by Monte Carlo methods. *J. Amer. Statist. Assoc.* 68: 66–74.

CARR, W. E. 1980. Fisher's exact test extended to more than two samples of equal size. *Technometrics* 22: 269–270. See also: 1981. Corrigendum. *Technometrics* 23: 320.

Literature Cited

CASAGRANDE, J. T., M. C. PIKE, and P. G. SMITH. 1978. An improved approximate formula for calculating sample sizes for comparing binomial distributions. *Biometrics* 34: 483–486.

CATTELL, R. B. 1978. Spearman, C. E., pp. 1036–1039. *In* Kruskal and Tanur (1978).

CAUDILL, S. B. 1988. Type I errors after preliminary tests for heteroscedasticity. *Statistician* 37: 65–68.

CHAPMAN, J.-A. W. 1976. A comparison of the X^2, $-2 \log R$, and multinomial probability criteria for significance tests when expected frequencies are small. *J. Amer. Statist. Assoc.* 71: 854–863.

CHATTERJEE, S., A. S. HADI, and B. PRICE. 2006. *Regression Analysis by Example*. 4th ed. John Wiley, Hoboken, NJ. 375 pp.

CHOW, B., J. E. MILLER, and P. C. DICKINSON. 1974. Extensions of Monte-Carlo comparison of some properties of two rank correlation coefficients in small samples. *J. Statist. Computa. Simula.* 3: 189–195.

CHRISTENSEN, R. 1990. *Log-Linear Models*. Springer-Verlag, New York. 408 pp.

CHURCH, J. D. and E. L. WIKE. 1976. The robustness of homogeneity of variance tests for asymmetric distributions: A Monte Carlo study. *Bull. Psychomet. Soc.* 7: 417–420.

CICCHITELLI, G. 1989. On the robustness of the one-sample *t* test. *J. Statist. Computa. Simula.* 32: 249–258.

CLEVELAND, W. S., C. L. MALLOWS, and J. E. McRAE. 1993. ATS methods: Nonparametric regression for non-Gaussian data. *J. Amer. Statist. Assoc.* 88: 821–835.

CLINCH, J. J. and H. J. KESELMAN. 1982. Parametric alternatives to the analysis of variance. *J. Educ. Statist.* 7: 207–214.

CLOPPER, C. J. and E. S. PEARSON. 1934. The use of confidence or fiducial limits illustrated in the case of the binomial. *Biometrika* 26: 404–413. [Cited in Agresti and Coull, 1998.]

COAKLEY, C. W. and M. HEISE, 1996. Versions of the sign test in the presence of ties. *Biometrics* 52: 1242–1251.

COCHRAN, W. G. 1937. The efficiencies of the binomial series tests of significance of a mean and of a correlation coefficient. *J. Roy. Statist. Soc.* 100: 69–73.

COCHRAN, W. G. 1942. The χ^2 correction for continuity. *Iowa State Coll. J. Sci.* 16: 421–436.

COCHRAN, W. G. 1947. Some consequences when the assumptions for analysis of variance are not satisfied. *Biometrics* 3: 22–38.

COCHRAN, W. G. 1950. The comparison of percentages in matched samples. *Biometrika* 37: 256–266.

COCHRAN, W. G. 1952. The χ^2 test for goodness of fit. *Ann. Math. Statist.* 23: 315–345.

COCHRAN, W. G. 1954. Some methods for strengthening the common χ^2 tests. *Biometrics* 10: 417–451.

COCHRAN, W. G. 1977. *Sampling Techniques*. 3rd. ed. John Wiley, New York. 428 pp.

COCHRAN, W. G. and G.M. COX. 1957. *Experimental Designs*. 2nd ed. John Wiley, New York. 617 pp.

COHEN, A. 1991. Dummy variables in stepwise regression. *Amer. Statist.* 45: 226–228.

COHEN, J. 1988. *Statistical Power Analysis for the Behavioral Sciences*. 2nd ed. Lawrence Erlbaum Associates, Hillsdale, N.J. 567 pp.

COHEN. J., P. COHEN, S. G. WEST, and L. S. AIKEN. 2003. *Applied Multiple Regression/Correlation Analysis for the Behavioral Sciences*. 3rd ed. Lawrence Applebaum Associates, Mahwah, NJ. 703 pp.

CONNETT, J. E., J. A. SMITH, and R. B. McHUGH. 1987. Sample size and power for pair-matched case-control studies. *Statist. Med.* 6: 53–59.

CONOVER, W. J. 1973. On the methods of handling ties in the Wilcoxon signed-rank test. *J. Amer. Statist. Assoc.* 68: 985–988.

CONOVER, W. J. 1974. Some reasons for not using the Yates continuity correction on 2×2 contingency tables. *J. Amer. Statist. Assoc.* 69: 374–376. *Also:* Comment, by C.F. Starmer, J. E. Grizzle, and P. K. Sen. *ibid.* 69: 376–378; Comment and a suggestion, by N. Mantel, *ibid.* 69: 378–380; Comment, by O. S. Miettinen, *ibid.* 69: 380–382; Rejoinder, by W. J. Conover, *ibid.* 69: 382.

Literature Cited

CONOVER, W. J. 1999. *Practical Nonparametric Statistics*. 3rd ed. John Wiley, New York. 584 pp.

CONOVER, W. J. and R. L. IMAN. 1976. On some alternative procedures using ranks for the analysis of experimental designs. *Cummunic. Statist. — Theor. Meth.* A5: 1349–1368.

CONOVER, W. J. and R. L. IMAN. 1981. Rank transformations as a bridge between parametric and nonparametric statistics. *Amer. Statist.* 85: 124–129.

COTTINGHAM, K. L., J. T. LENNON, and B. L. BROWN. 2005. Knowing where to draw the line: Designing more informative ecological experiments. *Front. Ecol. Environ.* 3: 145–152.

COWLES, S. M. and C. DAVIS. 1982. On the origins of the .05 level of statistical significance. *Amer. Psychol.* 37: 553–558.

COX, D. R. 1958. *Planning of Experiments*. John Wiley, New York. 308 pp.

CRAMER, E. M. 1972. Significance test and tests of models in multiple regression. *Amer. Statist.* 26(4): 26–30.

CRAMÉR, H. 1946. *Mathematical Methods of Statistics*. Princeton University Press, Princeton, N.J. 575 pp.

CRESSIE, N. and T. R. C. READ. 1989. Pearson's X^2 and the loglikelihood ratio statistic G^2: A comparative review. *Internat. Statist. Rev.* 57: 19–43.

CHATTERJEE, S., A. S. HADI, and B. PRICE. 2006. *Regression Analysis by Example*. 4th ed. Wiley, Hoboken, N.J. 375 pp.

CRITCHLOW, D. E. and M. A. FLIGNER. 1991. On distribution-free multiple comparisons in the one-way analysis of variance. *Communic. Statist. — Theor. Meth.* 20: 127–139.

CURETON, E. E. 1967. The normal approximation to the signed-rank sampling distribution when zero differences are present. *J. Amer. Statist. Assoc.* 62: 1068–1069.

D'AGOSTINO, R. B. 1970. Transformation to normality of the null distribution of g_1. *Biometrika* 57: 679–681.

D'AGOSTINO, R. B. 1986. Tests for the normal distribution, pp. 367–419. *In* R. B. D'Agostino and M. A. Stephens (eds.), *Goodness-of-fit Techniques*. Marcel Dekker, New York.

D'AGOSTINO, R. B., A. BELANGER, and R. B. D'AGOSTINO, JR. 1990. A suggestion for using powerful and informative tests of normality. *Amer. Statist.* 44: 316–321.

D'AGOSTINO, R. B., W. CHASE, and A. BELANGER. 1988. The appropriateness of some common procedures for testing the equality of two independent binomial populations. *Amer. Statist.* 42: 198–202.

D'AGOSTINO, R. B. and G. E. NOETHER. 1973. On the evaluation of the Kolmogorov statistic. *Amer. Statist.* 27: 81–82.

D'AGOSTINO, R. B. and E. S. PEARSON. 1973. Tests of departure from normality. Empirical results for the distribution of b_2 and $\sqrt{b_1}$. *Biometrika* 60: 613–622.

D'AGOSTINO, R. B. and G. L. TIETJEN. 1973. Approaches to the null distribution of $\sqrt{b_1}$. *Biometrika* 60: 169–173.

DALE, A. I. 1989. An early occurrence of the Poisson distribution. *Statist. Prob. Lett.* 7: 21–22.

DANIEL, W. W. 1990. Applied Nonparametric Statistics. 2nd ed. PWS-kent, Boston, MA. 635 pp.

DAPSON, R. W. 1980. Guidelines for statistical usage in age-estimation technics. *J. Wildlife Manage.* 44: 541–548.

DARLINGTON, R. B. 1990. *Regression and Linear Models*. McGraw-Hill, New York. 542 pp.

DARNELL, A. C. 1988. Harold Hotelling 1895–1973. *Statist. Sci.* 3: 57–62.

DAVENPORT, E. C., JR. and N. A. EL-SANHURRY. 1991. Phi/Phimax: Review and Synthesis. *Educ. Psychol. Meas.* 51: 821–828.

DAVENPORT, J. M. and J. T. WEBSTER. 1975. The Behrens-Fisher problem, an old solution revisited. *Metrika* 22: 47–54.

DAVID, F. N. 1962. *Games, Gods and Gambling*. Hafner Press, New York. 275 pp.

DAVID, H. A. 1995. First (?) occurrence of common terms in mathematical statistics. *Amer. Statist.* 49: 121–133.

Literature Cited

DAVID, H. A. 1998a. First (?) occurrence of common terms in probability and statistics — A second list with corrections. *Amer. Statist.* 52: 36–40.

DAVID, H. A. 1998b. Early sample measures of variability. *Statist. Sci.* 13: 368–377.

DAVID, H. A. 2005. Tables related to the normal distribution: A short history. *Amer. Statist.* 59: 309–311.

DAVID, H. A. 2006. The introduction of matrix algebra into statistics. *Amer. Statist.* 60: 162.

DAVID, H. A. and D. F. MORRISON. 2006. Samuel Stanley Wilks (1906–1964). *Amer. Statist.* 60: 46–49.

DAVID, H. A. and H. A. FULLER. 2007. Sir Maurice Kendall (1907–1983): A centenary appreciation. *Amer. Statist.* 61: 41–46.

DAY, R. W. and G. P. QUINN. 1989. Comparisons of treatments after an analysis of variance in ecology. *Ecol. Monogr.* 59: 433–463.

DE JONGE, C. and M. A. J. VAN MONTFORT. 1972. The null distribution of Spearman's S when $n = 12$. *Statist. Neerland.* 26: 15–17.

DEMPSTER, A. P. 1983. Reflections on W. G. Cochran, 1909–1980. *Intern. Statist. Rev.* 51: 321–322.

DESMOND, A. E. 2000. Francis Ysidro Edgeworth 1845–1926, pp. 78–87. *In* Houston, K. (ed.), *Creators of Mathematics: The Irish Connection*. University College Dublin Press, Dublin.

DESU, M. M. and D. RAGHAVARAO. 1990. *Sample Size Methodology*. Academic Press, Boston, MA. 135 pp.

DETRE, K. and C. WHITE. 1970. The comparison of two Poisson-distributed observations. *Biometrics* 26: 851–854.

DHARMADHIKARI, S. 1991. Bounds of quantiles: A comment on O'Cinneide. *Amer. Statist.* 45: 257–258.

DIJKSTRA, J. B. and S. P. J. WERTER. 1981. Testing the equality of several means when the population variances are unequal. *Communic. Statist. — Simula. Computa.* B10: 557–569.

DIXON, W. J. and F. J. MASSEY, JR. 1969. *Introduction to Statistical Analysis*. 3rd ed. McGraw-Hill, New York. 638 pp.

DOANE, D. P. 1976. Aesthetic frequency classifications. *Amer. Statist.* 30: 181–183.

DODGE, Y. 1996. A natural random number generator. *Internat. Statist. Rev.* 64: 329–344.

DONALDSON, T. S. 1968. Robustness of the F-test to errors of both kinds and the correlation between the numerator and denominator of the f-ratio. *J. Amer. Statist. Assoc.* 63: 660–676.

DONNER, A. and G. WELLS. 1986. A comparison of confidence interval methods for the interclass correlation coefficient. *Biometrics* 42: 401–412.

DRAPER, N. R. and W. G. HUNTER. 1969. Transformations: Some examples revisited. *Technometrics* 11: 23–40.

DRAPER, N. R. and H. SMITH. 1998. *Applied Regression Analysis*. 3rd ed. John Wiley, New York. 706 pp.

DUNCAN, D. B. 1951. A significance test for difference between ranked treatments in an analysis of variance. *Virginia J. Sci.* 2: 171–189.

DUNCAN, D. B. 1955. Multiple range and multiple F tests. *Biometrics* 11: 1–42.

DUNN, O. J. 1964. Multiple comparisons using rank sums. *Technometrics* 6: 241–252.

DUNN, O. J. and V. A. CLARK. 1987. *Applied Sciences: Analysis of Variance and Regression*. 2nd ed. John Wiley, New York. 445 pp.

DUNNETT, C. W. 1955. A multiple comparison procedure for comparing several treatments with a control. *J. Amer. Statist. Assoc.* 50: 1096–1121.

DUNNETT, C. W. 1964. New tables for multiple comparisons with a control. *Biometrics* 20: 482–491.

DUNNETT, C. W. 1970. Multiple comparison tests. *Biometrics* 26: 139–141.

Literature Cited

DUNNETT, C. W. 1980a. Pairwise multiple comparisons in the homogeneous variance, equal sample size case. *J. Amer. Statist. Assoc.* 75: 789–795.

DUNNETT, C. W. 1980b. Pairwise multiple comparisons in the unequal variance case. *J. Amer. Statist. Assoc.* 75: 796–800.

DUNNETT, C. W. 1982. Robust multiple comparisons. *Communic. Statist. — Theor. Meth.* 11: 2611–2629.

DUPONT, W. D. 1986. Sensitivity of Fisher's exact test to minor perturbations in 2×2 contingency tables. *Statist. in Med.* 5: 629–635.

DURAND, D. and J. A. GREENWOOD. 1958. Modifications of the Rayleigh test for uniformity in analysis of two-dimensional orientation data. *J. Geol.* 66: 229–238.

DWASS, M. 1960. Some k-sample rank-order tests, pp. 198–202. *In* I. Olkin, S. G. Ghurye, W. Hoeffding, W. G. Madow, and H. B. Mann. *Contributions to Probability and Statistics. Essays in Honor of Harold Hotelling.* Stanford University Press, Stanford, CA.

DYKE, G. 1995. Obituary: Frank Yates. *J. Roy. Statist. Soc.* Ser. A, 158: 333–338.

EASON, G., C. W. COLES, and G. GETTINBY. 1980. *Mathematics and Statistics for the Bio-Sciences.* Ellis Horwood, Chichester, England. 578 pp.

EBERHARDT, K. R. and M. A. FLIGNER. 1977. A comparison of two tests for equality of proportions. *Amer. Statist.* 31: 151–155.

EDGINGTON, E. S. 1961. Probability table for number of runs of signs of first differences in ordered series. *J. Amer. Statist. Assoc.* 56: 156–159.

EDWARDS, A. W. F. 1986. Are Mendel's results really too close? *Biol. Rev.* 61: 295–312.

EDWARDS, A. W. F. 1993. Mendel, Galton, and Fisher. *Austral. J. Statist.* 35: 129–140.

EELLS, W. C. 1926. A plea for a standard definition of the standard deviation. *J. Educ. Res.* 13: 45–52.

EINOT, I. and K. R. GABRIEL. 1975. A survey of the powers of several methods of multiple comparisons. *J. Amer. Statist. Assoc.* 70: 574–583.

EISENHART, C. 1947. The assumptions underlying the analysis of variance. *Biometrics* 3: 1–21.

EISENHART, C. 1968. Expression of the uncertainties of final results. *Science* 160: 1201–1204.

EISENHART, C. 1978. Gauss, Carl Friedrich, pp. 378–386. *In* Kruskal and Tanur (1978).

EISENHART, C. 1979. On the transition from "Student's" z to "Student's" t. *Amer. Statist.* 33: 6–10.

EVERITT, B. S. 1979. A Monte Carlo investigation of the robustness of Hotelling's one- and two-sample tests. *J. Amer. Statist. Assoc.* 74: 48–51.

EVERITT, B. S. 1992. *The Analysis of Contingency Tables.* 2nd ed. Chapman & Hall, New York. 164 pp.

EZEKIEL, M. 1930. *Methods of Correlation Analysis.* John Wiley, New York. 427 pp.

FAHOOME, G. 2002. Twenty nonparametric statistics and their large sample approximations. *J. Modern Appl. Statist. Meth.* 1: 248–268.

FELDMAN, S. E. and E. KLUGER. 1963. Short cut calculation of the Fisher-Yates "exact test." *Psychometrika* 28: 289–291.

FELLINGHAM, S. A. and D. J. STOKER. 1964. An approximation for the exact distribution of the Wilcoxon test for symmetry. *J. Amer. Statist. Assoc.* 59: 899–905.

FELTZ, C. J. 1998. Generalizations of the delta-corrected Kolmogorov-Smirnov goodness-of-fit test. *Austral. & New Zealand J. Statist.* 40: 407–413.

FELTZ, C. J. and G. E. MILLER. 1996. An asymptotic test for the equality of coefficients of variation from k populations. *Statist. in Med.* 15: 647–658.

FÉRON, R. 1978. Poisson, Siméon Denis, pp. 704–707. *In* Kruskal and Tanur (1978).

FIELLER, E. C., H. O. HARTLEY, and E. S. PEARSON. 1957. Tests for rank correlation coefficients. I. *Biometrika* 44: 470–481.

FIELLER, E. C., H. O. HARTLEY, and E. S. PEARSON. 1961. Tests for rank correlation coefficients. II. *Biometrika* 48: 29–40.

Literature Cited

FIENBERG, S. E. 1970. The analysis of multidimensional contingency tables. *Ecology* 51: 419–433.

FIENBERG, S. E. 1972. The analysis of incomplete multiway contingency tables. *Biometrics* 28: 177–202.

FIENBERG, S. E. 1980. *The Analysis of Cross-Classified Categorical Data.* 2nd ed. MIT Press, Cambridge, MA. 198 pp.

FISHER, N. I. 1993. *Statistical Analysis of Circular Data.* Cambridge University Press, Cambridge, England. 277 pp.

FISHER, N. I. and A. J. LEE. 1982. Nonparametric measures of angular-angular association. *Biometrika* 69: 315–321.

FISHER, N. I. and A. J. LEE. 1983. A correlation coefficient for circular data. *Biometrika* 70: 327–332.

FISHER, N. I., T. LEWIS, and B. J. J. EMBLETON. 1987. *Statistical Analysis of Spherical Data.* Cambridge University Press, Cambridge, England. 329 pp.

FISHER, R. A. 1918a. The correlation between relatives on the supposition of Mendelian inheritance. *Trans. Roy. Soc. Edinburgh* 52: 399–433.

FISHER, R. A. 1918b. The causes of human variability. *Eugenics Rev.* 10: 213–220.

FISHER, R. A. 1921. On the "probable error" of a coefficient of correlation deduced from a small sample. *Metron* 1: 3–32.

FISHER, R. A. 1922. On the interpretation of χ^2 from contingency tables and the calculation of *P. J. Royal Statist. Soc.* 85: 87–94.

FISHER, R. A. 1922a. The goodness of fit of regression formulæ, and the distribution of regression coefficients. *J. Roy. Statist. Soc.* 85: 597–612.

FISHER, R. A. 1925a. Applications of "Student's" distribution. *Metron* 5: 90–104.

FISHER, R. A. 1925b. *Statistical Methods for Research Workers.* [1st ed.] Oliver and Boyd, Edinburgh, Scotland. 239 pp.+ 6 tables.

FISHER, R. A. 1926. The arrangement of field experiments. *J. Ministry Agric.* 33. 503–513. [Cited in Sahai and Ageel, 2000: 488.]

FISHER, R. A. 1928. On a distribution yielding the error functions of several well known statistics, pp. 805–813. *In: Proc. Intern. Math. Congr., Toronto, Aug. 11–16, 1924,* Vol. II. University of Toronto Press, Toronto. [Cited in Eisenhart, 1979.]

FISHER, R. A. 1934. *Statistical Methods for Research Workers.* 5th ed. Oliver and Boyd, Edinburgh, Scotland.

FISHER, R. A. 1935. The logic of inductive inference. *J. Roy. Statist. Soc.* Ser. A, 98: 39–54.

FISHER, R. A. 1936. Has Mendel's work been rediscovered? *Ann. Sci.* 1: 115–137.

FISHER, R. A. 1939a. "Student." *Ann. Eugen.* 9: 1–9.

FISHER, R. A. 1939b. The comparison of samples with possibly unequal variances. *Ann. Eugen.* 9: 174–180.

FISHER, R. A. 1958. *Statistical Methods for Research Workers.* 13th ed. Hafner, New York. 356 pp.

FISHER, R. A. and F. YATES. 1963. *Statistical Tables for Biological, Agricultural, and Medical Research.* 6th ed. Hafner, New York. 146 pp.

FISZ, M. 1963. *Probability Theory and Mathematics Statistics.* 3rd ed. John Wiley, New York. 677 pp.

FIX, E. and J. L. HODGES, JR. 1955. Significance probabilities of the Wilcoxon test. *Ann. Math. Statist.* 26: 301–312.

FLEISS, J. L., E. LEVIN, and M. C. PAIK. 2003. *Statistical Methods for Rates and Proportions.* John Wiley, Hoboken, NJ. 760 pp.

FLEISS, J. L., A. TYTUN, and H. K. URY. 1980. A simple approximation for calculating sample sizes for comparing independent proportions. 3rd ed. *Biometrics* 36: 343–346.

FLIGNER, M. A. and G. E. POLICELLO II. 1981. Robust rank procedures for the Behrens-Fisher problem. *J. Amer. Statist. Assoc.* 76: 162–168.

Literature Cited

Fong, D. Y. T., C. W. Kwan, K. F. Lam, and K. S. L. Lam. 2003. Use of the sign test for the meadian in the presence of ties. *Amer. Statist.* 57: 237–240. Correction: 2005, *Amer. Statist.* 59: 119.

Franklin, L. A. 1988a. The complete exact distribution of Spearman's rho for $n = 12(1)18$. *J. Statist. Computa. Simula.* 29: 255–269.

Franklin, L. A. 1988b. A note on approximations and convergence in distribution for Spearman's rank correlation coefficient. *Communic. Statist. — Theor. Meth.* 17: 55–59.

Franklin, L. A. 1989. A note on the Edgeworth approximation to the distribution of Spearman's rho with a correction to Pearson's approximation. *Communic. Statist. — Simula. Computa.* 18: 245–252.

Freeman, M. F. and J. W. Tukey. 1950. Transformations related to the angular and the square root. *Ann. Math. Statist.* 21: 607–611.

Freund, R. J. 1971. Some observations on regressions with grouped data. *Amer. Statist.* 25(3): 29–30.

Friedman, M. 1937. The use of ranks to avoid the assumption of normality implicit in the analysis of variance. *J. Amer. Statist. Assoc.* 32: 675–701.

Friedman, M. 1940. A comparison of alternate tests of significance for the problem of m rankings. *Ann. Math. Statist.* 11: 86–92.

Friendly, M. 1994. Mosaic displays for multiway contingency tables. *J. Amer. Statist. Assoc.* 89: 190–200.

Friendly, M. 1995. Conceptual and visual models for categorical data. *Amer. Statist.* 49: 153–160.

Friendly, M. 1999. Extending mosaic displays: Marginal, conditional, and partial views of categorical data. *J. Comput. Graph. Statistics* 8: 373–395.

Friendly, M. 2002. A brief history of the mosaic display. *J. Comput. Graph. Statist.* 11: 89–107.

Frigge, M., D. C. Hoaglin, and B. Iglewicz. 1989. Some implications of the boxplot. *Amer. Statist.* 43: 50–54.

Fujino, Y. 1980. Approximate binomial confidence limits. *Biometrika* 67: 677–681.

Gabriel, K. R. and P. A. Lachenbruch. 1969. Nonparametric ANOVA in small samples: A Monte Carlo study of the adequacy of the asymptotic approximation. *Biometrics* 25: 593–596.

Gaito, J. 1960. Scale classification and statistics. *Psychol. Bull.* 67: 277–278.

Gaito, J. 1980. Measurement scales and statistics: Resurgence of an old misconception. *Psychol. Bull.* 87: 564–567.

Games, P. A. and J. F. Howell. 1976. Pairwise multiple comparison procedures with unequal n's and/or variances: A Monte Carlo study. *J. Educ. Statist.* 1: 113–125.

Games, P. A., H. J. Keselman, and J. C. Rogan. 1981. Simultaneous pairwise multiple comparison procedures for means when sample sizes are unequal. *Psychol. Bull.* 90: 594–598.

Games, P. A. and P. A. Lucas. 1966. Power of the analysis of variance of independent groups on nonnormal and normally transformed data. *Educ. Psychol. Meas.* 26: 311–327.

Gans, D. J. 1991. Preliminary test on variances. *Amer. Statist.* 45: 258.

Garside, G. R. and C. Mack. 1976. Actual type 1 error probabilities for various tests in the homogeneity case of the 2×2 contingency table. *Amer. Statist.* 30: 18–21.

Garson, G. D. 2006. *Logistic Regression.* www2.chass.ncsu.edu/garson/PA765/logistic.htm.

Gart, J. J. 1969a. An exact test for comparing matched proportions in crossover designs. *Biometrika* 56: 75–80.

Gart, J. J. 1969b. Graphically oriented tests of the Poisson distribution. *Bull. Intern. Statist. Inst.,* 37th Session, pp. 119–121.

Gartside, P. S. 1972. A study of methods for comparing several variances. *J. Amer. Statist. Assoc.* 67: 342–346.

Geary, R. C. and C. E. V. Leser. 1968. Significance tests in multiple regression. *Amer. Statist.* 22(1): 20–21.

Literature Cited

GEIRINGER, H. 1978. Von Mises, Richard, pp. 1229–1231. *In* Kruskal and Tanur (1978).

GEORGE, E. O. 1987. An approximation of F distribution by binomial probabilities. *Statist. Prob. Lett.* 5: 169–173.

GHENT, A. W. 1972. A method for exact testing of 2×2, 2×3, 3×3, and other contingency tables, employing binomial coefficients. *Amer. Midland Natur.* 88: 15–27.

GHENT, A. W. 1991. Insights into diversity and niche breadth analysis from exact small-sample tests of the equal abundance hypothesis. *Amer. Midland Natur.* 126: 213–255.

GHENT, A. W. 1993. An exact test and normal approximation for centrifugal and centripetal patterns in line and belt transects in ecological studies. *Amer. Midland Natur.* 130: 338–355.

GHENT, A. W. and J. H. ZAR. 1992. Runs of two kinds of elements on a circle: A redevelopment, with corrections, from the perspective of biological research. *Amer. Midland Natur.* 128: 377–396.

GIBBONS, J. D. and S. CHAKRABORTI. 2003. *Nonparametric Statistical Inference.* 4th ed. Marcel Dekker, New York. 645 pp.

GILL, J. L. 1971. Analysis of data with heterogeneous data: A review. *J. Dairy Sci.* 54: 369–373.

GIRDEN, E. R. 1992. *ANOVA: Repeated Measures.* Sage Publications, Newbury Park, CA, 77 pp.

GLANTZ, S. A. and B. K. SLINKER. 2001. *Primer of Applied Regression and Analysis of Variance.* 2nd ed. McGraw-Hill, New York. 949 pp.

GLASS, G. V. and K. D. HOPKINS. 1996. *Statistical Methods in Education and Psyhology.* Allyn and Bacon, Boston. 674 pp.

GLASS, G. V., P. D. PECKHAM, and J. R. SANDERS. 1972. Consequences of failure to meet assumptions underlying the fixed effects analysis of variance and covariance. *Rev. Educ. Res.* 42: 239–288.

GLEJSER, H. 1969. A new test for heteroscedasticity. *J. Amer. Statist. Assoc.* 64: 316–323.

GOODMAN, L. A. 1970. The multivariate analysis of qualitative data: Interactions among multiple classifications. *J. Amer. Statist. Assoc.* 65: 226–256.

GORMAN, J. W. and R. J. TOMAN. 1966. Selection of variables for fitting equations to data. *Technometrics* 8: 27–51.

GREENWOOD, J. A. and D. DURAND. 1955. The distribution of length and components of the sum of n random unit vectors. *Ann. Math. Statist.* 26: 233–246.

GRIZZLE, J. E. 1967. Continuity correction in the χ^2-test for 2×2 tables. *Amer. Statist.* 21(4): 28–32.

GREEN, S. B. 1991. How many subjects does it take to do a regression analysis? *Multivar. Behav. Res.* 26: 499–510.

GROENEVELD, R. A. and G. MEEDEN. 1984. Measuring skewness and kurtosis. *Statistician* 33: 391–399.

GUENTHER, W. C. 1964. *Analysis of Variance.* Prentice Hall, Englewood Cliffs, NJ. 199 pp.

GULLBERG, J. 1997. *Mathematics: From the Birth of Numbers.* W. W. Norton, New York. 1093 pp.

GUMBEL, E. J., J. A. GREENWOOD, and D. DURAND. 1953. The circular normal distribution: Theory and tables. *J. Amer. Statist. Assoc.* 48: 131–152.

GURLAND, J. and R. C. TRIPATHI. 1971. A simple approximation for unbiased estimation of the standard deviation. *Amer. Statist.* 25(4): 30–32.

HABER, M. 1980. A comparison of some continuity corrections for the chi-squared test on 2×2 tables. *J. Amer. Statist. Assoc.* 75: 510–515.

HABER, M. 1982. The continuity correction and statistical testing. *Intern. Statist. Rev.* 50: 135–144.

HABER, M. 1984. A comparison of tests for the hypothesis of no three-factors interaction $2 \times 2 \times 2$ contingency tables. *J. Statist. Comput. Simula* 20: 205–215.

HABER, M. 1986. An exact unconditional test for the 2×2 comparative trial. *Psychol. Bull.* 99: 129–132.

HABER, M. 1987. A comparison of some conditional and unconditional exact tests for 2×2 contingency tables. *Communic. Statist. – Simula. Computa.* 16: 999–1013.

HABER, M. 1990. Comments on "The test of homogeneity for 2×2 contingency tables: A review of and some personal opinions of the controversy" by G. Camilli. *Psychol. Bull.* 108: 146–149.

Literature Cited

HAFNER, K. 2001. A new way of verifying old and familiar sayings. *New York Times,* Feb. 1, 2001. Section G, p. 8.

HAHN, G. J. 1972. Simultaneous prediction intervals to contain the standard deviations or ranges of future samples from a normal distribution. *J. Amer. Statist. Assoc.* 67: 938–942.

HAHN, G. J. 1977. A prediction interval on the difference between two future sample means and its application to a claim of product superiority. *Technometrics* 19: 131–134.

HAHN, G. J. and W. Q. MEEKER. 1991. *Statistical Intervals: A Guide for Practitioners.* John Wiley, New York. 392 pp.

HAIGHT, F. A. 1967. *Handbook of the Poisson Distribution.* John Wiley, New York. 168 pp.

HAIR, J. F., JR., W. C. BLACK, B. J. BABIN, R. E. ANDERSON, and R. L. TATHAM. 2006. *Multivariate Data Analysis.* 6th ed. Prentice Hall, Upper Saddle River, NJ. 899 pp.

HALBERG, F. and J.-K. LEE. 1974. Glossary of selected chronobiologic terms, pp.XXXVII–L. *In* L. E. Scheving, F. Halberg, and J. E. Pauly (eds.), *Chronology.* Igaku Shoin, Tokyo.

HALD, T. N. 1981. Thiele's contributions to Statistics. *Internat. Statist. Rev.* 4: 1–20.

HALD, A. 1984. Nicolas Bernoulli's Theorem. *Internat. Statist. Rev.* 52: 93–99.

HAMAKER, H. C. 1978. Approximating the cumulative normal distribution and its inverse. *Appl. Statist.* 27: 76–77.

HAMILTON, D. 1987. Sometimes $R^2 > r_{yx_1}^2 + r_{yx_2}^2$. *Amer. Statist.* 41: 129–132. [See also Hamilton's (1988) reply to and reference to other writers on this topic: *Amer. Statist.* 42: 90–91.]

HAND, D. J. and C. C. TAYLOR. 1987. *Multivariate Analysis of Variance and Repeated Measures.* Chapman and Hall, London. 262 pp.

HÄRDLE, W. 1990. *Applied Nonparametric Regression.* Cambridge University Press, Cambridge, England. 333 pp.

HARDY, M. A. 1993. *Regression with Dummy Variables.* Sage Publications, Newbury Park, CA. 90 pp.

HARLOW, L. L., S. A. MULAIK, and J. H. STEIGER (eds.). 1997. *What If There Were No Significance Tests?* Lawrence Erlbaum Associates, Mahwah, NJ. 446 pp.

HARRIS, J. A. 1910. The arithmetic of the product moment method of calculating the coefficient of correlation. *Amer. Natur.* 44: 693–699.

HARRISON, D. and G. K. KANJI. 1988. The development of analysis of variance for circular data. *J. Appl. Statist.* 15: 197–223.

HARRISON, D., G. K. KANJI, and R. J. GADSDEN. 1986. Analysis of variance for circular data. *J. Appl. Statist.* 13: 123–138.

HARTER, H. L. 1957. Error rates and sample sizes for range tests in multiple comparisons. *Biometrics* 13: 511–536.

HARTER, H. L. 1960. Tables of range and studentized range. *Ann. Math. Statist.* 31: 1122–1147.

HARTER, H. L. 1970. *Order Statistics and Their Use in Testing and Estimation*, Vol. 1. Tests Based on Range and Studentized Range of Samples from a Normal Population. U.S. Government Printing Office, Washington, DC. 761 pp.

HARTER, H. L., H. J. KHAMIS, and R. E. LAMB. 1984. Modified Kolmogorov-Smirnov tests for goodness of fit. *Communic. Statist. — Simula. Computa.* 13: 293–323.

HARTIGAN, J. A. and B. KLEINER. 1981. Mosaics for contingency tables, pp. 268–273. *In* W. F. Eddy (ed.), *Computer Science and Statistics: Proceedings of the 13th Symposium on the Interface.* Springer-Verlag, New York.

HARTIGAN, J. A. and B. KLEINER. 1984. A mosaic of television ratings. *Amer. Statist.* 38: 32–35.

HARWELL, M. R., E. N. RUBINSTEIN, W. S. HAYES, and C. C. OLDS. 1992. Summarizing Monte Carlo results in methodological research: The one- and two-factor fixed effects ANOVA cases. *J. Educ. Statist.* 17: 315–339.

HASTINGS, C., JR. 1955. *Approximations for Digital Computers.* Princeton University Press, Princeton, NJ. 201 pp.

Literature Cited

HAUCK, W. W. and S. ANDERSON. 1986. A comparison of large-sample confidence interval methods for the difference of two binomial populations. *Amer. Statist.* 20: 318–322.

HAUCK, W. W., JR. and A. DONNER. 1977. Wald's test as applied to hypotheses in logit analysis. *J. Amer. Statist. Soc.* 72: 851–853.

HAVILAND, M. B. 1990. Yate's correction for continuity and the analysis of 2×2 contingency tables. *Statist. Med.* 9: 363–367. *Also:* Comment, by N. Mantel, *ibid.* 9: 369–370; S. W. Greenhouse, *ibid.* 9: 371–372; G. Barnard, *ibid.* 9: 373–375; R. B. D'Agostino, *ibid.* 9: 377–378; J. E. Overall, *ibid.* 9: 379–382; Rejoinder, by M. B. Haviland, *ibid.* 9: 383.

HAVLICEK, L. L. and N. L. PETERSON. 1974. Robustness of the t test: A guide for researchers on effect of violations of assumptions. *Psychol. Reports* 34: 1095–1114.

HAWKINS, D. M. 1980. A note on fitting a regression without an intercept term. *Amer. Statist.* 34(4): 233.

HAYS, W. L. 1994. *Statistics*, 5th ed. Harcourt Brace College Publishers, Fort Worth, TX. 1112 pp.

HAYTER, A. J. 1984. A proof of the conjecture that the Tukey-Kramer multiple comparisons procedure is conservative. *Ann. Statist.* 12: 61–75.

HAYTER, A. J. 1986. The maximum familywise error rate of Fisher's least significant difference test. *J. Amer. Statist. Assoc.* 81: 1000–1004.

HEALY, M. J. R. 1984. The use of R^2 as a measure of goodness of fit. *J. Roy. Statist. Soc.* Ser. A, 147: 608–609.

HENDERSON, D. A. and D. R. DENISON. 1989. Stepwise regression in social and psychological research. *Psychol. Reports* 64: 251–257.

HETTMANSPERGER, T. P. and J. W. McKEAN. 1998. *Robust Nonparametric Statistical Methods.* John Wiley, New York. 467 pp.

HEYDE, C. C. and E. SENETA. (eds.). 2001. *Statisticians of the Centuries.* Springer-Verlag, New York. 500 pp.

HICKS, C. R. 1982. *Fundamental Concepts in Design of Experiments,* 3rd ed. Holt, Rinehart, & Winston, New York. 425 pp.

HINES, W. G. S. 1996. Pragmatics of pooling in ANOVA tables. *Amer. Statist.* 50: 127–139.

HODGES, J. L., JR. and E. L. LEHMANN. 1956. The efficiency of some nonparametric competitors of the t-test. *Ann. Math. Statist.* 27: 324–335.

HODGES, J. L., JR. 1955. A bivariate sign test. *Ann. Math. Statist.* 26: 523–527.

HODGES, J. L. JR., P. H. RAMSEY, and S. WECHSLER. 1990. Improved significance probabilities of the Wilcoxon test. *J. Educ. Statist.* 15: 249–265.

HOENIG, J. M. and D. M. HEISEY. 2001. The abuse of power: The pervasive fallacy of power calculations for data analysis. *Amer. Statist.* 55: 19–24.

HOLLANDER, M. and J. SETHURAMAN. 1978. Testing for agreement between two groups of judges. *Biometrika* 65: 403–411.

HOLLANDER, M. and D. A. WOLFE. 1999. *Nonparametric Statistical Methods.* 3rd ed. John Wiley, New York. 787 pp.

HORNICK, C. W. and J. E. OVERALL. 1980. Evaluation of three sample-size formulae for 2×2 contingency tables. *J. Educ. Statist.* 5: 351–362.

HOSMANE, B. 1986. Improved likelihood ratio tests for the hypothesis of no three-factor interaction in two dimensional contingency tables. *Communic. Statist. — Theor. Meth.* 15: 1875–1888.

HOSMANE, B. 1987. An empirical investigation of chi-square tests for the hypothesis of no three-factor interaction in I x J x K contingency tables. *J. Statist. Computa. Simula.* 28: 167–178.

HOSMER, D. W., JR. and S. LEMESHOW. 2000. *Applied Logistic Regression.* 2nd ed. John Wiley, New York. 375 pp.

HOTELLING, H. 1931. The generalization of Student's ratio. *Ann. Math. Statist.* 2: 360–378.

Literature Cited

HOTELLING, H. 1951. A generalized *T* test and measure of generalized disperion, pp. 23–41. *In* Neyman, J. (ed.), *Proceedings of the Second Berkeley Symposium on Mathematical Statistics and Probability*. University of California Press, Berkeley.

HOTELLING, H. 1953. New light on the correlation coefficient and its transform. *J. Roy. Statist. Soc.* B15: 193–232.

HOTELLING, H. and M. R. PABST. 1936. Rank correlation and tests of significance involving no assumption of normality. *Ann. Math. Statist.* 7: 29–43.

HOTELLING, H. and L. M. SOLOMONS. 1932. The limits of a measure of skewness. *Ann. Math. Statist.* 3: 141–142.

HOWELL, D. C. 2007. *Statistical Methods for Psychology*. 6th ed. Thomson Wadsworth, Belmont, CA. 739 pp.

HSU, J. C. 1996. *Multiple Comparisons: Theory Methods*. Chapman and Hall, London. 277 pp.

HSU, P. L. 1938. Contributions to the theory of "Student's" *t*-test as applied to the problem of two samples. *Statist. Res. Mem.* 2: 1–24. [Cited in Ramsey, 1980.]

HUBBARD, R. and M. L. BAYARRI. 2003. Confusion over measures of evidence (*p*'s) versus errors (*α*'s) in classical statistical testing. *Amer. Statist.* 57: 171–178. *Also:* Discussion, by K. N. Berk, *ibid.,* 57: 178–179; M. A. Carlton, *ibid.,* 5: 179–181; Rejoinder, by R. Hubbard and M. L. Bayarri, *ibid.,* 181–182.

HUBER, P. 1964. Robust estimation of a location parameter. *Ann. Math. Statist.* 35: 73–101.

HUBER, P. 2004. *Robust Statistics*. John Wiley, New York. 308 pp.

HUBERTY, C. J. and S. A. MOURAD. 1980. Estimation in multiple correlation/prediction. *Educ. Psychol. Meas.* 40: 101–112.

HUCK, S. W. and B. H. LAYNE. 1974. Checking for proportional *n*'s in factorial ANOVA's. *Educ. Psychol. Meas.* 34: 281–287.

HUFF, D. 1954. *How to Lie with Statistics*. W. W. Norton, New York. 142 pp.

HUITEMA, B. E. 1974. Three multiple comparison procedures for contrasts among correlation coefficients. *Proc. Soc. Statist. Sect., Amer. Statist. Assoc.,* 1974, pp. 336–339.

HUITEMA, B. E. 1980. *The Analysis of Covariance and Alternatives*. John Wiley, New York. 445 pp.

HUMMEL, T. J. and J. R. SLIGO. 1971. Empirical comparisons of univariate and multivariate analysis of variance procedures. *Psychol. Bull.* 76: 49–57.

HURLBERT, S. H. 1984. Psudoreplication and the design of ecological field experiments. *Ecol. Monogr.* 54: 187–211. (Designated a "Citation Classic," 1993. *Current Contents* 24(12): 18.)

HURLBERT, S. H. 1990. Spatial distribution of the montane unicorn. *Oikos* 58: 257–271.

HUTCHESON, K. 1970. A test for comparing diversities based on the Shannon formula. *J. Theoret. Biol.* 29: 151–154.

HUTCHINSON, T. P. 1979. The validity of the chi-squared test when expected frequencies are small: A list of recent research references. *Communic. Statist. — Theor. Meth.* A8: 327–335.

HUTSON, A. D. 1999. Calculating nonparametric confidence intervals using fractional order statistics. *J. Appl. Statist.* 26: 343–353.

HUYNH, H. and FELDT, L. S. 1970. Conditions under which mean square ratios in repeated measurements designs have exact F-distributions. *J. Amer. Statist. Assoc.* 65: 1582–1589.

IMAN, R. L. 1974a. Use of a *t*-statistic as an approximation to the exact distribution of the Wilcoxon signed ranks test statistic. *Communic. Statist. — Theor. Meth.* A3: 795–806.

IMAN, R. L. 1974b. A power study of a rank transform for the two-way classification model when interaction may be present. *Can. J. Statist.* 2: 227–229.

IMAN, R. L. 1987. Tables of the exact quantiles of the top-down correlation coefficient for *n* = 3(1)14. *Communic. Statist. — Theor. Meth.* 16: 1513–1540.

IMAN, R. L. and W. J. CONOVER. 1978. Approximation of the critical region for Spearman's rho with and without ties present. *Communic. Statist. — Simula. Computa.* 7: 269–282.

IMAN, R. L. and W. J. CONOVER. 1979. The use of the rank transform in regression. *Technometrics* 21: 499–509.

Literature Cited

IMAN, R. L. and J. M. CONOVER. 1985. A measure of top-down correlation. Technical Report SAND85-0601, Sandia National Laboratories, Albuquerque, New Mexico. 44 pp.

IMAN, R. L. and W. J. CONOVER. 1987. A measure of top-down correlation. *Technometrics* 29: 351–357. Correction: *Technometrics* 31: 133 (1989).

IMAN, R. L. and J. M. DAVENPORT. 1976. New approximations to the exact distribution of the Kruskal-Wallis test statistic. *Communic. Statist. — Theor. Meth.* A5: 1335–1348.

IMAN, R. L. and J. M. DAVENPORT. 1980. Approximations of the critical region of the Friedman statistic. *Cummunic. Statist. — Theor. Meth.* A9: 571–595.

IMAN, R. L., S. C. HORA, and W. J. CONOVER. 1984. Comparison of asymptotically distribution-free procedures for the analysis of complete blocks. *J. Amer. Statist. Assoc.* 79: 674–685.

IMAN, R. L., D. QUADE, and D. A. ALEXANDER. 1975. Exact probability levels for the Kruskal-Wallis test, pp. 329–384. *In* H. L. Harter and D. B. Owen, *Selected Tables in Mathematical Statistics*, Volume III. American Mathematical Society, Providence, RI.

INTERNATIONAL BUSINESS MACHINES CORPORATION. 1968. *System/360 Scientific Subroutine Package (360A-CM-03X) Version III. Programmer's Manual.* 4th ed. White Plains, NY. 454 pp.

IRWIN, J. O. 1935. Tests of significance for differences between percentages based on small numbers. *Metron* 12: 83–94.

IRWIN, J. O. 1978. Gosset, William Sealy, pp. 409–413. *In* Kruskal and Tanur (1978).

IVES, K. H. and J. D. GIBBONS. 1967. A correlation measure for nominal data. *Amer. Statist.* 21(5): 16–17.

JACCARD, J., M. A. BECKER, and G. WOOD. 1984. Pairwise multiple comparison procedures: A review. *Psychol. Bull.* 96: 589–596.

JACQUES, J. A. and M. NORUSIS. 1973. Sampling requirements on the estimation of parameters in heteroscedastic linear regression. *Biometrics* 29: 771–780.

JAMMALAMADAKA, S. R. and A. SENGUPTA. 2001. *Topics in Circular Statistics*. World Scientific Publishing, Singapore. 322 pp.

JEYARATNAM, S. 1992. Confidence intervals for the correlation coefficient. *Statist. Prob. Lett.* 15: 389–393.

JOHNSON, M. E. and V. W. LOWE, JR. 1979. Bounds on the sample skewness and kurtosis. *Technometrics* 21: 377–378.

JOHNSON, B. R. and D. J. LEEMING. 1990. A study of the digits of π, e, and certain other irrational numbers. *Sankhyā: Indian J. Statist.* 52B: 183–189.

JOHNSON, R. A. and D. W. WICHERN. 2002. *Applied Multivariate Statistical Analysis*. 5th ed. Prentice Hall, Upper Saddle River, NJ. 767 pp.

JOLLIFFE, I. T. 1981. Runs test for detecting dependence between two variances. *Statistician* 30: 137–141.

JUPP, P. E. and K. V. MARDIA. 1980. A general correlation coefficient for directional data and related regression problems. *Biometrika* 67: 163–173.

KEMP, A. W. 1989. A note on Stirling's expansion for factorial n. *Statist. Prob. Lett.* 7: 21–22.

KEMPTHORNE, O. 1979. In dispraise of the exact test: Reactions. *J. Statist. Plan. Inference* 3: 199–213.

KENDALL, M. G. 1938. A new measure of rank correlation. *Biometrika* 30: 81–93.

KENDALL, M. G. 1943. *The Advanced Theory of Statistics*. Vol. I. Charles Griffin, London, England. [Cited in Eisenhart, 1979.]

KENDALL, M. G. 1945. The treatment of ties in ranking problems. *Biometrika* 33: 239–251.

KENDALL, M. G. 1962. *Rank Correlation Methods*. 3rd ed. Charles Griffin, London, England. 199 pp.

KENDALL, M. G. 1970. *Rank Correlation Methods*. 4th ed. Charles Griffin, London, England. 202 pp.

KENDALL, M. G. and B. BABINGTON SMITH. 1939. The problem of m rankings. *Ann. Math. Statist.* 10: 275–287.

Literature Cited

KENDALL, M. and J. D. GIBBONS. 1990. *Rank Correlation Methods*. 5th ed. Edward Arnold, London, England. 260 pp.

KENDALL, M. G. and A. STUART. 1966. *The Advanced Theory of Statistics*, Vol. 3. Hafner, New York. 552 pp.

KENDALL, M. G. and A. STUART. 1979. *The Advanced Theory of Statistics*, Vol. 2. 4th ed. Griffin, London, England. 748 pp.

KENNARD, R. W. 1971. A note on the C_p statistics. *Technometrics* 13: 899–900.

KENNEDY, J. J. 1992. *Analyzing Qualitative Data. Log-linear Analysis for Behavioral Research.* 2nd ed. Praeger, New York. 299 pp.

KEPNER, J. L. and D. H. ROBINSON. 1988. Nonparametric methods for detecting treatment effects in repeated-measures designs. *J. Amer. Statist. Assoc.* 83: 456–461.

KESELMAN, H. J., P. A. GAMES, and J. C. ROGAN. 1979. An addendum to "A comparison of the modified-Tukey and Scheffé methods of multiple comparisons for pairwise contrasts." *J. Amer. Statist. Assoc.* 74: 626–627.

KESELMAN, H. J., R. MURRAY, and J. C. ROGAN. 1976. Effect of very unequal group sizes on Tukey's multiple comparison test. *Educ. Psychol. Meas.* 36: 263–270.

KESELMAN, H. J. and J. C. ROGAN. 1977. The Tukey multiple comparison test: 1953–1976. *Psychol. Bull.* 84: 1050–1056.

KESELMAN, H. J. and J. C. ROGAN. 1978. A comparison of the modified-Tukey and Scheffé methods of multiple comparisons for pairwise contrasts. *J. Amer. Statist. Assoc.* 73: 47–52.

KESELMAN, H. J., J. C. ROGAN, and B. J. FEIR-WALSH. 1977. An evaluation of some non-parametric and parametric tests for location equality. *Brit. J. Math. Statist. Psychol.* 30: 213–221.

KESELMAN, H. J. and L. E. TOOTHAKER. 1974. Comparisons of Tukey's T-method and Scheffé's S-method for various numbers of all possible differences of averages contrasts under violation of assumptions. *Educ. Psychol. Meas.* 34: 511–519.

KESELMAN, H. J., L. E. TOOTHAKER, and M. SHOOTER. 1975. An evaluation of two unequal n_k forms of the Tukey multiple comparison statistic. *J. Amer. Statist. Assoc.* 70: 584–587.

KESELMAN, H. J., R. WILCOX, J. TAYLOR, and R. K. KOWALCHUK. 2000. Tests for mean equality that do not reuire honogeneity of variances: Do they really work? *Communic. Statist. — Simula.* 29: 875–895.

KEULS, M. 1952. The use of the "studentized range" in connection with an analysis of variance. *Euphytica* 1: 112–122.

KHAMIS, H. J. 1990. The δ-corrected Kolmogorov-Smirnov test for goodness of fit. *J. Statist. Plan. Infer.* 24: 317–335.

KHAMIS, H. J. 1993. A comparative study of the δ-corrected Kolmogorov-Smirnov test. *J. Appl. Statist.* 20: 401–421.

KHAMIS, H. J. 2000. The two-stage δ-corrected Kolmogorov-Smirnov test. *J. Appl. Statist.* 27: 439–450.

KIHLBERG, J. K., J. H. HERSON, and W. E. SCHUTZ. 1972. Square root transformation revisited. *J. Royal Statist. Soc. Ser. C. Appl. Statist.* 21: 76–81.

KIM, S.-H. and A. S. COHEN. 1998. On the Behrens-Fisher problem: A review. *J. Educ. Behav. Statist.* 23: 356–377.

KIRBY, W. 1974. Algebraic boundedness of sample statistics. *Water Resources Res.* 10: 220–222.

KIRBY, W. 1981. [Letter to the editor.] *Technometrics* 23: 215–216.

KIRK, R. E. 1995. *Experimental Design: Procedures for the Behavioral Sciences.* 3rd ed. Brooks/Cole, Pacific Grove, CA. 921 pp.

KLEINBAUM, D. G. and M. KLEIN. 2002. *Logistic Regression: A Self-Learning Text.* 2nd ed. Springer-Verlag, New York. 513 pp.

KNOKE, D. and P. J. BURKE. 1980. *Log-Linear Models.* Sage Publications, Beverly Hills, CA. 80 pp.

KOEHLER, K. J. 1986. Goodness-of-fit tests for log-linear models in sparse contingency tables. *J. Amer. Statist. Assoc.* 81: 483–492.

Literature Cited

KOEHLER, K. J. and K. LARNTZ. 1980. An empirical investigation of goodness-of-fit statistics for sparse multinomials. *J. Amer. Statist. Assoc.* 75: 336–344.

KOHR, R. L. and P. A. GAMES. 1974. Robustness of the analysis of variance, the Welch procedure, and a Box procedure to heterogeneous variances. *J. Exper. Educ.* 43: 61–69.

KOLMOGOROV, A. 1933. Sulla determinazione empirica di una legge di distribuzione. *Giornalle dell Instituto Italiano degli Attuari* 4: 1–11.

KORNBROT, D. E. 1990. The rank difference test: A new and meaningful alternative to the Wilcoxon signed ranks test for ordinal data. *Brit. J. Math. Statist. Psychol.* 43: 241–264.

KÖRNER, T. W. 1996. *The Pleasures of Counting.* Cambridge University Press, Cambridge, England. 534 pp.

KOWALSKI, C. J. 1972. On the effects of non-normality on the distribution of the sample product-moment correlation coefficient. *J. Roy. Statist. Soc.* C21: 1–12.

KRAMER, C. Y. 1956. Extension of multiple range tests to group means with unequal numbers of replications. *Biometrics* 12: 307–310.

KROLL, N. E. A. 1989. Testing independence in 2×2 contingency tables. *J. Educ. Statist.* 14: 47–79.

KRUSKAL, W. H. 1957. Historical notes on the Wilcoxon unpaired two-sample test. *J. Amer. Statist. Assoc.* 52: 356–360.

KRUSKAL, W. H. 1958. Ordinal measures of association. *J. Amer. Statist. Assoc.* 53: 814–861.

KRUSKAL, W. H. and J. M. TANUR (eds.). 1978. *International Encyclopedia of Statistics.* Free Press, New York. 1350 pp.

KRUSKAL, W. H. and W. A. WALLIS. 1952. Use of ranks in one-criterion analysis of variance. *J. Amer. Statist. Assoc.* 47: 583–621.

KRUTCHKOFF, R. G. 1988. One-way fixed effects analysis of variance when the error variances may be unequal. *J. Statist. Computa. Simula.* 30: 259–271.

KUIPER, N. H. 1960. Tests concerning random points on a circle. *Ned. Akad. Wetensch. Proc. Ser. A* 63: 38–47.

KULKARNI, P. M. and A. K. SHAH. 1995. Testing the equality of several binomial proportions to a prespecified standard. *Statist. Prob. Lett.* 25: 213–219.

KUTNER, M. H., C. J. NACHTSHEIM, and J. NETER. 2004. *Applied Linear Regression Models.* 4th ed. McGraw-Hall/Irwin, New York. 701 pp.

KVÅLSETH, T. O. 1985. Cautionary note about R^2. *Amer. Statist.* 39: 279–285.

LANCASTER, H. O. 1949. The combination of probabilities arising from data in discrete distributions. *Biometrika* 36: 370–382.

LANCASTER, H. O. 1969. *The Chi-Squared Distribution.* John Wiley, New York. 356 pp.

LANDRY, L. and Y. LEPAGE. 1992. Empirical behavior of some tests for normality. *Communic. Statist. — Simula.* 21: 971–999.

LAPLACE, P. S. 1812. *Theéorie Analytique de Probabilitès.* Coursier, London. [Cited in Agresti, 2002: 15.]

LARGEY, A. and J. E. SPENCER. 1996. *F*- and *t*-tests in multiple regression: The possibility of 'conflicting' outcomes. *Statistician* 45: 102–109.

LARNTZ, K. 1978. Small-sample comparisons of exact levels for chi-squared goodness-of-fit statistics. *J. Amer. Statist. Assoc.* 73: 253–263.

LAUBSCHER, N. F., F. W. STEFFANS, and E. M. DE LANGE. 1968. Exact critical values for Mood's distribution-free statistic for dispersion and its normal approximation. *Technometrics* 10: 447–507.

LAWAL, H. B. 1984. Comparisons of the X^2, Y^2, Freeman-Tukey and Williams's improved G^2 test statistics in small samples of one-way multinomials. *Biometrika* 71: 415–418.

LAWLEY, D. N. 1938. A generalization of Fisher's z test. *Biometrika* 30: 180–187.

LEADBETTER, M. R. 1988. Harald Cramér, 1893–1985. *Intern. Statist. Rev.* 56: 89–97.

Literature Cited

LEE, A. F. S. and N. S. FINEBERG. 1991. A fitted test for the Behrens-Fisher problem. *Communic. Statist. — Theor. Meth.* 20: 653–666.

LEE, A. F. S. and J. GURLAND. 1975. Size and power of tests for equality of means of two normal populations with unequal variances. *J. Amer. Statist. Assoc.* 70: 933–941.

LEHMANN, E. L. 1999. "Student" and small-sample theory. *Statist. Sci.* 14: 418–426.

LEHMANN, E. L. and C. REID. 1982. Jerzy Neyman 1894–1981. *Amer. Statist.* 36: 161–162.

LENTH, R. A. 2001. Some practical guidelines for effective sample size determination. *Amer. Statist.* 55: 187–193.

LEONHARDT, D. 2000. John Tukey, 85, statistician who coined 2 crucial words. *The New York Times*, July 28, 2000, Section A, p. 19.

LESLIE, P. H. 1955. A simple method of calculating the exact probability in 2×2 contingency tables with small marginal totals. *Biometrika* 42: 522–523.

LEVENE, H. 1952. On the power function of tests of randomness based on runs up and down. *Ann. Math. Statist.* 23: 34–56.

LEVENE, H. 1960. Robust tests for equality of variances, pp. 278–292. *In* Iolkin, S. G. Ghurye, W. Hoeffding, W. G. Madow, and H. B. Mann (eds.), *Contributions to Probability and Statistics: Essays in Honor of Harold Hotelling*. Stanford University Press, Stanford, CA.

LEVIN, B. and X. CHEN. 1999. Is the one-half continuity correction used once or twice to derive a well-known approximate sample size formula to compare two independent binomial distributions? *Amer. Statist.* 53: 62–66.

LEVIN, J. R., R. C. SERLIN, and L. WEBNE-BEHRMAN. 1989. Analysis of variance through simple correlation. *Amer. Statist.* 43: 32–34.

LEVY, K. J. 1975a. Some multiple range tests for variances. *Educ. Psychol. Meas.* 35: 599–604.

LEVY, K. J. 1975b. Comparing variances of several treatments with a control. *Educ. Psychol. Meas.* 35: 793–796.

LEVY, K. J. 1975c. An empirical comparison of several multiple range tests for variances. *J. Amer. Statist. Assoc.* 70: 180–183.

LEVY, K. J. 1976. A multiple range procedure for independent correlations. *Educ. Psychol. Meas.* 36: 27–31.

LEVY, K. J. 1978a. An empirical comparison of the ANOVA F-test with alternatives which are more robust against heterogeneity of variance. *J. Statist. Computa. Simula.* 8: 49–57.

LEVY, K. J. 1978b. An empirical study of the cube root test for homogeneity of variances with respect to the effects of a non-normality and power. *J. Statist. Computa. Simula.* 7: 71–78.

LEVY, K. J. 1978c. A priori contrasts under conditions of variance heterogeneity. *J. Exper. Educ.* 47: 42–45.

LEVY, K. J. 1979. Pairwise comparisons associated with the K independent sample median test. *Amer. Statist.* 33: 138–139.

LEWONTIN, R. C. 1966. On the measurement of relative variability. *Systematic Zool.* 15: 141–142.

LEYTON, M. K. 1968. Rapid calculation of exact probabilities for 2×3 contingency tables. *Biometrics* 24: 714–717.

LI, L. and W. R. SCHUCANY. 1975. Some properties of a test for concordance of two groups of rankings. *Biometrika* 62: 417–423.

LIDDELL, D. 1976. Practical tests of 2×2 contingency tables. *Statistician* 25: 295–304.

LIDDELL, D. 1980. Practical tests for comparative trials: a rejoinder to N. L. Johnson. *Statistician* 29: 205–207.

LIGHT, R. J. and B. H. MARGOLIN. 1971. Analysis of variance for categorical data. *J. Amer. Statist. Assoc.* 66: 534–544.

LIN, J.-T. 1988a. Approximating the cumulative chi-square distribution and its inverse. *Statistician* 37: 3–5.

LIN, J.-T. 1988b. Alternatives to Hamaker's approximations to the cumulative normal distribution and its inverse. *Statistician* 37: 413-414.

Literature Cited

LIN, L. I-K. 1989. A concordance correlation coefficient to evaluate reproducibility. *Biometrics* 45: 255–268.

LIN, L. I-K. 1992. Assay validation using the concordance correlation coefficient. *Biometrics* 48: 599–604.

LIN, L. I-K. 2000. A note on the concordance correlation coefficient. *Biometrics* 56: 324–325.

LIN, L. I-K. and V. CHINCILLI. 1996. Rejoinder to the letter to the editor from Atkinson and Klevill [Comment on the use of concordance correlation to assess the agreement between two variables]. *Biometrics* 53: 777–778.

LINDSAY, R. B. 1976. John William Strutt, Third Baron Rayleigh, pp. 100–107. *In* C. C. Gillispie (ed.), *Dictionary of Scientific Biography*, Vol. XIII. Charles Scribner's Sons, New York.

LING, R. F. 1974. Comparison of several algorithms for computing sample means and variances. *J. Amer. Statist. Assoc.* 69: 859–866.

LITTLE, R. J. A. and D. B. RUBIN. 2002. *Statistical Analysis with Missing Data.* John Wiley, Hoboken, NJ. 381 pp.

LIX, L. M., J. C. KESELMAN, and H. J. KESELMAN. 1996. Consequences of assumption violations revisited: A quantitative review of alternatives to the one-way analysis of variance *F* test. *Rev. Educ. Res.* 66: 579–619.

LLOYD, C. J. 1988. Doubling the one-sided *P*-value in testing independence in 2×2 tables against a two-sided alternative. *Statist. Med.* 7: 1297–3006.

LLOYD, M., J. H. ZAR, and J. R. KARR. 1968. On the calculation of information-theoretical measures of diversity. *Amer. Midland Natur.* 79: 257–272.

LOCKHART, R. A. and M. A. STEPHENS. 1985. Tests of fit for the von Mises distribution. *Biometrika* 72: 647–652.

LUDWIG, J. A. and J. F. REYNOLDS. 1988. *Statistical Ecology.* John Wiley, New York. 337 pp.

LUND, U. 1999. Least circular distance regression for directional data. *J. Appl. Statist.* 26: 723–733.

MAGURRAN, A. E. 2004. *Measuring Biological Diversity.* Blackwell Publishing, Malden, MA. 256 pp.

MAAG, U. R. 1966. A *k*-sample analogue of Watson's U^2 statistic. *Biometrika* 53: 579–583.

MACKINNON, W. J. 1964. Table for both the sign test and distribution-free confidence intervals of the median for sample sizes to 1,000. *J. Amer. Statist. Assoc.* 59: 935–956.

MAITY, A. and M. SHERMAN. 2006. The two-sample *T* test with one variance unknown. *Amer. Statist.* 60: 1163–1166.

MALLOWS, C. L. 1973. Some comments on C_p. *Technometrics* 15: 661–675.

MANN, H. B. and D. R. WHITNEY. 1947. On a test of whether one of two random variables is stochastically larger than the other. *Ann. Math. Statist.* 18: 50–60.

MANTEL, N. 1966. *F*-ratio probabilities from binomial tables. *Biometrics* 22: 404–407.

MANTEL, N. 1970. Why stepdown procedures in variable selection. *Technometrics* 12: 621–625.

MANTEL, N. 1974. Comment and a suggestion. [Re Conover, 1974.] *J. Amer. Statist. Assoc.* 69: 378–380.

MANTEL, N. and S. W. GREENHOUSE. 1968. What is the continuity correction? *Amer. Statist.* 22(5): 27–30.

MAOR, E. 1994. *e: The Story of a Number.* Princeton University Press, Princeton, NJ. 223 pp.

MARASCUILO, L. A. 1971. *Statistical Methods for Behavioral Science Research.* McGraw-Hill, New York. 578 pp.

MARASCUILO, L. A. and M. MCSWEENEY. 1967. Nonparametric post hoc comparisons for trend. *Psychol. Bull.* 67: 401–412.

MARASCUILO, L. A. and M. MCSWEENEY. 1977. *Nonparametric and Distribution-free Methods for the Social Sciences.* Brooks/Cole, Monterey, California. 556 pp.

MARCOULIDES, G. A. and S. L. HERSHBERGER. 1997. *Multivariate Statistical Methods: A First Course.* Lawrence Earlbaum Associates, Mahwah, NJ. 322 pp.

Literature Cited

MARDIA, K. V. 1967. A non-parametric test for the bivariate two-sample location problem. *J. Royal Statist. Soc.* B29: 320–342.

MARDIA, K. V. 1969. On the null distribution of a non-parametric test for the bivariate two-sample problem. *J. Royal Statist. Soc.* B31: 98–102.

MARDIA, K. V. 1970a. Measures of multivariate skewness and kurtosis with applications. *Biometrika* 57: 519–530.

MARDIA, K. V. 1970b. A bivariate non-parametric *c*-sample test. *J. Royal Statist. Soc.* B32: 74–87.

MARDIA, K. V. 1972a. *Statistics of Directional Data.* Academic Press, New York. 357 pp.

MARDIA, K. V. 1972b. A multisample uniform scores test on a circle and its parametric competitor. *J. Royal Statist. Soc.* B34: 102–113.

MARDIA, K. V. 1975. Statistics of directional data. (With discussion.) *J. Royal Statist. Soc.* B37: 349–393.

MARDIA, K. V. 1976. Linear-angular correlation coefficients and rhythmometry. *Biometrika* 63: 403–405.

MARDIA, K. V. 1981. Directional statistics in geosciences. *Communic. Statist. — Theor. Meth.* A10: 1523–1543.

MARDIA, K. V. 1990. Obituary: Professor B. L. Welch. *J. Roy. Statist. Soc.* Ser. A, 153: 253–254.

MARDIA, K. V. 1998. Obituary: Geoffrey Stuart Watston. *Statistician* 47: 701–702.

MARDIA, K. V. and P. E. JUPP. 2000. *Directional Statistics.* John Wiley, New York. 414 pp.

MARDIA, K. V. and B. D. SPURR. 1973. Multisample tests for multimodal and axial circular populations. *J. Roy. Statist. Soc.* B35: 422–436.

MARGOLIN, B. H. and R. J. LIGHT. 1974. An analysis of variance for categorical data, II: Small sample comparisons with chi square and other competitors. *J. Amer. Statist. Assoc.* 69: 755–764.

MARKOWSKI, C. A. and E. P. MARKOWSKI. 1990. Conditions for the effectiveness of a preliminary test of variance. *Amer. Statist.* 44: 322–326.

MARTIN, L., R. LEBLANC, and N. K. TOAN. 1993. Tables for the Friedman rank test. *Can. J. Statist.* 21: 39–43.

MARTÍN ANDRÉS, A. 1991. A review of classic non-asymptotic methods for comparing two proportions by means of independent samples. *Communic. Statist. — Simula. Computa.* 20: 551–583.

MARTÍN ANDRÉS, A. and I. HERRANZ TEJEDOR. 1995. Is Fisher's exact test very conservative? *Comp. Statist. Data Anal.* 19: 579–591.

MARTÍN ANDRÉS, A. and I. HERRANZ TEJEDOR. 2000. On the minimum expected quantity for the validity of the chi-squared test in 2×2 tables. *J. Appl. Statist.* 27: 807–820.

MARTÍN ANDRÉS, A., I. HERRANZ TEJEDOR, and J. D. LUNA DEL CASTILLO. 1992. Optimal correction for continuity in the chi-squared test in 2×2 tables (conditioned method). *Communic. Statist. — Simula.* 21: 1077–1101.

MARTÍN ANDRÉS, A., I. HERRANZ TEJEDOR, and A. SILVA MATO. 1995. The Wilcoxon, Spearman, Fisher, χ^2-, Student and Pearson tests and 2×2 tables. *Statistician* 44: 441–450.

MARTÍN ANDRÉS, A. and A. SILVA MATO. 1994. Choosing the optimal unconditioned test for comparing two independent proportions. *Comput. Statist. & Data Anal.* 17: 555–574.

MARTÍN ANDRÉS, A., A. SILVA MATO, and I. HERRANZ TEJEDOR. 1992. A critical review of asymptotic methods for comparing two proportions by means of independent samples. *Communic. Statist. — Simula. Computa.* 21: 551–586.

MARTÍN ANDRÉS, A. and J. M. TAPIA GARCIA. 2004. Optimal unconditional asymptotic test in 2×2 multinomial trials. *Communic. Statist.* 33: 83–97.

MASSEY, F. J., JR. 1951. The Kolmogorov-Smirnov test for goodness of fit. *J. Amer. Statist. Assoc.* 46: 68–78.

MAURAIS, J. and R. OUIMET. 1986. Exact critical values of Bartlett's test of homogeneity of variances for unequal sample sizes for two populations and power of the test. *Metrika* 33: 275–289.

Literature Cited

MAXWELL, A. E. 1970. Comparing the classification of subjects by two independent judges. *Brit. J. Psychiatry* 116: 651–655.

MAXWELL, S. E. 1980. Pairwise multiple comparisons in repeated measures designs. *J. Educ. Statist.* 5: 269–287.

MAXWELL, S. E. and H. D. DELANEY. 2004. *Designing Experiments and Analyzing Data. A Model Comparison Perspective.* 2nd ed. Lawrence Earlbaum Associates, Mahwah, NJ. 1079 pp.

McCORNACK. R. L. 1965. Extended tables of the Wilcoxon matched pair signed rank statistic. *J. Amer. Statist. Assoc.* 60: 864–871.

McCULLOUGH, B. D. 1998. Assessing the reliability of statistical software: Part I. *Amer. Statist.* 52: 358–366.

McCULLOUGH, B. D. 1999. Assessing the reliability of statistical software: Part II. *Amer. Statist.* 53: 149–159.

McCULLOCH, C. E. 1987. Tests for equality of variances with paired data. *Communic. Statist. — Theor. Meth.* 16: 1377–1391.

McGILL, R., J. W. TUKEY, and W. A. LARSEN. 1978. Variations of box plots. *Amer. Statist.* 32: 12–16.

McKAY, A. T. 1932. Distribution of the coefficient of variation and the extended "t" distribution. *J. Roy. Statist. Soc.* A95: 695–698.

McKEAN, J. W. and T. J. VIDMAR. 1994. A comparison of two rank-based methods for the analysis of linear models. *Amer. Statist.* 48: 220–229.

McNEMAR, Q. 1947. Note on the sampling error of the difference between correlated proportions or percentages. *Psychometrica* 12: 153–157.

MEAD, R., T. A. BANCROFT, and C. HAN. 1975. Power of analysis of variance test procedures for incompletely specified mixed models. *Ann. Statist.* 3: 797–808.

MEDDIS, R. 1984. *Statistics Using Ranks: A Unified Approach.* Basil Blackwell, Oxford, England. 449 pp.

MEE, W. M., A. K. SHAH, and J. J. LEFANTE. 1987. Comparing k independent sample means with a known standard. *J. Qual. Technol.* 19: 75–81.

MEEKER, W. Q., JR. and G. J. HAHN. 1980. Prediction intervals for the ratios of normal distribution sample variances and exponential distribution sample means. *Technometrics* 22: 357–366.

MEHTA, C. R. and J. F. HILTON. 1993. Exact power of conditional and unconditional tests: Going beyond the 2×2 contingency table. *Amer. Statist.* 47: 91–98.

MEHTA, J. S. and R. SRINIVASAN. 1970. On the Behrens-Fisher problem. *Biometrika* 57: 649–655.

MENARD, S. 2002. *Applied Logistic Analysis.* 2nd ed. Sage Publications, Thousand Oaks, CA. 111 pp.

MENDEL, G. 1865. Versuche über Planzen-Hybriden [Experiments on Plant Hybrids]. *Verhandlungen des naturforschenden Vereines in Brünn* 4, Abhandlungen, pp 3–47 (which appeared in 1866, and which has been published in English translation many times beginning in 1901, including as Mendel, 1933, and as pp. 1–48 in Stern and Sherwood, 1966).

MENDEL, G. 1933. *Experiments in Plant Hybridization.* Harvard University Press, Cambridge, MA. 353 pp.

MEYERS, L. S., G. GAMST, and A. J. GUARINO. 2006. *Applied Multivariate Research: Design and Interpretation.* Sage Publications, Thousand Oaks, CA. 722 pp.

MICKEY, R. M., O. J. DUNN, and V. A. CLARK. 2004. *Applied Statistics: Analysis of Variance and Regression.* John Wiley, New York. 448 pp.

MILLER, A. J. 1972. [Letter to the editor.] *Technometrics* 14: 507.

MILLER, G. E. 1991. Asymptotic test statistics for coefficients of variation. *Communic. Statist. — Theor. Meth.* 20: 2251–2262.

MILLER, G. E. and C. J. FELTZ. 1997. Asymptotic inference for coefficients of variation. *Communic. Statist. — Theor. Meth.* 26: 715–726.

MILLER, J. 2001. "Earliest Uses of Symbols for Variables": members.aol.com/jeff570/variables.html

Literature Cited

MILLER, J. 2004a. "Earliest Known Use of Some of the Words in Mathematics": members.aol.com/jeff570/mathword.html

MILLER, J. 2004b. "Earliest Uses of Symbols of Operation": members.aol.com/jeff570/operation.html

MILLER, J. 2004c. "Earliest Uses of Symbols in Probability and Statistics": members.aol.com/jeff570/stat.html

MILLER, L. H. 1956. Table of percentage points of Kolmogorov statistics. *J. Amer. Statist. Assoc.* 51: 111–121.

MILLER, R. G., JR. 1981. *Simultaneous Statistical Inference.* 2nd ed. McGraw-Hill, New York. 299 pp.

MILTON, R. C. 1964. An extended table of critical values for the Mann-Whitney (Wilcoxon) two-sample statistic. *J. Amer. Statist. Assoc.* 59: 925–934.

MOGULL, R. G. 1994. The one-sample runs test: A category of exception. *J. Exper. Behav. Statist.* 19: 296–303.

MOLENAAR, W. 1969a. How to poison Poisson (when approximating binomial tails). *Statist. Neerland.* 23: 19–40.

MOLENAAR, W. 1969b. Additional remark on "How to poison Poisson (when approximating binomial tails)." *Statist. Neerland.* 23: 241.

MONTGOMERY, D. C. 2005. *Design and Analysis of Experiments.* 6th ed. John Wiley, Hoboken, NJ. 643 pp.

MONTGOMERY, D. C., W. A. PECK, and G. G. VINING. 2006. *Introduction to Linear Regression Analysis.* John Wiley Interscience, New York to be 2006 edition. 612 pp.

MOOD, A. M. 1950. *Introduction to the Theory of Statistics.* McGraw-Hill, New York. 433 pp.

MOOD, A. M. 1954. On the asymptotic efficiency of certain non-parametric two-sample tests. *Ann. Math. Statist.* 25: 514–522.

MOORE, B. R. 1980. A modification of the Rayleigh test for vector data. *Biometrika* 67: 175–180.

MOORE, D. S. 1986. Tests of chi-squared type, pp. 63–95. *In* R. B. D'Agostino and M. A. Stephens (eds.), *Goodness-of-Fit Techniques.* Marcel Dekker, New York.

MOORS, J. J. A. 1986. The meaning of kurtosis: Darlington revisited. *Amer. Statist.* 40: 283–284.

MOORS, J. J. A. 1988. A quantile alternative for kurtosis. *Statistician* 37: 25–32.

MOSER, B. K. and G. R. STEVENS. 1992. Homogeneity of variance in the two-sample means test. *Amer. Statist.* 46: 19–21.

MOSIMANN, J. E. 1968. *Elementary Probablilty for the Biological Sciences.* Appleton-Century-Crofts, New York. 255 pp.

MUDDAPUR, M. V. 1988. A simple test for correlation coefficient in a bivariate normal population. *Sankyā: Indian J. Statist. Ser. B.* 50: 60–68.

MYERS, J. L. and A. D. WELL. 2003. *Research Design and Statistical Analysis.* 2nd ed. Lawrence Earlbaum Associates, Mahwah, NJ. 226, 760 pp.

NAGASENKER, P. B. 1984. On Bartlett's test for homogeneity of variances. *Biometrika* 71: 405–407.

NEAVE, H. R. and O. L. WORTHINGTON. 1988. *Distribution-Free Tests.* Unwin-Hyman, London, England. 430 pp.

NELSON, W., Y. L. TONG, J.-K. LEE, and F. HALBERG. 1979. Methods for cosinor-rhythmometry. *Chronobiologia* 6: 305–323.

NEMENYI, P. 1963. *Distribution-Free Multiple Comparisons.* State University of New York, Downstate Medical Center. [Cited in Wilcoxon and Wilcox (1964).]

NEWCOMBE, R. G. 1998a. Two-sided confidence intervals for the single proportion: Comparison of seven methods. *Statist. Med.* 17: 857–872.

NEWCOMBE, R. G. 1998b. Interval estimation for the difference between independent propotions: Comparison of eleven methods. *Statist. Med.* 17: 873–890.

Literature Cited

NEWMAN, D. 1939. The distribution of range in samples from a normal population, expressed in terms of an independent estimate of standard deviation. *Biometrika* 31: 20–30.

NEYMAN, J. 1959. Optimal asymptotic tests of composite hypothesis, pp. 213–234. *In* U. Grenander (ed.), *Probability and Statistics: The Harald Cramér Volume*. John Wiley, New York.

NEYMAN, J. 1967. R. A. Fisher: An appreciation. *Science* 156: 1456–1460.

NEYMAN, J. and E. S. PEARSON. 1928a. On the use and interpretation of certain test criteria for purposes of statistical inference. Part I. *Biometrika* 20A: 175–240.

NEYMAN, J. and E. S. PEARSON. 1928b. On the use and interpretation of certain test criteria for purposes of statistical inference. Part II. *Biometrika* 20A: 263–294.

NEYMAN, J. and E. S. PEARSON. 1931. On the problem of k samples. *Bull. Acad. Polon. Sci. Lett. Ser. A*, 3: 460–481.

NEYMAN, J. and E. S. PEARSON. 1933. On the problem of the most efficient tests of statistical hypotheses. *Phil. Trans. Roy. Soc. London. Ser. A*, 231: 239–337.

NOETHER, G. E. 1963. Note on the Kolmogorov statistic in the discrete case. *Metrika* 7: 115–116.

NOETHER, G. E. 1984. Nonparametrics: The early years—impressions and recollections. *Amer. Statist.* 38: 173–178.

NORRIS, R. C. and H. F. HJELM. 1961. Non-normality and product moment correlation. *J. Exp. Educ.* 29: 261–270.

NORTON, H. W. 1939. The 7 × 7 squares. *Ann. Eugen.* 9: 269–307.

NORWOOD, P. K., A. R. SAMPSON, K. McCARROLL, and R. STAUM. 1989. A multiple comparisons procedure for use in conjunction with the Benard-van Elteren test. *Biometrics* 45: 1175–1182.

O'BRIEN, P. C. 1976. A test for randomness. *Biometrics* 32: 391–401.

O'BRIEN, P. C. and P. J. DYCK. 1985. A runs test based on run lengths. *Biometrics* 41: 237–244.

O'BRIEN, R. G. and M. KAISER. 1985. MANOVA method for analyzing repeated measures designs: An extensive primer. *Psychol. Bull.* 97: 316–333.

O'CINNEIDE, C. A. 1990. The mean is within one standard deviation of any median. *Amer. Statist.* 44: 292–293.

O'CONNOR, J. J. and E. F. ROBERTSON. 1996. "Christopher Clavius." School of Mathematics and Statistics, St. Andrews University, Scotland: www-history.mcs.st-andrews.ac.uk/history/BiogIndex.html

O'CONNOR, J. J. and E. F. ROBERTSON. 1997. "Christian Kramp" (*ibid.*).

O'CONNOR, J. J. and E. F. ROBERTSON. 1998. "Charles Babbage" (*ibid.*).

O'CONNOR, J. J. and E. F. ROBERTSON. 2003. "John Venn" (*ibid.*).

ODEH, R. E. and J. O. EVANS. 1974. The percentage points of the normal distribution. *J. Royal Statist. Soc. Ser. C. Appl. Statist.* 23: 98–99.

OLDS, E. G. 1938. Distributions of sums of squares of rank differences for small numbers of individuals. *Ann. Math. Statist.* 9: 133–148.

OLSON, C. L. 1974. Comparative robustness of six tests on multivariate analysis of variance. *J. Amer. Statist. Assoc.* 69: 894–908.

OLSON, C. L. 1976. On choosing a test statistic in multivariate analysis of variance. *Psychol. Bull.* 83: 579–586.

OLSON, C. L. 1979. Practical considerations in choosing a MANOVA test statistic: A rejoinder to Stevens. *Psychol. Bull.* 86: 1350–1352.

OLSSON, U., F. DRASGOW, and N. J. DORANS. 1982. The polyserial correlation coefficient. *Biometrika* 47: 337–347.

OLVER, F. W. J. 1964. Bessel functions of integer order, pp. 355–433. *In* M. Abramowitz and I. Stegun (eds.), *Handbook of Mathematical Functions*, National Bureau of Standards, Washington, DC. (Also, Dover, New York, 1965.)

Literature Cited

O'Neill, M. E. and K. Mathews. 2000. A weighted least squares approach to Levene's test of homogeneity of variance. *Austral. and New Zeal. J. Statist.* 42: 81–100.

Ord, K. 1984. Marine George Kendall, 1907–1983. *Amer. Statist.* 38: 36–37.

Ostle, B. and L. C. Malone. 1988. *Statistics in Research*. 4th ed. Iowa State University Press, Ames, Iowa. 664 pp.

Otieno, B. S. and C. M. Anderson-Cook. 2003. A more efficient way of obtaining a unique median estimate for circular data. *J. Modern Appl. Statist. Meth.* 2: 168–176.

Otten, A. 1973. The null distribution of Spearman's S when $n = 13(1)16$. *Statist. Neerland.* 27: 19–20.

Overall, J. E., H. M. Rhoades and R. R. Starbuck. 1987. Small-sample tests for homogeneity of response probabilities in 2×2 contingency tables. *Psychol. Bull.* 98: 307–314.

Owen, D. B. 1962. *Handbook of Statistical Tables*. Addison-Wesley, Reading, MA. 580 pp.

Ozer, D. J. 1985. Correlation and the coefficient of determination. *Psychol. Bull.* 97: 307–315.

Page, W. and V. N. Murty. 1982. Nearness relations among measures of central tendency and dispersion: Part 1. *Two-Year Coll. Math. J.* 13: 315–327.

Pampel, F. C. 2000. *Logistic Regression: A Primer*. Sage Publications, Thousand Oaks, CA. 86 pp.

Papanastasiou, B. C. 2003. Greek letters in measurement and statistics. Is it all Greek to you?. *STATS* 36: 17–18.

Parshall, C. G. and J. D. Kromrey. 1996. Tests of independence in contingency tables with small samples: A comparison of statistical power, *Educ. Psychol. Meas.* 56: 26–44.

Patel, J. K. 1989. Prediction intervals — A review. *Communic. Statist. — Theor. Meth.* 18: 2393–2465.

Patil, K. D. 1975. Cochran's Q test: Exact distribution. *J. Amer. Statist. Assoc.* 70: 186–189.

Paul, S. R. 1988. Estimation of and testing significance for a common correlation coefficient. *Communic. Statist. — Theor. Meth.* 17: 39–53.

Paull, A. E. 1950. On a preliminary test for pooling mean squares in the analysis of variance. *Ann. Math. Statist.* 21: 539–556.

Pazer, H. L. and L. A. Swanson. 1972. *Modern Methods for Statistical Analysis*. Intext Educational Publishers, Scranton, PA. 483 pp.

Pearson, E. S. 1932. The analysis of variance in cases of nonnormal variation. *Biometrika* 23: 114–133.

Pearson, E. S. 1939. Student as a statistician. *Biometrika* 30: 210–250.

Pearson, E. S. 1947. The choice of statistical tests illustrated on the interpretation of data classed in a 2×2 table. *Biometrika* 34: 139–167.

Pearson, E. S. 1967. Studies in the history of probability and statistics. XVII. Some reflections on continuity in the development of mathematical statistics, 1885–1920. *Biometrika* 54: 341–355.

Pearson, E. S., R. B. D'Agostino, and K. O. Bowman. 1977. Tests for departure from normality. Comparison of powers. *Biometrika* 64: 231–246.

Pearson, E. S. and H. O. Hartley. 1951. Charts for the power function for analysis of variance tests, derived from the non-central F-distribution. *Biometrika* 38: 112–130.

Pearson, E. S. and H. O. Hartley. 1966. *Biometrika Tables for Statisticians*, Vol. 1. 3rd ed. Cambridge University Press, Cambridge, England. 264 pp.

Pearson, E. S. and H. O. Hartley. 1976. *Biometrika Tables for Statisticians*, Vol. 2, 2nd ed. Cambridge University Press, Cambridge, England. (1972, reprinted 1976.) 385 pp.

Pearson, E. S., R. L. Plackett, and G. A. Barnard. 1990. *'Student': A Statistical Biography of William Sealy Gossett*. Clarendon Press, Oxford, England. 142 pp.

PEARSON, E. S. and N. W. PLEASE. 1975. Relation between the shape of population distribution and the robustness of four simple test statistics. *Biometrika* 62: 223–241.

PEARSON, K. 1900. On a criterion that a given system of deviations from the probable in the case of a correlated system of variables is such that it can be reasonably supposed to have arisen in random sampling. *Phil. Mag. Ser. 5* 50: 157–175.

PEARSON, K. 1901. On the correlation of characters not quantitatively measured. *Philosoph. Trans. Roy. Soc. London* A195: 1–47. [Cited in Sheskin, 2004: 997.]

PEARSON, K. 1904. Mathematical contributions to the theory of evolution. XIII. On the theory of contingency and its relation to association and normal correlation. Draper's Co. Res. Mem., Biometric Ser. 1. 35 p. [Cited in Lancaster (1969).]

PEARSON, K. 1905. "Das Fehlergesetz und Seine Verallgemeinerungen durch Fechner und Pearson." A rejoinder. *Biometrika* 4: 169–212.

PEARSON, K. 1920. Notes on the history of correlation. *Biometrika* 13: 25–45.

PEARSON, K. 1924. I. Historical notes on the origin of the normal curve of errors. *Biometrika* 16: 402–404.

PEDHAZUR, E. J. 1997. *Multiple Regression in Behavioral Research: Explanation and Prediction.* Harcourt Brace, Fort Worth, TX. 1058 pp.

PETERS, W. S. 1987. *Counting for Something.* Springer-Verlag, New York. 275 pp.

PETRINOVICH, L. F. and HARDYCK, C. D. 1969. Error rates for multiple comparison methods: Some evidence concerning the frequency of erroneous conclusions. *Psychol. Bull.* 71: 43–54.

PETTITT, A. N. and M. A. STEPHENS. 1977. The Kolmogorov-Smirnov goodness-of-fit statistic with discrete and grouped data. *Technometrics* 19: 205–210.

PFANZAGL, J. 1978. Estimation: Confidence intervals and regions, pp. 259–267. *In* Kruskal and Tanur (1978).

PIELOU, E. C. 1966. The measurement of diversity in different types of biological collections. *J. Theoret. Biol.* 13: 131–144.

PIELOU, E. C. 1974. *Population and Community Ecology.* Gordon Breach, New York, 424 pp.

PIELOU, E. C. 1977. *Mathematical Ecology.* John Wiley, New York. 385 pp.

PILLAI, K. C. S. 1955. Some new test criteria in multivariate analysis. *Ann. Math. Statist.* 26: 117–121.

PIRIE, W. R. and M. A. HAMDAN. 1972. Some revised continuity corrections for discrete distributions. *Biometrics* 28: 693–701.

PITMAN, E. J. G. 1939. A note on normal correlation. *Biometrika* 31: 9–12.

PLACKETT, R. L. 1964. The continuity correlation in 2 × 2 tables. *Biometrika* 51: 327–338.

POLLAK, M. and J. COHEN. 1981. A comparison of the independent-samples *t*-test and the paired-samples *t*-test when the observations are nonnegatively correlated pairs. *J. Statist. Plan. Inf.* 5: 133–146.

POSTEN, H. O. 1992. Robustness of the two-sample *t*-test under violations of the homogeneity of variance assumption, part II. *Communic. Statist. — Theor. Meth.* 21: 2169–2184.

POSTEN, H. O., H. C. YEH, and D. B. OWEN. 1982. Robustness of the two-sample *t*-test under violations of the homogeneity of variance assumption. *Communic. Statist. — Theor. Meth.* 11: 109–126.

PRATT, J. W. 1959. Remarks on zeroes and ties in the Wilcoxon signed rank procedures. *J. Amer. Statist. Assoc.* 54: 655–667.

PRZYBOROWSKI, J. and H. WILENSKI. 1940. Homogeneity of results in testing samples from Poisson series. *Biometrika* 31: 313–323.

Literature Cited

QUADE, D. 1979. Using weighted rankings in the analysis of complete blocks with additive block effects. *J. Amer. Statist. Assoc.* 74: 680–683.

QUADE, D. and I. SALAMA. 1992. A survey of weighted rank correlation, pp. 213–224. *In* P. K. Sen and I. Salama (eds.), *Order Statistics and Nonparametrics: Theory and Applications*. Elsevier, New York.

QUENOUILLE, M. H. 1950. *Introductory Statistics*. Butterworth-Springer, London. [Cited in Thöni (1967: 18).]

QUENOUILLE, M. H. 1956. Notes on bias in estimation. *Biometrika* 43: 353–360.

QUINN, G. P. and M. J. KEOUGH. 2002. *Experimental Design and Data Analysis for Biologists*. Cambridge University Press, Cambridge, England. 537 pp.

RACTLIFFE, J. F. 1968. The effect on the t-distribution of non-normality in the sampled population. *J. Royal Statist. Soc. Ser. C. Appl. Statist.* 17: 42–48.

RADLOW, R. and E. F. ALF, JR. 1975. An alternate multinomial assessment of the accuracy of the χ^2 test of goodness of fit. *J. Amer. Statist. Assoc.* 70: 811–813.

RAFF, M. S. 1956. On approximating the point binomial. *J. Amer. Statist. Assoc.* 51: 293–303.

RAHE, A. J. 1974. Table of critical values for the Pratt matched pair signed rank statistic. *J. Amer. Statist. Assoc.* 69: 368–373.

RAHMAN, M. and Z. GOVINDARAJULU. 1997. A modification of the test of Shapiro and Wilk for normality. *J. Appl. Statist.* 24: 219–236.

RAMSEY, P. H. 1978. Power differences between pairwise multiple comparisons. *J. Amer. Statist. Assoc.* 73: 479–485.

RAMSEY, P. H. 1980. Exact Type I error rates for robustness of Student's t test with unequal variances. *J. Educ. Statist.* 5: 337–349.

RAMSEY, P. H. 1989. Determining the best trial for the 2×2 comparative trial. *J. Statist. Computa. Simula.* 34: 51–65.

RAMSEY, P. H. and P. P. RAMSEY. 1988. Evaluating the normal approximation to the binomial test. *J. Educ. Statist.* 13: 173–182.

RAMSEYER, G. C. and T.-K. TCHENG. 1973. The robustness of the studentized range statistic to violations of the normality and homogeneity of variance assumptions. *Amer. Educ. Res. J.* 10: 235–240.

RANNEY, G. B. and C. C. THIGPEN. 1981. The sample coefficient of determinationin simple linear regression. *Amer. Statist.* 35: 152–153.

RAO, C. R. 1992. R. A. Fisher: The founder of modern statistics. *Statist. Sci.* 7: 34–48.

RAO, C. R. and I. M. CHAKRAVARTI. 1956. Some small sample tests of significance for a Poisson distribution. *Biometrics* 12: 264–282.

RAO, C. R., S. K. MITRA, and R. A. MATTHAI. 1966. *Formulae and Tables for Statistical Work*. Statistical Publishing Society, Calcutta, India. 233 pp.

RAO, J. S. 1976. Some tests based on arc-lengths for the circle. *Sankhyā: Indian J. Statist. Ser. B.* 26: 329–338.

RAYLEIGH. 1919. On the problems of random variations and flights in one, two, and three dimensions. *Phil. Mag. Ser.* 6, 37: 321–347.

RENCHER, A. C. 1998. *Multivariate Statistical Inference and Applications*. John Wiley, New York. 559 pp.

RENCHER, A. C. 2002. *Methods of Multivariate Analysis*. John Wiley, New York. 708 pp.

RHOADES, H. M. and J. E. OVERALL. 1982. A sample size correction for Pearson chi-square in 2×2 contingency tables. *Psychol. Bull.* 91: 418–423.

RICHARDSON, J. T. E. 1990. Variants of chi-square for 2×2 contingency tables. *J. Math. Statist. Psychol.* 43: 309–326.

RICHARDSON, J. T. E. 1994. The analysis of 2×1 and 2×2 contingency tables: An historical review. *Statist. Meth. Med. Res.* 3: 107–133.

RODGERS, J. L. and W. L. NICEWANDER. 1988. Thirteen ways to look at the correlation coefficient. *Amer. Statist.* 42: 59–66.

Literature Cited

ROGAN, J. C. and H. J. KESELMAN. 1977. Is the ANOVA F-test robust to variance heterogeneity when sample sizes are equal?: An investigation via a coefficient of variation. *Amer. Educ. Res. J.* 14: 493–498.

ROSCOE, J. T. and J. A. BYARS. 1971. Sample size restraints commonly imposed on the use of the chi-square statistic. *J. Amer. Statist. Assoc.* 66: 755–759.

ROSS, G. J. S. and D. A. PREECE. 1985. The negative binomial distribution. *Statistician* 34: 323–336.

ROTHERY, P. 1979. A nonparametric measure of intraclass correlation. *Biometrika* 66: 629–639.

ROUANET, H. and D. LÉPINE. 1970. Comparison between treatments in a repeated-measurement design: ANOVA and multivariate methods. *Brit. J. Math. Statist. Psychol.* 23: 147–163.

ROUTLEDGE, R. D. 1990. When stepwise regression fails: correlated variables some of which are redundant. *Intern. J. Math. Educ. Sci. Technol.* 21: 403–410.

ROY, S. N. 1945. The individual sampling distribution of the maximum, minimum, and any intermediates of the p-statistics on the null-hypothesis. *Sankhyā* 7(2): 133–158.

ROY, S. N. 1953. On a heuristic method of test construction and its use in multivariate analysis. *Ann. Math. Statist.* 24: 220–238.

ROYSTON, E. 1956. Studies in the history of probability and statistics. III. A note on the history of the graphical presentation of data. *Biometrika* 43: 241–247.

ROYSTON, J. P. 1982a. An extension of Shapiro and Wilk's W test for normality to large samples. *Appl. Statist.* 31: 115–124.

ROYSTON, J. P. 1982b. The W test for normality. *Appl. Statist.* 31: 176–180.

ROYSTON, J. P. 1986. A remark on AS181. The W test for normality. *Appl. Statist.* 35: 232–234.

ROYSTON, J. P. 1989. Correcting the Shapiro-Wilk W for ties. *J. Statist. Comput. — Simulat.* 31: 237–249.

RUBIN, D. B. 1990. Comment: Neyman (1923) and causal inference in experiments and observational studies. *Statist. Sci.* 8: 472–480.

RUDAS, T. 1986. A Monte Carlo comparison of the small sample behaviour of the Pearson, the likelihood ratio and the Cressie-Read statistics. *J. Statist. Computa. Simula.* 24: 107–120.

RUSSELL, G. S. and D. J. LEVITIN. 1994. An expanded table of probability values for Rao's spacing test. Technical Report No. 94-9, Institute of Cognitive and Decision Sciences, University of Oregon, Eugene. 13 pp.

RUST, S. W. and M. A. FLIGNER. 1984. A modification of the Kruskal-Wallis statistic for the generalized Behrens-Fisher problem. *Communic. Statist. — Theor. Meth.* 13: 2013–2028.

RYAN, G. W. and S. D. LEADBETTER. 2002. On the misuse of confidence intervals for two means in testing for the significance of the difference between the means. *J. Modern Appl. Statist. Meth.* 1: 473–478.

RYAN, T. P. 1997. *Modern Regression Methods.* John Wiley, New York. 515 pp.

SAHAI, H. and M. I. AGEEL. 2000. *The Analysis of Variance: Fixed, Random and Mixed Models.* Birkhäser, Boston. 742 pp.

SAKODA, J. M. and B. H. COHEN. 1957. Exact probabilities for contingency tables using binomial coefficients. *Psychometrika* 22: 83–86.

SALAMA, I. and D. QUADE. 1982. A nonparametric comparison of two multiple regressions by means of a weighted measure of correlation. *Communic. Statist. — Theor. Meth.* A11: 1185–1195.

SATTERTHWAITE, F. E. 1946. An approximate distribution of estimates of variance components. *Biometrics Bull.* 2: 110–114.

SAVAGE, I. R. 1956. Contributions to the theory of rank order statistics — the two-sample case. *Ann. Math. Statist.* 27: 590–615.

SAVAGE, I. R. 1957. Non-parametric statistics. *J. Amer. Statist. Assoc.* 52: 331–334.

SAVAGE, L. J. 1976. On rereading R. A. Fisher. *Ann. Statist.* 4: 441–483. *Also:* Discussion by B. Efron, *ibid.* 4: 483–484; C. Eisenhart, *ibid.* 4: 484; B. de Finetti, *ibid.* 4: 485–488; F. A. S. Fraser, *ibid.* 4: 488–489; V. O. Godambe, 4: 490–492; I. J. Good, *ibid.* 4: 492–495; O. Kempthorne, *ibid.* 4: 495–497; S. M. Stigler, *ibid.* 4: 498–500.

Literature Cited

SAVILLE, D. J. 1990. Multiple comparison procedures: The practical solution. *Amer. Statist.* 44: 174–180.

SAWILOWSKY, S. S. 2002. Fermat, Schubert, Einstein, and Behrens-Fisher: The probable difference between two means when $\sigma_1^2 \neq \sigma_2^2$. *J. Modern Appl. Statist. Meth.* 1: 461–472.

SAWILOWSKY, S. S., C. C. BLAIR, and J. J. HIGGINS. 1989. An investigation of the Type I error and power properties of the rank transform procedure in factorial ANOVA. *J. Educ. Statist.* 14: 255–267.

SCHADER, M. and F. SCHMID. 1990. Charting small sample characteristics of asymptotic confidence intervals for the binomial proportion *p. Statist. Papers* 31: 251–264.

SCHEFFÉ, H. 1953. A method of judging all contrasts in the analysis of variance. *Biometrika* 40: 87–104.

SCHEFFÉ, H. 1959. *The Analysis of Variance.* John Wiley, New York. 477 pp.

SCHEFFÉ, H. 1970. Practical solutions of the Behrens-Fisher problem. *J. Amer. Statist. Assoc.* 65: 1501–1508.

SCHENKER, N. and J. F. GENTLEMAN. 2001. On judging the significance of differences by examining the overlap between confidence intervals. *Amer. Statist.* 55: 182–186.

SCHLITTGEN, R. 1979. Use of a median test for a generalized Behrens-Fisher problem. *Metrika* 26: 95–103.

SCHUCANY, W. R. and W. H. FRAWLEY. 1973. A rank test for two group concordance. *Psychometrika* 38: 249–258.

SCHWERTMAN, N. C. and R. A. MARTINEZ. 1994. Approximate Poisson confidence limits. *Communic. Statist. — Theor. Meth.* 23: 1507–1529.

SEAL, H. L. 1967. Studies in the history of probability and statistics. XV. The historical development of the Gauss linear model. *Biometrika* 54: 1–14.

SEAMAN, J. W., JR., S. C. WALLS, S. E. WISE, AND R. G. JAEGER. 1994. Caveat emptor: Rank transform methods and interaction. *Trends Ecol. Evol.* 9: 261–263.

SEBER, G. A. F. and A. J. LEE. 2003. *Linear Regression Analysis.* 2nd ed. John Wiley, New York. 557 pp.

SEBER, G. A. F. and C. J. WILD. 1989. *Nonlinear Regression.* John Wiley, New York. 768 pp.

SENDERS, V. L. 1958. *Measurement and Statistics.* Oxford University Press, New York. 594 pp.

SERLIN, R. C. and M. R. HARWELL. 1993. An empirical study of eight tests of partial correlation coefficients. *Communic. Statist. — Simula. Computa.* 22: 545–567.

SERLIN, R. C. and L. A. MARASCUILO. 1983. Planned and post hoc comparisons in tests of concordance and discordance for *G* groups of judges. *J. Educ. Statist.* 8: 187–205.

SHAFFER, J. P. 1977. Multiple comparison emphasizing selected contrasts: An extension and generalization of Dunnett's procedure. *Biometrics* 33: 293–303.

SHANNON, C. E. 1948. A mathematical theory of communication. *Bell System Tech. J.* 27: 379–423, 623–656.

SHAPIRO, S. S. 1986. *How to Test Normality and Other Distributional Assumptions.* Volume 3. *The ASQC Basic References in Quality Control: Statistical Techniques,* E. J. Dudewicz (ed.). American Society for Quality Control, Milwaukee, WI, 67 pp.

SHAPIRO, S. S. and M. B. WILK. 1965. An analysis of variance test for normality (complete samples). *Biometrika* 52: 591–611.

SHAPIRO, S. S., M. B. WILK, and H. J. CHEN. 1968. A comparative study of various tests for normality. *J. Amer. Statist. Assoc.* 63: 1343–1372.

SHAPIRO, S. S. and R. S. FRANCIA. 1972. An approximate analysis of variance test for normality. *J. Amer. Statist. Assoc.* 67: 215–216.

SHARMA, S. 1996. *Applied Multivariate Techniques.* John Wiley, New York. 493 pp.

SHEARER, P. R. 1973. Missing data in quantitative designs. *J. Royal Statist. Soc. Ser. C. Appl. Statist.* 22: 135–140.

SHESKIN, D. J. 2004. *Handbook of Parametric and Nonparametric Statistical Procedures.* 3rd ed. Chapman & Hall/CRC, Boca Raton, FL. 1193 pp.

Literature Cited

SICHEL, H. S. 1973. On a significance test for two Poisson variables. *J. Royal Statist. Soc. Ser. C. Appl. Statist.* 22: 50–58.

SIEGEL, S. and N. J. CASTELLAN JR. 1988. *Nonparametric Statistics for the Behavioral Sciences.* McGraw-Hill, New York. 399 pp.

SIEGEL, S. and J. TUKEY. 1960. A nonparametric sum of ranks procedure for relative spread in unpaired samples. *J. Amer. Statist. Assoc.* 55: 429–445. (Correction in 1991 in *J. Amer. Statist. Assoc.* 56: 1005.)

SIMONOFF, J. S. 2003. *Analyzing Categorical Data.* Springer, New York. 496 pp.

SIMPSON, G. G., A. ROE, and R. C. LEWONTIN. 1960. *Quantitative Zoology.* Harcourt, Brace, Jovanovich, New York. 44 pp.

SKILLINGS, J. H. and G. A. MACK. 1981. On the use of a Friedman-type statistic in balanced and unbalanced block designs. *Technometrics* 23: 171–177.

SMIRNOV, N. V. 1939a. Sur les écarts de la courbe de distribution empirique. *Recueil Mathématique N. S.* 6: 3–26. [Cited in Smirnov (1948).]

SMIRNOV, N. V. 1939b. On the estimation of the discrepancy between empirical curves of distribution for two independent samples. (In Russian.) *Bull. Moscow Univ. Intern. Ser. (Math.)* 2: 3–16. [Cited in Smirnov (1948).]

SMITH, D. E. 1951. *History of Mathematics.* Vol. I. Dover Publications, New York. 596 pp.

SMITH, D. E. 1953. *History of Mathematics.* Vol. II. Dover Publications, New York. 725 pp.

SMITH, H. 1936. The problem of comparing the results of two experiments with unequal means. *J. Council Sci. Industr. Res.* 9: 211–212. [Cited in Davenport and Webster (1975).]

SMITH, R. A. 1971. The effect of unequal group size on Tukey's HSD procedure. *Psychometrika* 36: 31–34.

SNEDECOR, G. W. 1934. *Calculation and Interpretation of Analysis of Variance and Covariance.* Collegiate Press, Ames, Iowa. 96 pp.

SNEDECOR, G. W. 1954. Biometry, its makers and concepts, pp. 3–10. *In* O. Kempthorne, T. A. Bancroft, J. W. Gowen, and J. L. Lush (eds.), *Statistics and Mathematics in Biology.* Iowa State College Press, Ames, Iowa.

SNEDECOR, G. W. and W. G. COCHRAN. 1989. *Statistical Methods.* 8th ed. Iowa State University Press, Ames, Iowa. 503 pp.

SOKAL, R. R. and F. J. ROHLF. 1995. *Biometry.* 3rd ed., W. H. Freeman & Co., New York. 887 pp.

SOMERVILLE, P. N. 1993. On the conservatism of the Tukey-Kramer multiple comparison procedure. *Statist. Prob. Lett.* 16: 343–345.

SOPER, H. E. 1914. Tables of Poisson's exponential binomial limit. *Biometrika* 10: 25–35.

SPEARMAN, C. 1904. The proof and measurement of association between two things. *Amer. J. Psychol.* 15: 72–101.

SPEARMAN, C. 1906. 'Footrule' for measuring correlation. *Brit. J. Psychol.* 2: 89–108.

SPRENT, P. and N. C. SMEETON. 2001. *Applied Nonparametric Statistics.* 3rd ed. Chapman & Hall/CRC, Boca Raton, FL. 461 pp.

SRIVASTAVA, A. B. L. 1958. Effect of non-normality on the power function of the *t*-test. *Biometrika* 45: 421–429.

SRIVASTAVA, A. B. L. 1959. Effects of non-normality on the power of the analysis of variance test. *Biometrika* 46: 114–122.

SRIVASTAVA, M. S. 2002. *Methods of Multivariate Statistics.* John Wiley, New York. 697 pp.

STARMER, C. F., J. E. GRIZZLE, and P. K. SEN. 1974. Some Reasons for not using the Yates continuity correction on 2×2 contingency tables: Comment. *J. Amer. Statist. Assoc.* 69: 376–378.

STEEL, R. G. D. 1959. A multiple comparison rank sum test: Treatments versus control. *Biometrics* 15: 560–572.

STEEL, R. G. D. 1960. A rank sum test for comparing all pairs of treatments. *Technometrics* 2: 197–207.

Literature Cited

STEEL, R. G. D. 1961a. Error rates in multiple comparisons. *Biometrics* 17: 326–328.

STEEL, R. G. D. 1961b. Some rank sum multiple comparison tests. *Biometrics* 17: 539–552.

STEEL, G. D., J. H. TORRIE, and D. A. DICKEY. 1997. *Principles and Procedures of Statistics. A Biometrical Approach.* 3rd ed. WCB/McGraw-Hill, Boston. 666 pp.

STEIGER, J. H. 1980. Tests for comparing elements in a correlation matrix. *Psychol. Bull.* 87: 245–251.

STELZL, I. 2000. What sample sizes are needed to get correct significance levels for log-linear models? — A Monte Carlo study using the SPSS-procedure "Hiloglinear." *Meth. Psychol. Res.* 5: 95–116.

STEPHENS, M. A. 1969a. Tests for randomness of directions against two circular alternatives. *J. Amer. Statist. Assoc.* 64: 280–289.

STEPHENS, M. A. 1969b. A goodness-of-fit statistic for the circle, with some comparisons. *Biometrika* 56: 161–168.

STEPHENS, M. A. 1972. Multisample tests for the von Mises distribution. *J. Amer. Statist. Assoc.* 67: 456–461.

STEPHENS, M. A. 1982. Use of the von Mises distribution to analyze continuous proportions. *Biometrika* 69: 197–203.

STEVENS, G. 1989. A nonparametric multiple comparison test for differences in scale parameters. *Metrika* 36: 91–106.

STEVENS, J. 2002. *Applied Multivariate Statistics for the Social Sciences.* 4th ed. Lawrence Earlbaum, Mahwah, NJ. 699 pp.

STEVENS, S. S. 1946. On the theory of scales of measurement. *Science* 103: 677–680.

STEVENS, S. S. 1968. Measurement, statistics, and the schemapiric view. *Science* 161: 849–856.

STEVENS, W. L. 1939. Distribution of groups in a sequence of alternatives. *Ann. Eugen.* 9: 10–17.

STIGLER, S. M. 1978. Francis Ysidro Edgeworth, statistician. *J. Roy. Statist. Soc. Ser. A,* 141: 287–322.

STIGLER, S. M. 1980. Stigler's law of eponymy. *Trans. N.Y. Acad. Sci. Ser. II.* 39: 147–157.

STIGLER, S. M. 1982. Poisson on the Poisson distribution. *Statist. Prob. Lett.* 1: 33–35.

STIGLER, S. M. 1986. *The History of Statistics: The Measurement of Uncertainty Before 1900.* Belknap Press, Cambridge, MA. 410 pp.

STIGLER, S. M. 1989. Francis Galton's account of the invention of correlation. *Statist. Sci.* 4: 73–86.

STIGLER, S. M. 1999. *Statistical on the Table. The History of Statistical Concepts and Methods.* Harvard University Press, Cambridge, MA. 488 pp.

STIGLINE, S. M. 2000. The problematic unity of biometrics. *Biometrics* 56: 653–658.

STOLINE, M. R. 1981. The status of multiple comparisons: Simultaneous estimation of all pairwise comparisons in one-way ANOVA designs. *Amer. Statist.* 35: 134–141.

STONEHOUSE, J. M. and G. J. FORRESTER. 1998. Robustness of *t* and *U* tests under combined assumption violations. *J. Appl. Statist.* 25: 63–74.

STORER, B. E. and C. KIM. 1990. Exact properties of some exact test statistics for comparing two binomial populations. *J. Amer. Statist. Assoc.* 85: 146–155.

STREET, D. J. 1990. Fisher's contributions to agricultural statistics. *Biometrics* 46: 937–945.

STRUIK, D. J. 1967. *A Concise History of Mathematics.* 3rd ed. Dover, NY. 299 pp.

STUART, A. 1984. Sir Maurice Kenall, 1907–1983. *J. Roy. Statist. Soc.* 147: 120–122.

STUDENT. 1908. The probable error of a mean. *Biometrika* 6: 1–25.

STUDENT. 1927. Errors of routine analysis. *Biometrika* 19: 151–164. [Cited in Winer, Brown, and Michels, 1991: 182.]

STUDIER, E. H., R. W. DAPSON, and R. E. BIGELOW. 1975. Analysis of polynomial functions for determining maximum or minimum conditions in biological systems. *Comp. Biochem. Physiol.* 52A: 19–20.

Literature Cited

SUTHERLAND, C. H. V. 1992. Coins and coinage: History of coinage: Origins; ancient Greek coins, pp. 530–534. *In* R. McHenry (general ed.), *The New Encyclopædia Britannica*, 15th ed., Vol. 16. Encyclopædia Britannica, Chicago, IL.

SUTTON, J. B. 1990. Values of the index of determination at the 5% significance level. *Statistician* 39: 461–463.

SWED, F. S. and C. EISENHART, 1943. Tables for testing randomness of grouping in a sequence of alternatives. *Ann. Math. Statist.* 14: 66–87.

SYMONDS, P. M. 1926. Variations of the product-moment (Pearson) coefficient of correlation. *J. Educ. Psychol.* 17: 458–469.

TABACHNICK, B. G. and L. S. FIDELL. 2001. *Using Multivariate Statistics*. 4th ed., Allyn and Bacon, Boston. 966 pp.

TAMHANE, A. C. 1979. A comparison of procedures for multiple comparisons. *J. Amer. Statist. Assoc.* 74: 471–480.

TAN, W. Y. 1982. Sampling distributions and robustness of t, F and variance-ratio in two samples and ANOVA models with respect to departure from normality. *Communic. Statist. — Theor. Meth.* 11: 2485–2511.

TAN, W. Y. and S. P. WONG. 1980. On approximating the null and nonnull distributions of the F ratio in unbalanced random-effects models from nonnormal universes. *J. Amer. Statist. Assoc.* 75: 655–662.

TATE, M. W. and S. M. BROWN. 1964. *Tables for Comparing Related-Sample Percentages and for the Median Test.* Graduate School of Education, University of Pennsylvania, Philadelphia, PA.

TATE, M. W. and S. M. BROWN. 1970. Note on the Cochran Q test. *J. Amer. Statist. Assoc.* 65: 155–160.

TATE, M. W. and L. A. HYER. 1973. Inaccuracy of the X^2 goodness of fit when expected frequencies are small. *J. Amer. Statist. Assoc.* 68: 836–841.

THIELE, T. N. 1897. *Elementær Iagttagelseslære.* Gyldendal, København, Denmark. 129 pp. [Cited in Hald (1981).]

THIELE, T. N. 1899. Om Iagttagelseslærens Halvinvarianter. *Overs. Vid. Sels. Forh.* 135–141. [Cited in Hald (1981).]

THODE, H. C., JR. 2002. *Testing for Normality.* Marcel Dekker, New York. 479 pp.

THOMAS, G. E. 1989. A note on correcting for ties with Spearman's ρ. *J. Statist. Computa. Simula.* 31: 37–40.

THÖNI, H. 1967. Transformation of Variables Used in the Analysis of Experimental and Observational Data. A Review. Tech. Rep. No. 7, Statistical Laboratory, Iowa State University, Ames, IA. 61 pp.

THORNBY, J. I. 1972. A robust test for linear regression. *Biometrics* 28: 533–543.

TIKU, M. L. 1965. Chi-square approximations for the distributions of goodness-of-fit statistics U_N^2 and W_N^2. *Biometrika* 52: 630–633.

TIKU, M. L. 1967. Tables of the power of the F-test. *J. Amer. Statist. Assoc.* 62: 525–539.

TIKU, M. L. 1972. More tables of the power of the F-test. *J. Amer. Statist. Assoc.* 67: 709–710.

TOLMAN, H. 1971. A simple method for obtaining unbiased estimates of population standard deviations. *Amer. Statist.* 25(1): 60.

TOMARKIN, A. J. and R. C. SERLIN. 1986. Comparison of ANOVA alternatives under variance heterogeneity and specific noncentrality structures. *Psychol. Bull.* 99: 90–99.

TOOTHAKER, L. E. 1991. *Multiple Comparisons for Researchers.* Sage Publications, Newbury Park, CA. 167 pp.

TOOTHAKER, L. E. and H. CHANG. 1980. On "The analysis of ranked data derived from completely randomized designs." *J. Educ. Statist.* 5: 169–176.

TOOTHAKER, L. E. and D. NEWMAN. 1994. Nonparametric competitors to the two-way ANOVA. *J. Educ. Behav. Statist.* 19: 237–273.

Literature Cited

TUKEY, J. W. 1949. One degree of freedom for non-additivity. *Biometrics* 5: 232–242.

TUKEY, J. W. 1953. *The problem of multiple comparisons. Department of Statistics*, Princeton University. (unpublished)

TUKEY, J. W. 1993. Where should multiple comparisons go next?, pp. 187–207. *In* Hoppe, F. M., *Multiple Comparisons. Selection, and Applications in Biometry*, Marcel Dekker, New York.

TWAIN, M. (S. L. CLEMENS). 1950. *Life on the Mississipi*. Harper & Row, New York. 526 pp.

TWEDDLE, I. 1984. Approximating n! Historical origins and error analysis. *Amer. J. Phys.* 52: 487–488.

UNISTAT LTD. 2003. *UNISTAT©: Statistical Package for MS Windows^{TM}*. Version 5.5. UNISTAT Ltd., London, England. 932 pp.

UPTON, G. J. G. 1976. More multisample tests for the von Mises distribution. *J. Amer. Statist. Assoc.* 71: 675–678.

UPTON, G. J. G. 1982. A comparison of alternative tests for the 2×2 comparative trial. *J. Roy. Statist. Soc., Ser. A*, 145: 86–105.

UPTON, G. J. G. 1986. Approximate confidence intervals for the mean direction of a von Mises distribution. *Biometrika* 73: 525–527.

UPTON, G. J. G. 1992. "Fisher's exact test." *J. Roy. Statist. Soc. Ser. A*, 155: 395–402.

UPTON, G. J. G. and B. FINGLETON. 1985. *Spatial Data Analysis by Example*. Volume 1. Point Pattern and Quantitative Data. John Wiley, New York. 410 pp.

UPTON, G. J. G. and B. FINGLETON. 1989. *Spatial Data Analysis by Example*. Volume 2. Categorical and Directional Data. John Wiley, New York. 416 pp.

URY, H. K. 1982. Comparing two proportions: Finding p_2 when p_1, n, α, and β are specified. *Statistician* 31: 245–250.

URY, H. K. and J. L. FLEISS. 1980. On approximate sample sizes for comparing two independent proportions with the use of Yates' correction. *Biometrics* 36: 347–351.

VAN ELTEREN, P. and G. E. NOETHER. 1959. The asymptotic efficiency of the χ^2-test for a balanced incomplete block design. *Biometrika* 46: 475–477.

VITTINGHOFF, E., D. V. GLIDDEN, S. C. SHIBOSKI, and C. E. McCULLOCH. 2005. *Regression Methods in Biostatistics: Linear, Logistic, Survival, and Repeated Measures*. Springer Sciences+Business Media, New York, 340 pp.

VOLLSET, S. E. 1993. Confidence intervals for a binomial proportion. *Statist. Med.* 12: 809–824.

VON EYE, A. and C. SCHUSTER. 1998. *Regression Analysis for Social Sciences*. Academic Press, San Diego, CA. 386 pp.

VON MISES, R. 1918. Über die "Ganzzahligkeit" der Atomgewichte und verwandte Fragen. *Physikal. Z.* 19: 490–500.

VON NEUMANN, J., R. H. KENT, H. R. BELLINSON, and B. I. HART. 1941. The mean square successive difference. *Ann. Math. Statist.* 12: 153–162.

VOSS, D. T. 1999. Resolving the mixed models controvery. *Amer. Statist.* 53: 352–356.

WALD, A. and J. WOLFOWITZ. 1940. On a test whether two samples are from the same population. *Ann. Math. Statist.* 11: 147–162.

WALKER, H. M. 1929. *Studies in the History of Statistical Method with Special Reference to Certain Educational Problems*. Williams and Wilkins, Baltimore, MD. 229 pp.

WALKER, H. M. 1934. Abraham de Moivre. *Scripta Math.* 3: 316–333.

WALKER, H. M. 1958. The contributions of Karl Pearson. *J. Amer. Statist. Assoc.* 53: 11–22.

WALLIS, W. A. 1939. The correlation ratio for ranked data. *J. Amer. Statist. Assoc.* 34: 533–538.

WALLIS, W. A. and G. H. MOORE. 1941. A significance test for time series analysis. *J. Amer. Statist. Assoc.* 36: 401–409.

WALLIS, W. A. and H. V. ROBERTS. 1956. *Statistics: A New Approach*. Free Press, Glencoe, IL. 646 pp.

Literature Cited

WALLRAFF, H. G. 1979. Goal-oriented and compass-oriented movements of displaced homing pigeons after confinement in differentially shielded aviaries. *Behav. Ecol. Sociobiol.* 5: 201–225.

WANG, F. T. and D. W. SCOTT. 1994. The L_1 method for robust non-parametric regression. *J. Amer. Statist. Assoc.* 89: 65–76.

WANG, Y. H. 2000. Fiducial intervals: What are they? *Amer. Statist.* 54: 105–111.

WANG, Y. Y. 1971. Probabilities of the Type I errors of the Welch tests for the Behrens-Fisher problem. *J. Amer. Statist. Assoc.* 66: 605–608.

WATSON, G. S. 1961. Goodness of fit tests on a circle. *Biometrika* 48: 109–114.

WATSON, G. S. 1962. Goodness of fit tests on a circle. II. *Biometrika* 49: 57–63.

WATSON, G. S. 1982. William Gemmell Cochran 1909–1980. *Ann. Statist.* 10: 1–10.

WATSON, G. S. 1983. *Statistics on Spheres.* John Wiley, New York. 238 pp.

WATSON, G. S. 1994. The mean is within one standard deviation from any median. *Amer. Statist.* 48: 268–269.

WATSON, G. S. 1995. U_n^2 test for uniformity of discrete distributions. *J. Appl. Statist.* 22: 273–276.

WATSON, G. S. and E. J. WILLIAMS. 1956. On the construction of significance tests on the circle and the sphere. *Biometrika* 43: 344–352.

WEHRHAHN, K. and J. OGAWA. 1978. On the use of the *t*-statistic in preliminary testing procedures for the Behrens-Fisher problem. *J. Statist. Plan. Inference* 2: 15–25.

WEINFURT, K. P. 1995. Multivariate analysis of variance, pp. 245–276. *In* Grimm, L. G. and P. R. Yarnold (eds.), *Reading and Understanding Multivariate Statistics.* American Psychological Association, Washington, DC.

WEISBERG, S. 2005. *Applied Linear Regression.* John Wiley, Hoboken, NJ. 303 pp.

WELCH, B. L. 1936. Specificiation of rules for rejecting too variable a product, with particular reference to an electric lamp problem. *J. Royal Statist. Soc.*, Suppl. 3: 29–48.

WELCH, B. L. 1938. The significance of the difference between two means when the population variances are unequal. *Biometrika* 29: 350–361.

WELCH, B. L. 1947. The generalization of "Student's" problem when several different population variances are involved. *Biometrika* 34: 28–35.

WELCH, B. L. 1951. On the comparison of several mean values: An alternate approach. *Biometrika* 38: 330–336.

WHEELER, S. and G. S. WATSON. 1964. A distribution-free two-sample test on the circle. *Biometrika* 51: 256–257.

WHITE, J. S. 1970. Tables of normal percentile points. *J. Amer. Statist. Assoc.* 65: 635–638.

WILCOXON, F. 1945. Individual comparisons by ranking methods. *Biometrics Bull.* 1: 80–83.

WILCOXON, F., S. K. KATTI, and R. A. WILCOX. 1970. Critical values and probability levels for the Wilcoxon rank sum test and the Wilcoxon signed rank test, pp. 171–259. *In* H. L. Harter and D. B. Owen (eds.), *Selected Tables in Mathematical Statistics. Vol. I.* Markham Publishing, Chicago.

WILCOXON, F. and R. A. WILCOX. 1964. *Some Rapid Approximate Statistical Procedures.* Lederle Laboratories, Pearl River, NY. 59 pp.

WILKINSON, L. and G. E. DALLAL. 1977. Accuracy of sample moments calculations among widely used statistical programs. *Amer. Statist.* 31: 128–131.

WILKS, S. S. 1932. Certain generalizations in the analysis of variance. *Biometrika* 24: 471–494.

WILKS, S. S. 1935. The likelihood test of independence in contingency tables. *Ann. Math. Statist.* 6: 190–196.

WILLIAMS, K. 1976. The failure of Pearson's goodness of fit statistic. *Statistician* 25: 49.

WILSON, E. B. 1927. Probable inference, the law of succession, and statistical inference. *J. Amer. Statist. Assoc.* 22: 209–212.

WILSON, E. B. 1941. The controlled experiment and the four-fold table. *Science* 93: 557–560.

WILSON, E. B. and M. M. HILFERTY. 1931. The distribution of chi-square. *Proc. Nat. Acad. Sci.*, Washington, DC. 17: 684–688.

Literature Cited

Windsor, C. P. 1948. Factorial analysis of a multiple dichotomy. *Human Biol.* 20: 195–204.

Winer, B. J., D. R. Brown, and K. M. Michels. 1991. *Statistical Procedures in Experimental Design.* 3rd ed. McGraw-Hill, New York. 1057 pp.

Winterbottom, A. 1979. A note on the derivation of Fisher's transformation of the correlation coefficient. *Amer. Statist.* 33: 142–143.

Wittkowski, K. M. 1998. Versions of the sign test in the presence of ties. *Biometrics* 54: 789–791.

Yates, F. 1934. Contingency tables involving small numbers and the χ^2 test. *J. Royal Statist. Soc. Suppl.* 1: 217–235.

Yates, F. 1964. Sir Ronald Fisher and the Design of Experiments. *Biometrics* 20: 307–321.

Yates, F. 1984. Tests of significance for 2×2 contingency tables. *J. Royal Statist. Soc., Ser. A*, 147: 426–449. *Also:* Discussion by G. A. Barnard, *ibid.* 147: 449–450; D. R. Cox *ibid.* 147: 451; G. J. G. Upton, *ibid.* 147: 451–452; I. D. Hill, *ibid.* 147: 452–453; M. S. Bartlett, *ibid.* 147: 453; M. Aitkin and J. P. Hinde, *ibid.* 147: 453–454; D. M. Grove, *ibid.* 147: 454–455; G. Jagger, *ibid.* 147: 455; R. S. Cormack, *ibid.* 147: 455; S. E. Fienberg, *ibid.* 147: 456; D. J. Finney, *ibid* 147: 456; M. J. R. Healy, *ibid.* 147: 456–457; F. D. K. Liddell, *ibid.* 147: 457; N. Mantel, *ibid.* 147: 457–458; J. A. Nelder, *ibid.* 147: 458; R. L. Plackett, *ibid.* 147: 458–462.

Young, L. C. 1941. On randomness in ordered sequences. *Ann. Math. Statist.* 12: 293–300.

Yuen, K. K. 1974. The two-sample trimmed *t* for unequal population variances. *Biometrika* 61: 165–170.

Yule, G. U. 1900. On the association of attributes in statistics. *Phil. Trans. Royal Soc. Ser. A* 94: 257.

Yule, G. U. 1912. On the methods of measuring the association between two attributes. *J. Royal Statist. Soc.* 75: 579–642.

Zabell, S. L. 2008. On Student's 1908 article "The probable error of a mean." *J. Amer. Statist. Assoc.* 103: 1–7. *Also* Comment, by S. M. Stigler, *ibid.* 103: 7–8; J. Aldreich, *ibid.* 103: 8–11; A. W. F. Edwards, *ibid.* 103: 11–13. E. Seneta, *ibid.* 103: 13–15; P. Diaconis and E. Lehmann, *ibid.* 103: 16–19. Rejoinder, by S. L. Zabell, *ibid.* 103: 19–20.

Zar, J. H. 1967. The effect of changes in units of measurement on least squares regression lines. *BioScience* 17: 818–819.

Zar, J. H. 1968. The effect of the choice of temperature scale on simple linear regression equations. *Ecology* 49: 1161.

Zar, J. H. 1972. Significance testing of the Spearman rank correlation coefficient. *J. Amer. Statist. Assoc.* 67: 578–580.

Zar, J. H. 1978. Approximations for the percentage points of the chi-squared distribution. *J. Royal Statist. Soc. Ser. C. Appl. Statist.* 27: 280–290.

Zar, J. H. 1984. *Biostatistical Analysis.* 2nd ed. Prentice Hall, Englewood Cliffs, NJ. 718 pp.

Zar, J. H. 1987. A fast and efficient algorithm for the Fisher exact test. *Behav. Res. Meth., Instrum., & Comput.* 19: 413–415.

Zelen, M. and N. C. Severo. 1964. Probability functions, pp. 925–995. *In* M. Abramowitz and I. Stegun (eds.), *Handbook of Mathematical Functions*, National Bureau of Standards, Washington, DC. (Also, Dover, New York, 1965.)

Zerbe, G. O. and D. E. Goldgar. 1980. Comparison of intraclass correlation coefficients with the ratio of two independent *F*-statistics. *Communic. Statist. — Theor. Meth.* A9: 1641–1655.

Zimmerman, D. W. 1987. Comparative power of Student *T* test and Mann-Whitney *U* test for unequal samples sizes and variances. *J. Exper. Educ.* 55: 171–179.

Zimmerman, D. W. 1994a. A note on the influence of outliers on parametric and nonparametric tests. *J. Gen. Psychol.* 12: 391–401.

Zimmerman, D. W. 1994b. A note on modified rank correlation. *J. Educ. Behav. Statist.* 19: 357–362.

Zimmerman, D. W. 1996. A note on homogeneity of variances of scores and ranks. *J. Exper. Educ.* 64: 351–362.

Zimmerman, D. W. 1997. A note on interpretation of the paired-samples *t* test. *J. Educ. Behav. Statist.* 22: 349–360.

Literature Cited

ZIMMERMAN, D. W. 1998. Invalidation of parametric and nonparametric tests by concurrent violation of two assumptions. *J. Exper. Educ.* 67: 55–68.

ZIMMERMAN, D. W. 2000. Statistical significance levels of nonparametric tests biased by heterogeneous variances of treatment groups. *J. Gen. Psychol.* 127: 354–364.

ZIMMERMAN, D. W. and B. D. ZUMBO. 1993. Rank transformations and the power of the Student t test and Welch t' test for non-normal populations with unequal variances. *Can. J. Exper. Psychol.* 47: 523–539.

This page intentionally left blank

Index

Page references followed by "f" indicate illustrated figures or photographs; followed by "t" indicates a table.